DISCRETE MATHEMATICS

DISCRETE MATHEMATICS

Gary Chartrand and Ping Zhang

Western Michigan University

WAVELAND

PRESS, INC.

Long Grove, Illinois

For information about this book, contact:
Waveland Press, Inc.
4180 IL Route 83, Suite 101
Long Grove, IL 60047-9580
(847) 634-0081
info@waveland.com
www.waveland.com

Table of Contents

PREFACE

THE EMERGENCE OF MATHEMATICS

Although the emergence of mathematics can be traced back to many regions of the world, there is an international mathematics that had its roots in Egypt and Babylonia and developed significantly in ancient Greece. The mathematics of Greece was translated into Arabic, as was some mathematics of India about the same time. This mathematics was later translated into Latin and became the mathematics of Western Europe and, centuries later, the mathematics of the world. Although mathematics developed in other parts of the world, particularly China, Japan and South India, this Asian mathematics did not have the impact that the international mathematics did.

Mathematical history has not followed the same path as other histories of the human spirit. Whether one is considering science, medicine or space exploration, nearly every human endeavor has seen its history marked by advances, corrections and reversals. Mathematics, however, has experienced only advancement. The theorems of yesterday are also theorems today. The accomplishments of the mathematicians of the past have become the foundation of present-day mathematics. With every new era in mathematics, a new peak has been created.

During the latter part of the 17th century and into the 18th century, much attention was paid to the development of calculus. During that period, there was great emphasis on real numbers and continuous mathematics. By the time the 19th century had arrived and early into the 20th century, logic and the emphasis placed on valid arguments to establish the truth of mathematical statements achieved ever-increasing importance.

THE ROAD TO DISCRETE MATHEMATICS

By the middle of the 20th century, another kind of mathematics, quite unlike the continuous mathematics of calculus, had become prominent: discrete mathematics. Instead of being concerned with real numbers, this mathematics was more involved with integers and other "similar" collections of objects, as well as with finite collections.

Discrete mathematics became primarily the study of the mathematical properties of sets, systems and structures having a "countable" collection of elements. The growing popularity of and interest in discrete mathematics was greatly influenced by its applications to areas such as computer science, engineering, communication and transportation as well as its emergence as a fascinating area of mathematics.

Like all mathematics, the validity and understanding of discrete mathematics rely on logic, sets, functions and methods of proof. More unique to discrete mathematics, however, are topics involving procedures (algorithms) for solving problems and the number of objects possessing certain properties of interest – the area of discrete mathematics called enumerative combinatorics. Also, discrete mathematics is concerned with the study of relations and graph theory, a growing area of discrete mathematics with a much shorter history.

In the middle of the 20th century, a beginning college student, even those with a strong mathematical background, would likely take courses in college algebra, trigonometry and analytic geometry, followed then by a sequence in calculus. At the same time, high school seniors were taking courses in Euclidean geometry, advanced algebra and trigonometry. As time went by, it became commonplace for beginning college students to take courses in precalculus and perhaps calculus, while high school students were taking courses dealing with functions, statistics and continuous mathematics (an introduction to calculus).

The passing years saw first the introduction of, then advances in and now the bombardment of technology into college courses and everyday life. Computer science became a major area of study, especially computer programming, development of software, data structures and analysis of algorithms. Discrete mathematics is the mathematics underlying computer science. Indeed, this mathematics has become important for many areas of study and has developed into a major area of mathematics of its own.

A COURSE IN DISCRETE MATHEMATICS

Developing an appreciation of and learning any area of mathematics requires a comprehension of the concepts occurring in the area (each of which requires a clear definition) and a knowledge of certain valid statements, each of which is either an axiom (whose truth is accepted without proof) or a theorem (whose truth can be established with the aid of concepts and other statements). This is best accomplished if we can understand proofs of theorems and are able to write some proofs of our own. We will be introduced to several methods of proof that can be used to establish the truth of theorems as well as ways to show that statements are false. These methods are based on logic, which allow us to use reasoning to show that a given statement is true or false. Although it is not our intention to go into any of this in great depth, it is our goal to present enough details and examples so that a sound introduction to proofs can be obtained. Developing a good understanding of proofs requires a great deal of practice and experience and comprehending how others prove theorems. These are the subjects of Chapters 1–3.

Chapter 4 is devoted to an important proof technique that can be used to establish the truth of special kinds of statements: mathematical induction. This proof technique is frequently employed in computer science. The related topic of recurrence is discussed in Chapter 4 as well.

Chapter 5 deals with concepts that are fundamental to all areas of mathematics: relations and functions. The important subject of equivalence relations is introduced in this chapter. Functions that are both one-to-one and onto are especially important and are emphasized in the chapter.

Algorithms are discussed in Chapter 6. The concept of algorithms is well known to and encountered often in computer science. There are problems in which it is convenient to have a solution consisting of an efficient step-by-step set of instructions from which a computer program can be written. The number of steps needed to use some algorithms to solve certain kinds of problems is analyzed. Particular emphasis is paid to search and sort algorithms.

The numbers encountered most often in discrete mathematics are the integers. Chapter 7 is devoted to the study of integers, particularly their divisibility properties. Primes and their properties are discussed, as is the important topic of congruence. The Division Algorithm and Euclidean Algorithm are introduced and illustrated in this chapter as well. Integer representations and a discussion of the bases 2, 8 and 16 occur here.

Chapters 8–10 deal with the area of discrete mathematics called enumerative combinatorics: the study of counting. The basic principles of counting are introduced in Chapter 8: multiplication, addition, pigeonhole, inclusion-exclusion. In Chapter 8, permutations and combinations are also introduced. These two concepts give rise to a wide variety of counting problems. Chapter 9 discusses the concept of Pascal triangles and the related binomial theorem and presents more advanced counting problems, some through the subject of generating functions. Chapter 10 gives an introduction to discrete probability. Random variables and expected values are also presented in this chapter.

There are many ways in which a pair of elements can be compared or related. Particularly when elements belong to the same set, there are numerous occurrences of this idea in mathematics. One of these relations has the same basic properties of the equality of numbers and is called an equivalence relation (discussed in Chapter 5). Another relation has the same basic properties of the "less than or equal to" comparison of two numbers and is called a partial ordering. Partially ordered sets are discussed in Chapter 11. This gives rise to the structures referred to as lattices and Boolean algebras, which are also discussed in Chapter 11.

Chapters 12–15 deal with the area of graph theory, a major subject within discrete mathematics. Chapter 12 introduces the fundamental concepts of graph theory and concepts involving paths and cycles. This gives rise to the study of Eulerian graphs and Hamiltonian graphs. Chapter 13 is devoted to the important class of graphs called trees. Minimum spanning trees are discussed and two algorithms, Kruskal's algorithm and Prim's algorithm, are presented and illustrated. Depth-first search and breadth-first search are also discussed in Chapter 13. One of the most famous problems in graph theory (in fact, in mathematics) was the Four Color Problem. This problem involves the concepts of planar graphs and graph colorings, which are the major topics in Chapter 14. Chapter 15 deals with directed graphs, the most studied class being the tournaments. Directed graphs are useful in the study of finite-state machines, another concept discussed in Chapter 15.

TO THE STUDENT

- **Prerequisites**

  A background in algebra and precalculus (and recalling the major topics and techniques in these areas) is an important prerequisite for a course in discrete mathematics. It would also be helpful if the student had taken a course or two in calculus. It is probably most important, however, that a student enters a course in discrete mathematics with a desire for learning and a goal of doing the best that he or she can do. This certainly means attending class regularly, being an attentive student and doing homework faithfully. Being a successful student in discrete mathematics takes a certain amount of dedication. Discrete mathematics, unlike much of calculus and precalculus, may not build on other courses the student has already taken. For most students, this is an exciting new subject whose topics will be encountered often again if there are more mathematics, computer science or engineering courses in a student's future.

- **Concepts and Theorems**

  As with all areas of mathematics, discrete mathematics deals with the understanding of concepts and theorems. The definitions of all concepts are presented and, if the concept is considered a major concept, it is prominently displayed. Almost all theorems are proved. Care has been taken to give clear proofs. Time will not permit your instructor to prove all theorems in class but, especially if your instructor emphasizes proofs in class, it is a good idea to read the proofs of any theorems discussed in class.

- **Examples**

  This textbook contains a large number of examples that illustrate the concepts, theorems and various methods introduced. These include methods for solving certain kinds of problems and methods for proving the truth of statements. Before attempting any exercises that have been assigned and as part of a study plan to prepare for a forthcoming quiz or exam, it is useful to read and understand the examples presented. Prior to reading the examples, it is useful to review the concepts and their definitions and theorems that the exercises deal with.

- **Exercises**

  Each chapter contains a large number of exercises. There are exercises at the end of each section and additional exercises at the end of each chapter. The exercises range from routine (which are often similar to examples presented in the text that require a basic understanding of the concepts or theorems presented) to moderately challenging (which will require some thought on the student's part) to more challenging and innovative (which require coming up with ideas for solving the problem).

  Some students spend little time, if any, reading a textbook. This is not a good idea. Only when an exercise is assigned that they don't know how to do (and no classmate or the instructor is available for help) do such students search through the book, looking primarily for examples that are similar to the exercise assigned. It is far better to read the textbook and understand the examples in the textbook before attempting to do the exercises assigned. The authors have spent much time and effort to make this textbook clear and informative (and hopefully interesting as well). It's good to take advantage of this.

Answers or hints to odd-numbered exercises (both section exercises and chapter exercises) are given at the end of the textbook.

- **Chapter Highlights**

At the end of each chapter are Chapter Highlights. This includes a list of the major concepts introduced in the chapter (arranged alphabetically) with a brief definition of each concept. Also included are the major results introduced in the chapter (arranged in the order presented). When the chapter has been completed, it is probably a good idea to review the concepts and theorems covered in class to refresh your memory. It is also good to review these again prior to a quiz or an exam on this material.

TO THE INSTRUCTOR

- **Construction of Chapters**

Each chapter is divided into sections (and sometimes even further into topics) for the purpose of making it easier to decide which portions of a chapter to discuss each class period. With few exceptions, proofs of the theorems are provided. Even if it is decided not to present a proof in class, proofs can be read by students. Each chapter contains many examples. The instructor could present some of these in class. If the instructor prefers not to do this, other examples can be given. Possibly exercises could be presented to serve as examples.

- **Concepts and Theorems**

Each chapter contains several concepts and theorems central to the chapter. The definition of each concept is presented and definitions of the major concepts are prominently displayed. Proofs of most theorems are given. Care has been taken to give clear proofs. Many theorems are preceded by a "Proof Strategy," which provides a discussion of a plan of what needs to be proved and how we might be able to do this. Many proofs of theorems are followed by a "Proof Analysis," which is a discussion that reflects on a proof just given and emphasizes certain details that might have gone unnoticed. Occasionally, an example is followed by an "Analysis," which serves a similar purpose.

- **Exercises**

Each section is followed by a set of exercises dealing with the material in that section. In addition, each chapter is followed by a supplemental set of exercises dealing with the material in that chapter. There are a number of exercises that range from routine to moderately challenging so that it can be determined if the students have an understanding of the concepts, examples and theorems presented in the section or chapter. In addition, there are more innovative exercises to challenge students to think for themselves and come up with ideas of their own.

- **Solutions Manual for Instructors**

There is a *Solutions Manual for Instructors* (prepared by the authors) that contains detailed solutions of all exercises in the textbook (both section exercises and chapter exercises), including complete proofs of all exercises requesting a proof (not simply hints) and often explanations of solutions of exercises having a numerical answer (not simply the answer).

TEACHING A COURSE FROM THIS TEXTBOOK

There is enough material in this textbook for a 2-semester sequence in discrete mathematics but considerably more material than one would attempt to teach in a 1-semester course. What portions of the textbook to be covered depends, to a great degree, on what the instructor's goals are. If the instructor wants to emphasize logic and proofs, the first four chapters can be covered in some detail. It is likely that some of Chapter 5 (Relations and Functions) would be covered. In many instances, algorithms are covered in Computer Science courses. If this is the case, Chapter 6 can be omitted. How each chapter in this text relies on other chapters is indicated in the diagram (digraph) below.

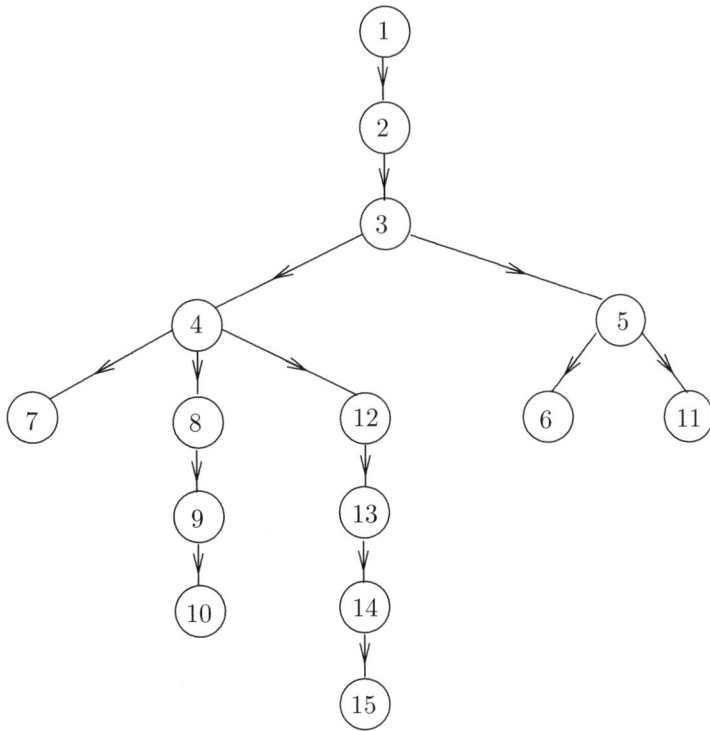

ACKNOWLEDGMENTS

The authors have taught discrete mathematics many times and have used various versions of notes that have led to this textbook. We thank many students and colleagues for their comments on and suggestions for these notes. This has had a profound impact on how this textbook was continually revised. The authors especially acknowledge and thank Neil Rowe, Publisher, Waveland Press, Inc., however, for his constant interest, support and communication involving this project.

Gary Chartrand and Ping Zhang

Chapter 0

What is Discrete Mathematics?

Before attempting to describe what discrete mathematics is or at least some of the topics it includes, let's first see what it isn't. It isn't continuous mathematics, which is studied in calculus. Calculus deals with the rate at which certain variables change. When a stone is dropped into water, the circumference of water ripples created by this expands continuously. When a ball is thrown in the air, its velocity changes continuously. If a clock has hour, minute and second hands, then these hands (often) revolve continuously.

When you were introduced to the whole numbers $0, 1, 2, 3, \ldots$ (the nonnegative integers) as a youngster and learned the rules of arithmetic involving them, you were being taught some discrete mathematics, albeit very elementary. The games and puzzles you learned and enjoyed were probably part of discrete mathematics as well. Digital clocks, which are so common these days, can be considered part of discrete mathematics.

While real numbers such as $\sqrt{2}$, π and e are so important to continuous mathematics, it's the nonnegative integers that mainly concern us in discrete mathematics. Functions such as x^2, $\sin x$ and e^x (where x is a real number) are of great importance to us in continuous mathematics, while n^2 and 2^n (where n is a positive integer) are important to us in discrete mathematics. The graphs of $y = x^2$ and $m = n^2$ are shown in Figure 1.

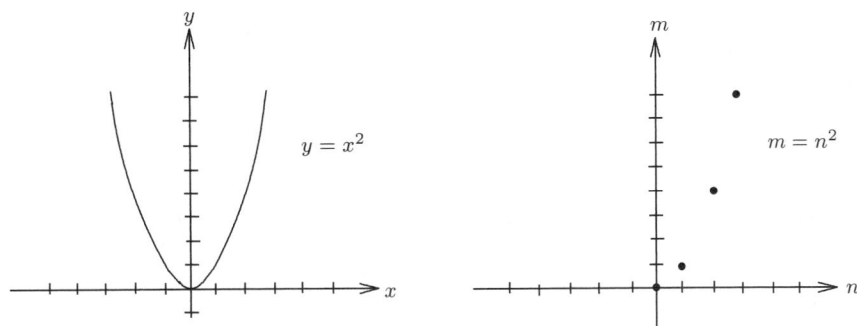

Figure 1: The graphs of $y = x^2$ and $m = n^2$

The exponential function $y = e^x$ is encountered often in calculus – perhaps not surprisingly since e is the base of the natural logarithm. This function also has the interesting property that its

1

derivative $\frac{dy}{dx} = e^x$ as well. So when $x = 1$, both the y-value and the slope of the tangent line to the graph of this function are e. While the importance of e^x may be clear in calculus, one might wonder what, if any, useful properties 2^n might have in discrete mathematics. Here we are considering the numbers $2^0, 2^1, 2^2, 2^3, \ldots$. Just as every positive integer can be expressed as sums of powers of 10, the same is true of powers of 2. For example,

$$432 = 10^2 + 10^2 + 10^2 + 10^2 + 10^1 + 10^1 + 10^1 + 10^0 + 10^0 = 2^8 + 2^7 + 2^5 + 2^4.$$

Suppose that a coin is flipped. Then there are two possibilities, namely heads (H) or tails (T). If the coin is flipped twice, there are $4 = 2^2$ possibilities: HH HT TH TT. If a coin is flipped three times, there are $8 = 2^3$ possibilities:

HHH HHT HTH HTT THH THT TTH TTT.

One might guess (correctly) that if a coin is flipped n times, then there are 2^n possibilities for the sequence of outcomes.

A concept that is encountered often in both continuous mathematics and discrete mathematics is that of a set. A *set* is a collection of objects, called the *elements* of the set. If a set S consists of the numbers 1 and 2, then we write this set as $S = \{1, 2\}$. A set T is called a *subset* of S if every element of T is an element of S. There are $4 = 2^2$ subsets of the set $S = \{1, 2\}$, namely

$$\{ \}, \{1\}, \{2\}, \{1, 2\}.$$

So one of the subset of S is $\{ \}$, called the *empty set*, and consists of no elements; while one of the subsets is S itself. There are $8 = 2^3$ subsets of the set $\{1, 2, 3\}$, namely

$$\{ \}, \{1\}, \{2\}, \{3\}, \{1, 2\}, \{1, 3\}, \{2, 3\}, \{1, 2, 3\}.$$

Here too one might correctly guess that there are 2^n subsets of a set with n elements.

A class of numbers that are encountered often are the primes, integers at least 2 whose only positive integer divisors are 1 and itself. The number 77 is the product of two different primes and its only positive integer divisors are the $4 = 2^2$ numbers $1, 7, 11, 77$. The number 30 is the product of three different primes and the $8 = 2^3$ positive integer divisors that divide it are

1, 2, 3, 5, 6, 10, 15, 30.

It turns out in this case as well that if a number A is the product of n different primes, then there are exactly 2^n different positive integers that divide A.

Suppose that in a gathering of people, every two people are either friends or strangers. Three people 1, 2, 3 happen to find themselves talking to one another. Then either one of these three people is friends with the other two or one of these people is strangers with the other two. Although this simple problem can be looked at in a variety of ways, one of these ways uses an area of discrete mathematics called graph theory, which we will visit in Chapters 12-15. Suppose that we represent each of the three people 1, 2 and 3 by a point (called a *vertex* in graph theory). We then join every two vertices by a line (called an *edge* in graph theory). If an edge joins two vertices representing two friends, we color this edge red (R) while if an edge joins two vertices representing two strangers we color this edge blue (B). There are $8 = 2^3$ different ways that the edges of this graph can be colored (see Figure 2). In every case, at least one of the three people is a friend of the other two or at least one of the three people is not acquainted with of the other two. If a graph has n edges and each edge is colored red and blue, then there are 2^n different ways that the edges of this graph can be colored.

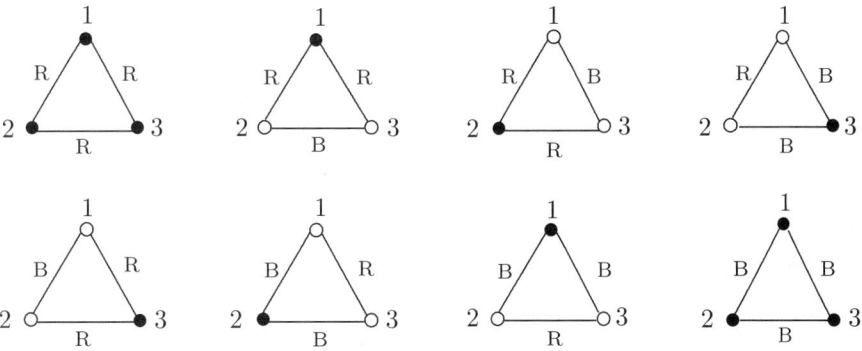

Figure 2: A graph model

There is a related problem, a version of which appeared in the 1953 Putnam Exam, a competitive exam that many undergraduates in the United States and Canada take. Suppose this time that we have a gathering of six people 1, 2, 3, 4, 5, 6 where, again, every two people are either friends or strangers. Then there is either some person who is friends with two others who are friends of each other or there is some person who is strangers with two others who are strangers of each other. Another way to state this fact is: There are three mutual friends or three mutual strangers. Yet another way to say this is: Every graph with six vertices, every two of which are joined by an edge, where each edge is colored red or blue, there is either a red triangle or a blue triangle. In this case, the graph has 15 edges. Since every edge can be colored either red or blue, there are $2^{15} = 32,768$ different colorings of this graph. Unlike the situation in Figure 2, where there are only 8 colorings of the graph, it is not practical to list all colorings of the graph and to check out each one individually to see that it has the desired property. There is, however, another approach – a more mathematical approach. Consider any coloring whatsoever of the six edges of the graph and consider any one of the six vertices, say vertex 1. Since there are five edges joining 1 to the other five vertices, three of these edges must be colored the same, say red (R). Suppose that these red edges join vertex 1 to the three vertices 2, 3 and 4 (see Figure 3).

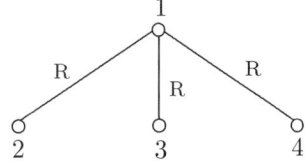

Figure 3: Another graph model

If any two of 2, 3 and 4 are joined by a red edge, a red triangle is produced; while if this does not occur, then there is a blue triangle with the vertices 2, 3 and 4.

Another modern application of discrete mathematics comes from the phrase:

Information is only a click away.

When some key words are typed into www.google.com, for example, the Google search engine goes into action and usually several web pages are listed (from a few to many). The web pages that are listed first are, in a certain sense, the most relevant. One of the best known and most influential methods (called an algorithm) for determining the relevance of web pages is the algorithm used by the Google search algorithm, which was developed by Sergey Brin and Larry Page, both graduate students at Stanford University at the time. (As of this writing, Page became CEO of Google in April 2011.) How relevant a certain web page P is to the key words typed at www.google.com can be determined by considering the web pages that have a link to P. For example, suppose that we create a certain web page A and this web page has a link to the web page B. This means that for the topic that web page A is dealing with, web page B is important. There may in fact be many pages that have a link to B, implying that it is common to believe that B is important. On the other hand, if only one web page, say C, has a link to B, then C is saying that B is the authoritative web page for this topic. In this manner, a "rank" can be assigned to each web page, determined by the ranks of the pages that have a link to it.

For this purpose, the World Wide Web can be represented by a huge graph (actually a "directed graph" in this case), where each web page is a vertex and there is a "directed" edge from vertex A to vertex B if there is a link from page A to page B. As an example, suppose that we have a very small internet, consisting of only four web pages:

www.A.com   www.B.com   www.C.com   www.D.com

Suppose that these appear as follows.

http://www.A.com	http://www.B.com
- - - - - - - - - -	- - - - - - - - - - -
- - - - - - - - - -	- - - - - - - - - - -
For more information	- - - - - - - - - - -
See:	For more information
http://www.B.com	See:
http://www.C.com	http://www.C.com
http://www.D.com	http://www.D.com

http://www.C.com	http://www.D.com
- - - - - - - - - -	- - - - - - - - - -
- - - - - - - - - -	- - - - - - - - - -
- - - - - - - - - -	For more information
For more information	See:
See:	http://www.A.com
http://www.A.com	http://www.C.com

The directed graph that represents this situation is shown in Figure 4.

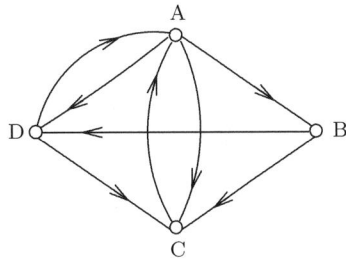

Figure 4: An internet directed graph

The directed graph is Figure 4 is now transformed into a "weighted" directed graph where each directed edge is assigned a number (a "weight"). Since vertex A has three outgoing directed edges, a weight of $\frac{1}{3}$ is assigned to each such directed edge to indicate that $\frac{1}{3}$ of its importance is passed on to the other three vertices (web pages). This gives us the weighted directed graph in Figure 5.

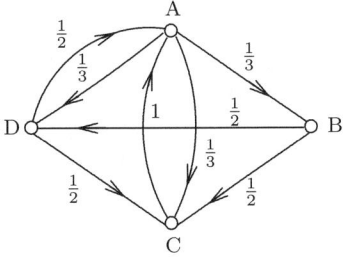

Figure 5: An internet weighted directed graph

The information supplied by Figure 5 can also be given in the table (the matrix)

$$\begin{bmatrix} 0 & 0 & 1 & \frac{1}{2} \\ \frac{1}{3} & 0 & 0 & 0 \\ \frac{1}{3} & \frac{1}{2} & 0 & \frac{1}{2} \\ \frac{1}{3} & \frac{1}{2} & 0 & 0 \end{bmatrix}$$

A deeper look at this information requires more mathematics, such as the subject of linear algebra.

While the mathematics that serves as the foundation of physics and engineering is continuous mathematics and this will always be important, it is discrete mathematics that is the foundation of computer science. Here we are concerned with digital logic: 1 or 0, true or false, yes or no, heads or tails, red or blue. While technology has moved from analog only to primarily digital, it has become increasingly important to understand not only the classical continuous mathematics but the more modern mathematics of the 21st century: discrete mathematics.

Chapter 1

Logic

While calculus is an area of continuous mathematics, in which the primary numbers of interest are the real numbers, discrete mathematics is a branch of mathematics in which the numbers encountered most often are the integers. In mathematics, it is not numbers that are most important to us, however. Every area of mathematics deals with concepts and topics that are relevant to that area. We are often interested in how these concepts are related to one another. These relationships are expressed as statements. It is not only important to understand what these statements are saying, it is essential to know that these statements are true and how to convince ourselves of this. Verifying that statements are true deals with the topic of proof. There are specific methods for proving that mathematical statements are true. In order to have an understanding of the various methods of proofs, it is necessary to have a knowledge of the concept of sets, which will be discussed in the next chapter, and a knowledge of logic, which is the subject of this chapter.

It is through logic that we are able to build on what we know. The famous physicist Albert Einstein once stated that logic can carry us only so far:

> *Logic will get you from A to B. Imagination will take you everywhere.*

Progress in every area is due to the imagination and creativity of people – often only a few people. There is certainly no formula as to how ideas occur to people. However, being aware of what is already known as well as learning to ask probing questions are important initial steps. The well-known writer J. K. Rowling (author of the *Harry Potter* novels) considered this topic herself and wrote:

> *Sometimes ideas just come to me. Other times I have to sweat and almost bleed to make ideas come. It's a mysterious process, but I hope I never find out exactly how it works.*

1.1   Statements

We are about to embark on a study of discrete mathematics. Throughout our study we will encounter many statements. It is not only essential that we understand what these statements are saying but how statements can be used. As expected, it is crucial that these statements are true. While there are instances when we might simply agree to accept the truth of certain statements, more often than not such statements must be shown to be true. If it is someone else who has verified the truth of a statement, then the method used to do this must be convincing. Sometimes, we may need to verify the truth of statements ourselves. At the base of all this is an understanding of what is meant by a statement.

In English grammar, sentences are divided into categories. In a **declarative sentence**, something is being declared or asserted; in an **interrogative sentence**, a question is being asked; in an **imperative sentence**, a command is given; while in an **exclamatory sentence**, an emotional expression is made.

Example 1.1

These four kinds of sentences are illustrated below.

1. Los Angeles is the capital of California. (declarative sentence)

2. Broadway is a street in New York City. (declarative sentence)

3. Which state capital has the largest population? (interrogative sentence)

4. Drive to Main Street and then turn right. (imperative sentence)

5. I can't believe it! (exclamatory sentence)

6. It was then that he arrived there. (declarative sentence)   ♦

We use the symbol ♦ to indicate the conclusion of a discussion of an example or of a solution of a problem. Although Example 1.1 presents four different types of sentences, it is the declarative sentences that will be of greatest interest to us.

Definition 1.2   *A* **statement** *is a declarative sentence that is either true or false but not both.*

Truth Values

The subject of logic deals with statements, what they say and the reasoning one uses to verify that they are true or to show that they are false. (As in many parts of mathematics, the terminology that is used is not always universal. For example, some authors refer to statements as **propositions**.) Every statement has a **truth value**. The truth value of a true statement is **true**, denoted by T; while the truth value of a false statement is **false**, denoted by F.

In Example 1.1, sentence 1 is a false statement and so the truth value of this statement is F. The capital of California is the city of Sacramento, not Los Angeles. Sentence 2 is a true statement as Broadway is a major street in New York City and the location of many theatres, and so the truth value of this statement is T. It should be noted that when the truth value of a statement is being determined, one may need to consider the context of the sentence. For example, it is conceivable that the capital could be moved to Los Angeles some day (unlikely) or that the street Broadway in New York City may be renamed some day (also unlikely).

A sentence that is interrogative, imperative or exclamatory does not have a truth value; thus none of these kinds of sentences are statements. Therefore, none of the sentences 3–5 are statements. Although sentence 6 is a declarative sentence, it is not a statement. Sentence 6 refers to a date (then), a person (he) and a place (there). Without knowing the date referred to, who the person is and what place is being referred to, we cannot determine whether sentence 6 is true or false and so this sentence is not a statement. That is, this sentence contains variables (namely, "then," "he" and "there") and without knowing the value of each variable, the sentence is not a statement.

Of course, none of the sentences 1–6 in Example 1.1 has anything to do with mathematics. Let's look at some that do.

Example 1.3

Classify the following mathematical sentences as declarative, interrogative, imperative or exclamatory. Which declarative sentences are statements and which statements are true?

1. Is it true that $\frac{-9}{3} + \frac{4}{2} = \frac{-9+4}{3+2}$?

2. Multiply the numbers 2/3 and 9/10.

3. $1.414 \neq \sqrt{2}$.

4. $\sqrt{(-3)^2} = -3$.

5. What a difficult calculus exam!

6. $3x + 1 = 7$.

Solution. Sentence 1 is interrogative, sentence 2 is imperative, sentences 3 and 4 are declarative, sentence 5 is exclamatory and sentence 6 is declarative. Sentences 3 and 4 are statements. Statement 3 is true as $\sqrt{2}$ does *not* equal 1.414. The number 1.414 is only an approximation of $\sqrt{2}$. Statement 4 is false as $\sqrt{(-3)^2} = \sqrt{9} = 3$. Sentence 6, although declarative, is not a statement. It is impossible to assign a truth value to it without knowing the value of x. If $x = 2$, then sentence 6 is a true statement; while for any number $x \neq 2$, sentence 6 is a false statement. ♦

Open Sentences

Many sentences encountered in mathematics contain variables. These kinds of sentences bring up additional questions.

Definition 1.4 *An **open sentence** is a declarative sentence containing one or more variables and whose truth or falseness depends on the values of these variables.*

Sentence 6 in Example 1.3 is therefore an open sentence, not a statement. In fact, sentence 6 in Example 1.1 is also an open sentence.

We often denote statements or open sentences by symbols, typically representing statements by P, Q, R or P_1, P_2, P_3 and so on.

Example 1.5

Consider the two statements

$$P_1 : |-2| = -2. \quad P_2 : (-1)^3 = -1.$$

Statement P_1 is false since the absolute value of -2 is 2, not -2. However, statement P_2 is true. ♦

An open sentence containing a variable x is typically represented by $P(x)$ or some similar notation. If x represents an object in a collection S of objects, then S is called the **domain** of x and $P(x)$ is referred to as an **open sentence over the domain** S. The domain S of a variable is often a collection of numbers of some type. (The collection S is called a *set* in mathematics, which is the main subject of Chapter 2.) If an open sentence P contains two variables x and y, where the domain of x is X and the domain of y is Y, then we typically write P as $P(x, y)$, where then x is an object of X and y is an object of Y. Of course, an open sentence may contain more than two variables.

Let's look at some open sentences.

Example 1.6

Consider the open sentence

$$P(x) : 3x - 9 = 0$$

where the element x represents a real number, that is, the domain of x is the collection of real numbers. Thus

$$P(3): \ 3 \cdot 3 - 9 = 0$$

is a true statement and

$$P(-3): \ 3(-3) - 9 = 0$$

is a false statement.   ♦

Example 1.7

Consider the open sentence

$$Q(a,b): \quad 3a + 5b \text{ is even}$$

where a and b are integers, that is, the domain of both a and b is the collection of integers. Therefore, the two statements

$$Q(1,-3): 3 \cdot 1 + 5 \cdot (-3) = -12 \text{ is even.} \quad Q(6,2): 3 \cdot 6 + 5 \cdot 2 = 28 \text{ is even.}$$

are true; while

$$Q(5,4): 3 \cdot 5 + 5 \cdot 4 = 35 \text{ is even}$$

is a false statement.   ♦

Example 1.8

We encountered the open sentence

$$\text{It was then that he arrived there.}$$

in Example 1.1. This open sentence could be expressed as

$$P(\text{then, he, there}): \quad \text{It was then that he arrived there.}$$

for example. The truth or falseness of $P(\text{then, he, there})$ depends on the values of the variables "then," "he" and "there." For example, the statement
$P(\text{on 20 July 1969, Neil Armstrong, on the moon}):$

$$\text{It was on 20 July 1969 that Neil Armstrong arrived on the moon.}$$

is true; while the statement
$P(\text{on 4 July 1776, Christopher Columbus, in Cleveland}):$

$$\text{It was on 4 July 1776 that Christopher Columbus arrived in Cleveland.}$$

is false.   ♦

Example 1.9

Each of the following is an open sentence, where n represents a positive integer. For which positive integers n are the following open sentences true statements?

(a) $P_1(n): \; |n+1| = 0.$

(b) $P_2(n): \; n^2 - 4 \leq 0.$

(c) $P_3(n): \; n^2 - n - 6 = 0.$

(d) $P_4(n): \; n + \frac{1}{n} \leq 2.$

Solution. Let n be a positive integer.

(a) Observe that $|n+1| = 0$ only when $n+1 = 0$ and $n+1 = 0$ only when $n = -1$. However, -1 is not a positive integer and so there is no positive integer n for which $P_1(n)$ is a true statement.

(b) Since $n^2 - 4 \le 0$ can be written as $n^2 \le 4$ and this inequality holds when $-2 \le n \le 2$, the only positive integers n for which $n^2 - 4 \le 0$ is true are therefore $n = 1$ and $n = 2$. Thus $n = 1$ and $n = 2$ are the only positive integers for which $P_2(n)$ is a true statement. That is, $P_2(1)$ and $P_2(2)$ are true, but $P_2(n)$ is false when n is any other positive integer.

(c) Since $n^2 - n - 6 = (n - 3)(n + 2)$, it follows that $n^2 - n - 6 = 0$ when $(n - 3)(n + 2) = 0$ and so $n = 3$ or $n = -2$. However, since n is a positive integer, $n^2 - n - 6 = 0$ only when $n = 3$. Therefore, $P_3(3)$ is true but $P_3(n)$ is false for every positive integer n different from 3.

(d) For $n = 1$, we have $n + \frac{1}{n} = 1 + 1 = 2$. When $n \ge 2$, however, $n + \frac{1}{n} > 2$. Therefore, the only positive integer n for which $n + \frac{1}{n} \le 2$ is $n = 1$. So $P_4(1)$ is true but $P_4(n)$ is false when n is an integer with $n \ge 2$.   ◆

Example 1.10

For an integer n, consider the open sentence

$$Q(n): \ 4n + 2 \text{ is even.}$$

Determine all those integers n for which $Q(n)$ is a true statement.

Solution. For every integer n, observe that the integer $4n + 2$ is even. Therefore, $Q(n)$ is a true statement for *every* integer n.   ◆

Open sentences that are true statements for every value of each variable in its domain will be of special interest to us. We will visit these sentences again soon and often.

Exercises for Section 1.1

1. Classify each of the following sentences as declarative, interrogative, imperative or exclamatory. Which declarative sentences are statements? Give the truth value for each statement.

   (a) Do you know who teaches this class?

   (b) What bad luck!

   (c) Clemson University is in South Carolina.

   (d) Please give me a copy of the syllabus.

   (e) She sold her book to what's-his-name.

   (f) South Dakota is south of North Carolina.

   (g) Can you believe that I have to take this class?

   (h) Tuition will decrease next year.

   (i) He told me that he understood mathematics.

2. Classify each of the following mathematical sentences as declarative, interrogative, imperative or exclamatory. Which declarative sentences are statements? Give the truth value for each statement.

   (a) Find a common denominator of 3/8 and 5/12.

   (b) $\pi = 22/7$.

   (c) $x^2 - 4 = 0$.

   (d) Is 5 a root of the equation $x^2 - 2x - 15 = 0$?

   (e) The Pythagorean Theorem is an astonishing theorem!

   (f) $\sqrt{2} < 1.5 < \sqrt{3}$.

(g) $\sqrt{n^2}$ is an integer.

3. Classify the following mathematical sentences as declarative, interrogative, imperative or exclamatory. Which declarative sentences are statements? Give the truth value for each statement.

   (a) The integer 2^3 is even.
   (b) The integer 0 is odd.
   (c) Is $5 \times 2 = 10$?
   (d) $x(x+1) = 12$.
   (e) Multiply $x + y$ by 7.
   (f) $7x$ is an odd integer.
   (g) What an impossible homework problem!

4. Each of the following is an open sentence, where n denotes an integer. For which integers n are the following sentences true statements?

   (a) $P_1(n)$: $|n - 2| = 3$.
   (b) $P_2(n)$: $2n^2 - 5n + 2 = 0$.
   (c) $P_3(n)$: $n^2 > 0$.
   (d) $P_4(n)$: $12/n^2$ is an integer.
   (e) $P_5(n)$: $n^3 = 2n$.
   (f) $P_6(n)$: $|n| + |n - 1| + |n - 2| = 3$.

5. For an integer n, consider the open sentence $P(n)$: $n^2 - 4n + 1$ is an odd integer.

   (a) Give an example of an integer n such that $P(n)$ is true.
   (b) Give an example of an integer n such that $P(n)$ is false.

6. For an integer n, consider the open sentence $P(n)$: $\left(\frac{1}{2}\right)^n$ is an integer.

   (a) Give an example of an integer n such that $P(n)$ is false.
   (b) Give an example of an integer n such that $P(n)$ is true.

7. For a real number x, consider the open sentence $P(x)$: $x(x - 5) = 6$. For which values of x is $P(x)$ a true statement?

8. For integers x and y, consider the open sentence $Q(x, y)$: $(x + 2)(y - 1)^3(2x - 1) = 0$. For which values of x and y is $Q(x, y)$ a true statement?

9. For integers x, y and z, consider the open sentence

$$R(x, y, z): \ (2x - 1)^2(z - 2)^2 + (3y + 1)^2(2z - 1)^2 + (3x - 2)^2(2y + 1)^2 \le 8.$$

   For which values of x, y and z is $R(x, y, z)$ a true statement?

10. For the following open sentences, each of variables x, y and z is 1, 2 or 3.

$$P(x, y): \ xy \text{ is even}, \quad Q(x, z): \ x + z \text{ is odd}, \quad R(y, z): \ 2^{y-2}2^{z-2} \text{ is odd}.$$

   Find distinct values of x, y and z such that all of the statements resulting from these open sentences are true.

11. For positive integers x and y, consider the following open sentences:

$$P(x,y): \ x^2y \geq 4, \ \ Q(x,y): \ x + 2y \leq 4, \ \ R(x,y): \ x^y > 1.$$

Find positive integer values for each of x and y such that all of the statements resulting from these open sentences are true but when the values of x and y are interchanged, all of the statements are false.

12. For each of the following statements, construct an open sentence $P(x)$, $P(x,y)$ or $P(x,y,z)$, where x, y and z are variables, such that for some choice of values of the variables we obtain the original statement and for a different choice of values of the variables, it is possible to obtain a statement whose truth value is different from that of the original statement.

  (a) The square of 3 is larger than the cube of 2.
  (b) The population of New York City is greater than the population of Wyoming.
  (c) The people of China ordinarily speak Chinese.
  (d) $4 \cdot 3 + 8 = 30$.
  (e) John Kennedy was President of the United States.

1.2 Negation, Conjunction and Disjunction

Much of our interest in numbers comes not only from the numbers themselves but from properties they possess through performing certain operations on them, such as addition, subtraction and multiplication. A similar comment can be made about statements. That is, our interest in statements is greatly influenced by the ways statements can be combined to produce new statements.

New statements can be obtained from given statements in a variety of ways. Furthermore, the truth values of these new statements depend on and are defined in terms of the truth values of the given statements. Some of the most common examples of this are described next.

Truth Tables

A table that gives all possible assignments of truth values for a statement or statements under consideration is called a **truth table**. The truth table for a single statement P is shown in Figure 1.1. This table has two rows, giving the two possible truth values that P can have, namely true (T) or false (F). If Q is a statement, then here too Q can either be true or false. The table of possible truth values for Q is also given in Figure 1.1. If we happen to be considering the two statements P and Q simultaneously, then there are $4 = 2^2$ possible combinations of truth values for P and Q, which we give in the (standard) order: TT, TF, FT, FF. This is also shown in Figure 1.1. Therefore, this table has four rows.

P
T
F

Q
T
F

P	Q
T	T
T	F
F	T
F	F

Figure 1.1: Truth tables giving the different assignments of truth values
for P and Q separately and for P and Q simultaneously

Example 1.11

Suppose that Janet is a college student who took a programming course last semester. She took a final exam at the end of the semeste. Consider the following two statements:

P: Janet studied for the exam.
Q: Janet received an A on the exam.

For these two statements, the four rows in the third table in Figure 1.1 correspond, respectively, to

(1) Janet studied for the exam and received an A on it.

(2) Janet studied for the exam but didn't receive an A on it.

(3) Janet didn't study for the exam but received an A on it.

(4) Janet didn't study for the exam and didn't receive an A on it.   ♦

If there were a third statement R to consider at the same time as P and Q, then there would be eight rows in the table that considers all $8 = 2^3$ combinations of truth values for P, Q and R (see Exercise 1).

Negation

Of course, a new number can (ordinarily) be obtained from a given number simply by taking the negative of the given number. There is a similar situation for statements.

Definition 1.12 *The **negation** of a statement P is the statement*

not P (*or* It is not the case that P)

written in symbols as $\sim P$.

Thus the **negation operator** $\sim$ produces a new statement from a single given statement.
    The truth table for the negation of a statement is shown in Figure 1.2. Therefore, for every statement P, the truth value of its negation $\sim P$ is opposite to that of P.

P	$\sim P$
T	F
F	T

Figure 1.2: The truth table for negation

Example 1.13

Let's consider statements 1 and 2 in Example 1.1, which we now denote by Q_1 and Q_2:

Q_1 : Los Angeles is the capital of California.
Q_2 : Broadway is a street in New York City.   (1.1)

The negation of Q_1 is

$\sim Q_1$: It is not the case that Los Angeles is the capital of California.

It sounds much better, however, if we say

$$\sim Q_1 : \text{ Los Angeles is } not \text{ the capital of California.}$$

The negation of Q_2 is

$$\sim Q_2 : \text{ Broadway is } not \text{ a street in New York City.}$$

Notice that Q_1 is false and $\sim Q_1$ is true, while Q_2 is true and $\sim Q_2$ is false. This was anticipated by the truth table in Figure 1.2. ♦

We can also take the negation of open sentences, although there is no truth value associated with these since in each case we are only creating a new open sentence.

Example 1.14

For a real number x, the negation of the open sentence $P(x) : (x-2)^2 > 0$ is

$$\sim P(x) : (x-2)^2 \le 0.$$

Although $P(x)$ is not a statement, it *is* a statement for each specific real number x. For example, both $P(3)$ and $P(2)$ are statements. In fact, $P(3)$ is true and $P(2)$ is false. As expected, $\sim P(3)$ is false and $\sim P(2)$ is true. ♦

Conjunction

New statements can also be created from two given statements in a variety of ways. We now look at two of the best known ways of doing this.

Definition 1.15 *For two statements P and Q, the* **conjunction** *of P and Q is the statement*

$$P \textbf{ and } Q$$

denoted by $P \wedge Q$.

The truth table for $P \wedge Q$ is given in Figure 1.3. Therefore, $P \wedge Q$ is true only when both P and Q are true.

P	Q	$P \wedge Q$
T	T	T
T	F	F
F	T	F
F	F	F

Figure 1.3: The truth table for conjunction

Example 1.16

For the two statements Q_1 and Q_2 in (1.1), the conjunction of Q_1 and Q_2 is the statement

$$Q_1 \wedge Q_2: \text{ Los Angeles is the capital of California } and$$
$$\text{Broadway is a street in New York City.}$$

According to the truth table in Figure 1.3, the statement $Q_1 \wedge Q_2$ is false since Q_1 is false (even though Q_2 is true).

The following statement Q_3 is true:

$$Q_3: \text{ Walt Disney World is located in Florida.}$$

Since Q_2 and Q_3 are both true, the conjunction of Q_2 and Q_3 is the following true statement:

$$Q_2 \wedge Q_3 : \quad \text{Broadway is a street in New York City and} \\ \text{Walt Disney World is located in Florida.} \quad \blacklozenge$$

Disjunction

A new statement can also be produced from two given statements by combining the given statements with the word "or."

Definition 1.17 *For two statements P and Q, the **disjunction** of P and Q is the statement*

$$P \text{ or } Q$$

denoted by $P \vee Q$.

The truth table for $P \vee Q$ is given in Figure 1.4. According to this truth table, $P \vee Q$ is false only when both P and Q are false. Therefore, $P \vee Q$ is true if at least one of P and Q is true.

P	Q	$P \vee Q$
T	T	T
T	F	T
F	T	T
F	F	F

Figure 1.4: The truth table for disjunction

The word "or" is used in everyday conversation in two ways. For example, if you tell one of your friends,

> "Before attending my discrete math class tomorrow, I'm stopping at the computer lab or going to the library."

and it turns out that you visit both the computer lab and the library, then what you said is true. What you said was not to be interpreted that if you went to one of these two places, then you were absolutely not going to the other.

There are instances, however, when the word "or" carries with it a different interpretation. Perhaps you've been saving money to buy a new car and you've decided to buy a Ford or a Toyota. There is certainly no intention that you will buy both.

Inclusive and Exclusive Or

Therefore, for two statements P and Q, sometimes P *or* Q has the intended interpretation that P and Q may both be true; while at other times P *or* Q has the intended interpretation that P and Q cannot both be true. In the first instance, P *or* Q is true if at least one of P and Q is true. This is referred to as the **inclusive or**. (In English usage, we often say or write "P and/or Q.") This is how the disjunction $P \vee Q$ is to be interpreted. Of course, if it's impossible for both P and Q to be true, then this has no effect on the truth value of $P \vee Q$. On the other hand, should our intended use of "or" in P *or* Q be that we only want P *or* Q to be true if *exactly one* of P and Q is true, then this is referred to as the **exclusive or** (as it excludes P and Q from both being true). It is ordinarily clear which use of "or" is intended. In mathematics, however, the single word "or" is understood to mean the "inclusive or." Indeed, the "exclusive or" is seldom encountered in mathematics.

Definition 1.18 *For two statements P and Q, the **exclusive or** of P and Q is the statement*

P or Q but not both

and is denoted by $P \oplus Q$.

The truth table for $P \oplus Q$ is given in Figure 1.5. Thus $P \oplus Q$ is true only when exactly one of P and Q is true.

P	Q	$P \oplus Q$
T	T	F
T	F	T
F	T	T
F	F	F

Figure 1.5: The truth table for exclusive or

As with negation, the conjunction, disjunction and exclusive or of two open sentences can also be formed. As is probably expected, the result of performing either of these operations produces yet another open sentence.

Example 1.19

For an integer n, consider the two open sentences

$$P(n): \ n^3 + 2n \text{ is even.} \quad Q(n): \ n^2 - 4 < 0.$$

Then the conjunction, disjunction and exclusive or of $P(n)$ and $Q(n)$ are the open sentences:

$P(n) \wedge Q(n)$: $n^3 + 2n$ is even *and* $n^2 - 4 < 0$.
$P(n) \vee Q(n)$: $n^3 + 2n$ is even *or* $n^2 - 4 < 0$.
$P(n) \oplus Q(n)$: $n^3 + 2n$ is even *or* $n^2 - 4 < 0$ *but not both.*

Now, $P(1) \wedge Q(1)$ is false, while $P(1) \vee Q(1)$ is true. On the other hand, both $P(3) \wedge Q(3)$ and $P(3) \vee Q(3)$ are false statements. Furthermore, $P(0) \oplus Q(0)$ is false, while $P(1) \oplus Q(1)$ and $P(2) \oplus Q(2)$ are both true. ◆

Example 1.20

For an integer n, consider the two open sentences

$$P(n): \ n^3 - n = 0. \quad Q(n): \ (3n - 2)^2 \geq 20.$$

(a) Determine all integers n for which $P(n)$ is a true statement.

(b) Determine all integers n for which $Q(n)$ is a true statement.

(c) Investigate the truth values of $P(n) \wedge Q(n)$, $P(n) \vee Q(n)$ and $P(n) \oplus Q(n)$ for various integers n.

Solution.

(a) Since $n^3 - n = n(n^2 - 1) = n(n - 1)(n + 1)$, it follows that $n^3 - n = 0$ when $n = -1$, $n = 0$ or $n = 1$. Therefore, $P(n)$ is a true statement when $n = -1$, $n = 0$ or $n = 1$ and is a false statement for all other integers.

(b) The only integers n for which $(3n - 2)^2 < 20$ are $n = 0$, $n = 1$ and $n = 2$. Therefore, $Q(n)$ is a false statement when $n = 0$, $n = 1$ or $n = 2$ and is a true statement for all other integers n.

(c) Since $P(-1)$ and $Q(-1)$ are both true, so is $P(-1) \wedge Q(-1)$. On the other hand, $P(1)$ is true and $Q(1)$ is false. Therefore, $P(1) \wedge Q(1)$ is false.

We have seen that the only integers n for which $Q(n)$ is false are $n = 0$, $n = 1$ and $n = 2$. However, $P(0)$ and $P(1)$ are both true. Therefore, $P(n) \vee Q(n)$ is a true statement for $n = 0$ and $n = 1$. Since $P(2)$ and $Q(2)$ are both false, $P(2) \vee Q(2)$ is false.

We have observed that $P(n)$ is true for $n = 0$ and $n = 1$, while $Q(n)$ is false for these values of n. Therefore, both $P(0) \oplus Q(0)$ and $P(1) \oplus Q(1)$ are true statements. Since $P(n)$ and $Q(n)$ are both true for $n = -1$ and both false for $n = 2$, it follows that $P(-1) \oplus Q(-1)$ and $P(2) \oplus Q(2)$ are false statements.   ◆

Compound Statements

The operations $\sim$, $\wedge$, $\vee$ and $\oplus$ are examples of **logical connectives**. We will see other logical connectives shortly. A **compound statement** is a statement constructed from one or more given statements (called **component statements** in this context) and one or more logical connectives. For example, $P \wedge Q$ and $P \vee Q$ can be considered as compound statements expressed in terms of the component statements P and Q.

If R and S are compound statements constructed from one or more logical connectives and from the two component statements P and Q, say, then once we know the truth value of each of P and Q, the truth value of each of R and S can be determined. If we should find that the truth values of R and S are the same for each combination of truth values of P and Q, then this is of special interest to us.

Definition 1.21 *Two compound statements R and S constructed from the same component statements are called* **logically equivalent** *if R and S have the same truth value for all combinations of truth values of their component statements. If R and S are logically equivalent, then we write $R \equiv S$; while if R and S are not logically equivalent, then we write $R \not\equiv S$.*

You might have noticed that $P \wedge Q$ and $Q \wedge P$ have the same truth value for each combination of truth values of the two statements P and Q. The same can be said for $P \vee Q$ and $Q \vee P$. The observations concerning $P \wedge Q$ and $Q \wedge P$ as well as $P \vee Q$ and $Q \vee P$ provide us with two laws of logic, which are verified by the truth tables shown in Figure 1.6. These two laws are expressed below.

Theorem 1.22 (Commutative Laws) *For every two statements P and Q,*

$$P \wedge Q \equiv Q \wedge P \text{ and } P \vee Q \equiv Q \vee P.$$

P	Q	$P \wedge Q$	$Q \wedge P$
T	T	**T**	**T**
T	F	**F**	**F**
F	T	**F**	**F**
F	F	**F**	**F**

P	Q	$Q \vee P$	$P \vee Q$
T	T	**T**	**T**
T	F	**T**	**T**
F	T	**T**	**T**
F	F	**F**	**F**

Figure 1.6: Verifying the Commutative Laws for Conjunction and Disjunction

It follows by Theorem 1.22 therefore that if $P(n)$ and $Q(n)$ are open sentences over some domain S, then for each value of n in S, the truth value of $P(n) \wedge Q(n)$ is the same as the truth value of $Q(n) \wedge P(n)$. A similar remark holds for $P(n) \vee Q(n)$ and $Q(n) \vee P(n)$.

Example 1.23

According to the commutative law of disjunction, there is no difference (in terms of logic) between saying (or writing)

$$n^3 + 2n \text{ is even or } n^2 - 4 < 0$$

and saying (or writing)

$$n^2 - 4 < 0 \text{ or } n^3 + 2n \text{ is even.} \qquad \blacklozenge$$

De Morgan's Laws

There are two extremely useful laws of logic that combine all three of the logical connectives negation, disjunction and conjunction. These laws, called De Morgan's Laws, are named for Augustus De Morgan, a famous mathematician of the mid-19th century. We will encounter him again but for the present, we are most interested in laws of logic associated with him.

Theorem 1.24 (**De Morgan's Laws**) *For every two statements P and Q,*

(a) $\sim (P \vee Q) \equiv (\sim P) \wedge (\sim Q)$.

(b) $\sim (P \wedge Q) \equiv (\sim P) \vee (\sim Q)$.

These laws can be verified quite simply by means of truth tables. The first of these laws is established in Figure 1.7. The entries in the columns headed by $\sim (P \vee Q)$ and $(\sim P) \wedge (\sim Q)$ are in bold to call attention to these columns, which establish the logical equivalence.

P	Q	$P \vee Q$	$\sim (P \vee Q)$	$\sim P$	$\sim Q$	$(\sim P) \wedge (\sim Q)$
T	T	T	**F**	F	F	**F**
T	F	T	**F**	F	T	**F**
F	T	T	**F**	T	F	**F**
F	F	F	**T**	T	T	**T**

Figure 1.7: Verifying De Morgan's Law: $\sim (P \vee Q) \equiv (\sim P) \wedge (\sim Q)$

Example 1.25

Use De Morgan's Laws to express the negation of the following.

(1) Either I get this job or I take another class.

(2) I'm eating dinner out and going to a movie.

Solution.

(1) I don't get this job and I don't take another class.

(2) I'm not eating dinner out or I'm not going to a movie. $\qquad \blacklozenge$

Example 1.26

For two integers a and b, consider the two sentences

$$P:\ a \text{ is odd.} \quad Q:\ b \text{ is odd.}$$

(1) State the disjunction $P \vee Q$ of P and Q.

(2) State the negation $\sim (P \vee Q)$ of $P \vee Q$ by using the phrase "it is not the case that."

(3) Use an appropriate De Morgan's Law to restate $\sim (P \vee Q)$.

(4) State the conjunction $P \wedge Q$ of P and Q.

(5) State the negation $\sim (P \wedge Q)$ of $P \wedge Q$ by using the phrase "it is not the case that."

(6) Use an appropriate De Morgan's Law to restate $\sim (P \wedge Q)$.

Solution.

(1) The disjunction of P and Q is

$$P \vee Q:\ a \text{ is odd } or \text{ } b \text{ is odd. (or: Either } a \text{ or } b \text{ is odd.)}$$

(2) The negation of $P \vee Q$ is

$$\sim (P \vee Q): \text{ It is not the case that } a \text{ is odd or } b \text{ is odd.}$$

(3) By De Morgan's Laws,

$$\sim (P \vee Q) \equiv (\sim P) \wedge (\sim Q):\ a \text{ is not odd } and \text{ } b \text{ is not odd.}$$
$$\text{(or: Both } a \text{ and } b \text{ are even.)}$$

(4) The conjunction of P and Q is

$$P \wedge Q:\ a \text{ is odd } and \text{ } b \text{ is odd. (or: Both } a \text{ and } b \text{ are odd.)}$$

(5) The negation of $P \wedge Q$ is

$$\sim (P \wedge Q): \text{ It is not the case that both } a \text{ and } b \text{ are odd.}$$

(6) By De Morgan's Laws,

$$\sim (P \wedge Q) \equiv (\sim P) \vee (\sim Q):\ a \text{ is not odd } or \text{ } b \text{ is not odd.}$$
$$\text{(or: Either } a \text{ or } b \text{ is even.)} \blacklozenge$$

Example 1.27

For a real number x, let $P(x):\ x^2 - 8x + 15 = 0$.

(a) Use the word "or" to describe those real numbers x for which $P(x)$ is true.

(b) Use De Morgan's Laws to describe those real numbers x for which $P(x)$ is false.

Solution. Let x be a real number. Observe that $x^2 - 8x + 15 = (x - 3)(x - 5)$.

(a) $P(x)$ is true when $x = 3$ *or* $x = 5$.

(b) $P(x)$ is false when $x \neq 3$ *and* $x \neq 5$. ♦

Not surprisingly, for any statement P, the statements P and $\sim (\sim P)$ are logically equivalent. As expected, this can be verified by means of a truth table (see Figure 1.8). This fact is stated below as a theorem.

P	$\sim P$	$\sim (\sim P)$
T	F	**T**
F	T	**F**

Figure 1.8: Verifying $P \equiv \sim (\sim P)$

Theorem 1.28 *For every statement P,*

$$P \equiv \sim (\sim P).$$

Example 1.29

Use De Morgan's Laws (Theorem 1.24) and Theorem 1.28 to express the negation of the following.

(1) Either he's not checking his email or he's not answering his email.

(2) This summer, she is neither buying an iPhone nor an iPad.

Solution.

(1) He checks and answers his email.

(2) The sentence above can also be written as follows: This summer, she is not buying an iPhone and not buying an iPad. Thus its negation is: This summer, she is either buying an iPhone or an iPad. ♦

We have seen that truth tables can be used to verify that two compound statements are logically equivalent. On the other hand, there are instances when we may be able to use some known facts to verify that two compound statements are logically equivalent.

Example 1.30

Let P and Q be two statements. Use De Morgan's Laws and Theorem 1.28 to verify that

(1) $\sim (P \vee (\sim Q)) \equiv (\sim P) \wedge Q$.

(2) $\sim ((\sim P) \wedge (\sim Q)) \equiv P \vee Q$.

Solution.

(1) By De Morgan's Law (Theorem 1.24(a)),

$$\sim (P \vee (\sim Q)) \equiv (\sim P) \wedge (\sim (\sim Q))$$

and by Theorem 1.28,

$$\sim (\sim Q) \equiv Q.$$

Therefore,
$$\sim (P \vee (\sim Q)) \equiv (\sim P) \wedge (\sim (\sim Q)) \equiv (\sim P) \wedge Q.$$

(2) Proceeding as in (1), we have

$$\sim ((\sim P) \wedge (\sim Q)) \quad \equiv \quad (\sim (\sim P)) \vee (\sim (\sim Q))$$
$$\text{by De Morgan's Law (Theorem 1.24(b))}$$
$$\equiv \quad P \vee Q \quad \text{by Theorem 1.28.} \blacklozenge$$

For every two real numbers a and b, the sum $a + b$ and product ab are also real numbers. For this reason, addition and multiplication are referred to as **binary operations** on the collection of real numbers. Both addition and multiplication are commutative. That is,

$$a + b = b + a \text{ and } ab = ba$$

for every two real numbers a and b. Conjunction and disjunction are also binary operations on a collection of statements. (The "exclusive or" is also a binary operation.) In fact, by Theorem 1.22 conjunction and disjunction are commutative since

$$P \wedge Q \equiv Q \wedge P \text{ and } P \vee Q \equiv Q \vee P$$

for every two statements P and Q.

Associative and Distributive Laws

Since addition and multiplication of real numbers are associative, that is,

$$a + (b + c) = (a + b) + c \text{ and } a(bc) = (ab)c$$

for all real numbers a, b and c, there is no confusion in writing $a + b + c$ and abc without parentheses. Furthermore, the distributive law holds, namely

$$a(b + c) = ab + ac$$

for all real numbers a, b and c. These properties of addition and multiplication of real numbers suggest logical equivalences of statements involving conjunction and disjunction.

Theorem 1.31 *Let P, Q and R be three statements. Then*

(a) (**Associative Laws**)

$P \vee (Q \vee R) \equiv (P \vee Q) \vee R$ *and* $P \wedge (Q \wedge R) \equiv (P \wedge Q) \wedge R.$

(b) (**Distributive Laws**)

$P \vee (Q \wedge R) \equiv (P \vee Q) \wedge (P \vee R)$ *and* $P \wedge (Q \vee R) \equiv (P \wedge Q) \vee (P \wedge R).$

So there are two distributive laws in this case. Each of the laws in Theorem 1.31 can be verified by truth tables (see Exercise 16). Here too, because of the associative laws, if P, Q and R are three statements, then $P \wedge Q \wedge R$ and $P \vee Q \vee R$ can be written without parentheses. In fact, for any integer $k \geq 3$ and any k statements $P_1, P_2, \ldots, P_k$, we can write $P_1 \wedge P_2 \wedge \cdots \wedge P_k$ as the **conjunction** of the statements and $P_1 \vee P_2 \vee \cdots \vee P_k$ as the **disjunction** of these k statements. The statement $P_1 \wedge P_2 \wedge \cdots \wedge P_k$ is true only when all of the statements $P_1, P_2, \ldots, P_k$ are true, while $P_1 \vee P_2 \vee \cdots \vee P_k$ is true only when at least one of the statements $P_1, P_2, \ldots, P_k$ is true.

Example 1.32

For real numbers x and y, consider the following open sentence

$$P(x, y): \quad x^2 + (y^2 - 4)^2 = 0.$$

Since $x^2 \geq 0$ and $(y^2 - 4)^2 \geq 0$ for all real numbers x and y, we can only have $x^2 + (y^2 - 4)^2 = 0$ when both $x^2 = 0$ and $(y^2 - 4)^2 = 0$. The equation $x^2 = 0$ has only one solution, namely $x = 0$; while $(y^2 - 4)^2 = 0$ has two solutions, namely $y = 2$ and $y = -2$. Therefore, $x^2 + (y^2 - 4)^2 = 0$ only when $x = 0$ and either $y = 2$ or $y = -2$. That is, $P(x, y)$ is true only when $x = 0$ and either $y = 2$ or $y = -2$. By the second distributive law in Theorem 1.31, $P(x, y)$ is true only when $x = 0$ and $y = 2$, or when $x = 0$ and $y = -2$. ◆

Exercises for Section 1.2

1. (a) Write a table that gives the different assignments of truth values for three statements P, Q and R if they are being considered simultaneously.

   (b) Maria has an interview for a job. For the statements P, Q and R below, write a sentence that corresponds to each row in the table in (a).

   P : Maria prepared for the interview.

   Q : Maria did well on the interview.

   R : Maria got the job.

2. State the negation of each of the following statements.

   (a) $\sqrt{3} > 1.7$.

   (b) The integer 0 is even.

   (c) The number 7 is not a root of the equation $x^2 - 7 = 0$.

3. For an integer n, consider the following open sentences

   $$P(n): \quad 2^n > 1. \quad Q(n): \quad |n^3 + 1| > 0.$$

   (a) Find an integer n such that $\sim P(n)$ is true.

   (b) Find an integer n such that $\sim Q(n)$ is true.

4. Consider the following two statements

   $$P: \quad 2^3 + 3^2 \text{ is even.} \quad Q: \quad 4 \cdot 3 \cdot 2 \cdot 1 < 5^2.$$

   State the conjunction and disjunction of P and Q. Are these compound statements true or false?

5. For an integer n, consider the following two open sentences

   $$P(n): \quad n^3 - n = 0. \quad Q(n): \quad n^2 \geq 2^n.$$

   (a) Find an integer n such that $P(n) \vee Q(n)$ is true but $P(n) \wedge Q(n)$ is false.

   (b) Find an integer n such that $P(n) \vee Q(n)$ is true but $P(n) \oplus Q(n)$ is false.

   (c) Does there exist an integer n such that $P(n) \wedge Q(n)$ and $P(n) \oplus Q(n)$ are both true?

   (d) Does there exist an integer n such that $P(n) \wedge Q(n)$ and $P(n) \oplus Q(n)$ are both false?

6. For an integer n, consider the following two open sentences

   $$P(n): \quad 2n^2 + 3n + 5 \text{ is even.} \quad Q(n): \quad (n + 1)(n - 3) < 0.$$

   (a) Give an example of an integer n such that both $P(n) \vee Q(n)$ and $P(n) \wedge Q(n)$ are true.

(b) Give an example of an integer n such that both $P(n) \vee Q(n)$ and $P(n) \oplus Q(n)$ are true.

(c) Does there exist an integer n such that $P(n) \vee Q(n)$ and $P(n) \oplus Q(n)$ are both false?

7. For an integer n, consider the following two open sentences

$$P(n): \ n^2 + n - 2 = 0. \quad Q(n): \ n^2 - n - 6 = 0.$$

Determine every integer n for which $P(n) \wedge Q(n)$ is true.

8. Let P and Q be statements. Construct a truth table for each of the following.
(a) $(\sim P) \vee Q$.   (b) $P \wedge (\sim Q)$.   (c) $(\sim P) \vee (P \wedge (\sim Q))$.

9. Consider the two statements

$$P: \ 15 \text{ is odd.} \quad Q: \ 21 \text{ is even.}$$

State each of the following in words and determine whether they are true or false.
(a) $\sim P$.   (b) $P \vee Q$.   (c) $P \wedge Q$.   (d) $(\sim P) \vee Q$.   (e) $P \wedge (\sim Q)$.

10. For two statements P and Q, verify that $P \oplus Q \equiv Q \oplus P$.

11. Verify the following De Morgan's Law by a truth table.
For two statements P and Q, $\sim (P \wedge Q) \equiv (\sim P) \vee (\sim Q)$.

12. Use De Morgan's Laws to state the negations of the following.

(a) Either $x < -3$ or $x > 3$.

(b) The integer a is odd and the integer b is even.

13. Use De Morgan's Laws to state the negations of the following.

(a) Either $x = 0$ or $y = 0$.

(b) The integers a and b are both nonnegative.

14. For two statements P and Q, use truth tables to verify the following.
(a) $\sim (P \vee Q) \not\equiv (\sim P) \vee (\sim Q)$.   (b) $\sim (P \wedge Q) \not\equiv (\sim P) \wedge (\sim Q)$.

15. For two nonzero integers a and b such that $a + b \neq 0$, consider the two open sentences

$$P: ab > 0. \quad Q: a + b > 0.$$

(a) State $P \wedge Q$.

(b) State $P \wedge (\sim Q)$.

(c) State $(\sim P) \wedge Q$.

(d) State $(\sim P) \wedge (\sim Q)$.

(e) State $\sim (P \wedge Q)$ using the phrase "it is not the case that."

(f) Use an appropriate De Morgan's Law to restate $\sim (P \wedge Q)$.

16. For statements P, Q and R, use truth tables to verify the following associative laws and distributive laws.

(a) $P \vee (Q \vee R) \equiv (P \vee Q) \vee R$.

(b) $P \wedge (Q \wedge R) \equiv (P \wedge Q) \wedge R$.

(c) $P \vee (Q \wedge R) \equiv (P \vee Q) \wedge (P \vee R)$.

(d) $P \wedge (Q \vee R) \equiv (P \wedge Q) \vee (P \wedge R)$.

17. For two statements P and Q, use truth tables to verify the following.

(a) $P \vee (P \wedge Q) \equiv P$. (b) $P \wedge (P \vee Q) \equiv P$.

18. Let P, Q and R be statements. Determine whether the following is true.

$$P \oplus (Q \oplus R) \equiv (P \oplus Q) \oplus R.$$

19. Let P, Q and R be statements. Determine whether the following is true.

$$P \wedge (Q \oplus R) \equiv (P \wedge Q) \oplus (P \wedge R).$$

20. Let P, Q and R be statements. Determine whether the following is true.

$$P \vee (Q \oplus R) \equiv (P \vee Q) \oplus (P \vee R).$$

21. For a real number a, consider the two sentences

$$P: \ a \leq 2. \quad Q: \ a > -2.$$

(a) State $P \vee Q$ and $P \wedge Q$.

(b) Use De Morgan's Laws to state $\sim (P \vee Q)$ and $\sim (P \wedge Q)$.

22. For integers x and y, let

$$Q(x, y): \ (x + 2)^2 + 2y^2 = 1.$$

(a) Complete the following: $Q(x, y)$ is a true statement
if $x =$ _____ and $y =$ _____
or $x =$ _____ and $y =$ _____.

(b) Use a distributive law to restate the answer in (a).

1.3 Implications

The compound statement of greatest interest to us is the implication. We will see that many theorems are expressed (or can be expressed) as implications.

Definition 1.33 *For two statements P and Q, the* **implication** $P \Rightarrow Q$ *is commonly written as*

If P, then Q.

An implication is also sometimes referred to as a **conditional**. *The statement P in the implication $P \Rightarrow Q$ is the* **hypothesis** *of $P \Rightarrow Q$, while Q is the* **conclusion** *of $P \Rightarrow Q$.*

Implications often occur in everyday communication:

(1) If he gets a promotion, then we'll all go out to dinner.

(2) If we score a touchdown, then we'll win the game.

(3) If this jacket is on sale, then I'll buy it.

(4) If he complains one more time, then I'll scream.

In (1) above, a condition is given under which we will go out to dinner, namely, if this person gets a promotion. If statement (1) is true and the individual gets a promotion, then we will all go out to dinner. This statement would be false if he got a promotion but we didn't go out to dinner. This is the only situation that would make statement (1) false. Consequently, if he doesn't get a promotion, then whether we go out to dinner or not, statement (1) would be true. Similar discussions can be made about implications (2)–(4) above.

For two statements P and Q, the truth table for $P \Rightarrow Q$ is defined in Figure 1.9. According to this truth table, we see that:

(i) if the hypothesis P is false, then $P \Rightarrow Q$ is true regardless of the truth value of Q;

(ii) if the conclusion Q is true, then $P \Rightarrow Q$ is true regardless of the truth value of P;

(iii) $P \Rightarrow Q$ is false only when the hypothesis P is true and the conclusion Q is false.

P	Q	$P \Rightarrow Q$
T	T	T
T	F	F
F	T	T
F	F	T

Figure 1.9: The truth table for implication

For the statements Q_1 and Q_2 in (1.1), the implication $Q_1 \Rightarrow Q_2$ is

$$Q_1 \Rightarrow Q_2 : \text{If Los Angeles is the capital of California,}$$
$$\text{then Broadway is a street in New York City.}$$

Since the statement Q_1 is a false statement and Q_2 is true, it follows from the third row of the truth table in Figure 1.9 that $Q_1 \Rightarrow Q_2$ is true. Of course, the explanation we just gave for why $Q_1 \Rightarrow Q_2$ is a true statement relies on the truth table for implication. It may seem that this is a peculiar way to define when an implication is true and when it is false. However, we might think of this truth table as describing conditions under which the contract "If P, then Q" holds. What this says is that if condition P is satisfied, then the contract specifies that condition Q must be satisfied as well. Therefore, the only time this contract will be broken is when P is satisfied but Q is not. This is precisely what the truth table for implication specifies. While there is no apparent connection between the statements Q_1 and Q_2 in the implication $Q_1 \Rightarrow Q_2$ above, there is ordinarily a connection between statements P and Q whether $P \Rightarrow Q$ is encountered in mathematics or, indeed, in everyday conversation.

Example 1.34

A man hires a neighbor to paint the living room of his house for $300. The neighbor says:

> *If you pay me an additional $50, then I will also paint your dining room.*   (1.2)

We now consider the following two statements:

P: You pay me an additional $50.
Q: I will also paint your dining room.

Let's consider the truth or falseness of the implication $P \Rightarrow Q$ in (1.2) for the possible combinations of truth values of P and Q.

Suppose that P and Q are both true. Then the man pays the neighbor an additional $50 and the neighbor paints the dining room. Since the neighbor did as he promised, the (verbal) contract has been fulfilled and so $P \Rightarrow Q$ is true, which agrees with the first row of the truth table in Figure 1.9.

Second, suppose that P is true and Q is false. In this case, the man pays the neighbor an additional $50 but the neighbor does not paint the dining room. The neighbor clearly did not do as he promised. Thus the contract is broken and $P \Rightarrow Q$ is false. This agrees with the second row of the truth table for $P \Rightarrow Q$.

Next, suppose that P is false and Q is true. Here the man does not pay the neighbor an additional $50, yet the neighbor paints the dining room. Why? Perhaps it was a gesture of friendship. In any case, the contract is not broken as nothing was promised if the man did not pay the neighbor an additional $50. This agrees with the third row of the truth table for $P \Rightarrow Q$.

Finally, suppose that P and Q are both false. In this case, the man does not pay the neighbor an additional $50 and the neighbor does not paint the dining room. The neighbor undoubtedly did exactly as the man expected. Here too nothing was promised if no additional money was paid. So here too the contract was not broken and $P \Rightarrow Q$ is true. This agrees with the fourth row of the truth table for $P \Rightarrow Q$. ♦

Let's now look at the truth and falseness of some specific implications from mathematics.

Example 1.35

Determine the truth value of each of the following implications.

(a) If $2 + 3 = 5$, then $4 + 6 = 10$.

(b) If $4 + 6 = 10$, then $5 + 7 = 14$.

(c) If $5 + 7 = 14$, then $6 + 9 = 15$.

(d) If $8 + 11 = 21$, then $12 + 14 = 28$.

Solution.

(a) "If $2 + 3 = 5$, then $4 + 6 = 10$" reduces to: If T, then T. This is a true implication according to the first row of the truth table in Figure 1.9.

(b) "If $4 + 6 = 10$, then $5 + 7 = 14$" reduces to: If T, then F. According to the second row of the truth table in Figure 1.9, this implication is false.

(c) "If $5 + 7 = 14$, then $6 + 9 = 15$" reduces to: If F, then T. By the third row of the truth table in Figure 1.9, this implication is true.

(d) "If $8 + 11 = 21$, then $12 + 14 = 28$" reduces to: If F, then F. This implication is true according to the fourth row of the truth table in Figure 1.9.

We could have also determined the truth of (a), (c) and (d) above more quickly. As we have mentioned, if P is false, then $P \Rightarrow Q$ is true regardless of the truth value of Q; while if Q is true, then $P \Rightarrow Q$ is true regardless of the truth value of P. Therefore, the implications in (a), (c) and (d) are true. Furthermore, as we mentioned, the only situation under which $P \Rightarrow Q$ is false is when P is true and Q is false. Therefore, (b) is false. ♦

We have mentioned that if P and Q are statements, then $P \Rightarrow Q$ is also a statement and therefore has a truth value (which, of course, depends on the truth values of P and Q). Since it is possible to consider the negation of an open sentence as well as the disjunction and conjunction of two open sentences, it should come as no surprise that we can consider $P \Rightarrow Q$ when P and Q are open sentences.

Example 1.36

For a real number x, consider the two open sentences

$$P(x):\ x - 2 = 0. \quad Q(x):\ x^2 - x - 2 = 0.$$

Investigate, for all real numbers x, the truth or falseness of the implication

$$P(x) \Rightarrow Q(x):\ \text{If } x - 2 = 0, \text{ then } x^2 - x - 2 = 0.$$

Solution. The only real number x for which $P(x)$ is true is $x = 2$, that is, $P(2)$ is true. Notice that $Q(2)$ is also true. Since $P(2)$ and $Q(2)$ are both true, it follows that $P(2) \Rightarrow Q(2)$ is a true statement according to the first row of the truth table in Figure 1.9. If we were to select a different real number x, then $P(x)$ would be false. In this case, by the third and fourth rows of the truth table in Figure 1.9, $P(x) \Rightarrow Q(x)$ is true regardless of the truth value of $Q(x)$ for the real number x under consideration. That is, $P(x) \Rightarrow Q(x)$ is true for *every* real number x. ♦

We will be especially interested in implications $P(x) \Rightarrow Q(x)$ that are true for every object in the domain of the variable x and will examine this relationship in detail in Chapter 3.

Example 1.37

For a real number x, consider

$$P(x):\ x^2 - 9 = 0. \quad Q(x):\ (x - 3)^2 = 0.$$

Investigate, for all real numbers x, the truth or falseness of the implication

$$P(x) \Rightarrow Q(x):\ \text{If } x^2 - 9 = 0, \text{ then } (x - 3)^2 = 0.$$

Solution. First observe that $P(x)$ is true only when $x = 3$ or $x = -3$. However, $Q(3)$ is true, while $Q(-3)$ is false. Thus $P(3) \Rightarrow Q(3)$ is true and $P(-3) \Rightarrow Q(-3)$ is false. If x is any real number that is different from 3 and -3, then $P(x)$ is false and so $P(x) \Rightarrow Q(x)$ is true. Therefore, $P(x) \Rightarrow Q(x)$ is true for every real number except for $x = -3$. ♦

Stating Implications in Words

We have already mentioned that for statements (or open sentences) P and Q, the implication $P \Rightarrow Q$ can be expressed in words as: If P, then Q. This implication can also be expressed in a variety of other ways. First, a slight variation of "If P, then Q" is "Q if P." However, $P \Rightarrow Q$ can also be expressed as "P implies Q." Indeed, common ways of expressing $P \Rightarrow Q$ in words are listed in Figure 1.10.

$P \Rightarrow Q$
If P, then Q
Q if P
P **implies** Q
P **only if** Q
P is **sufficient** for Q
Q is **necessary** for P

Figure 1.10: Expressing $P \Rightarrow Q$ in words

Perhaps the most surprising entry in Figure 1.10 to express the implication $P \Rightarrow Q$ in words is "P only if Q." Nevertheless, this is correct. For example, suppose that the Department of Mathematics has sign-up sheets for students who plan to take certain mathematics courses next semester. Above one sheet lying on a table is the following notice:

You should sign this sheet only if you plan to take discrete mathematics next semester.

What this means is that if you sign your name on the sheet, then you plan to take the course.

Example 1.38

Of course, if I obtained all A's last semester and took a math course last semester, then I must have received an A in math as well. Let's state this implication in the various ways indicated in Figure 1.10.

> If I obtained all A's, then I received an A in math.
> I received an A in math if I obtained all A's.
> Obtaining all A's implies that I received an A in math.
> I obtained all A's only if I received an A in math.
> Obtaining all A's is sufficient for receiving an A in math.
> Receiving an A in math is necessary for obtaining all A's.

In particular, the fourth sentence says that the only way I can obtain an A in all courses is to receive an A in math. ◆

Let's look at a mathematical example where an implication is involved and restate this implication in the manners suggested in Figure 1.10.

Example 1.39

For a real number x, let

$$P(x)\colon\ x = 5. \qquad Q(x)\colon\ x^2 = 25.$$

Express the implication $P(x) \Rightarrow Q(x)$ in words in a variety of ways.

Solution. According to the table in Figure 1.10, this implication can be expressed as follows:

> If $x = 5$, then $x^2 = 25$.
> $x^2 = 25$ if $x = 5$.
> $x = 5$ implies that $x^2 = 25$.
> $x = 5$ only if $x^2 = 25$.
> $x = 5$ is sufficient for $x^2 = 25$.
> $x^2 = 25$ is necessary for $x = 5$. ◆

Example 1.40

Express each of the following implications as an "If, then" sentence.

(a) $|a| = 2$ is sufficient for $a^4 = 16$.

(b) My team is going to the Super Bowl if they win their next game.

(c) I do well on an exam only if I studied for it.

(d) Not doing well this semester implies that I took too many courses.

(e) Doing well in math in necessary for me to have a good semester.

Solution.

(a) If $|a| = 2$, then $a^4 = 16$.

(b) If my team wins their next game, then they're going to the Super Bowl.

(c) If I do well on an exam, then I studied for it.

(d) If I don't do well this semester, then I took too many courses.

(e) If I had a good semester, then I did well in math.   ◆

Converse of an Implication

For two given statements P and Q, we can construct not only the implication $P \Rightarrow Q$ but the implication $Q \Rightarrow P$ as well.

Definition 1.41 *For statements (or open sentences) P and Q, the implication $Q \Rightarrow P$ is called the* **converse** *of $P \Rightarrow Q$.*

Let's look at the converses of the implications (1) – (4) stated at the beginning of this section.

(1') If we all go out to dinner, then he gets a promotion.

(2') If we win the game, then we score a touchdown.

(3') If I buy this jacket, then it is on sale.

(4') If I scream, then he will complain one more time.

Example 1.42

Recall once again the two statements Q_1 and Q_2 in (1.1):

$$Q_1: \quad \text{Los Angeles is the capital of California.}$$
$$Q_2: \quad \text{Broadway is a street in New York City.}$$

We have already noted that the implication

$$Q_1 \Rightarrow Q_2 : \quad \text{If Los Angeles is the capital of California,}$$
$$\text{then Broadway is a street in New York City.}$$

is a true statement. However, according to the truth table in Figure 1.9, its converse

$$Q_2 \Rightarrow Q_1 : \quad \text{If Broadway is a street in New York City,}$$
$$\text{then Los Angeles is the capital of California.}$$

is false because here the hypothesis Q_2 is true and the conclusion Q_1 is false.   ◆

Therefore, the implication $Q_2 \Rightarrow Q_1$ above is the converse of $Q_1 \Rightarrow Q_2$. In addition, $Q_1 \Rightarrow Q_2$ is the converse of $Q_2 \Rightarrow Q_1$. What this shows is that for two statements P and Q, the two implications $P \Rightarrow Q$ and $Q \Rightarrow P$ need not have the same truth value. If we think about this observation a bit, we should be able to convince ourselves that if P and Q are two statements for which $P \Rightarrow Q$ and $Q \Rightarrow P$ have the same truth value, then this truth value must be true (see Exercise 9).

Example 1.43

For a real number x, consider the two open sentences:

$$P(x): \quad (x-1)(x-2) = 0. \quad Q(x): \quad (x-2)(x-3) = 0.$$

Let's look at the implication $P(x) \Rightarrow Q(x)$ for $x = 1$, say. Then $P(1)$ is true, while $Q(1)$ is false. Hence $P(1) \Rightarrow Q(1)$ is a false statement. Let's consider the converse of $P(x) \Rightarrow Q(x)$ for $x = 3$ say. Then $Q(3)$ is true but $P(3)$ is false; so $Q(3) \Rightarrow P(3)$ is false. However, both $P(2) \Rightarrow Q(2)$ and $Q(2) \Rightarrow P(2)$ are true. ♦

Contrapositive of an Implication

We will have numerous opportunities to consider the converse of an implication. In fact, there will be many occasions when we will want to consider both an implication *and* its converse. For statements (or open sentences) P and Q, there is another implication that is related to $P \Rightarrow Q$ and in which we will be greatly interested.

Definition 1.44 *For two statements (or open sentences) P and Q, the* **contrapositive** *of the implication $P \Rightarrow Q$ is $(\sim Q) \Rightarrow (\sim P)$.*

Example 1.45

For the two statements Q_1 and Q_2 in (1.1), the contrapositive of $Q_1 \Rightarrow Q_2$ is

$$(\sim Q_2) \Rightarrow (\sim Q_1): \text{ If Broadway is } not \text{ a street in New York City,}$$
$$\text{then Los Angeles is } not \text{ the capital of California. } ♦$$

Just as $Q_1 \Rightarrow Q_2$ is a true statement, so too is $(\sim Q_2) \Rightarrow (\sim Q_1)$. This is, in fact, what happens for every true implication $P \Rightarrow Q$ formed from statements P and Q. This is the reason that the contrapositive of an implication interests us so much.

Theorem 1.46 *For every two statements P and Q,*

$$P \Rightarrow Q \ \equiv \ (\sim Q) \Rightarrow (\sim P).$$

That the two implications $P \Rightarrow Q$ and $(\sim Q) \Rightarrow (\sim P)$ are logically equivalent is verified in the truth table shown in Figure 1.11.

P	Q	$P \Rightarrow Q$	$\sim Q$	$\sim P$	$(\sim Q) \Rightarrow (\sim P)$
T	T	**T**	F	F	**T**
T	F	**F**	T	F	**F**
F	T	**T**	F	T	**T**
F	F	**T**	T	T	**T**

Figure 1.11: The logical equivalence of an implication and its contrapositive

Example 1.47

For an integer n, consider the two open sentences

$$P(n): \ 3n + 11 \text{ is even.} \quad Q(n): \ n \text{ is odd.}$$

Then for a specific integer n, the implication

$$P(n) \Rightarrow Q(n): \text{ If } 3n + 11 \text{ is even, then } n \text{ is odd.}$$

and the implication

$(\sim Q(n)) \Rightarrow (\sim P(n))$: If n is not odd, then $3n + 11$ is not even.

(or: If n is even, then $3n + 11$ is odd.) are both true or both false.   ♦

For statements P and Q, the implication $P \Rightarrow Q$ is also logically equivalent to a statement that does not involve an implication.

Theorem 1.48 *For every two statements P and Q,*

$$P \Rightarrow Q \ \equiv \ (\sim P) \vee Q.$$

That the statements $P \Rightarrow Q$ and $(\sim P) \vee Q$ are logically equivalent is verified in the truth table in Figure 1.12.

P	Q	$P \Rightarrow Q$	$\sim P$	$(\sim P) \vee Q$
T	T	**T**	F	**T**
T	F	**F**	F	**F**
F	T	**T**	T	**T**
F	F	**T**	T	**T**

Figure 1.12: Verifying $P \Rightarrow Q \ \equiv \ (\sim P) \vee Q$

Example 1.49

For the two statements

P: You don't love me.
Q: You must leave me.

the implication $P \Rightarrow Q$ is

$P \Rightarrow Q$: If you don't love me, then you must leave me.

According to Theorem 1.48, this implication is logically equivalent to

$(\sim P) \vee Q$: Either you love me or you must leave me.

which results in

$(\sim P) \vee Q$: Love me or leave me. ♦

There is now an immediate consequence of Theorem 1.48.

Theorem 1.50 *For every two statements P and Q,*

$$\sim (P \Rightarrow Q) \ \equiv \ P \wedge (\sim Q).$$

We could verify Theorem 1.50 by means of a truth table, but it can also be verified using the information we have:

$$
\begin{aligned}
\sim (P \Rightarrow Q) \ &\equiv \ \sim ((\sim P) \vee Q) \quad \text{by Theorem 1.48} \\
&\equiv \ (\sim (\sim P)) \wedge (\sim Q) \quad \text{by De Morgan's Law (Theorem 1.24(a))} \\
&\equiv \ P \wedge (\sim Q) \quad \text{by Theorem 1.28.}
\end{aligned}
$$

Therefore, $\sim (P \Rightarrow Q) \equiv P \wedge (\sim Q)$.

Example 1.51

For an integer n, consider the open sentences

$$P(n):\ n \text{ is even.} \qquad Q(n):\ n \text{ is not the sum of three odd integers.}$$

(a) State $P(n) \Rightarrow Q(n)$ in words.

(b) State $\sim (P(n) \Rightarrow Q(n))$ in words using the phrase "it is not the case that."

(c) Use Theorem 1.50 to restate $\sim (P(n) \Rightarrow Q(n))$ in words.

Solution.

(a) If n is even, then n is not the sum of three odd integers.

(b) It is not the case that if n is even, then n is not the sum of three odd integers.

(c) The even number n is the sum of three odd integers. ♦

Exercises for Section 1.3

1. Determine whether the following implications are true or false:

   (a) If 2 is an even integer, then $5 > 0$.
   (b) If 2 is an even integer, then $5 < 0$.
   (c) If 2 is an odd integer, then $5 > 0$.
   (d) If 2 is an odd integer, then $5 < 0$.

2. For a real number x, consider the open sentences

$$P(x):\ x^2 = x. \qquad Q(x):\ (x+1)^2 = 0.$$

   (a) Express the implication $P(x) \Rightarrow Q(x)$ in words.
   (b) Give an example of a real number a such that $P(a) \Rightarrow Q(a)$ is true.
   (c) Give an example of a real number b such that $P(b) \Rightarrow Q(b)$ is false.

3. For an integer n, consider the open sentences

$$P(n):\ 5n + 3 \text{ is odd.} \qquad Q(n):\ n \text{ is even.}$$

   (a) State $P(n) \Rightarrow Q(n)$ in words.
   (b) State $P(3) \Rightarrow Q(3)$ in words. Is this statement true or false?
   (c) State $P(2) \Rightarrow Q(2)$ in words. Is this statement true or false?

4. For an integer n, consider the open sentences

$$P(n):\ 2^{n-1} \cdot 3^n + 2^n \cdot 3^{n-1} \text{ is even.} \qquad Q(n):\ (n+1)(n+4)/2 \text{ is even.}$$

   For $n = 1, 2, 3$, determine whether the implication $P(n) \Rightarrow Q(n)$ is true or false.

5. Determine whether each of the following statements is true or false.

   (a) 0 is negative only if -1 is negative.
   (b) $|3| = -3$ is sufficient for $|-3| = 3$.

(c) $(-1)^2 = -1$ is necessary for $(-1)^3 = -1$.

6. Express each of the following implications as an "If, then" sentence.

   (a) Saving enough money is necessary for me to take a vacation.
   (b) I'm lucky if my car lasts through the winter.
   (c) Getting a tax refund is sufficient to make me happy.
   (d) I will go to the movies only if I have the time.
   (e) Being invited to the awards luncheon implies that I did well on the competition.

7. Express each of the implications below using (1) the phrase "only if" and (2) the word "sufficient."

   (a) If ab is odd, then a is odd. (b) If $x = -5$, then $x^2 = 25$.

8. Consider the implication: If x and y are even, then xy is even.

   (a) State the implication using the phrase "only if ."
   (b) State the implication using the word "sufficient ."
   (c) State the converse of the implication.
   (d) State the contrapositive of the implication.

9. For two statements P and Q, show that if $P \Rightarrow Q$ and $Q \Rightarrow P$ have the same truth value, then this truth value is true.

10. Consider the statements

$$P:\ \ 7 \text{ is an even integer.} \quad Q:\ \ 0 \text{ is a positive integer.}$$

   (a) Write the implication $P \Rightarrow Q$ in words and determine whether it is true or false.
   (b) Write the implication $Q \Rightarrow P$ in words and determine whether it is true or false.

11. Give an example of two open sentences $P(n)$ and $Q(n)$, where n is an integer and three integers a, b and c such that (1) $P(a) \Rightarrow Q(a)$ and $Q(b) \Rightarrow P(b)$ are both false and (2) $P(c) \Rightarrow Q(c)$ and $Q(c) \Rightarrow P(c)$ are both true.

12. State in words the converse and the contrapositive of the implication $P \Rightarrow Q$, where

$$P:\ \ 101 \text{ is even.} \quad Q:\ \ 110 \text{ is even.}$$

   Determine whether each of these three implications is true or false.

13. State in words the converse and the contrapositive of the implication $P(x) \Rightarrow Q(x)$, where

$$P(x):\ \ x = 2. \quad Q(x):\ \ |x| = 2.$$

14. Consider the implication: If ab is odd, then a is odd and b is odd.

   (a) State the converse of this implication.
   (b) State the contrapositive of this implication.

15. Consider the implication: If $7n - 8$ is odd, then n is even.

   (a) State the converse of this implication.
   (b) State the contrapositive of this implication.

16. Construct a truth table for $P \wedge (Q \Rightarrow (\sim P))$.

17. Construct a truth table for $(P \Rightarrow Q) \Rightarrow (\sim P)$.

18. The **inverse** of the implication of $P \Rightarrow Q$ is the implication $(\sim P) \Rightarrow (\sim Q)$.

    (a) Use a truth table to verify that $P \Rightarrow Q \not\equiv (\sim P) \Rightarrow (\sim Q)$.
    (b) Find another implication that is logically equivalent to $(\sim P) \Rightarrow (\sim Q)$ and verify your answer.

19. Let P, Q and R be three statements. Given that the implication $(Q \vee R) \Rightarrow (\sim P)$ is false and Q is false, determine the truth values of R and P.

20. Consider the following two statements:

$$P: \text{ Today is Saturday or Sunday.} \quad Q: \text{ I do not have class today.}$$

    State each of the following in words.
    (a) $P \Rightarrow Q$.   (b) $(\sim Q) \Rightarrow (\sim P)$.   (c) $(\sim P) \vee Q$.
    (d) $Q \Rightarrow P$.   (e) $(\sim Q) \vee P$.   (f) $(\sim P) \Rightarrow (\sim Q)$.

21. Consider the following two statements:

$$P: \text{ You drive over 70 miles per hour.} \quad Q: \text{ You receive a speeding ticket.}$$

    State each of the following in words.
    (a) $P \Rightarrow Q$.   (b) $(\sim Q) \Rightarrow (\sim P)$.   (c) $(\sim P) \vee Q$.
    (d) $Q \Rightarrow P$.   (e) $(\sim Q) \vee P$.   (f) $(\sim P) \Rightarrow (\sim Q)$.

22. Consider the statement

$$S: \text{ If I get an } A \text{ on the final exam, then I'll get an } A \text{ for my final grade.}$$

    (a) Assuming that S is true and I got an A on the final exam, does this mean that I'll get an A for my final grade?
    (b) Assuming that S is true and I got an A for my final grade, does this mean that I got an A on my final exam?
    (c) Assuming that S is true and I didn't get an A on my final exam, does this mean that I didn't get an A for my final grade?

23. For a real number x, consider the open sentences:

$$P(x): \ x = -2. \quad Q(x): \ x^2 = 4.$$

    State the following in words.

    (a) $\sim P(x)$ and $\sim Q(x)$.
    (b) $P(x) \Rightarrow Q(x)$ and $Q(x) \Rightarrow P(x)$.
    (c) $(\sim P(x)) \Rightarrow (\sim Q(x))$ and $(\sim Q(x)) \Rightarrow (\sim P(x))$.

24. For three statements P, Q and R, use truth tables to verify the following.

    (a) $(P \Rightarrow Q) \wedge (P \Rightarrow R) \equiv P \Rightarrow (Q \wedge R)$.
    (b) $(P \Rightarrow R) \wedge (Q \Rightarrow R) \equiv (P \vee Q) \Rightarrow R$.

(c) $(P \Rightarrow Q) \vee (P \Rightarrow R) \equiv P \Rightarrow (Q \vee R)$.

(d) $(P \Rightarrow R) \vee (Q \Rightarrow R) \equiv (P \wedge Q) \Rightarrow R$.

(e) $(P \Rightarrow Q) \wedge (Q \Rightarrow R) \equiv P \Rightarrow R$.

25. For two statements P and Q, use truth tables to verify the following.

(a) $P \vee Q \equiv (\sim P) \Rightarrow Q$.

(b) $P \wedge Q \equiv \sim (P \Rightarrow (\sim Q))$.

(c) $\sim (P \Rightarrow Q) \equiv P \wedge (\sim Q)$.

26. For an integer n, consider the open sentences

$$P(n): \ n \text{ is odd.} \quad Q(n): \ n \text{ is not the sum of two even integers.}$$

(a) Write $P(n) \Rightarrow Q(n)$ in words.

(b) Write $\sim (P(n) \Rightarrow Q(n))$ in words using the phrase "it is not the case that."

(c) Use Theorem 1.50 to rewrite $\sim (P(n) \Rightarrow Q(n))$ in words.

1.4   Biconditionals

There is a concert coming to Henry's town that he would like to attend but he thinks it's likely to be too expensive for his budget. In fact, he's decided that he will go if it costs him less than $25 and that's the only way he would go. So

$$\text{If it costs Henry less than \$25 to attend the concert, then he will go.} \tag{1.3}$$

Also,

$$\text{If Henry attends the concert, then it cost him less than \$25.} \tag{1.4}$$

In this case then:

$$\text{Henry will attend the concert if and only if it costs him less than \$25.} \tag{1.5}$$

If and only if

The sentence (1.5) is the conjunction of the implication (1.3) and its converse (1.4). In mathematics such sentences are encountered often. We will discuss these kinds of sentences in this section.

Definition 1.52 *For two statements P and Q, the **biconditional** of P and Q is the conjunction of the implication $P \Rightarrow Q$ and its converse $Q \Rightarrow P$. The biconditional of P and Q is denoted by $P \Leftrightarrow Q$. So $P \Leftrightarrow Q$ is the statement $(P \Rightarrow Q) \wedge (Q \Rightarrow P)$.*

The biconditional $P \Leftrightarrow Q$ is expressed in words as

$$P \text{ **if and only if** } Q.$$

or

$$P \text{ **is necessary and sufficient for** } Q.$$

P	Q	$P \Rightarrow Q$	$Q \Rightarrow P$	$P \Leftrightarrow Q$
T	T	T	T	**T**
T	F	F	T	**F**
F	T	T	F	**F**
F	F	T	T	**T**

Figure 1.13: The truth table for the biconditional

Because of the way $P \Leftrightarrow Q$ is defined, we can determine the truth table for this statement ourselves, as shown in Figure 1.13. Therefore, $P \Leftrightarrow Q$ is true provided that P and Q have the same truth value; that is, $P \Leftrightarrow Q$ is true only when P and Q are both true or are both false. Let's consider an example of a biconditional.

Example 1.53

For the statements

$$P: \ 3^2 > 2^3. \quad Q: \ 4^3 > 3^4.$$

the biconditional of these statements is

$$P \Leftrightarrow Q: \ 3^2 > 2^3 \text{ if and only if } 4^3 > 3^4.$$

Since P is true and Q is false, P and Q have different truth values and so $P \Leftrightarrow Q$ is false.   ◆

Example 1.54

Consider the following two statements:

$$P: \text{ I will receive an A on the exam.}$$
$$Q: \text{ I study for at least 10 hours.}$$

Then the biconditional of P and Q is

$P \Leftrightarrow Q$: I will receive an A on the exam if and only if I study for at least 10 hours.

The statement $P \Leftrightarrow Q$ is therefore true provided either

(a) I study for at least 10 hours and receive an A on the exam or

(b) I study for less than 10 hours and do not receive an A on the exam.   ◆

There is a logical explanation as to why $P \Leftrightarrow Q$ can be expressed in words as "P if and only if Q." Since $P \Leftrightarrow Q$ is defined as the statement $(P \Rightarrow Q) \wedge (Q \Rightarrow P)$ and the commutative law of conjunction holds (by Theorem 1.22), it follows that

$$(P \Rightarrow Q) \wedge (Q \Rightarrow P) \equiv (Q \Rightarrow P) \wedge (P \Rightarrow Q).$$

Recall that $Q \Rightarrow P$ can be expressed in words as "P if Q" and $P \Rightarrow Q$ can be expressed in words as "P only if Q." Therefore, we can express $P \Leftrightarrow Q$ in words as

$$P \text{ if } Q \text{ and } P \text{ only if } Q$$

or as

$$P \text{ if and only if } Q.$$

Many mathematicians abbreviate "if and only if" as "iff," which, of course, is not a word. However, since this abbreviation is so common, we should be aware of it. In a similar way, there is a logical explanation as to why $P \Leftrightarrow Q$ can be expressed as "P is necessary and sufficient for Q" (see Exercise 11). We can also consider $P \Leftrightarrow Q$ when P and Q are open sentences.

Example 1.55

For an integer n, consider the two open sentences

$$P(n): \ (n-1)^2 = 0. \quad Q(n): \ 7n - 3 = 0.$$

The biconditional of these two open sentences is

$$P(n) \Leftrightarrow Q(n): \ (n-1)^2 = 0 \text{ if and only if } 7n - 3 = 0.$$

Investigate the truth or falseness of $P(n) \Leftrightarrow Q(n)$ for various integers n.

Solution. First, $P(n)$ is true only when $n = 1$. Since $Q(1)$ is false, $P(1) \Leftrightarrow Q(1)$ is false. For an integer $n \neq 1$, $P(n)$ is false. However, $Q(n)$ is false for every integer n. Therefore, for $n \neq 1$, both implications $P(n)$ and $Q(n)$ are false and so $P(n) \Leftrightarrow Q(n)$ is true. $\blacklozenge$

We now look at two examples of biconditionals dealing with triangles.

Example 1.56

A triangle is *equilateral* if its three sides have the same length; while a triangle is *equiangular* if its three angles have the same measure (namely 60° or $\pi/3$ radians). The converse of the implication

If a triangle is equilateral, then it is equiangular.

is the implication

If a triangle is equiangular, then it is equilateral.

The conjunction of these two implications gives us the biconditional

A triangle is equilateral if and only if it is equiangular. $\blacklozenge$

Example 1.57

A triangle is a *right triangle* if one of its angles is a right angle (namely, 90° or $\pi/2$ radians). The well-known *Pythagorean theorem* can be stated as

If a, b and c are the lengths of the three sides of a right triangle, where $a \leq b \leq c$, then $a^2 + b^2 = c^2$.

Its converse is the implication:

If the lengths of the three sides of a triangle are a, b and c, where $a \leq b \leq c$ and $a^2 + b^2 = c^2$, then the triangle is a right triangle.

The conjunction of these two implications is a biconditional:

Let a, b and c be the lengths of the three sides of a triangle T, where $a \leq b \leq c$. Then T is a right triangle if and only if $a^2 + b^2 = c^2$. $\blacklozenge$

Example 1.58

Suppose that P and Q are two statements such that $P \Rightarrow Q$ is true and $Q \Rightarrow P$ is false. Investigate the truth or falseness of the biconditional P if and only if Q.

Solution. Since the biconditional

$$P \Leftrightarrow Q: \quad P \text{ if and only if } Q$$

is the conjunction of $P \Rightarrow Q$ and $Q \Rightarrow P$, it follows that $P \Leftrightarrow Q$ is true only when both $P \Rightarrow Q$ and $Q \Rightarrow P$ are true. Since this is not the case, $P \Leftrightarrow Q$ is false. ◆

Exercises for Section 1.4

1. Consider the statements

   $$P: \quad -5 \text{ is odd.} \quad Q: \quad 2^3 + 1 \text{ is even.}$$

   (a) Express the biconditional $P \Leftrightarrow Q$ using "if and only if" and "necessary and sufficient."
   (b) Determine whether $P \Leftrightarrow Q$ is true or false.

2. Determine, with explanation, whether the statement

   $$-1^2 > 0 \text{ if and only if } 7 \cdot 5 = 6^2 - 1.$$

   is true or false.

3. For an integer n and the open sentences

   $$P(n): \quad 3n^2 \text{ is even.} \quad Q(n): \quad n^3 \text{ is even.}$$

   express the biconditional $P(n) \Leftrightarrow Q(n)$ using "if and only if" and "necessary and sufficient."

4. For an integer n and the open sentences

   $$P(n): \quad n \text{ is odd.} \quad Q(n): \quad n^2 \text{ is odd.}$$

   state the biconditional $P(n) \Leftrightarrow Q(n)$ using "if and only if" and "necessary and sufficient."

5. For an integer n, consider the biconditional

   $$2^n > n^2 \text{ if and only if } 3n + 1 \text{ is even.}$$

   Determine an integer a for which the biconditional is true and an integer b for which the biconditional is false.

6. For an integer n, consider the open sentences

   $$P(n): \quad 5n + 7 \text{ is even.} \quad Q(n): \quad n \text{ is odd.}$$

   State $P(n) \Leftrightarrow Q(n)$ in words.

7. For integers a and b, consider the biconditional

   $$ab \text{ is even if and only if } a \text{ is even and } b \text{ is even.}$$

   (a) Give an example of two integers a and b for which this biconditional is true.
   (b) Give an example of two integers a and b for which this biconditional is false.

8. For integers a and b, consider the biconditional

   $$a + b \text{ is even if and only if } 3a \text{ and } 5b \text{ are even.}$$

(a) Give an example of two distinct integers a and b for which this biconditional is true.

(b) Give an example of two distinct integers a and b for which this biconditional is false.

9. For integers a, b and c, consider the biconditional

At least two of a, b and c are odd if and only if at least two of ab, ac and bc are odd.

(a) Give an example of integers a, b and c for which this biconditional is true.

(b) Give an example of integers a, b and c for which this biconditional is false.

10. For an integer n, consider the open sentences

$$P(n): \ n(n+1)(2n+1)/6 \text{ is even.} \quad Q(n): \ (n+1)^2(n+2)^2/4 \text{ is even.}$$

Determine whether the biconditionals $P(1) \Leftrightarrow Q(1)$ and $P(2) \Leftrightarrow Q(2)$ are true or false.

11. For two statements P and Q, give an explanation why the biconditional $P \Leftrightarrow Q$ can be expressed as "P is necessary and sufficient for Q."

12. For every two statements P and Q, use truth tables to verify the following.

(a) $P \Leftrightarrow Q \ \equiv \ (\sim P) \Leftrightarrow (\sim Q)$.

(b) $P \Leftrightarrow Q \ \equiv \ (P \wedge Q) \vee ((\sim P) \wedge (\sim Q))$.

(c) $\sim (P \Leftrightarrow Q) \ \equiv \ P \Leftrightarrow (\sim Q)$.

1.5   Tautologies and Contradictions

We have now seen the following logical connectives:

$\sim$	negation (not)
$\wedge$	conjunction (and)
$\vee$	disjunction (or)
$\oplus$	exclusive or
$\Rightarrow$	implication (implies)
$\Leftrightarrow$	biconditional (if and only if).

Recall that a compound statement is a statement constructed from one or more given statements (the component statements) and one or more logical connectives. Ordinarily, a compound statement is true for some combinations of truth values of its component statements and false for other combinations of truth values of its component statements – *but not always*.

Definition 1.59 *A compound statement is a* **tautology** *if it is true for all possible combinations of truth values of its component statements.*

Definition 1.60 *A compound statement is a* **contradiction** *if it is false for all possible combinations of the truth values of its component statements.*

Therefore, a compound statement S is a tautology if and only if its negation $\sim S$ is a contradiction. The simplest examples of a tautology and a contradiction are constructed from a single component statement.

For every statement P, the statement $P \vee (\sim P)$ is a tautology since certainly (exactly) one of P and $\sim P$ is true. In fact, because one of P and $\sim P$ is also false, the statement $P \wedge (\sim P)$ is a contradiction. These are verified in the truth table of Figure 1.14, where the entries in the columns headed by $P \vee (\sim P)$ and $P \wedge (\sim P)$ are in bold to call attention to the resulting tautology and contradiction, respectively.

P	$\sim P$	$P \vee (\sim P)$	$P \wedge (\sim P)$
T	F	**T**	**F**
F	T	**T**	**F**

Figure 1.14: A truth table for a tautology and a contradiction

Example 1.61

Since $P \vee (\sim P)$ is a tautology and $P \wedge (\sim P)$ is a contradiction for every statement P, if we let

$$P(n): \ n \text{ is even.}$$

where n is an integer, then

$$P(n) \vee (\sim P(n)): \ \ n \text{ is even or } n \text{ is odd.}$$

is a true statement for every integer n; while

$$P(n) \wedge (\sim P(n)): \ \ n \text{ is even and } n \text{ is odd.}$$

is a false statement for every integer n. ♦

Furthermore, for any two logically equivalent compound statements R and S, the biconditional $R \Leftrightarrow S$ is a tautology.

Example 1.62

For every two statements P and Q, the statement $(\sim Q) \vee (P \Rightarrow Q)$ is a tautology. This is verified in the truth table of Figure 1.15. ♦

P	Q	$\sim Q$	$P \Rightarrow Q$	$(\sim Q) \vee (P \Rightarrow Q)$
T	T	F	T	**T**
T	F	T	F	**T**
F	T	F	T	**T**
F	F	T	T	**T**

Figure 1.15: A truth table for the tautology in Example 1.62

Example 1.63

For every two statements P and Q, the statement $(P \wedge Q) \wedge (Q \Rightarrow (\sim P))$ is a contradiction (see the truth table in Figure 1.16). ♦

P	Q	$\sim P$	$P \wedge Q$	$Q \Rightarrow (\sim P)$	$(P \wedge Q) \wedge (Q \Rightarrow (\sim P))$
T	T	F	T	F	**F**
T	F	F	F	T	**F**
F	T	T	F	T	**F**
F	F	T	F	T	**F**

Figure 1.16: A truth table for the contradiction in Example 1.63

Modus Ponens and Modus Tollens

Example 1.64

Let P and Q be two statements. Show that $(P \wedge (P \Rightarrow Q)) \Rightarrow Q$ is a tautology.

Solution. That $(P \wedge (P \Rightarrow Q)) \Rightarrow Q$ is a tautology is verified in the truth table of Figure 1.17. ◆

P	Q	$P \Rightarrow Q$	$P \wedge (P \Rightarrow Q)$	$(P \wedge (P \Rightarrow Q)) \Rightarrow Q$
T	T	T	T	**T**
T	F	F	F	**T**
F	T	T	F	**T**
F	F	T	F	**T**

Figure 1.17: $(P \wedge (P \Rightarrow Q)) \Rightarrow Q$ is a tautology

The tautology

$$(P \wedge (P \Rightarrow Q)) \Rightarrow Q$$

considered in Example 1.64 is called **modus ponens** (mode that affirms the hypothesis) in logic. We will encounter this tautology in Chapter 3, as well as the tautology

$$((P \Rightarrow Q) \wedge (\sim Q)) \Rightarrow (\sim P),$$

which is called **modus tollens** (mode that denies the conclusion). (See Exercise 5.)

Exercises for Section 1.5

1. For a given statement P, what is the negation of the tautology $P \vee (\sim P)$?

2. Verify by a truth table that for every two statements P and Q, each of the following is a tautology.
   (a) $(P \wedge Q) \Rightarrow P$.   (b) $P \Rightarrow P \vee Q$.   (c) $(P \wedge Q) \Rightarrow (P \vee Q)$.

3. Verify by a truth table that for two statements P and Q, each of the following is a tautology.
   (a) $\sim (P \Rightarrow Q) \Rightarrow P$.   (b) $\sim (P \Rightarrow Q) \Rightarrow (\sim Q)$.

4. Verify by a truth table that for two statements P and Q, each of the following is a tautology.
   (a) $(\sim P) \Rightarrow (P \Rightarrow Q)$.   (b) $(P \wedge Q) \Rightarrow (P \Rightarrow Q)$.

5. For two statements P and Q, show that the compound statement modus tollens

$$((P \Rightarrow Q) \wedge (\sim Q)) \Rightarrow (\sim P)$$

   is a tautology.

6. Verify by a truth table that for two statements P and Q, the statement

$$(P \wedge (\sim Q)) \wedge (P \wedge Q)$$

   is a contradiction.

7. For statements P and Q, determine whether the compound statement

$$(P \wedge Q) \Leftrightarrow ((\sim P) \vee (\sim Q))$$

is a tautology, a contradiction or neither.

8. For statements P and Q, determine whether the compound statement

$$(P \wedge (\sim Q)) \Rightarrow (P \vee Q)$$

is a tautology, a contradiction or neither.

9. For statements P and Q, determine whether the compound statement

$$(P \vee Q) \Rightarrow (\sim P)$$

is a tautology, a contradiction or neither.

10. Let S and R be two compound statements with the same component statements. If S is a tautology and R is a contradiction, then what is the truth value of the following?

(a) $S \vee R$. (b) $S \wedge R$. (c) $S \Rightarrow R$. (d) $R \Rightarrow S$. (e) $S \Leftrightarrow R$.

11. Let R and S be two compound statements with the same component statements, where S is a contradiction. If the implication $(\sim R) \Rightarrow S$ is true, then what can be said about R?

12. Find a compound statement constructed from two component statements P and Q that is true for 50% of the combinations of truth values of P and Q.

13. Find a compound statement constructed from three component statements P, Q and R that is true for 50% of the combinations of truth values of P, Q and R.

1.6 Some Applications of Logic

The Scottish-born author Sir Arthur Conan Doyle is best known for his famous fictional detective Sherlock Holmes who would often use his skillful reasoning and astute observations to solve a criminal case he was investigating. He would announce how he arrived at his conclusion by uttering "It's elementary, my dear Watson." to his friend and associate Dr. John Watson. While Holmes used logic to solve difficut cases, our goal is to use logic to solve problems and to understand mathematical explanations. In this section we present a few examples of problems to see how logic can be used to find solutions.

Example 1.65

Your friend knows the truth value of a certain statement P but you don't. For every statement constructed from P, your friend promises always to give the correct truth value or always to give the incorrect truth value. That is, your friend will always be truthful or will always lie. Show that you can determine the truth value of P by giving two statements to your friend.

Solution. Notice that it wouldn't be useful to give the statement P itself to your friend, for, regardless of how your friend responded, we wouldn't know if it's the truth or a lie. First, give your friend a statement whose truth value is known to you. For example, give your friend the statement $P \vee (\sim P)$. Since this is a tautology and is therefore true, you will learn instantly whether your friend is always being truthful or is always lying. Once you know this, then give your friend the statement P. ♦

Example 1.66

Lieutenant Klumbo of the Los Angeles Police Department has spent the day interrogating four suspects in a murder case: Adams, Benjamin, Carter, Dickens. Each claims that he has an airtight alibi for the night the murder took place as each says that he was in Las Vegas that night (not in Los Angeles where the murder took place). At the end of the day, Lieutenant Klumbo has learned the following:

(1) If Adams went to Las Vegas, so did Benjamin.

(2) Benjamin and Carter did not both go to Las Vegas.

(3) If Carter went to Las Vegas, then Adams stayed in Los Angeles.

(4) Carter and Dickens did not both stay in Los Angeles.

(5) If Adams stayed in Los Angeles, so did Dickens.

(6) If Dickens went to Las Vegas, then Benjamin stayed in Los Angeles.

At the end of the day, Lieutenant Klumbo decides to release one of the four suspects. Which one and why?

Solution. Lieutenant Klumbo releases Carter.

(i) Suppose first that Benjamin went to Las Vegas. Then Carter did not go to Las Vegas. So Dickens did, implying that Benjamin stayed in Los Angeles and producing a contradiction.

(ii) Suppose next that Adams went to Las Vegas. Then Benjamin went to Las Vegas. By (i), a contradiction occurs.

(iii) Suppose next that Dickens went to Las Vegas. Then Benjamin stayed in Los Angeles and so did Adams. However, then Dickens stayed in Los Angeles as well, producing a contradiction.

Therefore, a contradiction is produced for each of Benjamin, Adams and Dickens. That is, each has been caught in a lie. There is no such problem for Carter, however. ♦

Example 1.67

The defendant in a trial has been accused of a robbery. During the trial, the prosecutor announces:

If the defendant committed the robbery, then he had an accomplice.

Upon hearing this, the defense attorney jumps to his feet and cries out:

That's a lie!

Determine which of the following is true.

(i) The jury in the trial should find the defendant guilty.

(ii) The jury in the trial should find the defendant not guilty.

(iii) The jury in the trial does not have enough information to reach a verdict.

Solution. (i) is true. If we let

P: The defendant committed the robbery. Q: He had an accomplice.

then the prosecutor states the implication $P \Rightarrow Q$. The defense attorney therefore proclaims that the implication $P \Rightarrow Q$ is false. The only way this can occur is if P is true and Q is false. However, since P is true, the defendant committed the robbery. Furthermore, since Q is false, he committed the robbery by himself. So the jury should find the defendant guilty (and the defendant should fire the defense attorney). ♦

Example 1.68

You are given two wicks (the wicks only, not as parts of candles), each of which takes one hour to burn completely and neither of which burns at a constant rate. How can these wicks be used to determine when 45 minutes has elapsed?

Solution. Light both ends of wick #1 and one end of wick #2 at the same time. When wick #1 is completely burned, 30 minutes have elapsed and at that instant light the other end of wick #2. When wick #2 is completely burned, 45 minutes have elapsed. ♦

Exercises for Section 1.6

1. You are given three wicks (the wicks only, not as parts of candles), each of which takes 80 minutes to burn completely but none of which burns at a constant rate. How can these wicks be used to determine when an hour and a half has elapsed?

2. There are three light switches on a wall, each of which operates one of three light bulbs in a room. This room can only be entered and seen from a doorway that is a few steps from the location of the light switches. All light switches are in the down position and so all light bulbs are dark. When one of the light switches is moved to the up position, one of the light bulbs is lit; however, we don't know which light switch operates which light bulb since the room cannot be seen from this location. Show that it is possible to determine the light switch that operates each light bulb by entering the room exactly once.

3. You are given a balance scale and eight coins, one of which is counterfeit and weighs less than the other coins. How can the counterfeit coin be determined in two weighings. [Problems of this nature will also be explored in more detail in Section 13.2, Chapter 13.]

4. After dividing a collection of coins into two groups, it is observed that 75 times the difference between the number of coins in each group equals the difference of the squares of these two numbers. How many coins are in the collection?

5. At a party, Charlie asks two friends to give him an example of an integer. He notices that the sum of one of these numbers and the square of the other gives the same result. Are these two numbers equal?

6. Figure 1.18 shows a 5×5 checkerboard whose 25 squares are labeled by $1, 2, \ldots, 25$. A coin placed on one of the squares can be moved to any adjacent square in the same row or column of the first square. For example, if a coin is placed on square 6, then the coin can be moved to square 1, 7 or 11. Determine whether the 25 squares can be listed in some order so that it is possible to move the coin from each square on the list to the next square on the list and then move the coin from the last square on the list back to the first square on the list.

1	2	3	4	5
6	7	8	9	10
11	12	13	14	15
16	17	18	19	20
21	22	23	24	25

Figure 1.18: The 5×5 checkerboard in Exercise 6

7. A puzzle consists of a 4×4 grid, some of whose squares are assigned one of the numbers 1, 2, 3, 4. A solution of the puzzle consists of assigning each blank square of the grid one of the numbers 1, 2, 3, 4 so that each of these numbers appears in every row, in every column and in each of the four 2×2 boxes of the grid. Determine which of the three puzzles of Figure 1.19 have solutions.

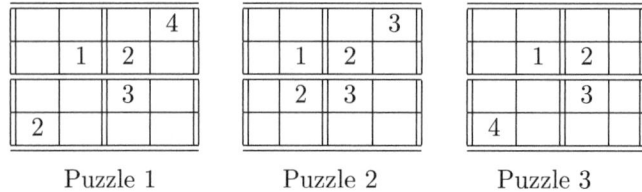

Figure 1.19: Three puzzles in Exercise 7

8. Every year some of Harry's friends give him a birthday party and do something that attempts to make him think. Five of Harry's friends enter the party one-by-one in the order: Kelly, Chuck, Barb, Lionel, Maria. Who would you expect to enter next: Robert, Daniel or Patrick?

9. Suppose that we are given a 3×6 checkerboard of Figure 1.20, where each square is red (R) or black B, and a collection of coins, each of which can be placed on a square of the checkerboard. Two squares are adjacent if they are in the same row or same column and there are no squares between them.

   (a) Is it possible to place coins on this checkerboard so that every black square has 1 coin on its adjacent squares and every red square has no coins on its adjacent squares?

   (b) Is it possible to place coins on the checkerboard, at most one coin on each square, so that every black square has an odd number of coins on its adjacent squares and every red square has an even number of coins on its adjacent squares?

R	B	R	B	R	B
B	R	B	R	B	R
R	B	R	B	R	B

Figure 1.20: The 3×6 checkerboard in Exercise 9

10. Figure 1.21 shows a building with nine rooms. Each room has a light switch. When the light switch is pressed, the light in the room turns on if it was originally off or turns off if it was originally on. The same thing happens to the rooms adjacent to the room (in the same row or column). So if all lights are off and the light switch in R_2 is pressed, the light goes on in the rooms R_1, R_2, R_3 and R_5. Suppose that all lights are off originally.

   (a) Show that it is possible to turn all the lights on.

   (b) What is the minimum number of light switches that must be pressed to turn all lights on?

R_1	R_2	R_3
R_4	R_5	R_6
R_7	R_8	R_9

Figure 1.21: A buildling with nine rooms in Exercise 10

Chapter 1 Highlights

Key Concepts

biconditional, $P \Leftrightarrow Q$: P if and only if Q. If P and Q are statements, then $P \Leftrightarrow Q$ is true only when P and Q are both true or are both false.

compound statement: a statement constructed from one or more statements called **component statements** and one or more logical connectives.

conclusion (of $P \Rightarrow Q$): the statement (or sentence) Q in the implication $P \Rightarrow Q$.

conjunction (of P and Q), $P \wedge Q$: P and Q. If P and Q are statements, then $P \wedge Q$ is true only when P and Q are both true.

contradiction: a compound statement that is false for all possible combinations of truth values of its component statements.

contrapositive (of $P \Rightarrow Q$): $(\sim Q) \Rightarrow (\sim P)$.

converse (of $P \Rightarrow Q$): $Q \Rightarrow P$.

disjunction (of P and Q), $P \vee Q$: P or Q. If P and Q are statements, then $P \vee Q$ is true only when at least one of P and Q is true.

exclusive or (of P and Q), $P \oplus Q$: P or Q but not both. If P and Q are statements, then $P \oplus Q$ is true if exactly one of P and Q is true.

hypothesis (of $P \Rightarrow Q$): the statement (or sentence) P in the implication $P \Rightarrow Q$.

implication, $P \Rightarrow Q$: if P, then Q, where P is the **hypothesis** of $P \Rightarrow Q$ and Q is the **conclusion** of $P \Rightarrow Q$. If P and Q are statements, then $P \Rightarrow Q$ is a statement that is true unless P is true and Q is false.

inclusive or: P or Q (same as disjunction).

logical connectives: $\sim$ (not), $\vee$ (or), $\oplus$ (exclusive or), $\wedge$ (and), $\Rightarrow$ (implies), $\Leftrightarrow$ (if and only if).

logically equivalent, $R \equiv S$: R and S have the same truth value for each combination of truth values of the component statements that form R and S.

modus ponens: the tautology $(P \wedge (P \Rightarrow Q)) \Rightarrow Q$.

modus tollens: the tautology $((P \Rightarrow Q) \wedge (\sim Q)) \Rightarrow (\sim P)$.

negation (of P), $\sim P$: not P. If P is a statement, then $\sim P$ is a statement whose truth value is opposite to that of P.

open sentence: a declarative sentence containing one or more variables and whose truth or falseness depends on the truth values of the variables.

statement: a declarative sentence that is true or false but not both.

tautology: a compound statement that is true for all possible combinations of truth values of its component statements.

truth table: a table displaying the relationships among the truth values of statements.

truth value: true T or false F.

Key Results

- For statements P, Q and R, the following hold:

  Commutative Laws: $P \vee Q \equiv Q \vee P$ and $P \wedge Q \equiv Q \wedge P$.

  De Morgan's Laws: $\sim (P \vee Q) \equiv (\sim P) \wedge (\sim Q)$ and $\sim (P \wedge Q) \equiv (\sim P) \vee (\sim Q)$.

  Associative Laws: $P \vee (Q \vee R) \equiv (P \vee Q) \vee R$ and $P \wedge (Q \wedge R) \equiv (P \wedge Q) \wedge R$.

  Distributive Laws: $P \vee (Q \wedge R) \equiv (P \vee Q) \wedge (P \vee R)$ and $P \wedge (Q \vee R) \equiv (P \wedge Q) \vee (P \wedge R)$.

  Logical Equivalences:

  $$P \equiv \sim (\sim P).$$
  $$P \Rightarrow Q \equiv (\sim Q) \Rightarrow (\sim P).$$
  $$P \Rightarrow Q \equiv (\sim P) \vee Q.$$
  $$\sim (P \Rightarrow Q) \equiv P \wedge (\sim Q).$$

Supplementary Exercises for Chapter 1

1. Classify each of the following declarative sentences as (1) an open sentence, (2) a true statement or (3) a false statement.

   (a) $(25)^2 = 2 \cdot 3 \cdot 100 + 25$. (b) $x > y$. (c) 0 is even.

   (d) $(\frac{1}{2})^2 > \frac{1}{2}$. (e) $3n + 5$ is even.

2. Let $P(x)$: The month x has 31 days. Determine the truth value for each of the following statements.

   (a) $P(\text{June})$.   (b) $P(\text{July})$.   (c) $P(\text{August})$.

3. The following is an open sentence where x is a positive integer:

   $$P(x) : (2x - 1)(x + 3) = 0.$$

   (a) For what values of x is $P(x)$ a true statement?

   (b) For what values of x is $P(x)$ a false statement?

4. Each of the following is an open sentence, where n represents an integer. For which integers n are the following open sentences true statements?

   (a) $Q_1(n) : |n + 4| > 0$. (b) $Q_2(n) : n^2 - 2n < 0$. (c) $Q_3(n) : n^2 - 3n = 4$.

   (d) $Q_4(n) : n + \frac{4}{n} \leq 4$. (e) $Q_5(n) : n^2 + 2^n \leq 8$.

5. For an integer n, consider the following open sentences:

   $$P(n): \ n^2 - 4n < 0. \quad Q(n): \ n^3 - n = 0.$$

   Determine all integers n for which the following are true.

   (a) $P(n) \wedge Q(n)$.   (b) $P(n) \vee Q(n)$.   (c) $P(n) \oplus Q(n)$.

6. For an integer n, consider the following open sentences:

   $$P(n): \ (n + 1)(n + 3) \text{ is odd.} \quad Q(n): \ 2n + 5 \text{ is even.}$$

(a) Give an example of an integer a such that $P(a) \Rightarrow Q(a)$ is true.

(b) Give an example of an integer b such that $P(b) \Rightarrow Q(b)$ is false.

7. For the statements

$$P:\ 4^2 \neq 2^4. \quad Q:\ 1+2+3 = 1 \cdot 2 \cdot 3.$$

state each of the following in words and determine whether they are true or false.

(a) $\sim P$.　(b) $P \vee Q$.　(c) $P \wedge Q$.　(d) $P \Rightarrow Q$.

8. State the negation of each of the following.

(a) Either x is an odd integer or y is an odd integer.

(b) Both x and y are odd integers.

(c) Neither x nor y is an odd integer.

9. For the sentences

$$P:\ ab \text{ is odd.} \quad Q:\ a \text{ is odd and } b \text{ is odd.}$$

state each of the following in words:

(a) $P \Rightarrow Q$ using the phrase "only if."

(b) $P \Rightarrow Q$ using the word "sufficient."

(c) $P \Rightarrow Q$ using the word "necessary."

(d) $\sim (P \Rightarrow Q)$.

(e) the converse of $P \Rightarrow Q$.

(f) the contrapositive of $P \Rightarrow Q$.

10. Each of the following statements is an implication $P \Rightarrow Q$. For each statement, indicate what P and Q are.

(a) I'm going to my class reunion only if I lose weight.

(b) To win a free $20 gift certificate, I must spend $100 at the store.

(c) To win the game, it is necessary that we score a touchdown.

(d) It is necessary to do research to be promoted to professor.

(e) I'll get an A on this exam if I'm lucky.

(f) All I need is a B on the final exam to get an A in the course.

11. Consider the statement S: If I get paid today, then I'm going to the party tonight.

(a) Assuming that S is true and that I got paid today, does this mean that I'm going to the party tonight?

(b) Assuming that S is true and that I went to the party tonight, does this mean that I got paid today?

(c) Assuming that S is true and that I didn't get paid today, does this mean that I don't go to the party tonight?

12. Consider the statement S: I'm going to the party tonight only if I get paid today.

(a) Assuming that S is true and that I got paid today, does this mean that I'm going to the party tonight?

(b) Assuming that S is true and that I went to the party tonight, does this mean that I got paid today?

(c) Assuming that S is true and that I didn't get paid today, does this mean that I'm not going to the party tonight?

13. Consider the following two statements

$$P: \text{I didn't get the job.} \quad Q: \text{I will go back to college.}$$

Express the following in words.
(a) $P \Rightarrow Q$.   (b) $(\sim P) \vee Q$.   (c) $(\sim P) \Rightarrow (\sim Q)$.   (d) $P \wedge Q$.

14. Consider the following two statements

$$P: \text{The rent doesn't increase.} \quad Q: \text{I will renew my lease.}$$

Express the following in words.
(a) $P \Rightarrow Q$.   (b) $(\sim P) \Rightarrow (\sim Q)$.   (c) $(\sim P) \vee Q$.   (d) $P \wedge Q$.

15. Consider the following two statements

$$P: \text{It is Friday.} \quad Q: \text{I don't go to class.}$$

Express the following in words.
(a) $P \Rightarrow Q$.   (b) $(\sim Q) \Rightarrow (\sim P)$.   (c) $P \wedge Q$.   (d) $P \vee (\sim Q)$.

16. State the negation of each of the following implications using Theorem 1.50.

(a) If you didn't vote, then you shouldn't complain.
(b) If you study discrete mathematics, then you should do well in this course.
(c) If $a > b$, then $a^2 > b^2$.

17. State the converse and contrapositive of each of the following implications.

(a) If you do every problem in this book, then you will get an A in this course.
(b) If $a + b$ is odd, then a is even and b is odd.
(c) If $a > 1$, then $a^2 > 1$.

18. For statements P and Q, construct a truth table for $((\sim P) \vee Q) \Rightarrow P$.

19. For an integer n, consider the following open sentences

$$P(n): 5n + 3 \text{ is even.} \quad Q(n): 4n + 3 \text{ is even.}$$

(a) Give an example of an integer a such that $P(a) \Rightarrow Q(a)$ is true.
(b) Give an example of an integer b such that $P(b) \Rightarrow Q(b)$ is false.

20. Consider the following statements

$$P: 2^3 + 1 \text{ is odd.} \quad Q: 3^2 + 1 \text{ is odd.}$$

Determine the truth value of $P \Leftrightarrow Q$.

21. Without using a truth table, show for statements P and Q that

$$\sim (P \vee ((\sim P) \wedge Q)) \equiv (\sim P) \wedge (\sim Q).$$

22. Each of the following is an open sentence over the domain S consisting of the three integers $1, 2$ and 3.

$P(n) : \frac{2^{n+1}+3^{n-1}}{5^{n-1}}$ is an integer.

$Q(n) : \frac{n^2+2n+3}{n^3+1} \geq 1$.

$R(n) : \frac{n^2+3n}{2}$ is an odd integer.

The integers a, b and c are the integers $1, 2$ and 3 in some order. What are the values of a, b and c if the implications

$$P(a) \Rightarrow Q(a), \quad Q(b) \Rightarrow R(b), \quad R(c) \Rightarrow P(c)$$

are true but all of their converses are false?

23. If we know the truth value of each of the following compound statements, would this give us any information about the truth value of P?

   (a) $((\sim P) \Rightarrow P) \Rightarrow (P \Rightarrow (\sim P))$.
   (b) $(P \vee (\sim P)) \Rightarrow P$.
   (c) $P \Rightarrow (P \Rightarrow (\sim P))$.

24. Consider the following statement:

   If Rob can get a car to drive, then he will take Sue to the movies.

   Which of the following is logically equivalent to this statement?

   (a) Either Rob won't get a car to drive or he will take Sue to the movies.
   (b) If Rob takes Sue to the movies, then he can get a car to drive.
   (c) If Rob doesn't take Sue to the movies, then Rob couldn't get a car to drive.
   (d) If Rob can't get a car to drive, then he won't take Sue to the movies.

25. Al's doctor has prescribed two medicines. In each case, Al is required to take one pill of each medication per day for 10 days. The pills are identical in shape and color. After taking the pills on the fifth day, Al accidentally places two pills from one bottle on a table containing a pill from the other bottle. He can't tell the pills apart. What should Al do? (Ask Marilyn, *Parade Magazine*, March 23, 2008, page 11.)

26. Each animal at a zoo is housed in a numbered cage. The number on the cage for four animals is given below.

   Bear 16;   Polar Bear 37;   Cougar: 24;   Elephant 34.

   From the information given above, what might be a logical number on the cage of the Giraffe?

27. Little Red Riding Hood is taking her customary walk through the woods one day to visit her grandmother. Her route requires her to cross seven bridges. On this particular day, she has made some cakes to take on her trip to give to her grandmother as a gift. The wolves in the woods learn of her plan and one wolf is stationed at each bridge. When Red arrives at a bridge, she is required to give the wolf there half of the cakes she has at that moment. In exchange for that, the wolf will return one of these cakes to her. How many cakes must Red have at the beginning of her walk in order to give her grandmother two cakes?

28. Let $S = \{1, 2, 3, 4, 5\}$. A puzzle consists of a 4×4 grid, some of whose squares are assigned one of the numbers from S. A solution of the puzzle consists of assigning each blank square of the grid an element of S so that no element of S appears twice in any row, in any column or in any of the four 2×2 boxes of the grid. Determine which of the two puzzles of Figure 1.22 have solutions.

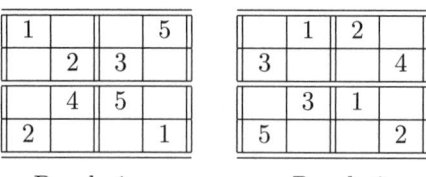

Figure 1.22: Two puzzles in Exercise 28

Chapter 2

Sets

One of the most fundamental concepts in mathematics is that of a set. The study of sets and their properties is a subject in itself, namely set theory. The history of set theory is unlike many mathematical subjects. The origins of many mathematical areas can often be traced back to a number of mathematicians whose ideas over a relatively short period of time created the subject. Set theory, however, is essentially the creation of a single individual: Georg Cantor. Although much of his work is beyond the scope of this book, we will nonetheless see some of his best known discoveries in Chapter 5 (Section 5.5). In this chapter, however, we will be content to introduce the basic concepts, notation and ideas of elementary set theory.

The importance of set theory is not only due to its usefulness in the foundation of discrete mathematics but to its usefulness in all of modern mathematics.

2.1 Sets and Subsets

For our purposes, a **set** is a collection of objects. These objects, although often numbers of some kind, have no restrictions on what they can be. The objects of a set are called the **elements** of the set. We will see many ways to describe sets. Regardless of how a set is described, what is most important is that its description should make it clear which elements it contains. If a set has just a few elements, then the set is often described simply by listing its elements between braces, separated by commas. For example, $\{x, y, z\}$ is a set consisting of the three elements x, y and z. Normally, we use a capital (upper case) letter to denote a set, such as A, B, C, S, X, Y or perhaps A_1, A_2, A_3 and so on. We could therefore denote the set above by $S = \{x, y, z\}$. Typically, a general element that belongs to a set is represented by a lower case letter. If a is an element of a set A, then we write $a \in A$. If b is not an element of A, we write $b \notin A$. Therefore, for the set $S = \{x, y, z\}$, $x \in S$ but $w \notin S$.

Two sets A and B are **equal**, written $A = B$, if they consist of exactly the same elements. That is, if $A = B$, then A and B are simply different names for the same set. If A and B are not equal, then we write $A \neq B$. If $A \neq B$, then at least one of these sets contains an element not belonging to the other set. For example, if $S = \{x, y, z\}$ and $T = \{w, x, y, z\}$, then $S \neq T$ since T contains an element, namely w, that does not belong to S.

If a set is described by listing its elements, then the order in which the elements are listed does not matter. For example,

$$S = \{x, y, z\} = \{y, x, z\} = \{z, y, x\}.$$

The set $A = \{1, 2, \ldots, 100\}$ consists of the first one hundred positive integers and $B = \{2, 4, 6, \ldots\}$ is the set of positive even integers. The three dots "$\ldots$", called an **ellipsis**, means "and so on up to" for the set A and "and so on" for the set B.

A set need not contain any elements. The set containing no elements is called the **empty set**. (This set is sometimes called the **null set** or **void set** as well.) The empty set is denoted by $\emptyset$. Thus $\emptyset = \{\ \}$. A **nonempty set** therefore contains at least one element. For a finite set A (a set with a finite number of elements), we write $|A|$ to denote the number of elements in A. This is called the **cardinality** of A. Therefore, the set $S = \{x, y, z\}$ that we encountered earlier has cardinality 3 and so $|S| = 3$.

Example 2.1

If A is the set of letters in the English alphabet and C is the set of playing cards in a standard deck of cards, then $|A| = 26$ and $|C| = 52$.   ◆

Example 2.2

Since the sets $\{\emptyset\}$ and $\emptyset$ do not consist of the same elements, $\{\emptyset\} \neq \emptyset$. (An empty box and a box containing an empty box are not the same thing.) The set $\{\emptyset\}$ has one element, namely the empty set $\emptyset$ and so $|\{\emptyset\}| = 1$. However, $\emptyset$ has no elements and so $|\emptyset| = 0$.   ◆

Example 2.3

The set $A = \{1, \{1, 3\}, \emptyset, a\}$ has four elements, two of which are sets, namely $\{1, 3\}$ and $\emptyset$. Since A has four elements, its cardinality $|A|$ equals 4.   ◆

Well-Known Sets of Numbers

Some of the best known sets are infinite sets of numbers (those containing an infinite number of elements) and many of these are denoted by a special symbol. For example, the set $\{\ldots, -2, -1, 0, 1, 2, \ldots\}$ of integers is often denoted by $\mathbb{Z}$. This symbol is used because Z is the first letter of the word *Zahlen* (the German word for numbers). The set of positive integers or **natural numbers** is denoted by $\mathbb{N}$. Thus $\mathbb{N} = \{1, 2, 3, \ldots\}$. (While historically 0 was not considered to be a natural number, that issue became clouded in later years. Nevertheless, we continue to describe the set of natural numbers by $\mathbb{N} = \{1, 2, 3, \ldots\}$.)

The set of real numbers is denoted by $\mathbb{R}$ and the set of positive real numbers is denoted by $\mathbb{R}^+$. Although it is not easy to describe precisely what a real number is (indeed, the definition of a real number is quite technical and involved), the real numbers consist of the rational numbers and irrational numbers.

A **rational number** is the ratio a/b of two integers a and b, where $b \neq 0$. The set of rational numbers is denoted by $\mathbb{Q}$ (for the first letter of the word *quotient*). An **irrational number** is a real number that is not rational. Examples of irrational numbers are $\sqrt{2}$, $\sqrt{3}$, $\sqrt[3]{2}$ and π. The number e, encountered throughout calculus, is also irrational. In particular,

$$0 \in \mathbb{Z} \text{ but } 0 \notin \mathbb{N},$$
$$\sqrt{2} \in \mathbb{R} \text{ but } \sqrt{2} \notin \mathbb{Q},$$
$$3/2 \in \mathbb{Q} \text{ but } 3/2 \notin \mathbb{Z}.$$

We now consider another useful and common way to describe the elements that belong to a set. For a given set S, let $P(x)$ denote some open sentence involving elements $x \in S$. Then the set

$$A = \{x \in S :\ P(x)\}$$

describes the set of all those elements $x \in S$ for which $P(x)$ is true. If the set S is clear or understood, then A may also be written as

$$A = \{x :\ P(x)\}.$$

In the description of the set A, the colon is considered to mean "such that", although some use a vertical line for this purpose. That is,

$$A = \{x \in S \mid P(x)\} = \{x \in S : P(x)\} \text{ or } A = \{x \mid P(x)\} = \{x : P(x)\}.$$

Example 2.4

The set

$$A = \{n \in \mathbb{Z} : n^2 \leq 4\}$$

consists of all those integers n such that $n^2 \leq 4$. Since the only integers whose squares are at most 4 are $0, \pm 1, \pm 2$, it follows that $A = \{0, \pm 1, \pm 2\}$. Therefore,

$$A = \{n \in \mathbb{Z} : n^2 \leq 4\} = \{n \in \mathbb{Z} \mid n^2 \leq 4\} = \{-2, -1, 0, 1, 2\}.$$

Also, $A = \{n \in \mathbb{Z} : |n| \leq 2\}$.   ◆

Example 2.5

Write each of the following sets by listing its elements.

$$A = \{x \in \mathbb{R} : x^2 - x - 6 = 0\} \text{ and } B = \{x \in \mathbb{R} : x^2 + 1 = 0\}.$$

Solution. The set A represents the set of (real number) solutions of the equation $x^2 - x - 6 = 0$. Since $x^2 - x - 6 = (x - 3)(x + 2) = 0$, it follows that $A = \{-2, 3\}$. Because the equation $x^2 + 1 = 0$ has no real number solutions, $B = \{ \ \} = \emptyset$.   ◆

Subsets

Definition 2.6 *A set A is called a* **subset** *of a set B, written $A \subseteq B$, if every element of A also belongs to B.*

For example, $\mathbb{N} \subseteq \mathbb{Z}$, $\mathbb{Z} \subseteq \mathbb{Q}$ and $\mathbb{Q} \subseteq \mathbb{R}$. In particular, $A \subseteq A$ for every set A. If A is not a subset of B, then we write $A \nsubseteq B$. If $A \nsubseteq B$, then there is some element of A that does not belong to B. For every set A, the empty set $\emptyset$ is also a subset of A. Let's see why this is true. If this weren't so, then $\emptyset \nsubseteq A$ would mean that there is some element in $\emptyset$ that is not in A. However, $\emptyset$ has no elements. Thus, as we claimed, $\emptyset \subseteq A$ for every set A.

Example 2.7

If $A = \{a, b, c\}$ and $B = \{a, b, c, d, e\}$, then $A \subseteq B$. For the sets $C = \{1, 2\}$ and $D = \{1, 3\}$, we have $C \nsubseteq D$ since $2 \in C$ but $2 \notin D$. Also, $D \nsubseteq C$.   ◆

Since two sets A and B are equal if they have precisely the same elements, this means that every element of A belongs to B and every element of B belongs to A. That is,

$$A = B \text{ if and only if } A \subseteq B \text{ and } B \subseteq A.$$

Definition 2.8 *A set A is a* **proper** *subset of a set B, written $A \subset B$, if $A \subseteq B$ but $A \neq B$.*

Therefore, $\mathbb{N} \subset \mathbb{Z}$, $\mathbb{Z} \subset \mathbb{Q}$ and $\mathbb{Q} \subset \mathbb{R}$. If A is a subset of a set B and A is not a proper subset of B, then necessarily $A = B$.

Example 2.9

For the sets $A = \{a, b, c\}$ and $C = \{a, b, c, d, e\}$, find all sets B such that $A \subset B \subset C$.

Solution. The only sets B for which $A \subset B \subset C$ are $\{a, b, c, d\}$ and $\{a, b, c, e\}$.   ◆

The use of the symbols $\subseteq$ and $\subset$ should remind you of the symbols $\leq$ and $<$ and the distinction between them. Nevertheless, there is some inconsistency among authors in the notation for set inclusion and proper set inclusion. For example, some write $A \subset B$ to mean A is a subset of B and $A \subsetneq B$ to mean A is a proper subset of B. Again, we write $A \subseteq B$ to mean A is a subset of B and $A \subset B$ to mean A is a proper subset of B.

Often when there is a discussion concerning sets, the sets involved are all subsets of some specified set, called the **universal set**, which is usually denoted by U. For example, if we are considering certain sets of integers, the universal set is $\mathbb{Z}$; while if we are considering certain sets of real numbers, the universal set is $\mathbb{R}$.

Venn Diagrams

It is occasionally convenient to draw a diagram in the plane, called a **Venn diagram**, to pictorially represent a set or sets under consideration. A rectangle is often drawn to represent the universal set. Within the rectangle, a closed curve (often a circle or ellipse) is drawn for each set being discussed. The location of regions inside or outside the curves indicates the sets to which an element may or may not belong. Venn diagrams are illustrated in the next example.

Example 2.10

Suppose that $U = \{1, 2, \ldots, 6\}$ is the universal set. Draw a Venn diagram for each of the following:

(a) the set $A = \{1, 2, 5, 6\}$.

(b) the set $B = \{1, 3, 6\}$.

(c) the two sets $C = \{1, 2, 4\}$ and $D = \{1, 2, 4, 5, 6\}$.

(d) the two sets $E = \{1, 2, 3\}$ and $F = \{3, 5, 6\}$.

Solution. See Figure 2.1.

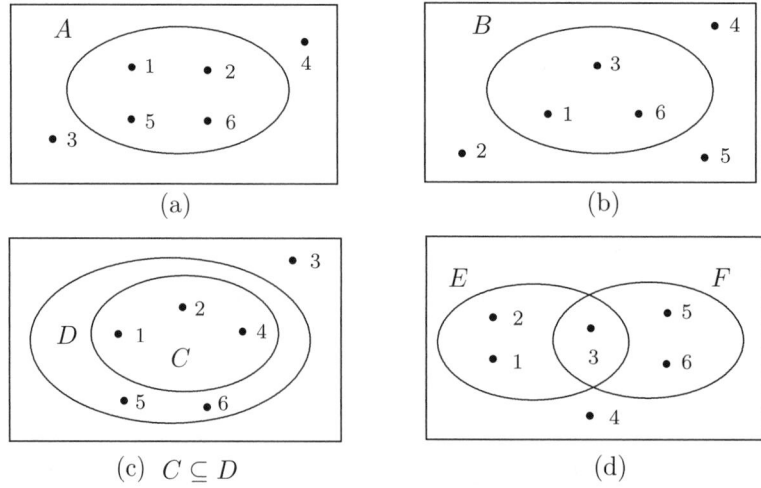

Figure 2.1: Venn diagrams for the sets in Example 2.10

(a) In this situation, the two elements 3 and 4 of U do not belong to A.

(b) Here the elements 2, 4 and 5 do not belong to B.

(c) Since every element of C belongs to D, it follows that $C \subseteq D$. In fact, $C \neq D$ and so $C \subset D$. The element 3 belongs to U but not to D.

(d) The elements 1, 2 and 3 belong to E. Since 1 and 2 do not belong to F, these elements are written inside E but outside of F. Since 5 and 6 belong to F but not to E, these two elements are drawn inside F but outside of E. The element 4 belongs to U but to neither E nor F. Only the element 3 belongs to both E and F. ♦

The standard and most appropriate way to draw a Venn diagram for two general sets A and B is shown in Figure 2.2 as it allows for the four possibilities:

- some elements may belong to both A and B,
- some elements may belong to A but not to B,
- some elements may belong to B but not to A and
- some elements may belong to neither A nor B.

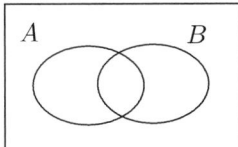

Figure 2.2: A Venn diagram for two general sets A and B

Example 2.11

A Venn diagram for two specific sets A and B is shown in Figure 2.3.

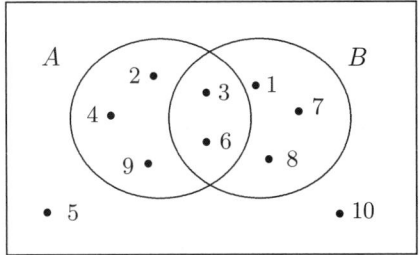

Figure 2.3: A Venn diagram for the two sets in Example 2.11

From this diagram, we can see that

- 3 and 6 belong to both sets A and B,
- 2, 4 and 9 belong to A but not to B,
- 1, 7 and 8 belong to B but not to A, while
- 5 and 10 belong to neither A nor B.

Therefore, $A = \{2, 3, 4, 6, 9\}$, $B = \{1, 3, 6, 7, 8\}$ and the universal set is $U = \{1, 2, \ldots, 10\}$. ♦

We mentioned that there are no restrictions on what the elements of a set can be. In fact, we have seen that an element of a set can itself be a set.

Example 2.12

(a) If $A_1 = \{1\}$ and $A_2 = \{\{1\}\}$, then each of A_1 and A_2 has a single element. So $|A_1| = |A_2| = 1$. The only element of A_1 is the number 1, while the only element of A_2 is the set $A_1 = \{1\}$, that is, $A_2 = \{A_1\}$. Notice that $A_1 \neq A_2$. Thus, $A_1 \in A_2$ and $A_1 \nsubseteq A_2$.

(b) The set $B = \{\{1, 2\}, \{2, 3\}\}$ has two elements and so $|B| = 2$. One element of B is the set $\{1, 2\}$, which contains 1 and 2, while the other element of B is the set $\{2, 3\}$ containing 2 and 3. However, none of the numbers 1, 2 and 3 belong to B.

(c) The set $C = \{1, 2, \{1, 2\}\}$ has three elements and so $|C| = 3$. Two of the elements of C are numbers, namely 1 and 2 and the other element of C is the set $\{1, 2\}$. Notice that $\{1, 2\} \subset C$ and $\{1, 2\} \in C$, that is, the set $\{1, 2\}$ is both a proper subset of C and an element of C.

(d) The set $D = \{0, \emptyset\}$ has cardinality 2. One element of D is the integer 0 and the other element of D is the empty set $\emptyset$. ◆

For a set A containing elements that are also sets, it can occasionally be confusing whether a given element belongs to A or whether a given set is a subset of A. Let's look at some examples of this.

Example 2.13

Let $A = \{\{1, \{1\}\}\}$, $B = \{1, 2, 3\}$, $C = \{1, \{\emptyset\}\}$, $D = \{1, 2, \{2\}\}$ and $E = \{\emptyset, \{0\}, 0\}$. Which of the following statements are true?

(a) $1 \in A$.

(b) $\{\emptyset\} \subseteq B$.

(c) $\{\emptyset\} \subseteq C$.

(d) $\{2\} \subseteq D$.

(e) $\{\emptyset\} \in E$.

Solution.

(a) The statement is false. The set A only has one element, namely the set $\{1, \{1\}\}$. Therefore, 1 is not an element of A.

(b) The statement is false. In order for $\{\emptyset\}$ to be a subset of B, $\emptyset$ must be an element of B. However, the only elements of B are 1, 2 and 3.

(c) The statement is false. The set C contains two elements, namely 1 and $\{\emptyset\}$. Since $\emptyset$ is not an element of C, the set $\{\emptyset\}$ is not a subset of C. However, $\{\emptyset\} \in C$.

(d) The statement is true. The elements of D are 1, 2 and $\{2\}$. Since 2 is an element of D, it follows that $\{2\} \subseteq D$. Observe that $\{2\} \in D$ as well.

(e) The statement is false. Since the elements of E are $\emptyset, \{0\}$ and 0, it follows that $\{\emptyset\}$ is not an element of E. However, $\{\emptyset\} \subseteq E$. ◆

Example 2.14

Give an example of three sets A, B and C such that $A \in B$, $A \subset B$ and $B \in C$.

Solution. We start with a simple example for A, say $A = \{1\}$. Since we want both $A \in B$ and $A \subset B$, this suggests choosing $B = \{A, 1\} = \{\{1\}, 1\}$. Since we are required to have $B \in C$, we choose $C = \{B\} = \{\{\{1\}, 1\}\}$. ◆

Power Sets

We have seen several sets for which some or all of their elements are sets. We next consider a set, all of whose elements are subsets of a given set.

Definition 2.15 *The set of all subsets of a set A is called the* **power set** *of A and is denoted by $\mathcal{P}(A)$. Thus*

$$\mathcal{P}(A) = \{B : B \subseteq A\}.$$

Since $A \subseteq A$ and $\emptyset \subseteq A$ for every set A, it follows that $A \in \mathcal{P}(A)$ and $\emptyset \in \mathcal{P}(A)$. Therefore, every nonempty set has at least two distinct subsets.

Example 2.16

Determine the power set of each of the sets $A = \{a, b\}$, $B = \{x, y, z\}$, $C = \{0, \{\emptyset\}\}$ and $D = \emptyset$.

Solution. The power sets of A, B, C and D are

$$
\begin{aligned}
\mathcal{P}(A) &= \{\emptyset, \{a\}, \{b\}, A\} \\
\mathcal{P}(B) &= \{\emptyset, \{x\}, \{y\}, \{z\}, \{x, y\}, \{x, z\}, \{y, z\}, B\} \\
\mathcal{P}(C) &= \{\emptyset, \{0\}, \{\{\emptyset\}\}, C\} \\
\mathcal{P}(D) &= \{\emptyset\}.
\end{aligned}
$$

Notice that even though D is empty, its power set $\mathcal{P}(D)$ is nonempty. ◆

Observe that in Example 2.16 we have listed the subsets of A, B and C systematically, first listing $\emptyset$, then the subsets with one element, the subsets with two elements and so on. You may have also noticed that each of the sets A and C in Example 2.16 has two elements, while both $\mathcal{P}(A)$ and $\mathcal{P}(C)$ have $4 = 2^2$ elements. Furthermore, the set B has three elements and $\mathcal{P}(B)$ has $8 = 2^3$ elements. This might very well lead you to suspect that if A has n elements for some nonnegative integer n, then $\mathcal{P}(A)$ has 2^n elements. This, in fact, is true and will be verified later. For the present however, we state this result for future use.

Theorem 2.17 *If A is a set with $|A| = n$, where n is a nonnegative integer, then*

$$|\mathcal{P}(A)| = 2^n.$$

Example 2.18

For $A = \{x \in \mathbb{Z} : |x| \leq 3\}$, how many elements are in $\mathcal{P}(A)$?

Solution. Since $A = \{-3, -2, -1, 0, 1, 2, 3\}$, it follows that $|A| = 7$. By Theorem 2.17, the number of elements in $\mathcal{P}(A)$ is $|\mathcal{P}(A)| = 2^7 = 128$. ◆

Exercises for Section 2.1

1. Write each of the following sets by listing its elements.

   (a) $A = \{n \in \mathbb{Z} : n^2 - n - 2 < 0\}$.

   (b) $B = \{x \in \mathbb{R} : x^3 = x\}$.

   (c) $C = \{n \in \mathbb{Z} : -4 < n \le 4\}$.

   (d) $D = \{n \in \mathbb{N} : n^3 < 100\}$.

   (e) $E = \{x \in \mathbb{R} : x^2 + 5 = 0\}$.

2. Let $S = \{-3, -2, -1, 0, 1, 2, 3, 4\}$. Describe each of the following subsets of S as $\{x \in S : P(x)\}$, where $P(x)$ is some open sentence involving elements $x \in S$.

   (a) $A = \{1, 2, 3, 4\}$. (b) $B = \{0, 1, 2, 3, 4\}$.

   (c) $C = \{-3, -2, -1\}$. (d) $D = \{-2, 0, 2, 4\}$.

3. Describe each of the following sets in symbols.

   (a) The set of all integers whose absolute value is at most 3.

   (b) The set of all positive integers whose square root is also an integer.

   (c) The set of all integers strictly between -3 and 5.

4. Describe each of the following sets in words.

   (a) $\{x \in \mathbb{R} : x^5 - 3x^3 - x^2 + 4x - 1 = 0\}$.

   (b) $\{n \in \mathbb{Z} : \frac{n}{3} \in \mathbb{Z}\}$.

   (c) $\{\ldots, -5, -3, -1, 1, 3, 5, \ldots\}$.

5. For each of the following sets, indicate whether 2 is an element of the set.

   (a) $A = \{x \in \mathbb{N} : x > 2\}$.

   (b) $B = \{x \in \mathbb{Z} : |x| \le 2\}$.

   (c) $C = \{x \in \mathbb{Q} : x(x - 2) = 0\}$.

   (d) $D = \{x \in \mathbb{R} - \mathbb{Q} : x(x - 2) = 0\}$

   (e) $E = \{\{2\}\}$.

6. For each of the following sets, indicate whether a is an element of the set.

   (a) $A = \{a, \{b\}\}$. (b) $B = \{\{a\}, b\}$.

   (c) $C = \{\{a, b\}\}$. (d) $D = \{\{a\}, \{b\}\}$.

7. Determine the cardinality of each of the following sets.

   (a) $A = \{a, b, c, d, e\}$.

   (b) $B = \{0, 2, 4, \ldots, 30\}$.

   (c) $C = \{35, 36, 37, \ldots, 50\}$.

   (d) $D = \{1, \{1\}, \{1, 2\}\}$.

   (e) $E = \{\emptyset, \{\emptyset\}\}$.

   (f) $F = \{\{2\}, \{2, 3, 4\}\}$.

8. Determine the cardinality of each of the following sets.

   (a) $A = \{\{\emptyset\}\}$. (b) $B = \{\emptyset, \{\emptyset\}\}$. (c) $C = \{\emptyset, \{\emptyset\}, \{\emptyset, \{\emptyset\}\}\}$.

9. Which of the following pairs of sets are equal?

   (a) $A_1 = \{1, 2, 3\}$ and $A_2 = \{2, 3, 1\}$.

   (b) $B_1 = \{1, 2\}$ and $B_2 = \{\{1\}, \{2\}\}$.

   (c) $C_1 = \{1, 2\}$ and $C_2 = \{\{1, 2\}\}$.

10. Which of the following sets are equal?

    (a) $A = \{n \in \mathbb{Z} : |n| < 2\}$.

    (b) $B = \{n \in \mathbb{Z} : n^2 < 4\}$.

    (c) $C = \{n \in \mathbb{Z} : n^3 - n = 0\}$.

    (d) $D = \{-1, 0, 1\}$.

    (e) $E = \{n \in \mathbb{Z} : n^2 \leq n\}$.

11. Determine whether each of the following statements is true or false.

    (a) $a \in \{a\}$.　(b) $\{a\} \in \{a\}$.　(c) $\{a\} \subseteq \{a\}$.

    (d) $\emptyset \in \{a\}$.　(e) $\emptyset \subseteq \{a\}$.　(f) $\{\emptyset\} \subseteq \{a\}$.

12. Let $A = \{x \in \mathbb{Z} : |x| \leq 2\}$ and $C = \{x \in \mathbb{Z} : -2 \leq x \leq 4\}$. Find all sets B such that $A \subset B \subset C$.

13. Draw a Venn diagram for two general sets C and D and shade the region that contains the elements of C that do not belong to D.

14. Draw a Venn diagram for two general sets C and D and shade the region that contains the elements of C that do not belong to D and the elements of D that do not belong to C.

15. Give an example of three sets A, B and C such that none of these three sets is a subset of either of the remaining two sets.

16. Give an example of three sets A, B and C such that $A \subseteq B$, $B \subseteq C$, $C \not\subseteq A$ and $C \subseteq B$.

17. Give an example of three sets A, B and C such that $A \subset B$, $A \subseteq C$ and $C \subseteq B$ and such that exactly two of the three sets A, B and C are equal.

18. Give an example of two sets A and B such that $A \in B$ and $A \subseteq B$.

19. Give examples of three sets A, B and C such that

    (a) $A \subseteq B \subset C$.

    (b) $A \in B$, $B \in C$ and $A \notin C$.

    (c) $A \in B$ and $A \subset C$.

20. Determine the power set of each of the following sets.

    (a) $\emptyset$.　(b) $\{\emptyset\}$.　(c) $\{0\}$.　(d) $\{1, 2\}$.　(e) $\{a, b, c\}$.

21. Determine the power set of each of the following sets.

    (a) $A = \{\emptyset, a\}$.　(b) $B = \{\emptyset, \{a\}\}$.　(c) $C = \mathcal{P}(\{a\})$.

22. Determine the power set of each of the following sets.

    (a) $A = \{0, \{0\}\}$.　　(b) $B = \{\emptyset, \{\emptyset\}\}$.　(c) $C = \{0, \emptyset\}$.

    (d) $D = \{0, \{0\}, \emptyset\}$.　　(e) $E = \{0, \emptyset, \{\emptyset\}\}$.

23. How many elements are in $\mathcal{P}(A)$ if $A = \{1, 2, 3, 4, 5\}$?

24. How many elements are in $\mathcal{P}(A)$ if $A = \{n \in \mathbb{Z} : |n| \le 5\}$?

25. Let $A = \{1, 2, \ldots, 10\}$. Give an example of a set B such that

    (a) B is a subset of $\mathcal{P}(A)$ and $|B| = 5$.

    (b) B is an element of $\mathcal{P}(A)$ and $|B| = 5$.

26. For $n \in \mathbb{N}$, two sets A and C have the property that $A \subset C$, $|A| = n$ and $|C| = n + 3$. How many sets B are there such that $A \subset B \subset C$?

2.2   Set Operations and Their Properties

Much of our interest in sets is because of properties they possess when combined to produce other sets. You may very well be aware of some of these properties. We begin with two of the most common ways of combining two given sets to form a new set.

Intersections and Unions

Definition 2.19 *Let A and B be two sets. The* **intersection** *$A \cap B$ of A and B is the set of elements belonging to both A and B. Thus*

$$A \cap B = \{x : \ x \in A \ and \ x \in B\}.$$

Definition 2.20 *Let A and B be two sets. The* **union** *$A \cup B$ of A and B is the set of elements belonging to at least one of A and B; that is,*

$$A \cup B = \{x : \ x \in A \ or \ x \in B\}.$$

Therefore, the word "or" in the definition of the union $A \cup B$ of A and B is the "inclusive or." Consequently, an element x belongs to $A \cup B$ if x belongs to both A and B or to exactly one of A and B. Because an element belongs to $A \cup B$ if it belongs to $A \cap B$, it follows that

$$A \cap B \subseteq A \cup B.$$

Venn diagrams for the intersection and union of two general sets A and B are shown in Figure 2.4. The shading indicates the portion of the diagram that represents the set being considered.

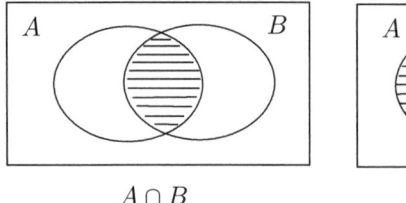

Figure 2.4: Venn diagrams for $A \cap B$ and $A \cup B$

Example 2.21

For the sets $C = \{1, 2, 4, 5\}$ and $D = \{1, 3, 5\}$,

$$C \cap D = \{1, 5\} \text{ and } C \cup D = \{1, 2, 3, 4, 5\}.$$

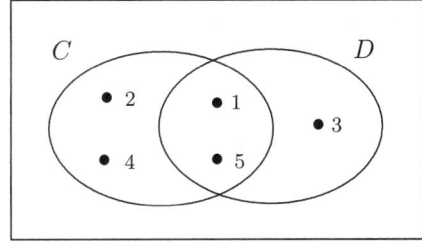

Figure 2.5: A Venn diagram for the sets C and D in Example 2.21

Figure 2.5 shows a Venn diagram for C and D. ◆

Intersection and union satisfy some familiar-sounding properties.

Theorem 2.22 *For every three sets A, B and C,*

(a) (**Commutative Laws**)

$A \cap B = B \cap A$ *and* $A \cup B = B \cup A$.

(b) (**Associative Laws**)

$(A \cap B) \cap C = A \cap (B \cap C)$ *and* $(A \cup B) \cup C = A \cup (B \cup C)$.

(c) (**Distributive Laws**)

$A \cap (B \cup C) = (A \cap B) \cup (A \cap C)$ *and* $A \cup (B \cap C) = (A \cup B) \cap (A \cup C)$.

The properties of intersection and union of sets stated in Theorem 2.22 are probably not surprising and are even suggested by Venn diagrams; however, none of these diagrams constitutes a proof of a property. Figure 2.6(a) shows a customary way of drawing a Venn diagram involving three general sets. See Figure 2.6(b) – (f) for the first of the two distributive laws in Theorem 2.22(c).

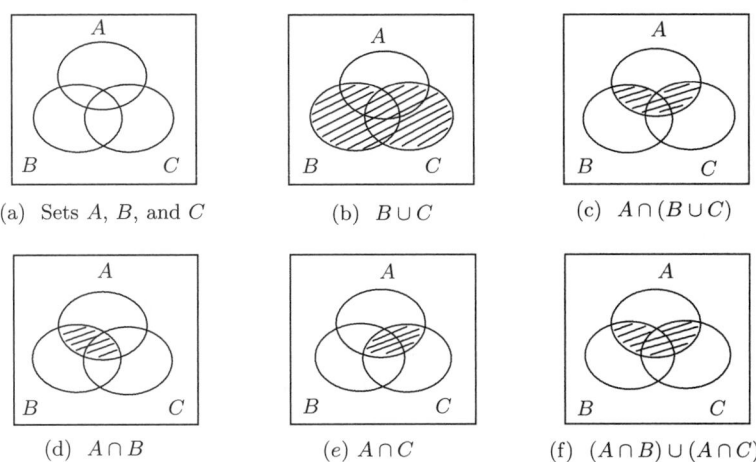

(a) Sets A, B, and C (b) $B \cup C$ (c) $A \cap (B \cup C)$

(d) $A \cap B$ (e) $A \cap C$ (f) $(A \cap B) \cup (A \cap C)$

Figure 2.6: Illustrating a distributive law

The properties of sets stated in Theorem 2.22 can be verified with the aid of the properties of logic that were presented in Chapter 1. For example, to establish that $A \cap B = B \cap A$ for every two sets A and B, we can show that $A \cap B$ and $B \cap A$ consist of exactly the same elements, that is, any element that belongs to one of $A \cap B$ and $B \cap A$ belongs to the other. Observe that

$$x \in A \cap B \;\equiv\; x \in A \text{ and } x \in B \quad \text{(definition of intersection of A and B)}$$

$$\equiv \quad x \in B \ \text{ and } \ x \in A \quad \text{(commutative law of conjunction)}$$
$$\equiv \quad x \in B \cap A \quad \text{(definition of intersection of } B \text{ and } A).$$

Strictly speaking, writing $A \cup B \cup C$ for three sets A, B and C is not defined, as union is only defined for two sets. There are two natural interpretations of $A \cup B \cup C$, namely $(A \cup B) \cup C$ and $A \cup (B \cup C)$. However, according to the associative law for union in Theorem 2.22(b), these two sets are equal. Consequently, there is no confusion in writing $A \cup B \cup C$ without parentheses. The same is true for $A \cap B \cap C$. Thus,

$$A \cap B \cap C \ = \ \{x : \ x \in A, x \in B \text{ and } x \in C\}$$
$$A \cup B \cup C \ = \ \{x : \ x \in A, x \in B \text{ or } x \in C\}.$$

More generally, for $n \geq 2$ sets $A_1, A_2, \ldots, A_n$, the **intersection** of these sets is

$$\bigcap_{i=1}^{n} A_i = A_1 \cap A_2 \cap \cdots \cap A_n = \{x : \ x \in A_i \text{ for } every \text{ } i \text{ with } 1 \leq i \leq n\};$$

while the **union** of these sets is

$$\bigcup_{i=1}^{n} A_i = A_1 \cup A_2 \cup \cdots \cup A_n = \{x : \ x \in A_i \text{ for } some \text{ } i \text{ with } 1 \leq i \leq n\}.$$

Example 2.23

For sets $A = \{1, 3, 4\}$, $B = \{3, 4, 6\}$ and $C = \{2, 3, 5\}$,

$$A \cap B \cap C = \{3\} \text{ and } A \cup B \cup C = \{1, 2, \ldots, 6\}. \ \blacklozenge$$

Example 2.24

Let $A_1 = \{1, 2\}$, $A_2 = \{2, 3\}$, $\ldots$, $A_{10} = \{10, 11\}$. That is, $A_i = \{i, i + 1\}$ for $i = 1, 2, \ldots, 10$. Then

$$A_1 \cap A_2 \cap \cdots \cap A_{10} = \emptyset \ \text{ and } \ A_1 \cup A_2 \cup \cdots \cup A_{10} = \{1, 2, \ldots, 11\}.$$

Note that for $1 \leq i \leq 9$, $A_i \cap A_{i+1} = \{i + 1\}$. $\blacklozenge$

Definition 2.25 *Two sets A and B are* **disjoint** *if they have no elements in common, that is, if $A \cap B = \emptyset$. A collection of sets is said to be* **pairwise disjoint** *if every two distinct sets in the collection are disjoint.*

Example 2.26

The sets $A = \{1, 2, 3\}$ and $B = \{4, 5, 6\}$ are disjoint, the set of even integers and the set of odd integers are disjoint and the set of rational numbers and the set of irrational numbers are disjoint. The set of nonnegative integers and the set of nonpositive integers are not disjoint, however, since 0 belongs to both sets. $\blacklozenge$

Example 2.27

The sets $A = \{1, 2, 3\}$, $B = \{4, 5\}$ and $C = \{6, 7, 8\}$ are pairwise disjoint since every two of the sets A, B and C are disjoint. The sets $A_1 = \{a, b\}$, $A_2 = \{a, c\}$ and $A_3 = \{b, c\}$ are not pairwise disjoint, however, since $A_1 \cap A_2 = \{a\} \neq \emptyset$, for example. On the other hand, $A_1 \cap A_2 \cap A_3 = \emptyset$. $\blacklozenge$

Difference and Symmetric Difference

Definition 2.28 *The **difference** $A - B$ of two sets A and B is defined as*

$$A - B = \{x :\ x \in A \ and \ x \notin B\}.$$

Some individuals denote the set $A - B$ by $A \smallsetminus B$.

Definition 2.29 *The **symmetric difference** $A \oplus B$ of two sets A and B is defined by*

$$A \oplus B = (A - B) \cup (B - A).$$

Venn diagrams for the difference and symmetric difference of two general sets are shown in Figure 2.7. While $A \oplus B = B \oplus A$ for every two sets A and B, this is not the case for $A - B$ and $B - A$. That is, there is no commutative law for set difference.

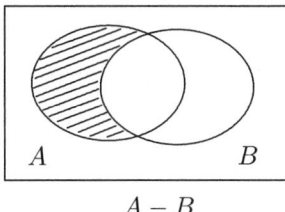 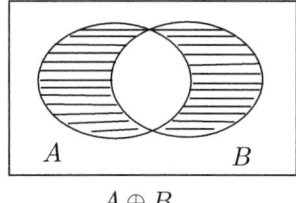

$$A - B \qquad\qquad A \oplus B$$

Figure 2.7: Venn diagrams for $A - B$ and $A \oplus B$

Therefore, for an element $x \in A \oplus B$, we must have $x \in A - B$ or $x \in B - A$. If $x \in A - B$, then $x \in A$ but $x \notin B$; while if $x \in B - A$, then $x \in B$ but $x \notin A$. In other words, if $x \in A \oplus B$, then x belongs to exactly one of A and B. That is, if $x \in A \oplus B$, then $x \in A$ *or* $x \in B$, where the use of the word "or" here is the "exclusive or." In symbols,

$$x \in A \oplus B \equiv (x \in A) \oplus (x \in B), \tag{2.1}$$

where the first occurrence of $\oplus$ in (2.1) denotes the symmetric difference of two sets and the second occurrence of $\oplus$ in (2.1) denotes the "exclusive or" of two statements. The Venn diagram for $A \oplus B$ in Figure 2.7 also illustrates the fact that

$$A \oplus B = (A \cup B) - (A \cap B).$$

Example 2.30

For the sets $A = \{i \in \mathbb{Z} :\ 1 \le i \le 10\}$ and $B = \{i \in \mathbb{Z} :\ 5 \le i \le 15\}$,

$$
\begin{aligned}
A - B &= \{i \in \mathbb{Z} :\ 1 \le i \le 4\} = \{1, 2, 3, 4\} \\
B - A &= \{i \in \mathbb{Z} :\ 11 \le i \le 15\} = \{11, 12, 13, 14, 15\} \\
A \cup B &= \{i \in \mathbb{Z} :\ 1 \le i \le 15\} = \{1, 2, \ldots, 15\} \\
A \cap B &= \{i \in \mathbb{Z} :\ 5 \le i \le 10\} = \{5, 6, 7, 8, 9, 10\} \\
A \oplus B &= (A - B) \cup (B - A) = (A \cup B) - (A \cap B) \\
&= \{1, 2, 3, 4, 11, 12, 13, 14, 15\}. \ \blacklozenge
\end{aligned}
$$

For general sets A, B and C, Venn diagrams for the sets $(A - B) \cap (A - C)$ and $A - (B \cup C)$ are shown in Figure 2.8. From the two Venn diagrams in Figure 2.8, it appears that $(A - B) \cap (A - C)$ and $A - (B \cup C)$ are equal sets. As we mentioned earlier, even though Venn diagrams may suggest that two sets are equal, these diagrams do not constitute a mathematical proof of this. Recall that two sets X and Y are equal if they contain exactly the same elements. We have also observed that

$$X = Y \text{ if and only if } X \subseteq Y \text{ and } Y \subseteq X. \tag{2.2}$$

Although methods of proof will not be discussed until Chapter 3, we can describe now how we would attempt to verify that two sets X and Y are equal by means of (2.2). In particular, it suffices to show that $X \subseteq Y$ and $Y \subseteq X$. To verify that $X \subseteq Y$, we show that each element of X belongs to Y. This can be accomplished by taking a general element of X, denoted by x say and showing that x must also belong to Y. To verify that $Y \subseteq X$, we show that a general element y belonging to Y must also belong to X. From this, it follows that $X = Y$.

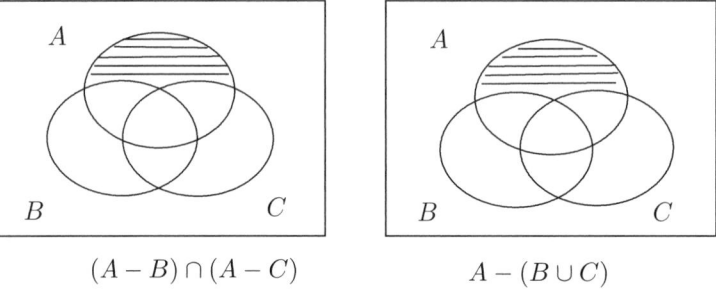

$$(A - B) \cap (A - C) \qquad\qquad A - (B \cup C)$$

Figure 2.8: Venn diagrams for $(A - B) \cap (A - C)$ and $A - (B \cup C)$

Example 2.31

Let A, B and C be sets. Show that

$$(A - B) \cap (A - C) = A - (B \cup C).$$

Solution. First, we show that $(A - B) \cap (A - C) \subseteq A - (B \cup C)$. Let $x \in (A - B) \cap (A - C)$. Then $x \in A - B$ and $x \in A - C$. Since $x \in A - B$, it follows that $x \in A$ and $x \notin B$. Since $x \in A - C$, it follows that $x \in A$ and $x \notin C$. Since $x \notin B$ and $x \notin C$, we have $x \notin B \cup C$. Thus $x \in A$ and $x \notin B \cup C$ and so $x \in A - (B \cup C)$. Therefore,

$$(A - B) \cap (A - C) \subseteq A - (B \cup C). \tag{2.3}$$

Next, we show that $A - (B \cup C) \subseteq (A - B) \cap (A - C)$. Let $y \in A - (B \cup C)$. Hence $y \in A$ and $y \notin B \cup C$. Since $y \notin B \cup C$, we can conclude that $y \notin B$ and $y \notin C$. Thus $y \in A - B$ and $y \in A - C$; so $y \in (A - B) \cap (A - C)$. Hence

$$A - (B \cup C) \subseteq (A - B) \cap (A - C). \tag{2.4}$$

Combining (2.3) and (2.4), we obtain

$$(A - B) \cap (A - C) = A - (B \cup C),$$

as desired.   ♦

Complement of a Set

Definition 2.32 *For a set A (which is therefore a subset of the universal set U being considered), the **complement** $\overline{A}$ of A is the set of elements (in the universal set) not belonging to A. That is,*

$$\overline{A} = \{x \in U :\ x \notin A\} = U - A.$$

The Venn diagram for the complement of a set is shown in Figure 2.9. Therefore,

$$A \cup \overline{A} = U \quad \text{and} \quad A \cap \overline{A} = \emptyset.$$

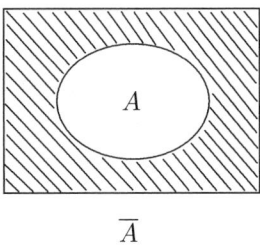

$$\overline{A}$$

Figure 2.9: Venn diagram for $\overline{A}$

Example 2.33

Let $\mathbb{Z}$ be the universal set and let E be the set of even integers. Then the complement $\overline{E}$ of E is the set O of odd integers. Furthermore, $E \cup \overline{E} = E \cup O = \mathbb{Z}$ and $E \cap \overline{E} = E \cap O = \emptyset$. ♦

Example 2.34

If $U = \mathbb{R}$, then the complement $\overline{\mathbb{Q}}$ of the set $\mathbb{Q}$ of all rational numbers is the set of irrational numbers. If $U = \mathbb{Z}$, then $\overline{\mathbb{N}} = \{n \in \mathbb{Z} : n \leq 0\}$. ♦

Two well-known properties of sets involving unions, intersections and complements are named after the British mathematician Augustus De Morgan, whom we encountered in Chapter 1.

Theorem 2.35 (De Morgan's Laws) *For two sets A and B,*

$$\overline{A \cup B} = \overline{A} \cap \overline{B} \quad \text{and} \quad \overline{A \cap B} = \overline{A} \cup \overline{B}.$$

Each law in Theorem 2.35 can be anticipated by considering Venn diagrams. De Morgan's Laws for the complement of the union and the intersection of two sets are consequences of De Morgan's Laws for the negation of the disjunction and of the conjunction of two statements. In order to see this, let A and B be two sets that are subsets of some universal set U. Then

$$
\begin{aligned}
x \in \overline{A \cup B} \;\; &\equiv \;\; \sim (x \in A \cup B) \equiv \; \sim (x \in A \text{ or } x \in B) \\
&\equiv \;\; \sim ((x \in A) \vee (x \in B)) \equiv \; (x \notin A) \wedge (x \notin B) \\
&\equiv \;\; (x \in \overline{A}) \wedge (x \in \overline{B}) \equiv \; x \in \overline{A} \cap \overline{B}.
\end{aligned}
$$

That is, $\overline{A \cup B} = \overline{A} \cap \overline{B}$. Furthermore,

$$
\begin{aligned}
x \in \overline{A \cap B} \;\; &\equiv \;\; \sim (x \in A \cap B) \equiv \; \sim (x \in A \text{ and } x \in B) \\
&\equiv \;\; \sim ((x \in A) \wedge (x \in B)) \equiv \; (x \notin A) \vee (x \notin B) \\
&\equiv \;\; (x \in \overline{A}) \vee (x \in \overline{B}) \equiv \; x \in \overline{A} \cup \overline{B}.
\end{aligned}
$$

Consequently, $\overline{A \cap B} = \overline{A} \cup \overline{B}$.

Example 2.36

For the sets $C = \{1, 2, 4, 5\}$ and $D = \{1, 3, 5\}$ and the universal set $U = \{1, 2, 3, 4, 5\}$, it follows that

$$C - D = \{2,4\},\ D - C = \{3\},\ C \oplus D = \{2,3,4\},\ \overline{C} = \{3\},\ \overline{D} = \{2,4\},$$
$$\overline{C} \cup \overline{D} = \{2,3,4\},\ C \cap D = \{1,5\},\ \overline{C \cap D} = \{2,3,4\}. \blacklozenge$$

Exercises for Section 2.2

1. For $A = \{1,3,5,7,9\}$, $B = \{2,3,4,6,7,8\}$ and the universal set $U = \{1,2,\ldots,10\}$, determine the following.

   (a) $A \cap B$. (b) $A \cup B$. (c) $A - B$. (d) $B - A$. (e) $A \oplus B$. (f) $\overline{A}$. (g) $\overline{B}$.

2. For $A = \{a,c,d\}$, $B = \{c,d,e,f\}$ and the universal set $U = \{a,b,c,d,e,f,g,h\}$, determine the following.

   (a) $A \cap B$. (b) $A \cup B$. (c) $A - B$. (d) $B - A$. (e) $A \oplus B$. (f) $\overline{A}$. (g) $\overline{B}$.

3. For the set $\mathbb{Q}$ of rational numbers, the set $\mathbb{I}$ of irrational numbers and the universal set $\mathbb{R}$ of real numbers, determine the following.

   (a) $\mathbb{Q} \cap \mathbb{I}$. (b) $\mathbb{Q} \cup \mathbb{I}$. (c) $\mathbb{Q} - \mathbb{I}$. (d) $\mathbb{I} - \mathbb{Q}$. (e) $\mathbb{Q} \oplus \mathbb{I}$. (f) $\overline{\mathbb{Q}}$. (g) $\overline{\mathbb{I}}$.

4. Let $U = \{1,2,3\}$ be the universal set and let $A = \{1,2\}$, $B = \{2,3\}$ and $C = \{1,3\}$. Determine the following.

   (a) $(A \cup B) - (B \cap C)$. (b) $\overline{A}$. (c) $\overline{B \cup C}$.

5. Let A, B and C be sets. Draw a Venn diagram for each of the following.

   (a) $A \cap \overline{B}$. (b) $A \cup (C - B)$. (c) $(A - B) \cap C$.

6. Let $U = \{1,2,\ldots,10\}$. For the sets $A = \{1,3,4,6,7,9,10\}$, $B = \{2,5,6,8,9\}$ and $C = \{1,3,4,5,9,10\}$, determine the following sets.

   (a) $A - \overline{(B - C)}$. (b) $\overline{(A - B)} - C$.

7. Illustrate the distributive law $A \cup (B \cap C) = (A \cup B) \cap (A \cup C)$ by drawing Venn diagrams.

8. For two sets A and B, which of the following sets are equal?

   $A \oplus B$, $(A - B) \oplus (B - A)$, $(A \cup B) \oplus (B \cap A)$.

9. Let $A = \{2,3,4\}$ and $B = \{3,4,5,6\}$, where $U = \{1,2,\ldots,7\}$ is the universal set. Determine the following.

   (a) $A \cap B$. (b) $A \cup B$. (c) $A - B$. (d) $B - A$. (e) $A \oplus B$.
   (f) $\overline{A}$. (g) $\overline{B}$. (h) $\overline{A \cap B}$. (i) $\overline{A \cup B}$. (j) $\overline{A} \cap \overline{B}$. (k) $\overline{A} \cup \overline{B}$.

10. Give examples of three sets A, B and C such that

    (a) $A \in B$, $A \subseteq C$ and $B \not\subseteq C$.
    (b) $B \in A$, $B \subset C$ and $A \cap C \neq \emptyset$.
    (c) $A \in B$, $B \subseteq C$ and $A \not\subseteq C$.

11. Let S be the set of faculty members in the Department of Mathematics at your university or college. Let A be the subset of S consisting of those members of S you have had as an instructor and let B be the subset of S consisting of those members of S with whom you have had a conversation. Describe the members of the following sets.

    (a) $A \cup B$. (b) $A \cap B$. (c) $A - B$. (d) $B - A$.
    (e) $A \oplus B$. (f) $\overline{A \cap B}$. (g) $\overline{A \cup B}$.

12. Let S be the set of students at your university or college who have been your classmate in one or more courses and let T be the set of students with whom you have studied. Describe each of the following sets in terms of S and T.

    (a) the set of students who have neither been your classmate nor with whom you have studied,

    (b) the set of students with whom you have studied but who have never been your classmate,

    (c) the set of your classmates with whom you have never studied,

    (d) the set of students with whom you have either studied or have been your classmate but not both,

    (e) the set of your classmates with whom you have studied.

13. (a) Illustrate the following by drawing Venn diagrams.
        For two sets A and B, $A - B = A \cap \overline{B}$.

    (b) For sets A and B, verify that $A - B = A \cap \overline{B}$.

14. For $n \in \mathbb{N}$, let $A_n = \left\{ \frac{n}{n+2} \right\}$ and $B_n = \left\{ \frac{1}{n} \right\}$. Determine $\left(\bigcup_{n=1}^{3} A_n \right) \cap \left(\bigcup_{n=1}^{3} B_n \right)$.

15. Let A and B be two sets. For each of the sets listed in (a)-(e) below, determine whether

    (1) the set is always a subset of A,

    (2) the set is always a subset of B,

    (3) A is always a subset of the set,

    (4) B is always a subset of the set,

    (5) none of the above.

    (a) $A \cap B$.   (b) $A \cup B$.   (c) $A - B$.   (d) $B - A$.   (e) $A \oplus B$.

16. For a universal set U, draw a Venn diagram for three sets A, B and C indicating the locations of the elements of U if A, B and C satisfy all of the conditions (a)–(h) and then determine A, B, C and U.

    (a) $A \cap B \cap C = \{2\}$,
    (b) $\overline{A} \cap \overline{B} \cap \overline{C} = \{3, 6, 12\}$,
    (c) $A \cap \overline{B} \cap \overline{C} = \{1, 9, 11\}$,
    (d) $B \cap \overline{A} \cap \overline{C} = \{4, 8\}$,
    (e) $C \cap \overline{A} \cap \overline{B} = \{10\}$,
    (f) $(B \cup C) \cap \overline{A} = \{4, 8, 10\}$,
    (g) $(A \cup C) \cap \overline{B} = \{1, 5, 7, 9, 10, 11\}$,
    (h) $(A \cup B) \cap \overline{C} = \{1, 4, 8, 9, 11\}$.

17. Use a distributive property of logic to verify the following.
    For any three sets A, B and C, $A \cap (B \cup C) = (A \cap B) \cup (A \cap C)$.

18. Let $A = \{0, \{0\}, \{0, \{0\}\}\}$.

    (a) Determine which of the following are elements of A: 0, $\{0\}$, $\{\{0\}\}$.

    (b) Determine $|A|$.

    (c) Determine which of the following are subsets of A: 0, $\{0\}$, $\{\{0\}\}$.
        For (d)–(i), determine the indicated sets.

(d) $\{0\} \cap A$.

(e) $\{\{0\}\} \cap A$.

(f) $\{\{\{0\}\}\} \cap A$.

(g) $\{0\} \cup A$.

(h) $\{\{0\}\} \cup A$.

(i) $\{\{\{0\}\}\} \cup A$.

19. Is it possible for two different sets A and B to have the property that $\mathcal{P}(A - B) = \mathcal{P}(B - A)$?

20. For a set A, the power set $\mathcal{P}(A)$ contains three distinct elements $\emptyset$, B and C.

(a) Determine another set belonging to $\mathcal{P}(A)$.

(b) Is there yet another set belonging to $\mathcal{P}(A)$?

21. For the universal set $U = \{1, 2, 3, 4\}$, give an example of four different subsets A, B, C and D of U each with at least three elements such that

$$0 < |A \cap B \cap C \cap D| < |A \cap B \cap C| < |A \cap B| < |A| = 4.$$

22. For sets A and B of integers, define $A + B = \{a + b : a \in A, b \in B\}$. If $|A| = |B| = 5$, how small and how large can $|A + B|$ be?

2.3   Cartesian Products of Sets

We have seen that new sets can be constructed from two given sets A and B in a variety of ways. There is another set that can be created from A and B but whose elements are of an entirely different nature.

For two elements a and b, we write (a, b) for the **ordered pair** in which a is the first element (or first coordinate) of the pair and b is the second element (second coordinate) of the pair. While $\{a, b\} = \{b, a\}$ for distinct elements a and b, this is not so for ordered pairs. In particular, $(a, b) = (c, d)$ if and only if $a = c$ and $b = d$. That is, if $a \neq b$, then $(a, b) \neq (b, a)$. This might remind you of the points $(1, 2)$ and $(2, 1)$ in the Cartesian plane, which are not same. Mentioning the Cartesian plane, in fact, leads us to the major concept of this section.

Definition 2.37   *For two sets A and B, the* **Cartesian product** *$A \times B$ of A and B is the set of all ordered pairs whose first coordinate belongs to A and whose second coordinate belongs to B. That is,*
$$A \times B = \{(a, b) :\ a \in A \ and \ b \in B\}.$$

Example 2.38

For the sets $A = \{1, 2\}$ and $B = \{x, y, z\}$, determine $A \times B$, $B \times A$, $A \times A$ and $B \times B$.

Solution. For these sets A and B,

$$
\begin{aligned}
A \times B &= \{(1, x), (1, y), (1, z), (2, x), (2, y), (2, z)\} \\
B \times A &= \{(x, 1), (x, 2), (y, 1), (y, 2), (z, 1), (z, 2)\} \\
A \times A &= \{(1, 1), (1, 2), (2, 1), (2, 2)\} \\
B \times B &= \{(x, x), (x, y), (x, z), (y, x), (y, y), (y, z), (z, x), (z, y), (z, z)\}. \ \blacklozenge
\end{aligned}
$$

You might have observed for the sets $A = \{1, 2\}$ and $B = \{x, y, z\}$ in Example 2.38 that

$$|A| = 2, \ |B| = 3 \text{ and } |A \times B| = 6.$$

Actually, for every two finite sets A and B,

$$|A \times B| = |A| \cdot |B| = |B| \cdot |A| = |B \times A|. \tag{2.5}$$

Example 2.38 also shows that $A \times B \neq B \times A$ for the sets $A = \{1, 2\}$ and $B = \{x, y, z\}$. Indeed, it is not unusual for $A \times B \neq B \times A$. (See Exercises 10 and 11.)

Example 2.39

For the sets $A = \{0, 1\}$ and $B = \{\emptyset, \{1\}, 2\}$, determine $A \times B$.

Solution. For these two sets A and B,

$$A \times B = \{(0, \emptyset), (0, \{1\}), (0, 2), (1, \emptyset), (1, \{1\}), (1, 2)\}. \ \blacklozenge$$

Example 2.40

The Cartesian product

$$\mathbb{R} \times \mathbb{R} = \{(x, y) : \ x, y \in \mathbb{R}\}$$

is the set of ordered pairs of real numbers. Therefore, $\mathbb{R} \times \mathbb{R}$ is the set of all points in the Cartesian plane. This set is sometimes denoted by $\mathbb{R}^2$ as well. $\blacklozenge$

The **Cartesian product** of $n \geq 2$ sets A_1, A_2, ..., A_n is denoted by $A_1 \times A_2 \times \cdots \times A_n$ and is defined by

$$A_1 \times A_2 \times \cdots \times A_n = \{(a_1, a_2, \ldots, a_n) : \ a_i \in A_i \text{ for } 1 \leq i \leq n\}.$$

The elements $(a_1, a_2, \ldots, a_n)$ are referred to as **ordered n-tuples**, where a_1 is the first element (or the first coordinate) of the n-tuple, a_2 is the second element and so on. So ordered 2-tuples are ordered pairs. Ordered 3-tuples are also called ordered triples. If $A_i = A$ for $1 \leq i \leq n$, then $A_1 \times A_2 \times \cdots \times A_n$ is also denoted by A^n.

Example 2.41

For $A = \{0, 1\}$, $B = \{1, 2\}$ and $C = \{1, 2, 3\}$, determine $A \times B \times C$.

Solution. The Cartesian product of A, B and C consists of all ordered triples (a, b, c), where $a \in A$, $b \in B$ and $c \in C$. Therefore,

$$\begin{aligned} A \times B \times C \ = \ & \{(0, 1, 1), (0, 1, 2), (0, 1, 3), (0, 2, 1), (0, 2, 2), (0, 2, 3), \\ & (1, 1, 1), (1, 1, 2), (1, 1, 3), (1, 2, 1), (1, 2, 2), (1, 2, 3)\}. \ \blacklozenge \end{aligned}$$

Exercises for Section 2.3

1. For the sets $A = \{a, b\}$ and $B = \{a, b, c\}$, determine $A \times B$, $B \times A$, $A^2 = A \times A$ and $B^2 = B \times B$.

2. For the set $A = \{0, \{1\}\}$, determine $A^2 = A \times A$.

3. For the sets $A = \{1, 2\}$, $B = \{2, 3\}$ and $C = \{3, 4\}$, determine the following.
   (a) $B \cap C$.   (b) $A \times (B \cap C)$.   (c) $A \times B$.   (d) $A \times C$.   (e) $(A \times B) \cap (A \times C)$.

4. For the set $A = \{1, 2, 3\}$, let $U = A^2 = A \times A$ be the universal set. For $B = \{1, 2\}$, determine $\overline{B \times B}$.

5. For $A = \{a, b\}$, determine $A \times \mathcal{P}(A)$.

6. For $A = \{1, 2\}$ and $B = \{\emptyset\}$, determine $\mathcal{P}(A) \times \mathcal{P}(B)$.

7. For $A = \{1, 2\}$ and $B = \{1\}$, determine $\mathcal{P}(A \times B)$.

8. For $A = \emptyset$ and $B = \{0\}$, determine $(A \times B) \cap \mathcal{P}(A \times B)$.

9. For $A = \{a, b, c\}$, $B = \{0, 1\}$ and $C = \{x, y\}$, determine

   (a) $A \times B \times C$.   (b) $C \times B \times A$.   (c) $B^3 = B \times B \times B$.

10. Show that if A and B are two nonempty sets such that $A \neq B$, then $A \times B \neq B \times A$.

11. Give an example of two distinct sets A and B such that $A \times B = B \times A$.

12. Verify, for sets A, B and C, that $(A \times B) \cap (A \times C) = A \times (B \cap C)$.

13. The universal set U has n elements for some positive integer n and A is a subset of U consisting of a single element of U. For what values of n can $\overline{A \times A} = \overline{A} \times \overline{A}$?

14. For two sets A and B of real numbers, the set $A \cdot B$ is defined by

$$A \cdot B = \{ab : \ a \in A, b \in B\}.$$

   Determine each of the following sets.

   (a) $A \cdot B$ for $A = \{\frac{1}{2}, 1, \sqrt{2}\}$ and $B = \{\sqrt{2}, 2, 4\}$.
   (b) $\mathbb{R} \cdot \mathbb{R}$.
   (c) $\mathbb{R} \cdot C$ where $C \subseteq \mathbb{R}$ with $|C| = 2$.

15. Let $A = \{0, 1\}$.

   (a) Give an example of four elements belonging to $(A \times A) \times (A \times A)$.
   (b) Determine $|(A \times A) \times (A \times A)|$.

2.4   Partitions

There are many instances when it is useful to divide a nonempty set A into nonempty subsets in such a way that each element of A belongs to exactly one of these subsets. For example, the set P of Americans who have served as president of the United States can be partitioned into the set P_1 of individuals who have been elected to exactly one term as president, the set P_2 of individuals who have been elected to exactly two terms as president, the set P_3 of individuals who have been elected to three or more terms as president and the set P_0 of individuals who have been elected to no terms as president.

   The set $\mathbb{Z}$ of integers can be divided into the set E of even integers and the set O of odd integers. Certainly, every integer belongs to exactly one of E and O. The set $\mathbb{R}$ of real numbers can be divided into the three non-overlapping subsets: the set $\mathbb{R}^+$ of positive real numbers, the set of negative real numbers (which we can denote by $\mathbb{R}^-$) and the set $\{0\}$ consisting only of 0. Similarly, $\mathbb{R}$ can be divided into the set $\mathbb{Q}$ of rational numbers and the set of irrational numbers. If we were considering the polynomial $x^2 - 4$, where $x \in \mathbb{R}$, we might wish to divide $\mathbb{R}$ into the sets A, B and C, where

$$
\begin{aligned}
A &= \{x \in \mathbb{R} : \ x^2 - 4 = 0\} \\
B &= \{x \in \mathbb{R} : \ x^2 - 4 < 0\} \\
C &= \{x \in \mathbb{R} : \ x^2 - 4 > 0\}.
\end{aligned}
$$

As it turns out,

$$
\begin{aligned}
A &= \{-2, 2\} \\
B &= \{x \in \mathbb{R} : \ -2 < x < 2\} \\
C &= \{x \in \mathbb{R} : \ x < -2 \text{ or } x > 2\}.
\end{aligned}
$$

What we have been discussing are various ways that a set might be "partitioned" into subsets.

Definition 2.42 *A **partition** of a nonempty set A is a collection of nonempty subsets of A such that every element of A belongs to exactly one of these subsets.*

A partition of a nonempty set A is therefore a collection of pairwise disjoint nonempty subsets of A whose union is A. Thus, if $\mathcal{P} = \{S_1, S_2, \ldots, S_k\}$, $k \geq 1$, is a partition of a nonempty set A, then

(1) every subset S_i is nonempty,

(2) every two different subsets S_i and S_j are disjoint and

(3) the union of all subsets in $\mathcal{P}$ is A.

Example 2.43

For the set $A = \{1, 2, \ldots, 10\}$ and the subsets

$$S_1 = \{1, 7, 8\}, \ S_2 = \{2, 4, 9\}, \ S_3 = \{3\}, \ S_4 = \{5, 6, 10\}$$

of A, the set $\mathcal{P} = \{S_1, S_2, S_3, S_4\}$ is a partition of A (see Figure 2.10).   ♦

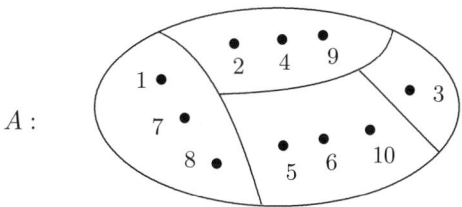

Figure 2.10: The partition $\mathcal{P}$ of the set $A = \{1, 2, \ldots, 10\}$ given in Example 2.43

Example 2.44

Let $A = \{1, 2, \ldots, 8\}$. Which of the following collections of subsets of A are partitions of A?

(a) $\mathcal{P}_1 = \{\{1, 4, 7, 8\}, \{3, 5, 6\}, \{2\}\}$.

(b) $\mathcal{P}_2 = \{\{1, 4\}, \{2, 8\}, \{3, 5, 7\}\}$.

(c) $\mathcal{P}_3 = \{\{1, 2, 4\}, \{3, 6, 8\}, \emptyset, \{5, 7\}\}$.

(d) $\mathcal{P}_4 = \{\{1, 7, 8\}, \{2, 5, 6\}, \{3, 4, 7\}\}$.

Solution.   Only $\mathcal{P}_1$ is a partition of A. The set $\mathcal{P}_2$ is not a partition of A since 6 belongs to no subset in $\mathcal{P}_2$. The set $\mathcal{P}_3$ is not a partition of A since all subsets in a partition are required to be nonempty. The set $\mathcal{P}_4$ is not a partition of A since 7 belongs to two different subsets in $\mathcal{P}_4$.   ♦

Example 2.45

List all of the partitions of the set $A = \{1, 2, 3\}$.

Solution. The partitions of A are

$$\mathcal{P}_1 = \{\{1, 2, 3\}\} = \{A\}, \ \ \mathcal{P}_2 = \{\{1\}, \{2, 3\}\}, \ \mathcal{P}_3 = \{\{2\}, \{1, 3\}\},$$
$$\mathcal{P}_4 = \{\{3\}, \{1, 2\}\}, \ \ \mathcal{P}_5 = \{\{1\}, \{2\}, \{3\}\}. \quad \blacklozenge$$

Exercises for Section 2.4

1. Which of the following are partitions of the set $A = \{a, b, c, d, e, f, g\}$? For each collection of subsets that is not a partition of A, explain your answer.

   (a) $\mathcal{P}_1 = \{\{a, c, e, g\}, \{b, f\}, \{d\}\}$.

   (b) $\mathcal{P}_2 = \{\{a, b, c, d\}, \{e, f\}\}$.

   (c) $\mathcal{P}_3 = \{\{a\}, \emptyset, \{b, c, d\}, \{e, f, g\}\}$.

   (d) $\mathcal{P}_4 = \{\{a, c, d\}, \{b, g\}, \{e\}, \{b, f\}\}$.

   (e) $\mathcal{P}_5 = \{\{a, b\}, \{c, d, e\}, \{f, g\}, \{f, g\}\}$.

   (f) $\mathcal{P}_6 = \{A\}$.

2. Let $A = \{1, 2, \ldots, 10\}$. Give an example of a partition $\mathcal{P}$ of A such that $|\mathcal{P}| = 3$.

3. Let $A = \{1, 2\}$ and $B = \{3, 4\}$.

   (a) Give an example of a partition of $A \times B$.

   (b) Give an example of a partition of the power set $\mathcal{P}(A)$ of A.

4. Let $\mathcal{P}$ be a partition of a set S containing elements a and b. Suppose that A is a set in $\mathcal{P}$ that contains a and B is a set in $\mathcal{P}$ that contains b.

   (a) If $b \in A$, then what is $A \cap B$?

   (b) If $b \notin A$, then what is $A \cap B$?

5. List all partitions of the set $A = \{a, b, c, d\}$.

6. Let $S = \{1, 2, \ldots, 15\}$.

   (a) Give an example of a partition $\mathcal{P}_1$ of S such that $\mathcal{P}_1$ consists of five subsets of S.

   (b) Give an example of a partition $\mathcal{P}_2$ of the partition $\mathcal{P}_1$ in (a) such that $\mathcal{P}_2$ consists of three subsets of $\mathcal{P}_1$.

7. Let $A = \{0, 1, \{0, 1\}\}$. Give an example of a partition S of the power set $\mathcal{P}(A)$ such that $|S| = 3$.

8. Let $A = \{1\}$ and $B = \{2\}$. Give an example of a partition S of the power set $\mathcal{P}(A \times B)$ such that $|S| > |A \times B|$.

9. Let $n \geq 3$ be an odd integer and let $k = (n + 1)/2$. Show that there is a partition of the set $A = \{1, 2, \ldots, n\}$ into k subsets $A_1, A_2, \ldots, A_k$ such that the sum of the integers belonging to each subset A_i $(1 \leq i \leq k)$ gives the same answer.

10. Let $A = \{1, 2, 3, 4\}$. Partition the power set $\mathcal{P}(A)$ of A into as many subsets as possible such that no two subsets have the same number of elements.

Chapter 2 Highlights

Key Concepts

cardinality (of a finite set A), $|A|$: the number of elements in A.

Cartesian product (of A and B), $A \times B$: the set of all ordered pairs whose first coordinate belongs to A and whose second coordinate belongs to B.

Cartesian product (of the n sets $A_1, A_2, \ldots, A_n$), $A_1 \times A_2 \times \cdots \times A_n$: the set of ordered n-tuples $(a_1, a_2, \ldots, a_n)$, where $a_i \in A_i$ for $1 \le i \le n$.

complement (of A), $\overline{A}$: the set of all elements (in the universal set) not belonging to A.

difference (of A and B), $A - B$: the set of all elements belonging to A but not to B.

disjoint sets: two sets having no elements in common.

element: an object belonging to a set.

empty set, $\emptyset$: the set containing no elements.

equal sets, $A = B$: A and B consist of the same elements.

intersection (of A and B), $A \cap B$: the set of all elements belonging to both A and B.

irrational number: a real number that is not rational.

$\mathbb{N}$: set of natural numbers (positive integers).

natural number: a positive integer.

ordered pair, (a, b): a pair of two (not necessarily distinct) elements where one is designated as the first element of the pair and the other is the second element.

pairwise disjoint sets: a collection of sets such that every two distinct sets are disjoint.

partition (of a nonempty set A): a collection of nonempty subsets of A such that every element of A belongs to exactly one of these subsets.

power set (of A), $\mathcal{P}(A)$: the set of all subsets of A.

proper subset (A is a proper subset of B), $A \subset B$: A is a subset of B and $A \ne B$.

$\mathbb{Q}$: the set of rational numbers.

$\mathbb{R}$: the set of real numbers, which consists of the rational numbers and irrational numbers.

$\mathbb{R}^+$: the set of positive real numbers.

$\mathbb{R}^-$: the set of negative real numbers.

rational number (fraction): a number of the form a/b, where $a, b \in \mathbb{Z}$ and $b \ne 0$.

set: a collection of objects.

subset (A is a subset of B), $A \subseteq B$: every element of A is an element of B.

symmetric difference (of A and B), $A \oplus B$: the set of elements belonging to A or B but not to both.

union (of A and B), $A \cup B$: the set of all elements belonging to at least one of A and B.

universal set: the set containing all objects under consideration.

Venn diagram: a graphical representation of sets.

$\mathbb{Z}$: the set of integers.

Key Results

- If A is a set with $|A| = n$, where n is a nonnegative integer, then $|\mathcal{P}(A)| = 2^n$.

- For any three sets A, B and C, the following laws hold:

  Commutative Laws : $A \cap B = B \cap A$ and $A \cup B = B \cup A$.

  Associative Laws : $(A \cap B) \cap C = A \cap (B \cap C)$ and $(A \cup B) \cup C = A \cup (B \cup C)$.

  Distributive Laws : $A \cap (B \cup C) = (A \cap B) \cup (A \cap C)$ and $A \cup (B \cap C) = (A \cup B) \cap (A \cup C)$.

  De Morgan's Laws : $\overline{A \cup B} = \overline{A} \cap \overline{B}$ and $\overline{A \cap B} = \overline{A} \cup \overline{B}$.

Supplementary Exercises for Chapter 2

1. Write each of the following sets by listing its elements.

   (a) $A = \{x \in \mathbb{Q} : x^4 - 4 = 0\}$.

   (b) $B = \left\{x \in \mathbb{R} : \frac{x+2}{x} = \frac{x}{x+2}\right\}$.

   (c) $C = \{x \in \mathbb{Q} : x^2 - (\pi + e)x + \pi e = 0\}$.

   (d) $D = \{n \in \mathbb{Z} : |2n - 1| + |2n + 1| \le 3\}$.

   (e) $E = \{x \in \mathbb{Z} : x^2 - 2x - 8 \le 0\}$.

2. For a certain set A, the power set of A is $\mathcal{P}(A) = \{\emptyset, \{0\}, B\}$, where B is a set. What is A?

3. Let A and B be sets. Show that if $A \ne B$, then $\mathcal{P}(A) \ne \mathcal{P}(B)$.

4. For $S = \{-3, -2, \ldots, 2\}$, list the elements of the set $A = \{n \in S : 0 < |n| < 2\}$.

5. For $S = \{1, 2, 3\}$, give an example of two sets A and B such that $|B| = |A| + 1$, where $A \subseteq S$ and $B \not\subseteq S$.

6. Give an example of three sets A, B and C such that $A \in B$, $A \notin C$ and $B \cap C = \emptyset$.

7. Determine the power set $\mathcal{P}(A)$ for $A = \{1, 3, 5\}$.

8. Let A be the set of students in a discrete mathematics course in your university and B the set of students majoring in computer science in your university. Describe the students in each of the following sets.

   (a) $A \cup B$.   (b) $A \cap B$.   (c) $A - B$.   (d) $B - A$.

   (e) $A \oplus B$.   (f) $\overline{A \cap B}$.   (g) $\overline{A \cup B}$.

9. Let C be the set of professors in a mathematics department who taught a calculus class and D the set of professors in a mathematics department who taught a discrete mathematics course. Describe each of the following sets in terms of C and D.

   (a) the set of professors in the mathematics department who taught both a calculus course and a discrete mathematics course.

   (b) the set of professors in the mathematics department who taught either a calculus course or a discrete mathematics course.

   (c) the set of professors in the mathematics department who taught neither a calculus course nor a discrete mathematics course.

   (d) the set of professors in the mathematics department who taught a calculus course but not a discrete mathematics course.

(e) the set of professors in the mathematics department who taught a discrete mathematics course but not a calculus course.

10. For $A = \{1, 2\}$, $B = \{-1, 0, 1\}$ and the universal set $U = \{-2, -1, 0, 1, 2\}$, determine

   (a) $A \cup B$.   (b) $A \cap B$.   (c) $A - B$.   (d) $\overline{B}$.   (e) $A \times B$.

11. For a set S, let $\mathcal{P}^*(S)$ be the set of all proper nonempty subsets of S. Find $\mathcal{P}^*(S)$ for each of the following. (a) $S = \{x, y\}$. (b) $S = \{1\}$.

12. For $S = \{\{1\}, \{1, 2\}, \{1, 2, 3\}\}$, find the power set $\mathcal{P}(S)$.

13. For the set $S = \{1, 2, \ldots, 6\}$, give an example of a partition $\mathcal{P}$ of S such that $|\mathcal{P}| = 3$.

14. Determine whether each of the following statements is true or false.

   (a) $0 \in \{\emptyset\}$.   (b) $0 \subseteq \{\emptyset\}$.   (c) $\emptyset \in \{0\}$.   (d) $\emptyset \subseteq \{0\}$.
   (e) $\{0\} \in \{0\}$.   (f) $\{0\} \subseteq \{0\}$.   (g) $\{\emptyset\} \in \{\emptyset\}$.   (h) $\{\emptyset\} \subseteq \{\{\emptyset\}\}$.

15. Give examples of two sets A and B such that

   (a) $A \in B$ and $A \subset B$.
   (b) $A \in B$ and $A \cap B = \emptyset$.
   (c) $A \in B$, $A \not\subseteq B$ and $A \cap B \neq \emptyset$.

16. Give examples of three sets A, B and C such that

   (a) $A \subseteq B \not\subseteq C$.
   (b) $A \subseteq B$, $B \in C$ and $A \cap C = \emptyset$.
   (c) $A \in B$, $A \subset B$ and $A \not\subseteq C$.
   (d) $A \in B$, $A \not\subseteq B$ and $B \in C$.

17. For the universal set $U = \{1, 2, \ldots, 10\}$, draw a Venn diagram for three sets A, B and C indicating the locations of the elements of U if A, B and C satisfy all of the conditions (a)-(h):

   (a) $A \cap B \cap C = \{5\}$,
   (b) $\overline{A \cup B \cup C} = \{10\}$,
   (c) $A - (B \cup C) = \{3\}$,
   (d) $B - (A \cup C) = \{4\}$,
   (e) $C - (A \cup B) = \{2\}$,
   (f) $(B \cup C) - A = \{1, 2, 4, 9\}$,
   (g) $(A \cup C) - B = \{2, 3, 7, 8\}$,
   (h) $(A \cup B) - C = \{3, 4, 6\}$.

18. Let $S = \{1, 2, \ldots, 10\}$. Give an example of

   (a) a set A of cardinality 2 such that $A \subseteq \mathcal{P}(S)$.
   (b) a set B of cardinality 2 such that $B \in \mathcal{P}(S)$.
   (c) a partition $\mathcal{P}$ of S such that $|\mathcal{P}| = 2$.

19. Let $S = \{a, b, c, d, e, f, g\}$. Give an example of

   (a) a set A such that $A \subseteq \mathcal{P}(S)$ with $|A| = 2$.
   (b) a set B such that $B \in \mathcal{P}(S)$ with $|B| = 3$.

(c) a partition $\mathcal{P}$ of S with $|\mathcal{P}| = 4$.

20. Let $A_i = \{1, 2, \ldots, i\}$ for $1 \le i \le 10$. Determine the following.

(a) $\displaystyle\bigcap_{i=1}^{10} A_i$.   (b) $\displaystyle\bigcup_{i=1}^{10} A_i$.

21. For the universal set $U = \{1, 2, \ldots, 8\}$, let $A = \{2, 3, 7\}$, $B = \{1, 3, 4, 7\}$ and $C = \{1, 5, 7, 8\}$. Determine the following.

(a) $A \cup B$.   (b) $A \oplus B$.   (c) $B - C$.

(d) $\overline{A} \cap B$.   (e) $A \cap B \cap C$.   (f) $|A - \overline{(B \cup C)}|$.

22. Let $S = \{1, 3, 5, \ldots, 15\}$. For each of the following sets, describe the set by listing its elements.

(a) $A = \{x \in S : x - 8 \in S\}$.

(b) $B = \{x \in S : x \ge 10\}$.

(c) $C = \{x \in S : |x - 5| \in S\}$.

23. Let $A = \{1, \{1\}\}$ and $B = \{1, 2, \{2\}\}$. Determine the following.

(a) $A \times B$.   (b) $B \times A$.   (c) $A \times A$.   (d) $\mathcal{P}(A)$.   (e) $\mathcal{P}(B)$.

24. For $A = \{1, 2, \ldots, 10\}$, let $\mathcal{P}$ be a collection of k nonempty subsets of A for some positive integer k. Is the following statement true or false? If every element of A does not belong to exactly $k - 1$ subsets in $\mathcal{P}$, then $\mathcal{P}$ is a partition of A.

25. Determine whether the following statements are true or false.

(a) $\{x\} \in \{\{x\}\}$.   (b) $\{(x, y)\} \subseteq \{x, y\}$.

(c) $\{x, \{x\}\} \subseteq \{\{x\}, \{\{x\}\}\}$.   (d) $\{x, \{y\}\} \in \{x, \{y\}\}$.

(e) $\{(x, \{x\})\} \subseteq \{x, \{x\}\}$.   (f) $x \in \{\{x\}, \{\{x\}\}\}$.

(g) $\{(\{x\}, \{y\})\} \subseteq \{\{x\}, \{y\}\}$.   (h) $\{x\} \in \{x, \{x, \{x\}\}\}$.

26. (a) For $A = \{1, 2, 3\}$, give an example of three subsets A_1, A_2 and A_3 of A such that $|A_i \cap A_j| = |i - j|$ for every pair i, j of distinct integers with $1 \le i, j \le 3$.

(b) For $B = \{1, 2, 3, 4, 5, 6\}$, give an example of four subsets B_1, B_2, B_3 and B_4 of B such that $|B_i \cap B_j| = |i - j|$ for every pair i, j of distinct integers with $1 \le i, j \le 4$.

27. (a) Let $A = \{1, 2, 3\}$. Give an example of three distinct subsets A_1, A_2 and A_3 of A such that $\mathcal{P} = \{A_1 \cap A_2, A_1 \cap A_3, A_2 \cap A_3\}$ is a partition of A.

(b) Let $B = \{1, 2, 3, 4, 5, 6\}$. Give an example of four distinct subsets B_1, B_2, B_3 and B_4 of B such that $\mathcal{P} = \{B_1 \cap B_2, B_1 \cap B_3, B_1 \cap B_4, B_2 \cap B_3, B_2 \cap B_4, B_3 \cap B_4\}$ is a partition of B.

Chapter 3

Methods of Proof

In a criminal trial, the goal of the prosecuting attorney is to prove, beyond a reasonable doubt, that the defendant is guilty of the crime with which he or she has been charged. If the defendant is found guilty, then it is still possible that he or she did not commit the crime. To test the safety and efficacy of a new drug, a pharmaceutical company conducts clinical trials of the drug on individuals with the purpose of proving that the drug is effective in treating a particular medical condition. If the drug is approved for use, then it is quite likely that there will be individuals having this condition but for whom the drug is not effective. While proof does not mean certainty in these situations, proof does mean certainty in mathematics.

Indeed, one of the major reasons that we discussed logic and sets in the two preceding chapters was to provide ourselves with a proper background so that we can learn how to read and understand proofs of theorems and, on occasion, be able to write correct proofs of our own. The goal of this chapter is to introduce some methods of proof. (Another method will be introduced in Chapter 4.)

In a proof of a theorem, we often employ definitions, assumptions, axioms (statements accepted without proof) and other theorems that are known to the readers. A proof ordinarily consists of a sequence of statements, each following logically from those precede it, and ending with the desired conclusion. These statements are normally written as English sentences, with enough clarity so that a person having an appropriate background will understand the proof. The amount of detail in a proof depends on the anticipated knowledge and background of the audience.

3.1 Quantified Statements

We have seen that each open sentence involves a variable (or variables) whose values are taken from a set called the domain of the variable. Furthermore, a statement can be formed from an open sentence by assigning to each variable in the open sentence a value from its domain. Statements can also be formed from open sentences by what is called a quantifier. There are two kinds of quantifiers: universal and existential. In this section we will see how to construct statements from open sentences with the aid of these quantifiers. In the following section, we will learn when such statements are true and be introduced to one method of verifying the truth of some statements constructed in this manner.

Universal Quantifiers

Suppose that $R(x)$ is an open sentence over a domain S. Then for each element $a \in S$, $R(a)$ is a statement. Each of the phrases "for all," "for every" and "for each" is referred to as a **universal**

quantifier and is denoted by the symbol $\forall$. The sentence

$$\forall x \in S, \ R(x) \tag{3.1}$$

is stated in words as:

For every $x \in S$, $R(x)$.

This sentence is, in fact, a statement and is called a **quantified statement**.

Example 3.1

For the open sentence

$$R(n): \ n^2 - n \text{ is even.}$$

over the domain $\mathbb{Z}$ of integers, the sentence

$$\forall n \in \mathbb{Z}, R(n): \ \text{For every integer } n, \ n^2 - n \text{ is even.}$$

is therefore a quantified statement.   ◆

The quantified statement (3.1) is often rephrased as an implication. For example, for an open sentence $R(x)$ over a domain S, the quantified statement $\forall x \in S, R(x)$ can be expressed as:

If $x \in S$, then $R(x)$.

Example 3.2

The quantified statement:

For every integer n, $n^2 - n$ is even.

in Example 3.1 can therefore be expressed as

If n is an integer, then $n^2 - n$ is even.   ◆

Example 3.3

For the open sentence

$$R(n): \ 5n + 3 \text{ is even.}$$

over the domain S of odd integers, the quantified statement

$$\forall n \in S, \ R(n)$$

can be expressed as

For every odd integer n, $5n + 3$ is even.

or as

If n is an odd integer, then $5n + 3$ is even.   ◆

If an open sentence $R(x)$ over a domain S is itself an implication, say $R(x)$ is the implication $P(x) \Rightarrow Q(x)$ for open sentences $P(x)$ and $Q(x)$, then the quantified statement

$$\forall x \in S, P(x) \Rightarrow Q(x)$$

can be expressed in a variety of ways:
- For every $x \in S$, if $P(x)$ then $Q(x)$.
- If $x \in S$, then $P(x)$ implies $Q(x)$.
- Let $x \in S$. If $P(x)$, then $Q(x)$.

Example 3.4

For the open sentences

$$P(n): \ 5n^2 - 2 \text{ is odd.} \quad Q(n): \ n \text{ is odd.}$$

where the domain of the variable n is $\mathbb{Z}$, the quantified statement

$$\forall n \in \mathbb{Z}, \ P(n) \Rightarrow Q(n)$$

can therefore be expressed in words in any of the following ways:
- For every integer n, if $5n^2 - 2$ is odd then n is odd.
- If n is an integer, then $5n^2 - 2$ is odd implies that n is odd.
- Let n be an integer. If $5n^2 - 2$ is odd, then n is odd.   ♦

If an open sentence $R(x)$ over a domain S is the biconditional $P(x) \Leftrightarrow Q(x)$ for open sentences $P(x)$ and $Q(x)$, then the quantified statement

$$\forall x \in S, \ P(x) \Leftrightarrow Q(x)$$

can be expressed in words in any of the following ways:
- For every $x \in S$, $P(x)$ if and only if $Q(x)$.
- Let $x \in S$. Then $P(x)$ if and only if $Q(x)$.
- Let $x \in S$. Then $P(x)$ is necessary and sufficent for $Q(x)$.

Example 3.5

For the open sentences

$$P(n): \ n^2 \text{ is even.} \quad Q(n): \ n \text{ is even.}$$

where the domain of the variable n is $\mathbb{Z}$, the quantified statement

$$\forall n \in \mathbb{Z}, \ P(n) \Leftrightarrow Q(n)$$

can be expressed in words in any of the following ways:
- For every integer n, n^2 is even if and only if n is even.
- Let n be an integer. Then n^2 is even if and only if n is even.
- Let $n \in \mathbb{Z}$. Then n^2 is even is a necessary and sufficent condition for n to be even.   ♦

Existential Quantifiers

There is another way to form a statement from an open sentence by means of quantification. Each of the phrases "there exists," "there is," "for some" or "for at least one" is called an **existential quantifier**, which is denoted in symbols by $\exists$. For an open sentence $Q(x)$ over a domain S, the sentence

$$\exists x \in S, \ Q(x)$$

is a quantified statement that can be expressed in any of the following ways:
- There exists $x \in S$ such that $Q(x)$.
- For some $x \in S$, $Q(x)$.
- For at least one $x \in S$, $Q(x)$.

Example 3.6

For the open sentence

$$Q(x): \quad x^4 + 2 = 2x^2.$$

over the domain $\mathbb{R}$ of real numbers, the quantified statement

$$\exists x \in \mathbb{R}, Q(x)$$

can be expressed in any of the following ways:
- There exists $x \in \mathbb{R}$ such that $x^4 + 2 = 2x^2$.
- For some real number x, $x^4 + 2 = 2x^2$.
- For at least one real number x, $x^4 + 2 = 2x^2$.

This statement can also be expressed as:

$$\text{The equation } x^4 + 2 = 2x^2 \text{ has a real number solution.} \qquad \blacklozenge$$

Negations of Quantified Statements

Let's now consider negations of quantified statements. For an open sentence $R(x)$ over a domain S, the negation of $\forall x \in S, R(x)$ is:

$$\sim (\forall x \in S, R(x)): \text{It is not the case that } R(x) \text{ for every } x \in S.$$

This says that:

$$\text{There exists } x \in S \text{ such that not } R(x).$$

In symbols:

$$\sim (\forall x \in S, R(x)) \;\equiv\; \exists x \in S, \sim R(x). \qquad (3.2)$$

The negation of $\exists x \in S, R(x)$ is:

$$\sim (\exists x \in S, R(x)): \text{There does not exist } x \in S \text{ such that } R(x).$$

This implies that:

$$\text{For every } x \in S, \text{ not } R(x).$$

Therefore,

$$\sim (\exists x \in S, R(x)) \;\equiv\; \forall x \in S, \sim R(x). \qquad (3.3)$$

In other words, the negation of a quantified statement is also a quantified statement, namely the one with the opposite quantifier applied to the negation of the original open sentence.

Example 3.7

State the negation of each of the following quantified statements.

(a) Everyone likes "The Wizard of Oz."

(b) There is a city whose population exceeds that of Mexico City.

Solution. The negation of the quantified statement in (a) is

$$\text{There is someone who does not like "The Wizard of Oz."}$$

(Notice that the negation of the quantified statement in (a) is *not*: No one likes "The Wizard of Oz.") The negation of the statement in (b) is

The population of every city is not more than the population of Mexico City.

This statement could also be expressed as

The population of no city exceeds that of Mexico City. ♦

Example 3.8

State the negation of each of the following quantified statements.

(a) For every integer n, the integer $2n^2 + 5n + 1$ is odd.

(b) There exists a real number x such that $x^2 < 0$.

Solution. The negation of the quantified statement in (a) is

There exists an integer n such that $2n^2 + 5n + 1$ is not odd.

Of course, the negation of the statement in (a) could also be expressed as

There exists an integer n such that $2n^2 + 5n + 1$ is even.

The negation of the quantified statement in (b) is

For every real number x, $x^2 \geq 0$.

The negation of the statement in (b) could also be written as

There is no real number x for which $x^2 < 0$. ♦

Example 3.9

State the negation of each of the following quantified statements.

(a) If n is an even integer, then $3n + 5$ is an even integer.

(b) There exists an integer n between $\sqrt{5}$ and e.

Solution. The quantified statement in (a) can also be expressed as

For every even integer n, $3n + 5$ is an even integer.

Consequently, the negation of this statement can be expressed as

There exists an even integer n such that $3n + 5$ is not even.

or

There exists an even integer n such that $3n + 5$ is odd.

The negation of the quantified statement in (b) can be written as

There exists no integer between $\sqrt{5}$ and e.

or

No integer is between $\sqrt{5}$ and e.

or

For every integer n, n does not lie between $\sqrt{5}$ and e.

or

$$\text{If } n \text{ is an integer, then } n \text{ is not between } \sqrt{5} \text{ and } e. \qquad \blacklozenge$$

There are occasions when a quantified statement may contain both types of quantifiers. For example, if $R(x, y)$ is an open sentence with variables x and y such that the domain of x is S and the domain of y is T, then the quantified statement

$$\forall x \in S, \exists y \in T, R(x, y) \tag{3.4}$$

can be expressed as

For every $x \in S$, there exists $y \in T$ such that $R(x, y)$.

On the other hand, the quantified statement

$$\exists x \in S, \forall y \in T, R(x, y) \tag{3.5}$$

can be expressed as

There exists $x \in S$ such that for every $y \in T$, $R(x, y)$.

The negation of the statement in (3.4) is

$$\begin{aligned}
\sim (\forall x \in S, \exists y \in T, R(x, y)) &\equiv \exists x \in S, \sim (\exists y \in T, R(x, y)) \\
&\equiv \exists x \in S, \forall y \in T, \sim R(x, y),
\end{aligned}$$

which can therefore be expressed as

There exists $x \in S$ such that for every $y \in T$, not $R(x, y)$.

The negation of the statement in (3.5) is

$$\begin{aligned}
\sim (\exists x \in S, \forall y \in T, R(x, y)) &\equiv \forall x \in S, \sim (\forall y \in T, R(x, y)) \\
&\equiv \forall x \in S, \exists y \in T, \sim R(x, y),
\end{aligned}$$

which can therefore be expressed as

For every $x \in S$, there exists $y \in T$ such that not $R(x, y)$.

Example 3.10

Express the following statements in symbols and express their negations, both in symbols and in words.

(a) For every real number x, there exists a real number y such that $x + y = 0$.

(b) There exists an integer a such that for every integer b, $ab = 0$.

Solution.

(a) In this case, the domain of both x and y is the set $\mathbb{R}$ of real numbers. If we let $P(x, y)$ be the open sentence

$$P(x, y): \quad x + y = 0.$$

then the statement in (a) can be expressed as:

$$\forall x \in \mathbb{R}, \exists y \in \mathbb{R}, P(x, y). \tag{3.6}$$

The negation of the statement in (3.6) is

$$\sim (\forall x \in \mathbb{R}, \exists y \in \mathbb{R}, P(x, y)) \quad \equiv \quad \exists x \in \mathbb{R}, \forall y \in \mathbb{R}, \sim P(x, y).$$

Therefore, in words, the negation of the statement in (a) is

There exists a real number x such that for all real numbers y, $x + y \neq 0$.

(b) Here the domain of both a and b is the set $\mathbb{Z}$ of integers. Let $Q(a, b)$ be the open sentence

$$Q(a, b): \quad ab = 0.$$

The statement in (b) can therefore be expressed as:

$$\exists\, a \in \mathbb{Z}, \forall\, b \in \mathbb{Z}, Q(a, b). \tag{3.7}$$

The negation, in symbols, of the statement in (3.7) is

$$\sim (\exists a \in \mathbb{Z}, \forall b \in \mathbb{Z}, Q(a, b)) \quad \equiv \quad \forall a \in \mathbb{Z}, \exists b \in \mathbb{Z}, \sim Q(a, b).$$

In words, the negation of the statement in (b) is:

For every integer a, there exists an integer b such that $ab \neq 0$.   ◆

Exercises for Section 3.1

1. For a real number x, consider the open sentence $P(x)$: $(x-1)^2 > 0$. State each of the following quantified statements in words.

   (a) $\forall x \in \mathbb{R}, P(x)$.   (b) $\exists x \in \mathbb{R}, P(x)$.

2. State the following as a quantified statement:

   If n is an integer, then $n^3 + n + 1$ is an odd integer.

3. For an odd integer n, let $P(n)$: $7n + 4$ is odd. Express the quantified statement $\forall n \in S, P(n)$ in words as an implication, where S is the set of odd integers.

4. Express the following in symbols:

   There is some integer n such that $|n - 2| + |n - 1| = |n|$.

5. Let S denote the set of odd integers and consider the open sentences

   $$P(n): n^2 + 1 \text{ is even.} \quad Q(n): n^2 \text{ is even.}$$

   State $\forall n \in S$, $P(n)$ and $\exists n \in S$, $Q(n)$ in words.

6. Describe a set S and define an open sentence $R(x)$ of your own for $x \in S$. Then state $\forall x \in S$, $R(x)$ and $\exists x \in S$, $R(x)$ in words.

7. State the negations of the following quantified statements.

   (a) For every set A, $A \cap \overline{A} = \emptyset$.

   (b) There exists a set A such that $\overline{A} \subseteq A$.

8. State the negations of the following quantified statements.

   (a) For every rational number r, the number $1/r$ is rational.

   (b) There exists a rational number r such that $r^2 = 2$.

9. State the negations of the following quantified statements.

(a) For every irrational number s, there exists a rational number r such that $rs = 0$.

(b) For every rational number r, there exists an irrational number s such that $rs = 0$.

(c) There exists an even integer a such that for every integer b, ab is even.

(d) There exists an odd integer a such that for every integer b, ab is odd.

10. Express the following quantified statements in symbols.

(a) For every integer a, there exists an integer b such that $|a - b| = 1$.

(b) There exists an integer a such that for every integer b, $|a - b| = 1$.

11. State the negations of the following quantified statements.

(a) For every even integer a, both a^2 and $a + 2$ are even.

(b) There exists some real number x such that $x + 3 > 0$.

12. State the negations of the following quantified statements.

(a) For every successful man, there is a woman who supports him.

(b) For every successful woman, there is a man who supports her.

(c) There exists a woman W such that for every man M, W is richer than M.

(d) There exists a woman W such that for every man M, W is older than M.

(e) For every movie, there is always someone who likes it.

13. State the negations of the following quantified statements.

(a) Once upon a time, there was a man who owned a singing frog.

(b) There exists an odd integer that is the sum of two odd integers.

(c) $\sqrt{2}$ is an irrational number.

(d) There is no smallest positive real number.

(e) There are infinitely many primes.

(f) Every rational number can be expressed as the sum of two rational numbers.

14. Consider the following quantified statement: For every even integer a and every odd integer b, $a + b$ is odd.

(a) Express this quantified statement in symbols.

(b) Express the negation of this quantified statement in symbols.

(c) Express the negation of this quantified statement in words.

15. Consider the following quantified statement: There exist an even integer a and an odd integer b such that $(a + 2)^2 + (b + 3)^2 = 0$.

(a) Express this quantified statement in symbols.

(b) Express the negation of this quantified statement in symbols.

(c) Express the negation of this quantified statement in words.

16. Consider the following quantified statement: For every real number x, there exists a positive real number y such that $y < x^2$.

(a) Express this quantified statement in symbols.

(b) Express the negation of this quantified statement in symbols.

(c) Express the negation of this quantified statement in words.

17. Consider the following quantified statement: There exists an integer a such that $ab \geq 0$ for every integer b.

    (a) Express this quantified statement in symbols.

    (b) Express the negation of this quantified statement in symbols.

    (c) Express the negation of this quantified statement in words.

18. State the negation of the quantified statement below.

    For every integer a, there exists an integer b such that $\left|\frac{a+1}{2} - b\right| \leq 1$.

19. State the negation of the quantified statement below.

    For every integer a, there exists an integer b such that $\left|\frac{2a+1}{2} - b\right| < \frac{1}{2}$.

3.2   Direct Proof

For an open sentence $R(x)$ over a domain S, we have seen two kinds of quantified statements that can be constructed from $R(x)$, namely one by means of the universal quantifier

$$\forall x \in S, R(x) \tag{3.8}$$

and the other by means of the existential quantifier

$$\exists x \in S, R(x). \tag{3.9}$$

Since each of (3.8) and (3.9) is a statement, each has a truth value.

- *The statement $\forall x \in S, R(x)$ is true if $R(x)$ is true for each $x \in S$. Therefore, $\forall x \in S, R(x)$ is false if $R(x)$ is false for at least one element $x \in S$.*

- *The statement $\exists x \in S, R(x)$ is true if there exists at least one element $x \in S$ for which $R(x)$ is true. Consequently, $\exists x \in S, R(x)$ is false if $R(x)$ is false for every element $x \in S$.*

We now illustrate these ideas.

Example 3.11

Let $S = \{3, 4, 5\}$ and let

$$R(x) : \quad \frac{x^2 + 5x + 4}{2} \quad \text{is even.}$$

be an open sentence over the domain S.

(a) State $R(x)$ for each $x \in S$ and determine its truth value.

(b) State $\forall x \in S, R(x)$ and determine its truth value.

(c) State $\exists x \in S, R(x)$ and determine its truth value.

Solution.

(a) The statements $R(x)$ for each $x \in S$ along with their truth values are

   $R(3)$: 14 is even. (a true statement)

   $R(4)$: 20 is even. (a true statement)

   $R(5)$: 27 is even. (a false statement)

(b) The quantified statement

$$\forall x \in S,\ R(x): \quad \text{For every } x \in S,\ \frac{x^2 + 5x + 4}{2} \text{ is even.}$$

is false since $R(5)$ is a false statement.

(c) The quantified statement

$$\exists x \in S,\ R(x): \quad \text{There exists } x \in S \text{ such that } \frac{x^2 + 5x + 4}{2} \text{ is even.}$$

is true since $R(3)$ (also $R(4)$) is a true statement.   ◆

But how does one establish that a statement such as (3.8) is true if the domain S contains more than just a few elements? The answer to this question is that we provide a *proof*.

Definition 3.12 *A **proof** of a statement is a presentation of a logical argument that demonstrates the truth of the statement.*

A proof of a statement R ordinarily consists of a sequence of statements, each following logically from those that precede it and which terminates with the desired conclusion that the statement R is true. Within the proof we typically make use of one or more of the following:

(1) definitions of concepts,

(2) axioms or principles that have been agreed upon,

(3) assumptions we may have made,

(4) previous theorems.

The manner in which a proof is written and the number of sentences in a proof can vary widely. Indeed, a proof is written for an intended audience and the person writing the proof must be aware of what the audience knows or is expected to know. Even a proof that is written correctly may not be persuasive to a person reading the proof. In order to have a successfully written proof, a person reading the proof must have a certain degree of conviction that the argument presented is convincing. A correct proof may not be convincing because (i) the writer has made some questionable assumptions about what a reader knows and is likely to recall or (ii) the presentation fails to include some details that the reader needs. While clear and well-written proofs are always preferred, short proofs are not necessarily better. We will see many examples of proofs in this and later chapters.

Learning to write proofs of your own is not an easy task. Although there is certainly no particular approach to accomplish this, there is little doubt that reading and understanding the proofs of others is a good way to begin. The first several proofs that you write on your own will no doubt mimic those you have seen. Eventually it will become necessary to create your own proofs as you encounter statements that are different from those you have previously seen. As you do this, you may write proofs that are unclear to others and that contain errors. Although our goal is always to write clear and correct proofs, you should not be overly concerned with this. After all, the only way to truly avoid making mistakes in proofs is to not even try writing proofs.

We begin by discussing some techniques that are used to prove quantified statements of the type

$$\forall x \in S, R(x) \tag{3.10}$$

are true, where $R(x)$ is an open sentence over a domain S. We have mentioned that statements of the type (3.10) are often expressed as

$$\text{For every } x \in S,\ R(x).$$

or

$$\text{If } x \in S, \text{ then } R(x).$$

Typically, $R(x)$ is an implication, that is,

$$R(x)\colon\ \ P(x) \Rightarrow Q(x)$$

for open sentences $P(x)$ and $Q(x)$ over a domain S. Thus, we are discussing how to prove that

$$\forall x \in S, P(x) \Rightarrow Q(x) \tag{3.11}$$

is true. We have seen that the statement (3.11) can be expressed in a number of ways, including

$$\text{For every } x \in S, \text{ if } P(x) \text{ then } Q(x).$$

or

$$\text{Let } x \in S. \text{ If } P(x), \text{ then } Q(x).$$

Probably the most common way to verify that a statement

$$\forall x \in S,\ P(x) \Rightarrow Q(x)$$

is true is by means of a proof technique called a **direct proof**. To show that $\forall x \in S,\ P(x) \Rightarrow Q(x)$ is true, we must show that $P(x) \Rightarrow Q(x)$ is true for every element $x \in S$. Ordinarily, it is not practical to show that $P(x) \Rightarrow Q(x)$ is true for every element $x \in S$ by considering each element in S individually. According to the truth table for implication, we know that $P(x) \Rightarrow Q(x)$ is a true statement for each element $x \in S$ for which $P(x)$ is false. Thus it remains only to show that $P(x) \Rightarrow Q(x)$ is a true statement for every element $x \in S$ for which $P(x)$ is true. Of course, in order for $P(x) \Rightarrow Q(x)$ to be true when $P(x)$ is true, $Q(x)$ must also be true. In a direct proof, we assume that $P(x)$ is a true statement for some arbitrary element $x \in S$ and show that $Q(x)$ must also be true. Let's summarize what is done in a direct proof.

> To prove that $\forall x \in S,\ P(x) \Rightarrow Q(x)$ is true by means of a direct proof, we begin by assuming that $P(x)$ is true for an arbitrary element $x \in S$ and then we show that $Q(x)$ is true.

In the case where we are attempting to prove that

$$\forall x \in S,\ R(x) \quad \text{or} \quad \text{If } x \in S, \text{ then } R(x).$$

is true by means of a direct proof, we assume that x is an arbitrary element in S and show that $R(x)$ is a true statement.

A few additional comments are useful when giving a direct proof of the quantified statement $\forall x \in S,\ P(x) \Rightarrow Q(x)$. We have mentioned that we assume that $P(x)$ is true for an arbitrary element $x \in S$ and that our goal is to show that $Q(x)$ is true for this element x. It is expected that someplace during the course of the proof we will make use of our assumption that $P(x)$ is true. We are certainly permitted to make use of other things during the proof. As mentioned above, we can make use of any axioms and definitions that have been agreed upon, as well as any theorems that are known (or should be known) to the reader.

Suppose that we have been successful in giving a direct proof of $\forall x \in S,\ P(x) \Rightarrow Q(x)$. *What do we now know that we didn't know before?* First, let's mention some things that we don't know. In particular, we do not know that $P(x)$ is true for an arbitrary element $x \in S$. We only assumed that $P(x)$ is true for an arbitrary element $x \in S$ in the course of giving the proof. At the end of the proof, we also do not know that $Q(x)$ is true for an arbitrary element $x \in S$. What, in fact, we know is that for each element $x \in S$, the implication

$$\text{If } P(x), \text{ then } Q(x).$$

is true – nothing more and nothing less.

Why are we interested in proving a statement such as

$$\forall x \in S,\ P(x) \Rightarrow Q(x)?$$

One answer to this question is that this is the only way to be absolutely certain that this statement is true. Another answer is that it's a challenge to the mind to be able to prove such a statement. Yet another answer is that once we have proved that a statement $\forall x \in S,\ P(x) \Rightarrow Q(x)$ is true, we can use this information. We then know that if we should ever encounter a situation where $P(a)$ is true for some element $a \in S$, then $Q(a)$ must be true as well. (What we are essentially using here is the tautology in Example 1.64 in Section 1.5 called modus ponens.) This may simply be interesting to know or perhaps it has other consequences, such as having some applications or may be useful in proving some other statements.

We are about to give examples of true mathematical statements and possible proofs (where we use a direct proof or some other proof technique that we will introduce later). We will refer to such examples as *results*. While some authors call every true mathematical statement a theorem, others refer to such statements as observations, facts, results, propositions or theorems, according to their importance. We will reserve the term *theorem*, however, for a true mathematical statement that is especially useful, interesting or significant. Of course, such an interpretation is necessarily quite subjective. For us then, results are examples, often presented to illustrate a proof technique. Therefore, the statement:

$$\text{Let } x \in \mathbb{R}. \text{ If } x - 2 = 0, \text{ then } x^2 - x - 2 = 0. \tag{3.12}$$

will be a result to us, not a theorem.

Examples of Direct Proof

We first illustrate the direct proof technique with examples involving real numbers, beginning with a direct proof of the statement (3.12). From time to time, when giving a proof of some result or theorem, we may precede the proof with a discussion of an idea or a plan we may have to construct a proof or we may follow a proof with a discussion of some ideas that were used in the proof in order to better understand the proof. We will refer to the first type of discussion as a **proof strategy** and the second type of discussion as a **proof analysis** (or simply as an **analysis** in the case of an example). We will conclude a proof strategy, a proof analysis or an analysis with the symbol ◆.

Result to Prove: Let x be a real number. If $x - 2 = 0$, then $x^2 - x - 2 = 0$.

Proof Strategy. In a direct proof, we begin by assuming that $x - 2 = 0$. Our goal is now to show that $x^2 - x - 2$ is also 0. Observe that $x^2 - x - 2$ can be factored as $(x - 2)(x + 1)$. Since we know that $x - 2 = 0$, we now see how a proof can be given.   ◆

Result 3.13 *Let x be a real number. If $x - 2 = 0$, then $x^2 - x - 2 = 0$.*

Proof. Assume that $x - 2 = 0$. Then $x^2 - x - 2 = (x - 2)(x + 1) = 0 \cdot (x + 1) = 0$.   ∎

Proof Analysis. The statement of Result 3.13 begins with the sentence "Let x be a real number." Again, this is informing us that the domain of the variable x is the set of real numbers. A proof of Result 3.13 could have also been given by assuming that $x - 2 = 0$ and so $x = 2$. Then $x^2 - x - 2 = 4 - 2 - 2 = 0$.   ◆

The symbol ∎ at the end of the proof of Result 3.13 indicates that the proof is complete. This symbol is commonly used for this purpose. Over the years, many wrote Q.E.D. to indicate the end of a proof. Some still do. Q.E.D. is an abbreviation for the Latin phrase "quod erat demonstrandum," whose English translation is "which was to be demonstrated."

A familiar property of real numbers is:

$$a^2 \geq 0 \text{ for every real number } a \text{ and } a^2 > 0 \text{ if } a \neq 0. \tag{3.13}$$

This is useful in the proof of the following result.

Result to Prove: Let x be a real number. If $(x-1)^2 \leq 0$, then $3x - 2x^2 \geq 0$.

Proof Strategy. In a direct proof, we would begin by assuming that $(x-1)^2 \leq 0$. However, by (3.13) we already know that $(x-1)^2 \geq 0$. This means that $(x-1)^2 = 0$ and so $x = 1$. Now that we know $x = 1$, we are able to conclude that $3x - 2x^2 \geq 0$. ◆

Result 3.14 *Let x be a real number. If $(x-1)^2 \leq 0$, then $3x - 2x^2 \geq 0$.*

Proof. Assume that $(x-1)^2 \leq 0$. Since $(x-1)^2 \geq 0$ as well, it follows that $(x-1)^2 = 0$. Therefore, $x - 1 = 0$ and so $x = 1$. Hence $3x - 2x^2 = 3 \cdot 1 - 2 \cdot 1^2 = 3 - 2 = 1 \geq 0$. ∎

In order to illustrate any proof technique in some detail, we need to deal with areas of mathematics with which we are all familiar. Probably one such area involves even and odd integers. Although we have encountered even and odd integers before and probably have a good understanding of these numbers, in order to be precise we now give the standard definitions of these two concepts.

Definition 3.15 *An integer n is **even** if $n = 2a$ for some integer a. An integer n is **odd** if $n = 2b + 1$ for some integer b.*

These definitions of even and odd integers are consequences of the fact that any even integer has a remainder of 0 when divided by 2 and an odd integer has a remainder of 1 when divided by 2. Since these are the only two possible remainders when an integer is divided by 2, every integer is either even or odd. This will be discussed in greater detail in Chapter 7. For example, $6, 0$ and -10 are even since $6 = 2 \cdot 3$, $0 = 2 \cdot 0$ and $-10 = 2 \cdot (-5)$, where, of course, $3, 0$ and -5 are integers. In addition, $5, 1$ and -7 are odd since $5 = 2 \cdot 2 + 1$, $1 = 2 \cdot 0 + 1$ and $-7 = 2 \cdot (-4) + 1$, where $2, 0$ and -4 are integers.

Before giving some examples, we make some assumptions, namely:

$$\text{If } a \text{ and } b \text{ are integers, then so too are } -a, a + b \text{ and } ab. \tag{3.14}$$

Although we are surely familiar with many properties of even and odd integers, such as the sum of two even integers is even and the product of two odd integers is odd, we do not allow ourselves to make any such assumptions about even and odd integers. We are required to *prove* such properties of even and odd integers. On the other hand, once we have proved any of these, we can use these properties. We now give an example of a direct proof of a result dealing with even and odd integers.

Result to Prove: If n is an odd integer, then $5n + 3$ is an even integer.

Proof Strategy. As always, we begin a direct proof with an assumption. In this case, we assume that n is an odd integer. By the definition of an odd integer, we can write $n = 2k + 1$ for some integer k. Our goal is to show that $5n + 3$ is an even integer, that is, to show that we can write $5n + 3$ as 2ℓ for some integer ℓ. Since $n = 2k + 1$, this suggests making this substitution for n in $5n + 3$. ◆

Result 3.16 *If n is an odd integer, then $5n + 3$ is an even integer.*

Proof. Let n be an odd integer. Therefore, $n = 2k + 1$ for some integer k. Then

$$5n + 3 = 5(2k + 1) + 3 = 10k + 5 + 3 = 10k + 8 = 2(5k + 4).$$

By the properties in (3.14), $5k + 4$ is an integer. Therefore, $5n + 3$ is an even integer. ∎

Proof Analysis. In the proof of Result 3.16, we used the fact that $5k$ is an integer since 5 and k are integers. Also, since $5k$ and 4 are integers, so is $5k + 4$. Finally, since $5n + 3$ is written as twice an integer, it follows that $5n + 3$ is an even integer. ♦

We now give another direct proof involving even and odd integers.

Result 3.17 *If n is an even integer, then $n^2 + 4n - 3$ is odd.*

Proof. Let n be an even integer. Therefore, $n = 2k$ for some integer k. Then

$$\begin{aligned} n^2 + 4n - 3 &= (2k)^2 + 4(2k) - 3 = 4k^2 + 8k - 3 \\ &= 4k^2 + 8k - 4 + 1 = 2(2k^2 + 4k - 2) + 1. \end{aligned}$$

Again by (3.14), $2k^2 + 4k - 2$ is an integer and so $n^2 + 4n - 3$ is odd. ∎

Proof Analysis. In the proof of Result 3.17, we wanted to show that $n^2 + 4n - 3$ is odd. We were able to show that $n^2 + 4n - 3$ could be written as $4k^2 + 8k - 3$ for some integer k. Since $4k^2 + 8k - 3 = 2(2k^2 + 4k - 1) - 1$, it may seem that this shows that $n^2 + 4n - 3$ is odd. But it doesn't. An integer m was defined to be odd if it can be written as $2b + 1$ for some integer b, not as $2b - 1$. It is essential that we have an understanding of all definitions. ♦

It is often the case that when a direct proof of

$$\text{For all } x \in S,\ P(x) \Rightarrow Q(x).$$

is to be given, the opening step where $P(x)$ is assumed to be true for an arbitrary element $x \in S$ is not included. This step is simply understood and accepted. Let's illustrate this in the following example.

Result 3.18 *If n is an even integer, then so is $5n^3$.*

Proof. Since n is even, $n = 2k$ for some integer k. Therefore,

$$5n^3 = 5(2k)^3 = 5(8k^3) = 40k^3 = 2(20k^3).$$

Since $20k^3$ is an integer, $5n^3$ is even. ∎

We have seen that statements can be constructed from open sentences involving two or more variables. Let $P(x, y)$ and $Q(x, y)$ be open sentences, where the domain of x is S and the domain of y is T. There are occasions when a direct proof can be used to show that the statement

$$\text{For } x \in S \text{ and } y \in T,\ P(x, y) \Rightarrow Q(x, y).$$

is true. In this case, we assume that $P(x, y)$ is true for an arbitrary element $x \in S$ and an arbitrary element $y \in T$ and we then show that $Q(x, y)$ is true.

Our next example involves rational numbers. Recall that a real number r is **rational** if $r = a/b$, where $a, b \in \mathbb{Z}$ and $b \neq 0$. We also use the fact:

The product of two real numbers is nonzero if and only if both numbers are nonzero.

The contrapositive of this biconditional is:

The product of two real numbers is 0 if and only if at least one of these two numbers is 0.

Result to Prove: If r and s are rational numbers, then $r + s$ is a rational number.

Proof Strategy. As usual, we begin by assuming that r and s are rational numbers. This means that $r = a/b$ and $s = c/d$, where $a, b, c, d \in \mathbb{Z}$ and $b, d \neq 0$. We now only need to add the two fractions a/b and c/d and show that the sum can be expressed as the ratio of two integers, where the denominator is nonzero. ♦

Result 3.19 *If r and s are rational numbers, then r + s is a rational number.*

Proof. Since r and s are rational numbers, $r = a/b$ and $s = c/d$, where $a, b, c, d \in \mathbb{Z}$ and $b, d \neq 0$. Therefore

$$r + s = \frac{a}{b} + \frac{c}{d} = \frac{ad + bc}{bd}.$$

Since $ad + bc$ and bd are integers and $bd \neq 0$, the number $r + s$ is rational. ∎

Analysis. It is convenient here to analyze the statement of Result 3.19, not the proof. We used the symbols r and s for convenience in stating the result. Certainly other symbols could have been used. Indeed, Result 3.19 could have been stated as

> *The sum of two rational numbers is rational.*

If this had been the statement of Result 3.19, then it would probably have been good to begin the proof by saying: Let r and s be two rational numbers. Then we have convenient symbols to use. The proof would then continue as written above. ◆

Exercises for Section 3.2

1. Let $S = \{2, 4, 6\}$ and let

$$R(n): \quad \frac{n^3 - n}{6} \text{ is even.}$$

be an open sentence over the domain S.

   (a) State $R(n)$ for each $n \in S$ and determine the truth value of each such statement.

   (b) State $\forall n \in S$, $R(n)$ in words and determine its truth value.

   (c) State $\exists n \in S$, $R(n)$ in words and determine its truth value.

2. Let $S = \{0, 3, 4\}$ and let

$$P(n): \quad \frac{n(n + 1)(2n + 1)}{6} \text{ is even.}$$

be an open sentence over the domain S.

   (a) State $P(n)$ for each $n \in S$ and determine the truth value of each such statement.

   (b) State $\forall n \in S$, $P(n)$ in words and determine its truth value.

   (c) State $\exists n \in S$, $P(n)$ in words and determine its truth value.

3. Let x be a real number. Prove that if $(x - 1)^2 = 0$, then $x^3 - 1 = 0$.

4. Let x be a real number. Prove that if $x^3 + x = 0$, then $x^2 - x \leq 0$.

5. Let x be a real number. Prove that if $(x - 2)^4 \leq 0$, then $9 - x^2 \geq 0$.

6. Let n be an integer. Prove that if $2n^2 + n - 1 = 0$, then $n^3 < 0$.

7. Prove that if n is an even integer, then $7n - 2$ is an even integer.

8. Prove that if n is an odd integer, then $3n + 10$ is an odd integer.

9. Prove that if n is an odd integer, then $7n^2 - 2n + 15$ is an even integer.

10. Let $S = \{2, 3\}$. Prove that if n is an odd integer and $n \in S$, then $3n + 1$ is even.

11. Let $S = \{n \in \mathbb{Z} : n \leq 0\}$. Prove that if n is a nonnegative integer and $n \in S$, then $2^{-n}(n^2 + 3)$ is odd.

12. Let $a, b \in \mathbb{Z}$. Prove that if a and b are odd, then $ab + a + b$ is odd.

13. Prove that if r and s are rational numbers, then $5r + 7s$ is a rational number.

14. Prove that if r and s are rational numbers, then $r - s$ is a rational number.

15. Prove that the product of two rational numbers is rational.

16. Prove that if a and b are positive integers, then $\frac{a}{b} + \frac{b}{a} \geq 2$.

17. Let P and Q be statements. Prove that if $P \wedge Q$ is true, then $P \vee Q$ is true.

18. (a) Prove the following:

    Result If a and b are even integers, then $a + b$ is even.

    (b) If c and d are even integers, do we know that $c + d$ is even?

    (c) For integers x and y, if we know that $x + y$ is even, do we know that x and y are even?

    (d) If a and b are integers that are not both even, do we know that $a + b$ is not even?

19. A triangle is *equilateral* if its three sides have the same length and a triangle is *isosceles* if at least two of its sides have the same length.

    (a) Prove the following:

    Result If T is an equilateral triangle, then T is isosceles.

    (b) If T' is an equilateral triangle, do we know that T' is isosceles?

    (c) If T_0 is an isosceles triangle, do we know that T_0 is equilateral?

    (d) If T is a triangle that is not equilateral, do we know that T is not isosceles?

3.3   Proof by Contrapositive

Recall, for two statements or open sentences P and Q, that the contrapositive of the implication $P \Rightarrow Q$ is the implication $(\sim Q) \Rightarrow (\sim P)$. Furthermore, by Theorem 1.46, for two statements P and Q, the implication $P \Rightarrow Q$ and its contrapositive $(\sim Q) \Rightarrow (\sim P)$ are logically equivalent, that is,

$$P \Rightarrow Q \quad \equiv \quad (\sim Q) \Rightarrow (\sim P).$$

Consequently, for every two statements P and Q,

$$P \Rightarrow Q \text{ and } (\sim Q) \Rightarrow (\sim P) \text{ are either both true or both false.}$$

This provides us with another method to prove that the statement

$$\forall x \in S, \, P(x) \Rightarrow Q(x)$$

is true. If we can verify that its contrapositive

$$\forall x \in S, \, (\sim Q(x)) \Rightarrow (\sim P(x))$$

is true using a direct proof, then we will have proved that $\forall x \in S, \, P(x) \Rightarrow Q(x)$ is true. This method of proof is called a **proof by contrapositive**. Let's summarize what is done in a proof by contrapositive.

> To prove that $\forall x \in S$, $P(x) \Rightarrow Q(x)$ is true using a proof by contrapositive, we begin by assuming that $Q(x)$ is false for an arbitrary element $x \in S$ and then we show that $P(x)$ is also false.

We now give an example of a proof by contrapositive.

Result 3.20 *Let n be an integer. If $7n + 3$ is an odd integer, then n is an even integer.*

Proof. Assume that n is not an even integer. Then n is an odd integer and so $n = 2k + 1$ for some integer k. Therefore,

$$7n + 3 = 7(2k + 1) + 3 = 14k + 10 = 2(7k + 5).$$

Since $7k + 5$ is an integer, $7n + 3$ is even. ∎

Proof Analysis. The contrapositive of the implication

> If $7n + 3$ is an odd integer, then n is an even integer.

is the implication

> If n is *not* an even integer, then $7n + 3$ is *not* an odd integer.　(3.15)

Since we were giving a proof by contrapositive, we proved the implication (3.15) by a direct proof and so we began by assuming that n is not an even integer. Thus n is an odd integer and so n can be expressed as $2k + 1$ for some integer k. It only remained then to show that $7n + 3$ is not odd. ♦

One question that might have occurred to you as we look back at the proof we gave in Result 3.20 is why we decided to use a proof by contrapositive rather than a direct proof. If we had decided to use a direct proof, then we would have started the proof by assuming that $7n + 3$ is an odd integer. So $7n + 3 = 2a + 1$ for some integer a. We are then required to show that n is an even integer. This probably suggests solving $7n + 3 = 2a + 1$ for n. When we do this, we get $n = (2a - 2)/7$. However, now what do we do? It's probably not clear. In fact, if we weren't already informed that n is an integer, we wouldn't know that n is an integer, much less an even integer. On the other hand, if we use a proof by contrapositive, then we begin by assuming that n is an odd integer. As we saw in the proof of Result 3.20, the argument goes quite simply after that. Therefore, the decision to use a proof by contrapositive was based on allowing us to make an initial assumption that leads to a clear and simple proof.

We consider another example.

Result 3.21 *Let n be an integer. If $5n + 2$ is an odd integer, then n is an odd integer.*

Proof. Assume that n is an even integer. Then $n = 2k$ for some integer k. Hence

$$5n + 2 = 5(2k) + 2 = 10k + 2 = 2(5k + 1).$$

Since $5k + 1$ is an integer, $5n + 2$ is even. ∎

We give one additional example where proof by contrapositive is the appropriate proof technique to use.

Result 3.22 *Let x be a real number. If $x^3 + 3x^2 + 2x + 1 \leq 0$, then $x < 0$.*

Proof. Assume that $x \geq 0$. Then $x^3 \geq 0$, $3x^2 \geq 0$ and $2x \geq 0$. Thus

$$x^3 + 3x^2 + 2x + 1 \geq 0 + 0 + 0 + 1 > 0. \quad \blacksquare$$

Proofs of Biconditionals

We next consider the quantified statement

$$\forall x \in S, \, P(x) \Leftrightarrow Q(x)$$

constructed from two open sentences $P(x)$ and $Q(x)$ over a domain S. Recall that the biconditional $P(x) \Leftrightarrow Q(x)$ is defined as $(P(x) \Rightarrow Q(x)) \wedge (Q(x) \Rightarrow P(x))$, which is the conjunction of an implication and its converse. Therefore, we have the following.

> To prove that $\forall x \in S, \, P(x) \Leftrightarrow Q(x)$ is true, we must prove that both $\forall x \in S, \, P(x) \Rightarrow Q(x)$ is true and $\forall x \in S, \, Q(x) \Rightarrow P(x)$ is true.

Result to Prove: Let n be an integer. Then $3n + 8$ is odd if and only if n is odd.

Proof Strategy. To prove this result, we must prove for every integer n that both "$3n + 8$ is odd if n is odd" and "$3n + 8$ is odd only if n is odd" are true. The first of these (the "if" part of the "if and only if") can also be expressed as:

> If n is odd, then $3n + 8$ is odd.

It is natural to use a direct proof to prove this implication. The converse of this implication is "$3n+8$ is odd only if n is odd" (the "only if" part of the "if and only if"), which can also be expressed as

> If $3n + 8$ is odd, then n is odd.

It would seem reasonable to use a proof by contrapositive to prove this implication.   ◆

Result 3.23 *Let n be an integer. Then $3n + 8$ is odd if and only if n is odd.*

Proof. First, we prove that if n is an odd integer, then $3n + 8$ is odd. Assume that n is odd. Then $n = 2k + 1$ for some integer k. So

$$3n + 8 = 3(2k + 1) + 8 = 6k + 11 = 2(3k + 5) + 1.$$

Since $3k + 5$ is an integer, $3n + 8$ is odd.

Next, we verify the converse, namely, if $3n + 8$ is odd, then n is odd. Assume that n is even. Then $n = 2\ell$ for some integer ℓ. Therefore,

$$3n + 8 = 3(2\ell) + 8 = 6\ell + 8 = 2(3\ell + 4).$$

Since $3\ell + 4$ is an integer, $3n + 8$ is even.   ∎

Proof Analysis. In the proof of Result 3.23, we saw that there were two implications to verify. When we verified that "if n is an odd integer, then $3n + 8$ is odd," we assumed that n is odd and so wrote that $n = 2k + 1$ for some integer k; while when we verified that "if $3n + 8$ is odd, then n is odd," we assumed that n is even and wrote that $n = 2\ell$ for some integer ℓ. Although we chose to use different symbols, namely k and ℓ, for the integers they represent since these integers are almost certainly different, we could have chosen the same symbol, say k, for each as two different implications were being verified in this proof.   ◆

The following example will be helpful to us soon. Consequently, we refer to it as a theorem.

Theorem to Prove: Let n be an integer. Then n^2 is even if and only if n is even.

Proof Strategy. Because the statement we wish to prove is a biconditional, we have two implications to prove, namely "n^2 is even if n is even" and "n^2 is even only if n is even." The implication "n^2 is even if n is even" (the "if" part of the "if and only if") can be written as

> If n is even, then n^2 is even.

We will prove this implication with a direct proof. Its converse is the implication "n^2 is even only if n is even" (the "only if" part of the "if and only if") and can be expressed as:

If n^2 is even, then n is even.

We will use a proof by contrapositive to prove this implication. ♦

Theorem 3.24 *Let n be an integer. Then n^2 is even if and only if n is even.*

Proof. Assume that n is even. Then $n = 2a$, where a is an integer. Therefore,

$$n^2 = (2a)^2 = 4a^2 = 2(2a^2).$$

Because $2a^2$ is an integer, n^2 is even.

Next, we verify the converse. Assume that n is odd. So $n = 2b + 1$, where $b \in \mathbb{Z}$. Then

$$n^2 = (2b + 1)^2 = 4b^2 + 4b + 1 = 2(2b^2 + 2b) + 1.$$

Since $2b^2 + 2b$ is an integer, n^2 is odd. ∎

Since the contrapositive of the implication "if n is even, then n^2 is even" is

if n^2 is odd, then n is odd.

while the contrapositive of the implication "if n^2 is even, then n is even" is

if n is odd, then n^2 is odd.

it follows that Theorem 3.24 can also be worded as follows.

Theorem 3.25 *Let n be an integer. Then n^2 is odd if and only if n is odd.*

Exercises for Section 3.3

1. Let n be an integer. Prove that if $5n + 7$ is odd, then n is even.

2. Let n be an integer. Prove that if $9n - 5$ is even, then n is odd.

3. Let $n \in S = \{1, 2, 3\}$. Prove that if $3n + 4$ is odd, then n is odd.

4. Let n be an integer. Prove that $3n - 11$ is odd if and only if n is even.

5. Let n be an integer. Prove that n^3 is even if and only if n is even.

6. Let n be an integer. Prove that n^4 is odd if and only if n is odd.

7. Let n be an integer. Prove that $2n^2 - n - 1 = 0$ if and only if $3n^2 - n - 2 = 0$.

8. Give a proof of

Let $n \in \mathbb{Z}$. Then $n - 3$ is even if and only if $n + 4$ is odd.

using

(a) two direct proofs.
(b) one direct proof and one proof by contrapositive.
(c) two proofs by contrapositive.

9. Let x be a real number. Prove that if $x^3 + 5x + 1 \leq 0$, then $x < 0$.

10. Let x and y be integers. Prove that if $x + y \geq 9$, then either $x \geq 5$ or $y \geq 5$.

11. Let x and y be integers. Prove that if $2x + 3y \geq 1$, then $x \geq 1$ or $y \geq 1$.

12. Let a, b and m be integers. Prove that if $2a + 3b \geq 12m + 1$, then $a \geq 3m + 1$ or $b \geq 2m + 1$.

13. Let a, b and c be nonnegative integers. Prove that if $a + 2b + 3c \geq 5$, then $a \geq 3$, $b \geq 2$ or $c \geq 1$.

14. Consider the following:

    Result Let $n \in \mathbb{Z}$. Then $5n + 7$ is even only if n is odd.

    Next, consider the following:

    Proof. Assume first that n is an odd integer. Then $n = 2a + 1$ for some integer a. Therefore,

    $$5n + 7 = 5(2a + 1) + 7 = 10a + 5 + 7 = 10a + 12 = 2(5a + 6).$$

    Since $5a + 6$ is an integer, $5n + 7$ is even.

    Next, assume that n is an even integer. Then $n = 2b$ for some integer b. Hence

    $$5n + 7 = 5(2b) + 7 = 10b + 7 = 2(5b + 3) + 1.$$

    Since $5b + 3$ is an integer, $5n + 7$ is odd.   ∎

    What is wrong with the preceding proof?

3.4   Proof by Cases

Suppose that $R(x)$ is an open sentence over a domain S and, as usual, we wish to prove that the quantified statement $\forall x \in S$, $R(x)$ is true. If we are using a direct proof, then we would begin with an arbitrary element x in S and attempt to show that $R(x)$ is true. There are some occasions when knowing that $x \in S$ doesn't seem to provide enough information to show that $R(x)$ is true. In such instances, it may be useful to consider a collection $\mathcal{P}$ of two or more subsets of S such that every element of S belongs to at least one of these subsets. Typically, $\mathcal{P}$ is a partition of S. The proof is then divided into cases, according to which subset in $\mathcal{P}$ the element x belongs. The fact that x belongs to a specific subset of S may provide just the additional information about x we need that allows us to give a proof.

For example, if we are attempting to prove that a statement concerning an integer n is true, then partitioning $\mathbb{Z}$ into the set of even integers and the set of odd integers may be a key step in obtaining a proof. Also, if we are attempting to prove some statement dealing with the absolute value $|x|$ of a real number x, then it may be useful to consider the partition of $\mathbb{R}$ into the set of positive real numbers, the set of negative real numbers and $\{0\}$ or perhaps the partition of $\mathbb{R}$ into $\{x \in \mathbb{R} : x \geq 0\}$ and $\{x \in \mathbb{R} : x < 0\}$.

In summary then, if we wish to prove some statement concerning an element x in a set S, then it might be useful to consider a partition $\mathcal{P}$ of S into two or more subsets of S. The proof can then be divided into **cases**, according to the particular subset in $\mathcal{P}$ to which x belongs. This is called a **proof by cases**. We now illustrate this approach.

Result to Prove: If n is an integer, then $n^2 - n$ is an even integer.

Proof Strategy. If we begin with an integer n, then there seems to be no way to show that $n^2 - n$ is an even integer. Certainly, n is even or n is odd. If n is even, then we can write $n = 2a$ for some integer a and showing that $n^2 - n$ is even is no longer difficult. Similarly, if n is odd, then we can write $n = 2b + 1$ for some integer b and again showing that $n^2 - n$ is even is straightforward.   ◆

Result 3.26 *If n is an integer, then $n^2 - n$ is an even integer.*

Proof. Let n be an integer. We consider two cases, according to whether n is even or n is odd.

Case 1. n is even. Then $n = 2a$ for some integer a. Therefore,

$$n^2 - n = (2a)^2 - (2a) = 4a^2 - 2a = 2(2a^2 - a).$$

Since $2a^2 - a$ is an integer, $n^2 - n$ is even.

Case 2. n is odd. Then $n = 2b + 1$ for some integer b. Therefore,

$$\begin{aligned} n^2 - n &= (2b+1)^2 - (2b+1) = (4b^2 + 4b + 1) - (2b+1) \\ &= 4b^2 + 2b = 2(2b^2 + b). \end{aligned}$$

Since $2b^2 + b$ is an integer, $n^2 - n$ is even. ■

Parity of Integers

Two integers m and n are said to be of the **same parity** if they are both even or both odd; otherwise, m and n are of **opposite parity**. For example, 3 and 7 are of the same parity, as are -2 and 10; while 3 and 8 are of opposite parity. Observe that every two integers are of the same parity or of opposite parity (but not both). Therefore, the negation of "two integers m and n are of the same parity" is "two integers m and n are of opposite parity." We now consider an example dealing with these ideas.

Result to Prove: Let m and n be two integers. Then $3m + 5n$ is even if and only if m and n are of the same parity.

Proof Strategy. There are two implications to prove here. One of these is "if m and n are of the same parity, then $3m + 5n$ is even." Either we want to use a direct proof or a proof by contrapositive. If we were to use a proof by contrapositive, then we would begin the proof by assuming that $3m + 5n$ is odd. We would then want to show that m and n are of opposite parity. But how would we do this? Using a direct proof seems to be the appropriate method. We would then begin the proof by assuming that m and n are of the same parity. Since this means that either m and n are both even or m and n are both odd, it immediately suggests a proof by cases.

The second implication is "if $3m + 5n$ is even, then m and n are of the same parity." Here the reasonable proof technique is a proof by contrapositive, where we would begin by assuming that m and n are of opposite parity and divide the proof into the two cases (1) m is even and n is odd and (2) m is odd and n is even. ♦

Result 3.27 *Let m and n be two integers. Then $3m + 5n$ is even if and only if m and n are of the same parity.*

Proof. First, we show that if m and n are of the same parity, then $3m + 5n$ is even. We use a direct proof to do this. Assume that m and n are of the same parity. We consider two cases.

Case 1. m and n are even. Therefore, $m = 2a$ and $n = 2b$ for integers a and b. So

$$3m + 5n = 3(2a) + 5(2b) = 6a + 10b = 2(3a + 5b).$$

Since $3a + 5b$ is an integer, $3m + 5n$ is even.

Case 2. m and n are odd. Then $m = 2a + 1$ and $n = 2b + 1$, where $a, b \in \mathbb{Z}$. Hence

$$\begin{aligned} 3m + 5n &= 3(2a+1) + 5(2b+1) = 6a + 3 + 10b + 5 \\ &= 6a + 10b + 8 = 2(3a + 5b + 4). \end{aligned}$$

Since $3a + 5b + 4$ is an integer, $3m + 5n$ is even.

Next, we verify the converse, that is, if $3m + 5n$ is even, then m and n are of the same parity. We establish this using a proof by contrapositive. Assume that m and n are of opposite parity. We consider two cases.

Case 1. m is even and n is odd. So $m = 2a$ and $n = 2b + 1$ for integers a and b. Therefore,

$$3m + 5n = 3(2a) + 5(2b + 1) = 6a + 10b + 5 = 2(3a + 5b + 2) + 1.$$

Since $3a + 5b + 2$ is an integer, $3m + 5n$ is odd.

Case 2. m is odd and n is even. Thus $m = 2a + 1$ and $n = 2b$ for integers a and b. Therefore,

$$3m + 5n = 3(2a + 1) + 5(2b) = 6a + 3 + 10b = 2(3a + 5b + 1) + 1.$$

Since $3a + 5b + 1$ is an integer, $3m + 5n$ is odd.   ■

Proof Analysis. In the preceding example, we first used a direct proof to show that "if m and n are of the same parity, then $3m + 5n$ is even" and considered two cases. We then used a proof by contrapositive to show that "if $3m + 5n$ is even, then m and n are of the same parity" and considered two cases here as well. The first implication we verified in the proof of Result 3.27 was "if m and n are of the same parity, then $3m + 5n$ is even." In Case 1, when m and n are both even, we wrote $m = 2a$ and $n = 2b$ for integers a and b. In Case 2, when m and n are both odd, we wrote $m = 2a + 1$ and $n = 2b + 1$ for integers a and b. Although it was perfectly permissible to use a and b in these two cases since the two cases are independent of each other, it would be incorrect to write $m = 2a$ and $n = 2a$ for some integer a in Case 1, say, for this would imply that $m = n$, which would be incorrect, since the statement of Result 3.27 did not say that m and n are two *equal* integers.   ♦

Without Loss of Generality

From time to time there are occasions when the proofs of two cases are so similar that including proofs of both cases is repetitive. When this occurs, we ordinarily choose to prove only one case, saying that we are doing this **without loss of generality** (that is, nothing is lost by considering only one of the cases). We present one additional example to illustrate this idea.

Result 3.28 *Let m and n be integers. Then mn is odd if and only if m and n are both odd.*

Proof. First, we prove that if m and n are odd, then mn is odd. Let m and n be odd integers. Then $m = 2k + 1$ and $n = 2\ell + 1$ for integers k and ℓ. Then

$$
\begin{aligned}
mn &= (2k + 1)(2\ell + 1) = 4k\ell + 2k + 2\ell + 1 \\
   &= 2(2k\ell + k + \ell) + 1.
\end{aligned}
$$

Since $2k\ell + k + \ell$ is an integer, mn is odd.

Next, we verify the converse, namely if mn is odd, then m and n are both odd. We use a proof by contrapositive. Suppose that m is even or n is even. Without loss of generality, we may assume that m is even. Then $m = 2k$ for some integer k. Thus

$$mn = (2k)n = 2(kn).$$

Since kn is an integer, mn is even.   ■

Proof Analysis. In Result 3.28, when proving that if mn is odd, then m and n are both odd, we used a proof by contrapositive. We began, therefore, by assuming that it is not the case that m and n are odd. By De Morgan's Law (Theorem 1.24), this means that m is even or n is even. Thus, there are two cases, namely

Case 1: *m is even.*   Case 2: *n is even.*

(If m and n are both even, this situation would be covered by either case.) Since the proofs of these two cases are very similar, we used the phrase "without loss of generality" (some abbreviate this as WLOG or WOLOG) to indicate this and chose to prove only one of them. (See also Exercise 12.) ♦

In Example 2.31 of Chapter 2 we verified the following:

Let A, B and C be sets. Then $(A - B) \cap (A - C) = A - (B \cup C)$.

We would now refer to this example as a result and a solution to the example as a proof of the result. Verifying equality of sets often involves proofs by cases. We give an example of this.

Result 3.29 *Let A, B and C be sets. Then*

$$(A - B) \cup (A - C) = A - (B \cap C).$$

Proof. First, we show that $(A - B) \cup (A - C) \subseteq A - (B \cap C)$. Let $x \in (A - B) \cup (A - C)$. Then $x \in A - B$ or $x \in A - C$. We consider these two cases.

Case 1. $x \in A - B$. Then $x \in A$ and $x \notin B$. Since $x \notin B$, it follows that $x \notin B \cap C$. Because $x \in A$ and $x \notin B \cap C$, we have $x \in A - (B \cap C)$. Hence $(A - B) \cup (A - C) \subseteq A - (B \cap C)$.

Case 2. $x \in A - C$. Then $x \in A$ and $x \notin C$. Because $x \notin C$, we have $x \notin B \cap C$. So $x \in A$ and $x \notin B \cap C$. Therefore, $x \in A - (B \cap C)$. Consequently, $(A - B) \cup (A - C) \subseteq A - (B \cap C)$.

Next, we show that $A - (B \cap C) \subseteq (A - B) \cup (A - C)$. Let $y \in A - (B \cap C)$. Hence $y \in A$ and $y \notin B \cap C$. Since $y \notin B \cap C$, either $y \notin B$ or $y \notin C$. We consider these two cases.

Case 1. $y \notin B$. Since $y \in A$, we have $y \in A - B$. Therefore, $y \in (A - B) \cup (A - C)$ and so $A - (B \cap C) \subseteq (A - B) \cup (A - C)$.

Case 2. $y \notin C$. Since $y \in A$, it follows that $y \in A - C$. Thus $y \in (A - B) \cup (A - C)$ and so $A - (B \cap C) \subseteq (A - B) \cup (A - C)$.

Therefore, $(A - B) \cup (A - C) = A - (B \cap C)$. ∎

As in the proof of Result 3.28, a proof of Result 3.29 could also be given using "without loss of generality" (in fact, using it twice). See Exercise 13.

Exercises for Section 3.4

1. Prove that if n is an integer, then $n^2 - 3n + 5$ is odd.

2. Prove that if n is an integer, then $n^3 - n$ is even.

3. Let $n \in \mathbb{Z}$. Prove that if $n^2 + n = 0$, then $\frac{2^n + 3^n}{12^n}$ is even.

4. Prove that if n is an integer, then $3n + 1$ and $5n + 2$ are of opposite parity.

5. (a) Let m and n be two integers. Prove that $m + n$ is even if and only if m and n are of the same parity.

   (b) Let $S = \{a, b, c, d\}$ be a set of four integers. Prove that if m pairs of distinct integers of S are of the same parity and the remaining n pairs of distinct integers of S are of opposite parity, then m and n are of the same parity.

6. Let m and n be two integers. Prove that $3m - n$ is even if and only if m and n are of the same parity.

7. Let m and n be two integers. Prove that $7m + 3n$ is odd if and only if m and n are of opposite parity.

8. Let m and n be two integers. Prove that if m and n are of opposite parity, then $2m + n$ and $3m - 4n$ are of opposite parity.

9. Let m and n be two integers. Prove that mn^2 is odd if and only if m and n are odd.

10. Let m and n be two integers. Prove that mn and $m + n$ are both even if and only if m and n are both even.

11. Give a proof of

$$\text{Let } n \in \mathbb{Z}. \text{ If } |2n - 1| \leq 5, \text{ then } n \leq 3 \text{ and } n \geq -2.$$

using

   (a) a direct proof.

   (b) a proof by contrapositive.

12. In the proof of Result 3.28 it was proved, for integers m and n, that if mn is odd, then m and n are both odd. This was done using a proof by contrapositive. In the proof that was given, it was assumed, without loss of generality, that m is even. Give a proof of this implication, using a proof by cases namely:

   Case 1. m is even .   Case 2. n is even.

13. Give a proof of Result 3.29 where "without loss of generality" is used twice.

14. Use a proof by cases (as in the proof of Result 3.29) to prove the following distributive law for sets:

   Let A, B and C be sets. Then $A \cap (B \cup C) = (A \cap B) \cup (A \cap C)$.

15. Prove the following:

   Let A and B be sets. Then $(A - B) \cup (B - A) = (A \cup B) - (A \cap B)$.

16. Let $S = \{0, 1, 3, 4\}$. Prove that if $x \in S$, then $x^2 - 4x + 3 = 0$ or $x^2 - 4x + 3 = 3$.

3.5   Counterexamples

For an open sentence $R(x)$ over a domain S, we have seen that the statement $\forall x \in S, R(x)$ is true if $R(x)$ is true for every element $x \in S$. Recall also that

$$\sim (\forall x \in S, R(x)) \equiv \exists x \in S, \sim R(x)$$

and so that $\forall x \in S, R(x)$ is false if $R(x)$ is a false statement for at least one element $a \in S$. Often, $R(x)$ is an implication $P(x) \Rightarrow Q(x)$ for open sentences $P(x)$ and $Q(x)$ over S. We have described two methods (direct proof and proof by contrapositive) that can be used to prove that a statement $\forall x \in S, P(x) \Rightarrow Q(x)$ is true. To show that $\forall x \in S, P(x) \Rightarrow Q(x)$ is false, however, we need only show that there exists some element $a \in S$ such that $P(a) \Rightarrow Q(a)$ is false. Such an element $a \in S$ is called a **counterexample** for (or of) the statement $\forall x \in S, P(x) \Rightarrow Q(x)$. Therefore, a counterexample of $\forall x \in S, P(x) \Rightarrow Q(x)$ is an element $a \in S$ for which $P(a)$ is true and $Q(a)$ is false. Such an element is said to **disprove** the statement. We now illustrate this.

Example 3.30 *Disprove the following: Let n be an integer. If $4n + 5$ is odd, then n is even.*

Solution. The integer $n = 1$ is a counterexample since $4n + 5 = 4 \cdot 1 + 5 = 9$ is odd and $n = 1$ is not even. ♦

Analysis. Let's reflect on Example 3.30 and the solution we gave. Since we were asked to disprove

$$\text{Let } n \text{ be an integer. If } 4n + 5 \text{ is odd, then } n \text{ is even.}$$

we needed to present a counterexample, which, in this case, is an integer n for which "$4n + 5$ is odd" is true and "n is even" is false. We saw that the integer $n = 1$ has this property. (Actually, every odd integer n is a counterexample here.) ♦

Example 3.31 *Disprove the following: Let $m, n \in \mathbb{Z}$. The integer $3m + 7n$ is even if and only if m and n are both even.*

Solution. The integers $m = 1$ and $n = 1$ are not both even (in fact, $m = 1$ and $n = 1$ are both odd) but $3m + 7n = 3 \cdot 1 + 7 \cdot 1 = 10$ is even. So $m = 1$ and $n = 1$ is a counterexample. ♦

Analysis. The biconditional in Example 3.31 is the conjunction of two implications, namely

$$\text{Let } m, n \in \mathbb{Z}. \text{ If } 3m + 7n \text{ is even, then } m \text{ and } n \text{ are both even.} \tag{3.16}$$

and

$$\text{Let } m, n \in \mathbb{Z}. \text{ If } m \text{ and } n \text{ are both even, then } 3m + 7n \text{ is even.} \tag{3.17}$$

For the statement in Example 3.31 to be false, we need to show that at least one of the statements (3.16) and (3.17) is false. Even though statement (3.17) is true, statement (3.16) is false. A counterexample that disproves statement (3.16) is what we gave in the solution of Example 3.31. ♦

Example 3.32 *Disprove: If x is a real number, then $(x^2 - 2)^2 > 0$.*

Solution. The number $x = \sqrt{2}$ is a counterexample since

$$(x^2 - 2)^2 = \left((\sqrt{2})^2 - 2 \right)^2 = (2 - 2)^2 = 0. ♦$$

The number $x = -\sqrt{2}$ would have also been a counterexample in Example 3.32. The following example shows the importance of the domain involved in a quantified statement.

Example 3.33

Consider the two open sentences

$$P(x): \ |x| = 1. \quad Q(x): \ x^3 = 1.$$

and the statement

$$\forall x \in S, \ P(x) \Rightarrow Q(x): \ \text{Let } x \in S. \text{ If } |x| = 1, \text{ then } x^3 = 1. \tag{3.18}$$

for two different choices for the domain S of x.

First, let $S = \mathbb{R}^+$ be the set of positive real numbers. Then the implication is true since $x = 1$ is the only positive real number such that $|x| = 1$. Since $1^3 = 1$, we have proved that the statement (3.18) is true.

Second, let $S = \mathbb{R}$ be the set of real numbers. Then the implication is false. Note that there are two real numbers x for which $|x| = 1$, namely, $x = 1$ and $x = -1$. If $x = 1$, then $x^3 = 1^3 = 1$. However, if $x = -1$, then $x^3 = (-1)^3 = -1 \neq 1$. Therefore, the implication $P(-1) \Rightarrow Q(-1)$ is false and $x = -1$ is a counterexample of the statement (3.18). ♦

Exercises for Section 3.5

1. Disprove: If n is an even integer, then $3n + 2$ is odd.

2. Disprove: Let m and n be integers. The integer mn is even if and only if m and n are of opposite parity.

3. Disprove: Let $x, y \in \mathbb{R}$. If $|x| < |y|$, then $x < y$.

4. Disprove: For a real number x, $x^2 - x - 2 > 0$ if and only if $-1 < x < 1$.

5. Disprove: Let A, B and C be sets. If $A \cap B = A \cap C$, then $B = C$.

6. Disprove: Let A, B and C be sets. If $A \cup B = A \cup C$, then $B = C$.

7. Disprove: Let A and B be sets. If $A \cup B \neq \emptyset$, then $A \neq \emptyset$ and $B \neq \emptyset$.

8. Disprove: For every two sets A and B, $\mathcal{P}(A \cup B) = \mathcal{P}(A) \cup \mathcal{P}(B)$.

9. Disprove: For every three sets A, B and C, $A \cup (B - C) = (A \cup B) - (A \cup C)$.

10. Disprove: For every $n \in \mathbb{N}$, there exists $m \in \mathbb{N}$ such that $n < m < n^2$.

11. Disprove: Every positive integer can be expressed as the sum of two positive integers.

12. Prove or disprove the following:

    (a) Let n be a positive integer. If $n^2 - 4 = 0$, then $n - 2 = 0$.
    (b) Let n be a negative integer. If $n^2 - 4 = 0$, then $n - 2 = 0$.

13. Prove or disprove the following:

    (a) Let x be an integer. If $2x - 1 = 0$, then $2x^2 - 3x - 2 = 0$.
    (b) Let x be a rational number. If $2x - 1 = 0$, then $2x^2 - 3x - 2 = 0$.

14. Prove or disprove the following:

    (a) Let $S = \{1, 4, 5, 8\}$. If $n \in S$, then $(n^2 - n)/2$ is an even integer.
    (b) Let $S = \{1, 4, 5, 6, 8\}$. If $n \in S$, then $(n^2 - n)/2$ is an even integer.

15. Prove or disprove the following:

    Let n be an integer. Then $n^2 + n$ is even if and only if n is even.

16. Prove or disprove the following:

    Let a and b be two integers. Then $2a + 3b$ is odd if and only if a is even and b is odd.

17. Prove or disprove the following: Let $a, b \in \mathbb{Z}$. Then $3ab$ is even if and only if a and b are even.

3.6 Existence Proofs

We have been investigating the truth or falseness of quantified statements of the type $\forall x \in S$, $R(x)$, where $R(x)$ is an open sentence over a domain S. We now turn to quantified statements of the type $\exists x \in S$, $R(x)$. Recall that a statement $\exists x \in S$, $R(x)$ is true if $R(x)$ is true for at least one element $x \in S$. In this case, a proof requires showing only that there is some element $a \in S$ for which $R(a)$ is true. This is called an **existence proof**.

Result 3.34 *There exists an integer n such that $2 - n^2 > 0$.*

Proof. Let $n = 1$. Then $2 - n^2 = 2 - 1^2 = 1 > 0$. ∎

In the proof of Result 3.34, it was not particularly difficult to notice that $n = 1$ had the desired property that $2 - n^2 > 0$. Had we selected $n = -1$ or $n = 0$, this too would have resulted in a successful proof.

Result 3.35 *There exist integers m and n of the same parity such that*

$$(m - 1)^2 + (n - 4)^2 \le 1.$$

Proof. Let $m = 2$ and $n = 4$. Then

$$(m - 1)^2 + (n - 4)^2 = (2 - 1)^2 + (4 - 4)^2 = 1 + 0 \le 1,$$

as desired. ∎

A correct proof of Result 3.35 could have also been given by selecting (1) $m = 0$, $n = 4$, (2) $m = 1$, $n = 5$ or (3) $m = 1$, $n = 3$.

Result 3.36 *There exists a real number x such that $x^2 - 8 = 0$.*

Proof. Let $x = \sqrt{8}$. Then $x^2 - 8 = (\sqrt{8})^2 - 8 = 8 - 8 = 0$. ∎

A proof of Result 3.36 consists only of finding a real number solution of the equation $x^2 - 8 = 0$. This equation has two solutions of course, namely $\sqrt{8}$ and $-\sqrt{8}$.

For the next result, it is useful to recall that we mentioned that $\sqrt{2}$ is an irrational number. Although we have never verified that fact, we will in the next section (in Theorem 3.49).

Result 3.37 *There exist rational numbers a and b such that a^b is irrational.*

Proof. If $a = 2$ and $b = 1/2$, then $a^b = 2^{\frac{1}{2}} = \sqrt{2}$, which is irrational. ∎

To give a proof of Result 3.37, it was useful to recall that for every positive real number a, the number $a^{\frac{1}{2}}$ is $\sqrt{a}$. Result 3.37 suggests another question and illustrates an important point. An existence statement can be verified, either by giving an appropriate example or by showing an appropriate example must exist even if a specific example can't be found.

Result 3.38 *There exist irrational numbers a and b such that a^b is rational.*

Proof. Consider the number $\sqrt{2}^{\sqrt{2}}$. Of course, this number is either rational or irrational. We consider these possibilities separately.

Case 1. $\sqrt{2}^{\sqrt{2}}$ is rational. Letting $a = b = \sqrt{2}$ verifies that the statement is true.

Case 2. $\sqrt{2}^{\sqrt{2}}$ is irrational. In this case, consider the number obtained by raising the (irrational) number $\sqrt{2}^{\sqrt{2}}$ to the (irrational) power $\sqrt{2}$; that is, consider a^b, where $a = \sqrt{2}^{\sqrt{2}}$ and $b = \sqrt{2}$. Observe that

$$a^b = \left(\sqrt{2}^{\sqrt{2}}\right)^{\sqrt{2}} = \sqrt{2}^{\sqrt{2} \cdot \sqrt{2}} = \sqrt{2}^2 = 2,$$

which is rational and verifies that the statement is true.   ∎

Proof Analysis.   The proof of Result 3.38 may seem unsatisfactory to you since we still don't know two *specific* irrational numbers a and b such that a^b is rational. We actually do know a bit more, namely, either (1) $\sqrt{2}^{\sqrt{2}}$ is rational or (2) $\sqrt{2}^{\sqrt{2}}$ is irrational and $\left(\sqrt{2}^{\sqrt{2}}\right)^{\sqrt{2}}$ is rational. (It has actually been proved elsewhere that $\sqrt{2}^{\sqrt{2}}$ is an irrational number. Hence there *are* two known irrational numbers a and b such that a^b is rational.)   ♦

Recall that a statement $\exists x \in S, R(x)$ is false only when $R(x)$ is a false statement for every $x \in S$. Establishing such a fact requires a different kind of argument.

Example 3.39

Disprove the following statement:

$$\text{There exists } x \in \mathbb{R} \text{ such that } x^4 + 2 = 2x^2. \tag{3.19}$$

Before providing a solution for Example 3.39, let's consider what we are being asked to do. We are being asked to show that the statement is false. If there did exist a real number x such that $x^4 + 2 = 2x^2$, then such a real number x would be a solution of this equation. However, we are being asked to show that the statement (3.19) is false, that is, we are being asked to show that the equation $x^4 + 2 = 2x^2$ has no real number solutions. This is logically equivalent to the statement

$$\text{For every real number } x, \ x^4 + 2 \neq 2x^2.$$

Observing that the equation $x^4 + 2 = 2x^2$ can be written as $x^4 - 2x^2 + 2 = 0$ suggests a method for solving this problem.

Solution of Example 3.39. Let $x \in \mathbb{R}$. Since

$$x^4 - 2x^2 + 2 = (x^4 - 2x^2 + 1) + 1 = (x^2 - 1)^2 + 1 \geq 0 + 1 = 1 > 0,$$

it follows that $x^4 - 2x^2 + 2 \neq 0$ and so $x^4 + 2 \neq 2x^2$ for every real number x.   ♦

There are instances when we encounter quantified statements containing both universal and existential quantifiers. For example, suppose that $R(x,y)$ is an open sentence containing variables x and y, where the domain of x is S and the domain of y is T. Then the quantified statement

$$\forall x \in S, \exists y \in T, R(x,y)$$

can be expressed as:

$$\text{For every } x \in S, \text{ there exists } y \in T \text{ such that } R(x,y);$$

while the quantified statement

$$\exists x \in S, \forall y \in T, R(x,y)$$

can be expressed as:

$$\text{There exists } x \in S \text{ such that for every } y \in T, \ R(x,y).$$

We now illustrate both kinds of quantified statements, along with a proof of each.

Result to Prove: For every irrational number s, there exists a rational number r such that rs is irrational.

Proof Strategy.   The statement we wish to prove can also be expressed as an implication.

If s is an irrational number, then there exists a rational number r such that rs is irrational.

If we were to give a direct proof of this statement, then we would begin with an arbitrary irrational number s. Our goal then is to show that there is some rational number r such that rs is irrational. The number $r = 1$ has this property. ♦

Result 3.40 *For every irrational number s, there exists a rational number r such that rs is irrational.*

Proof. Let s be an irrational number. Consider the rational number $r = 1$. Thus $rs = 1 \cdot s = s$ is irrational. ∎

Result to Prove: There exists a rational number r such that for every irrational number s, the number rs is rational.

Proof Strategy. A proof in this case consists of finding a rational number r having the property that rs is rational for every irrational number s. Observe that $r = 0$ has this property. ♦

Result 3.41 *There exists a rational number r such that for every irrational number s, the number rs is rational.*

Proof. Let $r = 0$. Then for each irrational number s, it follows that $rs = 0 \cdot s = 0$ is rational. ∎

Recall that the negation of $\forall x \in S, \exists y \in T, R(x,y)$ is

$$\sim (\forall x \in S, \exists y \in T, R(x,y)) \quad\equiv\quad \exists x \in S, \forall y \in T, \sim R(x,y).$$

Therefore, to show that the quantified statement $\forall x \in S, \exists y \in T, R(x,y)$ is false, we must show that there exists some $x \in S$ such that for every $y \in T$, $R(x,y)$ is false.

Example 3.42

Disprove the following statement: For every positive integer n, there exists a negative integer m such that $n + m = 1$.

Solution. Consider the positive integer $n = 1$. Then for every negative integer m,

$$n + m = 1 + m \leq 1 + (-1) = 0.$$

Therefore, $n = 1$ is a counterexample. ♦

We have seen that the negation of $\exists x \in S, \forall y \in T, R(x,y)$ is

$$\sim (\exists x \in S, \forall y \in T, R(x,y)) \quad\equiv\quad \forall x \in S, \exists y \in T, \sim R(x,y).$$

Therefore, to show that the quantified statement $\exists x \in S, \forall y \in T, R(x,y)$ is false, we must show that for every element $x \in S$, there exists an element $y \in T$ such that $R(x,y)$ is false.

Example 3.43

Disprove the following statement: There exists a positive integer n such that for every negative integer m, $n + m < 0$.

Solution. We show that no positive integer has this property. Let n be a positive integer. Then $m = -n$ is a negative integer and $n + m = n + (-n) = 0$. ♦

Exercises for Section 3.6

1. Prove that there exists an integer n such that $100 - n^2 < 0$.

2. Prove that there exist integers m and n of opposite parity such that

$$(m-2)^2 + (n-6)^2 \le 1.$$

3. Prove that there exists a real number x such that $x^2 = 5$.

4. Prove that for every two rational numbers a and b with $a < b$, there exists a rational number r such that $a < r < b$.

5. Prove that there exist an irrational number a and a rational number b such that a^b is rational.

6. Prove that there exist an irrational number a and a rational number b such that a^b is irrational.

7. Prove that there exists a real number x such that $x^4 - x^2 - 2 = 0$.

8. Prove that there exists an integer n such that $4n^2 - 8n + 3 < 0$.

9. Prove that there exist integers m and n of the same parity such that

$$(m-2)^2 - (n-3)^3 = 1.$$

10. Disprove: There exists a real number x such that $x^4 - x^2 + 2 = 0$.

11. Prove that there exist distinct irrational numbers a and b such that a^b is rational.

12. Prove or disprove: For every integer b, there exists a positive integer a such that $|a - |b|| \le 1$.

13. Prove or disprove each of the following.

    (a) There exist distinct rational numbers a and b such that $(a-1)(b-1) = 1$.
    (b) There exist distinct rational numbers a and b such that $\frac{1}{a} + \frac{1}{b} = 1$.

14. Prove or disprove the following:

    There exist distinct positive integers a and b such that $\frac{a}{b} + \frac{3b}{a}$ is an integer.

15. Prove or disprove: For every nonempty set A, there exists a set B such that $A \cup B = \emptyset$.

16. Prove or disprove: Let S be a nonempty set. For every proper subset A of S, there exists a nonempty subset B of S such that $A \cup B = S$ and $A \cap B = \emptyset$.

3.7   Proof by Contradiction

Let R be a mathematical statement that we would like to show is true. If R is expressed as $\forall x \in S$, $P(x) \Rightarrow Q(x)$ for open sentences $P(x)$ and $Q(x)$ over a domain S, then we have already described two methods we might use to verify the truth of R, namely a direct proof and a proof by contrapositive. We now describe yet a third method. Suppose, by assuming that R is false, we are able to deduce a statement that contradicts some known fact or some assumption we made in the proof. If we denote the fact or assumption by P, then what we have deduced is $\sim P$. Thus we now have produced the contradiction $P \wedge (\sim P)$, which we also denote by C. We therefore have the truth of the implication $(\sim R) \Rightarrow C$. However, since C is false and $(\sim R) \Rightarrow C$ is true, it follows (by the truth table for implication) that $\sim R$ is false and so R is true, which is the desired outcome. This method of proof is called a **proof by contradiction**. Let's summarize what is done in a proof by contradiction.

> *To verify that a statement R is true using a proof by contradiction, we assume that R is false and, from this, obtain a contradiction. If R is expressed in terms of an implication, say $R: \forall x \in S$, $P(x) \Rightarrow Q(x)$, then we begin a proof by contradiction by assuming that R is false, that is, by assuming that $\forall x \in S$, $P(x) \Rightarrow Q(x)$ is false. We have seen that this means there exists some element $x \in S$ for which $P(x)$ is true and $Q(x)$ is false. We then attempt to produce a contradiction from this.*

We now give some illustrations of proof by contradiction.

Result to Prove: There is no smallest positive real number.

Proof Strategy. Since we are proving this statement by contradiction, we begin by assuming that the statement is false. Hence we assume that there exists a smallest positive real number. Once we know a certain object exists, it is often helpful to give this object a name. So suppose that this smallest positive real number is denoted by r. Our goal now is to contradict some assumption or fact. This is always the difficult part of the proof. In this case, a contradiction would be produced if we can find a positive real number that is smaller than r. That is a key to this proof. ♦

Result 3.44 *There is no smallest positive real number.*

Proof. Assume, to the contrary, that there exists a smallest positive real number, say r. Since $0 < r/2 < r$, it follows that $r/2$ is a positive real number that is smaller than r, which is a contradiction. ∎

Proof Analysis. In the proof of Result 3.44, we wrote $0 < r/2 < r$ and assumed that these inequalities would be clear to the reader. If it was felt that these inequalities needed to be justified, then we could write (for the positive real number r) that $0 < r < r + r = 2r$ and so

$$0 < r < 2r. \tag{3.20}$$

Dividing the inequality in (3.20) by the positive number 2 gives us the desired $0 < r/2 < r$. ♦

Another comment about the proof of Result 3.44 may be useful here. Since we were giving a proof by contradiction, we started by assuming that the statement is false. In the proof, we used the phrase *to the contrary*. This is a common signal that is used to indicate that we have assumed the statement we wish to prove is false and that a proof by contradiction is being given.

Result 3.45 *No even integer can be expressed as the sum of an even integer and an odd integer.*

Proof. Suppose that the statement is false. Then there exists an even integer n that can be expressed as the sum of an even integer x and an odd integer y. So $x = 2a$ and $y = 2b + 1$, where $a, b \in \mathbb{Z}$. Therefore,
$$n = x + y = (2a) + (2b + 1) = 2(a + b) + 1.$$
Since $a + b$ is an integer, n is odd. This contradicts the assumption that n is even. ∎

Proof Analysis. Once again, a proof by contradiction was used as the proof technique to verify Result 3.45. As expected, we assumed that the statement is false, that is, we assumed that there is some even integer that is the sum of an even integer and an odd integer. Notice that in the proof, each of these integers was represented by a symbol (namely, n, x and y, respectively). This allowed us to write a simpler and clearer proof.

A different "proof" for Result 3.45 might have occurred to you. Suppose that x is an even integer and y is an odd integer. Then $x = 2a$ and $y = 2b + 1$ for integers a and b. Therefore,

$$x + y = (2a) + (2b + 1) = 2(a + b) + 1.$$

Since $a + b$ is an integer, $x + y$ is odd. This is not a proof of Result 3.45, however. What this proves is that the sum of an even integer and an odd integer is odd. This is not what Result 3.45 says. What Result 3.45 says is that for any even integer n under consideration, we cannot express n as the sum of an even integer and an odd integer. ♦

Result 3.46 *The sum of a rational number and an irrational number is irrational.*

Proof. Assume, to the contrary, that there exist a rational number r and an irrational number s such that $t = r + s$ is rational. Since r and t are rational, $r = a/b$ and $t = c/d$, where $a, b, c, d \in \mathbb{Z}$ and $b, d \neq 0$. Therefore,

$$\frac{c}{d} = t = r + s = \frac{a}{b} + s;$$

so

$$s = \frac{c}{d} - \frac{a}{b} = \frac{bc - ad}{bd}.$$

Since $bc - ad$ and bd are integers and $bd \neq 0$, it follows that s is a rational number, which contradicts our assumption that s is irrational.   ∎

One question that often occurs with any statement we wish to prove is:

> *How do we know which proof technique to use when attempting to verify the truth of a statement?*

Unfortunately, the answer is: We *don't* know, at least we don't always know. It is often unclear even for experienced mathematicians to determine which proof technique to use. In fact, sometimes no technique seems to work. Having just said this, however, there is something that the statements in Results 3.44 – 3.46 have in common. All are *negative-sounding*. That may be less obvious in Result 3.46 but that result deals with irrational numbers, which are real numbers that are *not* rational. Proof by contradiction is the common method of proof in such instances.

We now illustrate proof by contradiction with two examples where this would not be the expected method of proof. These examples are meant only to show that there is not always just one technique that can be used to verify the truth of a given statement.

Result 3.47 *If n is an odd integer, then $3n - 11$ is even.*

Proof. Assume, to the contrary, that there exists an odd integer n such that $3n - 11$ is also odd. Since n is odd, $n = 2k + 1$ for some integer k. Therefore,

$$3n - 11 = 3(2k + 1) - 11 = 6k - 8 = 2(3k - 4).$$

Since $3k - 4$ is an integer, $3n - 11$ is even. This, however, contradicts our assumption that $3n - 11$ is odd.   ∎

Result 3.48 *Let n be an integer. If $5n + 1$ is odd, then n is even.*

Proof. Assume, to the contrary, that there exists an integer n such that $5n + 1$ is odd and n is odd. Since n is odd, $n = 2k + 1$ for some integer k. Therefore,

$$5n + 1 = 5(2k + 1) + 1 = 10k + 5 + 1 = 2(5k + 3).$$

Since $5k + 3$ is an integer, $5n + 1$ is even. This, however, contradicts our assumption that $5n + 1$ is odd.   ∎

We have mentioned a few times that $\sqrt{2}$ is an irrational number. The proof of this fact is one of the best known examples of a proof by contradiction. In the proof it is useful to recall Theorem 3.24:

> *Let n be an integer. Then n^2 is even if and only if n is even.*

One other comment is useful. A fraction a/b is *reduced to lowest terms* if a and b contain no common integer factors greater than 1. For example, $3/2$ is reduced to lowest terms while $6/4$ is not as 6 and 4 have 2 as a common factor.

Theorem 3.49   *The real number $\sqrt{2}$ is irrational.*

Proof. Assume, to the contrary, that $\sqrt{2}$ is rational. Then $\sqrt{2} = a/b$, where $a, b \in \mathbb{Z}$ and $b \neq 0$. We may further assume that a/b has been reduced to lowest terms. Squaring both sides, we obtain $2 = a^2/b^2$ and so $a^2 = 2b^2$. Since b^2 is an integer, a^2 is even. By Theorem 3.24, a is even as well and so $a = 2c$ for some integer c. Thus, $a^2 = (2c)^2 = 4c^2 = 2b^2$ and so $b^2 = 2c^2$. Since c^2 is an integer, b^2 is even. Again, by Theorem 3.24, b is even. Therefore, a and b are both even and so contain 2 as a common factor, which contradicts our assumption that a/b has been reduced to lowest terms. ■

Proof Analysis. The subtle aspect of this proof is noticing that our assumption that a/b has been reduced to lowest terms has resulted in a contradiction. Perhaps even more subtle is observing that it might be worthwhile to express a/b in lowest terms. As we mentioned, when giving a proof by contradiction, it is often difficult to know where a contradiction will occur and what this contradiction might be. ♦

Exercises for Section 3.7

1. Prove that there is no largest negative rational number.

2. Prove that 100 cannot be expressed as the sum of three odd integers.

3. Prove that 101 cannot be expressed as the sum of two even integers.

4. Prove that no odd integer can be expressed as the sum of three even integers.

5. Prove that if a is a rational number and b is an irrational number, then $a - b$ is irrational.

6. (a) Disprove: The product of a rational number and an irrational number is irrational.

    (b) Prove that the product of a nonzero rational number and an irrational number is irrational.

7. Prove that there is no smallest positive irrational number.

8. Use a proof by contradiction to prove that if n is an even integer, then $7n + 9$ is odd.

9. Use a proof by contradiction to prove the following: Let n be an integer. If $3n + 14$ is even, then n is even.

10. Prove that if a and b are positive real numbers, then $\sqrt{a} + \sqrt{b} \neq \sqrt{a + b}$.

11. Suppose that you are given the fact: *Let n be an integer. Then $n^2 = 3a$ for some integer a if and only if $n = 3b$ for some integer b.* Use this fact to show that the real number $\sqrt{3}$ is irrational.

12. Prove that $\sqrt{2} + \sqrt{3}$ is an irrational number.

13. Prove that $\sqrt{6}$ is an irrational number. [Hint: Use the fact that if $m, n \in \mathbb{Z}$ such that mn is even, then m is even or n is even.]

14. Let $a, b, c \in \mathbb{R}$. Prove that if $a + b$, $a + c$ and $b + c$ are all rational, then a, b and c are all rational.

15. Prove that there do not exist three distinct positive real numbers a, b and c such that two of the three numbers $\sqrt{a + b}$, $\sqrt{a + c}$ and $\sqrt{b + c}$ are equal.

Chapter 3 Highlights

Key Concepts

cases, proof by: in a proof by cases of some statement concerning an element x in a set S, we consider a collection $\mathcal{P}$ of two or more subsets of S and divide the proof into cases, according to the particular subset in $\mathcal{P}$ to which x belongs.

contradiction, proof by: to verify a statement R using a proof by contradiction, we assume that R is false and, from this, produce a contradiction; to verify a statement $\forall x \in S, P(x) \Rightarrow Q(x)$ using a proof by contradiction, we assume there exists $x \in S$ such that $P(x)$ is true and $Q(x)$ is false and then produce a contradiction from this.

contrapositive, proof by: in a proof by contrapositive of $\forall x \in S, P(x) \Rightarrow Q(x)$, we assume that $Q(x)$ is false for an arbitrary $x \in S$ and show that $P(x)$ is false.

counterexample (of $\forall x \in S, P(x) \Rightarrow Q(x)$): an element $a \in S$ for which $P(a) \Rightarrow Q(a)$ is false, that is, $P(a)$ is true and $Q(a)$ is false.

direct proof: in a direct proof of $\forall x \in S, P(x) \Rightarrow Q(x)$, we assume that $P(x)$ is true for an arbitrary $x \in S$ and show that $Q(x)$ is true.

existence proof: in an existence proof of $\exists x \in S, R(x)$, we show that there is an element $a \in S$ for which $R(a)$ is true.

existential quantified statement, $\exists x \in S, Q(x)$: there exists $x \in S$ such that $Q(x)$.

existential quantifier $\exists$: "there exists," "there is," "for some," "for at least one."

proof by cases: see **cases, proof by**.

proof by contradiction: see **contradiction, proof by**.

proof by contrapositive: see **contrapositive, proof by**.

quantified statement: an existential quantified statement or a universal quantified statement.

universal quantified statement, $\forall x \in S, P(x)$: for all $x \in S$, $P(x)$.

universal quantifier $\forall$: "for all," "for every," "for each."

Key Results

- For an open sentence $R(x)$ over a domain S,

  $\sim (\forall x \in S, R(x)) \equiv \exists x \in S, \sim R(x)$.

  $\sim (\exists x \in S, R(x)) \equiv \forall x \in S, \sim R(x)$.

- For an open sentence $R(x, y)$ containing variables x and y, where the domain of x is S and the domain of y is T,

  $\sim (\forall x \in S, \exists y \in T, R(x, y)) \equiv \exists x \in S, \forall y \in T, \sim R(x, y)$.

  $\sim (\exists x \in S, \forall y \in T, R(x, y)) \equiv \forall x \in S, \exists y \in T, \sim R(x, y)$.

Supplementary Exercises for Chapter 3

1. Let S be the set of positive even integers and for an even integer n, let

$$P(n): \ 2^{n-2} \text{ is an even integer.}$$

State each of the following in words.

(a) $\forall n \in S, P(n)$. (b) $\exists n \in S, P(n)$. (c) $\sim (\forall n \in S, P(n))$. (d) $\sim (\exists n \in S, P(n))$.

2. Let $S = \{0, 1, 4, 5\}$ and let

$$P(n): \quad \frac{n^2 - 5n + 6}{2} \text{ is odd.}$$

be an open sentence over the domain S. Prove that $\forall n \in S, P(n)$ is true.

3. Let m and n be two integers. Prove that $mn + m$ is odd if and only if m is odd and n is even.

4. Let n be an integer. Prove that if $11n - 9$ is even, then n is odd.

5. Let x be a real number. Prove that if $(x^2 - 1)^2 = 0$, then $x^4 - x^2 = 0$.

6. Disprove: If r and s are irrational numbers, then rs is irrational.

7. Prove that if r and s are rational numbers and $s \neq 0$, then r/s is rational.

8. Let x and y be real numbers. Prove that if $(x - 3)^2 + (y - 4)^2 = 0$, then $x^2 + y^2 = 25$.

9. Let n be an integer. Prove that $5n + 1$ is even if and only if n is odd.

10. Let a and b be integers. Prove that if a and b are of the same parity, then $5a - 3b$ is even.

11. Disprove the following by providing a counterexample in each case.

   (a) If a and b are of opposite parity, then $2a + 3b$ is odd.
   (b) If x is a real number, then $(x + 1)^2 > 0$.

12. Prove that there exist two integers a and b such that $a + b > ab$.

13. Prove that if n is an even integer, then $5n - 7$ is an odd integer using
   (a) a direct proof. (b) a proof by contrapositive. (c) a proof by contradiction.

14. Prove for every integer n that if $3n + 5$ is odd, then n is an even integer using
   (a) a direct proof. (b) a proof by contrapositive. (c) a proof by contradiction.

15. Give a proof of

   Let $n \in \mathbb{Z}$. Then n is odd if and only if $7 - n$ is even.

   using

   (a) two direct proofs.
   (b) one direct proof and one proof by contrapositive.
   (c) two proofs by contrapositive.

16. Prove that if r is a rational number and s is an irrational number, then $2r - 3s$ is an irrational number using a proof by contradiction.

17. Prove that 10 cannot be expressed as the sum of an odd integer and two even integers.

18. Prove that 100 cannot be written as the sum of three integers, an even number of which are even.

19. Prove that there exists no positive integer x such that $x < x^2 < 2x$.

20. Let m and n be integers. Prove that if $m + n \geq 10$, then $m \geq 5$ or $n \geq 5$.

21. Let m and n be integers. Prove that if $mn = 1$, then either $m = n = 1$ or $m = n = -1$.

22. Let m and n be integers. Prove that if $m^2 = n^2$, then either $m = n$ or $m = -n$.

23. For a real number x, let $|x|$ be the absolute value of x, that is, $|x| = x$ if $x \geq 0$ and $|x| = -x$ if $x < 0$. Let a and b be real numbers. Prove that

    (a) $|ab| = |a||b|$. (b) $|a + b| \leq |a| + |b|$.

24. For $n \in \mathbb{Z}$, consider the following: $P(n)$: $n^2 < 4$. $Q(n)$: $n^3 = n$.

    State the following in words.

    (a) $\forall n \in \mathbb{Z}, P(n)$.

    (b) $\exists n \in \mathbb{Z}, Q(n)$.

    (c) $\sim (\forall n \in \mathbb{Z}, P(n))$.

    (d) $\sim (\exists n \in \mathbb{Z}, Q(n))$.

    (e) The converse of $P(n) \Rightarrow Q(n)$.

    (f) The contrapositive of $P(n) \Rightarrow Q(n)$.

25. Prove that if a is an odd integer and b is an even integer, then $3a + 5b - 4$ is odd.

26. Let $x \in \mathbb{R}$. Prove that if $x^3 = x$, then $x^2 < 2$.

27. For two real numbers a and b, $\min(a, b)$ denotes the smaller of a and b; while $\max(a, b)$ denotes the larger of a and b. So $\min(3, 5) = 3$ and $\max(3, 5) = 5$, while $\min(4, 4) = \max(4, 4) = 4$. For two real numbers a and b, let $m = \min(a, b)$ and $M = \max(a, b)$. Let r and s be real numbers. Prove that if $r \leq m$, then $r \leq a$ and $r \leq b$; while if $s \geq M$, then $s \geq a$ and $s \geq b$.

28. Let a and b be real numbers. Prove that $\min(a, b) + \max(a, b) = a + b$ (see Exercise 27).

29. Let a and b be real numbers. Prove that $\min(a, b) \leq \frac{a+b}{2} \leq \max(a, b)$ (see Exercise 27).

30. Let $x \in \mathbb{R}$. Prove that if $|x - 3| < 2$, then $1 < x < 5$.

31. Prove that there exist a rational number a and an irrational number b such that a^b is rational.

32. Prove that there exist a rational number a and an irrational number b such that a^b is irrational.

33. Prove that $\sqrt[3]{2}$ is irrational.

34. Prove that $\sqrt[3]{3}$ is irrational. (Use the fact that if n is an integer such that $n^3 = 3a$ for some integer a, then $n = 3b$ for some integer b.)

35. In the proof of Result 3.38, we began by considering $\sqrt{2}^{\sqrt{2}}$. Give a new proof of Result 3.38 by first considering $\sqrt[3]{2}^{\sqrt[3]{3}}$ (Recall from Exercises 33 and 34 that $\sqrt[3]{2}$ and $\sqrt[3]{3}$ are irrational.)

36. Prove the following De Morgan's law for sets:

    For every two sets A and B, $\overline{A \cup B} = \overline{A} \cap \overline{B}$.

37. It is known that $\sqrt{3}$ and $\sqrt{5}$ are irrational. Prove that $\sqrt{3} + \sqrt{5}$ is irrational.

38. Prove for every integer n that there exist two integers a and b of opposite parity such that $an + b$ is an odd integer.

39. Let a and b be two integers of opposite parity and let $n \in \mathbb{Z}$. Prove that $an + b$ and $a + bn$ are of opposite parity if and only if n is even.

Chapter 4

Mathematical Induction

For an open sentence $R(x)$ over a domain S, we have now seen three proof techniques that can be used to verify the truth of the quantified statement

$$\forall x \in S,\ R(x): \quad \text{For all } x \in S,\ R(x).$$

These three proof techniques are direct proof, proof by contrapositive and proof by contradiction. If $S = \mathbb{N}$, then there is another valuable method of proof that may be successful to prove that the quantified statement $\forall x \in S,\ R(x)$ is true. This method is called mathematical induction or simply induction and is typically used to prove quantified statements of the type $\forall n \in \mathbb{N},\ P(n)$, where $P(n)$ then is an open sentence over the domain $\mathbb{N}$.

There have been attempts to prove quantified statements of the type $\forall n \in \mathbb{N},\ P(n)$ by induction-like arguments that go back many centuries. About a thousand years ago, Abu al-Karaji, who lived near Baghdad much of his life, attempted to give an argument for generating the rows in the Pascal triangle (see Chapter 9, Section 9.1). He gave an argument for $n = 1$ and an argument for $n = 2$ based on his argument for $n = 1$ and an argument for $n = 3$ based on what he did for $n = 2$. He continued this a bit further and then stated that this process could be continued indefinitely. Blaise Pascal himself was not aware of what al-Karaji had done but during the middle of the 17th century Pascal wrote the following:

> *Even though this proposition may have an infinite number of cases, I shall give a very short proof of it assuming two lemmas. The first, which is self evident, is that the proposition is valid for the second row. The second is that if the proposition is valid for any row then it must necessarily be valid for the following row. From this it can be seen that it is necessarily valid for all rows; for it is valid for the second row by the first lemma; then by the second lemma it must be true for the third row and hence for the fourth row and so on to infinity.*

The method of proof that Pascal was attempting to use was induction and, although his argument lacks precision in that he was essentially trying to talk his way through a proof, he was on the right track. In this chapter we introduce the powerful method of proof by mathematical induction.

4.1 The Principle of Mathematical Induction

Suppose that $P(n)$ is an open sentence over the domain $\mathbb{N}$ of positive integers and we wish to prove that the quantified statement

$$\forall n \in \mathbb{N}, P(n) \tag{4.1}$$

is true. For the statement (4.1) to be true, $P(n)$ must be a true statement for each positive integer n, that is, all of the statements $P(1), P(2), P(3), \ldots$ must be true. Of course, if we can show that $P(n)$

is true for an arbitrary positive integer n, then we have verified (4.1) by means of a direct proof. On the other hand, if we can produce a contradiction by assuming that there exists a positive integer n such that $P(n)$ is false, then we have verified (4.1) using a proof by contradiction. However, there are occasions when neither of these techniques may be successful. Certainly, it is not practical to verify the truth of the statements $P(1)$, $P(2)$, $P(3)$, ... one by one. Fortunately, there is another method of proof that allows us to establish the truth of (4.1) by verifying that $P(n)$ is true when $n = 1$ and establishing the truth of a certain implication. This method is a consequence of the following principle.

The Principle of Mathematical Induction *The statement*

$$\forall n \in \mathbb{N}, \ P(n): \ \textit{For every positive integer } n, \ P(n).$$

is true if

(1) $P(1)$ *is true and*

(2) *the statement* $\forall k \in \mathbb{N}, P(k) \Rightarrow P(k+1)$ *is true.*

The Principle of Mathematical Induction therefore states that all of the statements $P(1)$, $P(2)$, $P(3)$, ... are true if we can show that $P(1)$ is true and prove the following:

$$\text{Let } k \in \mathbb{N}. \text{ If } P(k), \text{ then } P(k+1). \tag{4.2}$$

A proof of the statement $\forall n \in \mathbb{N}, P(n)$ using the Principle of Mathematical Induction is called an **induction proof**, a **proof by mathematical induction** or simply a **proof by induction**. Step (1) is often called the **base** or **basis step** (or the **anchor**) of the induction proof, while step (2) is often called the **inductive step** of the induction proof. In (4.2), $P(k)$ is often referred to as the **inductive hypothesis** or **induction hypothesis**. Although a proof of (4.2) could be given by any method, a direct proof is often used. In such a case, we assume that $P(k)$ is a true statement for an arbitrary positive integer k. Our goal then is to show that under this assumption, $P(k+1)$ is a true statement.

Suppose that we have verified the truth of statements (1) and (2). Then, of course, we know that $P(1)$ is a true statement by (1) and that

$$\text{If } P(1), \text{ then } P(2).$$

is true by (2). Since $P(1)$ *is* true, it follows that $P(2)$ is true. By the inductive step again,

$$\text{If } P(2), \text{ then } P(3).$$

is true. Since $P(2)$ is true, it follows that $P(3)$ is true. By similar reasoning, $P(4)$ is true, $P(5)$ is true and so on. Consequently, it is not surprising that verifying (1) and (2) in the Principle of Mathematical Induction establishes the truth of all of the statements $P(1)$, $P(2)$, $P(3)$, ... However, simply saying the informal phrase "and so on" does not constitute a formal proof.

The Principle of Mathematical Induction is often accepted as a fundamental property of the positive integers and is therefore taken as an axiom. Thus, we accept its truth without proof. Let's see how the Principle of Mathematical Induction works in practice.

Result to Prove: For every positive integer n,

$$\frac{1}{1 \cdot 2} + \frac{1}{2 \cdot 3} + \frac{1}{3 \cdot 4} + \cdots + \frac{1}{n(n+1)} = \frac{n}{n+1}. \tag{4.3}$$

Proof Strategy. What we wish to prove here is that for each positive integer n, there is a simple formula for the sum of the fractions $\frac{1}{1 \cdot 2}$, $\frac{1}{2 \cdot 3}$, $\frac{1}{3 \cdot 4}$, ..., $\frac{1}{n(n+1)}$; namely, the sum of these fractions is $\frac{n}{n+1}$. In this case, let $P(n)$ denote the open sentence (4.3), that is,

$$P(n): \ \frac{1}{1 \cdot 2} + \frac{1}{2 \cdot 3} + \frac{1}{3 \cdot 4} + \cdots + \frac{1}{n(n+1)} = \frac{n}{n+1}$$

where n is a positive integer. Let's consider $P(n)$ for a few values of n. For $n = 1, 2, 3, 4$, for example, the statements $P(n)$ are given below:

$$P(1): \quad \frac{1}{1 \cdot 2} = \frac{1}{2} = \frac{1}{1+1}$$

$$P(2): \quad \frac{1}{1 \cdot 2} + \frac{1}{2 \cdot 3} = \frac{1}{2} + \frac{1}{6} = \frac{2}{3} = \frac{2}{2+1}$$

$$P(3): \quad \frac{1}{1 \cdot 2} + \frac{1}{2 \cdot 3} + \frac{1}{3 \cdot 4} = \frac{1}{2} + \frac{1}{6} + \frac{1}{12} = \frac{3}{4} = \frac{3}{3+1}$$

$$P(4): \quad \frac{1}{1 \cdot 2} + \frac{1}{2 \cdot 3} + \frac{1}{3 \cdot 4} + \frac{1}{4 \cdot 5} = \frac{1}{2} + \frac{1}{6} + \frac{1}{12} + \frac{1}{20} = \frac{4}{5} = \frac{4}{4+1}.$$

Also, if $k - 1, k, k + 1$ and $k + 2$ are positive integers, then $P(k - 1)$, $P(k)$, $P(k + 1)$ and $P(k + 2)$ are given below:

$$P(k-1): \quad \frac{1}{1 \cdot 2} + \frac{1}{2 \cdot 3} + \cdots + \frac{1}{(k-1)k} = \frac{k-1}{k}$$

$$P(k): \quad \frac{1}{1 \cdot 2} + \frac{1}{2 \cdot 3} + \cdots + \frac{1}{k(k+1)} = \frac{k}{k+1}$$

$$P(k+1): \quad \frac{1}{1 \cdot 2} + \frac{1}{2 \cdot 3} + \cdots + \frac{1}{(k+1)(k+2)} = \frac{k+1}{k+2}$$

$$P(k+2): \quad \frac{1}{1 \cdot 2} + \frac{1}{2 \cdot 3} + \cdots + \frac{1}{(k+2)(k+3)} = \frac{k+2}{k+3}.$$

By the Principle of Mathematical Induction, the statement (4.3) can be shown to be true for every positive integer n if

(1) the truth of the statement $P(1)$ can be established and

(2) the truth of the statement:

$$\text{Let } k \in \mathbb{N}. \text{ If } P(k), \text{ then } P(k+1).$$

can be established.

Consequently, we can prove that the statement (4.3) is true if we can (1) show that

$$\frac{1}{1 \cdot 2} = \frac{1}{1+1}$$

and (2) verify the following implication for an arbitrary $k \in \mathbb{N}$:
If

$$\frac{1}{1 \cdot 2} + \frac{1}{2 \cdot 3} + \cdots + \frac{1}{k(k+1)} = \frac{k}{k+1},$$

then

$$\frac{1}{1 \cdot 2} + \frac{1}{2 \cdot 3} + \cdots + \frac{1}{(k+1)(k+2)} = \frac{k+1}{k+2}.$$

Let's now see how a proof of (4.3) can be given using the Principle of Mathematical Induction. ◆

When giving a proof by the Principle of Mathematical Induction, it is customary for the writer of the proof to inform the reader that a proof will be given using the Principle of Mathematical Induction, which is typically expressed by saying that mathematical induction (or, more simply, induction) is being used or employed.

Result 4.1 *For every positive integer n,*

$$\frac{1}{1 \cdot 2} + \frac{1}{2 \cdot 3} + \frac{1}{3 \cdot 4} + \cdots + \frac{1}{n(n+1)} = \frac{n}{n+1}.$$

Proof. We proceed by induction. First we verify that the statement is true for $n = 1$. Here we need only observe that $\dfrac{1}{1 \cdot 2} = \dfrac{1}{1+1}$ to establish the base step. Next, we verify the inductive step. Assume that

$$\frac{1}{1 \cdot 2} + \frac{1}{2 \cdot 3} + \cdots + \frac{1}{k(k+1)} = \frac{k}{k+1},$$

where k is a positive integer. We show that

$$\frac{1}{1 \cdot 2} + \frac{1}{2 \cdot 3} + \cdots + \frac{1}{(k+1)(k+2)} = \frac{k+1}{k+2}.$$

Observe that

$$\frac{1}{1 \cdot 2} + \frac{1}{2 \cdot 3} + \cdots + \frac{1}{(k+1)(k+2)} = \left[\frac{1}{1 \cdot 2} + \frac{1}{2 \cdot 3} + \cdots + \frac{1}{k(k+1)} \right] + \frac{1}{(k+1)(k+2)}.$$

From our assumption of the inductive hypothesis above that

$$\frac{1}{1 \cdot 2} + \frac{1}{2 \cdot 3} + \cdots + \frac{1}{k(k+1)} = \frac{k}{k+1},$$

we have

$$\begin{aligned}
\frac{1}{1 \cdot 2} + \frac{1}{2 \cdot 3} + \cdots + \frac{1}{k(k+1)} + \frac{1}{(k+1)(k+2)} &= \frac{k}{k+1} + \frac{1}{(k+1)(k+2)} \\
&= \frac{k(k+2)+1}{(k+1)(k+2)} = \frac{k^2 + 2k + 1}{(k+1)(k+2)} \\
&= \frac{(k+1)^2}{(k+1)(k+2)} = \frac{k+1}{k+2}.
\end{aligned}$$

By the Principle of Mathematical Induction, it then follows that

$$\frac{1}{1 \cdot 2} + \frac{1}{2 \cdot 3} + \frac{1}{3 \cdot 4} + \cdots + \frac{1}{n(n+1)} = \frac{n}{n+1}$$

for every positive integer n. ∎

As we mentioned in Result 4.1, the Principle of Mathematical Induction is used to show that there is a formula for the sum of the numbers

$$\frac{1}{1 \cdot 2}, \ \frac{1}{2 \cdot 3}, \ \ldots, \ \frac{1}{n(n+1)},$$

where n is a positive integer. Actually, there are many such formulas that give the sum of numbers of a certain type and these formulas are often verified by using the Principle of Mathematical Induction. We now consider additional examples of this type. Suppose that we are interested in obtaining a formula for the sum of the first n positive integers, where $n \in \mathbb{N}$. For $1 \leq n \leq 5$, observe that

$$\begin{aligned}
1 &= 1 \\
1 + 2 &= 3 \\
1 + 2 + 3 &= 6 \\
1 + 2 + 3 + 4 &= 10 \\
1 + 2 + 3 + 4 + 5 &= 15.
\end{aligned}$$

This may or may not suggest a formula for the sum $1 + 2 + \cdots + n$. However, if we write

$$A = 1 + 2 + \cdots + n$$

and then write the terms of this sum in reverse order, we also have

$$A = n + (n-1) + (n-2) + \cdots + 1.$$

Adding these two expressions for A, namely

$$
\begin{array}{ccccccccccccc}
A = & 1 & + & 2 & + & 3 & + & \cdots & + & (n-2) & + & (n-1) & + & n \\
A = & n & + & (n-1) & + & (n-2) & + & \cdots & + & 3 & + & 2 & + & 1
\end{array}
$$

we obtain

$$2A = (n+1) + (n+1) + \cdots + (n+1).$$

Since there are n terms in this sum, each equal to $n+1$, we have

$$2A = n(n+1) \text{ and so } A = n(n+1)/2,$$

that is,

$$1 + 2 + \cdots + n = \frac{n(n+1)}{2}. \tag{4.4}$$

Therefore, we have obtained and verified a formula for the sum of the first n positive integers. The story goes that when the brilliant mathematician Carl Friedrich Gauss was a student in elementary school, the children in his class were given the tedious task of adding the first 100 positive integers. Supposedly, Gauss quickly noticed that $1 + 2 + \cdots + 100$ equals the 50 sums $1 + 100$, $2 + 99$, ..., $50 + 51$, each of which is 101 and so $1 + 2 + \cdots + 100 = 50(101) = 5050$, thereby verifying the formula (4.4) for $n = 100$.

We give an alternative proof of (4.4), this time using the Principle of Mathematical Induction.

Result 4.2 *For every positive integer n,*

$$1 + 2 + \cdots + n = \frac{n(n+1)}{2}.$$

Proof. We use induction. Since $1 = \frac{1(1+1)}{2}$, the formula holds for $n = 1$. Assume that

$$1 + 2 + 3 + \cdots + k = \frac{k(k+1)}{2}$$

for a positive integer k. We show that

$$1 + 2 + 3 + \cdots + (k+1) = \frac{(k+1)(k+2)}{2}.$$

Observe that

$$
\begin{aligned}
1 + 2 + 3 + \cdots + (k+1) &= (1 + 2 + 3 + \cdots + k) + (k+1) \\
&= \frac{k(k+1)}{2} + (k+1) = \frac{k(k+1) + 2(k+1)}{2} \\
&= \frac{(k+1)(k+2)}{2}.
\end{aligned}
$$

By the Principle of Mathematical Induction, it follows that

$$1 + 2 + 3 + \cdots + n = \frac{n(n+1)}{2}$$

for every positive integer n. ∎

Proof Analysis. A remark about the proof of Result 4.2 may be useful. Within the proof, we assumed that $1 + 2 + 3 + \cdots + k = k(k+1)/2$ for a positive integer k and we wanted to show that

$1 + 2 + 3 + \cdots + (k + 1) = (k + 1)(k + 2)/2$. In general, to show that two expressions are equal, we often begin with one of these expressions and show that it is equal to the other expression. The expression we begin with is ordinarily chosen to be the one that seems easier to simplify.   ♦

A common attempted (but *incorrect*) proof of Result 4.2 might contain something like the following:

$$
\begin{aligned}
1 + 2 + \cdots + (k + 1) &= \frac{(k + 1)(k + 2)}{2} \qquad (4.5) \\
(1 + 2 + \cdots + k) + (k + 1) &= \frac{(k + 1)(k + 2)}{2} \\
\frac{k(k + 1)}{2} + (k + 1) &= \frac{(k + 1)(k + 2)}{2} \\
\frac{k(k + 1) + 2(k + 1)}{2} &= \frac{(k + 1)(k + 2)}{2} \\
\frac{(k + 1)(k + 2)}{2} &= \frac{(k + 1)(k + 2)}{2}. \qquad (4.6)
\end{aligned}
$$

There are several difficulties with this so-called "proof." Actually, it starts in (4.5) with what we want to show. We can't write this as we don't know that these two quantities are equal. The way this "proof" is written, there is one equality after another, each supposedly following from the equality immediately preceding it. However, we started with two quantities that we didn't know were equal (even though we wrote that they were equal) and from this we eventually deduced the equality in (4.6). However, the equality in (4.6) is totally obvious. No explanation is required. To make a correct proof from this, however, what can be done is to start with the leftmost expression in (4.5) and proceed down the expressions on the left of each equality to finally obtain the desired result.

The formula presented in Result 4.2 occurs surprisingly often and sometimes in unexpected places. We give one such example of this now.

Example 4.3

Suppose that we have 10 rolls of coins, numbered from 1 to 10. Each roll consists of 10 coins and all 100 coins look identical, but they are not. In 9 of the 10 rolls, each coin in the roll weighs 10 grams but in the other roll of (counterfeit) coins, each coin weighs 11 grams. We have a scale on which coins can be weighed. How can we determine which roll contains the counterfeit coins if we are only allowed to use the scale and we are only permitted one weighing of coins?

Solution. Of course, if a roll of standard coins is weighed, then its weight will be 100 grams; while if the roll of counterfeit coins is weighed, then it will weigh 110 grams and we will know that these are the counterfeit coins. However, there is only one chance out of 10 that we will select the roll of counterfeit coins to weigh. On the other hand, if we select one coin from roll #1, two coins from roll #2 and so on, up to all 10 coins from roll #10 and weigh this collection of coins, then we are weighing

$$1 + 2 + \cdots + 10 = \frac{10 \cdot 11}{2} = 55$$

coins (by the formula presented in Result 4.2). Not all of these coins weigh 10 grams, so the weight of these 55 coins is more than $55 \cdot 10 = 550$ grams. If roll #k ($1 \leq k \leq 10$) is the one containing the counterfeit coins, then the weight of these 55 coins will be $550 + k$ grams. Therefore, the weight of these 55 coins minus 550 is the number of the roll containing the counterfeit coins.   ♦

What Result 4.2 gives us, of course, is a formula for the sum of the first n positive integers. We now consider the sum of the first n odd positive integers. Observe that

$$1 = 1$$
$$1 + 3 = 4$$
$$1 + 3 + 5 = 9$$
$$1 + 3 + 5 + 7 = 16.$$

Based on these examples, it would be reasonable to guess that the sum of the first n odd positive integers is n^2. This would be a good guess.

Result 4.4 *For every positive integer n,*

$$1 + 3 + 5 + \cdots + (2n - 1) = n^2.$$

Proof. We proceed by induction. Since $2 \cdot 1 - 1 = 1^2$, the formula holds for $n = 1$. Assume that

$$1 + 3 + 5 + \cdots + (2k - 1) = k^2$$

for a positive integer k. We show that $1 + 3 + 5 + \cdots + (2k + 1) = (k + 1)^2$. Observe that

$$
\begin{aligned}
1 + 3 + 5 + \cdots + (2k + 1) &= [1 + 3 + 5 + \cdots + (2k - 1)] + (2k + 1) \\
&= k^2 + (2k + 1) = (k + 1)^2.
\end{aligned}
$$

By the Principle of Mathematical Induction, the formula holds for every positive integer n. ∎

There are other formulas for sums of integers that are similar to those we have seen. The next two formulas involve the sum of the squares and the sum of the cubes of the first n positive integers. The proofs too are similar to those we have seen and so are not presented here, but are left instead as exercises (see Exercises 7 and 8).

Result 4.5 *For every positive integer n,*

$$1^2 + 2^2 + \cdots + n^2 = \frac{n(n + 1)(2n + 1)}{6}.$$

Result 4.6 *For every positive integer n,*

$$1^3 + 2^3 + 3^3 + \cdots + n^3 = \frac{n^2(n + 1)^2}{4}.$$

We now present and verify a formula for the sum of powers of 2.

Result 4.7 *For every positive integer n,*

$$1 + 2 + 2^2 + \cdots + 2^n = 2^{n+1} - 1.$$

Proof. We proceed by induction. Since $1 + 2 = 2^2 - 1$, the formula holds for $n = 1$. Assume that

$$1 + 2 + 2^2 + \cdots + 2^k = 2^{k+1} - 1$$

for a positive integer k. We show that $1 + 2 + 2^2 + \cdots + 2^{k+1} = 2^{k+2} - 1$. Observe that

$$
\begin{aligned}
1 + 2 + 2^2 + \cdots + 2^{k+1} &= (1 + 2 + 2^2 + \cdots + 2^k) + 2^{k+1} \\
&= (2^{k+1} - 1) + 2^{k+1} = 2 \cdot 2^{k+1} - 1 \\
&= 2^{k+2} - 1,
\end{aligned}
$$

as desired. By the Principle of Mathematical Induction, the formula holds for every positive integer n. ∎

When giving a proof using the Principle of Mathematical Induction, it is essential that we establish both the base step and the inductive step, despite the fact that in the examples we have seen, verifying the base step may seem quite trivial and may appear to add little to the proof. For example, consider the expression $3n^2 - n + 3$, where $n \in \mathbb{N}$. Suppose that we believed that $3n^2 - n + 3$ is even for every positive integer n and that we attempt to give an induction proof of this, beginning with the inductive step. First, we assume that $3k^2 - k + 3$ is even for a positive integer k. Therefore, $3k^2 - k + 3 = 2\ell$, where $\ell \in \mathbb{Z}$. We then show that $3(k+1)^2 - (k+1) + 3$ is even as well. Observe that

$$
\begin{aligned}
3(k+1)^2 - (k+1) + 3 &= 3(k^2 + 2k + 1) - k - 1 + 3 \\
&= 3k^2 + 6k + 3 - k + 2 \\
&= (3k^2 - k + 3) + 6k + 2 = 2\ell + 2(3k + 1) \\
&= 2(\ell + 3k + 1).
\end{aligned}
$$

Since $\ell + 3k + 1$ is an integer, $3(k+1)^2 - (k+1) + 3$ is even and the inductive step has been verified. However, it is not true that $3n^2 - n + 3$ is even for every positive integer n. When $n = 1$, for example,

$$
3n^2 - n + 3 = 3 \cdot 1^2 - 1 + 3 = 5,
$$

which, of course, is odd. In fact, $3n^2 - n + 3$ is never even. The reason that our "induction proof" failed is that the base case was never established.

Exercises for Section 4.1

1. Prove that $\frac{1}{2 \cdot 3} + \frac{1}{3 \cdot 4} + \cdots + \frac{1}{(n+1)(n+2)} = \frac{n}{2n+4}$ for every positive integer n.

2. Prove that $\frac{1}{1 \cdot 3} + \frac{1}{3 \cdot 5} + \frac{1}{5 \cdot 7} + \cdots + \frac{1}{(2n-1)(2n+1)} = \frac{n}{2n+1}$ for every positive integer n.

3. Prove that $\frac{1}{1 \cdot 4} + \frac{1}{4 \cdot 7} + \frac{1}{7 \cdot 10} + \cdots + \frac{1}{(3n-2)(3n+1)} = \frac{n}{3n+1}$ for every positive integer n.

4. Find a formula for $2 + 4 + 6 + \cdots + 2n$ for every positive integer n and then verify your formula by the Principle of Mathematical Induction.

5. Prove that $1 + 5 + 9 + \cdots + (4n - 3) = n(2n - 1)$ for every positive integer n.

6. Prove that $2 + 5 + 8 + \cdots + (3n - 1) = n(3n + 1)/2$ for every positive integer n.

7. Prove the result in Result 4.5: For every positive integer n, $1^2 + 2^2 + \cdots + n^2 = \frac{n(n+1)(2n+1)}{6}$.

8. Prove the result in Result 4.6: For every positive integer n, $1^3 + 2^3 + 3^3 + \cdots + n^3 = \frac{n^2(n+1)^2}{4}$.

9. Prove that $1 + 3 + 3^2 + \cdots + 3^n = (3^{n+1} - 1)/2$ for every positive integer n.

10. Let $r \geq 2$ be an integer. Prove that $1 + r + r^2 + \cdots + r^n = \frac{r^{n+1} - 1}{r - 1}$ for every positive integer n.

11. Prove that $1 \cdot 2 + 2 \cdot 3 + 3 \cdot 4 + \cdots + n(n + 1) = \frac{n(n+1)(n+2)}{3}$ for every positive integer n.

12. (a) Let $k \in \mathbb{N}$. Prove that if $k^2 + k + 5$ is even, then $(k + 1)^2 + (k + 1) + 5$ is even.

    (b) What is the value of $k^2 + k + 5$ when $k = 7$?

    (c) What information do (a) and (b) provide?

13. Consider the expression $(n + 2)(n + 3)/2$, where $n \in \mathbb{N}$.

    (a) Is $(n + 2)(n + 3)/2$ even when $n = 1$?

    (b) Attempt to prove the following: Let $k \in \mathbb{N}$. If $(k+2)(k+3)/2$ is even, then $(k+3)(k+4)/2$ is even.

(c) What information do (a) and (b) give you?

14. Square tiles, one foot on each side, are to be placed on a square section of a floor, 10 feet on each side. This floor is divided into 10 strips denoted by s_1, s_2, ..., s_{10}, where each strip s_k ($1 \le k \le 10$) consists of $2k - 1$ squares (see Figure 4.1, where s_3 and s_4 are also shown). For each k ($1 \le k \le 10$), a total of $11 - k$ tiles are placed on each square in the strip s_k. For example, a total of $11 - 6 = 5$ tiles are placed on the square whose lower left corner has coordinates $(5, 3)$.

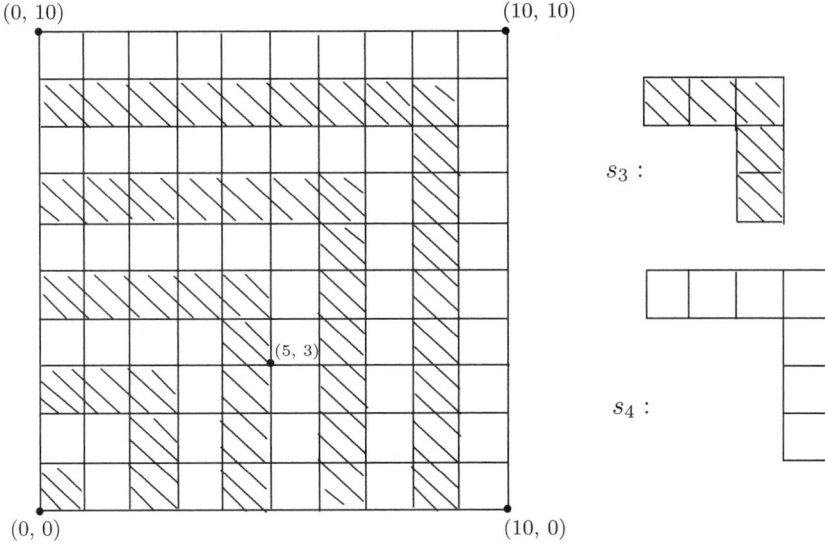

Figure 4.1: Illustrating Exercise 14

(a) How many tiles are placed on the square in the strip s_1?

(b) How many tiles are placed on each of the three squares in the strip s_2?

(c) How many tiles are placed on each square in the strip s_{10}?

(d) Use a formula from this section to find the total number of tiles placed on the floor.

(e) Use this situation to develop another formula that is equal to the formula in (d).

4.2 Additional Examples of Induction Proofs

In the preceding section, all of the results concerned establishing formulas for sums of numbers. This is only a small sample of the kinds of statements that can be verified with the aid of the Principle of Mathematical Induction. In this section we look at other types of examples. Before doing this, however, we mention that there are several variations and generalizations of the Principle of Mathematical Induction. We have seen that the Principle of Mathematical Induction can be used to prove that statements of the form

$$\forall n \in \mathbb{N}, \ P(n)$$

are true. There are numerous occasions when $\mathbb{N}$ is not the appropriate domain. Indeed, if the domain of the variable n in an open sentence $P(n)$ is the set

$$S = \{i \in \mathbb{Z} : \ i \ge m\} = \{m, m + 1, m + 2, \ldots\}$$

for some fixed integer m, then there is a more general principle that corresponds to this. Because this principle is so similar in statement to the Principle of Mathematical Induction, it is ordinarily referred to by the same name.

The Principle of Mathematical Induction *For a fixed integer m, let*

$$S = \{i \in \mathbb{Z} : \ i \geq m\}.$$

Then the statement

$$\forall n \in S, P(n) : \ \textit{For every integer } n \geq m, \ P(n).$$

is true if

(1) $P(m)$ *is true and*

(2) *the statement $\forall k \in S, P(k) \Rightarrow P(k+1)$ is true.*

When $m = 1$, this principle is exactly the Principle of Mathematical Induction that we stated in Section 4.1. To use the Principle of Mathematical Induction stated above, we need to show that the hypothesis of this principle is satisfied. This means that we must show that the basis step (1) and the inductive step (2) are satisfied.

For a positive integer n, the integer $n!$ (n **factorial**) is defined by

$$n! = n(n-1)(n-2) \cdots 3 \cdot 2 \cdot 1$$

and for every positive integer n,

$$(n+1)! = (n+1)n!. \tag{4.7}$$

One would certainly think that $n! = n(n-1)(n-2) \cdots 3 \cdot 2 \cdot 1$ is greater than $2 \cdot 2 \cdot 2 \cdots 2$ (n factors 2) if n is large enough, that is, $n! > 2^n$. The table below compares the values of $n!$ and 2^n for several positive integers n.

n	$n!$	2^n
1	1	2
2	2	4
3	6	8
4	24	16
5	120	32

Observe that $2^n > n!$ when $1 \leq n \leq 3$. However, $n! > 2^n$ for $n = 4$ and $n = 5$. There is good reason therefore to believe that $n! > 2^n$ for *every* integer $n \geq 4$. This is indeed the case, which we prove by the Principle of Mathematical Induction (where $m = 4$ here).

Result to Prove: For every integer $n \geq 4$, $n! > 2^n$.

Proof Strategy. In order to give an induction proof of this result, we must first show that $n! > 2^n$ when $n = 4$. However, $4! = 24$ and $2^4 = 16$ while $24 > 16$; so the desired inequality is obvious. Next, we need to verify the inductive step. That is, we need to prove that if $k! > 2^k$, where k is some integer with $k \geq 4$, then $(k+1)! > 2^{k+1}$. So we begin by assuming that $k! > 2^k$. Typically, to establish that some inequality holds, we start with one side and work towards the other. In this case, this means that we would start with $(k+1)!$ and show that it is greater than 2^{k+1} (recall that $k \geq 4$) or start with 2^{k+1} and show that it is less than $(k+1)!$. If we begin with $(k+1)!$ and use (4.7), then we have $(k+1)! = (k+1)k!$. Furthermore, from our assumption, we know that $k! > 2^k$. Multiplying this inequality by (the positive integer) $k+1$, we have

$$(k+1)k! > (k+1) \cdot 2^k.$$

We need to recall that our goal is to show that $(k+1)! > 2^{k+1}$. By (4.7), $(k+1)! = (k+1)k!$. If we can show that $(k+1) \cdot 2^k > 2^{k+1}$, then we will have what we want. Since $k \geq 4$, it follows that $k+1 \geq 5$. But $5 > 2$ and so

$$(k+1) \cdot 2^k \geq 5 \cdot 2^k > 2 \cdot 2^k = 2^{k+1}.$$

It only remains to put these pieces together and write a careful proof. ♦

Result 4.8 *For every integer $n \geq 4$, $n! > 2^n$.*

Proof. We use induction. Since $4! = 24 > 16 = 2^4$, the inequality holds for $n = 4$. Suppose that $k! > 2^k$ for an integer $k \geq 4$. We show that $(k+1)! > 2^{k+1}$. Observe that

$$(k+1)! = (k+1)k! > (k+1) \cdot 2^k \geq (4+1)2^k = 5 \cdot 2^k > 2 \cdot 2^k = 2^{k+1}.$$

Therefore, $(k+1)! > 2^{k+1}$. By the Principle of Mathematical Induction, $n! > 2^n$ for every integer $n \geq 4$. ∎

Next we compare values of 2^n and n for some positive integers n. Consider the following table.

n	2^n
1	2
2	4
3	8

After comparing n and 2^n for only $1 \leq n \leq 3$ and observing that $2^n > n$ for these values of n, it seems as if $2^n > n$ for *every* positive integer n. We attempt to prove this. If we use the Principle of Mathematical Induction, then m would have the value 1, which means that this is also an application of the Principle of Mathematical Induction as stated in Section 4.1.

Result 4.9 *For every positive integer n, $2^n > n$.*

Proof. We proceed by induction. Since $2^1 > 1$, the inequality holds for $n = 1$. Assume that $2^k > k$ for a positive integer k. We show that $2^{k+1} > k+1$. Observe that

$$2^{k+1} = 2 \cdot 2^k > 2k = k + k \geq k + 1.$$

Therefore, $2^{k+1} > k+1$. By the Principle of Mathematical Induction, $2^n > n$ for every positive integer n. ∎

We now consider some applications of the Principle of Mathematical Induction of an entirely different nature. There are often known properties of two objects of a certain type that hold for more than two objects of this type. In such instances, a proof by induction is a common proof technique. We now give two examples of this.

We previously mentioned the following fundamental property of real numbers:

If a and b are real numbers such that $ab = 0$, then either $a = 0$ or $b = 0$. (4.8)

This property also holds for products of more than two real numbers.

Result 4.10 *If $a_1, a_2, \ldots, a_n$ are $n \geq 2$ real numbers such that $a_1 a_2 \cdots a_n = 0$, then $a_i = 0$ for some integer i $(1 \leq i \leq n)$.*

Proof. We use induction. By the well-known property of real numbers described in (4.8), the statement is true for $n = 2$. That is, if a and b are two real numbers such that $ab = 0$, then $a = 0$ or $b = 0$. Assume that:

If $a_1, a_2, \ldots, a_k$ are any $k \geq 2$ real numbers whose product is 0, then $a_i = 0$ for some integer i with $1 \leq i \leq k$.

We wish to show the statement is true in the case of $k+1$ numbers, that is:

If $b_1, b_2, \ldots, b_{k+1}$ are $k+1$ real numbers such that $b_1 b_2 \cdots b_{k+1} = 0$, then $b_i = 0$ for some integer i $(1 \leq i \leq k+1)$.

Let $b_1, b_2, \ldots, b_{k+1}$ be $k+1$ real numbers such that $b_1 b_2 \cdots b_{k+1} = 0$. We show that $b_i = 0$ for some integer i $(1 \leq i \leq k+1)$. Let $b = b_1 b_2 \cdots b_k$. Then

$$b_1 b_2 \cdots b_{k+1} = (b_1 b_2 \cdots b_k) b_{k+1} = b b_{k+1} = 0.$$

Therefore, by (4.8) either $b = 0$ or $b_{k+1} = 0$. If $b_{k+1} = 0$, then we have the desired conclusion. On the other hand, if $b = b_1 b_2 \cdots b_k = 0$, then, since b is the product of k real numbers, it follows by the inductive hypothesis that $b_i = 0$ for some integer i $(1 \leq i \leq k)$. In any case, $b_i = 0$ for some integer i $(1 \leq i \leq k+1)$. By the Principle of Mathematical Induction, the result is true. ∎

Proof Analysis. A few comments about Result 4.10 and its proof are in order. First, in the statement of Result 4.10 the symbols $a_1, a_2, \ldots, a_n$ are only used to help us state the result. In fact, this result could have been stated as follows:

Whenever the product of $n \geq 2$ real numbers is 0, at least one of these numbers is 0.

Consequently, when verifying the basis step $(n = 2)$, the two numbers involved need not be called a_1 and a_2. Indeed, it seemed more natural to denote these two numbers more simply by a and b. In the inductive step, where we assumed that the statement is true for an integer $k \geq 2$ and then showed that the statement is true for the integer $k+1$, it was not necessary and would probably have been misleading to use $a_1, a_2, \ldots, a_k$ and then $a_1, a_2, \ldots, a_{k+1}$, for this might have incorrectly suggested that the k numbers must form a subset of the $k+1$ numbers. ♦

One of De Morgan's laws for sets states the following:

$$\text{If } A \text{ and } B \text{ are any two sets, then } \overline{A \cup B} = \overline{A} \cap \overline{B}. \tag{4.9}$$

We now show that there is a more general property which is a consequence of (4.9).

Result 4.11 *For any $n \geq 2$ sets $A_1, A_2, \ldots, A_n$,*

$$\overline{A_1 \cup A_2 \cup \cdots \cup A_n} = \overline{A_1} \cap \overline{A_2} \cap \cdots \cap \overline{A_n}.$$

Proof. We proceed by induction. By De Morgan's law, if A and B are any two sets, then

$$\overline{A \cup B} = \overline{A} \cap \overline{B}.$$

Hence the statement is true for $n = 2$. Assume, for any k sets $A_1, A_2, \ldots, A_k$, where $k \geq 2$, that

$$\overline{A_1 \cup A_2 \cup \cdots \cup A_k} = \overline{A_1} \cap \overline{A_2} \cap \cdots \cap \overline{A_k}.$$

Now let $B_1, B_2, \ldots, B_{k+1}$ be any $k+1$ sets. We show that

$$\overline{B_1 \cup B_2 \cup \cdots \cup B_{k+1}} = \overline{B_1} \cap \overline{B_2} \cap \cdots \cap \overline{B_{k+1}}.$$

Let $B = B_1 \cup B_2 \cup \cdots \cup B_k$. Observe that

$$
\begin{aligned}
\overline{B_1 \cup B_2 \cup \cdots \cup B_{k+1}} &= \overline{(B_1 \cup B_2 \cup \cdots \cup B_k) \cup B_{k+1}} = \overline{B \cup B_{k+1}} \\
&= \overline{B} \cap \overline{B_{k+1}} = \overline{(B_1 \cup B_2 \cup \cdots \cup B_k)} \cap \overline{B_{k+1}} \\
&= (\overline{B_1} \cap \overline{B_2} \cap \cdots \cap \overline{B_k}) \cap \overline{B_{k+1}} \\
&= \overline{B_1} \cap \overline{B_2} \cap \cdots \cap \overline{B_{k+1}}.
\end{aligned}
$$

The result then follows by the Principle of Mathematical Induction. ∎

We now use induction to verify a theorem (Theorem 2.17) that we encountered in Chapter 2, Section 2.1.

Theorem 4.12 *Let n be a nonnegative integer. If A is a set with $|A| = n$, then the cardinality of its power set is $|\mathcal{P}(A)| = 2^n$.*

Proof. We use induction. The power set of the empty set $\emptyset$ is $\mathcal{P}(\emptyset) = \{\emptyset\}$. Since

$$|\mathcal{P}(\emptyset)| = 1 = 2^0,$$

the statement is true for $n = 0$. Assume the following:

For a nonnegative integer k, the power set of every set with k elements has 2^k elements.

Let A be a set with $k + 1$ elements, say

$$A = \{a_1, a_2, \ldots, a_{k+1}\}.$$

We show that $|\mathcal{P}(A)| = 2^{k+1}$. Now each subset of A either contains a_{k+1} or it doesn't. We determine the number of subsets of each type.

First, we determine the number of subsets of A not containing a_{k+1}. Let

$$B = \{a_1, a_2, \ldots, a_k\}$$

be the subset of A obtained by removing a_{k+1} from A. The subsets of A not containing a_{k+1} are precisely the subsets of B. Since $|B| = k$, it follows by the inductive hypothesis that there are 2^k such subsets.

Next, we determine the number of subsets of A that contain a_{k+1}. Every subset of A containing a_{k+1} can be expressed as $C \cup \{a_{k+1}\}$, where $C \subseteq B$. Thus, the number of subsets of A containing a_{k+1} is the number of subsets C of B. Since there are 2^k subsets of B, there are 2^k sets $C \cup \{a_{k+1}\}$, where $C \subseteq B$.

Hence the total number of subsets of A is

$$2^k + 2^k = 2 \cdot 2^k = 2^{k+1}.$$

The result then follows by the Principle of Mathematical Induction. ∎

Exercises for Section 4.2

1. Use induction to prove that $2^n > n$ for every nonnegative integer n.

2. Prove that $n! > 3^n$ for every integer $n \geq 7$.

3. Prove that $n! > n^2$ for every integer $n \geq 4$.

4. Prove that $3^n > n$ for every positive integer n.

5. Prove by induction that $n^2 \geq 2n + 3$ for every integer $n \geq 3$.

6. Prove by induction that $n^2 > n + 1$ for every integer $n \geq 2$.

7. Prove that $2^n > n^2$ for every integer $n \geq 5$.

8. Let S be a set of real numbers having the property that if $a, b \in S$, then $a \cdot b \in S$. Prove that for every $n \geq 2$ elements $a_1, a_2, \ldots, a_n$ in S, their product $a_1 a_2 \cdots a_n \in S$.

9. Prove by induction that $n^2 + n + 1$ is odd for every nonnegative integer n.

10. Prove by induction that $n^2 - 3n$ is even for every nonnegative integer n.

11. Let $x \in \mathbb{Z}$. Prove for every integer $n \geq 2$ that x^n is even if and only if x is even.

12. Prove for any $n \geq 2$ sets $A_1, A_2, \ldots, A_n$ that $\overline{A_1 \cap A_2 \cap \cdots \cap A_n} = \overline{A_1} \cup \overline{A_2} \cup \cdots \cup \overline{A_n}$.

13. Prove the following generalization of a distributive law for sets: For any set A and any $n \geq 2$ sets $B_1, B_2, \ldots, B_n$, $A \cap (B_1 \cup B_2 \cup \cdots \cup B_n) = (A \cap B_1) \cup (A \cap B_2) \cup \cdots \cup (A \cap B_n)$.

14. Prove that $\frac{1}{\sqrt{1}} + \frac{1}{\sqrt{2}} + \cdots + \frac{1}{\sqrt{n}} > \sqrt{n+1}$ for every integer $n \geq 3$.

15. Prove for every positive integer n that $1 + \frac{1}{\sqrt{2}} + \frac{1}{\sqrt{3}} + \cdots + \frac{1}{\sqrt{n}} \leq 2\sqrt{n}$.

16. Prove for every positive integer n that $2! \cdot 4! \cdot 6! \cdots (2n)! \geq ((n+1)!)^n$.

17. For an integer $n \geq 3$, an n-gon is an n-sided polygon. Thus a 3-gon is a triangle, while a 4-gon is a quadrilateral. It is well known that the sum of the interior angles of each triangle is 180°. Use induction to prove that for every integer $n \geq 3$, the sum of the interior angles of every n-gon is $(n-2) \cdot 180^o$.

4.3   Sequences

Many induction problems that occur in discrete mathematics involve the concept of sequences. You may very well have seen sequences in calculus, which are encountered during the study of infinite series. For our purposes, a **sequence** is a listing of elements of some set. A sequence can be infinite or finite but we are more interested in infinite sequences here. Typically, the elements of a sequence belong to the set $\mathbb{Z}$ or $\mathbb{R}$. An infinite sequence is often denoted by $a_1, a_2, a_3, \ldots$ or $\{a_n\}$. The element a_1 is then called the first term of the sequence, a_2 the second term and so on. The *nth term* of the sequence is therefore a_n.

Example 4.13

Consider the sequence $a_1, a_2, a_3, \ldots$ or $\{a_n\}$, where

$$a_n = \frac{(-1)^n}{n} \text{ for } n \in \mathbb{N}.$$

This sequence can also be expressed as

$$-1, \ \frac{1}{2}, \ \frac{-1}{3}, \ \frac{1}{4}, \ \ldots, \text{ or as } \left\{ \frac{(-1)^n}{n} \right\}.$$

Thus -1 is the first term of the sequence and $\frac{1}{2}$ is the second term. The *nth* term is therefore $\frac{(-1)^n}{n}$. ♦

While the numbers $-1, \frac{1}{2}, \frac{-1}{3}, \frac{1}{4}$ are certainly the first four terms of the sequence $\left\{ \frac{(-1)^n}{n} \right\}$, this is not the only sequence having these numbers as its first four terms. We consider another example.

Example 4.14

Determine the first three terms of the sequence $\{a_n\}$, where

$$a_n = n^3 - 6n^2 + 12n - 6 \text{ for } n \in \mathbb{N}.$$

Solution. The first three terms of this sequence are $a_1 = 1$, $a_2 = 2$, $a_3 = 3$.   ♦

If we had asked for a sequence $\{a_n\}$ whose first three terms are 1, 2, 3, we almost certainly would not have expected $a_n = n^3 - 6n^2 + 12n - 6$ as a response. The obvious answer would have been $a_n = n$. Example 4.14 serves to illustrate that there is no single answer to such a question, but there may be an obvious and expected answer. Notice that if $a_n = n^3 - 6n^2 + 12n - 6$, then $a_4 = 10$. Since

$$a_n = n^3 - 6n^2 + 12n - 6 = (n-1)(n-2)(n-3) + n,$$

it is not surprising that $a_n = n$ for $n = 1, 2, 3$. Indeed, if we had defined

$$a_n = (n-1)(n-2)(n-3)(n-4) + n$$

for $n \in \mathbb{N}$, then $a_n = n$ for $n = 1, 2, 3, 4$, while $a_5 = 29$.

We now consider a few common sequences.

Example 4.15

(a) $1, 2, 3, 4, \ldots$ is a sequence whose nth term is n.

(b) $1, 4, 9, 16, \ldots$ is a sequence whose nth term is n^2.

(c) $1, 8, 27, 64, \ldots$ is a sequence whose nth term is n^3.

The sequences (a)-(c) are all *polynomial sequences* of the form $\{n^k\}$ for a fixed positive integer k.

(d) $2, 4, 8, 16, \ldots$ is a sequence whose nth term is 2^n.

A **geometric sequence** is a sequence in which the ratio of every two consecutive terms a_n and a_{n+1} is some constant r, that is, $a_{n+1}/a_n = r$ for each $n \in \mathbb{N}$. Hence the sequence in (e) is geometric with $r = 2$.

(e) $4, 7, 10, 13, \ldots$ is a sequence whose nth term is $3n + 1$.

An **arithmetic** (pronounced ar-ith-MET-ic) **sequence** is a sequence in which the difference of every two consecutive terms a_n and a_{n+1} is some constant k, that is, $a_{n+1} - a_n = k$ for each $n \in \mathbb{N}$. Hence the sequence in (f) is an arithmetic sequence with $k = 3$. ◆

Example 4.16

Consider the sequence $a_0, a_1, a_2, \ldots$, where

$$a_0 = -\tfrac{1}{3}, a_1 = \tfrac{2}{5}, a_2 = -\tfrac{4}{7} \text{ and } a_3 = \tfrac{8}{9}.$$

Determine the nth term of a sequence $\{a_n\}$ whose first four terms are those given above.

Solution. Because of the alternating sign, the expression for a_n contains either $(-1)^n$ or $(-1)^{n+1}$. A quick check tells us that $(-1)^{n+1}$ is the correct choice. Since the numerators are $1, 2, 4, 8, \cdots$, we see that the numerator is 2^n. Since the denominators are $3, 5, 7, 9, \ldots$, the terms are increasing by 2 and so a_n contains $2n + k$ in the denominator for some constant k. We see that $k = 3$ is the correct choice and so

$$a_n = (-1)^{n+1} \frac{2^n}{2n+3}.$$ ◆

Binary Strings

The sequences described above are all infinite sequences as each such sequence contains an infinite number of terms. We now consider some finite sequences. A finite sequence is also referred to as a **string**. The number of terms in a string is its **length**. Thus a string of length n can be expressed as $a_1, a_2, \ldots, a_n$ or $(a_1, a_2, \ldots, a_n)$ or $a_1 a_2 \ldots a_n$, where a_i belongs to some set S $(1 \leq i \leq n)$.

A **bit** (or **binary digit**) is a 0 or 1. When each term of a string is a bit, $S = \{0, 1\}$ and the string is called a **binary string** or a **bit string**. An n-**bit string** is a string of length n where each term is 0 or 1. Hence $(0, 0, 1, 0, 1)$ is a 5-bit string. This string can also be expressed more simply as $0, 0, 1, 0, 1$ or as 00101.

The following is an application of bit strings.

Example 4.17

There are $2^4 = 16$ subsets of a set S with 4 elements. These subsets can be represented as binary strings of length 4. First, select an ordering of the elements of S such as $S = \{a_1, a_2, a_3, a_4\}$. Each subset A of S can be represented as a 4-bit string whose ith term is 1 if $a_i \in A$ and whose ith term is 0 if $a_i \notin A$. For the subset $A = \{a_2, a_4\}$, for example, the first term of the 4-bit string that represents A is 0 because $a_1 \notin A$. For this subset A of S, the 2nd term of the 4-bit string representing A is 1 because $a_2 \in A$. Continuing in this manner, we see that $A = \{a_2, a_4\}$ can be represented by 0101. Therefore, the set S itself is represented by 1111 and the empty set $\emptyset$ is represented by 0000. The table below gives the representations of all subsets of S.   ♦

$\emptyset$	$\{a_1\}$	$\{a_2\}$	$\{a_3\}$
0000	1000	0100	0010
$\{a_4\}$	$\{a_1, a_2\}$	$\{a_1, a_3\}$	$\{a_1, a_4\}$
0001	1100	1010	1001
$\{a_2, a_3\}$	$\{a_2, a_4\}$	$\{a_3, a_4\}$	$\{a_1, a_2, a_3\}$
0110	0101	0011	1110
$\{a_1, a_2, a_4\}$	$\{a_1, a_3, a_4\}$	$\{a_2, a_3, a_4\}$	$\{a_1, a_2, a_3, a_4\}$
1101	1011	0111	1111

Binary representations of the subsets of $\{a_1, a_2, a_3, a_4\}$ in Example 4.17

All of the sequences that we have encountered thus far were explicitly defined by a closed-form expression. That is, the general term of the sequence can be determined simply by evaluating a given expression for an appropriate integer. For example, the finite sequence $a_1, a_2, \ldots, a_{10}$ where

$$a_n = (n^2 - 1)^3 + 3n \text{ for } n = 1, 2, \ldots, 10$$

and the infinite sequence $\{b_n\}$ where

$$b_n = \left\lfloor \frac{n-10}{3} \right\rfloor \text{ for } n \in \mathbb{N}$$

are explicitly defined.

Recursively Defined Sequences

Some sequences are defined in an entirely different way. A sequence is said to be recursively defined if one or more initial terms of the sequence are specified, while later terms in the sequence are defined according to the values of terms preceding them. More specifically, we have the following definition.

Definition 4.18 *A sequence $a_1, a_2, a_3, \ldots$ of real numbers is said to be **recursively defined** if:*

(1) *For some fixed positive integer t, the terms $a_1, a_2, \ldots, a_t$ are given.*

(2) *For each integer $n > t$, a_n is defined in terms of one or more of $a_1, a_2, \ldots, a_{n-1}$.*

*Here, $a_1, a_2, \ldots, a_t$ are called the **initial values** of $\{a_n\}$; while the relation that defines a_n in terms of $a_1, a_2, \ldots, a_{n-1}$ for $n > t$ is called the **recurrence relation** for $\{a_n\}$.*

The first term of a sequence $\{a_n\}$ or $\{b_n\}$ need not have subscript 1. For example, $a_0, a_1, a_2, \ldots$ is an infinite sequence as is $b_3, b_4, b_5, \ldots$. We now look at two examples of recursively defined sequences.

Example 4.19

A sequence $a_1, a_2, a_3, \ldots$ is defined recursively by

$$a_1 = 1 \text{ and } a_n = \tfrac{n}{n-1}a_{n-1} \text{ for } n \geq 2.$$

Determine a_2, a_3, a_4 and a_5.

Solution. By definition $a_1 = 1$. By the recurrence relation,

$$a_2 = \tfrac{2}{1}a_1 = 2 \cdot 1 = 2, \; a_3 = \tfrac{3}{2}a_2 = \tfrac{3}{2} \cdot 2 = 3,$$
$$a_4 = \tfrac{4}{3}a_3 = \tfrac{4}{3} \cdot 3 = 4 \text{ and } a_5 = \tfrac{5}{4}a_4 = \tfrac{5}{4} \cdot 4 = 5.$$

Based on these values, we might be led to believe that $a_n = n$ for every positive integer n. This, in fact, is true. ♦

Result 4.20 *A sequence $a_1, a_2, a_3, \ldots$ is defined recursively by*

$$a_1 = 1 \text{ and } a_n = \tfrac{n}{n-1}a_{n-1} \text{ for } n \geq 2.$$

Then $a_n = n$ for every positive integer n.

Proof. We proceed by induction. Since $a_1 = 1$, the formula holds for $n = 1$. Assume that $a_k = k$ for a positive integer k. We show that $a_{k+1} = k + 1$. Since $k + 1 \geq 2$, it follows that

$$a_{k+1} = \frac{k+1}{k}a_k = \frac{k+1}{k} \cdot k = k + 1.$$

By the Principle of Mathematical Induction, $a_n = n$ for every $n \in \mathbb{N}$. ∎

In the following example, there are two initial values.

Example 4.21

A sequence $a_1, a_2, a_3, \ldots$ is defined recursively by

$$a_1 = 2, \; a_2 = 3 \text{ and } a_n = 2a_{n-1} - a_{n-2} \text{ for } n \geq 3.$$

Determine a_1, a_2, a_3, a_4 and a_5.

Solution. First, the initial values $a_1 = 2$ and $a_2 = 3$ are given. By the recurrence relation,

$$a_3 = 2a_2 - a_1 = 2 \cdot 3 - 2 = 4,$$
$$a_4 = 2a_3 - a_2 = 2 \cdot 4 - 3 = 5 \text{ and}$$
$$a_5 = 2a_4 - a_3 = 2 \cdot 5 - 4 = 6.$$

Based on these values, we might be led to believe that $a_n = n + 1$ for every $n \in \mathbb{N}$. This will be verified in the next section. ♦

Let's look at some familiar sequences that can also be defined recursively.

Example 4.22

For a fixed positive real number a and a nonnegative integer n, the nth power of a is ordinarily defined by

$$a^n = \begin{cases} 1 & \text{if } n = 0 \\ a \cdot a \cdots a \; (n \text{ factors } a) & \text{if } n \in \mathbb{N}. \end{cases}$$

The definition of a^n, where $n \in \{0\} \cup \mathbb{N}$, can be expressed as a sequence $p_0, p_1, p_2, \ldots$ defined by

$$p_n = \begin{cases} 1 & \text{if } n = 0 \\ a \cdot a \cdots a \; (n \text{ factors } a) & \text{if } n \in \mathbb{N}. \end{cases}$$

That is, $p_n = a^n$. This definition, though common, may seem a bit awkward when $n \in \mathbb{N}$. The number a^n can also be defined recursively by

$$p_n = \begin{cases} 1 & \text{if } n = 0 \\ a \cdot p_{n-1} & \text{if } n \in \mathbb{N}. \end{cases}$$

Therefore, once a^{n-1} has been defined, a^n is then defined as the product of a and a^{n-1} when $n \geq 1$. In particular, $p_0 = 1$, $p_1 = a \cdot p_0 = a \cdot 1 = a$, $p_2 = a \cdot p_1 = a \cdot a = a^2$ and $p_3 = a \cdot p_2 = a \cdot a^2 = a^3$. ♦

Example 4.23

For a positive integer n, we have defined $n!$ (n factorial) by

$$n! = n(n-1)(n-2)\cdots 3 \cdot 2 \cdot 1.$$

Thus $1! = 1$, $2! = 2 \cdot 1 = 2$, $3! = 3 \cdot 2 \cdot 1 = 6$ and $4! = 4 \cdot 3 \cdot 2 \cdot 1 = 24$. For $n = 0$, $n!$ is defined as $n! = 0! = 1$. Therefore, for a nonnegative integer n, the number n factorial $n!$ is defined by

$$n! = \begin{cases} 1 & \text{if } n = 0 \\ n(n-1)\cdots 3 \cdot 2 \cdot 1 & \text{if } n \in \mathbb{N}. \end{cases}$$

The definition of $n!$, where $n \in \{0\} \cup \mathbb{N}$, can also be expressed as a sequence $f_0, f_1, f_2, \ldots$ defined by

$$f_n = \begin{cases} 1 & \text{if } n = 0 \\ n(n-1)\cdots 3 \cdot 2 \cdot 1 & \text{if } n \in \mathbb{N}. \end{cases}$$

Although this is the most common way to define $n!$, here too the definition of $n!$ when $n \in \mathbb{N}$ may seem somewhat awkward. There is also a way to define $n!$ recursively. For a nonnegative integer n,

$$f_n = \begin{cases} 1 & \text{if } n = 0 \\ nf_{n-1} & \text{if } n \in \mathbb{N}. \end{cases}$$

In this case, $f_n = n!$, where $n! = n \cdot (n-1)!$ when $n \in \mathbb{N}$. Thus $f_0 = 1$, $f_1 = 1 \cdot f_0 = 1 \cdot 1 = 1$, $f_2 = 2 \cdot f_1 = 2 \cdot 1 = 2$ and $f_3 = 3 \cdot f_2 = 3 \cdot 2 = 6$. ♦

Fibonacci Numbers

The best known and most studied recursively defined sequence is one that results from certain positive integers called the Fibonacci numbers.

Example 4.24

The **Fibonacci sequence** $F_1, F_2, F_3, \ldots$ is defined recursively by

$$F_n = \begin{cases} 1 & \text{if } n = 1, 2 \\ F_{n-2} + F_{n-1} & \text{if } n \geq 3. \end{cases}$$

The numbers F_1, F_2, F_3, ... are referred to as **Fibonacci numbers**. Therefore, $F_1 = 1$ and $F_2 = 1$ are the initial values here and since the recurrence relation is $F_n = F_{n-2} + F_{n-1}$ for $n \geq 3$, it follows that

$$\begin{aligned} F_3 &= F_1 + F_2 = 1 + 1 = 2 \\ F_4 &= F_2 + F_3 = 1 + 2 = 3 \\ F_5 &= F_3 + F_4 = 2 + 3 = 5. \end{aligned}$$

Consequently, the first few Fibonacci numbers are

$$1, \ 1, \ 2, \ 3, \ 5, \ 8, \ 13, \ 21, \ 34, \ 55, \ 89.$$ ♦

Fibonacci numbers occur often and sometimes unexpectedly.

Example 4.25

For a positive integer n, let s_n be the number of n-bit strings having no two consecutive 0s.

(a) Determine s_1, s_2 and s_3.

(b) Give a recursive definition of s_n for $n \geq 1$.

(c) Use (b) to determine s_i for $1 \leq i \leq 6$.

Solution.

(a) Certainly, the two 1-bit strings 0 and 1 do not have two consecutive 0s and so $s_1 = 2$. The 2-bit string 00 is the only one of the four 2-bit strings with two consecutive 0s. Thus $s_2 = 3$. The eight 3-bit strings are

$$000 \quad 001 \quad 010 \quad 011 \quad 100 \quad 101 \quad 110 \quad 111.$$

Since 5 of these do not have two consecutive 0s, it follows that $s_3 = 5$.

(b) We have noted in (a) that $s_1 = 2$ and $s_2 = 3$. For $n \geq 3$, an n-bit string with no two consecutive 0s either (1) has 1 as the last bit and so the first $n - 1$ bits give an $(n - 1)$-bit string having no two consecutive 0s or (2) has 10 as the last two bits and so the first $n - 2$ bits give an $(n - 2)$-bit string having no two consecutive 0s. Consequently, for $n \geq 3$, $s_n = s_{n-1} + s_{n-2}$ and so the recursive definition of s_n for $n \geq 1$ is:

$$s_1 = 2, \ s_2 = 3 \text{ and } s_n = s_{n-1} + s_{n-2} \text{ for } n \geq 3.$$

(c) This gives us $s_1 = 2$, $s_2 = 3$, $s_3 = s_2 + s_1 = 3 + 2 = 5$, $s_4 = s_3 + s_2 = 5 + 3 = 8$, $s_5 = s_4 + s_3 = 8 + 5 = 13$, $s_6 = s_5 + s_4 = 13 + 8 = 21$, all of which, of course, are Fibonacci numbers. ◆

There are several relationships involving products of Fibonacci numbers. We present one of these next. Another is given in Exercise 23.

Example 4.26 *For every integer $n \geq 2$,*

$$F_{n-1}F_{n+1} = F_n^2 + (-1)^n.$$

Proof. We proceed by induction. When $n = 2$, we have

$$F_1 F_3 = 1 \cdot 2 = 1^2 + 1 = F_2^2 + (-1)^2.$$

Thus the formula holds for $n = 2$. Assume for an arbitrary integer $k \geq 2$ that

$$F_{k-1}F_{k+1} = F_k^2 + (-1)^k.$$

Thus

$$F_k^2 = F_{k-1}F_{k+1} - (-1)^k. \tag{4.10}$$

We show that

$$F_k F_{k+2} = F_{k+1}^2 + (-1)^{k+1}.$$

Using (4.10) and the recursion relation for the Fibonacci numbers, we obtain

$$
\begin{aligned}
F_k F_{k+2} &= F_k(F_k + F_{k+1}) = F_k^2 + F_k F_{k+1} \\
&= [F_{k-1}F_{k+1} - (-1)^k] + F_k F_{k+1} \\
&= F_{k-1}F_{k+1} + F_k F_{k+1} + (-1)^{k+1} \\
&= (F_{k-1} + F_k)F_{k+1} + (-1)^{k+1} \\
&= F_{k+1}F_{k+1} + (-1)^{k+1} = F_{k+1}^2 + (-1)^{k+1}.
\end{aligned}
$$

Thus, $F_k F_{k+2} = F_{k+1}^2 + (-1)^{k+1}$. By the Principle of Mathematical Induction,

$$F_{n-1} F_{n+1} = F_n^2 + (-1)^n$$

for every integer $n \geq 2$.   ∎

The origin of Fibonacci numbers is rather curious. One of the great European mathematicians of the middle ages was Leonardo da Pisa, born in 1175 in Pisa, Italy, known for its famous Leaning Tower. Leonardo called himself Fibonacci. The name Fibonacci is a shortening of "filius Bonacci", which means "son of Bonaccio". His father's name was Guglielmo Bonaccio. (Bonacci is the plural of Bonaccio.) Fibonacci traveled a great deal in his early years about the Mediterranean coast and returned to Pisa in 1200. With the knowledge he had acquired during his travels, he wrote the book *Liber Abaci*, in which he introduced the decimal number system to the Latin-speaking world. In Chapter 12 of his book, he stated the following problem:

> *How Many Pairs of Rabbits Are Created by One Pair in One Year?*
>
> *A certain man had one pair of rabbits together in a certain enclosed place and one wishes to know how many are created from the pair in one year when it is the nature of them in a single month to bear a single pair and in the second month those born to bear also.*

Solution. First, it may be necessary to clarify some aspects of Fibonacci's Rabbit Problem. It is assumed that at the beginning of the year (the beginning of Month #1), there is a pair of rabbits (one male and one female). At the beginning of Month #2, the two rabbits have developed into adult rabbits and can bear other rabbits. At the beginning of Month #3, another pair of rabbits (one male and one female) are born. Therefore, at the beginning of Month #3, there are two pairs of rabbits (one adult pair and one juvenile pair). At the beginning of Month #4, the adult rabbits bear another pair of rabbits (a new juvenile pair) and the previous juvenile pair becomes an adult pair. Thus, at the beginning of Month #4, there are two adult pairs of rabbits and one juvenile pair, or three pairs in all.

Assuming that the rabbits do not die during the year, the number r_n of pairs of rabbits at the beginning of Month #n ($3 \leq n \leq 13$) is the number r_{n-1} of pairs of rabbits there are at the beginning of Month #$(n-1)$ plus the number of pairs of rabbits born at the beginning of Month #n. The number of pairs of rabbits born at that time equals the number of pairs of adult rabbits there are at the beginning of Month #$(n-1)$. This is the number r_{n-2} of pairs of rabbits at the beginning of Month #$(n-2)$. Therefore,

$$r_n = r_{n-2} + r_{n-1}.$$

This information is given in the table below.

Begining of Month #n	1	2	3	4	5	6	7	8	9	10	11	12	13
No. of pairs of rabbits	1	1	2	3	5	8	13	21	34	55	89	144	233
No. of pairs of adult rabbits	0	1	1	2	3	5	8	13	21	34	55	89	144

The numbers $r_1, r_2, \ldots, r_{13}$ are therefore $1, 1, 2, \ldots, 233$, which are Fibonacci numbers and so the number of rabbits at the end of the year is 233 pairs.   ♦

Although Fibonacci is understood to be neither the originator of this problem nor the numbers that bear his name, he was the one to make these numbers known. It was the French mathematician Edouard Lucas who gave the name Fibonacci numbers to this sequence of numbers. Indeed, there is a closely related sequence of numbers named after Lucas – the **Lucas numbers**:

$$2, 1, 3, 4, 7, 11, 18, 29, 47, 76, \ldots.$$

Fibonacci died in the 1240s. There is a statue honoring him at the Leaning Tower of Pisa.

If a sequence is defined recursively, then both an initial value (or values) must be specified and a recurrence relation must be defined. If either of these two is changed, then a different sequence is produced. We illustrate this remark with the Fibonacci numbers.

Example 4.27

A sequence $L_1, L_2, L_3, \ldots$ is defined recursively by

$$L_1 = 2, \ L_2 = 1 \text{ and } L_n = L_{n-2} + L_{n-1} \text{ for } n \geq 3.$$

Determine L_3, L_4 and L_5.

Solution. This recursively defined sequence has the same recurrence relation as the Fibonacci numbers but the first initial value is different. By the recurrence relation,

$$
\begin{aligned}
L_3 &= L_1 + L_2 = 2 + 1 = 3 \\
L_4 &= L_2 + L_3 = 1 + 3 = 4 \\
L_5 &= L_3 + L_4 = 3 + 4 = 7.
\end{aligned}
$$

The numbers $L_1, L_2, L_3, \ldots$ are in fact the Lucas numbers we referred to earlier. ◆

We have previously mentioned in Chapter 2 (Theorem 2.17 in Section 2.1) that the number of subsets of an n-element set is 2^n. The number of subsets of an n-element set can be defined recursively.

Example 4.28

For a nonnegative integer n, let s_n be the number of subsets of an n-element set.

(a) What are s_0, s_1 and s_2?

(b) Give a recursive definition of s_n for $n \geq 0$.

Solution.

(a) Since the only 0-element set is $\emptyset$ and the only subset of $\emptyset$ is $\emptyset$, it follows that $s_0 = 1$. The subsets of a 1-element set $\{a\}$ are $\emptyset$ and $\{a\}$, so $s_1 = 2$. Also, the subsets of a 2-element set $\{x, y\}$ are $\emptyset$, $\{x\}$, $\{y\}$ and $\{x, y\}$ and so $s_2 = 4$. Therefore, $s_0 = 1$, $s_1 = 2$ and $s_2 = 4$.

(b) First, $s_0 = 1$. Suppose that s_{n-1} $(n - 1 \geq 0)$ is the number of subsets of an $(n - 1)$-element set. Let A be an n-element set, say $A = \{a_1, a_2, \ldots, a_n\}$. Every subset of A either contains a_n or it does not. A subset of A that does not contain a_n is a subset of $B = \{a_1, a_2, \ldots, a_{n-1}\}$. There are s_{n-1} subsets of B. A subset of A that contains a_n can be expressed as $C \cup \{a_n\}$, where $C \subseteq B$. Since there are s_{n-1} choices for C, there are s_{n-1} subsets of A containing a_n. Hence there are $s_{n-1} + s_{n-1} = 2s_{n-1}$ subsets of A. Therefore, $s_0 = 1$ and $s_n = 2s_{n-1}$ for $n \geq 1$. ◆

In all of the examples of recurrence relations that we have seen, we computed the first few terms of the sequence being considered. In many instances, it would probably be useful if there were an explicit formula for the nth term of the sequence. The question of finding such a formula and verifying the correctness of the formula is important and will be discussed further in Chapter 9.

Exercises for Section 4.3

1. Write out the first four terms of the sequence $\{a_n\}$ for which $a_n = (-1)^{n+1} \frac{2n+3}{3 \cdot 2^{n-1}}$, where $n \in \mathbb{N}$.

2. Write out the first four terms of the sequence $\{a_n\}$ for which $a_n = \frac{n^2}{2n-1}$, where $n \in \mathbb{N} \cup \{0\}$.

3. Write out the first four terms of the sequence $\left\{ \frac{(-1)^{3n}}{2^{2n}} \right\}$, where $n \in \mathbb{N}$.

4. Let $n \in \mathbb{N}$. For n even, $\lceil n/2 \rceil = \lfloor n/2 \rfloor = n/2$ and for n odd, $\lceil n/2 \rceil = (n+1)/2$ and $\lfloor n/2 \rfloor = (n-1)/2$. Write out the first four terms of the following:

   (a) $\{\lceil n/2 \rceil\}$.   (b) $\{\lfloor n/2 \rfloor\}$.   (c) $\{\lceil n/2 \rceil - \lfloor n/2 \rfloor\}$.

5. Write out the first four terms of each of the following sequences whose nth term, $n \in \mathbb{N}$, is

   (a) 3.   (b) $2^n + (-2)^n$.   (c) $3^n - 2^n$.

6. Find an expression for the nth term a_n, where $n \in \mathbb{N}$, of a sequence whose first four terms are

   (a) $1, -1, 1, -1, \ldots$.   (b) $0, 2, 0, 2, \ldots$.   (c) $1, 3, 5, 7, \ldots$.

   (d) $2, 8, 18, 32, \ldots$.   (e) $2, 4, 12, 48, \ldots$.   (f) $6, 18, 54, 162, \ldots$.

   (g) $1, 4, 7, 10, \ldots$.   (h) $-3, 1, 5, 9, \ldots$.   (i) $1, 2, 5, 10, \ldots$.   (j) $5, 5, 5, 5, \ldots$.

7. Find the nth terms of three different sequences beginning with 1, 2, 4, where $n \in \mathbb{N}$.

8. Find the nth terms of three different sequences beginning with 1, 3, 9, where $n \in \mathbb{N}$.

9. Give an example of

   (a) a string of length 5 whose elements belong to $\{a, b, c, d, e\}$.

   (b) a bit string of length 6.

   (c) a bit string of length n for some $n \in \mathbb{N}$ such that the sum of its terms is $\lceil n/2 \rceil$.

10. For $X = \{x_1, x_2, x_3, x_4, x_5\}$, proceed as in Example 4.17 to determine

    (a) which 5-bit string corresponds to each of the subsets below.
        $X_1 = \{x_1, x_4\}$, $X_2 = \{x_2, x_4, x_5\}$ and $X_3 = \{x_3, x_5\}$

    (b) which subset of X corresponds to each of the 5-bit strings below.
        $s_1 = 00000$, $s_2 = 01001$ and $s_3 = 11111$.

11. In a children's game, a child must place either a black coin or a red coin on each of the four squares shown below.

1	2
3	4

    Explain how each placement of four coins corresponds to some 4-bit string.

12. A sequence $A_1, A_2, A_3, \ldots$ of sets is defined recursively by $A_1 = \{1\}$ and $A_n = ((A_1 \cup A_2 \cup \cdots \cup A_{n-1}) - A_{n-1}) \cup \{n\}$ for $n \geq 2$. Determine the sets A_2, A_3, A_4, A_5 and A_6.

13. A sequence $a_1, a_2, a_3, \ldots$ is defined recursively by $a_1 = 2$ and $a_n = 2a_{n-1} - 1$ for $n \geq 2$.

    (a) Determine a_2, a_3, a_4 and a_5.

    (b) Based on the values obtained in (a), make a guess for a formula for a_n for every positive integer n and use induction to verify that your guest is correct.

14. A sequence $a_1, a_2, a_3, \ldots$ is defined recursively by $a_1 = 3$ and $a_n = 2a_{n-1} + 1$ for $n \geq 2$.

    (a) Determine a_2, a_3, a_4 and a_5.

    (b) Based on the values obtained in (a), make a guess for a formula for a_n for every positive integer n and use induction to verify that your guest is correct.

15. A sequence $a_1, a_2, a_3, \ldots$ is defined recursively by $a_1 = 1$, $a_2 = 2$ and $a_n = 2a_{n-1} - a_{n-2}$ for $n \geq 3$. Determine a_3, a_4, a_5 and a_6.

16. For a positive integer n, let s_n be the number of n-bit strings that do not contain a 1 immediately followed by 0.

    (a) Determine s_1, s_2 and s_3.

    (b) Give a recursive definition of s_n for $n \geq 1$.

17. For a positive integer n, let s_n be the number of n-bit strings having no three consecutive 0s.

    (a) Determine s_1, s_2 and s_3.

    (b) Give a recursive definition of s_n for $n \geq 1$.

18. A sequence $a_1, a_2, a_3, \ldots$ is defined recursively by $a_1 = 2$, $a_2 = 1$, $a_3 = 3$ and $a_n = a_{n-3} + a_{n-2} + a_{n-1}$ for $n \geq 4$.

    (a) Determine a_4, a_5 and a_6.

    (b) Suppose that the first three terms of this sequence were defined instead as $a_1 = 1$, $a_2 = 2$, $a_3 = 3$ but the recurrence relation was not changed. In this case, what are a_4, a_5 and a_6?

    (c) Suppose that the initial values of the original sequence were not changed but the recurrence relation is given instead by $a_n = a_{n-3} + a_{n-1}$ for $n \geq 4$. What are a_4, a_5 and a_6 in this case?

19. For a positive integer n, let s_n be the number of 2-element subsets of an n-element set.

    (a) What are s_1, s_2 and s_3?

    (b) Give a recursive definition of s_n for $n \geq 1$.

    (c) Use the recursive definition in (b) to compute s_4.

20. A sequence $a_1, a_2, a_3, \cdots$ is defined recursively by $a_1 = 1$, $a_2 = 2$ and

$$a_n = \begin{cases} 2 & \text{if } a_{n-1}a_{n-2} = 1 \\ 1 & \text{if } a_{n-1}a_{n-2} \neq 1 \end{cases}$$

for $n \geq 3$.

    (a) Determine a_n for $3 \leq n \leq 9$.

    (b) Prove that if $a_{n-1}a_{n+1} = 1$ where $n \geq 2$, then $a_n = 2$.

21. Use induction to show the following for Fibonacci numbers: $F_1 + F_2 + \cdots + F_n = F_{n+2} - 1$ for every positive integer n.

22. Use induction to show the following for Fibonacci numbers: $F_2 + F_4 + \cdots + F_{2n} = F_{2n+1} - 1$ for every positive integer n.

23. Use induction to show the following for Fibonacci numbers: $F_{n-1}F_{n+2} = F_nF_{n+1} + (-1)^n$ for every integer $n \geq 2$.

4.4 The Strong Principle of Mathematical Induction

We have already mentioned that there are several variations of the Principle of Mathematical Induction. Perhaps the best known and most useful of these is called the Strong Principle of Mathematical Induction. This principle is also known by several other names, including the Strong Form of Mathematical Induction, the Second Principle of Mathematical Induction and the Alternate Form of Mathematical Induction.

The Strong Principle of Mathematical Induction *The statement*

$$\forall n \in \mathbb{N}, \; P(n): \quad \text{For every positive integer } n, \; P(n).$$

is true if

(1) *$P(1)$ is true and*

(2) *the statement $\forall k \in \mathbb{N}, P(1) \wedge P(2) \wedge \cdots \wedge P(k) \Rightarrow P(k+1)$ is true.*

The Strong Principle of Mathematical Induction then states that all of the statements $P(1)$, $P(2)$, $P(3)$, ... are true if it can be shown that (1) $P(1)$ is true and (2) the statement

Let $k \in \mathbb{N}$. If $P(i)$ for each $i \in \mathbb{N}$ with $1 \le i \le k$, then $P(k+1)$.

is true. As before, (1) is called the **base** or **basis step** of this induction and (2) is the **inductive step**. If a direct proof is used to verify the inductive step, we then assume for an arbitrary positive integer k that $P(i)$ is a true statement for *every* integer i with $1 \le i \le k$ (which is called the **induction** or **inductive hypothesis**) and show that $P(k+1)$ is a true statement. Assuming that $P(i)$ is a true statement for $1 \le i \le k$ means then that we are assuming the following k statements are true: $P(1)$, $P(2)$, ..., $P(k)$. The difference between using the Principle of Mathematical Induction and using the Strong Principle of Mathematical Induction to prove that $\forall n \in \mathbb{N}, \; P(n)$ is true lies in their respective inductive steps. When using a direct proof to prove the inductive step in the Strong Principle of Mathematical Induction, we are permitted to assume that all of the statements $P(1)$, $P(2)$, ..., $P(k)$ are true (rather than only assuming that $P(k)$ is true, as in the Principle of Mathematical Induction). In each situation, we need to show that $P(k+1)$ is true. That is, when attempting to show that $P(k+1)$ is true in the inductive step in the Strong Principle of Mathematical Induction, we are allowed to assume more (which may be just what is needed).

We asked earlier whether the nth term of a recursively defined sequence may also be expressed in a closed form (that is, by a formula). When there is a proposed formula, the anticipated formula can often be verified either by the Principle of Mathematical Induction (examples of which are shown in Section 4.3) or perhaps by the Strong Principle of Mathematical Induction. We first consider another example where the Principle of Mathematical Induction is used.

Result 4.29 *If $\{a_n\}$ is a sequence defined recursively by*

$$a_1 = 3 \text{ and } a_n = 2a_{n-1} + 1 \text{ for } n \ge 2,$$

then $a_n = 2^{n+1} - 1$ for every positive integer n.

Proof. We use the Principle of Mathematical Induction. Since $a_1 = 2^{1+1} - 1 = 3$, the formula holds for $n = 1$. Assume that $a_k = 2^{k+1} - 1$ for a positive integer k. We show that

$$a_{k+1} = 2^{(k+1)+1} - 1 = 2^{k+2} - 1.$$

Observe that

$$a_{k+1} = 2a_k + 1 = 2(2^{k+1} - 1) + 1 = 2^{k+2} - 1.$$

It then follows by the Principle of Mathematical Induction that $a_n = 2^{n+1} - 1$ for every positive integer n. ∎

In the previous result, there was only one initial value. In Example 4.21 (in Section 4.3), we encountered a recursively defined sequence in which there are two initial values. We now visit this example again.

Result 4.30 *A sequence $a_1, a_2, a_3, \ldots$ is defined recursively by*

$$a_1 = 2, a_2 = 3 \text{ and } a_n = 2a_{n-1} - a_{n-2} \text{ for } n \ge 3.$$

Then $a_n = n + 1$ for every positive integer n.

Proof. We proceed by the Strong Principle of Mathematical Induction. Since $a_1 = 2 = 1 + 1$, the formula holds for $n = 1$. Assume, for a positive integer k, that $a_i = i + 1$ for every integer i with $1 \le i \le k$. We show that $a_{k+1} = k + 2$. First, observe that when $k = 1$, $a_{k+1} = a_{1+1} = a_2 = 3 = 1 + 2$ and so the formula holds. Hence we may assume that $k \ge 2$. Since $k + 1 \ge 3$, it follows by the recurrence relation that

$$a_{k+1} = 2a_k - a_{k-1} = 2(k+1) - [(k-1) + 1] = 2k - 2 - k = k + 2.$$

It therefore follows by the Strong Principle of Mathematical Induction that $a_n = n + 1$ for each positive integer n. ∎

Proof Analysis. In the proof of Result 4.30, we wanted to show that $a_{k+1} = k + 2$. In the proof we used the fact that we know that $a_n = 2a_{n-1} - a_{n-2}$. However, we only know this when $n \ge 3$. So we could only write that $a_{k+1} = 2a_k - a_{k-1}$ when $k + 1 \ge 3$, that is when $k \ge 2$. The situation when $k = 1$ had to be handled separately.

We now consider two additional examples where the Strong Principle of Mathematical Induction is an appropriate technique.

Example 4.31

A sequence $a_1, a_2, a_3, \ldots$ is defined recursively by

$$a_1 = 1, a_2 = 4 \text{ and } a_n = 2a_{n-1} - a_{n-2} + 2 \text{ for } n \ge 3.$$

(a) Determine a_3, a_4 and a_5.

(b) Conjecture a formula for a_n for each positive integer n.

Solution.

(a) By the recurrence relation,

$$
\begin{aligned}
a_3 &= 2a_2 - a_1 + 2 = 2 \cdot 4 - 1 + 2 = 9 \\
a_4 &= 2a_3 - a_2 + 2 = 2 \cdot 9 - 4 + 2 = 16 \\
a_5 &= 2a_4 - a_3 + 2 = 2 \cdot 16 - 9 + 2 = 25.
\end{aligned}
$$

(b) Based on the initial values a_1 and a_2 and the values a_3, a_4 and a_5 determined in (a), a reasonable conjecture would be that $a_n = n^2$ for each positive integer n. ◆

We now show that the conjecture made in Example 4.31 is, in fact, correct.

Result to Prove: A sequence $a_1, a_2, a_3, \ldots$ is defined recursively by

$$a_1 = 1, a_2 = 4 \text{ and } a_n = 2a_{n-1} - a_{n-2} + 2 \text{ for } n \ge 3.$$

Then $a_n = n^2$ for each positive integer n.

Proof Strategy. We attempt to prove this result by the Strong Principle of Mathematical Induction. Because $a_1 = 1 = 1^2$, the formula holds for $n = 1$. This is the basis step. We now verify the inductive step. Thus we assume that $a_i = i^2$ for every integer i with $1 \le i \le k$, where k is an arbitrary positive integer and then show that $a_{k+1} = (k+1)^2$.

From the way in which the sequence $a_1, a_2, a_3, \ldots$ is defined, we can write $a_{k+1} = 2a_k - a_{k-1} + 2$ *provided* that $k + 1 \ge 3$, that is, provided that $k \ge 2$. However, it is possible that $k = 1$. If $k = 1$, then $k + 1 = 2$ and $a_{k+1} = a_2 = 4 = 2^2$. Thus $a_{k+1} = (k+1)^2$ when $k = 1$. On the other hand, if $k \ge 2$, then $k + 1 \ge 3$ and, as we just mentioned, $a_{k+1} = 2a_k - a_{k-1} + 2$. From the inductive hypothesis, $a_i = i^2$ for each integer i with $1 \leq i \leq k$ (where now $k \geq 2$). In particular, $a_k = k^2$ and $a_{k-1} = (k-1)^2$. Thus

$$a_{k+1} = 2a_k - a_{k-1} + 2 = 2k^2 - (k-1)^2 + 2 = (k+1)^2.$$

This is what we wanted to show.   ♦

We are now prepared to give a more concise proof of this result.

Result 4.32 *A sequence* $a_1, a_2, a_3, \ldots$ *is defined recursively by*

$$a_1 = 1, a_2 = 4 \text{ and } a_n = 2a_{n-1} - a_{n-2} + 2 \text{ for } n \geq 3.$$

Then $a_n = n^2$ *for each positive integer* n.

Proof. We employ the Strong Principle of Mathematical Induction. Since $a_1 = 1 = 1^2$, the formula holds for $n = 1$. Assume that $a_i = i^2$ for every integer i with $1 \leq i \leq k$, where k is an arbitrary positive integer. We show that $a_{k+1} = (k+1)^2$. When $k = 1$, we have $a_{k+1} = a_2 = 4 = 2^2$ and so $a_{k+1} = (k+1)^2$ for $k = 1$. Hence we may assume that $k \geq 2$ and so $k + 1 \geq 3$. Observe that

$$\begin{aligned} a_{k+1} &= 2a_k - a_{k-1} + 2 = 2k^2 - (k-1)^2 + 2 \\ &= 2k^2 - (k^2 - 2k + 1) + 2 = k^2 + 2k + 1 = (k+1)^2. \end{aligned}$$

Therefore, $a_{k+1} = (k+1)^2$ for every positive integer k. By the Strong Principle of Mathematical Induction, $a_n = n^2$ for each positive integer n.   ■

Proof Analysis. Let's take another look at the proof we just gave of Result 4.32. In the conclusion of the inductive step, we were required to show that $a_{k+1} = (k+1)^2$ where $k \geq 1$. Recall that we first verified that $a_{k+1} = (k+1)^2$ when $k = 1$. As in the proofs just preceding this, the reason for this was because we could only write $a_{k+1} = 2a_k - a_{k-1} + 2$ when $k + 1 \geq 3$ (from the recurrence relation on the definition of the sequence $a_1, a_2, a_3, \ldots$). We then showed that $a_{k+1} = (k+1)^2$ where $k \geq 2$. We could have actually divided the proof into two cases at that time (namely, *Case* 1 : $k = 1$ and *Case* 2 : $k \geq 2$) but because the case $k = 1$ could be handled so quickly, we decided not to give a proof by cases.

Another logical question that could be asked about the proof of Result 4.31 is why the Strong Principle of Mathematical Induction was used rather than the standard Principle of Mathematical Induction. This is actually quite easy to answer. Suppose that we had attempted to give a proof of Result 4.32 using the Principle of Mathematical Induction. In that case, we would need to verify the inductive step by assuming that $a_k = k^2$ for a positive integer k and show that $a_{k+1} = (k+1)^2$. Notice that we don't know that $a_{k-1} = (k-1)^2$. For $k \geq 2$, however, $a_{k+1} = 2a_k - a_{k-1} + 2$. Although we can write $a_k = k^2$, we have no simplified expression for a_{k-1} and we are unable to give a proof.   ♦

In Example 4.31, we were asked to conjecture (guess) a formula for a_n. Although in this case, it doesn't appear too difficult to make the right guess, it is still possible that we may have guessed incorrectly or that we can't come up with a good guess. In Section 9.5 we will visit a topic that will allow us to determine a formula (or closed form) for a_n in many instances when the sequence $\{a_n\}$ is defined recursively.

Result 4.33 *A sequence* $a_1, a_2, a_3, \ldots$ *is defined recursively by*

$$a_1 = 1, a_2 = 3 \text{ and } a_n = 3a_{n-1} - 2a_{n-2} - 2 \text{ for } n \geq 3.$$

Then $a_n = 2n - 1$ *for every positive integer* n.

Proof. We use the Strong Principle of Mathematical Induction. Since $a_1 = 1 = 2 \cdot 1 - 1$, the formula holds for $n = 1$. Assume that $a_i = 2i - 1$ for every integer i with $1 \leq i \leq k$, where k is an arbitrary positive integer. We show that

$$a_{k+1} = 2(k+1) - 1 = 2k + 1.$$

When $k = 1$, we have $a_{k+1} = a_2 = 2 \cdot 1 + 1 = 3$ and so $a_{k+1} = 2k + 1$ when $k = 1$. When $k \geq 2$, we have $k + 1 \geq 3$ and so

$$
\begin{aligned}
a_{k+1} &= 3a_k - 2a_{k-1} - 2 = 3(2k - 1) - 2(2k - 3) - 2 \\
&= 6k - 3 - 4k + 6 - 2 = 2k + 1.
\end{aligned}
$$

Thus, $a_{k+1} = 2k + 1$ for every positive integer k. By the Strong Principle of Mathematical Induction, $a_n = 2n - 1$ for every positive integer n. $\blacksquare$

We now consider a recursively defined sequence in which there are three initial values.

Result 4.34 *A sequence $a_1, a_2, a_3, \ldots$ is defined recursively by*

$$a_1 = 1, a_2 = 2, \ a_3 = 4 \ and \ a_n = a_{n-1} + a_{n-2} + 2a_{n-3} \ for \ n \geq 4.$$

Then $a_n = 2^{n-1}$ for each positive integer n.

Proof. We use the Strong Principle of Mathematical Induction. Since $a_1 = 1 = 2^{1-1}$, the formula holds for $n = 1$. Assume that $a_i = 2^{i-1}$ for every integer i with $1 \leq i \leq k$, where k is a positive integer. We show that $a_{k+1} = 2^{(k+1)-1} = 2^k$. When $k = 1$ and $k = 2$, we have $a_{k+1} = a_2 = 2^1 = 2$ and $a_{k+1} = a_3 = 2^2 = 4$, respectively. Therefore, $a_{k+1} = 2^k$ when $k = 1$ and $k = 2$. When $k \geq 3$,

$$
\begin{aligned}
a_{k+1} &= a_k + a_{k-1} + 2a_{k-2} = 2^{k-1} + 2^{k-2} + 2 \cdot 2^{k-3} \\
&= 2^{k-1} + 2^{k-2} + 2^{k-2} = 2^{k-1} + 2 \cdot 2^{k-2} \\
&= 2^{k-1} + 2^{k-1} = 2 \cdot 2^{k-1} = 2^k.
\end{aligned}
$$

Therefore, $a_{k+1} = 2^k$ for every positive integer k. It then follows by the Strong Principle of Mathematical Induction that $a_n = 2^{n-1}$ for each positive integer n. $\blacksquare$

There is also a more general form of the Strong Principle of Mathematical Induction. Again, because of the minor differences in their statements, we refer to this as well as the Strong Principle of Mathematical Induction.

The Strong Principle of Mathematical Induction *For a fixed integer m, let*

$$S = \{i \in \mathbb{Z} : \ i \geq m\}.$$

Then the statement

$$\forall n \in S, \ P(n) : \ For \ every \ integer \ n \geq m, \ P(n).$$

is true if

(1) *$P(m)$ is true and*

(2) *the statement $\forall k \in S, P(m) \wedge P(m+1) \wedge \cdots \wedge P(k) \Rightarrow P(k+1)$ is true.*

We return to the Fibonacci numbers again to establish a relationship that can be verified with the aid of the the Strong Principle of Mathematical Induction (with $m = 2$).

Result 4.35 *For every integer $n \geq 2$, $F_n \leq 2F_{n-1}$.*

Proof. We proceed by induction. Since $F_2 = 1 \leq 2 \cdot 1 = 2F_1$, the inequality holds for $n = 2$. Assume, for an arbitrary integer $k \geq 2$, that $F_i \leq 2F_{i-1}$ for every integer i with $2 \leq i \leq k$. We show that $F_{k+1} \leq 2F_k$. When $k = 2$, we have $F_3 = 2 \leq 2 \cdot 1 = 2F_2$. Thus $F_{k+1} \leq 2F_k$ when $k = 2$. Hence we may assume that $k \geq 3$. In this case,

$$F_{k+1} = F_{k-1} + F_k \leq 2F_{k-2} + 2F_{k-1} = 2(F_{k-2} + F_{k-1}) = 2F_k$$

and so $F_{k+1} \leq 2F_k$.

Thus, by the Strong Principle of Mathematical Induction, $F_n \leq 2F_{n-1}$ for every integer $n \geq 2$. ∎

With the assistance of Result 4.35, an upper bound for Fibonacci numbers can be established using the Principle of Mathematical Induction.

Corollary 4.36 *For every positive integer n, $F_n \leq 2^n$.*

Proof. We proceed by induction. Since $F_1 = 1 \leq 2^1$, the inequality holds for $n = 1$. Assume that $F_k \leq 2^k$ for an arbitrary positive integer k. We show that $F_{k+1} \leq 2^{k+1}$. By Result 4.35,

$$F_{k+1} \leq 2F_k \leq 2 \cdot 2^k = 2^{k+1}.$$

Thus $F_{k+1} \leq 2^{k+1}$. By the Principle of Mathematical Induction, $F_n \leq 2^n$ for every positive integer n.
                                                                                            ∎

Exercises for Section 4.4

1. A sequence $\{a_n\}$ is defined recursively by $a_1 = 1$ and $a_n = 2a_{n-1}$ for $n \geq 2$.

   (a) Determine a_2, a_3, a_4 and a_5.

   (b) Conjecture a formula for a_n for each positive integer n.

   (c) Verify the conjecture in (b).

2. A sequence $a_1, a_2, a_3, \ldots$ of integers is defined recursively by $a_1 = 0, a_2 = 2$ and $a_n = 2a_{n-1} - a_{n-2} + 2$ for $n \geq 3$.

   (a) Determine a_3, a_4 and a_5.

   (b) Conjecture a formula for a_n for each positive integer n.

   (c) Verify the conjecture in (b).

3. A sequence $a_1, a_2, a_3, \ldots$ of integers is defined recursively by $a_1 = 4$, $a_2 = 7$ and $a_n = 2a_{n-1} - a_{n-2} + 2$ for $n \geq 3$. Prove that $a_n = n^2 + 3$ for every positive integer n.

4. A sequence $\{a_n\}$ is defined recursively by $a_1 = 1, a_2 = 4$ and $a_n = -a_{n-1} + 2a_{n-2} + 6n - 7$ for $n \geq 3$. Prove that $a_n = n^2$ for every positive integer n.

5. A sequence $\{a_n\}$ is defined recursively by $a_1 = 1, a_2 = 4, a_3 = 9$ and $a_n = a_{n-1} - a_{n-2} + a_{n-3} + 2(2n - 3)$ for $n \geq 4$.

   (a) Determine a_4, a_5 and a_6.

   (b) Conjecture a formula for a_n for each positive integer n.

   (c) Verify your conjecture in (b).

6. A sequence $\{a_n\}$ is defined recursively with recurrence relation $a_n = 2a_{n-1} - a_{n-2}$ for $n \geq 3$.
   (a) For $a_1 = 1$ and $a_2 = 3$,
   (i) determine a_3, a_4 and a_5.
   (ii) conjecture a formula for a_n for each positive integer n.
   (iii) verify the conjecture in (b).
   (b) Repeat (a) for $a_1 = 1$ and $a_2 = 2$.   (c) Repeat (a) for $a_1 = 1$ and $a_2 = 1$.

7. A sequence $a_1, a_2, a_3, \ldots$ is defined recursively by $a_1 = 1$, $a_2 = -1$ and $a_n = a_{n-1} + 2a_{n-2}$ for $n \geq 3$. Conjecture a formula for a_n for $n \in \mathbb{N}$ and verify that your conjecture is correct.

8. In Example 4.36, we saw that nth Fibonacci number $F_n \leq 2^n$. Prove that $F_n \leq \left(\frac{5}{3}\right)^n$ for every positive integer n.

9. Find a positive constant c such that $c < \frac{5}{3}$ and $F_n \leq c^n$ for every positive integer n, where the F_n is the nth Fibonacci number. Verify your answer.

10. Prove that the nth Fibonacci number is $F_n = \frac{1}{\sqrt{5}}\left[\left(\frac{1+\sqrt{5}}{2}\right)^n - \left(\frac{1-\sqrt{5}}{2}\right)^n\right]$ for every positive integer n.

11. The following "proof" proposes to prove that $2n = 0$ for every nonnegative integer n. Discover the error.

    "Proof." We use the Strong Principle of Mathematical Induction. Since $2 \cdot 0 = 0$, the statement is true for $n = 0$. Assume, for an arbitrary nonnegative integer k, that $2i = 0$ for every integer i with $0 \leq i \leq k$. We show that $2(k+1) = 0$. Let i and j be integers such that $0 \leq i \leq k$ and $0 \leq j \leq k$ and $i + j = k + 1$. By the induction hypothesis, $2i = 0$ and $2j = 0$. Thus $2(k+1) = 2(i+j) = 2i + 2j = 0 + 0 = 0$.

    By the Strong Principle of Mathematical Induction, $2n = 0$ for every nonnegative integer n. ∎

12. Figure 4.2(i) shows a 2×3 board and Figure 4.2(ii) shows a standard domino. There are three ways that the six squares of the board can be covered by three dominos. This is shown in Figure 4.2(iii). Let s_n denote the number of ways that a $2 \times n$ board can be covered by n dominos. Thus $s_3 = 3$.

    (a) Compute s_1, s_2, s_4 and s_5.

    (b) Make a conjecture for s_n for every positive integer n and prove that this conjecture is correct.

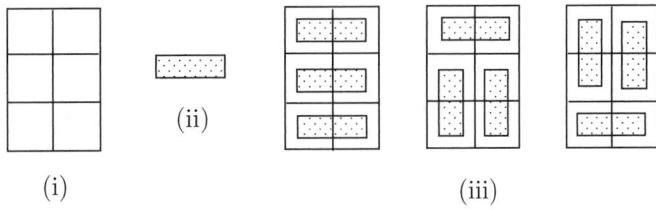

Figure 4.2: Illustrating Exercise 12

Chapter 4 Highlights

Key Concepts

arithmetic sequence: a sequence $\{a_n\}$ in which the difference of every two consecutive terms a_n and a_{n+1} is some constant k, that is, $a_{n+1} - a_n = k$.

base (or **basis step, anchor**)): the step in an induction proof that establishes the truth of a statement for a specific value.

bit: 0 or 1.

bit string: a finite sequence of 0s and 1s.

Fibonacci numbers F_n: $F_1 = 1$, $F_2 = 1$ and $F_n = F_{n-2} + F_{n-1}$ for $n \geq 3$.

geometric sequence: a sequence $\{a_n\}$ in which the ratio of every two consecutive terms a_n and a_{n+1} is some constant r, that is, $a_{n+1}/a_n = r$.

induction proof: a proof (often of statements of the type $\forall n \in \mathbb{N}, P(n)$) using the Principle of Mathematical Induction or the Strong Principle of Mathematical Induction.

inductive hypothesis (or **induction hypothesis**): the hypothesis of the implication in the inductive step of an induction proof.

inductive step: the step in an induction proof that establishes the truth of a certain required statement expressed as an implication.

initial values: some specified values for a recursively defined sequence.

length (of a string): the number of terms in the string.

mathematical induction: a proof technique that is a consequence of the Principle of Mathematical Induction or the Strong Principle of Mathematical Induction.

recurrence relation: a relation that defines a_n ($n \in \mathbb{N}$) in terms of $a_1, a_2, \ldots, a_{n-1}$, where the sequence $\{a_n\}$ is defined recursively.

recursively defined sequence: a sequence defined in terms of initial values and a recurrence relation.

sequence: a listing (finite or infinite) of elements of some set.

string: a finite sequence.

Key Results

- **The Principle of Mathematical Induction** The statement

$$\forall n \in \mathbb{N}, \ P(n): \ \text{for every positive integer } n, \ P(n).$$

  is true if (1) $P(1)$ is true and (2) the statement $\forall k \in \mathbb{N}, P(k) \Rightarrow P(k+1)$ is true.

- **The Principle of Mathematical Induction** For a fixed integer m, let $S = \{i \in \mathbb{Z} : i \geq m\}$. Then the statement

$$\forall n \in S, P(n): \ \text{For every integer } n \geq m, \ P(n).$$

  is true if (1) $P(m)$ is true and (2) the statement $\forall k \in S, P(k) \Rightarrow P(k+1)$ is true.

- **The Strong Principle of Mathematical Induction** The statement

$$\forall n \in \mathbb{N}, \ P(n): \ \text{for every positive integer } n, \ P(n).$$

  is true if (1) $P(1)$ is true and (2) the statement $\forall k \in \mathbb{N}, P(1) \wedge P(2) \wedge \cdots \wedge P(k) \Rightarrow P(k+1)$ is true.

- **The Strong Principle of Mathematical Induction** For a fixed integer m, let $S = \{i \in \mathbb{Z} : i \geq m\}$. Then the statement

$$\forall n \in S, \ P(n): \ \text{For every integer } n \geq m, \ P(n).$$

  is true if (1) $P(m)$ is true and (2) the statement $\forall k \in S, P(m) \wedge P(m+1) \wedge \cdots \wedge P(k) \Rightarrow P(k+1)$ is true.

Supplementary Exercises for Chapter 4

1. Prove that $1 + 3 + 6 + \cdots + \frac{n(n+1)}{2} = \frac{n(n+1)(n+2)}{6}$ for every positive integer n.

2. Find a formula for $1 + 4 + 7 + \cdots + (3n - 2)$ for positive integers n, and then verify your formula by induction.

3. Prove that $1^2 + 3^2 + \cdots + (2n - 1)^2 = \frac{n(2n-1)(2n+1)}{3}$ for every positive integer n.

4. Prove that $1 \cdot 2 \cdot 3 + 2 \cdot 3 \cdot 4 + \cdots + n \cdot (n + 1) \cdot (n + 2) = \frac{n(n+1)(n+2)(n+3)}{4}$ for every positive integer n.

5. Prove that $1(1!) + 2(2!) + \cdots + n(n!) = (n + 1)! - 1$ for every positive integer n.

6. Prove that $2^n > n^2 + n + 1$ for every integer $n \geq 5$.

7. Prove in two ways that $2^{n+1} < 1 + (n + 1)2^n$ for every positive integer n.

8. Prove that $3^n > n^2$ for every positive integer n.

9. Prove that $2^n > n^3$ for every integer $n \geq 10$.

10. Prove that $4^n > n^3$ for every positive integer n.

11. Prove by induction that $n^n > n!$ for every integer $n \geq 2$.

12. Prove that $1 + \frac{1}{4} + \frac{1}{9} + \cdots + \frac{1}{n^2} \leq 2 - \frac{1}{n}$ for every positive integer n.

13. Prove that $\frac{1}{\sqrt[3]{1}} + \frac{1}{\sqrt[3]{2}} + \cdots + \frac{1}{\sqrt[3]{n}} > (n + 1)^{2/3}$ for every integer $n \geq 4$.

14. Use induction to prove that $n^2 + n$ is even for every nonnegative integer n.

15. Use induction to prove that for every real number $x > -1$ and every positive integer n, $(1 + x)^n \geq 1 + nx$.

16. Let a be a real number. Use induction to prove that $\sum_{i=0}^{n}(a + i) = \frac{1}{2}(n + 1)(2a + n)$ for every nonnegative integer n.

17. Let a and b be two real numbers. Use induction to prove that $\sum_{i=0}^{n}(a + ib) = \frac{1}{2}(n + 1)(2a + nb)$ for every nonnegative integer n.

18. A sequence $\{a_n\}$ is defined recursively by $a_1 = 5, a_2 = 7$ and $a_n = 3a_{n-1} - 2a_{n-2} - 2$ for $n \geq 3$. Prove that $a_n = 2n + 3$ for every positive integer n.

19. A sequence $a_0, a_1, a_2, \ldots$ of integers is defined recursively by $a_0 = 0$ and $a_n = 3a_{n-1} + 3^{n-1}$ for $n \geq 1$. Prove that $a_n = n3^{n-1}$ for every nonnegative integer n.

20. A sequence $\{a_n\}$ is defined recursively by $a_1 = 2, a_2 = 4$ and $a_n = a_{n-1} + 2a_{n-1}$ for $n \geq 3$. Prove that $a_n = 2^n$ for every positive integer n.

21. A sequence $\{a_n\}$ is defined recursively by $a_1 = 2, a_2 = 6$ and $a_n = 2a_{n-1} - a_{n-2} + 2$ for $n \geq 3$. Prove that $a_n = n(n + 1)$ for every positive integer n.

22. A sequence $\{a_n\}$ is defined recursively by $a_1 = 2, a_2 = 2$ and $a_n = a_{n-2}a_{n-1}$ for $n \geq 3$.

    (a) Determine a_n for $3 \leq n \leq 6$.

    (b) Conjecture a formula for a_n for $n \geq 1$.

    (c) Prove that the conjecture in (b) is correct.

23. Let $\alpha = \left(1 + \sqrt{5}\right)/2$. Prove that the nth Fibonacci number $F_n > \alpha^{n-2}$ for every integer $n \geq 3$.

24. (a) Show that there do not exist nonnegative integers x and y such that $7 = 3x + 5y$.

    (b) Use induction to prove that for every integer $n \geq 8$, there exist nonnegative integers x and y such that $n = 3x + 5y$.

25. Let A be a nonempty set of real numbers. A number $m \in A$ is a **least element** of A if $x \geq m$ for every $x \in A$. For example, every finite nonempty set of real numbers and $\mathbb{N}$ have a least element, while $\mathbb{Z}$ and the open interval $(0, 1)$ of real numbers do not have a least element. A nonempty set S of real numbers is said to be **well-ordered** if every nonempty subset of S has a least element.

    (a) Show that if S is a nonempty set of real numbers and S does not have a least element, then S is not well-ordered.

    (b) Show that the closed interval $[0, 1]$ of real numbers is not well-ordered (thereby showing that a set with a least element may not be well-ordered).

26. The **Well-Ordering Principle** states: *The set $\mathbb{N}$ of positive integers is well-ordered.* (See Exercise 25.)

    Prove that the Well-Ordering Principle is true if and only if the Principle of Mathematical Induction is true.

27. Let $S = \{1, 2, \ldots, m\}$, where $m \in \mathbb{N}$. The following is called the Principle of Finite Induction.

    The Principle of Finite Induction *Let $P(n)$ be an open sentence over the domain S. Then the statement $\forall n \in S$, $P(n)$ is true if*

    (1) $P(1)$ *is a true statement and*

    (2) $\forall k \in S - \{m\}$, $P(k) \Rightarrow P(k+1)$ *is true.*

    Let $S = \{1, 2, \ldots, 20\}$. The sum of the elements in S is $1 + 2 + \cdots + 20 = 210$. Use the Principle of Finite Induction to prove that for each integer n with $1 \leq n \leq 210$, there exists a subset of S, the sum of whose elements is n.

28. Prove for every positive integer n and the Fibonacci numbers $F_1, F_2, \ldots$ that $F_{n+6} = 4F_{n+3} + F_n$.

29. Prove for every positive integer n that

$$\sqrt{2 + \sqrt{2 + \sqrt{2 + \cdots + \sqrt{2}}}} = 2\cos\frac{\pi}{2^{n+1}},$$

where the number 2 occurs n times in the expression on the left. [Note: A well-known identity from trigonometry states that $\cos 2x = 2\cos^2 x - 1$ for every real number x.]

Chapter 5

Relations and Functions

There will be a number of instances when we will be dealing with two sets A and B (possibly $A = B$) where typically some of the elements of A have a connection with some of the elements of B in some manner. Depending on the requirements of such a connection, we obtain two concepts encountered often in mathematics: relations and functions.

5.1 Relations

In mathematics an object in one set can be related to an object in another set in variety of ways. For example, an integer a can be related to an integer b if $a + b$ is even or if a and b have the same remainder when divided by 3. A straight line ℓ can be related to the real number m if m is the slope of ℓ. Recall that the Cartesian product $A \times B$ of two sets A and B is the set of all ordered pairs (a, b), where $a \in A$ and $b \in B$. That is,

$$A \times B = \{(a, b) : a \in A, b \in B\}.$$

For example, if $A = \{0, 1\}$ and $B = \{1, 2, 3\}$, then

$$A \times B = \{(0, 1), (0, 2), (0, 3), (1, 1), (1, 2), (1, 3)\}.$$

This brings us to the concept of a *relation* in mathematics.

Definition 5.1 *A relation R **from a set** A **to a set** B is a subset of $A \times B$. In addition, R is said to be a relation on $A \times B$. If $(a, b) \in R$, then a is said to be **related** to b; while if $(a, b) \notin R$, then a is not related to b. If $(a, b) \in R$, we also write $a \mathrel{R} b$; while if $(a, b) \notin R$, we write $a \mathrel{\not R} b$.*

Let's look at a few examples of relations.

Example 5.2 *Let $A = \{\sqrt{2}, e, 3, \pi\}$ and $B = \{1, 2, 3, 4\}$. An element $a \in A$ is said to be related to an element $b \in B$ if $|a - b| < 1$. Which elements of A are related to which elements of B?*

Solution. First, recall that $\sqrt{2} \approx 1.414$, $e \approx 2.718$ and $\pi \approx 3.14$. Denote this relation by R. That is, for $a \in A$ and $b \in B$, it follows that $a \mathrel{R} b$ if $|a - b| < 1$. In this case,

$$\sqrt{2} \mathrel{R} 1, \ \sqrt{2} \mathrel{R} 2, \ e \mathrel{R} 2, \ e \mathrel{R} 3, \ 3 \mathrel{R} 3, \ \pi \mathrel{R} 3 \text{ and } \pi \mathrel{R} 4.$$

Observe that if a and b are both integers, then $a \mathrel{R} b$ can only occur if $a = b$. The elements of A related to the elements of B can also be described by writing

$$R = \{(\sqrt{2}, 1), (\sqrt{2}, 2), (e, 2), (e, 3), (3, 3), (\pi, 3), (\pi, 4)\}. \ \blacklozenge$$

Example 5.3

For the sets $A = \{0, 1\}$ and $B = \{1, 2, 3\}$, suppose that

$$R = \{(0, 2), (0, 3), (1, 2)\}$$

is a relation from A to B. Therefore, $0 \; R \; 2$, $0 \; R \; 3$ and $1 \; R \; 2$. Since 0 is not related to 1 and 1 is not related to 3, we can also indicate this by writing $0 \; R\!\!\!/ \; 1$ and $1 \; R\!\!\!/ \; 3$.   ♦

Example 5.4

Let $\mathbb{N}$ be the set of natural numbers (positive integers) and let $\mathbb{N}^-$ denote the set of negative integers. A relation R from $\mathbb{N}$ to $\mathbb{N}^-$ is defined by $a \; R \; b$ if $a + b \in \mathbb{N}$. Give examples of some pairs of elements that are related by R and some that are not.

Solution. First, $5 \; R \; (-2)$ since $5 + (-2) = 3 \in \mathbb{N}$. Also, $5 \; R \; (-4)$ since $5 + (-4) = 1 \in \mathbb{N}$. On the other hand, because $2 + (-2) = 0 \notin \mathbb{N}$ and $3 + (-5) = -2 \notin \mathbb{N}$, it follows that $2 \; R\!\!\!/ \; (-2)$ and $3 \; R\!\!\!/ \; (-5)$.   ♦

If A and B are finite sets and R is a relation from A to B, then R can be represented by a diagram.

Example 5.5

For the sets $A = \{x, y, z\}$ and $B = \{0, 1, 2, 3\}$,

$$R = \{(x, 1), (x, 2), (y, 1), (z, 0), (z, 3)\}$$

is a relation from A to B. The relation R can also be described by the diagram in Figure 5.1.   ♦

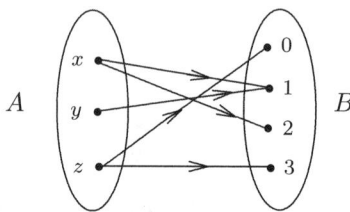

Figure 5.1: Representation of the relation in Example 5.5

One type of a relation that we will encounter often is defined from some set to the same set.

Definition 5.6 *A **relation** R **on a set** S is a relation from S to S. That is, R is a relation on a set S if R is a subset of $S \times S$.*

We have seen that if a set A has n elements, then there are 2^n subsets of A. Thus if a set S has three elements, for example, then $S \times S$ has 9 elements and there are $2^9 = 512$ different subsets of $S \times S$ and so there are 512 possible relations on S.

Example 5.7

Listed below are five (of the 512 possible) relations on the set $S = \{x, y, z\}$.

$$R_1 = \{(x, y), (x, z), (z, z)\}$$

$$R_2 = \{(x, x), (x, y), (y, z)\}$$

$R_3 = \{(x,x),(y,y),(z,z)\}$

$R_4 = \emptyset$ (the **empty relation**)

$R_5 = S \times S = \{(x,x),(x,y),(x,z),(y,x),(y,y),(y,z),(z,x),(z,y),(z,z)\}.$ ♦

There are several properties that a relation defined on a set can have that will be of particular interest to us. We now consider three of the best known properties.

Definition 5.8 *Let R be a relation defined on a nonempty set S. Then R is*

(1) **reflexive** *if $a\,R\,a$ for all $a \in S$; that is, if $a \in S$, then $(a,a) \in R$;*

(2) **symmetric** *if whenever $a\,R\,b$, then $b\,R\,a$ for all $a,b \in S$; that is, if $(a,b) \in R$, then $(b,a) \in R$;*

(3) **transitive** *if whenever $a\,R\,b$ and $b\,R\,c$, then $a\,R\,c$ for all $a,b,c \in S$; that is, if $(a,b) \in R$ and $(b,c) \in R$, then $(a,c) \in R$.*

In the definition of a transitive relation, it is not necessary that a,b and c be distinct. In particular, if $c = a$, then the definition says if $(a,b) \in R$ and $(b,a) \in R$, then (a,a) must be in R. Furthermore, if R is a transitive relation and $(a,b) \in R$ and $(b,a) \in R$, then $(b,a) \in R$ and $(a,b) \in R$ and so the ordered pair (b,b) must be in R as well.

A relation R defined on a set S is therefore

(1') **not reflexive** if $(x,x) \notin R$ for *some* $x \in S$;

(2') **not symmetric** if $(x,y) \in R$ but $(y,x) \notin R$ for *some* pair x,y of distinct elements of S;

(3') **not transitive** if $(x,y) \in R$ and $(y,z) \in R$ but $(x,z) \notin R$ for *some* $x,y,z \in S$.

Before we see some examples of relations that have some, none or all of the properties (1)–(3), there is one additional fact that is useful for us to recall. For a false statement P and any statement Q, the implication $P \Rightarrow Q$ is true. Consequently, if we have the empty relation R defined on a nonempty set S, then R is symmetric since for all $a,b \in S$ the statement "$(a,b) \in R$" is false. Thus, the implication "if $(a,b) \in R$, then $(b,a) \in R$" is true for all $a,b \in S$. By the same reasoning, if R is the empty relation on S, then R is transitive. Indeed, if for all $a,b,c \in S$ the statement "$(a,b) \in R$ and $(b,c) \in R$" is false, then R is transitive as the implication "if $(a,b) \in R$ and $(b,c) \in R$, then $(a,c) \in R$" is true for all $a,b,c \in S$. However, the empty relation on a nonempty set S is not reflexive since $(a,a) \notin R$ for all $a \in S$.

Example 5.9

Of the five relations in Example 5.7 only R_3 and R_5 are reflexive. All of the relations R_3, R_4 and R_5 are symmetric. The relation R_1 is not symmetric since, for example, $(x,y) \in R_1$ but $(y,x) \notin R_1$. All of the relations R_1, R_3, R_4 and R_5 are transitive. Since $(x,y) \in R_2$ and $(y,z) \in R_2$ but $(x,z) \notin R_2$, the relation R_2 is not transitive. Let's see why the relation R_1 *is* transitive. In order for R_1 to be transitive, the statement

$$\text{If } (a,b) \in R_1 \text{ and } (b,c) \in R_1, \text{ then } (a,c) \in R_1. \tag{5.1}$$

must be true for all $a,b,c \in S$. We consider all choices of $a,b,c \in S$ for which $(a,b) \in R_1$ and $(b,c) \in R_1$. There are three ordered pairs $(a,b) \in R_1$, namely, (x,y), (x,z) and (z,z). Suppose first that we consider $(a,b) = (x,y)$. Then $b = y$. However then, there is no ordered pair $(b,c) \in R_1$. If $(a,b) = (x,z)$, then $b = z$. The only ordered pair $(b,c) \in R_1$ where $b = z$ is $(b,c) = (z,z) \in R_1$. That is, $(x,z) \in R_1$ and $(z,z) \in R_1$ (so $a = x$ and $b = c = z$). Since $(a,c) = (x,z) \in R_1$, the implication in (5.1) is true when $(a,b) = (x,z)$ and $(b,c) = (z,z)$. The only other ordered pair $(a,b) \in R_1$ is $(a,b) = (z,z)$. In this case, the only ordered pair $(b,c) \in R_1$ is $(b,c) = (z,z)$; so $a = b = c = z$. But then $(a,c) = (z,z) \in R_1$, so (5.1) is true when $(a,b) = (z,z)$ and $(b,c) = (z,z)$. Consequently, the implication in (5.1) is true for all $a, b, c \in S$ and so R_1 is transitive. ◆

We now consider a number of other relations defined on a set and determine which of the properties reflexive, symmetric and transitive they may possess.

Example 5.10

Let $S = \{1, 2, 3, 4\}$. The relation

$$R = \{(1,1), (1,2), (1,4), (2,1), (2,2), (2,3), (3,1), (3,2), (3,3), (4,1)\}$$

is defined on the set S. Which of the properties reflexive, symmetric and transitive does R possess?

Solution. The relation R is not reflexive since $(4,4) \notin R$. The relation R is not symmetric since $(3,1) \in R$ but $(1,3) \notin R$. Furthermore, the relation R is not transitive since, for example, $(1,2) \in R$ and $(2,3) \in R$ but $(1,3) \notin R$. Also, $(4,1) \in R$ and $(1,4) \in R$ but $(4,4) \notin R$. ◆

Example 5.11

A relation R is defined on the set $\mathbb{N}$ of positive integers by $a \ R \ b$ if $a < b$. Which of the properties reflexive, symmetric and transitive does R possess?

Solution. Since $a < a$ is not true for any positive integer a, it follows that $a \not R a$ for every $a \in \mathbb{N}$ and the relation R is not reflexive. Suppose that a and b are positive integers such that $a \ R \ b$. Then $a < b$. Certainly, $b < a$ is not true; so $b \not R a$ and R is not symmetric. On the other hand, if $a, b, c \in \mathbb{N}$ such that $a \ R \ b$ and $b \ R \ c$, then $a < b$ and $b < c$. Thus $a < c$ and so $a \ R \ c$. Therefore, R is transitive. ◆

Example 5.12

A relation R is defined on the set $\mathbb{Z}$ of integers by $a \ R \ b$ if $ab \geq 0$. Which of the properties reflexive, symmetric and transitive does R possess?

Solution. Let $a \in \mathbb{Z}$. Since $a \cdot a = a^2 \geq 0$, it follows that $a \ R \ a$ for all $a \in \mathbb{Z}$ and so R is reflexive. Suppose that $a, b \in \mathbb{Z}$ such that $a \ R \ b$. Then $ab \geq 0$. Since $ba = ab$, it follows that $ba \geq 0$ and so $b \ R \ a$. Therefore, R is symmetric.

Since $2 \cdot 0 \geq 0$ and $0 \cdot (-3) \geq 0$, it follows that $2 \ R \ 0$ and $0 \ R \ (-3)$. However $2 \not R (-3)$ since $2 \cdot (-3) = -6 < 0$. Therefore, R is not transitive. ◆

Next, we look at an example of a relation having a geometric flavor.

Example 5.13

A relation R is defined on the set $\mathbb{R}$ of real numbers by $x \ R \ y$ if $2x + y \geq 0$. (That is, $x \ R \ y$ if (x, y) is a point in the Euclidean plane that lies on or to the right of the line $y = -2x$. See Figure 5.2.)

(a) Give an example of two real numbers a and b such that $a \ R \ b$ and two real numbers c and d such that $c \not R d$.

(b) Is R reflexive?

(c) Is R symmetric?

(d) Is R transitive?

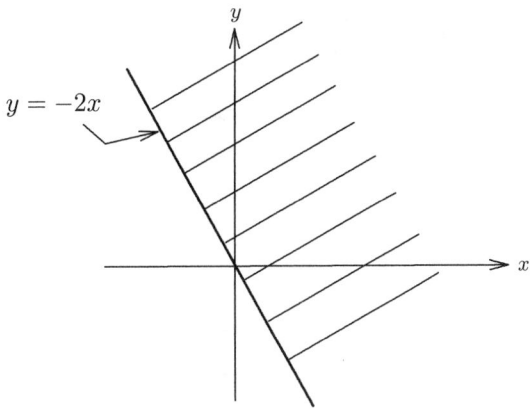

Figure 5.2: The line $y = -2x$ in the Euclidean plane

Solution.

(a) Since $2 \cdot 3 + 1 = 7 \geq 0$, it follows that $3 \; R \; 1$. Also, since $2 \cdot (-3) + (-1) = -7 < 0$, it follows that $(-3) \; \not{R} \; (-1)$.

(b) Since $2(-1) + (-1) = -3 < 0$, it follows that $(-1) \; \not{R} \; (-1)$ and so R is not reflexive.

(c) Since $2 \cdot 3 + (-6) = 0$, it follows that $3 \; R \; (-6)$. However, $2(-6) + 3 = -9 < 0$ and so $(-6) \; \not{R} \; 3$. Therefore, R is not symmetric.

(d) Since $2(-1) + 2 \geq 0$, it follows that $(-1) \; R \; 2$. Also, since $2 \cdot 2 + (-4) \geq 0$, it follows that $2 \; R \; (-4)$. However, $(-1) \; \not{R} \; (-4)$ because $2(-1) + (-4) = -6 < 0$. Therefore, R is not transitive. ◆

Exercises for Section 5.1

1. Let $A = \{1, 2, 3\}$ and $B = \{4, 5, 6\}$. Give an example of a relation R from A to B such that 1 is related to 5 but 2 is not related to 4.

2. Let $A = \{w, x, y\}$ and $B = \{x, y, z\}$.

   (a) Determine $A \times B$.

   (b) Give an example of a relation R from A to B such that $|R| = 4$.

   (c) Give an example of a relation R on A such that $|R| = 5$.

3. For $A = \{a, b, c, d\}$ and $B = \{x, y, z\}$, the set $R = \{(a, y), (a, z), (b, y), (c, x), (c, z)\}$ of ordered pairs is a relation from A to B. Describe this relation by a diagram.

4. For $A = \{1, 2, 3\}$ and $B = \{2, 3, 4, 5\}$, a relation R from A to B is described by the diagram in Figure 5.3. Express R as a set of ordered pairs.

5. Let $A = \{1, 2, 4, 5\}$ and $B = \{2, 3, 5\}$. A relation R from A to B is defined by $a \; R \; b$ if $a + b$ is a prime. Express R as a set of ordered pairs.

6. Let $S = \{1, 2, \ldots, 9\}$. A relation R is defined on S by $a \; R \; b$ if there exists a prime number with one or more digits whose first digit is a and whose last digit is b. Give examples of six pairs $a, b \in S$ for which $a \; R \; b$ and six pairs $a, b \in S$ for which $a \; \not{R} \; b$.

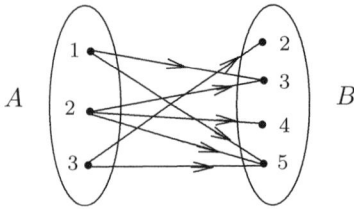

Figure 5.3: The relation R in Exercise 4

7. The following are relations on the set $S = \{1, 2, 3, 4\}$. Which of the properties reflexive, symmetric and transitive does each relation possess?

   (a) $R_1 = \{(1,2), (2,1), (3,4), (4,3)\}$.   (b) $R_2 = \{(1,2), (1,3), (1,4)\}$.

   (c) $R_3 = \{(1,1), (1,2), (2,2), (2,3), (3,3), (3,4), (4,4), (4,1)\}$.

   (d) $R_4 = \{(1,2), (3,4)\}$.   (e) $R_5 = \{(1,1), (2,2), (3,3), (3,1), (4,4)\}$.

8. Let S be a nonempty set and let $R = S \times S$. Which of the properties reflexive, symmetric and transitive does R possess?

9. A relation R is defined on the set $\mathbb{R}$ of real numbers by $x \, R \, y$ if $x - y \geq 0$, that is, $x \, R \, y$ if the point (x, y) in the plane lies on or to the right of the line $y = x$.

   (a) Give an example of two real numbers a and b such that $a \, R \, b$ and two real numbers c and d such that $c \, \cancel{R} \, d$.

   (b) Which of the properties reflexive, symmetric and transitive does R possess?

10. The following are relations on the set $\mathbb{R}$ of real numbers. Which of the properties reflexive, symmetric and transitive does each relation below possess?

    (a) $x \, R_1 \, y$ if $|x - y| \leq 1$.   (b) $x \, R_2 \, y$ if $y \leq 2x + 1$.

    (c) $x \, R_3 \, y$ if $y = x^2$.   (d) $x \, R_4 \, y$ if $x^2 + y^2 = 9$.

11. For the set $S = \{1, 2, 3, 4\}$, let $\mathcal{P}(S)$ be its power set. Which of the properties reflexive, symmetric and transitive does each relation below on $\mathcal{P}(S)$ possess?

    (a) $X \, R_1 \, Y$ if $X \subseteq Y$.   (b) $X \, R_2 \, Y$ if $X \cap Y = \emptyset$.

    (c) $X \, R_3 \, Y$ if $|X| \leq |Y|$.   (d) $X \, R_4 \, Y$ if $|X \cap Y| = 1$.

5.2   Equivalence Relations

Probably the best known relation on the set of integers is the equals relation, that is, the relation R defined on $\mathbb{Z}$ by $a \, R \, b$ if $a = b$. Certainly (1) $a = a$ for every $a \in \mathbb{Z}$, (2) if $a = b$, then $b = a$ for all $a, b \in \mathbb{Z}$ and (3) if $a = b$ and $b = c$, then $a = c$ for all $a, b, c \in \mathbb{Z}$. In other words, the equals relation is reflexive, symmetric and transitive. Many of the relations that we encounter in mathematics and computer science mimic the equals relation in that they are also reflexive, symmetric and transitive. Relations possessing all three of these properties are called equivalence relations.

Definition 5.14 *A relation R on a nonempty set is an* **equivalence relation** *if R is reflexive, symmetric and transitive.*

Consequently, the equals relation defined on the set of integers (or on any nonempty subset of real numbers) is an equivalence relation. We now consider some other examples of equivalence relations.

Example 5.15

A relation R is defined on the set $\mathcal{L}$ of straight lines in the Euclidean plane by $\ell_1 \, R \, \ell_2$ if two lines ℓ_1 and ℓ_2 coincide or are parallel. (See Figure 5.4.) Explain why R is an equivalence relation.

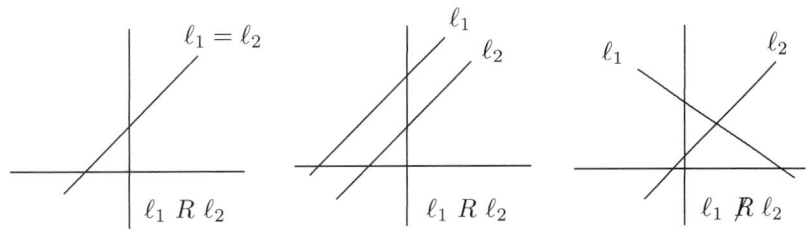

Figure 5.4: The equivalence relation R in Example 5.15

Solution. Certainly every line coincides with itself. Thus $\ell \, R \, \ell$ for every line ℓ and R is therefore reflexive. Suppose that $\ell_1 \, R \, \ell_2$ for lines ℓ_1 and ℓ_2. Then ℓ_1 and ℓ_2 coincide or ℓ_1 and ℓ_2 are parallel. Hence ℓ_2 and ℓ_1 coincide or ℓ_2 and ℓ_1 are parallel. Thus $\ell_2 \, R \, \ell_1$ and R is symmetric.

If $\ell_1 \, R \, \ell_2$ and $\ell_2 \, R \, \ell_3$ for lines ℓ_1, ℓ_2 and ℓ_3, then (1) ℓ_1 and ℓ_2 coincide or ℓ_1 and ℓ_2 are parallel and (2) ℓ_2 and ℓ_3 coincide or ℓ_2 and ℓ_3 are parallel. Thus ℓ_1 and ℓ_3 coincide or ℓ_1 and ℓ_3 are parallel. Hence $\ell_1 \, R \, \ell_3$ and R is transitive. Therefore, R is an equivalence relation on $\mathcal{L}$. ◆

Example 5.16 *A relation R is defined on $\mathbb{N} \times \mathbb{N}$ by $(a, b) \, R \, (c, d)$ if $ad = bc$. Show that R is an equivalence relation.*

Proof. Let $(a, b) \in \mathbb{N} \times \mathbb{N}$. Since $ab = ba$, it follows that $(a, b) \, R \, (a, b)$ and so R is reflexive. Let $(a, b), (c, d) \in \mathbb{N} \times \mathbb{N}$ such that $(a, b) \, R \, (c, d)$. Then $ad = bc$. Thus $cb = da$, which implies that $(c, d) \, R \, (a, b)$ and so R is symmetric.

Let $(a, b), (c, d), (e, f) \in \mathbb{N} \times \mathbb{N}$ such that $(a, b) \, R \, (c, d)$ and $(c, d) \, R \, (e, f)$. Hence $ad = bc$ and $cf = de$. We show that $(a, b) \, R \, (e, f)$. Since $ad = bc$ and $cf = de$, it follows that $a = bc/d$ and $f = de/c$. Hence

$$af = \left(\frac{bc}{d}\right)\left(\frac{de}{c}\right) = \frac{bcde}{dc} = be,$$

which implies that $(a, b) \, R \, (e, f)$ and so R is transitive. Therefore, R is an equivalence relation. ◆

Proof Analysis. The transitive property in Example 5.16 could also be verified by observing that

$$adf = (ad)f = (bc)f = b(cf) = b(de) = bde.$$

Thus $adf = bde$. Dividing by d, we obtain $af = be$. ◆

If we think of the ordered pair (a, b) in Example 5.16 as representing the positive rational number $\frac{a}{b}$, then two fractions $\frac{a}{b}$ and $\frac{c}{d}$ are related if $ad = bc$, that is, if $\frac{a}{b} = \frac{c}{d}$. Hence $\frac{a}{b}$ and $\frac{c}{d}$ are related if and only if these numbers are equal. Not surprisingly then, this relation is an equivalence relation.

Equivalence Classes

If R is an equivalence relation that is defined on a set A, then with each element of A, there is an associated subset of A that will be of special interest to us.

Definition 5.17 *Let R be an equivalence relation on a set A. For $a \in A$, the **equivalence class** $[a]$ is defined by*

$$[a] = \{x \in A : \ x \, R \, a\}.$$

Hence, for an equivalence relation R on a set A and an element $a \in A$, the equivalence class $[a]$ is the set of all elements of A that are related to a by R. So we can think of $[a]$ as the set consisting of all of the relatives of a. Since every equivalence relation is reflexive, $a \, R \, a$ and so $a \in [a]$. Therefore, $[a] \neq \emptyset$ for every element $a \in A$. Let's look at some examples of equivalence classes.

Example 5.18

Let $S = \{1, 2, 3, 4, 5, 6\}$. The relation

$$
\begin{aligned}
R \;=\; & \{(1,1),(2,2),(3,3),(4,4),(5,5),(6,6),(1,4), \\
& (2,3),(2,6),(3,2),(3,6),(4,1),(6,2),(6,3)\}
\end{aligned}
$$

on S is an equivalence relation. Therefore, with each element of S, there is an associated equivalence class. Since $(1,1),(4,1) \in R$ but $(2,1),(3,1),(5,1),(6,1) \notin R$, it follows that $[1] = \{1,4\}$. In this case, the equivalence classes are

$$[1] = \{1,4\}, \; [2] = \{2,3,6\}, \; [3] = \{2,3,6\},$$
$$[4] = \{1,4\}, \; [5] = \{5\} \text{ and } [6] = \{2,3,6\}.$$

Notice here that $[1] = [4]$ and $[2] = [3] = [6]$.   ♦

Example 5.19

In Example 5.15, a relation R is defined on the set $\mathcal{L}$ of straight lines in the Euclidean plane by $\ell_1 \, R \, \ell_2$ if ℓ_1 and ℓ_2 coincide or ℓ_1 and ℓ_2 are parallel, which was shown to be an equivalence relation. For a straight line ℓ, the equivalence class $[\ell]$ consists of ℓ and all straight lines parallel to ℓ. For example, if ℓ is the straight line with equation $y = 2x + 1$, then $[\ell]$ consists of all lines with slope 2; while if ℓ' is the straight line with equation $x = 3$, then $[\ell']$ consists of all vertical lines.   ♦

Example 5.20

In Example 5.16, a relation R is defined on the Cartesian product $\mathbb{N} \times \mathbb{N}$ such that $(a,b) \, R \, (c,d)$ if $ad = bc$. Since R is an equivalence relation, there is an equivalence class associated with each element of $\mathbb{N} \times \mathbb{N}$. For example,

$$
\begin{aligned}
[(1,2)] \;&=\; \{(x,y) \in \mathbb{N} \times \mathbb{N} : \; (x,y) \, R \, (1,2)\} \\
&=\; \{(x,y) \in \mathbb{N} \times \mathbb{N} : \; 2x = y\} \\
&=\; \{(1,2),(2,4),(3,6),\ldots\}.
\end{aligned}
$$

For positive integers a, b, c and d, we have noted that

$$ad = bc \text{ if and only if } \frac{a}{b} = \frac{c}{d}.$$

Again, if we think of the ordered pair (a,b) as representing the positive rational number $\frac{a}{b}$, then the equivalence class $[(a,b)]$ is the set of all ordered pairs (c,d) such that $\frac{a}{b} = \frac{c}{d}$.   ♦

Result to Prove: A relation R is defined on $\mathbb{Z}$ by $a \, R \, b$ if $a + b$ is even. Then R is an equivalence relation.

Proof Strategy. For the reflexive property, we are required to show that if $a \in \mathbb{Z}$, then $a \, R \, a$. Since $a \, R \, a$ if $a + a$ is even, we need to show that if $a \in \mathbb{Z}$, then $a + a = 2a$ is even. This is obvious. For the symmetric property, we must show that if $a \, R \, b$, where $a, b \in \mathbb{Z}$, then $b \, R \, a$. Suppose that $a \, R \, b$. Then $a + b$ is even. However, $b + a = a + b$. Thus $b + a$ is even and so $b \, R \, a$.

For the transitive property, we are required to show that if $a \, R \, b$ and $b \, R \, c$, where $a, b, c \in \mathbb{Z}$, then $a \, R \, c$. Now $a \, R \, c$ if $a + c$ is even. If we use a direct proof, then we begin by assuming that $a \, R \, b$ and $b \, R \, c$, with the goal to show that $a + c$ is even. Because $a \, R \, b$ and $b \, R \, c$, it follows that both $a + b$ and $b + c$ are even. One way to work $a + c$ into the discussion is to add $a + b$ and $b + c$. Since $a + b$ and $b + c$ are even, $a + b = 2x$ and $b + c = 2y$ for some integers x and y. Adding these, we get $(a + b) + (b + c) = 2x + 2y$ and so $a + c = 2x + 2y - 2b$. The remainder of the proof is straightforward.   ♦

Example 5.21 *A relation R is defined on $\mathbb{Z}$ by $a\ R\ b$ if $a + b$ is even.*

(a) *Show that R is an equivalence relation.*

(b) *Describe the equivalence classes $[0]$, $[1]$, $[-3]$ and $[4]$.*

Solution.

(a) **Proof.** Let $a \in \mathbb{Z}$. Then $a + a = 2a$ is an even integer and so $a\ R\ a$. Thus R is reflexive. Assume next that $a\ R\ b$, where $a, b \in \mathbb{Z}$. Then $a + b$ is even. Since $b + a = a + b$, it follows that $b + a$ is even. Therefore, $b\ R\ a$ and R is symmetric.

Finally, assume that $a\ R\ b$ and $b\ R\ c$, where $a, b, c \in \mathbb{Z}$. Hence $a + b$ and $b + c$ are both even and so $a + b = 2x$ and $b + c = 2y$ for some integers x and y. Adding these two equations, we obtain

$$(a + b) + (b + c) = 2x + 2y,$$

which implies that

$$a + c = 2x + 2y - 2b = 2(x + y - b).$$

Since $x + y - b$ is an integer, $a + c$ is even. Therefore, $a\ R\ c$ and R is transitive. ∎

(b) The equivalence classes are

$$\begin{aligned}
[0] &= \{x \in \mathbb{Z} :\ x\ R\ 0\} = \{x \in \mathbb{Z} :\ x + 0 \text{ is even}\} \\
&= \{x \in \mathbb{Z} :\ x \text{ is even}\} = \{\ldots, -4, -2, 0, 2, 4, \ldots\} \\
[1] &= \{x \in \mathbb{Z} :\ x\ R\ 1\} = \{x \in \mathbb{Z} :\ x + 1 \text{ is even}\} \\
&= \{x \in \mathbb{Z} :\ x \text{ is odd}\} = \{\ldots, -5, -3, -1, 1, 3, 5, \ldots\} \\
[-3] &= \{x \in \mathbb{Z} :\ x\ R\ (-3)\} = \{x \in \mathbb{Z} :\ x - 3 \text{ is even}\} \\
&= \{x \in \mathbb{Z} :\ x \text{ is odd}\} = \{\ldots, -5, -3, -1, 1, 3, 5, \ldots\} \\
[4] &= \{x \in \mathbb{Z} :\ x\ R\ 4\} = \{x \in \mathbb{Z} :\ x + 4 \text{ is even}\} \\
&= \{x \in \mathbb{Z} :\ x \text{ is even}\} = \{\ldots, -4, -2, 0, 2, 4, \ldots\}.
\end{aligned}$$

Consequently, in this case $[0] = [4]$ and $[1] = [-3]$. ◆

If the equivalence classes determined in the preceding examples are examined, then it might be observed that in each example every two equivalence classes are either equal or disjoint. It turns out that this is true in general. In order to establish this fact, we begin by presenting a necessary and sufficient condition for two equivalence classes to be equal.

Before we verify the next theorem, however, let's recall that two sets C and D can be shown to be equal if we can establish that both $C \subseteq D$ and $D \subseteq C$. Recall also that to verify that $C \subseteq D$, for example, we can select an arbitrary element $x \in C$ and show that $x \in D$ as well.

Theorem 5.22 *Let R be an equivalence relation on a nonempty set A and let a and b be elements of A. Then*

$$[a] = [b] \text{ if and only if } a\ R\ b.$$

Proof. Assume first that $a\ R\ b$. We verify that $[a] = [b]$. To establish that the sets $[a]$ and $[b]$ are equal, we show that $[a] \subseteq [b]$ and $[b] \subseteq [a]$, beginning with the first of these. Let $x \in [a]$. Then $x\ R\ a$. Since $a\ R\ b$, it follows by the transitive property that $x\ R\ b$. Thus $x \in [b]$ and so $[a] \subseteq [b]$. Next, we show that $[b] \subseteq [a]$. Let $y \in [b]$. Then $y\ R\ b$. Since $a\ R\ b$ and R is symmetric, it follows that $b\ R\ a$. Because $y\ R\ b$ and $b\ R\ a$, it follows by the transitive property that $y\ R\ a$. Therefore $y \in [a]$ and so $[b] \subseteq [a]$. Therefore, $[a] = [b]$.

To establish the converse, we begin by assuming that $[a] = [b]$. We show that $a\ R\ b$. Since $[a] = [b]$ and $a \in [a]$, it follows that $a \in [b]$. Hence $a\ R\ b$. ∎

Recall that a partition $\mathcal{P}$ of a nonempty set A is a collection of nonempty subsets of A such that every element of A belongs to exactly one of these subsets. We now show that the set of distinct equivalence classes of a nonempty set resulting from an equivalence relation on the set always produces a partition of the set.

Theorem 5.23   *Let R be an equivalence relation defined on a nonempty set A. If $\mathcal{P}$ is the set of all distinct equivalence classes of A resulting from R, then $\mathcal{P}$ is a partition of A.*

Proof. We already observed that each equivalence class is nonempty and every element of A belongs to at least one equivalence class. It remains only to show that every element of A belongs to exactly one equivalence class in $\mathcal{P}$. Assume, to the contrary, that x belongs to two distinct equivalence classes in $\mathcal{P}$, say $[a]$ and $[b]$. Since $x \in [a]$ and $x \in [b]$, it follows that $x \mathrel{R} a$ and $x \mathrel{R} b$. Because R is symmetric, $a \mathrel{R} x$. Thus $a \mathrel{R} x$ and $x \mathrel{R} b$. Since R is transitive, $a \mathrel{R} b$. By Theorem 5.22, it follows that $[a] = [b]$, which is a contradiction. Thus, as claimed, $\mathcal{P}$ is a partition of A.   ∎

As a consequence of Theorem 5.23, we have the following corollary.

Corollary 5.24   *Let R be an equivalence relation defined on a nonempty set A. If $[a]$ and $[b]$ are equivalence classes of A resulting from R, then either $[a] = [b]$ or $[a] \cap [b] = \emptyset$.*

From what we saw in Theorem 5.22, we can expand on Corollary 5.24 a bit more. In particular, if $a \mathrel{R} b$, then $[a] = [b]$; while if $a \not\mathrel{R} b$, then $[a] \cap [b] = \emptyset$. Also, if $[a] \cap [b] \neq \emptyset$, then $[a] = [b]$.

We now consider two additional examples of equivalence relations and the equivalence classes resulting from them.

Result to Prove: A relation R is defined on $\mathbb{Z}$ by $a \mathrel{R} b$ if $3a - 7b$ is even. Then R is an equivalence relation.

Proof Strategy.   The first step is to show that R is reflexive. In order to do this, we must show that every integer is related to itself. So let $x \in \mathbb{Z}$. According to the definition of R, $x \mathrel{R} x$ if $3x - 7x$ is even. However, $3x - 7x = -4x = 2(-2x)$ is even.

To show that R is symmetric, we need to show that whenever $x \mathrel{R} y$, we also have $y \mathrel{R} x$. Assume that $x \mathrel{R} y$. Then $3x - 7y$ is even. We need to show that $3y - 7x$ is even. Since $3x - 7y$ is even, $3x - 7y = 2k$ for some integer k. There are several ways to proceed. One way is to begin with $3y - 7x$ and write this as $3y - 3x - 4x$. Since $3x = 7y + 2k$, we can make this substitution in $3y - 3x - 4x$. Simplifying the algebra gives us the desired result.

To show that R is transitive, we are required to show that if $x \mathrel{R} y$ and $y \mathrel{R} z$, then $x \mathrel{R} z$. Assume that $x \mathrel{R} y$ and $y \mathrel{R} z$. So $3x - 7y$ and $3y - 7z$ are even. Our goal is to show that $3x - 7z$ is even. Since $3x - 7y$ and $3y - 7z$ are even, $3x - 7y = 2k$ and $3y - 7z = 2\ell$ for some integers k and ℓ. Notice that if we add these two equations, the expression $3x - 7z$ appears in the sum.   ◆

Example 5.25

A relation R is defined on $\mathbb{Z}$ by $a \mathrel{R} b$ if $3a - 7b$ is even.

(a) Prove that R is an equivalence relation.

(b) Describe the distinct equivalence classes resulting from R and show that the set of all distinct equivalence classes produces a partition of $\mathbb{Z}$.

Solution.

(a) **Proof.** Let $x \in \mathbb{Z}$. Since $3x - 7x = -4x = 2(-2x)$ and $-2x$ is an integer, $3x - 7x$ is even. Thus $x \mathrel{R} x$ and R is reflexive.

Next, we show that R is symmetric. Let $x \mathrel{R} y$, where $x, y \in \mathbb{Z}$. Thus $3x - 7y$ is even and so $3x - 7y = 2k$ for some integer k. Therefore, $3x = 7y + 2k$. Observe that

$$
\begin{aligned}
3y - 7x &= 3y - 3x - 4x = 3y - (7y + 2k) - 4x \\
&= -4y - 2k - 4x = 2(-2y - k - 2x).
\end{aligned}
$$

Since $-2y - k - 2x$ is an integer, $3y - 7x$ is even. So $y \mathrel{R} x$ and R is symmetric.

Finally, we show that R is transitive. Assume that $x \mathrel{R} y$ and $y \mathrel{R} z$, where $x, y, z \in \mathbf{Z}$. Then $3x - 7y$ and $3y - 7z$ are even. So $3x - 7y = 2k$ and $3y - 7z = 2\ell$, where $k, \ell \in \mathbf{Z}$. Adding these two equations, we obtain

$$(3x - 7y) + (3y - 7z) = 3x - 4y - 7z = 2k + 2\ell$$

and so

$$3x - 7z = 2k + 2\ell + 4y = 2(k + \ell + 2y).$$

Since $k + \ell + 2y$ is an integer, $3x - 7z$ is even. Therefore, $x \mathrel{R} z$ and R is transitive. ∎

(b) We begin with the element 0 of $\mathbb{Z}$. The equivalence class $[0]$ is

$$
\begin{aligned}
[0] &= \{x \in \mathbb{Z} : x \mathrel{R} 0\} = \{x \in \mathbb{Z} : 3x \text{ is even}\} \\
&= \{x \in \mathbb{Z} : x \text{ is even}\} = \{\ldots, -4, -2, 0, 2, 4, \ldots\}.
\end{aligned}
$$

Thus, $[0]$ is the set of all even integers. To find another equivalence class, we consider an integer not in $[0]$, say 1. Then

$$
\begin{aligned}
[1] &= \{x \in \mathbb{Z} : x \mathrel{R} 1\} = \{x \in \mathbb{Z} : 3x - 7 \cdot 1 \text{ is even}\} \\
&= \{x \in \mathbb{Z} : 3x \text{ is odd}\} = \{x \in \mathbb{Z} : x \text{ is odd}\} \\
&= \{\ldots, -5, -3, -1, 1, 3, 5, \ldots\}.
\end{aligned}
$$

Observe that $\{[0], [1]\}$ is a partition of $\mathbb{Z}$, which implies that these are the only two distinct equivalence classes in this case. ◆

Exercises for Section 5.2

1. A relation R is defined on $\mathbb{N} \times \mathbb{N}$ by $(a, b) \mathrel{R} (c, d)$ if $a + d = b + c$.

   (a) Show that R is an equivalence relation.

   (b) Describe the equivalence classes $[(3, 1)]$, $[(5, 5)]$, and $[(4, 7)]$.

2. Let $S = \{1, 2, 3, 4, 5, 6, 7\}$. The relation

   $$
   \begin{aligned}
   R = \ &\{(1, 1), (1, 3), (1, 4), (2, 2), (3, 1), (3, 3), (3, 4), (4, 1), \\
   &(4, 3), (4, 4), (5, 5), (5, 7), (6, 6), (7, 5), (7, 7)\}
   \end{aligned}
   $$

   on S is an equivalence relation. Determine the distinct equivalence classes.

3. Let R be an equivalence relation on the set $S = \{a, b, c, d, e, f\}$. If the distinct equivalence classes are $\{a, d\}$, $\{b, f\}$ and $\{c, e\}$, what is R?

4. Let R be an equivalence relation on the set $S = \{u, v, w, x, y, z\}$ having the following properties: (a) $u \in [x] \cap [y]$, (b) $v \notin [x] \cap [y]$, (c) $z \in [v]$ and (d) $w \notin [u] \cup [z]$. What is R?

5. An equivalence relation R on the set $S = \{1, 2, 3, 4, 5, 6\}$ results in three distinct equivalence classes. Given that (a) $3 \in [4] \cap [5]$, (b) $[2] \cap [6] = \emptyset$ and (c) $1 \in [3]$, what is R?

6. Let $\mathbb{R}^*$ denote the set of nonzero real numbers. Define a relation R on $\mathbb{R}^*$ by $a \mathrel{R} b$ if $ab > 0$.

   (a) Determine whether R is an equivalence relation.

   (b) If R is an equivalence relation, then describe the distinct equivalence classes.

7. Let R be a relation defined on $\mathbb{Z}$ by $a\,R\,b$ if $a + b = 0$ or $a - b = 0$.

   (a) Determine whether R is an equivalence relation.

   (b) If R is an equivalence relation, then describe the distinct equivalence classes.

8. A relation R is defined on the set $S = \{-7, -5, -4, -1, 3, 4, 9\}$ by $a\,R\,b$ if $a + 3b$ is even.

   (a) Show that R is an equivalence relation.

   (b) Describe the distinct equivalence classes resulting from R.

9. A relation R is defined on the set $\mathbb{Z}$ of integers by $a\,R\,b$ if $11a - 5b$ is even.

   (a) Show that R is an equivalence relation.

   (b) Describe the distinct equivalence classes resulting from R.

10. A relation R is defined on $\mathbb{Z} \times \mathbb{Z}$ by $(a, b)\,R\,(c, d)$ if $a + b + c + d$ is even.

    (a) Show that R is an equivalence relation.

    (b) Describe the distinct equivalence classes resulting from R.

11. A relation R is defined on $\mathbb{Z} \times \mathbb{Z}$ by $(a, b)\,R\,(c, d)$ if $abcd$ is even. Is R an equivalence relation?

12. Let $S = \{x, y, z\}$. A relation R on S has the following four properties:

    (1) z is related to at least one element of S,

    (2) y is related every element to which z is related,

    (3) x is related to every element to which either z or y is related,

    (4) R is symmetric.

    Which of the following is true?

    (a) R is an equivalence relation.

    (b) It is impossible to determine whether R is an equivalence relation.

13. Does there exist an example of an equivalence relation R on the set $S = \{a, b, c, d, e, f\}$ such that (1) no two distinct equivalence classes have the same number of elements and (2) $a\,R\,b$, $b\,R\,c$, $c\,R\,d$, $d\,R\,e$ and $e\,R\,f$?

14. A relation R is defined on the set $\mathbb{R}^+$ of positive real numbers by $a\,R\,b$ if the arithmetic mean (the average) of a and b equals the geometric mean of a and b, that is, if $\frac{a+b}{2} = \sqrt{ab}$.

    (a) Prove that R is an equivalence relation.

    (b) Describe the distinct equivalence classes resulting from R.

15. Let S be a nonempty set and let $\mathcal{P} = \{S_1, S_2, \ldots, S_k\}$ be a partition of S, where $k \geq 1$. Define a relation R on S by $a\,R\,b$ if $a, b \in S_i$ for some i with $1 \leq i \leq k$.

    (a) Prove that R is an equivalence relation.

    (b) Describe the distinct equivalence classes resulting from R.

16. Give an example of an equivalence relation R on the set $A = \{1, 2, 3, 4, 5, 6, 7\}$ with $\mathcal{P}$ the set of equivalence classes such that the following four properties are satisfied:

    (1) $|\mathcal{P}| = 3$,   (2) there exists no set $S \in \mathcal{P}$ such that $|S| = 3$,

    (2) $3\,R\,4$ but $3\,R\,5$ and (4) there exists a set $T \in \mathcal{P}$ such that $1, 7 \in T$.

17. Prove or disprove: Let A be a finite set with $|A| = n \geq 2$. There exists no equivalence relation R on A with $|R| = n + 1$.

5.3 Functions

In a relation from a set A to a set B, an element of A may be related to all, some or no elements of B. The situation when each element of A is related to exactly one element of B is of great interest and importance. These types of relations have a familiar name and will be discussed in this section.

Definition 5.26 *Let A and B be two nonempty sets. A **function** f from A to B is a relation from A to B that associates with each element of A a unique element of B. A function f from A to B is denoted by $f : A \to B$.*

A function $f : A \to B$ can therefore also be described as a subset of $A \times B$ such that for every element a of A there is exactly one ordered pair in f in which a is the first coordinate. Hence if g is a subset of $A \times B$ in which either

(i) some element a' of A is not the first coordinate of any ordered pair in g or

(ii) some element a'' of A is the first coordinate of more than one ordered pair in g,

then g is *not* a function from A to B.

Example 5.27

For the two sets $A = \{a_1, a_2, a_3, a_4\}$ and $B = \{b_1, b_2, b_3\}$, let f be the function from A to B that assigns

- to a_1 the element b_2,
- to a_2 the element b_3,
- to a_3 the element b_3,
- to a_4 the element b_2.

Therefore, f consists of the ordered pairs (a_1, b_2), (a_2, b_3), (a_3, b_3) and (a_4, b_2), that is,

$$f = \{(a_1, b_2), (a_2, b_3), (a_3, b_3), (a_4, b_2)\}. \qquad \blacklozenge$$

There is some common terminology and notation related to functions that we should be aware of. Suppose that $f : A \to B$. If $b \in B$ is the unique element assigned to $a \in A$ by f, then we write $b = f(a)$ and say that "b is f of a" and that b is the **image** of a (under f). For the function f in Example 5.27, $f(a_1) = b_2$ and so b_2 is the image of a_1.

Definition 5.28 *If $f : A \to B$ is a function from a set A to a set B, then A is called the **domain** of f and B is the **codomain** of f. The **range** $f(A)$ is the set of images of the elements of A, namely,*

$$f(A) = \{f(a) : \ a \in A\}.$$

For the function f in Example 5.27, the domain of f is therefore $A = \{a_1, a_2, a_3, a_4\}$ and the codomain of f is $B = \{b_1, b_2, b_3\}$. The range of f is $f(A) = \{b_2, b_3\}$.

More generally, we have the following concept.

Definition 5.29 *For a function f from a set A to a set B and a subset X of A, the **image** of X under f is the set*

$$f(X) = \{f(x) : \ x \in X\}.$$

For every subset X of A, $f(X) \subseteq B$. If $X = A$, then $f(X) = f(A)$ is the range of f. If $X = \{a_2, a_3\}$ in Example 5.27, then $f(X) = \{b_3\}$; while if $Y = \{a_3, a_4\}$, then $f(Y) = \{b_2, b_3\}$.

Representations of Functions

It is often convenient to visualize a function by means of a diagram. Suppose that f is a function from A to B. Then Figure 5.5 provides a pictorial representation of such a function. If $f(a) = b$, then this is represented by drawing an arrow from the element $a \in A$ to its image $b \in B$.

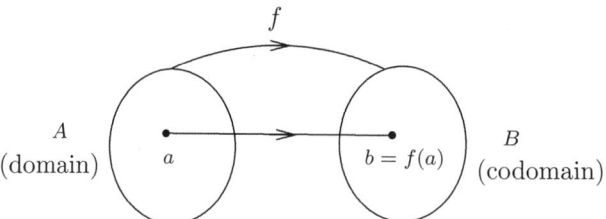

Figure 5.5: A function $f : A \to B$

Example 5.30

For the sets $A = \{1,2,3,4\}$ and $B = \{a,b,c,d,e\}$, the diagram in Figure 5.6 describes a function $f : A \to B$. The domain of f is A, the codomain of f is B and the range of f is $f(A) = \{a,b,d\}$. The function f can also be expressed as

$$f = \{(1,b),(2,a),(3,b),(4,d)\}$$

and also by writing

$$f(1) = b, \; f(2) = a, \; f(3) = b, \; f(4) = d.$$

Observe that neither c nor e is the image of any element of A. Also, if $A' = \{1,2,3\}$, then $f(A') = \{a,b\}$.

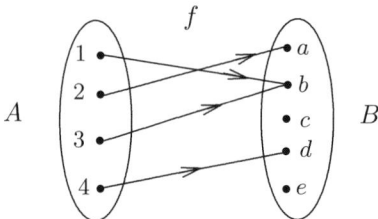

Figure 5.6: The function $f : A \to B$ in Example 5.30

For the sets $A = \{1,2,3,4\}$ and $B = \{a,b,c,d,e\}$ above, if

$$g = \{(1,c),(2,e),(3,a)\} \text{ and } h = \{(1,a),(2,b),(3,c),(3,e),(4,b)\},$$

then neither g nor h is a function from A to B. The reason that g is not a function from A to B is because the element 4 in A has no image in B; while h is not a function from A to B because the element 3 in A has more than one image in B, namely, c and e are both images of 3. On the other hand, if we were to let $A' = \{1,2,3\}$, then g *is* a function from the set A' to B.   ♦

Example 5.31

For each real number x, let $f(x)$ denote any real number y such that (x,y) lies on the circle $x^2 + y^2 = 25$. Is f a function from the set $\mathbb{R}$ of real numbers to $\mathbb{R}$?

Solution. No, f fails to be a function from $\mathbb{R}$ to $\mathbb{R}$ for two reasons. First, consider $x = 6$. There is no real number y such that $x^2 + y^2 = 36 + y^2 = 25$. This says that f is not defined for $x = 6$; that is, f associates no real number with 6 and so there is no number $f(6)$. Next, consider $x = 3$. There are two real numbers y for which $x^2 + y^2 = 9 + y^2 = 25$, namely $y = 4$ and $y = -4$. That is, $f(3) = 4$ and $f(3) = -4$ both satisfy the requirement. Therefore, f associates two distinct real numbers with 3. In fact, only for $x = 5$ and $x = -5$ is there a unique real number y (namely $y = 0$) such that $x^2 + y^2 = 25$. ♦

In general, for two nonempty sets A and B, there are many functions from A to B. Two functions $f : A \to B$ and $g : A \to B$ are said to be **equal** if $f(a) = g(a)$ for every element $a \in A$.

Example 5.32

Let $A = \{a, b\}$ and $B = \{0, 1\}$. Determine all functions from A to B.

Solution. The functions from A to B are

$$f_1 = \{(a, 0), (b, 0)\}, \quad f_2 = \{(a, 0), (b, 1)\},$$
$$f_3 = \{(a, 1), (b, 0)\}, \quad f_4 = \{(a, 1), (b, 1)\}.$$

Therefore, there are four such functions. ♦

In Chapter 8 we will describe how to determine the number of different functions from one finite set to another finite set.

For two nonempty sets A and B of real numbers and a function $f : A \to B$, the **graph** of f is the set of points (x, y) in the Cartesian plane for which $x \in A$, $y \in B$ and $y = f(x)$.

Example 5.33

For each real number x, let $f(x)$ be the real number $x^2 - 4x + 1$. Then f can be considered as a function both of whose domain and codomain is the set $\mathbb{R}$ of real numbers that associates with each real number x the unique real number $f(x) = x^2 - 4x + 1$. This is a function that you might very well have encountered in calculus or precalculus. In fact, the graph of this function is a parabola, which is shown in Figure 5.7. For each point (x, y) on this graph, $y = f(x)$. In particular, $(4, 1)$ is a point on the graph because $1 = f(4) = 4^2 - 4 \cdot 4 + 1$. The range of f is $\{y \in \mathbb{R} : y \geq -3\}$. ♦

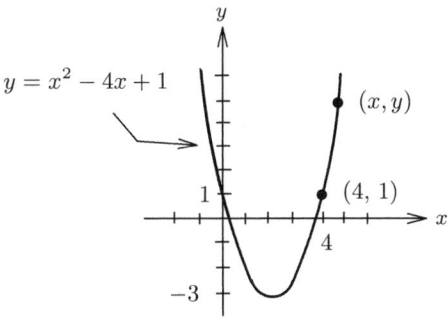

Figure 5.7: The graph of the function $y = f(x) = x^2 - 4x + 1$ in Example 5.33

Analysis. We noted that the range of f is $\{y \in \mathbb{R} : y \geq -3\}$. From the graph of $y = f(x) = x^2 - 4x + 1$ in Figure 5.7, it certainly *appears* that this is the range (the set of images or y-values). However, since this relies on the accuracy of the graph and our interpretation of the graph, it is not clear if our answer for the range is correct. Even a graphing calculator cannot help us to give a precise answer here. If we were to look at this function in another way, however, we can see that our answer for the range is accurate. Observe, by completing the square, that we can write

$$f(x) = x^2 - 4x + 1 = (x^2 - 4x + 4) - 3 = (x - 2)^2 - 3.$$

Since $(x - 2)^2 \geq 0$, it follows that $(x - 2)^2 - 3 \geq 0 - 3 = -3$ and so $f(x) \geq -3$. (See Exercise 14.) ♦

Although the functions that one considers in calculus, such as that in Example 5.33, appear to be given by some sort of formula, this is not necessary. In fact, as we have seen, if $f : A \to B$, then A and B need not even be sets of numbers. This is illustrated again in the next example.

Example 5.34

Let $A = \{a, b, c, d\}$ and $B = \{u, v, w, x, y, z\}$. If f assigns to both a and c the element w, assigns to b the element v, and assigns to d the element y, then f is a function from A to B for which $f(a) = w$, $f(b) = v$, $f(c) = w$ and $f(d) = y$. Therefore, the function f can be also expressed as a set of four ordered pairs, namely

$$f = \{(a, w), (b, v), (c, w), (d, y)\}.$$

This function can also be described by the diagram in Figure 5.8(a) and the graph in Figure 5.8(b), which in this case is a set of four points. Furthermore, the domain of f is A, the codomain of f is B and the range of f is $f(A) = \{v, w, y\}$. ♦

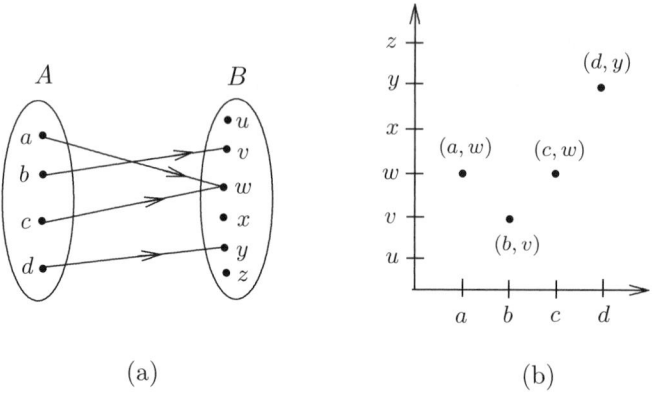

(a)   (b)

Figure 5.8: Representing the function of Example 5.34
by a diagram and a graph

Common Functions

The examples we've seen so far show that functions can be defined or described in a variety of ways. We now consider some of the most common types of functions.

Example 5.35

For a nonempty set A, the function $f : A \to A$ defined by $f(a) = a$ for every $a \in A$ is called the **identity function** on A. That is, the identity function on A assigns to each element of A that same element. For $A = \{1, 2, 3\}$, for example, the identity function on A is $f = \{(1, 1), (2, 2), (3, 3)\}$. The graphs of both this function and the identity function on the set $\mathbb{R}$ of real numbers are shown in Figures 5.9(a) and 5.9(b), respectively. ♦

Example 5.36

The function $f : \mathbb{R} \to \mathbb{R}$ defined by

$$f(x) = |x| = \begin{cases} x & \text{if } x \geq 0 \\ -x & \text{if } x < 0 \end{cases}$$

is the **absolute value function**, which assigns to each real number x its absolute value $|x|$. According to the definition then, the absolute value function assigns to each nonnegative real number the number itself; while it assigns to each negative real number the negative of that number. So the image of every real number under the absolute value function is a nonnegative real number. For

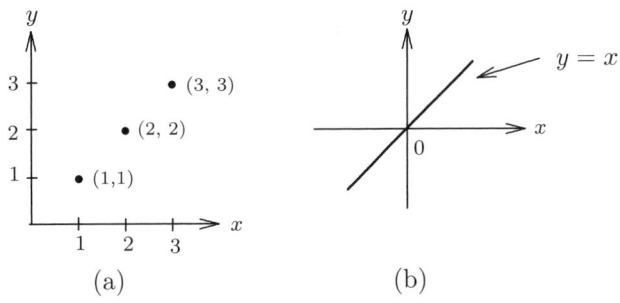

(a)   (b)

Figure 5.9: The graphs of the identity functions on $\{1, 2, 3\}$ and on $\mathbb{R}$

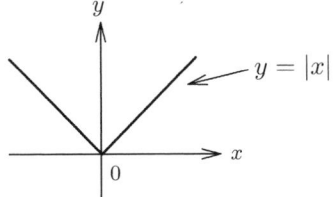

Figure 5.10: The graph of the absolute value function on $\mathbb{R}$

example, $f(5) = |5| = 5$, $f(0) = |0| = 0$ and $f(-2.5) = |-2.5| = -(-2.5) = 2.5$. The graph of the absolute value function on $\mathbb{R}$ is shown in Figure 5.10.   ♦

There are occasions when we are interested in integers that are near some given real number, namely those obtained by rounding up or rounding down the real number to the nearest integer. For a real number r, the **ceiling** $\lceil r \rceil$ of r is the smallest integer that is greater than or equal to r; while the **floor** $\lfloor r \rfloor$ of r is the greatest integer less than or equal to r. Therefore, $\lceil 6.4 \rceil = 7$ and $\lfloor 6.4 \rfloor = 6$. Also, $\lceil 4 \rceil = \lfloor 4 \rfloor = 4$, $\lceil -2 \rceil = \lfloor -2 \rfloor = -2$, $\lceil -3.5 \rceil = -3$ and $\lfloor -3.5 \rfloor = -4$. Both the ceiling and the floor produce a function.

Example 5.37

The function $f : \mathbb{R} \to \mathbb{Z}$ defined by $f(x) = \lceil x \rceil$ is the **ceiling function** and the function $g : \mathbb{R} \to \mathbb{Z}$ defined by $g(x) = \lfloor x \rfloor$ is the **floor function**. (See also Exercise 4 in Section 4.3.) The graphs of these two functions are shown in Figure 5.11.   ♦

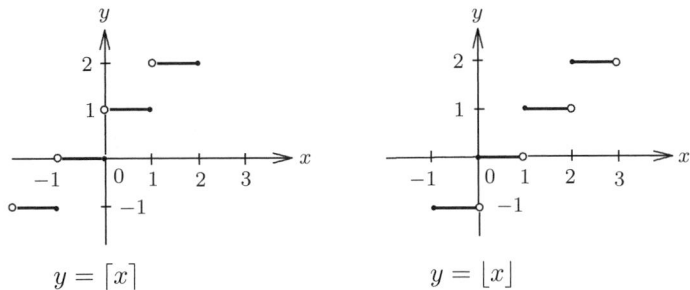

$y = \lceil x \rceil$ $y = \lfloor x \rfloor$

Figure 5.11: The graphs of the ceiling and floor functions

Example 5.38

If 100 CD ROMs are to be given to the 15 students in a computer science class so that the distribution among the students is as equal as possible, then each student must receive either $\lceil 100/15 \rceil = 7$ CD ROMs or $\lfloor 100/15 \rfloor = 6$ CD ROMs. (In fact, 10 students must receive 7 CD ROMs with the other 5 students receiving 6 CD ROMs.)   ◆

Two other common types of functions that are encountered in discrete mathematics are exponential and logarithmic functions. Let $a \in \mathbb{R}^+$ such that $a \neq 1$, where, recall, $\mathbb{R}^+$ is the set of positive real numbers. If $b \in \mathbb{R}$ and $a^b = c$, then this equality can also be expressed by writing $\log_a c = b$. That is,

$$\log_a c = b \text{ if and only if } a^b = c.$$

Necessarily, $c \in \mathbb{R}^+$. In each case, a is called the **base**. In calculus, the most common number to use as a base is the number e (an irrational number whose approximate value is 2.718), while 2 is a common base in computer science. Therefore,

$$y = \log_e x \text{ if and only if } x = e^y,$$

where $\log_e x$ is often written as $\ln x$. Also,

$$y = \log_2 x \text{ if and only if } x = 2^y.$$

Example 5.39

(a) The function $f : \mathbb{R} \to \mathbb{R}^+$ defined by $f(x) = 2^x$ is an **exponential function**. For example, $f(-3) = 2^{-3} = \frac{1}{2^3} = \frac{1}{8}$, $f(0) = 2^0 = 1$ and $f(2) = 2^2 = 4$.

(b) The function $g : \mathbb{R}^+ \to \mathbb{R}$ defined by $g(x) = \log_2 x$ is a **logarithmic function**. For example, $g\left(\frac{1}{8}\right) = \log_2\left(\frac{1}{8}\right) = \log_2\left(2^{-3}\right) = -3$, $g(1) = \log_2(1) = \log_2\left(2^0\right) = 0$ and $g(4) = \log_2(4) = \log_2\left(2^2\right) = 2$.

The graphs of these two functions are shown in Figure 5.12.   ◆

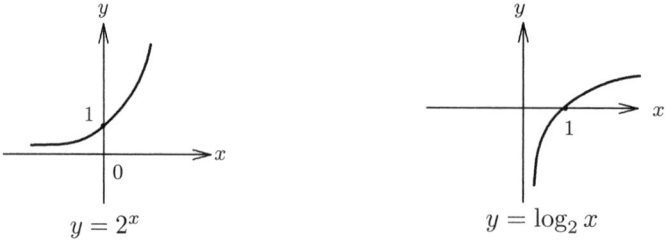

$$y = 2^x \qquad\qquad y = \log_2 x$$

Figure 5.12: The graphs of an exponential function and a logarithmic function

Example 5.40

(a) The function $f : \mathbb{N} \to \mathbb{R}$ defined by $f(n) = \frac{(-1)^{n+1}}{n}$ is an infinite sequence, which can be expressed as $f(1), f(2), f(3), \ldots$ or as $1, -\frac{1}{2}, \frac{1}{3}, -\frac{1}{4}, \ldots$ or as $\{a_n\}$, where $a_n = \frac{(-1)^{n+1}}{n}$ for $n \in \mathbb{N}$.

(b) The function $f : \mathbb{N} \cup \{0\} \to \mathbb{R}$ defined by $f(n) = \frac{n+1}{2n+1}$ is also an infinite sequence, which can be expressed as $f(0), f(1), f(2), \ldots$ or as $1, \frac{2}{3}, \frac{3}{5}, \ldots$ or as $\{a_n\}$, where $a_n = \frac{n+1}{2n+1}$ for $n \in \mathbb{N} \cup \{0\}$.

(c) For $n \in \mathbb{N}$, the function $f : \{1, 2, \ldots, n\} \to \{0, 1\}$ defined by

$$f(n) = \begin{cases} 0 & \text{if } n \text{ is even} \\ 1 & \text{if } n \text{ is odd} \end{cases}$$

is a finite sequence (or an n-bit string in this case). For $n = 5$, this sequence (5-bit string) can be expressed as $f(1), f(2), f(3), f(4), f(5)$ or as $1, 0, 1, 0, 1$ or as 10101.

Composition of Functions

Just as there are many ways of combining two numbers to produce a new number, many ways of combining two statements to produce a new statement and many ways of combining two sets to produce a new set, there are also many ways of combining two functions to produce a new function. We consider one of the best known ways of doing this – provided the domains and the codomains of the two given functions satisfy certain conditions.

Definition 5.41 *Let A, B and C be sets and suppose that $f : A \to B$ and $g : B \to C$ are two functions. The **composition** $g \circ f$ of f and g is the function from A to C defined by*

$$(g \circ f)(a) = g(f(a)) \text{ for } a \in A. \tag{5.2}$$

Therefore, in the definition above of the composition $g \circ f$ of two functions f and g, the domain of g is the codomain of f. The domain then of the composition $g \circ f$ is the domain of f, while the codomain of $g \circ f$ is the codomain of g. To determine the image of an element $a \in A$ by the function $g \circ f$, the definition in (5.2) tells us to first determine the image $f(a) \in B$ of a under f and then determine the image $g(f(a)) \in C$ of $f(a)$ under g. This is the element $(g \circ f)(a)$. Notice that in the composition $g \circ f$ we first deal with f and then with g. A diagram illustrating the composition of two such functions f and g is shown in Figure 5.13.

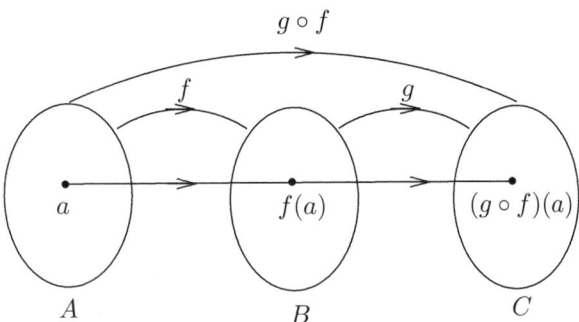

Figure 5.13: The composition $g \circ f$ of functions f and g

We now consider two examples of composition of functions.

Example 5.42

Let $A = \{1, 2, 3\}$, $B = \{a, b, c, d\}$ and $C = \{x, y, z\}$ and let $f : A \to B$ and $g : B \to C$ be functions, where

$$f = \{(1, c), (2, b), (3, a)\} \text{ and } g = \{(a, y), (b, x), (c, x), (d, z)\}.$$

Then

$$(g \circ f)(1) = g(f(1)) = g(c) = x.$$

Similarly, $(g \circ f)(2) = x$ and $(g \circ f)(3) = y$. Therefore,

$$g \circ f = \{(1, x), (2, x), (3, y)\}. \qquad \blacklozenge$$

Strictly speaking, when dealing with the composition $g \circ f$ of two functions f and g, it is not necessary that the domain of g be the same as the codomain of f. All that is required is that the range of f be a subset of the domain of g. Thus, for the functions f and g in Example 5.42, we could let

$$A = \{1, 2, 3\}, \; B = \{a, b, c, d\}, \; B' = \{a, b, c\}, \text{ and } C = \{x, y, z\},$$

and let $f : A \to B$ and $g : B' \to C$ be functions, where

$$f = \{(1, c), (2, b), (3, a)\} \text{ and } g = \{(a, y), (b, x), (c, x)\}.$$

Then

$$g \circ f = \{(1, x), (2, x), (3, y)\},$$

as we obtained in Example 5.42.

The Chain Rule from differential calculus also involves composition of functions. This concerns determining the derivative $(g \circ f)'$ of the composition $g \circ f$ of two functions f and g whenever we know the derivatives f' and g' of f and g, respectively. Although we are not concerned with derivatives here, let us consider an example of composition that might have occurred in calculus.

Example 5.43

Let $f : \mathbb{R} \to \mathbb{R}$ and $g : \mathbb{R} \to \mathbb{R}$, where $f(x) = \sin x$ and $g(x) = x^2$. Determine $(f \circ g)(x)$ and $(g \circ f)(x)$ for $x \in \mathbb{R}$.

Solution. First notice that because the domain and codomain are the same set (namely the set $\mathbb{R}$ of real numbers), both $f \circ g$ and $g \circ f$ are defined in this case. For $x \in \mathbb{R}$,

$$\begin{aligned} (f \circ g)(x) &= f(g(x)) = f(x^2) = \sin(x^2) \text{ and} \\ (g \circ f)(x) &= g(f(x)) = g(\sin x) = (\sin x)^2 = \sin^2 x. \; \blacklozenge \end{aligned}$$

Example 5.43 illustrates the fact that even when $f \circ g$ and $g \circ f$ are both defined, they need not be equal.

Example 5.44

Let $X = \{s_1, s_2, \ldots, s_7\}$ denote a set of 7 students taking an advanced computer science class, let $Y = \{A, B, C, D, F\}$ be the set of possible grades that a student in the class could earn and let $Z = \{0, 1, 2, \ldots, 7\}$. The function $f : X \to Y$ assigns to each student the grade the student earns in the class and $g : Y \to Z$ assigns to each grade the number of students receiving this grade. Suppose that

$$\begin{aligned} f &= \{(s_1, B), (s_2, B), (s_3, C), (s_4, A), (s_5, C), (s_6, B), (s_7, A)\} \\ g &= \{(A, 2), (B, 3), (C, 2), (D, 0), (F, 0)\}. \end{aligned}$$

The composition $g \circ f : X \to Z$ is therefore defined. For example,

$$(g \circ f)(s_1) = g(f(s_1)) = g(B) = 3,$$

which means that s_1 is one of three students who received the same grade, namely B, in the computer science class. In general,

$$g \circ f = \{(s_1, 3), (s_2, 3), (s_3, 2), (s_4, 2), (s_5, 2), (s_6, 3), (s_7, 2)\}. \qquad \blacklozenge$$

Exercises for Section 5.3

1. Let A be the set of all nonempty subsets of $B = \{1, 2, 3\}$. For each f described below, determine whether f is a function from A to B. For $S \in A$,

   (a) $f(S) = |S|$.

   (b) $f(S) = |S| - 1$.

   (c) $f(S) = 1/|S|$.

   (d) $f(S)$ is the sum of the elements in S.

   (e) $f(S)$ is the largest element in S.

   (f) $f(S)$ is the absolute value of the difference of the largest element and the smallest element in S.

2. For two nonempty sets A and B, define f_i, $i = 1, 2, 3, 4$, as shown in Figure 5.14. Determine whether f_i $(i = 1, 2, 3, 4)$ is a function from A to B. If f_i is not a function, explain.

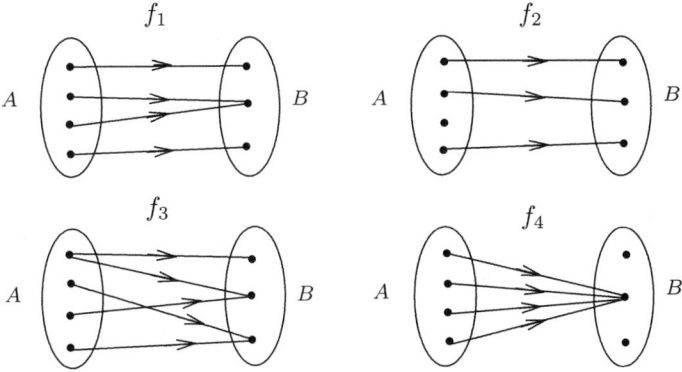

Figure 5.14: Diagrams for Exercise 6

3. Let $A = \{a, b, c, d, e\}$ and $B = \{x, y, z\}$ and let $f = \{(a, x), (b, x), (c, z), (d, x), (e, z)\}$ be a function from A to B.

   (a) Determine the domain, codomain and range of f.

   (b) Determine the image of d.

   (c) Determine whether y is an image.

   (d) Determine $f(X)$ where $X = \{a, c, d\}$.

   (e) Give an example of a function g from B to A.

4. For two nonempty sets A and B, let $f : A \to B$ be a function. For a nonempty subset X of A, suppose that $f(X) = C \subset B$. Furthermore, suppose that there is some element $y \in A - X$ such that $f(X \cup \{y\}) = C$. Complete the sentence:

   There is some element $x \in X$ such that _____.

5. Let $f : A \to B$ be a function. Prove that if $A_1, A_2 \subseteq A$, then $f(A_1 \cup A_2) = f(A_1) \cup f(A_2)$.

6. Let $f : A \to B$ be a function. Prove that if $A_1, A_2 \subseteq A$, then $f(A_1 \cap A_2) \subseteq f(A_1) \cap f(A_2)$.

7. Prove that there exist nonempty sets A and B, a function $f : A \to B$ and subsets A_1 and A_2 of A such that $f(A_1 \cap A_2) \neq f(A_1) \cap f(A_2)$.

8. For the sets $A = \{1, 2, 3, 4\}$ and $B = \{x, y, z\}$, give an example of a function $g : A \to B$ and a function $h : B \to A$.

9. Let $A = \{a, b\}$ and $B = \{0, 1, 2\}$. Determine all functions from A to B.

10. Let $A = \{a, b, c\}$ and $B = \{0, 1\}$. Determine all functions from A to B.

11. The graph of $x = y^2$ is a parabola, which is drawn in Figure 5.15. For each $x \in \mathbb{R}$, let $y = f(x)$ be a real number such that (x, y) lies on the graph. Is f a function from $\mathbb{R}$ to $\mathbb{R}$?

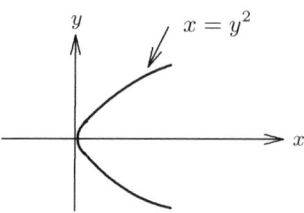

Figure 5.15: The graph of $x = y^2$ in Exercise 11

12. Let $S = \{-2, -1, 0, 1, 2\}$ and let $f = \{(x, y) \in S \times S : |x| + |y| = 2\}$. Is f a function from S to S?

13. A function $f : \mathbb{R} \to \mathbb{R}$ is defined by $f(x) = x(4 - x)$. Find the domain, codomain and range of f.

14. We have mentioned that if $f : \mathbb{R} \to \mathbb{R}$ is a function defined by $f(x) = x^2 - 4x + 1$ for $x \in \mathbb{R}$, then the range of f is $S = \{y \in \mathbb{R} : y \geq -3\}$. Show that every number $a \in S$ is the image of some real number x. [Hint: Consider $x^2 - 4x + 1 = a$ and use the quadratic formula.]

15. Let $A = \{-3, -2, \ldots, 3\}$ and $B = \{-10, -9, \ldots, 10\}$. A function $f : A \to B$ is defined by

$$f(n) = \begin{cases} n^2 & \text{if } n \in \{-1, 0, 1\} \\ -n & \text{if } n \in \{-2, 2\} \\ n^2 - 9 & \text{if } n \in \{-3, 3\}. \end{cases}$$

(a) Write f as a set of ordered pairs.

(b) What is the domain, codomain and range of f?

(c) What is $f(X)$ where $X = \{-3, -2, -1\}$?

16. Describe the identity function on the set $S_1 = \{t, u, v\}$ and on the set $S_2 = \mathcal{P}(A)$, where $A = \{1, 2\}$.

17. Draw the graphs of the following functions:

(a) $f : \mathbb{R} \to \mathbb{R}$ defined by $f(x) = |x - 2|$, where $x \in \mathbb{R}$.

(b) $f : \mathbb{R} \to \mathbb{R}$ defined by $f(x) = -2^x$, where $x \in \mathbb{R}$.

18. Find the following values:
   (a) $\lceil \frac{3}{5} \rceil$, $\lceil -\frac{3}{5} \rceil$, $\lfloor \frac{3}{5} \rfloor$, $\lfloor -\frac{3}{5} \rfloor$.   (b) $\lceil \frac{7}{5} \rceil$, $\lceil -\frac{7}{5} \rceil$, $\lfloor \frac{7}{5} \rfloor$, $\lfloor -\frac{7}{5} \rfloor$.   (c) $\lceil 3 \rceil$, $\lceil -3 \rceil$, $\lfloor 3 \rfloor$, $\lfloor -3 \rfloor$.

19. Let $f : \mathbb{Z} \to \mathbb{Z}$ be defined by $f(x) = \lfloor \frac{x^2}{3} \rfloor$ for each $x \in \mathbb{Z}$. Determine $f(S)$ for each of the following.

   (a) $S = \{0, 1, 2, 3\}$.   (b) $S = \{-2, -1, 0, 1, 2, 3\}$.
   (c) $S = \{1, 3, 5, 7\}$.   (d) $S = \{2, 4, 6, 8\}$.

20. Let $f : \mathbb{Z} \to \mathbb{Z}$ be defined as indicated below. For $S = \{-2, -1, 0, 1, 2\}$, determine $f(S)$ for each of the following.

    (a) $f(x) = 2$.   (b) $f(x) = 2x + 1$.   (c) $f(x) = x^2 + 1$.

    (d) $f(x) = |2x - 1|$.   (e) $f(x) = \lceil \frac{x}{3} \rceil$.

21. Let $f : \mathbb{R}^+ \to \mathbb{R}$, $g : \mathbb{R}^+ \to \mathbb{R}$ and $h : \mathbb{R}^+ \to \mathbb{R}$ be defined by $f(x) = \lceil x \rceil$, $g(x) = \lfloor x \rfloor$ and $h(x) = \log_2 x$ for $x \in \mathbb{R}^+$. Let $S = \{\frac{1}{2}, \sqrt{2}\}$. Compute the following.

    (a) $f(x)$ for each $x \in S$.   (b) $g(x)$ for each $x \in S$.   (c) $h(x)$ for each $x \in S$.

22. For every real number x, show the following.

    (a) $x \leq \lceil x \rceil < x + 1$.   (b) $\lfloor x \rfloor \leq x < \lfloor x \rfloor + 1$.

23. Let $x \in \mathbb{R}$. Prove that $\lceil x \rceil + \lfloor x \rfloor = 2x$ if and only if $x \in \mathbb{Z}$ or $2x \in \mathbb{Z}$.

24. Let $\ell, n \in \mathbb{N}$, where $n \geq 3$. Show that $\left\lceil \sqrt{n/2} \right\rceil = \ell$ if and only if $2(\ell - 1)^2 + 1 \leq n \leq 2\ell^2$. (See Exercise 22.)

25. (a) Is f a function from $\mathbb{R}$ to $\mathbb{R}$ if $f(x) = \frac{1}{x^2 - 1}$ for each $x \in \mathbb{R}$?

     (b) Is f a function from $\mathbb{Z}$ to $\mathbb{Z}$ if $f(n) = \frac{1}{n^2 + 1}$ for each $n \in \mathbb{Z}$?

     (c) Is f a function from $\mathbb{Z}$ to $\mathbb{N}$ if $f(n) = |-\sqrt{n^2}|$ for each $n \in \mathbb{Z}$?

26. For appropriate sets A and B, determine the range of a function $f : A \to B$ that assigns

    (a) to each integer the sum of that integer and its negative.

    (b) to each pair of integers the sum of twice of the first integer and three times the second integer.

    (c) to each real number the cube root of that number.

    (d) to each rational number two times that number.

    (e) to each positive 2-digit integer the first digit of that integer.

27. Suppose that $f : \mathbb{Z} \to \mathbb{Z}$ is a function with the property that $f(a + b) = f(a) + f(b)$ for every two integers a and b. Prove that if $f(c)$ is even for some odd integer c, then $f(c)$ is even for every integer c.

28. Let $A = \{1, 2, 3\}$, $B = \{1, 2, 3, 4, 5\}$ and $C = \{1, 2, 3, 4\}$. Also let $f : A \to B$ and $g : B \to C$, where $f = \{(1, 4), (2, 5), (3, 1)\}$ and $g = \{(1, 3), (2, 3), (3, 2), (4, 4), (5, 1)\}$,

    (a) Determine $(g \circ f)(1)$, $(g \circ f)(2)$ and $(g \circ f)(3)$.   (b) Determine $g \circ f$.

29. Let $A = \{a, b, c, d\}$ and $B = \{w, x, y, z\}$. Consider the functions $f : A \to A$ and $g : A \to B$, where $f = \{(a, c), (b, a), (c, a), (d, b)\}$ and $g = \{(a, z), (b, w), (c, y), (d, x)\}$. Determine the following. (a) $(g \circ f)(d)$.   (b) $g \circ f$.   (c) $f \circ f$.

30. Let $A = \{1, 2, 3, 4\}$ and $B = \{a, b, c, d\}$. Consider the functions $f : A \to B$ and $g : B \to A$, where $f = \{(1, a), (2, a), (3, c), (4, d)\}$ and $g = \{(a, 1), (b, 3), (c, 2), (d, 1)\}$. Determine the following. (a) $(g \circ f)(1)$ and $(f \circ g)(b)$.   (b) $g \circ f$ and $f \circ g$.

31. Two functions $f : \mathbb{R} \to \mathbb{R}$ and $g : \mathbb{R} \to \mathbb{R}$ are defined by $f(x) = 3x^2 + 1$ and $g(x) = 5x - 3$ for all $x \in \mathbb{R}$. Determine the following. (a) $(g \circ f)(0)$ and $(f \circ g)(0)$.   (b) $g \circ f$ and $f \circ g$.

32. Let $f : \mathbb{R} \to [1, \infty)$ and $g : [1, \infty) \to \mathbb{R}$ be defined by $f(x) = |x| + 1$ and $g = \sqrt{x - 1}$.

    (a) Determine $(g \circ f)(-4)$.   (b) Determine $(f \circ g)(5)$.

5.4   Bijective Functions

There are functions from a set A to a set B that often have one or both of the following properties:

(1) each element of B is the image of at most one element of A;

(2) each element of B is the image of at least one element of A.

These properties are encountered so often that familiarity with them is needed. We begin with functions that satisfy property (1).

One-to-one Functions

Definition 5.45 *For two nonempty sets A and B, a function $f : A \to B$ is said to be* **one-to-one** *if every two distinct elements of A have distinct images in B, that is, if $a, b \in A$ and $a \neq b$, then $f(a) \neq f(b)$.*

A one-to-one function is also referred to as an **injective function** or an **injection**. We consider two functions, only one of which is one-to-one.

Example 5.46

Let $A = \{a, b, c\}$ and $B = \{w, x, y, z\}$. Consider the functions $f : A \to B$ and $g : A \to B$ defined by

$$f = \{(a, x), (b, z), (c, w)\} \text{ and } g = \{(a, w), (b, y), (c, w)\}.$$

These functions are also described by the diagrams in Figure 5.16. For the function f, distinct elements of A have distinct images in B. Therefore, f is one-to-one. On the other hand, $g(a) = g(c) = w$ and so a and c do not have distinct images in B. Therefore, g is not one-to-one.   ♦

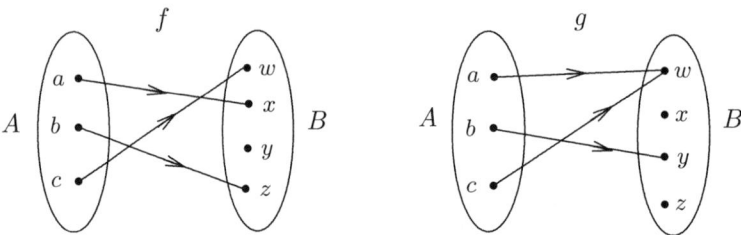

Figure 5.16: A one-to-one function and a function that is not one-to-one

Example 5.47

Let A be a nonempty set and let $f : A \to A$ be the function defined by $f(a) = a$ for every element $a \in A$. Thus f is the identify function on A. This function f is one-to-one for if x and y are any two distinct elements of A, then their images $f(x) = x$ and $f(y) = y$ are also distinct.   ♦

For two nonempty finite sets A and B, suppose that $f : A \to B$ is a one-to-one function. Then different elements of A must have different images in B. Therefore, if A has n elements, then the elements of A have n images in B. Consequently, the set B must contain at least n elements. Therefore, we have the following.

$$\text{If } f : A \to B \text{ is one-to-one, then we must have } |B| \geq |A|. \tag{5.3}$$

The contrapositive of the implication in (5.3) can therefore be stated as follows.

If $|B| < |A|$, then there is no one-to-one function from A to B.

In other words, if the codomain of a function has fewer elements than its domain, then the function is not one-to-one. In particular, if A is a finite nonempty set and B is a nonempty proper subset of A, then there is no one-to-one function $f : A \to B$. Perhaps surprisingly, for an infinite set A it *is* possible to have one-to-one functions from A to a proper subset of A.

Example 5.48

Let A be the set of integers (that is, $A = \mathbb{Z}$) and let B be the set of even integers. Therefore, B is an infinite proper subset of A. The function $f : A \to B$ defined by $f(n) = 2n$ for every integer n is one-to-one for if $a, b \in \mathbb{Z}$ and $a \neq b$, then $f(a) = 2a \neq 2b = f(b)$.   ◆

Although verifying that the function in Example 5.48 is one-to-one is rather straightforward, it may be difficult to show that other kinds of functions are one-to-one. For the purpose of showing a function is one-to-one, it is often helpful to consider the definition of a one-to-one function from another perspective.

First, let us restate the definition of a one-to-one function. A function $f : A \to B$ is **one-to-one** if whenever $a, b \in A$ and $a \neq b$, then $f(a) \neq f(b)$. The contrapositive of this implication is the following.

A function $f : A \to B$ is one-to-one if whenever $a, b \in A$ and $f(a) = f(b)$, then $a = b$.

Let's see how this version of the definition of one-to-one function can be used.

Result 5.49 *Let $f : \mathbb{R} \to \mathbb{R}$ be a function defined by $f(x) = 5x - 3$ for $x \in \mathbb{R}$. Then f is one-to-one.*

Proof. We begin with the assumption that $f(a) = f(b)$, where $a, b \in \mathbb{R}$ and attempt to show that $a = b$. Let $f(a) = f(b)$, where $a, b \in \mathbb{R}$. Then $5a - 3 = 5b - 3$. Adding 3 to both sides and then dividing by 5, we obtain $a = b$. Thus f is one-to-one.   ∎

A function $f : A \to B$ is therefore *not* one-to-one if there are distinct elements a and b of A such that $f(a) = f(b)$. Indeed, this is the way to show that a function f is not one-to-one, namely, by giving an example of two distinct elements a and b of A for which $f(a) = f(b)$.

Example 5.50

Show that the following functions are not one-to-one.

(a) $f : \mathbb{R} \to \mathbb{R}$ is defined by $f(x) = x^2 + 1$ for $x \in \mathbb{R}$.

(b) $g : \mathbb{Z} \to \mathbb{Z}$ is defined by $f(n) = \lceil n/2 \rceil$ for $n \in \mathbb{Z}$.

(c) $h : \mathbb{R} \to \mathbb{R}$ is defined by $h(x) = x^2 - 3x + 1$ for $x \in \mathbb{R}$.

Solution.

(a) The function f is not one-to-one since $f(1) = f(-1) = 2$ for example.

(b) The function g is not one-to-one since $g(1) = g(2) = 1$ for example.

(c) The function h is not one-to-one because $h(0) = h(3) = 1$ for example.   ◆

Analysis. It probably wasn't too difficult to find counterexamples in Example 5.50(a) and Example 5.50(b). In the case of Example 5.50(c), if we write

$$h(x) = x^2 - 3x + 1 = x(x - 3) + 1,$$

we see that $x(x-3) = 0$ has $x = 0$ or $x = 3$ as solutions and so $h(0) = h(3) = 1$. ♦

Onto Functions

While a function from a set A to a set B is one-to-one if every element of B is the image of at most one element of A, we now consider those functions having the property that every element of B is the image of at least one element of A.

Definition 5.51 *Let A and B be two nonempty sets. A function $f : A \to B$ is called **onto** if every element of B is the image of some element of A.*

In other words, $f : A \to B$ is onto if for every element $b \in B$, there is some element $a \in A$ whose image is b, that is, $f(a) = b$. Equivalently, f is onto provided the range of f is $f(A) = B$. A function that is onto is also called a **surjective function** or a **surjection**. We now consider two functions, one of which is onto and the other is not.

Example 5.52

Let $A = \{1, 2, 3, 4, 5\}$ and $B = \{a, b, c, d\}$. The functions $f : A \to B$ and $g : A \to B$ defined by

$$f = \{(1, b), (2, d), (3, d), (4, a), (5, c)\} \text{ and } g = \{(1, a), (2, a), (3, c), (4, c), (5, d)\}.$$

Diagrams for these functions are also shown in Figure 5.17. For the function f, every element of B is the image of some element of A. Thus f is onto. In the case of the function g, the element b is not the image of any element of A. Consequently, g is not onto. ♦

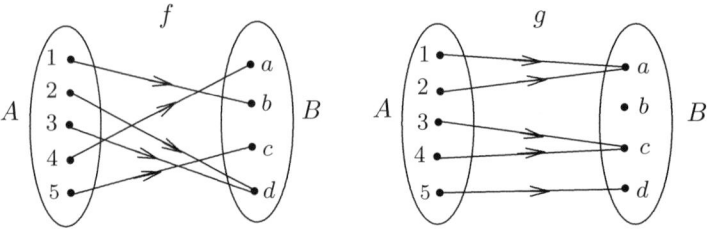

Figure 5.17: The functions f and g in Example 5.52

According to the definition, a function $f : A \to B$ is onto if every element in B is an image of some element of A. If A has n elements, then at most n elements of B can be images of A. Thus if f is onto, then B can contain at most n elements. Therefore, we have the following:

$$\text{If } f : A \to B \text{ is onto, then } |B| \le |A|. \tag{5.4}$$

Consequently, we have the following, which is the contrapositive of the implication (5.4):

$$\text{If } |A| < |B|, \text{ then there is no onto function from } A \text{ to } B.$$

We now consider a more complex example of an onto function.

Result to Prove: The function $f : \mathbb{R} \to \mathbb{R}$ defined by $f(x) = 4x - 9$ for $x \in \mathbb{R}$ is onto.

Proof Strategy: In order to show that f is onto, it is necessary to show that every real number is the image of some real number. Certainly, -1 and 7 are images since $f(2) = -1$ and $f(4) = 7$. However, *every* real number must be shown to be an image, not just -1 and 7. Before showing that f is onto, let us show that one additional real number, say π, is the image of some real number. We need to exhibit a real number x such that $f(x) = \pi$. Since $f(x) = 4x - 9$, we must have $4x - 9 = \pi$ for some real number x. Solving for x, we see that we must have $x = \frac{\pi + 9}{4}$. It remains to show that $f\left(\frac{\pi + 9}{4}\right)$ is indeed π. Thus

$$f\left(\tfrac{\pi+9}{4}\right) = 4\left(\tfrac{\pi+9}{4}\right) - 9 = \pi,$$

as desired. This observation leads us to a proof that f is onto. ♦

Result 5.53 *The function $f : \mathbb{R} \to \mathbb{R}$ defined by $f(x) = 4x - 9$ for $x \in \mathbb{R}$ is onto.*

Proof. Since we need to show that every real number is the image of some real number, we begin with an arbitrary real number, say r. We show that there is a real number x such that $f(x) = r$. Since r is a real number, so is $(r + 9)/4$. Observe that

$$f\left(\tfrac{r+9}{4}\right) = 4\left(\tfrac{r+9}{4}\right) - 9 = r.$$

Thus r is the image of $\frac{r+9}{4}$ and so f is onto. ∎

Proof Analysis. The crucial step in the solution of Example 5.53 was to consider $f\left(\tfrac{r+9}{4}\right)$. But where did $\frac{r+9}{4}$ come from? We started with an arbitrary real number r, which we wanted to show is the image of some real number x. For this to occur, $f(x)$ must equal r. However, since $f(x) = 4x - 9$, we must have $4x - 9 = r$. Solving for x, we obtain $x = \frac{r+9}{4}$ and that is why we decided to consider $f\left(\tfrac{r+9}{4}\right)$. The algebra used to obtain the number $\frac{r+9}{4}$ is not part of the justification that f is onto, however. ♦

Example 5.54

Determine whether the function $f : \mathbb{R} \to \mathbb{R}$ defined by

$$f(x) = x^2 - 2x + 5$$

for $x \in \mathbb{R}$ is onto.

Solution. Observe that

$$x^2 - 2x + 5 = (x^2 - 2x + 1) + 4 = (x - 1)^2 + 4.$$

Since $(x - 1)^2 \geq 0$, it follows that $f(x) = (x - 1)^2 + 4 \geq 4$. Therefore, no real number less than 4 can be an image and so f is not onto. ♦

Analysis. We could draw the graph of the function defined by $y = f(x) = x^2 - 2x + 5$, as in Figure 5.18, where it now certainly *appears* that every value of y (every image) is at least 4.

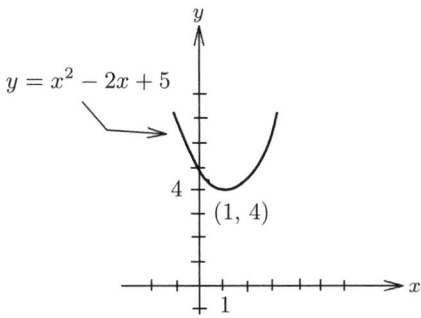

Figure 5.18: The graph of the function f in Example 5.54

While drawing the graph of the function does not constitute a solution, it may very well suggest a solution. Now select a number less than 4, say 3. If f is onto, then there is a real number x such that $f(x) = 3$ and so $x^2 - 2x + 5 = 3$ or $x^2 - 2x + 2 = 0$. Solving for x by the quadratic formula, we obtain

$$x = \frac{2 \pm \sqrt{4 - 8}}{2} = 1 \pm \sqrt{-1}.$$

Since neither $1 + \sqrt{-1}$ nor $1 - \sqrt{-1}$ is a real number, it follows that 3 is not an image of any real number and so f is not onto. This is an alternative explanation as to why f is not onto. ♦

Example 5.55

Determine whether the function $f : \mathbb{Z} \to \mathbb{Z}$ defined by $f(n) = 3n$ for $n \in \mathbb{Z}$ is one-to-one and onto.

Solution. First we show that f is one-to-one. Let $f(a) = f(b)$ for integers a and b. Then $3a = 3b$. Dividing by 3, we obtain $a = b$ and so f is one-to-one.

The function f is not onto however, for suppose it were. Then there would exist an integer n such that $f(n) = 1$ and so $3n = 1$. But this says that $n = 1/3$, which, of course, is not an integer. ♦

Analysis. Another explanation as to why the function f in Example 5.55 is not onto is to observe that all images must be multiples of 3. Since not all integers are multiples of 3, the function f is not onto.   ♦

Example 5.56

Suppose that in a computer science class with 25 students, the instructor gives a 20-point quiz periodically. Therefore, there are 21 possible scores on each quiz, namely, $0, 1, 2, \ldots, 20$. Let A be the set of the 25 students in class and let $B = \{0, 1, 2, \ldots, 20\}$. Then with each student $a \in A$ there is associated an integer $b \in B$, namely b is the score that a received on the quiz. This describes a function $f : A \to B$, where $b = f(a)$ if b is the score of a on the quiz. When the graded quiz is returned to the class, a student might ask: Did anyone get a perfect score on the quiz? If the answer is "no," then the function f is not onto. If the answer is "yes," then the function f still might not be onto. In fact, it is hoped that this function is not onto, as no one would want any student to receive a score of 0 on the quiz. Since $|A| > |B|$, that is, since there are more students than there are possible scores, it is impossible for this function to be one-to-one. So there must be at least two students who received the same quiz score.   ♦

Bijective Functions

If f is a function from a set A to a set B that is both one-to-one and onto, then every element of B is the image of *at most* one element of A and the image of *at least* one element of A. That is, if f is both one-to-one and onto, then every element of B is the image of *exactly one* element of A. Functions that are both injective and surjective are especially important in mathematics.

Definition 5.57 *A function that is one-to-one and onto is called a* **bijective function**, *a* **bijection** *or a* **one-to-one correspondence**.

Let's see an example of this.

Example 5.58

For $A = \{1, 2, 3, 4\}$ and $B = \{w, x, y, z\}$, the function $f : A \to B$ defined by

$$f = \{(1, y), (2, w), (3, z), (4, x)\}$$

is bijective. Two diagrams of this function are shown in Figure 5.19.   ♦

Suppose that A and B are finite nonempty sets. We stated that a function $f : A \to B$ is called bijective if it is both one-to-one and onto. Since f is one-to-one implies that $|A| \leq |B|$ and f is onto implies that $|A| \geq |B|$, it follows that $|A| = |B|$. A bijective function $f : A \to B$ pairs off the elements of A and B. For the function f in Example 5.58, this property is displayed more prominently in the second diagram in Figure 5.19.

Result 5.59 *The function $f : \mathbb{R}^+ \to \mathbb{R}^+$ defined by $f(x) = \sqrt{x}$ for $x \in \mathbb{R}^+$ is bijective.*

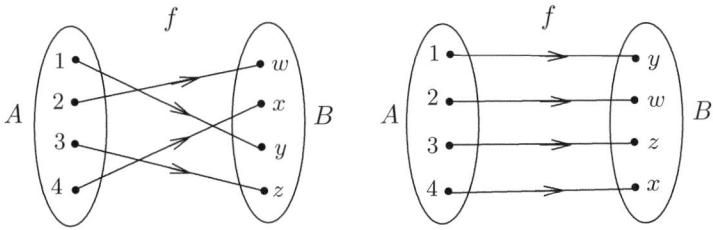

Figure 5.19: The function f in Example 5.58

Proof. First, we show that f is one-to-one. Let $f(a) = f(b)$, where $a, b \in \mathbb{R}^+$. Then $\sqrt{a} = \sqrt{b}$. Squaring both sides, we obtain $a = b$. Therefore, f is one-to-one. Next, we show that f is onto. Let r be a positive real number. Then r^2 is a positive real number and $f(r^2) = \sqrt{r^2} = r$; so f is onto. Thus f is bijective. ∎

Proof Analysis. Let's look at how the function f in Result 5.59 was shown to be onto. We began by selecting an arbitrary element in the codomain $\mathbb{R}^+$ of positive real numbers. This element was denoted by r. In order to show that f is onto, we needed to find an element x in the domain of f such that $f(x) = r$. Since the domain is also the set $\mathbb{R}^+$, we are seeking a positive real number x such that $f(x) = r$. However, since $f(x) = \sqrt{x}$, it follows that $\sqrt{x} = r$ and so $x = r^2$. This is why we decided to consider $f(r^2)$. ◆

The function f in Result 5.59 might suggest the consideration of other functions.

Example 5.60

Determine whether the function $g : \mathbb{R}^+ \to \mathbb{R}$ defined by $g(x) = \sqrt{x}$ for all $x \in \mathbb{R}^+$ is bijective.

Solution. Although this function g is one-to-one (the same argument used in the proof of Result 5.59 to show that f is one-to-one can be used here), g is not onto. Since $\sqrt{x} > 0$ for every $x \in \mathbb{R}^+$, no real number $y \le 0$ can be the image of any positive real number and so g is not onto. ◆

Hence by changing the codomain of the function in Result 5.59, we arrived at a function with a different property. Suppose we now change $\sqrt{x}$ to $\sqrt[3]{x}$.

Example 5.61

Determine whether the function $h : \mathbb{R}^+ \to \mathbb{R}$ defined by $h(x) = \sqrt[3]{x}$ for all $x \in \mathbb{R}^+$ is bijective.

Solution. In this case as well, $\sqrt[3]{x} > 0$ for every $x \in \mathbb{R}^+$. Thus no real number $y \le 0$ can be the image of any positive real number and so h is not onto. Therefore, h is not bijective. This function is one-to-one, however. ◆

Now we change the domain of the function in Example 5.61.

Example 5.62

Determine whether the function $\phi : \mathbb{R} \to \mathbb{R}$ defined by $\phi(x) = \sqrt[3]{x}$ for all $x \in \mathbb{R}$ is bijective.

Solution. In this case, we see that ϕ is indeed bijective.

Proof. First, we show that ϕ is one-to-one. Let $\phi(a) = \phi(b)$, where $a, b \in \mathbb{R}$. Then $\sqrt[3]{a} = \sqrt[3]{b}$. Cubing both sides, we obtain $a = b$. Therefore, ϕ is one-to-one. Next, we show that ϕ is onto. Let $r \in \mathbb{R}$. Then $r^3 \in \mathbb{R}$ and $\phi(r^3) = \sqrt[3]{r^3} = r$. So ϕ is onto. Therefore, ϕ is bijective. ∎

For a nonempty set A, a bijective function f from A to A is called a **permutation** on (or of) A. Thus, the function f in Result 5.59 is a permutation on $\mathbb{R}^+$, while the function ϕ in Example 5.62 is a permutation on $\mathbb{R}$.

Example 5.63

Give an example of a permutation on the set $S = \{1, 2, 3, 4\}$.

Solution. Since $g = \{(1, 2), (2, 1), (3, 4), (4, 3)\}$ is a bijective function from S to S, it follows that g is a permutation on S. ♦

Actually, the identity function on every nonempty set A is a permutation on A. Therefore, for the set $S = \{1, 2, 3, 4\}$ in Example 5.63,

$$f = \{(1, 1), (2, 2), (3, 3), (4, 4)\}$$

is also a permutation of S. There are *many* other permutations of the set S. The number of such permutations will be discussed in Chapter 9.

Compositions of Bijective Functions

If f and g are two functions for which the composition $g \circ f$ is defined, then $g \circ f$ is one-to-one whenever f and g are one-to-one. Furthermore, $g \circ f$ is onto whenever f and g are onto. These two facts are especially important in the study of composition of functions.

Theorem 5.64 *Let A, B and C be nonempty sets and let $f : A \to B$ and $g : B \to C$ be two functions.*

(a) *If f and g are one-to-one, then so is $g \circ f$.*

(b) *If f and g are onto, then so is $g \circ f$.*

Proof. Let $f : A \to B$ and $g : B \to C$ be one-to-one functions. Suppose that

$$(g \circ f)(a_1) = (g \circ f)(a_2),$$

where $a_1, a_2 \in A$. By definition,

$$g(f(a_1)) = g(f(a_2)).$$

Since g is one-to-one, it follows that $f(a_1) = f(a_2)$. Moreover, since f is one-to-one, it follows that $a_1 = a_2$. Therefore, $g \circ f$ is one-to-one.

Next suppose that $f : A \to B$ and $g : B \to C$ are onto. Let $c \in C$. We show that there is some element in A whose image under $g \circ f$ is c. Since g is onto, there exists $b \in B$ such that $g(b) = c$. On the other hand, since f is onto, it follows that there exists $a \in A$ such that $f(a) = b$. Therefore,

$$(g \circ f)(a) = g(f(a)) = g(b) = c.$$

This implies that the image of a under the function $g \circ f$ is c. Therefore, $g \circ f$ is onto. ∎

Proof Analysis. Looking once more at the proof of Theorem 5.64 and how it was shown that $g \circ f$ is one-to-one, we see that we started, as expected, with the assumption that $(g \circ f)(a_1) = (g \circ f)(a_2)$ for elements a_1 and a_2 in A. From this, it follows that $g(f(a_1)) = g(f(a_2))$. The elements $f(a_1)$ and $f(a_2)$ belong to B. Since g is one-to-one, whenever $g(b_1) = g(b_2)$ for $b_1, b_2 \in B$, we have $b_1 = b_2$. Applying this to $g(f(a_1)) = g(f(a_2))$ gives us $f(a_1) = f(a_2)$. ♦

A result that is a consequence of some theorem is referred to as a **corollary** of the theorem. Since bijective functions are both one-to-one and onto, we have an immediate corollary of Theorem 5.64.

Corollary 5.65 *Let A, B and C be nonempty sets and let $f : A \to B$ and $g : B \to C$ be two functions. If f and g are bijective, then so is $g \circ f$.*

Inverse Functions

For nonempty sets A and B, we have seen that a bijective function $f : A \to B$ pairs off the elements of A with the elements of B. For example, for

$$A = \{a,\, b,\, c,\, d,\, e\} \text{ and } B = \{v,\, w,\, x,\, y,\, z\},$$

the function $f : A \to B$ given by

$$f = \{(a,x),\, (b,v),\, (c,w),\, (d,z),\, (e,y)\}$$

is bijective. The diagram in Figure 5.20(a) shows this function. The elements of B are rearranged in Figure 5.20(b) to illustrate the bijective character of f more clearly. The **inverse function** (or the **inverse**) f^{-1} of f is obtained from f by replacing each ordered pair (r,s) of f by (s,r). A diagram of f^{-1} can therefore be obtained by reversing the directions of the arrows in Figure 5.20(b). This is shown in Figure 5.20(c). The function $f^{-1} : B \to A$ is bijective as well. In fact,

$$f^{-1} = \{(x,a),\, (v,b),\, (w,c),\, (z,d),\, (y,e)\}.$$

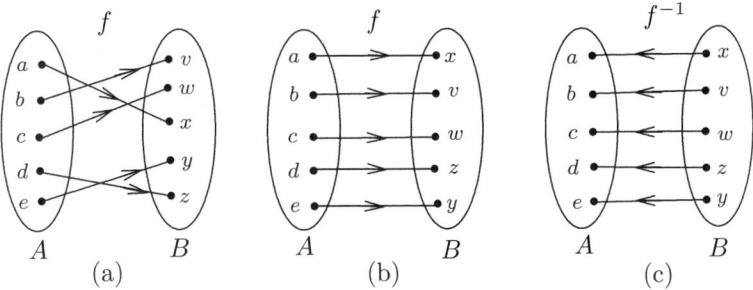

Figure 5.20: The inverse of a bijective function

Suppose that $f : A \to B$ is a function that is not bijective. We show that replacing each ordered pair (r,s) of f by (s,r) does not produce a function from B to A, for suppose that it did, resulting in a function $g : B \to A$. Since f is not bijective, either f is not one-to-one or f is not onto. If f is not one-to-one, then A contains distinct elements a_1 and a_2 such that $f(a_1) = f(a_2) = b$ for some $b \in B$. However then (b,a_1) and (b,a_2) are both ordered pairs in g, implying that b has two distinct images. (See Figure 5.21(a).) This is impossible since g is a function. On the other hand, if f is not onto, then there is some element $c \in B$ that is not the image of any element in A. This implies that there is no element $a \in A$ such that (a,c) is in f and therefore no element $a \in A$ such that (c,a) is in g. Thus c has no image in A. (See Figure 5.21(b)). This is impossible since g is a function from B to A. These observations are summarized in the theorem below.

Theorem 5.66 *Let A and B be nonempty sets. A function $f : A \to B$ has an inverse function $f^{-1} : B \to A$ if and only if f is bijective. Moreover, if f is bijective, then so is f^{-1}.*

Example 5.67

At an awards ceremony five honor students receive gifts for their accomplishments, namely

- Abraham receives a calculator,
- Stephen receives a gift certificate,
- Charlene receives a cell phone,
- Debra receives an original painting,
- Lawrence receives an iPod.

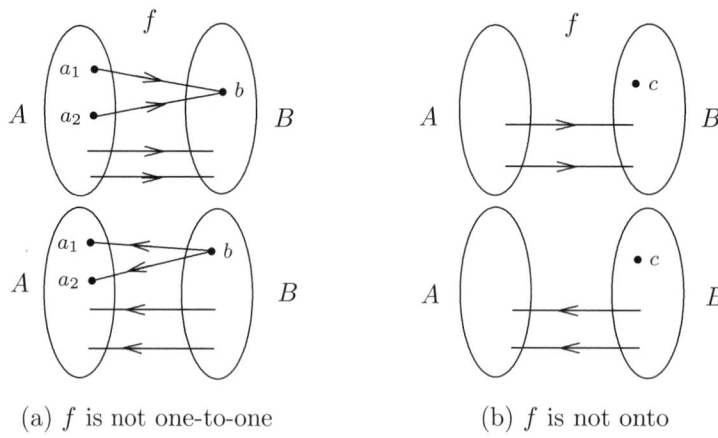

(a) f is not one-to-one   (b) f is not onto

Figure 5.21: Functions that are not bijective

If we let

$$A = \{\text{Abraham, Stephen, Charlene, Debra, Lawrence}\}$$
$$B = \{\text{calculator, certificate, cell phone, painting, iPod}\},$$

then the distribution of gifts describes a function $f : A \to B$, namely

$$f = \{(\text{Abraham, calculator}), (\text{Stephen, certificate}),$$
$$(\text{Charlene, cell phone}), (\text{Debra, painting}), (\text{Lawrence, iPod})\}.$$

Since f is bijective, the inverse $f^{-1} : B \to A$ exists, namely

$$f^{-1} = \{(\text{calculator, Abraham}), (\text{certificate, Stephen}),$$
$$(\text{cell phone, Charlene}), (\text{painting, Debra}), (\text{iPod, Lawrence})\}.$$

While the function f tells us that Stephen received a gift certificate, for example, the function f^{-1} tells us that the gift certificate was awarded to Stephen.   ◆

Let's consider another example of a bijective function.

Example 5.68

Show that the function $f : \mathbb{R} \to \mathbb{R}$ defined by $f(x) = 3x + 2$ for $x \in \mathbb{R}$ is bijective. Determine $f^{-1}(x)$ for $x \in \mathbb{R}$.

Solution. Let $f(a) = f(b)$, where $a, b \in \mathbb{R}$. Then $3a + 2 = 3b + 2$. Subtracting 2 from both sides and dividing by 3 give us $a = b$. Therefore, f is one-to-one. Next, let r be a real number. Then $\frac{r-2}{3}$ is a real number. Furthermore,

$$f\left(\tfrac{r-2}{3}\right) = 3\left(\tfrac{r-2}{3}\right) + 2 = r,$$

and so f is onto. Hence f is bijective.

Since f is a bijective function, f has an inverse function f^{-1}. Let $x \in \mathbb{R}$ and suppose that the image of x under f is y. Since $f(x) = y$, it follows that $y = 3x + 2$. Therefore, in f^{-1} the image of y is x. See Figure 5.22. Because $x = \frac{y-2}{3}$, it follows that the image of y in f^{-1} is $\frac{y-2}{3}$. Therefore, in f^{-1}, the image of a real number a is $\frac{a-2}{3}$ and in fact, the image of a real number x is $\frac{x-2}{3}$. Since the image of x is $f^{-1}(x)$, it follows that $f^{-1}(x) = \frac{x-2}{3}$.   ◆

Analysis. The solution of Example 5.68 suggests how to determine $f^{-1}(x)$ more quickly in this case. For the function $f : \mathbb{R} \to \mathbb{R}$ defined by $f(x) = 3x + 2$, which is bijective, we write $y = 3x + 2$. Solving for x, we obtain $x = (y - 2)/3$. Interchanging x and y, we obtain $y = (x - 2)/3$ and so $f^{-1}(x) = (x - 2)/3$.   ◆

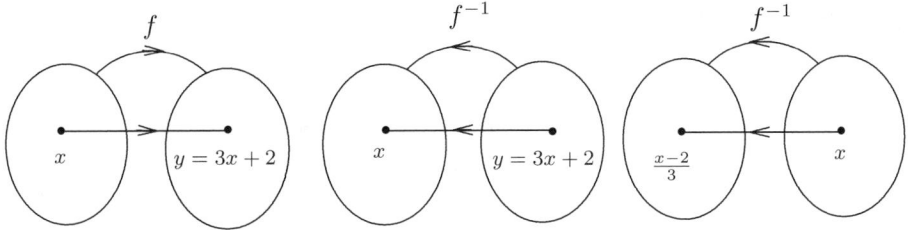

Figure 5.22: Determining the inverse of a function

Example 5.69

The function $f : \mathbb{R} \to \mathbb{R}$ defined by $f(x) = x^7 - 4$ for $x \in \mathbb{R}$ is known to be bijective. Determine $f^{-1}(x)$ for $x \in \mathbb{R}$.

Solution. Let $y = x^7 - 4$. Solving for x, we first obtain $x^7 = y + 4$ and so $x = \sqrt[7]{y + 4}$. Interchanging x and y, we obtain $y = \sqrt[7]{x + 4}$ and therefore $f^{-1}(x) = \sqrt[7]{x + 4}$. ♦

Even knowing that a function $f : \mathbb{R} \to \mathbb{R}$ is bijective does not necessarily mean that an expression for $f^{-1}(x)$ can be determined however. For example, the function $f : \mathbb{R} \to \mathbb{R}$ defined by $f(x) = x^7 + 2x^3 + 3x - 4$ is also bijective, but an algebraic expression for $f^{-1}(x)$ cannot be found.

Suppose that f is a bijective function from a set A to a set B and $f(a) = b$, where $a \in A$ and $b \in B$. Since f is bijective, f has an inverse function f^{-1} from B to A which is also bijective. Furthermore, $f^{-1}(b) = a$. Since $f : A \to B$ and $f^{-1} : B \to A$, the composition $f^{-1} \circ f : A \to A$ exists and

$$(f^{-1} \circ f)(a) = f^{-1}(f(a)) = f^{-1}(b) = a.$$

That is, $f^{-1} \circ f$ is the identity function on A. (See Figure 5.23.) Also, $f^{-1} : B \to A$ and $f : A \to B$, so the composition $f \circ f^{-1} : B \to B$ exists and

$$(f \circ f^{-1})(b) = f(f^{-1}(b)) = f(a) = b.$$

Therefore, $f \circ f^{-1}$ is the identity function on B. These observations are summarized below.

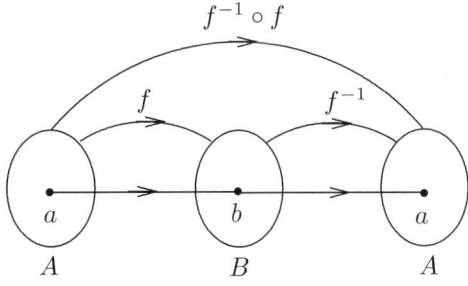

Figure 5.23: The composition of a function and its inverse

Theorem 5.70 *For nonempty sets A and B, let $f : A \to B$ be a bijective function. Then*

(a) *$f^{-1} \circ f$ is the identity function on A and*

(b) *$f \circ f^{-1}$ is the identity function on B.*

Exercises for Section 5.4

1. Let $A = \{1, 2, 3, 4\}$ and $B = \{1, 2, 3, 4, 5\}$. Give an example of

   (a) a function $f : A \to B$ that is not one-to-one but where $f(1) \neq f(2)$.

   (b) a one-to-one function $g : A \to B$, where $g(1) \neq 3 \neq g(2)$.

   (c) a one-to-one function $h : A \to B$ such that $h(n) > n$ for every $n \in A$.

2. Determine, with explanation, whether the following functions are one-to-one.

   (a) $f : \mathbb{Z} \to \mathbb{Z}$ is defined by $f(n) = 4n + 1$ for $n \in \mathbb{Z}$.

   (b) $f : \mathbb{R}^+ \to \mathbb{R}^+$ is defined by $f(x) = x^2$ for $x \in \mathbb{R}^+$.

   (c) $f : \mathbb{R} \to \mathbb{Z}$ is defined by $f(x) = \lceil x \rceil$ for $x \in \mathbb{R}$.

   (d) $f : \mathbb{R} \to \mathbb{R}$ is defined by $f(x) = x^2 + 2x + 1$ for $x \in \mathbb{R}$.

3. Let $A = \{a, b, c, d\}$ and $B = \{a, b, c\}$. Give an example of

   (a) a function $f : A \to B$ that is not onto but where $f(a) \neq f(b)$.

   (b) a function $g : A \to B$ that is onto but where $g(a) = g(b)$.

   (c) a function $h : A \to B$ that is onto but $h(x) \neq x$ for each $x \in B$.

4. Show that $f : \mathbb{R} \to \mathbb{R}$ defined by $f(x) = 2x + 1$ for $x \in \mathbb{R}$ is onto.

5. Determine, with explanation, whether the following functions are onto.

   (a) $f : \mathbb{Z} \to \mathbb{Z}$ is defined by $f(n) = 4n + 1$ for $n \in \mathbb{Z}$.

   (b) $f : \mathbb{R}^+ \to \mathbb{R}^+$ is defined by $f(x) = x^2$ for $x \in \mathbb{R}^+$.

   (c) $f : \mathbb{R} \to \mathbb{Z}$ is defined by $f(x) = \lceil x \rceil$ for $x \in \mathbb{R}$.

   (d) $f : \mathbb{R} \to \mathbb{R}$ is defined by $f(x) = x^2 + 2x + 1$ for $x \in \mathbb{R}$.

6. For each function f_i $(1 \leq i \leq 4)$ in Figure 5.24, determine whether f_i is (a) one-to-one and (b) onto.

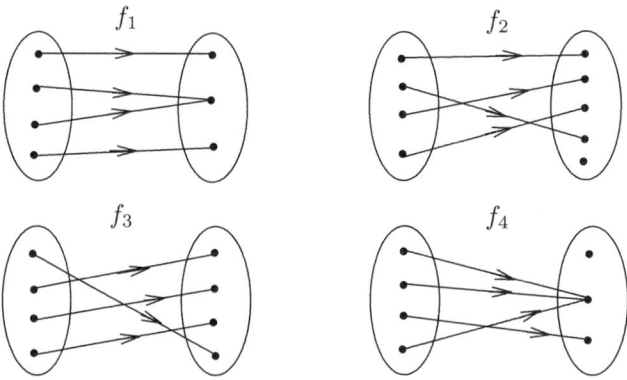

Figure 5.24: Functions f_i $(1 \leq i \leq 4)$ in Exercise 6

7. Let $A = \{a, b, c, d\}$ and $B = \{q, r, s, t\}$. Give an example, if such an example exists, of a function $f : A \to B$ that is

   (a) one-to-one but not onto.   (b) onto but not one-to-one.

   (c) one-to-one and onto.   (d) neither one-to-one nor onto.

   In each case where the function f fails to have a particular property, give an explanation.

8. Let $A = \{a, b, c, d\}$ and $B = \{1, 2, 3\}$. Does there exist a function from A to B that is one-to-one?

9. Let $A = \{a, b, c\}$ and $B = \{1, 2, 3, 4\}$. Does there exist a function from A to B that is onto?

10. Give an example of two finite sets A and B and two functions $f : A \to B$ and $g : B \to A$ such that f is one-to-one but not onto and g is onto but not one-to-one. Explain why f is not onto and why g is not one-to-one.

11. A function $f : \mathbb{Z} \to \mathbb{Z}$ is defined by $f(n) = 2n$. Determine with explanation whether f is (a) one-to-one, (b) onto.

12. A function $f : \mathbb{Z} \to \mathbb{Z}$ is defined by $f(n) = n - 5$. Determine with explanation whether f is (a) one-to-one, (b) onto.

13. A function $f : \mathbb{Z} \to \mathbb{Z}$ is defined by $f(n) = 5n + 1$. Determine with explanation whether f is (a) one-to-one, (b) onto.

14. (a) A function $f : \mathbb{Z} \to \mathbb{Z}$ is defined by $f(x) = 3x + 5$. Determine with explanation whether f is (i) one-to-one, (ii) onto.

   (b) A function $g : \mathbb{R} \to \mathbb{R}$ is defined by $g(x) = 3x + 5$. Determine with explanation whether g is (i) one-to-one, (ii) onto.

15. Let $f : \mathbb{R} - \{0\} \to \mathbb{R} - \{0\}$ be defined by $f(x) = \frac{1}{2x}$. Determine with explanation whether f is (a) one-to-one, (b) onto.

16. Determine with explanation which of the following functions are bijective functions from $\mathbb{R}$ to $\mathbb{R}$. (a) $f(x) = 5x - 1$. (b) $f(x) = x^2 + x + 1$. (c) $f(x) = 2x^3 + 3$. (d) $f(x) = x^4 + 3x^2 + 1$.

17. Let f be a permutation on $\mathbb{N}$. Prove that if there exists no integer n such that $f(n) < n$, then f is the identity function on $\mathbb{N}$.

18. Each of the following is a function from $\mathbb{N} \times \mathbb{Z}$ to $\mathbb{Z}$. Which of these are onto?

   (a) $f(a, b) = 2a + b$.   (b) $f(a, b) = b$.   (c) $f(a, b) = 2^a b$.
   (d) $f(a, b) = |a| - |b|$.   (e) $f(a, b) = a + 10$.

19. Each of the following is a function from $\mathbb{N} \times \mathbb{N}$ to $\mathbb{N}$. Which of these are one-to-one?

   (a) $f(a, b) = 2a + b$.   (b) $f(a, b) = b$.   (c) $f(a, b) = a^2 + b^2$.
   (d) $f(a, b) = a^b$.   (e) $f(a, b) = 2^a 3^b$.

20. Let $f : A \to B$ be a function and let A_1 and A_2 be subsets of A. Prove that if f is one-to-one, then $f(A_1 \cap A_2) = f(A_1) \cap f(A_2)$.

21. Let $f : A \to B$ be a function for nonempty sets A and B. Prove or disprove the following.

   (a) If, for every two nonempty subsets A_1 and A_2 of A, $f(A_1) = f(A_2)$ implies that $A_1 = A_2$, then f is one-to-one.

   (b) If, for every nonempty subset B_1 of B, there exists a subset A_1 of A such that $f(A_1) = B_1$, then f is onto.

22. For nonempty sets A, B and C, let $f : A \to B$ and $g : B \to C$ be functions. Prove that if $g \circ f$ is one-to-one, then f is one-to-one.

23. For nonempty sets A, B and C, let $f : A \to B$ and $g : B \to C$ be functions. Disprove each of the following.

   (a) If $g \circ f$ is one-to-one, then g is one-to-one.

(b) If g is onto, then $g \circ f$ is onto.

(c) If g is one-to-one, then $g \circ f$ is one-to-one.

24. Prove or disprove each of the following.

(a) There exist functions $f : A \to B$ and $g : B \to C$ such that f is not one-to-one and $g \circ f : A \to C$ is one-to-one.

(b) There exist functions $f : A \to B$ and $g : B \to C$ such that f is not onto and $g \circ f : A \to C$ is onto.

25. For a function $f : \mathbb{Z} \to \mathbb{Z}$, let $f^2 = f \circ f$. Show that if f^2 is injective, then f is injective.

26. Let f, g and h be functions from $\mathbb{R}$ to $\mathbb{R}$ defined by $f(x) = e^x$, $g(x) = x^3$ and $h(x) = 3x$ for each $x \in \mathbb{R}$. Determine each of the following:

(a) $(g \circ f)(x)$.   (b) $(f \circ g)(x)$.   (c) $(h \circ f)(x)$.   (d) $(f \circ h)(x)$.

(e) a composition of functions that results in e^{3x^3}.

(f) a composition of functions that results in $3e^{x^3}$.

27. Figure 5.25 shows a map consisting of four cities c_1, c_2, c_3, c_4 with a road between each pair of cities. Let $S = \{c_1, c_2, c_3, c_4\}$. The symbols r, b and g represent the colors red, blue and green, respectively. So there are two red roads, two blue roads and two green roads. For $i = 1, 2, 3, 4$, the function $f : S \to S$ is defined by $f(c_i) = c_i$, that is, we start in city c_i but do not drive anyplace. The function $f_r : S \to S$ is defined as $f_r(c_i) = c_j$ if we start in city c_i and go to city c_j using the red road. The functions f_b and f_g are defined similarly. Determine each of the following:

(a) $(f_b \circ f_r)(c_1)$ and $f_g(c_1)$.   (b) $f_g(f_r(f_g(c_2)))$ and $f_r(c_2)$.   (c) $(f_b \circ f_b)(c_i)$ for $i = 1, 2, 3, 4$.

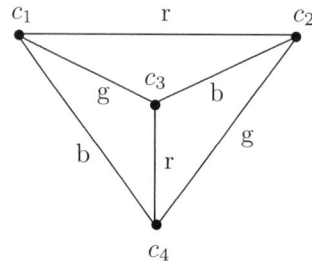

Figure 5.25: The map in Example 27

28. For $A = \{1, 2, 3\}$, let $f : A \to A$ be the function defined by $f = \{(1, 2), (2, 3), (3, 3)\}$. Determine, with explanation, whether f has an inverse function from A to A.

29. For $A = \{a, b, c\}$ and $B = \{w, x, y, z\}$, let $f : A \to B$ be the function defined by $f(a) = y$, $f(b) = z$ and $f(c) = w$. Determine, with explanation,

(a) whether f is one-to-one.   (b) whether f has an inverse function from B to A.

30. For $A = \{p, q, r, s\}$ and $B = \{t, u, v, w\}$, the function $f : A \to B$ defined by $f = \{(p, u), (q, w), (r, t), (s, v)\}$ is a bijective function. Determine $f^{-1} : B \to A$.

31. Show that the function $f : \mathbb{R} \to \mathbb{R}$ defined by $f(x) = 5x - 7$ is bijective and determine $f^{-1}(x)$ for $x \in \mathbb{R}$.

32. Show that the function $f : \mathbb{R} \to \mathbb{R}$ defined by $f(x) = 2 - x^3$ is bijective and determine $f^{-1}(x)$ for $x \in \mathbb{R}$.

33. Let $A = \{1, 2, 3, 4\}$ and $B = \{a, b, c, d, e\}$. The function $f : A \to B$ defined by $f(1) = c$, $f(2) = a$, $f(3) = b$, $f(4) = e$ is one-to-one but not onto. Since f is not bijective, f does not have an inverse from B to A. Show that there exists some set C and a bijective function $g : C \to A$ such that $g(y) = x$ if and only if $f(x) = y$ for every $x \in A$.

34. Let $A = \{-4, -3, -2, -1, 0, 1, 2, 3, 4\}$ and $B = \{0, 1, 4, 9, 16\}$. The function $f : A \to B$ is defined by $f(x) = x^2$ for each $x \in A$. Since f is not bijective, f does not have an inverse from B to A. Show that there exists some set C and a bijective function $g : B \to C$ such that $g(y) = x$ if and only if $f(x) = y$ for every $y \in B$.

35. Let $A = \{1, 2, 3\}$, $B = \{a, b, c\}$ and $C = \{x, y, z\}$. The functions $f : A \to B$ and $g : B \to C$ defined by $f = \{(1, c), (2, a), (3, b)\}$ and $g = \{(a, y), (b, z), (c, x)\}$ are bijective.

    (a) Determine $g \circ f : A \to C$ and $(g \circ f)^{-1} : C \to A$.

    (b) Determine $f^{-1} : B \to A$, $g^{-1} : C \to B$ and $f^{-1} \circ g^{-1} : C \to A$.

36. For sets A, B and C, let $f : A \to B$ and $g : B \to C$ be functions that are both one-to-one and onto. Does the function $g \circ f$ necessarily have an inverse function?

37. Give an example of a finite set S and a bijection $f : S \to S$ such that all of the elements $f(a), (f \circ f)(a), f^{-1}(a)$ are distinct for each $a \in S$.

5.5 Cardinalities of Sets

With the aid of functions, we can take a new and expanded look at what is meant by the number of elements in a set (especially an infinite set). We defined the **cardinality** of a finite set A as the number of elements in A and denoted this number by $|A|$. So if $C = \{c\}$ and $D = \{w, x, y, z\}$, then $|C| = 1$ and $|D| = 4$. Furthermore, $|\emptyset| = 0$. Of course, two finite sets A and B have the same cardinality if they have the same number of elements. For example, if $A = \{a, b, c\}$, then any set B having the same cardinality as A must have three elements. Historically, the cardinality of infinite sets was a difficult idea to understand. Prior to 1874 all infinite sets were considered to be of the same size. However, in 1874 Georg Cantor wrote a famous article in which he explained why there are different kinds of infinity. Before describing why this is the case, let us return to finite sets.

Sets Having the Same Cardinality

The sets $A = \{a, b, c\}$ and $B = \{x, y, z\}$ both have three elements and so $|A| = |B|$. We can construct a bijective function from A to B, say

$$f = \{(a, x), (b, y), (c, z)\}.$$

That is, f pairs off a and x, b and y and c and z. Indeed, for any two finite nonempty sets A and B, $|A| = |B|$ if and only if the elements of A and B can be paired off, that is, $|A| = |B|$ if and only if there exists a bijective function from A to B. Indeed, for finite nonempty sets A and B, we could have defined A and B to have the same cardinality if there exists a bijective function from A to B. This suggests a definition for A and B to have the same cardinality whether A and B are finite sets or not.

Definition 5.71 *Two nonempty sets A and B (finite or infinite) are defined to have the **same cardinality**, written $|A| = |B|$, if there exists a bijective function from A to B.*

While there is nothing new about this definition if A and B are finite, for infinite sets A and B, we now have, for the first time, a definition for $|A| = |B|$. Before considering some consequences of this definition for infinite sets, we make a few observations. Suppose that A and B are infinite sets having the same cardinality. Then there exists a bijective function f from A to B. By Theorem 5.66, f has an inverse function f^{-1} from B to A, which is also bijective. Therefore, to determine whether two infinite sets A and B have the same cardinality, we need to establish the existence of a bijective function either from A to B *or* from B to A. Also, suppose that A, B and C are infinite sets such that $|A| = |B|$ and $|B| = |C|$. Certainly, if these sets are finite, then we can immediately conclude that $|A| = |C|$. Because $|A| = |B|$ and $|B| = |C|$, there exist bijective functions $f : A \to B$ and $g : B \to C$. By Corollary 5.65, the composition $g \circ f : A \to C$ is also bijective and so, by definition, $|A| = |C|$. So if $|A| = |B|$ and $|B| = |C|$, then we can conclude that $|A| = |C|$ whether the sets involved are finite or infinite.

Denumerable Sets

We now consider an important class of infinite sets.

Definition 5.72 *A set A is called* **denumerable** *if $|A| = |\mathbb{N}|$.*

Consequently, a set A is denumerable if it has the same number of elements as the set of positive integers. That is, A is denumerable if there exists a bijective functions from A to $\mathbb{N}$ (or a bijective function from $\mathbb{N}$ to A). For example, the set $\mathbb{N}$ itself is denumerable since the identity function $f : \mathbb{N} \to \mathbb{N}$ defined by $f(n) = n$ for each $n \in \mathbb{N}$ is bijective. The following example of a denumerable set may be unexpected, however.

Result 5.73 *The set of positive even integers is denumerable.*

Solution. It suffices to show that there exists a bijective function from $\mathbb{N}$ to the set A of positive even integers. Consider the function $f : \mathbb{N} \to A$ defined by $f(n) = 2n$ for $n \in \mathbb{N}$. First, we show that f is one-to-one. Assume that $f(a) = f(b)$, where $a, b \in \mathbb{N}$. Then $2a = 2b$ and so $a = b$. Thus f is one-to-one. Next, we show that f is onto. Let $m \in A$. Then $m = 2k$ for some positive integer k. Since $f(k) = 2k = m$, the function f is onto. Because f is bijective, the set of positive even integers is denumerable.   ♦

The previous example illustrates a rather curious fact, namely, it is possible for a set and a proper subset to have the same cardinality. This cannot occur with finite sets, however.

We now provide another example of a denumerable set.

Theorem 5.74 *The function $f : \mathbb{N} \to \mathbb{Z}$ defined by*

$$
\begin{array}{ccccccccc}
1 & 2 & 3 & 4 & 5 & \cdots & n & & \cdots \\
\downarrow & \downarrow & \downarrow & \downarrow & \downarrow & \cdots & \downarrow & & \cdots \\
0 & 1 & -1 & 2 & -2 & \cdots & (-1)^n \lfloor n/2 \rfloor & & \cdots
\end{array}
$$

or

$$f(n) = (-1)^n \lfloor n/2 \rfloor$$

for each $n \in \mathbb{N}$ is bijective.

Proof. First we show that f is one-to-one. Assume that $f(a) = f(b)$, where $a, b \in \mathbb{N}$. Observe that 1 is the only positive integer whose image under f is 0. Hence if $f(a) = f(b) = 0$, then $a = b = 1$. Thus, we may assume that $f(a) = f(b) \neq 0$. We consider two cases.

Case 1. $f(a) = f(b) > 0$. Then a and b are both even, say $a = 2x$ and $b = 2y$, where $x, y \in \mathbb{N}$. Thus $f(a) = x$ and $f(b) = y$. Since $f(a) = f(b)$, it follows that $x = y$ and so $a = 2x = 2y = b$.

Case 2. $f(a) = f(b) < 0$. Then a and b are both odd, say $a = 2x + 1$ and $b = 2y + 1$, where $x, y \in \mathbb{N}$. Thus $f(a) = -x$ and $f(b) = -y$. Since $f(a) = f(b)$, it follows that $x = y$ and so $a = 2x + 1 = 2y + 1 = b$.

Hence f is one-to-one. Next, we show that f is onto. Let $n \in \mathbb{Z}$. If $n \in \mathbb{N}$, then $f(2n) = n$. If $n \leq 0$, then $f(-2n + 1) = n$. Thus f is onto. ∎

As a consequence of this result, we have the following.

Corollary 5.75 *The set of integers is denumerable.*

We now know that $|\mathbb{N}| = |\mathbb{Z}|$, that is, the set of positive integers and the set of integers have the same number of elements. Although this is probably counter-intuitive, when dealing with cardinalities it is essential to rely on the definition, not on our intuition.

Denote the set of positive rational numbers by $\mathbb{Q}^+$. There are certainly many positive rational numbers that are not positive integers, so we probably wouldn't expect that $|\mathbb{Q}^+| = |\mathbb{N}|$ but this is exactly what happens.

Theorem 5.76 *The set of positive rational numbers is denumerable.*

Proof. Certainly, every positive rational number can be expressed as the ratio a/b of two positive integers a and b. In fact, this can be done in many ways. For example,

$$\frac{2}{3} = \frac{4}{6} = \frac{6}{9} = \frac{8}{12} = \cdots.$$

Figure 5.26 shows a table whose rows and columns are both labeled by $1, 2, 3, \ldots$. In the entry whose column is labeled by a and whose row is labeled by b, we place the rational number a/b.

	1	2	3	4	5	$\cdots$
1	$\frac{1}{1}$	$\frac{2}{1}$	$\frac{3}{1}$	$\frac{4}{1}$	$\frac{5}{1}$	$\cdots$
2	$\frac{1}{2}$	$\frac{2}{2}$	$\frac{3}{2}$	$\frac{4}{2}$	$\frac{5}{2}$	$\cdots$
3	$\frac{1}{3}$	$\frac{2}{3}$	$\frac{3}{3}$	$\frac{4}{3}$	$\frac{5}{3}$	$\cdots$
4	$\frac{1}{4}$	$\frac{2}{4}$	$\frac{3}{4}$	$\frac{4}{4}$	$\frac{5}{4}$	$\cdots$
5	$\frac{1}{5}$	$\frac{2}{5}$	$\frac{3}{5}$	$\frac{4}{5}$	$\frac{5}{5}$	$\cdots$
$\vdots$	$\vdots$	$\vdots$	$\vdots$	$\vdots$	$\vdots$	$\vdots$

Figure 5.26: A table of positive rational numbers

Every positive rational number a/b, where $a, b \in \mathbb{N}$, appears in this table, in fact, appears infinitely often since

$$\frac{a}{b} = \frac{2a}{2b} = \frac{3a}{3b} = \cdots.$$

We use this table to show that there is a bijective function $f : \mathbb{N} \to \mathbb{Q}^+$. Starting in the upper lefthand corner of the table, we draw diagonals directed from the lower left to the upper right (as shown in Figure 5.27).

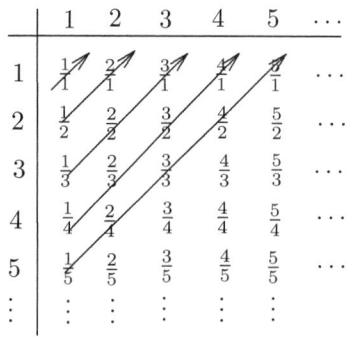

Figure 5.27: Illustrating a bijective function $f : \mathbb{N} \to \mathbb{Q}^+$

The first diagonal contains only one number (namely $1 = 1/1$), the second diagonal contains two numbers (first $1/2$ and then $2 = 2/1$) and so on. We now construct a bijective function f from $\mathbb{N}$ to $\mathbb{Q}^+$. In particular, $f(1) = 1$, $f(2) = 1/2$ and $f(3) = 2$. In general, for $n \in \mathbb{N}$, $f(n)$ is the nth number encountered in this manner that has not been previously encountered. For example, $f(4) = 1/3$ but $f(5)$ is not $2/2 = 1$ since $f(1)$ has already been defined to be 1. The next number after $2/2 = 1$ is $3/1 = 3$. Since the number 3 has not yet been encountered, we define $f(5) = 3$. The images of the first 11 positive integers are therefore:

$$
\begin{array}{ccccccccccc}
1 & 2 & 3 & 4 & 5 & 6 & 7 & 8 & 9 & 10 & 11 \\
\downarrow & \downarrow & \downarrow & \downarrow & \downarrow & \downarrow & \downarrow & \downarrow & \downarrow & \downarrow & \downarrow \\
1 & \frac{1}{2} & 2 & \frac{1}{3} & 3 & \frac{1}{4} & \frac{2}{3} & \frac{3}{2} & 4 & \frac{1}{5} & 5
\end{array} \cdots
$$

From the manner in which this function f is defined, f is bijective and so $|\mathbb{N}| = |\mathbb{Q}^+|$. Therefore, $\mathbb{Q}^+$ is denumerable. ∎

Not only is the set of positive rational numbers denumerable, the set of *all* rational numbers is denumerable.

Theorem 5.77 *The set $\mathbb{Q}$ of rational numbers is denumerable.*

Proof. According to Theorem 5.76, $\mathbb{Q}^+$ is denumerable and so there is a bijective function $f : \mathbb{N} \to \mathbb{Q}^+$. If q_n denotes the image of n under f, then $\mathbb{Q}^+ = \{q_1, q_2, q_3, \ldots\}$ and so $\mathbb{Q} = \{0, q_1, -q_1, q_2, -q_2, \ldots\}$. In fact, this says that the function $g : \mathbb{N} \to \mathbb{Q}$ whose images of $\mathbb{N}$ shown below

$$
\begin{array}{ccccccc}
1 & 2 & 3 & 4 & 5 & 6 & 7 \\
\downarrow & \downarrow & \downarrow & \downarrow & \downarrow & \downarrow & \downarrow \\
0 & q_1 & -q_1 & q_2 & -q_2 & q_3 & -q_3
\end{array} \cdots
$$

is bijective. Therefore, $|\mathbb{N}| = |\mathbb{Q}|$ and so $\mathbb{Q}$ is denumerable. ∎

At this time, it is useful to make an observation. First, if a set A is denumerable, then it is infinite and there is a bijective function $f : \mathbb{N} \to A$. Thus, every element of A appears exactly once in the sequence: $f(1), f(2), f(3), \ldots$. If we denote $f(n)$ by a_n for each positive integer n, then we are saying that the elements of A can be listed as $a_1, a_2, a_3, \ldots$. Indeed, if we have a set B whose elements can be listed as $b_1, b_2, b_3, \ldots$, then the function $g : \mathbb{N} \to B$ defined by $g(n) = b_n$ is bijective, $|\mathbb{N}| = |B|$ and B is denumerable. In summary:

> A set A is denumerable if and only if it is possible to list the elements of A as $a_1, a_2, a_3, \ldots$, that is, there is an infinite sequence $\{a_n\}$ in which every element of A appears exactly once.

Another useful result concerning denumerable sets is the following, which we state without proof.

Theorem 5.78 *Every infinite subset of a denumerable set is denumerable.*

Since we know that the set $\mathbb{Z}$ of integers and the set $\mathbb{Q}$ of rational numbers are denumerable, it follows by Theorem 5.78 that all of the following sets are denumerable.

(1) the set of even integers,

(2) the set of odd integers,

(3) the set $S = \{n \geq m : n \in \mathbb{Z}\}$ for a fixed integer m,

(4) the set $T = \{\frac{a}{b} : a \text{ and } b \text{ are odd integers}\}$,

(5) the set $U = \{\frac{1}{2^m 3^n} : m, n \in \mathbb{Z}\}$.

By Example 5.77, we are saying that $\mathbb{N}$ and $\mathbb{Q}$ have the same number of elements. Since every infinite set that we have seen is denumerable, it might appear that every two infinite sets have the same cardinality and that when it comes to sets, there is only one kind of "infinity." We have already stated that this is not the case, however. We next show that the closed interval

$$[0, 1] = \{x \in \mathbb{R} : 0 \leq x \leq 1\}$$

of real numbers is not denumerable. Before doing this however, we need to describe some characteristics of real numbers with which you may already be familiar. Of course, we know that every real number is either rational or irrational. Furthermore, every real number has a decimal expansion. It is known that irrational numbers have non-repeating decimal expansions. For example, we have seen that $\sqrt{2}$ is irrational. The decimal expansion of $\sqrt{2}$ begins with $1.42421\cdots$. This expansion is infinite and there is never a place in the decimal expansion that is periodic (repeats itself) from some point on. On the other hand, the decimal expansion of a rational number always repeats itself from some point on. For example,

$$\frac{6}{55} = 0.1090909\cdots$$

and the "09" continues to repeat; while

$$\frac{1}{7} = 0.142857142857142857\cdots$$

where "142857" continues to repeat.

Some rational numbers have two decimal expansions. For example,

$$\frac{1}{5} = 0.2 \quad (\text{or } \frac{1}{5} = 0.2000\cdots) \quad \text{and} \quad \frac{1}{5} = 0.1999\cdots.$$

Also,

$$\frac{3}{4} = 0.75000\cdots \text{ and } \frac{3}{4} = 0.74999\cdots.$$

Those rational numbers that have two decimal expansions have one decimal expansion where there is a last nonzero digit d followed only by 0s. The other decimal expansion of such a number is obtained by replacing d by $d-1$ and followed only by 9s. In fact, every rational number that is reduced to lowest terms and such that the only primes that divide its denominator are 2 or 5 has two decimal expansions; otherwise, it has only one decimal expansion. The number $1/3 = 0.333\cdots$ and the numbers $6/55$ and $1/7$ therefore have only one decimal expansion; while $1/40$ and $3/125$ have two decimal expansions.

We are now prepared to show that the closed interval $[0, 1]$ of real numbers is not denumerable. The nature of the statement suggests that we should use a proof by contradiction.

Theorem 5.79 *The closed interval $[0, 1]$ of real numbers is not denumerable.*

Proof. Assume, to the contrary, that $[0, 1]$ is denumerable. Then there exists a bijective function $f : \mathbb{N} \to [0, 1]$. For each $n \in \mathbb{N}$, let $f(n) = a_n \in [0, 1]$. Furthermore, for each $r \in [0, 1]$, there exists exactly one $m \in \mathbb{N}$ such that $f(m) = a_m = r$.

For $n \in \mathbb{N}$, suppose that the decimal expansion of a_n is $0.a_{n1}a_{n2}a_{n3} \cdots$, where if a_n is a rational number with two decimal expansions, then we select that decimal expansion containing all 9s from some point on. Hence

$$f(1) \;=\; a_1 = 0.a_{11}a_{12}a_{13} \cdots$$
$$f(2) \;=\; a_2 = 0.a_{21}a_{22}a_{23} \cdots$$
$$f(3) \;=\; a_3 = 0.a_{31}a_{32}a_{33} \cdots$$

and so on. Consider the number $b \in [0, 1]$ such that

$$b = 0.b_1 b_2 b_3 \cdots,$$

where

$$b_n = \left\{ \begin{array}{ll} 2 & \text{if } a_{nn} = 1 \\ 1 & \text{if } a_{nn} \neq 1. \end{array} \right.$$

For example, if $a_1 = 0.111 \cdots$, $a_2 = 0.02999 \cdots$, $a_3 = 0.141414 \cdots$ and $a_4 = .135167781 \cdots$, then the first four digits in the decimal expansion of b are $2, 1, 2, 2$, that is, $b = 0.2122 \cdots$. Since there are no 0s and no 9s in the decimal expansion of b, it follows that $0.b_1 b_2 b_3 \cdots$ is not the alternative decimal expansion of any rational number. Because the first digit in the decimal expansions of b and a_1 are different, $b \neq a_1$. Since the second digit in the decimal expansions of b and a_2 are different, $b \neq a_2$. In general, $b \neq a_n$ for all $n \in \mathbb{N}$. Consequently, there is no positive integer m such that $f(m) = b$ and so f is not onto. Thus f is not bijective, producing a contradiction. ∎

As a consequence of Theorem 5.79, $\mathbb{N}$ and $[0,1]$ have different cardinalities and so there are at least two different sizes of infinite sets.

Countable and Uncountable Sets

Definition 5.80 *A set that is either finite or denumerable is called* **countable**. *A denumerable set is also called* **countably infinite**. *A set that is not countable is called* **uncountable**.

Therefore, $\{1, 2, 3\}$, $\mathbb{N}$, $\mathbb{Z}$ and $\mathbb{Q}$ are countable sets, while $[0, 1]$ is uncountable. The following result is a consequence of Theorem 5.78.

Theorem 5.81 *Every set that contains an uncountable subset is itself uncountable.*

Proof. Assume, to the contrary, that there exists a countable set A that contains an uncountable subset B. Since B is uncountable, B is infinite. Furthermore, since $B \subseteq A$, the set A is infinite. Because A is infinite and countable, A is denumerable. Thus B is an infinite subset of a denumerable set. By Theorem 5.78, B is denumerable. This, however, produces a contradiction. ∎

Since the closed interval $[0, 1]$ is uncountable, we have the following consequence of Theorem 5.81.

Corollary 5.82 *The set $\mathbb{R}$ of real numbers is uncountable.*

You may very well have encountered complex numbers in your study of mathematics. A **complex number** is a number that can be expressed as $a + bi$ for some real numbers a and b and where $i = \sqrt{-1}$. The set of complex numbers is often denoted by $\mathbb{C}$. Since the number b can be 0 in the definition of a complex number, every real number is complex and so $\mathbb{R} \subseteq \mathbb{C}$. Since $\mathbb{R}$ is uncountable, we have the following.

Corollary 5.83 *The set $\mathbb{C}$ of complex numbers is uncountable.*

There are many other results involving cardinalities of sets. One example is the following.

Theorem 5.84 *If A and B are disjoint denumerable sets, then $A \cup B$ is denumerable.*

Proof. Since A and B are denumerable, the sets A and B can be expressed as

$$A = \{a_1, a_2, a_3, \ldots\} \text{ and } B = \{b_1, b_2, b_3, \ldots\}.$$

The function $f : \mathbb{N} \to A \cup B$ defined by

$$
\begin{array}{cccccccc}
1 & 2 & 3 & 4 & 5 & 6 & \cdots \\
\downarrow & \downarrow & \downarrow & \downarrow & \downarrow & \downarrow & \cdots \\
a_1 & b_1 & a_2 & b_2 & a_3 & b_3 & \cdots
\end{array}
$$

is bijective. Therefore, $A \cup B$ is denumerable.   ∎

An interesting consequence of the preceding theorem is the following.

Theorem 5.85 *The set of irrational numbers is uncountable.*

Proof. Denote the set of irrational numbers by $\mathbb{I}$. Assume, to the contrary, that $\mathbb{I}$ is denumerable. Since $\mathbb{Q}$ and $\mathbb{I}$ are disjoint denumerable sets, $\mathbb{Q} \cup \mathbb{I}$ is denumerable by Theorem 5.84. Since $\mathbb{Q} \cup \mathbb{I} = \mathbb{R}$, it follows that $\mathbb{R}$ is denumerable. This, however, contradicts Corollary 5.82.   ♦

We have mentioned what it means for two sets A and B to have the same cardinality. We now introduce the idea of inequalities with cardinal numbers. For two nonempty sets A and B, we say that $|A| \leq |B|$ if there exists a one-to-one function $f : A \to B$. (Of course, if there exists a one-to-one function f from A to B that is also onto, then $|A| = |B|$.) We say that $|A| < |B|$ if $|A| \leq |B|$ and $|A| \neq |B|$, that is, $|A| < |B|$ if there exists a one-to-one function f from A to B but no bijective function from A to B. Trivially, if $A = \{a, b, c\}$ and $B = \{w, x, y, z\}$, then $|A| < |B|$. Also $|\mathbb{Z}| < |\mathbb{R}|$.

The following theorem shows that for every set A, there exists a set B such that $|A| < |B|$. In particular, we can always take $B = \mathcal{P}(A)$. The proof of this theorem is somewhat subtle, however.

Theorem 5.86 *Every set has a smaller cardinality than its power set, that is,*

$$|A| < |\mathcal{P}(A)|$$

for every set A.

Proof. First, if $A = \emptyset$, then $\mathcal{P}(A) = \{\emptyset\}$. Thus $|A| = 0 < 1 = |\mathcal{P}(A)|$. Hence we may assume that $A \neq \emptyset$. Consider the function $f : A \to \mathcal{P}(A)$ defined by

$$f(a) = \{a\} \text{ for each } a \in A.$$

We show that f is one-to-one. Assume that $f(a) = f(b)$. Then $\{a\} = \{b\}$. This, however, implies that $a = b$ and so f is one-to-one. Thus $|A| \leq |\mathcal{P}(A)|$.

To show that $|A| < |\mathcal{P}(A)|$, it remains to prove that there is no bijective function from A to $\mathcal{P}(A)$. We use a proof by contradiction to verify this. Assume, to the contrary, that there exists a bijective function $g : A \to \mathcal{P}(A)$. For each $x \in A$, let $g(x) = A_x$. Thus $A_x \subseteq A$ for each $x \in A$. Moreover, for each $x \in A$, there are two possibilities, namely either $x \in A_x$ or $x \notin A_x$.

We now define a subset B of A as follows:

$$B = \{x \in A : \ x \notin A_x\}.$$

Since $B \subseteq A$ and $g : A \to \mathcal{P}(A)$ is a bijective function, there exists an element $y \in A$ such that $g(y) = B$. Since $g(y) = A_y$, this means that $B = A_y$. Consider the element y. If $y \in B$, then $y \notin A_y = B$, which is, of course, impossible. Thus $y \notin B = A_y$. However, by the definition of B, it follows that $y \in B$, which again is a contradiction.   ∎

The preceding theorem then has the following corollary.

Corollary 5.87 *There is no set of largest cardinality.*

Consequently, as far as cardinality of infinite sets is concerned, there are infinitely many kinds of infinite sets.

Exercises for Section 5.5

1. Let E denote the set of even integers and O the set of odd integers. Show that $|E| = |O|$.

2. Prove or disprove the following: If A and B are denumerable sets, then $|A| = |B|$.

3. Let A denote a denumerable set and let B be a nonempty finite set. Show that if A and B are disjoint, then $A \cup B$ is a denumerable set.

4. Prove that if A and B are denumerable sets, then $A \times B$ is denumerable.

5. Let $C = \{a + b\sqrt{2} : a, b \in \mathbb{Q}\}$. Prove that C is denumerable.

6. (a) Prove that the function $f : \mathbb{R} - \{1\} \to \mathbb{R} - \{2\}$ defined by $f(x) = \frac{2x}{x-1}$ is bijective.
   (b) Show that $|\mathbb{R} - \{1\}| = |\mathbb{R} - \{2\}|$.

7. Determine whether each of the following is true or false.

   (a) If A and B are nonempty sets with $A \subseteq B$, then $|A| \leq |B|$.
   (b) If A and B are nonempty sets with $A \subseteq B$ and $B \subseteq A$, then $|A| = |B|$.
   (c) If A and B are nonempty sets with $A \subset B$, then $|A| < |B|$.
   (d) If A and B are disjoint nonempty sets, then $|A| < |A \cup B|$.

8. Prove or disprove: If A is a denumerable subset of $\mathbb{R}$, then A consists of rational numbers.

9. Prove or disprove: If A is a nonempty subset of a denumerable set, then A is denumerable.

10. Prove or disprove: There is no set having more elements than the set $\mathbb{R}$ of real numbers.

11. Let A be a denumerable set. Determine, with explanation, whether each of the following is true or false.

    (a) The set of 0-element subsets of A is countable.
    (b) The set of 1-element subsets of A is countable.
    (c) The set of 2-element subsets of A is countable.
    (d) The set of all subsets of A is countable.

12. Prove or disprove: The set $S = \{(a, b) : a, b \in \mathbb{R}\}$ of all points in the plane is uncountable.

13. Let A be a denumerable subset of an uncountable set C. Prove or disprove: If B is a set such that $A \subseteq B \subset C$, then B is denumerable.

14. Prove or disprove: If A is a denumerable set and we continue to add new elements to A, then we will eventually arrive at a set B where $|A| < |B|$.

15. Prove or disprove: If S is a set containing the four distinct elements a, b, c and d, then $|S - \{a, b\}| = |S - \{c, d\}|$.

Chapter 5 Highlights

Key Concepts

bijective function (**bijection**): a function that is one-to-one and onto.

$\mathbb{C}$: the set of complex numbers.

cardinalities of two sets, comparing: $|A| = |B|$: two sets A and B have the same cardinality if there exists a bijective function from A to B; $|A| \leq |B|$: there exists a one-to-one function from the set A to the set B; $|A| < |B|$: $|A| \leq |B|$ and $|A| \neq |B|$ or there exists a one-to-one function from the set A to the set B but no bijective function from A to B.

ceiling function $\lceil x \rceil$: the smallest integer greater than or equal to x.

codomain (of f): the set B where f is a function from A to B.

complex number: $a + bi$ for some real numbers a and b and where $i = \sqrt{-1}$.

composition (of f and g), $g \circ f$: for functions $f : A \to B$ and $g : B \to C$, the composition $g \circ f : A \to C$ is the function defined by $(g \circ f)(x) = g(f(x))$ for each $x \in A$.

corollary: a consequence of another result.

countable set: a set that is either finite or denumerable.

denumerable set (or **countably infinite set**): a set A such that $|A| = |\mathbb{N}|$.

domain (of f): the set A where f is a function from A to B.

equivalence class (resulting from an equivalence relation R on A), $[a]$: the set of all elements in A that are related to a by R.

equivalence relation: a reflexive, symmetric and transitive relation on a set.

floor function $\lfloor x \rfloor$: the largest integer less than or equal to x.

function (from A to B): an assignment of a unique element of B to each element of A.

graph (of a function f from A to B, where $A, B \subseteq \mathbb{R}$): the set of points (x, y) in the Cartesian plane for which $x \in A$, $y \in B$ and $y = f(x)$.

image, b is an image of a under f: $b = f(a)$.

injective function (**injection** or one-to-one function): a function from A to B such that distinct elements of A have distinct images in B.

inverse (of a bijective function $f : A \to B$), f^{-1}: a bijective function from B to A such that $(r, s) \in f$ if and only if $(s, r) \in f^{-1}$.

one-to-one correspondence: a bijective function.

one-to-one function (injective function): a function from A to B such that distinct elements of A have distinct images in B.

onto function (surjective function): a function from A to B such that every element of B is the image of some element of A.

permutation (on a nonempty set A): a bijective function from A to A.

range (of f): the set of all images of f.

reflexive: a relation R on a set S is reflexive if $(a, a) \in R$ for all $a \in S$.

related to, $a \, R \, b$: a is related to b by the relation R if $(a, b) \in R$. $a \, \mathcal{R} \, b$: a is not related to b by the relation R if $(a, b) \notin R$.

relation (from a set A to a set B): a subset of $A \times B$.

relation (on a set S): a relation from S to S or a subset of $S \times S$.

surjective function (**surjection** or onto function): a function from A to B such that every element of B is the image of some element of A.

symmetric: a relation R on a set S is symmetric if $(a, b) \in R$, then $(b, a) \in R$ for all $a, b \in S$.

transitive: a relation R on a set S is transitive if $(a, b) \in R$ and $(b, c) \in R$, then $(a, c) \in R$ for all $a, b, c \in S$.

uncountable set: a set that is not countable.

Key Results

- Let R be an equivalence relation on a nonempty set A and let a and b be elements of A. Then $[a] = [b]$ if and only if $a \, R \, b$.

- Let R be an equivalence relation defined on a nonempty set A. If $\mathcal{P}$ is the set of distinct equivalence classes of A resulting from R, then $\mathcal{P}$ is a partition of A.

- Let R be an equivalence relation defined on a nonempty set A. If $[a]$ and $[b]$ are equivalence classes of A resulting from R, then either $[a] = [b]$ or $[a] \cap [b] = \emptyset$.

- Let A, B and C be nonempty sets and $f : A \to B$ and $g : B \to C$ two functions. Then the following hold:

    If f and g are one-to-one, then so is $g \circ f$.

    If f and g are onto, then so is $g \circ f$.

    If f and g are bijective, then so is $g \circ f$.

- A function $f : A \to B$ has an inverse $f^{-1} : B \to A$ if and only if f is bijective. In this case, f^{-1} is also bijective.

- If $f : A \to B$ is a bijective function, then $f^{-1} \circ f$ is the identity function on A and $f \circ f^{-1}$ is the identity function on B.

- The set $\mathbb{Z}$ of integers is denumerable.

- The set $\mathbb{Q}$ of rational numbers is denumerable.

- Every infinite subset of a denumerable set is denumerable.

- Every set that contains an uncountable subset is itself uncountable.

- The set $\mathbb{R}$ of real numbers is uncountable.

- The set $\mathbb{C}$ of complex numbers is uncountable.

- If A and B are disjoint denumerable sets, then $A \cup B$ is denumerable.

- The set of irrational numbers is uncountable.

- For every set A, $|A| < |\mathcal{P}(A)|$.

- There is no set of largest cardinality.

Supplementary Exercises for Chapter 5

1. For an integer $n \geq 2$, let $S_n = \{1, 2, \ldots, n\}$ and let T_n be the set of all 2-element subsets of S_n. For $A, B \in T_n$, define the relation R_n on T_n by $A\ R_n\ B$ if $A \cap B = \emptyset$. For $n = 4$, list all elements of R_n.

2. Let $A = \{a, b\}$. How many relations on A are there?

3. Let $A = \{1, 2, 3, 4\}$. For the relation $R = \{(1,1), (2,2), (2,3), (3,2), (4,4)\}$ on A, determine which of the properties reflexive, symmetric, transitive R possesses.

4. Let $A = \{1, 2, 3\}$ and let $R = \{(1,3)\}$ be a relation on A. Which of the properties reflexive, symmetric, transitive does R possess?

5. Let R be the relation defined on $\mathbb{Z}$ by $a\ R\ b$ if $a = b$ or $a = 2b$.

   (a) Give an example of two integers that are related by R and two integers that are not.

   (b) Which of the properties reflexive, symmetric, transitive does R possess?

6. A relation R is defined on $\mathbb{N}$ by $a\ R\ b$ if $b = a^n$ for some $n \in \mathbb{N}$.

   (a) Give an example of two positive integers that are related by R and two positive integers that are not.

   (b) Which of the properties reflexive, symmetric, transitive does R possess?

7. A relation R is defined on $\mathbb{R}$ by $a\ R\ b$ if $ab \leq 0$.

   (a) Give an example of two real numbers that are related by R and two real numbers that are not.

   (b) Which of the properties reflexive, symmetric, transitive does R possess?

8. A relation R is defined on $\mathbb{Z}$ by $a\ R\ b$ if $|a - b| \in \{0\} \cup \{2^n : n \in \mathbb{Z}, n \geq 0\}$.

   (a) Give an example of two integers that are related by R and two integers that are not.

   (b) Which of the properties reflexive, symmetric, transitive does R possess?

9. Let $S = \{1, 2, 3, 4, 5\}$.

   (a) Give an example of an equivalence relation R on S that results in four distinct equivalence classes. What are these equivalence classes for your example?

   (b) Give an example of an equivalence relation R' on S that results in three distinct equivalence classes. What are these equivalence classes for your example?

10. Let $S = \{1, 2, 3, 4, 5\}$. Give an example of an equivalence relation R on S having the property that for every 2-element subset A of S, there are two distinct equivalence classes, both of which have a nonempty intersection with A.

11. A relation R on a nonempty set A is defined to be **circular** if whenever $a\ R\ b$ and $b\ R\ c$, then $c\ R\ a$ for all $a, b, c \in A$. Prove that a relation R on A is an equivalence relation if and only if R is reflexive and circular.

12. A relation R is defined on $\mathbb{Z}$ by $a\ R\ b$ if $5a - b$ is even.

   (a) Prove that R is an equivalence relation.

   (b) Describe the distinct equivalence classes resulting from R.

13. A relation R is defined on $\mathbb{N}$ by $a\ R\ b$ if $a^2 + b^2$ is even.

(a) Show that R is an equivalence relation.

(b) Describe the distinct equivalence classes resulting from R.

14. A relation R is defined on the set $\mathbb{Z}$ of integers by $a\ R\ b$ if $|a| = |b|$.

(a) Show that R is an equivalence relation.

(b) Describe the distinct equivalence classes resulting from R.

15. Let S be a nonempty set. Prove that if R_1 and R_2 are equivalence relations on S, then $R_1 \cap R_2$ is also an equivalence relation on S.

16. Let S be a nonempty set. Show that if R_1 and R_2 are equivalence relations on S, then $R_1 \cup R_2$ need not be an equivalence relation on S.

17. Let $A = \{0, 1, 3, 4\}$ and $B = \{-2, -1, 0, 1, 2\}$.

(a) Give an example of a function $f : A \rightarrow B$. (b) What is the range of f?

18. Let $S = \{n \in \mathbb{Z} : -n \in \mathbb{N}\}$.

(a) Give an example of a function $g : \mathbb{N} \rightarrow S$. (b) What is the range of g?

19. Let $A = \{1, 3, 4\}$ and $B = \{-2, -1, 0, 1, 4\}$. For $a \in A$, define $f(a) = b$, where $|b| = a$. Is f a function from A to B?

20. Let $A = \{-3, -1, 0, 2\}$ and $B = \{0, 1, 4, 9\}$. Let $f = \{(x, y) \in A \times B : y = x^2\}$. Is f a function from A to B?

21. Let A denote the set of all real numbers x such that (x, y) lies on the graph of $x = y^2 - y$. If (x, y) lies on the graph, then let $y = f(x)$. Is f a function from A to $\mathbb{R}$?

22. Consider the function $f : A \rightarrow B$, where $A = \{1, 2, 3\}$ and $B = \{a, b, c, d\}$, shown in the diagram of Figure 5.28. (a) Is f one-to-one? (b) Is f onto?

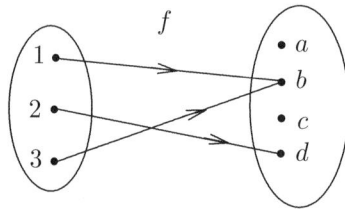

Figure 5.28: The diagram of the function f in Exercise 22

23. Define $f : \mathbb{N} \rightarrow \mathbb{N}$ by $f(n) = \lceil n/3 \rceil$. (a) Is f one-to-one? (b) Is f onto?

24. Let $f : \mathbf{R} \rightarrow \mathbf{R}$ be defined by $f(x) = 5x + 3$.

(a) Show that f is one-to-one.

(b) Show that f is onto.

(c) Find $f^{-1}(x)$ for $x \in \mathbb{R}$.

25. Let $f : A \rightarrow B$ be a function. For a subset B_1 of B, the **inverse image** $f^{-1}(B_1)$ of B_1 is defined as $f^{-1}(B_1) = \{a \in A : f(a) \in B_1\}$. For every two subsets B_1 and B_2 of B, prove the following.

(a) $f^{-1}(B_1 \cup B_2) = f^{-1}(B_1) \cup f^{-1}(B_2)$.

(b) $f^{-1}(B_1 \cap B_2) = f^{-1}(B_1) \cap f^{-1}(B_2)$.

26. Give an example of a function $f : \mathbb{N} \to \mathbb{N}$ that is

   (a) one-to-one and onto.

   (b) one-to-one but not onto.

   (c) onto but not one-to-one.

   (d) neither one-to-one nor onto.

27. Determine which of the following functions are bijective functions from $\mathbb{R}$ to $\mathbb{R}$.

   (a) $f(x) = -2x + 7$.

   (b) $f(x) = x(x - 5) + 1$.

   (c) $f(x) = x^3 - x$.

   (d) $f(x) = x^4 + 2x^2 + 3$.

28. A function $f : \mathbb{Z} \to \mathbb{Z}$ is defined by

$$f(n) = \begin{cases} n + 1 & \text{if } n \geq 0 \\ n - 1 & \text{if } n < 0 \text{ and } n \neq -10 \\ n & \text{if } n = -10. \end{cases}$$

   (a) Is f one-to-one? (b) Is f onto?

29. Show that there is a function $f : \mathbb{R} \to \mathbb{R}$ that is

   (a) onto but not one-to-one.

   (b) one-to-one but not onto.

30. Let $f : \mathbf{R} \to \mathbf{R}$ and $g : \mathbf{R} \to \mathbf{R}$ be defined by $f(x) = 3x + 1$ and $g(x) = x^2 - 2$. Determine $(f \circ g)(2)$ and $(g \circ f)(2)$.

31. Show that the function $f : \mathbb{R} \to \mathbb{R}$ defined by $f(x) = ax + b$, where $a, b \in \mathbb{R}$ and $a \neq 0$ is bijective.

32. Prove that the function $f : \mathbb{R} - \{3\} \to \mathbb{R} - \{1\}$ defined by $f(x) = \frac{x}{x-3}$ is bijective.

33. Prove that the function $f : \mathbb{R} - \{5\} \to \mathbb{R} - \{1\}$ defined by $f(x) = \frac{x+1}{x-5}$ is bijective.

34. Let A be the set of people who go to a particular movie theater on a particular day, $B = \mathbb{N}$ (the set of positive integers) and $C = \{6, 7, 10\}$. A function $f : A \to B$ associates with person $a \in A$ his or her age in years, while g associates with a person of a particular age, the cost of a movie ticket. In particular,

$$g(n) = \begin{cases} 6 & \text{if } n < 12 \\ 10 & \text{if } 12 \leq n < 60 \\ 7 & \text{if } n \geq 60. \end{cases}$$

   (a) For each $a \in A$, what does $(g \circ f)(a)$ represent?

   (b) If $A = \{a_1, a_2, \ldots, a_n\}$, then express in words what the sum

$$\sum_{i=1}^{n} (g \circ f)(a_i) = (g \circ f)(a_1) + (g \circ f)(a_2) + \cdots + (g \circ f)(a_n)$$

   represents.

35. Let $f : \mathbb{R}^+ \to \mathbb{R}$ be a function defined by $f(x) = \ln x$ for all $x \in \mathbb{R}^+$. Prove that the nth derivative of $f(x)$ is given by $f^{(n)}(x) = \frac{(-1)^{n+1}(n-1)!}{x^n}$ for every positive integer n.

36. Let $f : \mathbb{R} \to \mathbb{R}$ be a function defined by $f(x) = xe^{-x}$ for all $x \in \mathbb{R}$. Prove that the nth derivative of $f(x)$ is given by $f^{(n)}(x) = (-1)^n e^{-x}(x - n)$ for every positive integer n

37. Show that the set A of negative integers is denumerable.

38. Show that the set $\{q \in \mathbb{Q} : q \le 0\}$ is denumerable.

39. Prove that if A and B are disjoint countable sets, then $A \cup B$ is countable.

40. Determine, with explanation, whether the following is true or false. If A and B are disjoint sets such that A is countable and B is uncountable, then $A \cup B$ is uncountable.

41. Determine, with explanation, whether the following is true or false. If A and B are denumerable sets, then $A \cap B$ is denumerable.

42. Determine, with explanation, whether the following is true or false. If A is an infinite set and B is an infinite subset of A, then $|A - B| < |A|$.

43. Let $\mathbb{I}$ denote the set of irrational numbers.

   (a) Prove that if $n \in \mathbb{N}$ and $r \in \mathbb{I}$, then nr is irrational.

   (b) Determine, with explanation, whether the following is true or false. If S is an infinite subset of $\mathbb{I}$, then S is uncountable.

44. Give an example of a set D containing the set $\mathbb{C}$ of complex numbers as a proper subset. Is D countable or uncountable?

Chapter 6

Algorithms and Complexity

At this point a number of the fundamental concepts that occur throughout discrete mathematics have been discussed and proof techniques have been introduced to establish the truth of certain mathematical statements. Having laid the foundation of discrete mathematics, we now proceed more deeply into the subject with our emphasis changing to problem-solving. With each problem that we encounter, various questions can arise, often including one or more of the following:

- (i) Does the problem have a solution?
- (ii) If the problem does have a solution, how many solutions are there?
- (iii) Among all of the solutions of the problem, is there a best (or optimal) solution?
- (iv) Is there a procedure that can be used to find a solution to the problem?

Although all of these questions are important, our primary interest in this chapter will be with question (iv).

6.1 What is an Algorithm?

Suppose that we are given a list of four real numbers. Then it shouldn't be particularly difficult to:

(1) find the largest number on the list,

(2) add the numbers on the list,

(3) determine whether a specific number is on the list,

(4) determine whether some number appears twice on the list.

But what if the list contains a thousand or perhaps a million numbers? None of the problems (1) – (4) seems nearly as easy to solve now. Certainly none of these problems can be solved by observation for such a large list. Indeed, these problems seem better suited to computer-aided solutions. This suggests writing a computer program in each case that will solve the problem. Before writing such a program, however, what is needed is a plan – a detailed strategy for solving the problem.

Definition 6.1 *An **algorithm** is a procedure for solving a problem that consists of input, output and a finite sequence of steps that converts the input to the output.*

What we want to do therefore is to write algorithms that will solve the problems mentioned above for lists containing a great many numbers. In the process of doing this, we also want to learn techniques for writing algorithms that solve other problems as well. The word "algorithm" comes from the word "algorizm" which emanated from the name of the 9th century Persian mathematician Abu Abdullah Muhammad ibn Musa al-Khwarizm. Although the original use of the word "algorithm" dealt with performing arithmetic on the integers, the meaning of this word has expanded to include procedures for solving a wide variety of problems.

Examples of algorithms occurred long before the 9th century, however. In the next chapter, we will see an algorithm due to the famous geometer Euclid (325 BC–265 BC) for finding the greatest common divisor of two positive integers. It is believed that the first person to develop a computer algorithm was Augusta Ada Byron (1815–1852), the Countess of Lovelace and daughter of the illustrious poet Lord Byron. Tutored in mathematics by Augustus De Morgan, she suggested to Charles Babbage (1792–1871) a plan for how a computing machine might calculate certain numbers (called Bernoulli numbers). In her honor, a programming language developed by the United States Department of Defense was named "Ada" in the 1970s.

There are several ways to describe an algorithm. One way is simply to write the steps of the algorithm in sentences using English and, for many purposes, this is a useful method. Another method is to describe the steps using some computer language. A disadvantage of doing this, however, is that the algorithm would probably be difficult to understand if the individual reading this was not familiar with this language. Another common way to specify an algorithm is by means of what is called **pseudocode**, an artificial language that resembles (but still provides the precision of) the code (programs) of some familiar computer languages. This method is preferred by many computer scientists and mathematicians and, much of the time, this is the method we will use. However an algorithm is written, it is important that it is understandable and that it accomplishes what is intended.

To illustrate how an algorithm might be written using pseudocode, we present an algorithm that solves problem (1), that is, an algorithm that describes a method for finding the largest number in a list a, b, c, d of four numbers. We denote the largest of these four numbers by $\max(a, b, c, d)$. So $\max(9, 14, 7, 17) = 17$ and $\max(11, 16, 16, 13) = 16$.

Algorithm 6.2 *Find the Maximum Number in a List of Four Numbers.*

This algorithm finds the largest of four numbers a, b, c, d.

> **Input**: Four numbers a, b, c, d.
> **Output**: $x = \max(a, b, c, d)$
> 1. $x := a$ [x is assigned the value a]
> 2. **if** $b > x$ **then** $x := b$ [if $b > x$, then x is assigned the value b]
> 3. **if** $c > x$ **then** $x := c$ [if $c > x$, then x is assigned the value c]
> 4. **if** $d > x$ **then** $x := d$ [if $d > x$, then x is assigned the value d]
> 5. **output** x

Anything written within brackets [ ] is a comment. It is informational only; it is not a step to be performed. We will provide comments about many of the steps in the first few algorithms that we present.

We now make a few comments about Algorithm 6.2. The input for this algorithm is four numbers a, b, c, d. As we indicated, our goal is to find the largest of these numbers, which we denote by $x = \max(a, b, c, d)$. Step 1 tells us to assign x the value of a. First, as suggested above, the symbol := represents assignment. That is, $x := a$ means that x is initially assigned the value a. According to Step 2, if $b > x$ (that is, if $b > a$), then we replace x by b. On the other hand, if it is not true that $b > x$, then the conclusion of this step is not executed and we proceed to Step 3 and so on.

Let's see how this algorithm works in practice.

Example 6.3

Use Algorithm 6.2 to find the largest of the four numbers $3, 5, 2, 7$.

Solution. In this case, $a = 3, b = 5, c = 2, d = 7$. Thus x is assigned the value 3 in Step 1. That is, after Step 1 is performed, $x = 3$. In Step 2, we see whether $b > x$, that is, whether $5 > 3$. It is. So x is now assigned the number 5. In Step 3, we check whether $c > x$, that is, whether $2 > 5$. It is not. Consequently, we move on to Step 4. In this step, we determine whether $d > x$, that is, whether $7 > 5$. Since this is true, x is assigned 7. Hence after Step 4, $x = 7$. Step 5 outputs the value of x, namely 7, which is the maximum of the four numbers.   ♦

What Algorithm 6.2 does in Example 6.3 is illustrated by proceeding downward through the table below.

	x	output
$a = 3$	3	
$b = 5$	5	
$c = 2$	5	
$d = 7$	7	$x = 7$

When we write an algorithm using pseudocode, the standard arithmetic operators $+$, $-$, $*$ or juxtaposition (for multiplication) and $/$ (for division) are used, as well as the comparative logical operators $=$, $\neq$, $<$, $\leq$, $>$ and $\geq$. (Observe that $=$ is the equals operator, while $:=$ denotes assignment.) It is often useful to introduce loops into algorithms. Two common examples of loops are the for loop and the conditional while loop.

For integers a and b with $a \leq b$, a **for loop** has the following appearance, according to whether there is a single step or two or more steps to perform within the loop.

> **for** $i = a$ **to** b **do**
>   this step

> **for** $i = a$ **to** b **do**
>   **begin**
>   these steps
>   **end**

When a for loop is first encountered, i is assigned the integer a and the step on the next line is executed. If there is more than one step to be executed, then these steps are written between **begin** and **end** and all such steps are executed in the order listed. Once the step or steps are executed, the value of i is increased by 1 to $a + 1$ and the step or steps are executed again. The value of i is then increased by 1 again, this time to $a + 2$ and, once again, the step or steps are executed. This continues as long as $i \leq b$. When i reaches the value b, the step or steps are executed for the last time. When $i > b$, the loop terminates and the algorithm proceeds to the step that immediately follows the for loop.

For a statement P, a **while loop** has the following appearance, according to whether there is a single step to perform within the loop or two or more steps.

> **while** P **do**
>   this step

> **while** P **do**
>   **begin**
>   these steps
>   **end**

If a while loop is encountered in an algorithm and there is a single step on the next line, then this step is executed provided the statement P is true. If P is true and there is more than one step to execute, then the steps written between **begin** and **end** are executed in the order listed. If P is true and after the step or steps within the while loop have been executed, we return to the while loop and check once again to determine whether P is a true or false statement. If P is true, then we proceed as above. On the other hand, if the while loop is encountered and the statement P is false, then the algorithm proceeds immediately to the next step in the algorithm that follows the while loop. Ordinarily, there is more than one step within a while loop and the statement P is adjusted at some stage.

Indeed, what is accomplished in a for loop can also be accomplished in the following while loop, where a and b are integers with $a \leq b$.

> $i := a$
> **while** $i \leq b$ **do**
>   **begin**
>   these steps
> $i := i + 1$
>   **end**

We now use a for loop to write a more general algorithm that finds the largest number in a potentially long list of numbers.

Algorithm 6.4 *Find the Maximum Number in a List of Numbers.*

This algorithm finds the largest number in a list $s : a_1, a_2, \ldots, a_n$ of n numbers.

> **Input**: A positive integer n and a sequence $s : a_1, a_2, \ldots, a_n$ of n numbers.
> **Output**: $x = \max(s)$
> 1. $x := a_1$ [x is assigned the first number in the sequence and is the temporary
>   maximum]
> 2. **for** $i := 2$ **to** n **do**   [i is sequentially assigned the integers from 2 to n]
> 3.   **if** $a_i > x$ **then** $x := a_i$ [if $a_i > x$, then a value larger than x has been found
>   and x is replaced by a_i]
> 4. **output** x

What Algorithm 6.4 Does

1. Sets the temporary maximum equal to the first number on the list.

2. Compares the next number on the list with the temporary maximum and, if this number is larger, sets the temporary maximum equal to this number.

3. Repeats the previous step if there are more numbers remaining on the list.

4. Stops when the list has been exhausted and outputs the temporary maximum, which is the maximum number on the list.

We now find the largest number in a list of four numbers using the more general Algorithm 6.4.

Example 6.5

Illustrate Algorithm 6.4 for the sequence $s : 8, 11, 11, 9$.

Solution. In this case, $a_1 = 8$, $a_2 = 11$, $a_3 = 11$, $a_4 = 9$. The numbers in the sequence s are input, as is the length $n = 4$ of this sequence. In Step 1, x is assigned the number 8. Step 2 is a for loop and i is initially assigned the value 2. Because $a_2 = 11 > 8 = x$, the value 11 is assigned to x in Step 3. We return to the for loop in Step 2 and i is increased to 3. Because $a_3 = 11 > 11$ is not true, the conclusion of Step 3 is not executed. We return to Step 2, where i is increased to $4 = n$. Since $a_4 = 9 > 11 = x$ is not true and $i = n$, we proceed to Step 4, which outputs x. Because $x = 11$, this is the maximum of the numbers in the sequence s.   ♦

What Algorithm 6.4 does in Example 6.5 is shown in the table below.

i	a_i	x	output
		8	
2	11	11	
3	11	11	
4	9	11	$x = 11$

We now consider a general algorithm for solving problem (2); that is, rather than adding four numbers, the algorithm will find the sum of the numbers in any finite list of numbers.

Algorithm 6.6 *Determine the Sum of the Numbers in a Sequence.*

This algorithm finds the sum of the numbers in a sequence $s : a_1, a_2, \ldots, a_n$ of n numbers.

> **Input**: A positive integer n and a sequence $s : a_1, a_2, \ldots, a_n$ of n numbers.

Output: $x = \displaystyle\sum_{i=1}^{n} a_i.$

1. $x := 0$ [x is assigned the value 0, which is the temporary sum]
2. **for** $i := 1$ **to** n **do** [i is assigned the integers from 1 to n]
3. $x := x + a_i$ [a_i is added to the current value of x to provide the new value
                         of the temporary sum x]
4. **output** x

What Algorithm 6.6 Does

1. Sets the temporary sum equal to 0.

2. Adds the first (or next) number in the sequence to the temporary sum and this new sum becomes the temporary sum.

3. Repeats the previous step if there are more numbers remaining on the list.

4. Stops when the list has been exhausted and outputs the temporary sum, which is the sum of all numbers on the list.

We now use this algorithm to add four numbers.

Example 6.7

Illustrate Algorithm 6.6 for the sequence $s : 3, 7, 8, 4$.

Solution. In this case, $n = 4$ and $a_1 = 3$, $a_2 = 7$, $a_3 = 8$, $a_4 = 4$. In Step 1, x is assigned the value 0. Step 2 is a for loop in which i is initially assigned the value 1. In Step 3, x is assigned the number $0 + a_1 = 0 + 3 = 3$. We now return to Step 2, where i is increased to 2. In Step 3, x is now replaced by $3 + a_2 = 3 + 7 = 10$. Once again, we return to Step 2, where i is increased to 3. Now x is assigned the number $10 + a_3 = 10 + 8 = 18$ in Step 3. We return to Step 2, where i is increased to $4 = n$. In Step 3, x is assigned the number $18 + a_4 = 18 + 4 = 22$. Since the for loop is complete, we proceed to Step 4 and $x = 22$ (the sum of the four numbers in s) is output.   ◆

How Algorithm 6.6 is implemented in Example 6.7 is shown in the table below.

i	a_i	x	output
		0	
1	3	3	
2	7	10	
3	8	18	
4	4	22	$x = 22$

We now present a method for solving a problem that is more general than problem (3), that is, an algorithm to determine whether some specific number appears on a given finite list of numbers. A while loop is introduced in the following algorithm.

Algorithm 6.8 *Determine Whether a Specified Number Appears in a Sequence.*

This algorithm determines whether a specified number k is one of the numbers in a list $s : a_1, a_2, \ldots, a_n$ of n numbers.

Input: A positive integer n, a number k and a sequence $s : a_1, a_2, \ldots, a_n$ of n numbers.
Output: k "is in the sequence" or k "is not in the sequence" as appropriate.

1. $i := 1$ [i is assigned the number 1]
2. **while** $i \leq n$ **do** [provided $i \leq n$, Steps 3 and 4 are performed]
     begin
3.   **if** $a_i = k$ **then output** k "is in the sequence" and $i := n + 1$ [it is determined
                                whether $a_i = k$]
4. $i := i + 1$ [i is increased by 1]
     end
5. **if** $i = n + 1$ **then output** k "is not in the sequence"

What Algorithm 6.8 Does

1. Sets an index i equal to 1.

2. Compares the index i to the size n of the sequence and, provided $i \leq n$, determines whether
   the first (or next) number in the sequence is the number of interest (and, if so, outputs this
   information and increases the index to 1 plus the size); otherwise, it increases the index by 1.
   Setting $i = n + 1$ results in an exit from the while loop.

3. Repeats the previous step provided the index does not exceed the size of the sequence.

4. Outputs the information that the number of interest is not in the sequence if the index equals
   1 plus the size.

Let's see what Algorithm 6.8 does when $k = 10$ and $s : 8, 14, 10, 9$.

Example 6.9

Illustrate Algorithm 6.8 for $k = 10$ and $s : 8, 14, 10, 9$.

Solution. Here $k = 10$, $n = 4$ and $a_1 = 8$, $a_2 = 14$, $a_3 = 10$, $a_4 = 9$. In Step 1, i is assigned the
value 1. Step 2 is a while loop. Since $1 = i \leq n = 4$, the two steps (Steps 3 and 4) between **begin**
and **end** are performed. Since $a_1 \neq 10$, the conclusion of Step 3 is not executed. In Step 4, i is
increased to 2. Because $2 \leq 4$, Steps 3 and 4 are performed again. Since $a_2 \neq 10$, we move on to
Step 4, where i becomes 3. Because $3 \leq 4$, Steps 3 and 4 are performed once again. Now $a_3 = 10$
and so 10 *is in the sequence* is the output. Also, i is assigned the number 5. In Step 4, i is increased
to 6. Since $6 \leq 4$ is not true, we leave the while loop and proceed to Step 5. Because $i \neq 5$, the
algorithm ends and we are left with the output: 10 *is in the sequence.*   ♦

The table below illustrates how Algorithm 6.8 is applied to the sequence in Example 6.9.

k	i	a_i	output
10	1	8	
10	2	14	
10	3	10	10 is in the sequence
10	6	9	

We next see what Algorithm 6.8 does if the specified number k does not appear in the sequence
under consideration.

Example 6.10

Illustrate Algorithm 6.8 for $k = 10$ and $s : 8, 14, 9$.

Solution. In this case, $k = 10$, $n = 3$ and $a_1 = 8$, $a_2 = 14$, $a_3 = 9$. In Step 1, i is assigned the
number 1. Step 2 is a while loop. Since $1 = i \leq n = 3$, we perform Steps 3 and 4. Since $a_1 \neq 10$,
the conclusion of Step 3 is not executed. In Step 4, i is increased to 2. Because $2 \leq 3$, we perform
Steps 3 and 4. Since $a_2 \neq 10$, we move on to Step 4, where i becomes 3. Because $3 \leq 3$, we again
perform Steps 3 and 4. Since $a_3 \neq 10$, we move on to Step 4, where i is increased to 4. Since $4 \leq 3$

is false, we leave the while loop and go to Step 5. Because $4 = i = n + 1$, the following is output: 10 *is not in the sequence.* ◆

The table below shows the consequence of applying Algorithm 6.8 to the sequence in Example 6.10.

k	i	a_i	output
10	1	8	
10	2	14	
10	3	9	
			10 is not in the sequence

We now turn to the final problem (4) in our original list of problems. In this problem, we are asked whether some number appears two or more times in a given sequence of numbers. Let's suppose that we not only want to know whether a number appears twice in a sequence but which number has this property if this should occur. In the algorithm described below, a "nested while loop" appears. Just as the two innermost parentheses are paired off in $(\cdots(\cdots)\cdots)$, so too are the innermost **begin** and **end** paired off when two ends follow two begins.

Algorithm 6.11 *Determine Whether Some Number Appears Twice in a Sequence.*

This algorithm determines whether a sequence $s : a_1, a_2, \ldots, a_n$ of n numbers contains two equal terms.

Input: A positive integer n and a sequence $s : a_1, a_2, \ldots, a_n$ of n numbers.
Output: a_i "appears twice in the sequence" or "no term appears twice in the sequence" as appropriate.

1. $i := 0$
2. **while** $i \leq n - 1$ **do**
   begin
3. $i := i + 1$
4. $j := i + 1$
5.   **while** $j \leq n$ **do**
          begin
6.   **if** $a_i = a_j$ **then output** a_i "appears twice in the sequence" and
                             $i := n + 1$ and $j := n$
7. $j := j + 1$
          end
       end
8.   **if** $i = n$ **then output** "no term appears twice in the sequence"

We now apply this algorithm for a particular sequence.

Example 6.12

Illustrate Algorithm 6.11 for the sequence $s : 11, 9, 13, 9$.

Solution. For this sequence, $n = 4$ and $a_1 = 11$, $a_2 = 9$, $a_3 = 13$, $a_4 = 9$. In Step 1, i is assigned the value 0. Since $0 = i \leq n - 1 = 3$, all of Steps 3 – 5 are performed. By Steps 3 and 4, we have $i = 1$ and $j = 2$. Step 5 is a while loop that lies within the while loop at Step 2. Because $2 = j \leq n = 4$, we perform Steps 6 and 7 within the while loop at Step 5. Because $a_1 \neq a_2$, the conclusion of Step 6 is not performed. By Step 7, j is increased to 3. The while loop at Step 5 is still in progress. Since $3 = j \leq n = 4$, Steps 6 and 7 are performed. Because $a_1 \neq a_3$, the conclusion of Step 6 is not performed and we move to Step 7 where j is increased to 4. The while loop at Step 5 is still in progress. Because $4 = j \leq n = 4$, Steps 6 and 7 are performed. Because $a_1 \neq a_4$, we move to Step 7 where j is increased to 5. Because $5 = j \leq n = 4$ is not true, the while loop at Step 5 is completed.

Returning to the while loop at Step 2, we have $1 = i \leq n-1 = 3$ and so Steps 3 – 5 are performed. In Steps 3 and 4, i becomes 2 and j is assigned the value 3. We next encounter the while loop at Step 5. Because $3 = j \leq n = 4$, Steps 6 and 7 are performed. Since $a_2 \neq a_3$, we move to Step 7, where j is increased to 4. The while loop at Step 5 is still in progress. Because $4 = j \leq n = 4$, Steps 6 and 7 are performed. Since $a_2 = a_4$, the following is output: 9 *appears twice in the sequence* and i is assigned the value $n+1 = 5$ and j is assigned $n = 4$. In Step 7, j is increased to 5. Because $5 = j \leq n = 4$ is not true, the while loop at Step 5 is completed. Returning to the while loop at Step 2, we find that $5 = i \leq n-1 = 3$ is not true and the conclusions in Steps 3–5 are not performed. Now the while loop at Step 2 is complete and we move to Step 8. Because $5 = i = n = 4$ is false, the conclusion of Step 8 is not performed and the algorithm is completed.   ♦

The table below shows the consequence of applying Algorithm 6.11 to the sequence in Example 6.12.

i	j	a_i	a_j	output
0				
1	2	11	9	
1	3	11	13	
1	4	11	9	
2	3	9	13	
2	4	9	9	9 appears twice in the sequence
5	4			
5	5			

There are several variations of Algorithm 6.11. Some of these are explored in the exercises.

Exercises for Section 6.1

1. Illustrate Algorithm 6.4 for the sequence $s : 3, 5, 4, 5$.

2. Illustrate Algorithm 6.4 for the sequence $s : 7, 2, -1, 4, -8$.

3. (a) Write an algorithm to determine the smallest number in a list a, b, c of three numbers.

   (b) Give a step by step description of what this algorithm does for the sequence $8, 6, 4$.

   (c) Give a step by step description of what this algorithm does for the sequence $7, 5, 5$.

4. (a) Write an algorithm to determine the minimum number in a sequence $s : a_1, a_2, \ldots, a_n$ of n numbers.

   (b) Give a step by step description of what this algorithm does for the sequence $s : 6, 3, 5, 3, 7$.

5. (a) For a sequence s: $a_1, a_2, \ldots, a_n$ of $n \geq 2$ distinct numbers, write an algorithm to determine the next-to-largest number in the sequence.

   (b) Give a step by step description of what this algorithm does for the sequence $s : 7, 10, 12, 15$.

6. (a) Write an algorithm to determine the middle number (not the smallest or the largest) of three distinct numbers a, b, c.

   (b) Give a step by step description of what this algorithm does for the sequence $7, 6, 4$.

7. For an odd integer $n \geq 3$ and a sequence s: $a_1, a_2, \ldots, a_n$ of n distinct integers, consider the following algorithm (where the output is not given).

   Input: An odd positive integer n and a sequence s: $a_1, a_2, \ldots, a_n$ of n distinct integers.

   Output: ?

1. **for** $i := 1$ **to** n **do**
2. $a_{n+i} := a_i$
3. $x := a_1$
4. $r := 0$
5. $j := 1$
6. **while** $r \neq (n-1)/2$ **do**
     begin
7. $r := 0$
8.   **for** $i := j+1$ **to** $j+n-1$ **do**
9.   **if** $a_i > a_j$ **then** $r := r+1$
10.   **if** $r = (n-1)/2$ **then** $x := a_j$
11. $j := j+1$
     end
12. **output** x

(a) What is output for the sequence s : 23, 21, 11, 19, 7?

(b) For a sequence of n distinct integers, where $n \geq 3$ is odd, what is output? Justify your answer.

8. Write an algorithm to determine $\displaystyle\sum_{i=1}^{n-1} |a_i - a_{i+1}|$ for a sequence s : a_1, a_2, ..., a_n of $n \geq 2$ numbers.

9. Illustrate Algorithm 6.8 for $k = 11$ and s : $9, 10, 14, 11$.

10. Illustrate Algorithm 6.8 for $k = 17$ and s : $16, 14, 18$.

11. Write an algorithm that determines whether a sequence s : $a_1, a_2, \ldots, a_n$ of n numbers contains any negative numbers.

12. Write an algorithm that determines how many negative numbers there are in a sequence s : $a_1, a_2, \ldots, a_n$ of n numbers.

13. For a given number k and a sequence s : $a_1, a_2, \ldots, a_n$ of n distinct numbers, write an algorithm to determine whether k is a term in the sequence and, if so, which term in the sequence it is.

14. For a given number k and a sequence s : $a_1, a_2, \ldots, a_n$ of n numbers, write an algorithm that determines whether any term in s is within 1 of k.

15. Illustrate Algorithm 6.11 for the sequence s : $1, 2, 3$.

6.2   Growth of Functions

Ordinarily, there are several algorithms that give a solution to a particular problem. Given two algorithms that provide a solution to the same problem, it is natural to ask whether one of these algorithms might be preferred over the other. In order to study this question, we describe a common method of comparing the growth of two functions f and g defined on the set $\mathbb{N}$ of positive integers and whose values are positive real numbers, that is, whose codomain is the set $\mathbb{R}^+$.

Big-O of a Function

Definition 6.13 *A function $f : \mathbb{N} \to \mathbb{R}^+$ is **big-O** (or **big-oh**) of a function $g : \mathbb{N} \to \mathbb{R}^+$, written $f = O(g)$ or $f(n) = O(g(n))$, if there exist a positive constant C and positive integer k such that*

$$f(n) \leq Cg(n) \text{ for every integer } n \geq k. \tag{6.1}$$

Suppose that the graphs of two functions $y = f(x)$ and $y = g(x)$ are given in Figure 6.1(a), where $f : \mathbb{R}^+ \to \mathbb{R}^+$ and $g : \mathbb{R}^+ \to \mathbb{R}^+$. If there exist some positive constant C and positive integer k such that $f(x) \leq Cg(x)$ for every real number $x \geq k$ (see Figure 6.1(b)), then $f(n) \leq Cg(n)$ for every integer $n \geq k$ and $f(n) = O(g(n))$.

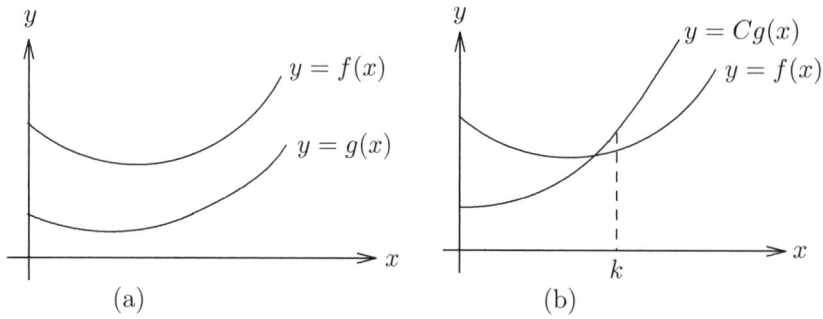

Figure 6.1: The graphs of functions f, g and Cg

If $f(n) = O(g(n))$, then for large values of n,

f grows no faster than a constant times g.

Observe that if $f(n) \leq Cg(n)$ for all $n \geq k$, then $f(n) \leq C'g(n)$ for all $n \geq k'$ for each constant $C' \geq C$ and each integer $k' \geq k$. Therefore, the constants C and k in (6.1) are not unique. In fact, if C is chosen large enough, then $f(n) \leq Cg(n)$ for *every* positive integers n (that is, k can be chosen to be 1). (See Exercise 1.)

Example 6.14

Let $f : \mathbb{N} \to \mathbb{R}^+$ and $g : \mathbb{N} \to \mathbb{R}^+$ be two functions defined by $f(n) = 2n^2$ and $g(n) = n^3$ for all $n \in \mathbb{N}$. Show that $f = O(g)$.

Solution. Since $n^2 \leq n^3$ whenever $n \geq 1$, it follows that $2n^2 \leq 2n^3$ for $n \geq 1$. Therefore, for every integer $n \geq 1$, $2n^2 \leq 2(n^3)$. Hence $f = O(g)$, where $C = 2$ and $k = 1$ in the definition.   ♦

Example 6.15

Let $f : \mathbb{N} \to \mathbb{R}^+$ and $g : \mathbb{N} \to \mathbb{R}^+$ be two functions defined by $f(n) = 3n^2 + 6$ and $g(n) = n^3 + n$ for all $n \in \mathbb{N}$. Show that $f = O(g)$.

Solution. Certainly $n^2 \leq n^3$ for $n \geq 1$ and so $3n^2 \leq 3n^3$ for $n \geq 1$. Also, $6 \leq 3n$ for $n \geq 2$. Therefore, for every integer $n \geq 2$,
$$3n^2 + 6 \leq 3(n^3 + n).$$

Hence $f = O(g)$, where $C = 3$ and $k = 2$ in the definition.   ♦

For two functions $f : \mathbb{N} \to \mathbb{R}^+$ and $g : \mathbb{N} \to \mathbb{R}^+$, the function f **is not big-O** of g and we write $f \neq O(g)$ or $f(n) \neq O(g(n))$ if there exist *no* positive constant C and positive integer k for which

$$f(n) \leq Cg(n) \text{ for every integer } n \geq k.$$

While $f = O(g)$ for the functions f and g defined in Examples 6.14 and 6.15, $g \neq O(f)$. We verify this for Example 6.15, while this is left as an exercise for Example 6.14 (see Exercise 2).

Example 6.16

Let $f : \mathbb{N} \to \mathbb{R}^+$ and $g : \mathbb{N} \to \mathbb{R}^+$ be two functions defined by $f(n) = 3n^2 + 6$ and $g(n) = n^3 + n$ for all $n \in \mathbb{N}$. Show that $g \neq O(f)$.

Solution. We use a proof by contradiction. Assume, to the contrary, that $g = O(f)$. Then there exist a positive constant C and a positive integer k such that

$$n^3 + n \leq C(3n^2 + 6) \text{ for } n \geq k. \tag{6.2}$$

Dividing both sides of the inequality in (6.2) by n^3, we have

$$1 + \frac{1}{n^2} \leq \frac{3C}{n} + \frac{6C}{n^3} \text{ for } n \geq k. \tag{6.3}$$

Let n be an integer that is greater than all of k, $6C$ and 2, that is, $n > \max(k, 6C, 2)$. Then

$$\frac{1}{n} < \frac{1}{6C} \text{ and } \frac{1}{n} < \frac{1}{2}.$$

So

$$\frac{3C}{n} = 3C \cdot \frac{1}{n} < 3C \cdot \frac{1}{6C} = \frac{1}{2}$$

and

$$\frac{6C}{n^3} = \frac{6C}{n^2} \cdot \frac{1}{n} < \frac{6C}{n^2} \cdot \frac{1}{6C} = \frac{1}{n^2} < \frac{1}{4}.$$

Thus the right side of (6.3) is

$$\frac{3C}{n} + \frac{6C}{n^3} < \frac{1}{2} + \frac{1}{4} = \frac{3}{4};$$

while the left side of (6.3) is $1 + \frac{1}{n^2} > 1$, which is a contradiction. ◆

In Examples 6.15 and 6.16, f is a polynomial function of degree 2 and g is a polynomial function of degree 3 and we showed that $f = O(g)$ but $g \neq O(f)$. In fact, the following holds:

If f and g are two polynomial functions where the degree of g is greater than that of f, then $f = O(g)$ but $g \neq O(f)$.

We consider another example to illustrate this fact.

Example 6.17

Let the functions $f : \mathbb{N} \to \mathbb{R}^+$ and $g : \mathbb{N} \to \mathbb{R}^+$ be defined by

$$f(n) = n \text{ and } g(n) = n^2 \text{ for all } n \in \mathbb{N}.$$

Then $f = O(g)$ but $g \neq O(f)$.

Solution. Observe that $f = O(g)$ since $n \leq n^2$ for every positive integer n (and so we may take $C = 1$ and $k = 1$ in the definition). On the other hand $g \neq O(f)$, for suppose, to the contrary, that $g = O(f)$. Then there exist a positive constant C and a positive integer k such that

$$n^2 \leq Cn \text{ for all } n \geq k.$$

Dividing the inequality $n^2 \leq Cn$ by n, we have that $n \leq C$ for all $n \geq k$. However, $n \leq C$ is not satisfied if we choose n to be an integer that is both greater than C and greater than k. Hence $n = O(n^2)$ but $n^2 \neq O(n)$.   ◆

We saw in Chapter 5 that

(1) $2^n \geq n$ for all $n \geq 1$,

(2) $2^n \geq n^2$ for all $n \geq 4$ and

(3) $n! \geq 2^n$ for all $n \geq 4$.

Consequently,

$$n = O(2^n), \ n^2 = O(2^n), \ \text{and} \ \ 2^n = O(n!).$$

However,

$$2^n \neq O(n), \ 2^n \neq O(n^2), \ \text{and} \ \ n! \neq O(2^n).$$

Several familiar functions are listed below in increasing order. That is, each function is big-O of every function listed to its right, while no function is big-O of a function listed to its left. We write $\log n$ for $\log_2 n$.

$$1 \qquad \log n \qquad n \qquad n \log n \qquad n^2 \qquad 2^n \qquad n!$$

The graphs of the functions $f_1(x) = 1$, $f_2(x) = \log x$, $f_3(x) = x$, $f_4(x) = x \log x$, $f_5(x) = x^2$ and $f_6(x) = 2^x$ are given in the Figure 6.2. Notice that the scales on the two axes are different. Observe that $x^2 > 2^x$ when $2 < x < 4$. For the one integer value 3 in this range, $3^2 > 2^3$. However, $2^x \geq x^2$ when $x \geq 4$. Thus $2^n \geq n^2$ for every integer $n \geq 4$. Taking $C = 1$ and $k = 4$ in the definition, we see that $n^2 = O(2^n)$.

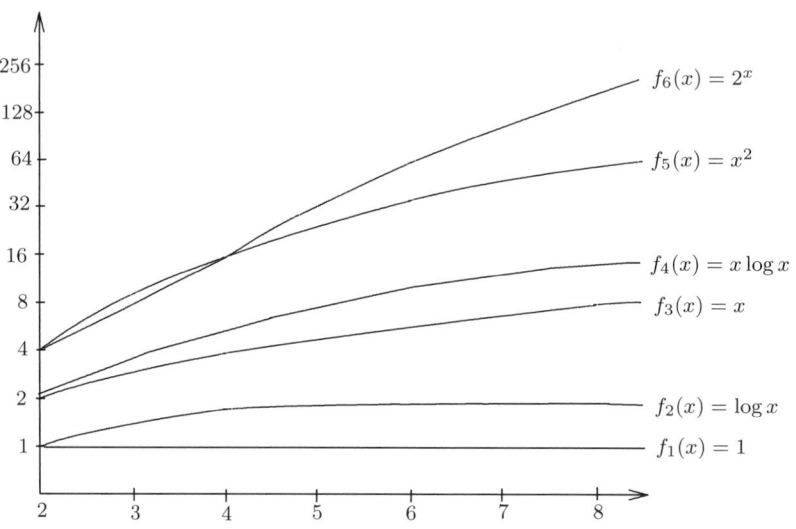

Figure 6.2: The graphs of functions

Big-Theta of a Function

If $f : \mathbb{N} \to \mathbb{R}^+$ and $g : \mathbb{N} \to \mathbb{R}^+$ are two functions for which $f = O(g)$, then we know that $f(n) \leq Cg(n)$ for some positive constant C and sufficiently large positive integers n (that is, $Cg(n)$

is an upper bound for $f(n)$ for sufficiently large positive integers n). Therefore, we know that f grows no faster than a constant times g for sufficiently large n, but this provides us with no clear indication on how to compare the growth of f and g. We now describe a concept that deals with this.

Definition 6.18 *A function $f : \mathbb{N} \to \mathbb{R}^+$ is **big-theta** of a function $g : \mathbb{N} \to \mathbb{R}^+$, written $f = \Theta(g)$ or $f(n) = \Theta(g(n))$, if there exist positive constants C_1 and C_2 and a positive integer k such that*

$$C_1 g(n) \leq f(n) \leq C_2 g(n) \text{ for every integer } n \geq k.$$

When $f = \Theta(g)$, we say that

$$f \text{ and } g \text{ grow at the same rate.}$$

Example 6.19

Let $f : \mathbb{N} \to \mathbb{R}^+$ and $g : \mathbb{N} \to \mathbb{R}^+$ be two functions defined by $f(n) = 2n^2 + 6$ and $g(n) = 3n^2 + 3n$ for all $n \in \mathbb{N}$. Show that $f = \Theta(g)$.

Solution. Observe that for each integer $n \geq 3$,

$$2n^2 + 6 \leq 2n^2 + 2n = \tfrac{2}{3}(3n^2 + 3n).$$

Furthermore, for each integer $n \geq 1$,

$$2n^2 + 6 = n^2 + n^2 + 6 \geq n^2 + n + 6 \geq n^2 + n = \frac{1}{3}(3n^2 + 3n).$$

Therefore, for each integer $n \geq 3$,

$$\frac{1}{3}(3n^2 + 3n) \leq 2n^2 + 6 \leq \frac{2}{3}(3n^2 + 3n),$$

or

$$\frac{1}{3}g(n) \leq f(n) \leq \frac{2}{3}g(n).$$

Thus $f = \Theta(g)$, where $C_1 = \frac{1}{3}$, $C_2 = \frac{2}{3}$ and $k = 3$ in the definition. ♦

According to Example 6.19 then, $2n^2 + 6 = \Theta(3n^2 + 3n)$. In fact, the following holds:

If f and g are two polynomial functions of the same degree, then $f = \Theta(g)$.

There is a close connection between the big-O and big-theta concepts.

Theorem 6.20 *Let $f : \mathbb{N} \to \mathbb{R}^+$ and $g : \mathbb{N} \to \mathbb{R}^+$ be two functions. Then $f = \Theta(g)$ if and only if $f = O(g)$ and $g = O(f)$.*

Proof. Assume first that $f = \Theta(g)$. Then there exist positive constants C_1 and C_2 and a positive integer k such that

$$C_1 g(n) \leq f(n) \leq C_2 g(n) \text{ for every integer } n \geq k.$$

In particular, $f(n) \leq C_2 g(n)$ for every integer $n \geq k$. So $f = O(g)$. Since $C_1 g(n) \leq f(n)$ for every integer $n \geq k$ and $C_1 > 0$, it follows that $g(n) \leq \left(\frac{1}{C_1}\right) f(n)$ for every integer $n \geq k$. Hence $g = O(f)$.

Next, we verify the converse. Assume that $f = O(g)$ and $g = O(f)$. Since $f = O(g)$, there exists a positive constant C' and a positive integer k' such that

$$f(n) \leq C' g(n) \text{ for every integer } n \geq k'.$$

Because $g = O(f)$, there exists positive constant C'' and a positive integer k'' such that

$$g(n) \leq C'' f(n) \text{ for every integer } n \geq k''.$$

Thus,

$$f(n) \geq \left(\tfrac{1}{C''}\right) g(n) \text{ for every integer } n \geq k''.$$

Let $k = \max(k',\ k'')$. Thus for every integer $n \geq k$,

$$\left(\tfrac{1}{C''}\right) g(n) \leq f(n) \leq C' g(n)$$

and so $f = \Theta(g)$.   ∎

Let's consider an illustration of the previous theorem.

Example 6.21

Let the functions $f : \mathbb{N} \to \mathbb{R}^+$ and $g : \mathbb{N} \to \mathbb{R}^+$ be defined by

$$f(n) = \frac{n^2 + 3n + 2}{2} \text{ and } g(n) = n^2$$

for all $n \in \mathbb{N}$. Verify that $f = \Theta(g)$.

Solution. Since

$$\frac{n^2 + 3n + 2}{2} \leq \frac{n^2 + 3n^2 + 2n^2}{2} = 3n^2$$

for every positive integer n, it follows that $f = O(g)$. On the other hand,

$$n^2 \leq n^2 + 3n + 2 = 2\left(\frac{n^2 + 3n + 2}{2}\right)$$

for all $n \in \mathbb{N}$. It follows that $g = O(f)$.

Since $f = O(g)$ and $g = O(f)$, it then follows by Theorem 6.20 that $f = \Theta(g)$.   ◆

The following is a fundamental property of the big-theta concept and can be considered as a consequence of Theorem 6.20 (see Exercise 13).

Theorem 6.22 *Let $f : \mathbb{N} \to \mathbb{R}^+$ and $g : \mathbb{N} \to \mathbb{R}^+$ be two functions. If $f = \Theta(g)$, then $g = \Theta(f)$.*

There is also a connection between the big-theta concept and limits.

Theorem 6.23 *Let $f : \mathbb{N} \to \mathbb{R}^+$ and $g : \mathbb{N} \to \mathbb{R}^+$ be two functions. If $\lim\limits_{n \to \infty} \dfrac{f(n)}{g(n)} = c$ for some positive real number c, then $f = \Theta(g)$.*

The converse of Theorem 6.23 is not true. For example, consider the functions $f : \mathbb{N} \to \mathbb{R}^+$ and $g : \mathbb{N} \to \mathbb{R}^+$ defined by $f(n) = n(2 + \sin n)$ and $g(n) = n$. Since $-1 \leq \sin n \leq 1$ for every positive integer n, it follows that $n \leq n(2 + \sin n) \leq 3n$, that is,

$$1 \cdot g(n) \leq f(n) \leq 3g(n)$$

for every positive integer n. Hence $f = \Theta(g)$. However, it turns out that

$$\lim_{n \to \infty} \frac{f(n)}{g(n)} = \lim_{n \to \infty} \frac{n(2 + \sin n)}{n} = \lim_{n \to \infty} (2 + \sin n)$$

does not exist.

Although we've only been concerned with functions having domain $\mathbb{N}$ and codomain $\mathbb{R}^+$ as these are the functions that we will encounter later in the chapter, it is possible to define and study the growth of functions having different domains and codomains.

Exercises for Section 6.2

1. For functions $f : \mathbb{N} \to \mathbb{R}^+$ and $g : \mathbb{N} \to \mathbb{R}^+$, $f = O(g)$ if there is a positive constant C and a positive integer k such that $f(n) \leq Cg(n)$ for every integer $n \geq k$. Show that there is a positive constant C' such that $f(n) \leq C'g(n)$ for every positive integer n.

2. For the functions $f : \mathbb{N} \to \mathbb{R}^+$ and $g : \mathbb{N} \to \mathbb{R}^+$ in Example 6.14 defined by $f(n) = 2n^2$ and $g(n) = n^3$ for all $n \in \mathbb{N}$, it was shown that $f = O(g)$. Show that $g \neq O(f)$.

3. Let $f : \mathbb{N} \to \mathbb{R}^+$ and $g : \mathbb{N} \to \mathbb{R}^+$ be functions defined by $f(n) = 5n + 7$ and $g(n) = n^2$ for all $n \in \mathbb{N}$. Show that $f = O(g)$ but $g \neq O(f)$.

4. Let $f(n) = 2n^2 + 7n - 1$. Show that $f = O(n^3)$.

5. Show that $\log n = O(n)$. [Hint: Recall that $2^n \geq n$ for every positive integer n.]

6. For which of the following is $f(n) = O(n)$?

   (a) $f(n) = 100$.　(b) $f(n) = 2n + 5$.　(c) $f(n) = \lfloor n/2 \rfloor$.

   (d) $f(n) = \lceil n/2 \rceil$.　(e) $f(n) = n^2 + 3n + 2$.　(f) $f(n) = n \log n$.

7. For which of the following is $f(n) = O(n^2)$?

   (a) $f(n) = 2n + 5$.　(b) $f(n) = \lfloor n/2 \rfloor$.　(c) $f(n) = n^2 + 3n + 2$.

   (d) $f(n) = n \log n$.　(e) $f(n) = n^2 \log n$.　(f) $f(n) = 2^n$.

8. For the function f defined by $f(n) = \frac{n^2+1}{n+1}$ for each $n \in \mathbb{N}$, show that $f(n) = O(n)$.

9. Let $f : \mathbb{N} \to \mathbb{R}^+$ and $g : \mathbb{N} \to \mathbb{R}^+$ be two functions defined by $f(n) = 2n + 1$ and $g(n) = n$ for all $n \in \mathbb{N}$. Show that $f = \Theta(g)$.

10. Let $f : \mathbb{N} \to \mathbb{R}^+$ and $g : \mathbb{N} \to \mathbb{R}^+$ be functions defined by $f(n) = n^2 + 3n$ and $g(n) = n^2 + 5$ for all $n \in \mathbb{N}$. Show that $f = \Theta(g)$.

11. Let $f : \mathbb{N} \to \mathbb{R}^+$ and $g : \mathbb{N} \to \mathbb{R}^+$ be functions defined by $f(n) = n^2 + 4n + 1$ and $g(n) = n^2 + 4$ for all $n \in \mathbb{N}$. Show that $f = \Theta(g)$.

12. Let $f : \mathbb{N} \to \mathbb{R}^+$ be a function. Prove that $f = \Theta(f)$.

13. Prove Theorem 6.22: *Let $f : \mathbb{N} \to \mathbb{R}^+$ and $g : \mathbb{N} \to \mathbb{R}^+$ be two functions. If $f = \Theta(g)$, then $g = \Theta(f)$.*

14. Let $f : \mathbb{N} \to \mathbb{R}^+$, $g : \mathbb{N} \to \mathbb{R}^+$ and $h : \mathbb{N} \to \mathbb{R}^+$ be three functions. Prove that if $f = \Theta(g)$ and $g = \Theta(h)$, then $f = \Theta(h)$.

15. Let f and g be two functions defined by $f(n) = \frac{1}{2}n^2 + 5n + 1$ and $g(n) = 2n^2 + 3$. Show that $f(n) = \Theta(g(n))$.

16. For functions $f : \mathbb{N} \to \mathbb{R}^+$ and $g : \mathbb{N} \to \mathbb{R}^+$, we say that f is **big-omega** of g, written as $f = \Omega(g)$ or $f(n) = \Omega(g(n))$, if there exist a positive constant C and a positive integer k such that $f(n) \geq Cg(n)$ for every integer $n \geq k$.

    (a) Prove that $f = \Theta(g)$ if and only if $f = O(g)$ and $f = \Omega(g)$.

    (b) Prove that $f = \Omega(g)$ if and only if $g = O(f)$.

6.3   Analysis of Algorithms

If we have a relatively complex computational problem to solve, then we have seen that a logical first step is to write an algorithm. This gives us a detailed step by step plan for solving the problem. It is always a good idea to try out this procedure on a small data set to convince ourselves that the algorithm is doing as we planned. The next step is to use the algorithm as a roadmap to guide us in writing a computer program. After doing this, it would probably be wise to run the program with some data sets for which the answer is known. Of course, one would expect that the larger the size of the data, the longer it would take to run the program. But what if we run the program and find that it is taking an inordinately long time to obtain a solution? Then this is unacceptable and all of our effort seems to have been wasted. It would certainly have been useful to know this could happen before we went to the trouble of writing the computer program and running it with data.

Complexity of an Algorithm

As we noted earlier, there are often many algorithms for solving a particular problem. Quite possibly, some of these algorithms are better than others. But what do we mean by this? It is simply too complicated and too time-consuming (and too expensive) to compare every pair of computer programs by running them with the same data set. Furthermore, we might reach a different conclusion if we change the data. Rather than consider two different programs that have been written to solve a particular problem, it makes more sense to study instead what led to the programs, namely, their respective algorithms. The idea then is to have a method for comparing two algorithms. By an **analysis of an algorithm**, we normally refer to a process for obtaining estimates of the time and space needed to execute the algorithm. The **complexity of an algorithm** is the amount of space and time needed to execute the algorithm. The **space complexity** of an algorithm concerns a study of computer memory and the data structures employed, which is not a subject of this book.

The **time complexity** of an algorithm concerns a study of the time required to solve the problem using the algorithm as a function of the size of the input. Good estimates of the time needed to execute a program are often extraordinarily difficult to obtain, for these depend on such things as the computer being used, how the data is represented, what software is being used and how the program is being translated into machine language. Satisfactory estimates can usually be obtained, however, by considering the time complexity of the underlying algorithm. To determine a measure of the speed of an algorithm, we count the number of certain kinds of operations performed by the algorithm. This ordinarily provides a reasonable estimate of the speed of the algorithm and gives us a mechanism for comparing two algorithms. If the major operations of an algorithm are comparative operations, then the time complexity of an algorithm is the number of comparisons made when the algorithm is used to solve a problem. On the other hand, if the major operations are arithmetic (addition, multiplication and so on), then the time complexity of an algorithm is the number of these operations.

We now determine the time complexity of some algorithms, beginning with Algorithm 6.4, where the maximum number in a sequence of n numbers is found.

Example 6.24

Determine the time complexity of Algorithm 6.4.

Algorithm 6.4:   Find the largest number in a list $s : a_1, a_2, \ldots, a_n$ of n numbers.

   Input: A positive integer n and a sequence $s : a_1, a_2, \ldots, a_n$ of n numbers.
   Output: $x = \max(s)$

    1. $x := a_1$ [x is assigned the first number in the sequence]
    2. **for** $i := 2$ **to** n **do** [i is sequentially assigned the integers from 2 to n]
    3.   **if** $a_i > x$ **then** $x := a_i$ [if $a_i > x$, then a value larger than x has been
                                     found and x is replaced by a_i]

    4. **output** x

Solution. A comparison occurs in Step 3 for each of the $n - 1$ values of i ($i = 2, 3, \ldots, n$). Thus there are $n - 1$ comparisons and the time complexity of Algorithm 6.4 is $n - 1$.   ◆

It is ordinarily the case that we are not concerned with the exact number of comparisons or arithmetic operations when computing the time complexity of an algorithm. For instance, in Example 6.24, we are not counting any arithmetic operations or assignments done within the algorithm. That is, the number of comparisons is only meant to give us an estimate or an idea of how much time it would take to use this algorithm on a data set of size n. To be sure, the time required to solve a problem using a particular algorithm does not depend only on the number of operations used in the algorithm. It depends also on the computer and software being used. However running a program that implements an algorithm to solve the problem on a different (faster) computer or with different (improved) software typically results in a time improvement that is a constant multiple of the earlier time used and depends little on n. This suggests the advantage of using the big-O notation to discuss time complexity. Hence what we wish to do is to express the time complexity as $O(f(n))$ or, better yet, $\Theta(f(n))$ for some (preferably common) function $f : \mathbb{N} \to \mathbb{R}^+$.

If the time complexity of an algorithm can be expressed as $\Theta(f(n))$ for some polynomial $f(n)$, then we say that the algorithm has **polynomial time complexity**. Consequently, the time complexity of such an algorithm can be expressed as $\Theta(n^p)$ for a nonnegative integer p. If $p = 0$, that is, if the time complexity is $O(1)$, then there are only a finite number of operations being counted in the algorithm and so the algorithm has **constant time complexity**. An algorithm with time complexity $\Theta(n)$ has **linear time complexity** and one with time complexity $\Theta(n^2)$ has **quadratic time complexity**. So Algorithm 6.4 has linear time complexity as $n - 1 = \Theta(n)$.

Suppose that two algorithms A and B have time complexities $O(f(n))$ and $O(g(n))$, respectively. If $f(n) = O(g(n))$ but $g(n) \neq O(f(n))$, then Algorithm A is said to be **more efficient** than Algorithm B. If $f(n) = \Theta(g(n))$, then f and g are considered to have the **same degree of efficiency**.

Next, we determine the time complexity of Algorithm 6.6, where the sum of the numbers in a sequence of size n is computed.

Example 6.25

Determine the time complexity of Algorithm 6.6.

Algorithm 6.6: Find the sum of the numbers in a sequence $s : a_1, a_2, \ldots, a_n$ of n numbers.

    Input: A positive integer n and a sequence $s : a_1, a_2, \ldots, a_n$ of n numbers.

    Output: $x = \displaystyle\sum_{i=1}^{n} a_i$.

    1. $x := a_1$ [x is assigned the value a_1]
    2. **for** $i := 2$ **to** n **do** [i is assigned the integers from 2 to n]
    3. $x := x + a_i$ [a_i is added to the current value of x to provide the new value of x]
    4. **output** x

Solution. This is not an algorithm that involves comparisons. Since addition is the key operation, we interpret the time complexity in this case as the number of additions. There is an addition for each value of i ($i = 2, 3, \ldots, n$). Thus there are $n - 1$ additions and the time complexity of Algorithm 6.6 is $n - 1 = \Theta(n)$.   ◆

Analyzing Algorithm 6.8, which determines whether some number k appears in a sequence of n numbers, introduces something we have not yet seen.

Algorithm 6.8: Determine whether a specified number k is one of the numbers in a list $s : a_1, a_2,$ $\ldots, a_n$ of n numbers.

> **Input**: A positive integer n, a number k and a sequence $s : a_1, a_2, \ldots, a_n$ of n numbers.
> **Output**: k "is in the sequence" or k "is not in the sequence" as appropriate.
> 1. $i := 1$ [i is assigned the number 1]
> 2. **while** $i \leq n$ **do** [provided $i \leq n$, Steps 3 and 4 are performed]
> > **begin**
> 3.   **if** $a_i = k$ **then output** k "is in the sequence" and $i := n + 1$ [it is determined
> > > whether $a_i = k$]
> 4. $i := i + 1$ [i is increased by 1]
> > **end**
> 5. **if** $i = n + 1$ **then output** k "is not in the sequence"

In Step 1, i is assigned the value 1. In Step 2, a comparison is made since 1 is compared with n. In Step 3, another comparison is made as it is determined whether $a_1 = k$. Suppose, in fact, that $a_1 = k$. Then i is assigned the value $n+1$. In Step 4, i is assigned the value $n+2$. In Step 2, $n+2$ is compared with n. Since $n + 2 \leq n$ is not true, we leave the while loop and move on to Step 5 where $i = n + 2$ is compared with $n + 1$. This is the final step of Algorithm 6.8. A total of 4 comparisons have been made. What we have just described is the *minimum* number of comparisons that can be made when Algorithm 6.8 is executed. The number of comparisons made when Algorithm 6.8 is used can vary widely depending on where in the sequence the number k appears or, in fact, whether k appears in the sequence at all.

The maximum number of comparisons that can occur when a problem is solved using an algorithm is called the **worst case time complexity** of the algorithm.

Example 6.26

Determine the worst case time complexity of Algorithm 6.8.

Solution. The maximum number of comparisons in Algorithm 6.8 occurs when k is the last term of the sequence or when k does not appear in the sequence. Suppose that one of these situations occurs. In Step 2, i and n are compared for $i = 1, 2, \ldots, n+1$. This is a total of $n + 1$ comparisons. In Step 3, a_i and k are compared for $i = 1, 2, \ldots, n$. This is a total of n comparisons. Whether $a_n = k$ or not, i is compared with $n + 1$ in Step 5. Hence the total number of comparisons is $(n + 1) + n + 1 = 2n + 2 = \Theta(n)$, which is the worst case time complexity of Algorithm 6.8.   ♦

Without additional information, it is impossible to know how likely it is that a given number k might appear in a sequence of n distinct numbers. However, if we know beforehand that k appears somewhere in the sequence, then it is equally likely that k is any one of the n numbers. The average number of comparisons needed to determine where k appears in the sequence if k is any one of the n terms is called the **average case time complexity** of this algorithm. This gives a user of the algorithm a better estimate of how many comparisons are likely to be needed if the algorithm is used (and if the number k is in the sequence). Of course, the worst case complexity tells a user that the number of comparisons will never be greater than a certain number ($2n + 2$ for Algorithm 6.8). We next determine the average case time complexity of Algorithm 6.8. For this purpose, it will be useful for us to recall the formula:

$$1 + 2 + \cdots + n = \frac{n(n + 1)}{2} \tag{6.4}$$

for every positive integer n, which we first encountered in Chapter 4.

Example 6.27

Determine the average case time complexity of Algorithm 6.8.

Solution. Suppose that $a_t = k$ where $1 \leq t \leq n$. In Step 2, i and n are compared for $i = 1, 2, \ldots, t$ and in Step 3, a_i and k are compared for $i = 1, 2, \ldots, t$. In Step 3, when $a_t = k$, i is changed from t to $n + 1$. In Step 4, i is increased to $n + 2$. In Step 2, $n + 2$ and n are compared and the while loop is exited. In Step 5, $n + 2$ and $n + 1$ are compared and the algorithm ends. Hence the total number of comparisons made when $a_t = k$ is $t + t + 1 + 1 = 2t + 2$. Since t can be any one of $1, 2, \ldots, n$, the average number of comparisons made by Algorithm 6.8 is

$$\frac{\sum_{t=1}^{n}(2t+2)}{n} = \frac{4 + 6 + 8 + \cdots + (2n+2)}{n}$$
$$= \frac{2[2 + 3 + 4 + \cdots + (n+1)]}{n}.$$

By the formula in (6.4),

$$1 + 2 + \cdots + (n+1) = \frac{(n+1)(n+2)}{2}$$

and so

$$2 + 3 + 4 + \cdots + (n+1) = \frac{(n+1)(n+2)}{2} - 1.$$

Therefore,

$$\frac{4 + 6 + 8 + \cdots + (2n+2)}{n} = \frac{2\left[\frac{(n+1)(n+2)}{2} - 1\right]}{n} = \frac{(n+1)(n+2) - 2}{n}$$
$$= \frac{n^2 + 3n}{n} = n + 3.$$

Thus the average number of comparisons made when Algorithm 6.8 is used (knowing that k appears in the sequence) is $n + 3 = \Theta(n)$, which is the average case time complexity of Algorithm 6.8. ◆

For a sequence $s : a_1, a_2, \ldots, a_n$ of $n \geq 2$ numbers, we now consider an algorithm that computes the sum $\sum_{1 \leq i < j \leq n} a_i a_j$. For example, $\sum_{1 \leq i < j \leq 3} a_i a_j = a_1 a_2 + a_1 a_3 + a_2 a_3$.

Algorithm 6.28 *Determine the Sum of the Products of Every Two Numbers in a Finite Sequence.*

This algorithm finds the sum of the products of every two numbers in a sequence $s : a_1, a_2, \ldots, a_n$ of $n \geq 2$ numbers.

Input: A positive integer $n \geq 2$ and a sequence $s : a_1, a_2, \ldots, a_n$ of n numbers.
Output: $x = \sum_{1 \leq i < j \leq n} a_i a_j = a_1 a_2 + a_1 a_3 + \cdots + a_{n-1} a_n$.

1. $x := 0$
2. **for** $i := 1$ **to** $n - 1$ **do**
   begin
3. $t := 0$
4.   **for** $j := i + 1$ **to** n **do**
       begin
5. $r := a_i a_j$
6. $t := t + r$
       end
7. $x := x + t$
   end
8. **output** x

Let's see what Algorithm 6.28 does for a sequence of length 3.

Example 6.29

Illustrate Algorithm 6.28 for the sequence $s : a_1, a_2, a_3$.

Solution. In Step 1, x is assigned the value 0. Step 2 is a for loop. We proceed through this loop. In Step 3, t is assigned the value 0. Step 4 is a for loop within the for loop at Step 2 with initial value $j = 2$. Steps 5 and 6 are performed. In Step 5, r is assigned $a_1 a_2$ and in Step 6, t is assigned $0 + a_1 a_2 = a_1 a_2$. We return to the for loop at Step 4 and j is increased to $3 = n$. We proceed through the for loop at Step 4 again. In Step 5, r is assigned the value $a_1 a_3$. In Step 6, t is assigned $a_1 a_2 + a_1 a_3$. Since $j = 3 = n$, the for loop at Step 4 has been executed and so we return to the for loop at Step 2 and so Step 7 is performed, where x is assigned the value $0 + a_1 a_2 + a_1 a_3 = a_1 a_2 + a_1 a_3$. We now return to the for loop at Step 2 with $i = 2 = n - 1$. In Step 3, t is assigned 0. Since $j = 3 = n$, we proceed through the for loop at Step 4. By Step 5, r is assigned $a_2 a_3$ and by Step 6, t is assigned $0 + a_2 a_3 = a_2 a_3$. This completes the execution of the for loop at Step 4. In Step 7, x is assigned the number $a_1 a_2 + a_1 a_3 + a_2 a_3$. Then we proceed to Step 8, where $x = a_1 a_2 + a_1 a_3 + a_2 a_3$ is output. ◆

A step by step account of how the values of variables and indexes are affected when Algorithm 6.28 is applied in Example 6.29 is shown in Figure 6.3.

x	i	t	j	r	output
0					
0	1				
0	1	0			
0	1	0	2		
0	1	0	2	$a_1 a_2$	
0	1	$a_1 a_2$	2	$a_1 a_2$	
0	1	$a_1 a_2$	3	$a_1 a_2$	
0	1	$a_1 a_2$	3	$a_1 a_3$	
0	1	$a_1 a_2 + a_1 a_3$	3	$a_1 a_3$	
$a_1 a_2 + a_1 a_3$	1	$a_1 a_2 + a_1 a_3$	3	$a_1 a_3$	
$a_1 a_2 + a_1 a_3$	2	$a_1 a_2 + a_1 a_3$	3	$a_1 a_3$	
$a_1 a_2 + a_1 a_3$	2	0	3	$a_1 a_3$	
$a_1 a_2 + a_1 a_3$	2	0	3	$a_1 a_3$	
$a_1 a_2 + a_1 a_3$	2	0	3	$a_2 a_3$	
$a_1 a_2 + a_1 a_3$	2	$a_2 a_3$	3	$a_2 a_3$	
$a_1 a_2 + a_1 a_3 + a_2 a_3$	2	$a_2 a_3$	3	$a_2 a_3$	$a_1 a_2 + a_1 a_3 + a_2 a_3$

Figure 6.3: Apply Algorithm 6.28 to Example 6.29

The main reason for presenting Algorithm 6.28 is so we can discuss its time complexity.

Example 6.30

Determine the time complexity of Algorithm 6.28.

Solution. For each of the $n - i$ values of j ($j = i + 1, i + 2, \ldots, n$), there is one multiplication in Step 5 and one addition in Step 6. Since there are $n - i$ values of j for each i ($i = 1, 2, \ldots, n - 1$), there are $2(n - i)$ arithmetic operations for each such integer i. Thus the total number of arithmetic operations within the for loop at Step 4 is

$$
\begin{aligned}
\sum_{i=1}^{n-1} 2(n - i) &= 2(n - 1) + 2(n - 2) + \cdots + 2 \cdot 1 \\
&= 2[1 + 2 + \cdots + (n - 1)] \\
&= \frac{2n(n - 1)}{2} = n^2 - n.
\end{aligned}
$$

There is an addition in Step 7 for each i ($i = 1, 2, \ldots, n-1$) within the for loop at Step 2. Thus the time complexity of Algorithm 6.28 is $(n^2 - n) + (n - 1) = n^2 - 1 = \Theta(n^2)$. ♦

Exercises for Section 6.3

1. For a sequence $s : a_1, a_2, \ldots, a_n$ of $n \geq 3$ distinct numbers, write an algorithm with time complexity $O(1)$ that finds a number x in s that is neither the maximum nor the minimum of the terms in s. Show that the time complexity of this algorithm is $O(1)$.

2. For a sequence $s : a_1, a_2, \ldots, a_n$ of n numbers, write an algorithm that computes $a_1 a_2 \cdots a_n + a_1 a_2 \cdots a_{n-1} + \cdots + a_1 a_2 + a_1$. Show that this algorithm has time complexity $\Theta(f(n))$ for some common function f.

3. For a sequence $s : a_1, a_2, \ldots, a_n$ of n numbers, write an algorithm that computes $(a_1 + a_2 + \cdots + a_n)(a_1 + a_2 + \cdots + a_{n-1}) \cdots (a_1 + a_2)a_1$. Show that this algorithm has time complexity $\Theta(f(n))$ for some common function f.

4. For a sequence $s : a_1, a_2, \ldots, a_n$ of $n \geq 3$ numbers, write an algorithm that computes $a_1 a_2 + a_2 a_3 + a_3 a_4 + \cdots + a_{n-1} a_n + a_n a_1$. Show that this algorithm has time complexity $\Theta(f(n))$ for some common function f.

5. Write an algorithm that computes the sum of the absolute values of the difference of every two numbers in a sequence of $n \geq 2$ numbers. Show, with verification, that this algorithm has time complexity $\Theta(f(n))$ for some common function f.

6. For a sequence $s : a_1, a_2, \ldots, a_n$ of $n \geq 2$ numbers, write an algorithm that computes the number $a_1 a_2^2 a_3^3 \cdots a_n^n$. Show, with verification, that this algorithm has time complexity $\Theta(f(n))$ for some common function f.

7. For a real number a and nonnegative integer n, define a^n recursively by $a^0 = 1$ and $a^n = a \cdot a^{n-1}$ for $n \geq 1$.

   (a) Write an algorithm that computes a^n for a real number a and nonnegative integer n.

   (b) What is the time complexity of the algorithm in (a)?

8. (a) For a positive integer n and $n+1$ constants $c_n, c_{n-1}, \ldots, c_1, c_0$ and a real number a, write an algorithm that uses the algorithm in Exercise 7 to compute a^k, $0 \leq k \leq n$ and computes $f(a)$, where f is the polynomial function defined by $f(x) = c_n x^n + c_{n-1} x^{n-1} + \cdots + c_1 x + c_0$ for each $x \in \mathbb{R}$.

   (b) Determine the time complexity of the algorithm in (a), where a^k, $0 \leq k \leq n$, is computed by the algorithm in Exercise 7.

9. For a positive integer n and $n+1$ constants $c_n, c_{n-1}, \ldots, c_1, c_0$ and a real number a, **Horner's algorithm** computes $f(a)$ for a polynomial function f defined by $f(x) = c_n x^n + c_{n-1} x^{n-1} + \cdots + c_1 x + c_0$ for each $x \in \mathbb{R}$ by using the fact that

$$f(x) = (\cdots (c_n x + c_{n-1}) x + c_{n-2}) x + \cdots + c_1) x + c_0.$$

   (a) Determine the time complexity of Horner's algorithm.

   (b) How do the efficiences of Horner's algorithm and the algorithm in Exercise 8 compare?

6.4   Searching and Sorting

In Algorithm 6.8, a procedure is described for determining whether a number k appears in a sequence $s : a_1, a_2, \ldots, a_n$. What this algorithm does is to proceed through the list term by term, beginning with a_1 and then a_2 and so on until either k is found somewhere in the list or it is discovered that k is not in the list. Because of the way this algorithm proceeds through the list searching for k, Algorithm 6.8 is called a linear search algorithm. We saw in Example 6.26 that the worst case time complexity of Algorithm 6.8 is $\Theta(n)$, that is, the worst case time complexity of the Linear Search Algorithm is linear. This algorithm is repeated below for future reference.

Algorithm 6.31   (Linear Search Algorithm)   *This algorithm determines whether a specified number k appears in a sequence $s : a_1, a_2, \ldots, a_n$ of n numbers.*

> **Input**: A positive integer n, a number k and a sequence $s : a_1, a_2, \ldots, a_n$ of n numbers.
> **Output**: k "is in the sequence" or k "is not in the sequence" as appropriate.
> 1. $i := 1$
> 2. **while** $i \leq n$ **do**
>   **begin**
> 3.   **if** $a_i = k$ **then output** k "is in the sequence" and $i := n + 1$
> 4. $i := i + 1$
>   **end**
> 5. **if** $i = n + 1$ **then output** k "is not in the sequence"

The Binary Search Algorithm

If the sequence $s : a_1, a_2, \ldots, a_n$ of n numbers is **increasing** (that is, $a_1 < a_2 < \cdots < a_n$), then, in the worst case, considerably fewer comparisons are needed to determine whether a given number k appears in the sequence. A sequence $s : a_1, a_2, \ldots, a_n$ of n numbers is **nondecreasing** if $a_1 \leq a_2 \leq \cdots \leq a_n$. For example, if we were to search for the word *traverse* in a dictionary, we would surely not start at the beginning of the dictionary and proceed word by word (or even page by page) until locating the word. The search algorithm we are about to describe is the binary search algorithm. The idea here is to divide the sequence into halves (approximately), selecting a middle number in the sequence, say a_p. There are then three possibilities, namely, $a_p = k$, $a_p > k$, or $a_p < k$.

(1) If $a_p = k$, then we know that k appears in the sequence and where k appears in the sequence and the search is complete.

(2) If $a_p > k$, then k can only appear in the subsequence $a_1, a_2, \ldots, a_{p-1}$ of s. Thus we repeat this procedure with the subsequence $a_1, a_2, \ldots, a_{p-1}$.

(3) If $a_p < k$, then k can only appear in the subsequence $a_{p+1}, a_2, \ldots, a_n$ of s. Thus we repeat this procedure with the subsequence $a_{p+1}, a_2, \ldots, a_n$.

In the likely event that the middle number a_p is not the number k, the number k (if it appears in the sequence at all) must now appear in a sequence whose length is approximately half of that of the original sequence. We then search for k in the smaller sequence in the same way, namely, by dividing the smaller sequence into two halves and so on.

Algorithm 6.32   (Binary Search Algorithm)   *This algorithm determines whether a specified number k is one of the numbers in an increasing sequence $s : a_1, a_2, \ldots, a_n$ of n distinct numbers.*

> **Input**: A positive integer n, a number k and an increasing sequence $s : a_1, a_2, \ldots, a_n$ of n numbers.

Output: k "is in position" p or k "is not in the sequence" as appropriate.
  1. $a := 1$ [a represents the first position of the sublist being considered – initially $a = 1$]
  2. $b := n$ [b represents the last position of the sublist being considered – initially $b = n$]
  3. **while** $a \leq b$ **do**
     begin
  4. $p := \lfloor (a + b)/2 \rfloor$
  5.   **if** $a_p = k$ **then output** k "is in position" p and $a := b + 2$
  6.   **if** $a_p > k$ **then** $b := p - 1$
  7.   **if** $a_p < k$ **then** $a := p + 1$
  8.   **if** $a = b + 1$ **then output** k "is not in the sequence"
     end

Let's see what the Binary Search Algorithm does in practice.

Example 6.33

Use the Binary Search Algorithm to determine whether the number 20 appears in the sequence
$s : 3, 7, 8, 11, 12, 15, 16, 18, 20, 22$.

Solution. In this case, $k = 20$, $n = 10$ and $a_1 = 3$, $a_2 = 7$, ..., $a_{10} = 22$. By Steps 1 and 2, a and
b are assigned the numbers 1 and 10, respectively. Step 3 is a while loop. Since $1 = a \leq b = 10$,
Steps 4 – 8 within the while loop are executed. In Step 4, p is assigned the number $\lfloor (1 + 10)/2 \rfloor = 5$.
Since $a_5 \neq k$, the conclusions of Step 5 are not performed. Since $a_5 = 12 > 20$ is false, we move
immediately to Step 6. Because $a_5 = 12 < 20 = k$, it follows that a is assigned the number 6.
(We will now consider the subsequence $a_6, a_7, \ldots, a_{10}$.) Since $6 = a \neq b + 1 = 11$, the conclusion
of Step 8 is not performed. Since $6 = a \leq b = 10$, we continue through the while loop at Step 3
again. In Step 4, p is assigned the value $\lfloor (6 + 10)/2 \rfloor = 8$. In Steps 5 – 7, only $18 = a_8 < k = 20$ is
true and a is assigned the value $8 + 1 = 9$. (We will now consider the subsequence a_9, a_{10}.) Since
$9 = a \leq b = 10$, we proceed through the while loop at Step 3 again. In Step 4, p is assigned the
value $\lfloor (9 + 10)/2 \rfloor = 9$. In Step 5, $a_9 = 20 = k$ is true, so 20 *appears at position* 9 is output and a
is assigned the value $10 + 2 = 12$. None of the conclusions in Steps 6 – 8 are performed. Because
$a \leq b$ is false, the steps within the while loop at Step 3 are not performed and the algorithm ends.
What the Binary Search Algorithm does to the sequence s is shown in Figure 6.4. ◆

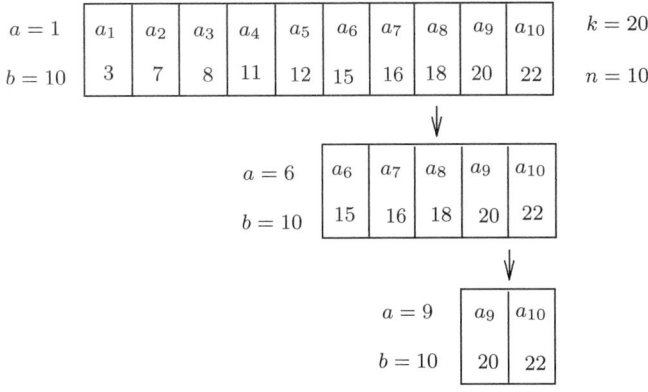

"20 is in position 9" is output

Figure 6.4: Applying the Binary Search Algorithm to Example 6.33

    Applying the Binary Search Algorithm to our example in which we were searching for the word
traverse in the dictionary, we would begin by opening the dictionary to the middle page (or there-
abouts) and probably see that we were on a page where the words begin with L, say. We know
that *traverse* comes after this, so we look for the middle page in the second half of the dictionary

(possibly arriving at a page whose words begin with R). This procedure is continued until *traverse* is found. If we don't find it, then the natural conclusion is that we don't have a correct spelling of it. Of course, in this dictionary example, our experience would tell us to open to a page that is later in the dictionary. For a sequence of numbers, we wouldn't know this, however, without additional information. We consider another example.

Example 6.34

Use the Binary Search Algorithm to determine whether the number 10 appears in the sequence $s : 3, 7, 8, 11, 12, 15, 16, 18, 20, 22$.

Solution. In this case, $k = 10$, $n = 10$ and $a_1 = 3$, $a_2 = 7$, ..., $a_{10} = 22$. By Steps 1 and 2, a and b are assigned the numbers 1 and 10, respectively. Step 3 is a while loop. Since $1 = a \leq b = 10$, Steps 4 – 8 are executed. In Step 4, p is assigned the number $\lfloor (1 + 10)/2 \rfloor = 5$. Among the Steps 5 – 7, only $a_5 > k$ is true and so b is assigned the number $5 - 1 = 4$. (We will now consider the subsequence a_1, a_2, a_3, a_4.) Since $1 = a \leq b = 4$, we proceed through the while loop at Step 3 again. In Step 4, p is assigned the value $\lfloor (1 + 4)/2 \rfloor = 2$. Among Steps 5 – 7, only $a_2 < k = 10$ is true and a is assigned the value 3. (We will now consider the subsequence a_3, a_4.) Since $3 = a \leq b = 4$ is true, we proceed through the while loop at Step 3 once again. At Step 4, p is assigned the value $\lfloor (3 + 4)/2 \rfloor = 3$. Of the Steps 5 – 8, only $a_3 < k$ is true and a is assigned the value $3 + 1 = 4$. (The subsequence under consideration now consists of the single term a_4.) Because $4 = a \leq b = 4$, we proceed through the while loop at Step 3 yet again. In Step 4, p is assigned the value $\lfloor (4 + 4)/2 \rfloor = 4$. Of Steps 5 – 7, only $a_4 > k$ is true and b is assigned $4 - 1 = 3$. In Step 8, $a = b + 1$ and 10 *is not in the sequence* is output. Because $a \leq b$ is false, the steps within the while loop at Step 3 are not executed and the algorithm ends. What the Binary Search Algorithm does to the sequence s is shown in Figure 6.5. ♦

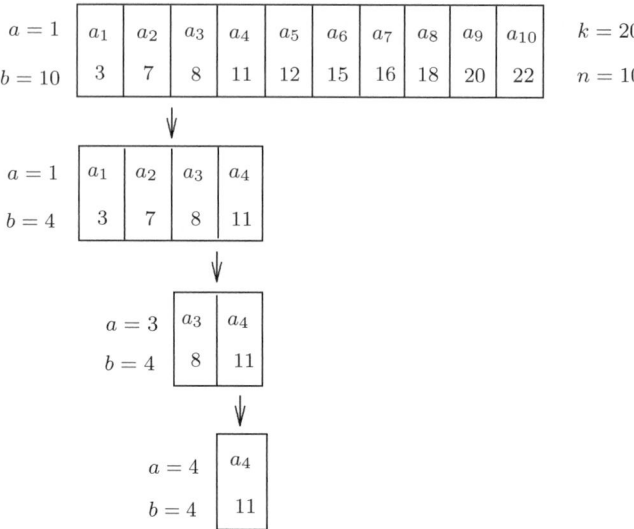

"$k = 10$ is not in the sequence" is output

Figure 6.5: Applying the Binary Search Algorithm to Example 6.34

Although it may appear that the number of comparisions made in the previous examples is not an improvement over the Linear Search Algorithm (and this is true), the number of comparisions is a major improvement if n is large.

Example 6.35

Determine the worst case time complexity of the Binary Search Algorithm.

Solution. In the worst case, the number k does not appears in the sequence. Suppose first that the number n of terms in the sequence s is a power of 2, say $n = 2^j$ for some positive integer j. Initially, then $a = 1$ and $b = n = 2^j$. After the initial comparison is made in Step 3 (namely that $1 \leq n$), a comparison is made in each of the Steps 5 – 8. Thus 5 comparisons have been made. When we return to the while loop at Step 3, a subsequence with 2^{j-1} terms is being considered and 5 comparisons are made again. After $j + 1$ passes through the while loop at Step 3, $a = b + 1$. There is only one additional comparison in Step 3, where $a \leq b$ is false. Hence the number of comparisons in the algorithm is $5(j + 1) + 1 = 5j + 6$. Since $j = \log n$, the worst case time complexity is $O(\log n)$. If n is not a power of 2, then $n < 2^j < 2n$ for some positive integer j and then the number of comparisons in the algorithm is no more than

$$5j + 6 < 5\log(2n) + 6 = 5\log n + 5\log 2 + 6,$$

that is, the worst case time complexity of the Binary Search Algorithm in any case is $O(\log n)$. ◆

As we already observed, $\log n < n$ for every positive integer n and so using the Binary Search Algorithm is more efficient than using the the Linear Search Algorithm. Of course, to use the Binary Search Algorithm we need to have numbers in the sequence s listed in increasing order (although decreasing order is perfectly acceptable as well). Even though this is not an unusual occurrence, there is no guarantee that the sequence under consideration has this desired property. Therefore, it may be useful to arrange or sort the numbers in the sequence so that an increasing (or nondecreasing) sequence results. There are many sorting algorithms. We describe three of them.

The Bubble Sort Algorithm

The first sorting algorithm we present is called the Bubble Sort Algorithm, so named because when this algorithm is applied to a list of numbers, given in a column, the largest numbers successively flow to the bottom. When using this algorithm, it is useful at times to interchange the values of two variables. For example, if a and b are two variables, then their values can be interchanged by performing the following steps, where a new variable t (for temporary) is introduced for this purpose:

$$
\begin{aligned}
t &:= a \\
a &:= b \\
b &:= t
\end{aligned}
$$

In the first step above, t takes on the value of a so that this value is not lost in the second step, where a takes on the value of b. The result of these three steps is referred to as the **swap of a and b**, which we denote by **swap**(a, b).

Algorithm 6.36 (**Bubble Sort Algorithm**) *This algorithm sorts the terms in a sequence s : $a_1, a_2, \ldots, a_n$ of n numbers so that the terms of s are reordered from smallest to largest.*

> **Input**: A positive integer n and a sequence $s : a_1, a_2, \ldots, a_n$ of n numbers.
> **Output**: A nondecreasing sequence whose terms are those of s.
> 1. **for** $j := 1$ **to** $n - 1$ **do**
> 2. **for** $i := 1$ **to** $n - j$ **do**
> 3.   **if** $a_i > a_{i+1}$ **then swap**(a_i, a_{i+1}).
> 4. **output** s

Let's see how the Bubble Sort Algorithm works with a specific example.

Example 6.37

Apply the Bubble Sort Algorithm to the sequence $s : 17, 8, 11, 5$.

Solution. In this sequence, $n = 4$ and $a_1 = 17$, $a_2 = 8$, $a_3 = 11$, $a_4 = 5$. In Step 1 of Algorithm 6.36, j is initially assigned the value 1. While $j = 1$, the variable i takes on the values $1, 2, 3$. When $i = 1$, a_1 and a_2 (17 and 8) are swapped; when $i = 2$, the current a_2 and a_3 (17 and 11) are swapped; and when $i = 3$, the current a_3 and a_4 (17 and 5) are swapped. At this time, a_4 is the maximum term of s, namely 17. Then j is increased to 2 at Step 1. While $j = 2$, the variable i takes on the values 1 and 2. When $i = 1$, the current a_1 and a_2 (8 and 11) are not swapped; and when $i = 2$, the current a_2 and a_3 (11 and 5) are swapped. Then j is increased to 3. Since $i = 1$, the current terms a_1 and a_2 (8 and 5) are swapped. This completes the reordering. What the algorithm does to the sequence s is shown in Figure 6.6.   ♦

original sequence	$i = 1$ $j = 1$	$i = 2$ $j = 1$	$i = 3$ $j = 1$	$i = 1$ $j = 2$	$i = 2$ $j = 2$	$i = 1$ $j = 3$	final sequence
17	**8**	8	8	**8**	8	**5**	5
8	**17**	**11**	11	**11**	**5**	**8**	8
11	11	**17**	**5**	5	**11**	11	11
5	5	5	**17**	17	17	17	17

Figure 6.6: Applying the Bubble Sort Algorithm to a sequence of numbers

Example 6.38

Determine the time complexity of the Bubble Sort Algorithm.

Solution. For each integer j ($1 \le j \le n - 1$), there are $n - j$ comparisons, one for each integer i with $1 \le i \le n - j$. Thus the total number of comparisons is

$$\sum_{j=1}^{n-1}(n - j) = 2[1 + 2 + \cdots + (n - 1)]$$

$$= 2\frac{n(n - 1)}{2} = n^2 - n = \Theta(n^2). \quad ♦$$

The Insertion Sort Algorithm

Although the Bubble Sort Algorithm provides a relatively simple method for sorting the numbers in a list, this may not be the first algorithm one would think of if there was a sequence to sort. For example, suppose that you had taken ten quizzes in a certain course and wanted to sort them from lowest to highest (or highest to lowest). How might you do this? A procedure that a number of people might use is the Insertion Sort Algorithm. Before formally stating this algorithm, we give an example to illustrate it.

Example 6.39

Use the Insertion Sort Algorithm to sort the list $s : 5, 2, 4, 6, 1, 3$ from smallest to largest.

Solution. We start with the second number 2 and determine whether it should precede or follow 5. Since $2 < 5$, the integer 2 should precede 5. So 2 is inserted before 5. Next we look at 4 and where it should be placed in the sequence that precedes it. Since $2 < 4 < 5$, the integer 4 is inserted between 2 and 5. We then continue in this manner. Figure 6.7 illustrates how the list changes as we proceed through the Insertion Sort Algorithm.   ♦

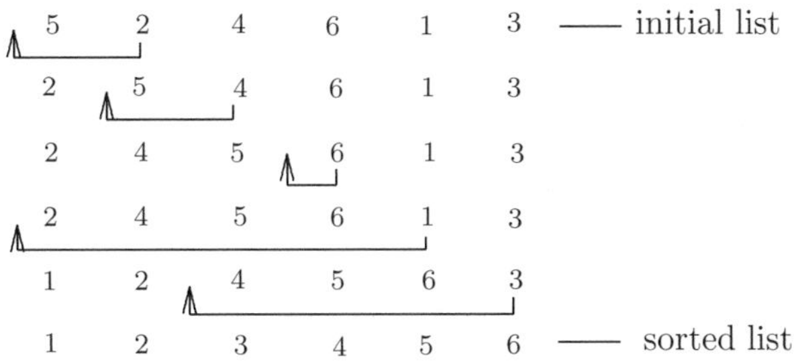
5 2 4 6 1 3 — initial list
2 5 4 6 1 3
2 4 5 6 1 3
2 4 5 6 1 3
1 2 4 5 6 3
1 2 3 4 5 6 — sorted list

Figure 6.7: Sorting a list by the Insertion Sort Algorithm

Before giving a formal statement of the Insertion Sort Algorithm, it is useful to introduce the **insertion** command, which we denote by **insert**$(a_j \to a_i)$ for $1 \le i < j \le n$ and which moves the current term a_j to where the current term a_i is located, with all the terms $a_i, a_{i+1}, \ldots, a_{j-1}$ moving to the right one place. The terms are then relabeled, as shown in Figure 6.8.

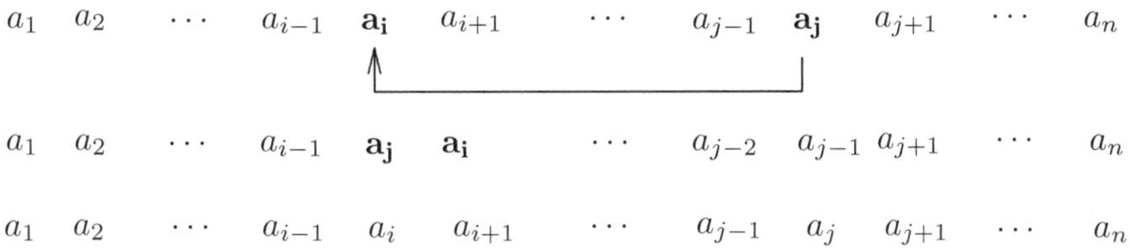
a_1 a_2 ⋯ a_{i-1} $\mathbf{a_i}$ a_{i+1} ⋯ a_{j-1} $\mathbf{a_j}$ a_{j+1} ⋯ a_n
a_1 a_2 ⋯ a_{i-1} $\mathbf{a_j}$ $\mathbf{a_i}$ ⋯ a_{j-2} a_{j-1} a_{j+1} ⋯ a_n
a_1 a_2 ⋯ a_{i-1} a_i a_{i+1} ⋯ a_{j-1} a_j a_{j+1} ⋯ a_n

Figure 6.8: Illustrating **insert**$(a_j \to a_i)$

The insertion **insert**$(a_j \to a_i)$ can be accomplished with the following steps:

$t := a_j$
for $k := 1$ **to** $j - i$ **do**
  $a_{j-k+1} := a_{j-k}$
$a_i := t$

We now state the Insertion Sort Algorithm.

Algorithm 6.40 (**Insertion Sort Algorithm**) *This algorithm sorts the terms in a sequence* $s : a_1, a_2, \ldots, a_n$ *of* n *numbers so that the terms are reordered from smallest to largest.*

Input: A positive integer n and a sequence $s : a_1, a_2, \ldots, a_n$ of n numbers.
Output: A nondecreasing sequence whose terms are the terms of s.
1. **for** $j := 2$ **to** n **do**
2.   **for** $i := 1$ **to** $j - 1$ **do**
3.   **if** $a_j \le a_i$ **then insert**$(a_j \to a_i)$ and $i := j - 1$
4. **output** s

Example 6.41

Determine the worst case time complexity of the Insertion Sort Algorithm.

Solution. For each value of i $(i = 1, 2, \ldots, j-1)$, there is a possible comparison in Step 3. Hence there may be as many as $j - 1$ comparisons. Since j has the values $2, 3, \ldots, n$, the number of comparisons can be as many as

$$\sum_{j=2}^{n}(j-1) = 1 + 2 + \cdots + (n-1) = \frac{n(n-1)}{2} = \Theta(n^2). \ \blacklozenge$$

We give one additional example using the Insertion Sort Algorithm.

Example 6.42

Sort the list $s : 15, 11, 9, 12, 10$ from smallest to largest using the Insertion Sort Algorithm.

Solution. The sorting of the elements in s is described in Figure 6.9.   ◆

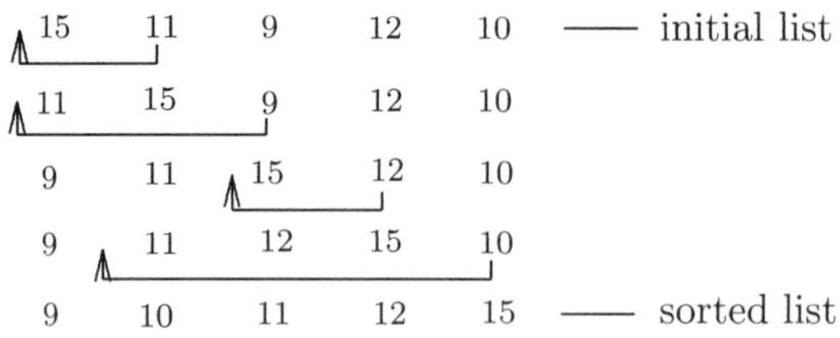

Figure 6.9: Sorting the list in Example 6.42 by the Insertion Sort Algorithm

The Merge Sort Algorithm

A more efficient algorithm for sorting the numbers in a list is the Merge Sort Algorithm – so named because in the process of using this algorithm, pairs of sorted lists are merged into larger sorted lists until the final desired sorted list has been obtained. In order to illustrate how this is done, suppose that we wish to merge two sorted lists $s_1 : 2, 4, 5, 6, 8, 9$, and $s_2 : 1, 3, 7$ into a single sorted list. This can be accomplished by comparing the first terms of the two lists. The smaller of these two numbers is removed from the relevant list and becomes the first term of the desired list. Figure 6.10 describes this in its entirety.

List 1	List 2	Comparision	Merged List
2 4 5 6 8 9	1 3 7	$1 < 2$	1
2 4 5 6 8 9	3 7	$2 < 3$	1 2
4 5 6 8 9	3 7	$3 < 4$	1 2 3
4 5 6 8 9	7	$4 < 7$	1 2 3 4
5 6 8 9	7	$5 < 7$	1 2 3 4 5
6 8 9	7	$6 < 7$	1 2 3 4 5 6
8 9	7	$7 < 8$	1 2 3 4 5 6 7
8 9	$\emptyset$		1 2 3 4 5 6 7 8 9

Figure 6.10: Merging two sorted lists

This procedure gives rise to the following theorem.

Theorem 6.43 *Suppose that $s_1 : a_1, a_2, \ldots, a_m$ and $s_2 : b_1, b_2, \ldots, b_n$ are two sorted lists with m and n numbers. Then these two lists can be merged into a single sorted list using at most $m + n - 1$ steps.*

With the aid of Theorem 6.43, we can now describe the **Merge Sort Algorithm**. Let s be a list of n numbers. Then $2^{k-1} < n \leq 2^k$ for some positive integer k. For simplicity, assume first that $n = 2^k$. This list is divided into two lists, each of size 2^{k-1}. Then each of these lists is further divided into two lists, each of size 2^{k-2}. We continue this until arriving at 2^k lists, with one element each. Proceeding in reverse order now, the two elements that occur in each of the 2^{k-1} pairs are compared and listed in nondecreasing order. In each case, $2^1 - 1 = 1$ comparison is needed to accomplish this. The total number of comparisons needed to do this is $2^{k-1}(2^1 - 1)$. The two pairs of two elements that occur in each of 2^{k-2} sets of 2^2 elements are now merged into a sorted list. By Theorem 6.43, each such merge requires at most $2^2 - 1 = 3$ comparisons. The total number of comparisons needed to do this is at most $2^{k-2}(2^2-1)$. Thus the total number of comparisons needed to produce a sorted list of size n is at most

$$
\begin{aligned}
2^{k-1}(2^1-1)+2^{k-2}(2^2-1)+\cdots+2^1(2^{k-1}-1) &= \sum_{i=1}^{k} 2^{i-1}(2^{k-i+1}-1) \\
&= \sum_{i=1}^{k}(2^k-2^{i-1}) \\
&= \sum_{i=1}^{k} 2^k - \sum_{i=1}^{k} 2^{i-1} \\
&= k2^k - (2^k-1) \\
&= n\log n - n + 1.
\end{aligned}
$$

(See Result 5.7 in Section 5.1.) Therefore, the time complexity of the Merge Sort Algorithm is $O(n\log n)$. If n is not a power of 2, then $n < 2^k < 2n$ for some positive integer k. Thus $k < \log n + \log 2$ and the number of comparisons needed to sort a list of size n using the Merge Sort Algorithm is at most

$$
\begin{aligned}
k2^k - 2^k + 1 &< (\log n + \log 2)2n - n + 1 \\
&= 2n\log n + (2\log 2 - 1)n + 1 \\
&= \Theta(n\log n).
\end{aligned}
$$

Hence in any case, the time complexity of the Merge Sort Algorithm is $O(n\log n)$.

Example 6.44

Use the Merge Sort Algorithm to arrange the sequence $s: 6, 7, 1, 4, 8, 2, 5, 3$ in nondecreasing order.

Solution. The sequence s of length 8 is divided into two subsequences of length 4, namely $6, 7, 1, 4$ and $8, 2, 5, 3$. Each of those sequences is then divided into two subsequences of length 2. In particular, $6, 7, 1, 4$ is divided into $6, 7$ and $1, 4$; while $8, 2, 5, 3$ is divided into $8, 2$ and $5, 3$. Each of the four sequences of length 2 is then divided into two subsequences of length 1. All of this is described in the table shown in Figure 6.11.

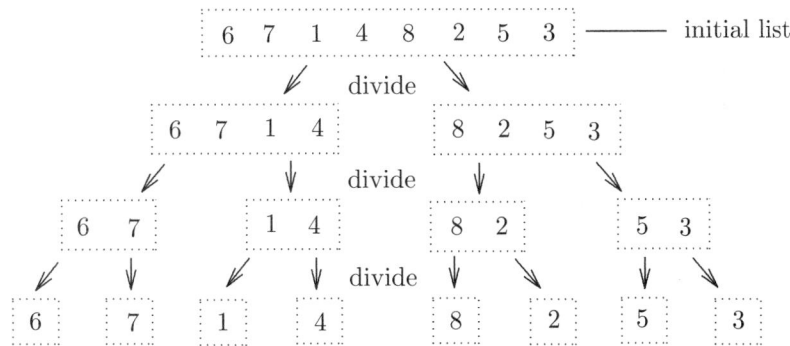

Figure 6.11: Dividing a list by the Merge Sort Algorithm in Example 6.44

The process is then reversed. The first two sequences of length 1 are merged into a sorted sequence of length 2. This is then done with the third and fouth sequences of length 1, the fifth and sixth sequences of length 1 and seventh and eighth sequences of length 1, producing four sorted sequences of length 2 as shown in Figure 6.12.

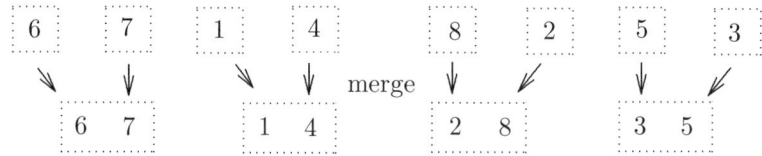

Figure 6.12: Merging the sequences of length 1 in Example 6.44
to produce sorted sequences of length 2 by the Merge Sort Algorithm

The first two sorted sequences of length 2 are merged into a sorted sequence of length 4, as are
the third and fouth sorted sequences of length 2. Finally, the two sorted sequences of length 4 are
merged into a sorted sequence of length 8, which gives the desired result shown in Figure 6.13.

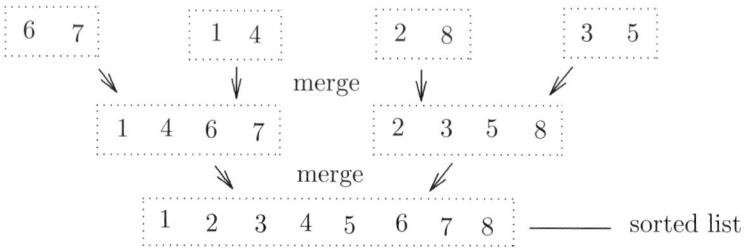

Figure 6.13: Merging the sequences of length 2 in Example 6.44
to produce the final sorted list by the Merge Sort Algorithm

The tables below show how the Merge Sort Algorithm transforms the first two rows in Figure 6.13.

List 1	List 2	Comparison	Merged List
6 7	1 4	$1 < 6$	1
6 7	4	$4 < 6$	1 4
6 7	$\emptyset$		1 4 6 7

List 1	List 2	Comparison	Merged List
2 8	3 5	$2 < 3$	2
8	3 5	$3 < 8$	2 3
8	5	$5 < 8$	2 3 5
8	$\emptyset$		2 3 5 8

List 1	List 2	Comparison	Merged List
1 4 6 7	2 3 5 8	$1 < 2$	1
4 6 7	2 3 5 8	$2 < 4$	1 2
4 6 7	3 5 8	$3 < 4$	1 2 3
4 6 7	5 8	$4 < 5$	1 2 3 4
6 7	5 8	$5 < 6$	1 2 3 4 5
6 7	8	$6 < 8$	1 2 3 4 5 6
7	8	$7 < 8$	1 2 3 4 5 6 7
$\emptyset$	8		1 2 3 4 5 6 7 8

We now give one additional example using the Merge Sort Algorithm.

Example 6.45

Use the Merge Sort Algorithm to arrange the sequence $s : 3, 8, 4, 2, 1, 7, 6, 5$ in nondecreasing order.

Solution. The sorting of the elements in s is described in Figure 6.14.

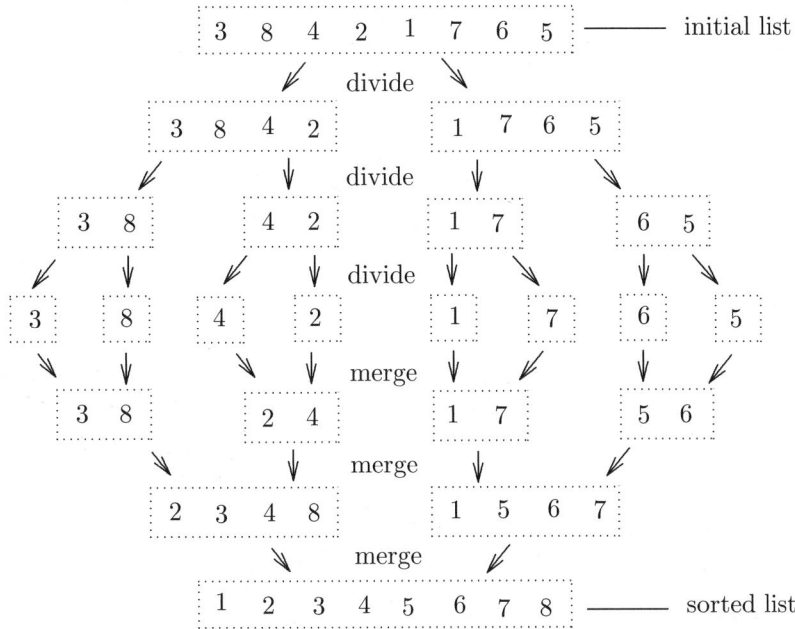

Figure 6.14: Sorting a list by the Merge Sort Algorithm in Example 6.45

Exercises for Section 6.4

1. Use the Binary Search Algorithm to determine whether the number 17 appears in the sequence $s : 3, 6, 7, 8, 15, 17, 19, 23, 24$. Give a step by step description of what the algorithm does in this case.

2. Use the Binary Search Algorithm to determine whether the number 11 appears in the sequence $s : 2, 4, 5, 7, 9, 10, 12, 14, 15, 16, 19, 23$. Give a step by step description of what the algorithm does in this case.

3. Use the Bubble Sort Algorithm on the sequence $s : 11, 9, 10, 8$ so that the terms are reordered from smallest to largest. Give a table that presents a step by step description of what the algorithm does in this case.

4. Use the Bubble Sort Algorithm on the sequence $s : 3, 9, 7, 6, 5$ so that the terms are reordered from smallest to largest. Give a table that presents a step by step description of what the algorithm does in this case.

5. Write a new Bubble Sort Algorithm that sorts the terms of a sequence $s : a_1, a_2, \ldots, a_n$ of n numbers so that the terms of s are reordered from largest to smallest.

6. Sort the list $s : 6, 4, 2, 1, 3, 5$ by the Insertion Sort Algorithm. Give a diagram that presents a step by step description of what the algorithm does in this case.

7. Sort the list $s : 6, 4, 2, 5, 3, 1$ by the Insertion Sort Algorithm. Give a diagram that presents a step by step description of what the algorithm does in this case.

8. Write a new Insertion Sort Algorithm that sorts the terms of a sequence $s : a_1, a_2, \ldots, a_n$ of n numbers so that the terms of s are reordered from largest to smallest.

9. Indicate how the two sorted lists $s_1 : 1, 3, 4, 7, 8, 9$ and $s_2 : 2, 5, 6$ are merged into a single sorted list by the Merge Sort Algorithm.

10. Sort the list $s : 7, 4, 3, 1, 2, 8, 6, 5$ by the Merge Sort Algorithm. Give a step by step description of what the algorithm does in this case.

Chapter 6 Highlights

Key Concepts

algorithm: a procedure for solving a problem that consists of input, output and a finite sequence of steps that converts the input to the output.

average case time complexity (of an algorithm): the average number of comparisons or arithmetic operations needed when a problem is solved using the algorithm.

big-O (f is big-O of g or f is **big-oh** of g), $f = O(g)$: there exist a positive constant C and a positive integer k such that $f(n) \leq Cg(n)$ for every integer $n \geq k$.

big-theta (f is big-theta of g), $f = \Theta(g)$: there exist positive constants C_1 and C_2 and a positive integer k such that $C_1 g(n) \leq f(n) \leq C_2 g(n)$ for every integer $n \geq k$.

Binary Search Algorithm: an algorithm that determines whether a specified number is one of the numbers in a sequence of distinct numbers by successively dividing the sequence in half.

Bubble Sort Algorithm: an algorithm that sorts the terms in a sequence of numbers listed in a column, so that the largest numbers successively flow to the bottom, resulting in a nondecreasing sequence.

complexity of an algorithm: the amount of space and time needed to execute the algorithm.

constant time complexity (of an algorithm): time complexity of the algorithm is $O(1)$.

Insertion Sort Algorithm: an algorithm that sorts the terms in a sequence of numbers so that the terms are successively inserted in a place in the sequence until a nondecreasing sequence results.

Linear Search Algorithm: an algorithm that determines whether a specified number appears in a sequence by successively proceeding through the sequence term by term.

linear time complexity (of an algorithm): time complexity of the algorithm is $O(n)$.

Merge Sort Algorithm: an algorithm that sorts the terms in a sequence of numbers by successively dividing the sequence and the resulting subsequences in half and then merging the sorted subsequences until a nondecreasing sequence results.

polynomial time complexity (of an algorithm): time complexity of the algorithm is $\Theta(f(n))$ for some polynomial $f(n)$.

quadratic time complexity (of an algorithm): time complexity of the algorithm is $O(n^2)$.

space complexity: the amount of space required by a given algorithm to solve a problem.

time complexity: the amount of time required by a given algorithm to solve a problem.

worst case time complexity (of an algorithm): the maximum number of comparisons or arithmetic operations that can occur when a problem is solved using the algorithm.

Key Results

- If f and g are two polynomial functions where the degree of g is greater than that of f, then $f = O(g)$ but $g \neq O(f)$.

- If f and g are two polynomial functions of the same degree, then $f = \Theta(g)$.

- Let $f : \mathbb{N} \to \mathbb{R}^+$ and $g : \mathbb{N} \to \mathbb{R}^+$ be two functions. Then the following hold:

  (a) $f = \Theta(g)$ if and only if $f = O(g)$ and $g = O(f)$.

  (b) If $f = \Theta(g)$, then $g = \Theta(f)$.

  (c) If $\lim\limits_{n \to \infty} \dfrac{f(n)}{g(n)} = c$ for some positive real number c, then $f = \Theta(g)$.

- Suppose that $s_1 : a_1, a_2, \ldots, a_m$ and $s_2 : b_1, b_2, \ldots, b_n$ are two sorted lists with m and n numbers. Then these two lists can be merged into a single sorted list using at most $m + n - 1$ steps.

Supplementary Exercises for Chapter 6

1. (a) For a sequence $s : a_1, a_2, \ldots, a_n$ of $n \geq 2$ integers, write an algorithm that determines whether some number occurs twice in s and, if so, where in the sequence this occurs.

   (b) Illustrate the algorithm in (a) for the sequence $s : 6, 7, 5, 7$.

2. (a) For a sequence $s : a_1, a_2, \ldots, a_n$ of n numbers, write an algorithm that determines whether there are two terms a_i and a_j (with $i \neq j$) such that $|a_i - a_j| \leq 1$.

   (b) Illustrate the algorithm in (a) for the sequence $s : 3.8, 5.2, 1.9, 2.9$.

3. (a) Give an example of a problem suggested by Algorithm 6.11.

   (b) Write an algorithm that solves the problem given in (a).

4. Give an example of two functions $f : \mathbb{N} \to \mathbb{R}^+$ and $g : \mathbb{N} \to \mathbb{R}^+$ such that $f \neq O(g)$ and $g \neq O(f)$.

5. Show that $n \log n = O(n^2)$.

6. For the function f defined by $f(n) = n^4 + 2n^2 + 4$ for each $n \in \mathbb{N}$, show that $f(n) = O(n^4)$.

7. For the function f defined by $f(n) = \frac{2n^3 + n}{n + 2}$ for each $n \in \mathbb{N}$, show that $f(n) = O(n^2)$.

8. For each of the following functions f, determine the smallest nonnegative integer k such that $f(n) = O(n^k)$.

   (a) $f(n) = n^2 + n \log n$.   (b) $f(n) = n^2 + (\log n)^3$.

   (c) $f(n) = \frac{2n^3 + n + 1}{n^2 + 1}$.   (d) $f(n) = (n^2 + 1)(n^3 + 1)$.

9. Let f and g be the functions defined by $f(n) = n^2 + 3n + 1$ and $g(n) = n^3$ for each $n \in \mathbb{N}$. Show that $f(n) = O(g(n))$ but $g(n) \neq O(f(n))$.

10. Let f be the function defined by $f(n) = n^3$ for each $n \in \mathbb{N}$. For which of the following functions g is $f(n) = O(g(n))$?

    (a) $g(n) = n^2 + 1$.   (b) $g(n) = n^3 + 1$.   (c) $g(n) = \frac{1}{2}n^3$.

    (d) $g(n) = \frac{1}{2}n^4$.   (e) $g(n) = 2^n$.   (f) $g(n) = n^2 \log n$.

11. Let f and g be the functions defined by $f(n) = 2^n$ and $g(n) = 3^n$ for each $n \in \mathbb{N}$. Show that $f(n) = O(g(n))$ but $f(n) \neq \Theta(g(n))$.

12. Let $f : \mathbb{N} \to \mathbb{R}^+$ and $g : \mathbb{N} \to \mathbb{R}^+$ be defined by $f(n) = 2n^3 + n + 10$ and $g(n) = n^3 + 4n^2 + 1$ for $n \in \mathbb{N}$. Show that $f = \Theta(g)$

13. For a sequence $s : a_1, a_2, \ldots, a_n$ of n numbers, write an algorithm that determines two numbers a_i and a_j, $i \neq j$, in s for which $|a_i - a_j|$ is minimum. Then show that your algorithm has time complexity $\Theta(f(n))$ for some common function f.

14. A sequence $s : a_1, a_2, \ldots, a_n$ of n distinct numbers is increasing if $a_1 < a_2 < \cdots < a_n$. A **subsequence** of s is a sequence $s' : a_j, a_{j+1}, \ldots, a_k$, where $1 \leq j \leq k \leq n$. For a sequence $s : a_1, a_2, \ldots, a_n$ of n distinct numbers, write an algorithm that determines the maximun length of an increasing subsequence of s.

15. For an increasing sequence $s : a_1, a_2, \ldots, a_n$ of n (distinct) integers, write an algorithm that inserts an integer a in an appropriate position into $a_1, a_2, \ldots, a_n$ so that the resulting sequence s' of length $n + 1$ is nondecreasing.

16. Write an algorithm that locates the first occurrence of a largest element in a list of integers, where the integers in the list are not necessaily distinct.

17. Write an algorithm that counts the number of 1s in an n-bit string.

18. Write an algorithm that finds the first term in a sequence of integers that is equal to some previous term in the sequence.

19. Write an algorithm that finds each term in a sequence of integers that is greater than the sum of all previous terms in the sequence.

20. Write an algorithm that finds the first term in a sequence of integers that is less than the immediately preceding term of the sequence.

Chapter 7

Integers

The branch of mathematics that is concerned with properties of the integers has traditionally been called number theory. Since integers play such an important role in discrete mathematics, it is essential to have an understanding of the fundamentals of number theory. This is the goal of the current chapter.

Although number theory was a favorite topic among the ancient Greeks, it was the book *Disquitiones Arithmeticae* in 1801 by Carl Friedrich Gauss that is considered by many as the beginning of modern number theory. Indeed, Gauss wrote:

> *Mathematics is the queen of the sciences and number theory is the queen of Mathematics.*

The German mathematician Leopold Kronecker (1823–1891) wrote:

> *God invented the integers; all else is the work of man.*

Thus the importance of integers in mathematics was recognized centuries ago.

Despite the great interest in number theory by famous mathematicians, this area includes many results and problems that can be understood by amateur mathematicians and non-mathematicians alike. One topic of continuing interest in number theory has been prime numbers, their properties and their distribution among the integers.

Although number theory has long been considered a rather obscure and abstract area of mathematics, its practical importance has only grown in recent decades because of its connections with certain aspects of computer science. One of these connections deals with the intriguing area of cryptography, which we will visit in this chapter.

7.1 Divisibility Properties

We have mentioned that an integer n is even if $n = 2k$ for some integer k, that is, n is even if n can be expressed as the product of 2 and some other integer. Integers that can or cannot be expressed as a product of integers in certain ways are important to our study.

For integers a and b with $a \neq 0$, we say that a **divides** b if $b = ac$ for some integer c. If a divides b, then we indicate this by writing $a \mid b$. Therefore, an integer n is even if and only if $2 \mid n$. If $a \mid b$, then a is called a **factor** or a **divisor** of b and b is called a **multiple** of a. Therefore, if n is even, then 2 is a divisor of n and n is a multiple of 2. If a does not divide b, then we write $a \nmid b$. For any two given integers a and b, $a \mid b$ is a *statement*. For example, $12 \mid 52$ is a false statement; while $12 \mid 72$ is a true statement. The statement $12 \mid 72$ has no numerical value, although $72/12 = 6$ of course. There is a major difference between the meaning of a vertical line and a slanted line.

Example 7.1

Determine which of the following statements are true:
  (a) 3 | 12,   (b) −7 | 14,   (c) 5 | 0,   (d) 8 | 20,   (e) 0 | 0.

Solution. Since $12 = 3 \cdot 4$, it follows that $3 \mid 12$. Also, $-7 \mid 14$ and $5 \mid 0$ as $14 = (-7)(-2)$ and $0 = 5 \cdot 0$. However, $8 \nmid 20$ since there is no integer c such that $20 = 8c$. Writing $0 \mid 0$ has no meaning since $a \neq 0$ in the definition of $a \mid b$. In summary, (a)–(c) are true; while (d) and (e) are false, but for different reasons.   ♦

There are several useful divisibility properties of integers. We now state and prove some of the most fundamental of these. Before doing this, however, there is an important fact that should be emphasized. If a and b are integers with $a \neq 0$ and we wish to show that $a \mid b$, then what we are required to show is that b can be written in a certain manner. In particular, we need to show that there is some integer c such that $b = ac$. The proofs of these kinds of results are generally direct proofs.

Theorem to Prove: Let a, b and c be integers with $a \neq 0$. If $a \mid b$ and $a \mid c$, then $a \mid (b+c)$.

Proof Strategy. In a direct proof of this theorem, we begin by assuming that $a \mid b$ and $a \mid c$. This means that there are integers r and s such that $b = ar$ and $c = as$. Since we want to show that $a \mid (b+c)$, we are required to show that $b+c$ can be expressed as the product of a and some integer. This suggests considering the integer $b+c$. Substituting the expressions ar and as for b and c, respectively, leads us to a proof.   ♦

Theorem 7.2   *Let a, b and c be integers with $a \neq 0$. If $a \mid b$ and $a \mid c$, then $a \mid (b+c)$.*

Proof. Assume that $a \mid b$ and $a \mid c$. Hence $b = ar$ and $c = as$ for some integers r and s. Therefore,

$$b + c = ar + as = a(r+s).$$

Since $r + s$ is an integer, $a \mid (b+c)$.   ■

A similar proof can be given of the next theorem.

Theorem 7.3   *Let a and b be integers with $a \neq 0$. If $a \mid b$, then $a \mid bx$ for every integer x.*

Proof. Assume that $a \mid b$. Then $b = ar$ for some integer r. Then

$$bx = (ar)x = a(rx).$$

Since rx is an integer, $a \mid bx$.   ■

We now look at an even more general result.

Theorem 7.4   *Let a, b and c be integers with $a \neq 0$. If $a \mid b$ and $a \mid c$, then $a \mid (bx + cy)$ for every two integers x and y.*

Proof. Assume that $a \mid b$ and $a \mid c$. Then $b = ar$ and $c = as$ for some integers r and s. Then

$$bx + cy = (ar)x + (as)y = a(rx) + a(sy) = a(rx + sy).$$

Since $rx + sy$ is an integer, $a \mid (bx + cy)$.   ■

Note that if $y = 0$ in Theorem 7.4, then we obtain Theorem 7.3; while if we let $x = 1$ and $y = 1$ in Theorem 7.4, then we obtain Theorem 7.2. For this reason, Theorem 7.4 is a more general result, and Theorems 7.2 and 7.3 are both corollaries of Theorem 7.4. The next theorem gives another divisibility property of integers.

Theorem 7.5   *Let a, b and c be integers with $a \neq 0$ and $b \neq 0$. If $a \mid b$ and $b \mid c$, then $a \mid c$.*

Proof. Assume that $a \mid b$ and $b \mid c$. Then $b = ar$ and $c = bs$ for some integers r and s. Therefore,

$$c = bs = (ar)s = a(rs).$$

Since rs is an integer, $a \mid c$. $\blacksquare$

We now turn to some examples involving divisors of integers and see how mathematical induction can be applied to verify certain divisibility properties. Recall that

$$(a + b)^3 = a^3 + 3a^2b + 3ab^2 + b^3.$$

Result to Prove: For every nonnegative integer n,

$$3 \mid \left(n^3 - n \right).$$

Proof Strategy. Since we have decided to prove this result by induction and the statement deals with nonnegative integers n, our first step (the basis step) is to show (or simply observe) that $3 \mid (0^3 - 0)$. To verify the inductive step, we assume that $3 \mid \left(k^3 - k \right)$ for a nonnegative integer k and show that $3 \mid \left[(k + 1)^3 - (k + 1) \right]$. Since $3 \mid \left(k^3 - k \right)$, it follows that $k^3 - k = 3\ell$ for some integer ℓ. Our goal is to show that we can express $(k + 1)^3 - (k + 1)$ as the product of 3 and an integer. $\blacklozenge$

Result 7.6 *For every nonnegative integer n,*

$$3 \mid \left(n^3 - n \right).$$

Proof. We proceed by induction. Since $3 \mid (0^3 - 0)$, the statement is true for $n = 0$. Assume that $3 \mid \left(k^3 - k \right)$, where k is a nonnegative integer. Then $k^3 - k = 3\ell$ for some integer ℓ. We show that 3 divides $(k + 1)^3 - (k + 1)$. Observe that

$$
\begin{aligned}
(k + 1)^3 - (k + 1) &= k^3 + 3k^2 + 3k + 1 - k - 1 = (k^3 - k) + 3k^2 + 3k \\
&= 3\ell + 3k^2 + 3k = 3(\ell + k^2 + k).
\end{aligned}
$$

Since $\ell + k^2 + k$ is an integer, 3 divides $(k+1)^3 - (k+1)$. By the Principle of Mathematical Induction, $3 \mid \left(n^3 - n \right)$ for every nonnegative integer n. $\blacksquare$

Result 7.7 *For every nonnegative integer n,*

$$4 \mid (5^n - 1).$$

Proof. We use induction. Since $4 \mid (5^0 - 1)$, the statement is true when $n = 0$. Assume that $4 \mid \left(5^k - 1 \right)$ for a nonnegative integer k. Then $5^k - 1 = 4\ell$ for some integer ℓ and so $5^k = 4\ell + 1$. We show that $4 \mid \left(5^{k+1} - 1 \right)$. Observe that

$$5^{k+1} - 1 = 5 \cdot 5^k - 1 = 5(4\ell + 1) - 1 = 20\ell + 4 = 4(5\ell + 1).$$

Since $5\ell + 1$ is an integer, $4 \mid \left(5^{k+1} - 1 \right)$. By the Principle of Mathematical Induction, $4 \mid (5^n - 1)$ for every nonnegative integer n. $\blacksquare$

Exercises for Section 7.1

1. For each pair a, b of integers, determine whether $a \mid b$. If $a \mid b$, then find an integer c such that $b = ac$.

   (a) $a = 7$ and $b = -70$.　(b) $a = 16$ and $b = -40$.　(c) $a = 1$ and $b = 10$.

   (d) $a = 8$ and $b = -8$.　(e) $a = 14$ and $b = 0$.　(f) $a = 0$ and $b = 14$.

2. Give an example of

   (a) three positive integers that are multiples of 3.

   (b) three positive integers that are multiples of 5.

   (c) a positive integer having 2, 3 and 5 as factors.

   (d) a positive integer exceeding 10 that 4 divides but 8 does not.

   (e) a positive integer that is a multiple of both 3 and 4.

   (f) a positive integer that is a multiple of both 4 and 6.

3. Let a and b be integers with $a \neq 0$. Prove that if $a \mid b$, then $a \mid (-b)$ and $(-a) \mid b$.

4. Let a, b and c be integers with $a \neq 0$. Prove that if $a \mid b$ and $a \mid (b+c)$, then $a \mid c$.

5. Let a, b and c be integers with $a \neq 0$ and $c \neq 0$. Prove that $ac \mid bc$ if and only if $a \mid b$.

6. Let a and b be integers with $a \neq 0$ and $b \neq 0$. Prove that if $a \mid b$, then $|a| \leq |b|$.

7. Disprove the following: Let a and b be integers with $a \neq 0$ and $b \neq 0$. If $a \mid b$ and $b \mid a$, then $a = b$.

8. Let a and b be integers such that $3 \mid a$. Prove that if $3 \nmid (a+b)$, then $3 \nmid b$.

9. Let $a, b, c \in \mathbb{Z}$ and $a \neq 0$. Show that if $a \nmid bc$, then $a \nmid b$ and $a \nmid c$.

10. (a) Show that if an integer a is a multiple of 36, then a is a multiple of 12.

    (b) The result in (a) is also a consequence of a theorem in the text. Which theorem?

11. Prove that $3 \mid (4n^3 + 5n)$ for every nonnegative integer n.

12. Prove that $3 \mid (2^{2n} - 1)$ for every nonnegative integer n.

13. Prove that $7 \mid (2^{3n} - 1)$ for every nonnegative integer n.

14. Prove that $4 \mid (7^n - 3^n)$ for every nonnegative integer n.

15. In Result 7.6, it was shown that $3 \mid (n^3 - n)$ for every nonnegative integer n. Use Result 7.6 to prove that $3 \mid (n^3 - n)$ for *every integer* n.

16. Prove that $4 \mid (3^{2n-1} + 1)$ for every positive integer n.

17. Prove that $133 \mid (11^{n+1} + 12^{2n-1})$ for every positive integer n.

18. (a) Let n be a positive integer. Show that every integer m with $1 \leq m \leq 2n$ can be expressed as $2^{\ell}k$, where ℓ is a nonnegative integer and k is an odd integer with $1 \leq k < 2n$.

    (b) Prove for every positive integer n and every subset S of $\{1, 2, \ldots, 2n\}$ with $|S| = n + 1$ that there exist distinct integers $a, b \in S$ such that $a \mid b$.

7.2 Primes

The integers that are of greatest interest to us, both historically and in terms of their properties, are the primes.

Definition 7.8 *A* **prime** *is an integer $p \geq 2$ whose only positive integer divisors are 1 and p.*

The 15 smallest primes are

$$2 \quad 3 \quad 5 \quad 7 \quad 11 \quad 13 \quad 17 \quad 19 \quad 23 \quad 29 \quad 31 \quad 37 \quad 41 \quad 43 \quad 47.$$

The Fundamental Theorem of Arithmetic

One of the most important properties of primes lies in the following famous theorem. After developing some properties of primes, we will present a proof of this theorem in Section 7.6.

Theorem 7.9 (The Fundamental Theorem of Arithmetic) *Every integer $n \geq 2$ is either prime or can be expressed as a product of (not necessarily distinct) primes, that is,*

$$n = p_1 p_2 \cdots p_k,$$

where $p_1, p_2, \ldots, p_k$ are primes. This factorization is unique except possibly for the order in which the primes appear.

Let's look at some illustrations of Theorem 7.9.

Example 7.10

Express the following integers as products of primes.

(a) 30.   (b) 12.   (c) 100.   (d) 54.   (e) 605.   (f) 216.

Solution.

(a) $30 = 2 \cdot 3 \cdot 5$.

(b) $12 = 2 \cdot 2 \cdot 3 = 2^2 \cdot 3$.

(c) $100 = 2 \cdot 2 \cdot 5 \cdot 5 = 2^2 \cdot 5^2$.

(d) $54 = 2 \cdot 3 \cdot 3 \cdot 3 = 2 \cdot 3^3$.

(e) $605 = 5 \cdot 11 \cdot 11 = 5 \cdot 11^2$.

(f) $216 = 2 \cdot 2 \cdot 2 \cdot 3 \cdot 3 \cdot 3 = 2^3 \cdot 3^3$.   ◆

There are some instances when one can tell rather quickly whether a certain prime p divides an integer n. This is probably the case with the integers in Example 7.10. As we have noted, 2 divides n if and only if n is even. Of course, it is easy to determine whether an integer n is even: the last digit of n must be even. Actually, $4 = 2^2$ divides n if and only if 4 divides the number formed from the last two digits of n and so on. For example, $4 \mid 3212$ because $4 \mid 12$, but $8 \nmid 3212$ since $8 \nmid 212$. The prime 3 divides an integer n if and only if 3 divides the sum of the digits of n. For example, $3 \mid 5241$ since $3 \mid (5 + 2 + 4 + 1)$. The same is true of $9 = 3^2$, that is, 9 divides n if and only if 9 divides the sum of the digits of n. Since $9 \nmid (5 + 2 + 4 + 1)$, it follows that $9 \nmid 5241$. On the other hand, $9 \mid 747$. This property does not hold for $27 = 3^3$, however; for example, certainly $27 \mid 27$ but $27 \nmid (2 + 7)$.

It is easy to determine whether the prime 5 divides an integer n, as all we need to do is to check whether the last digit of n is 5 or 0. While there is no simple way to determine whether 7 divides an integer n, there is a method for determining whether 11 divides an integer n. If we were to add alternating digits of n, getting a say and then add the remaining digits, obtaining b, then $11 \mid n$ if and only if $11 \mid (a - b)$. For example, suppose that $n = 8,241,739$. Then $a = 8 + 4 + 7 + 9 = 28$ and $b = 2 + 1 + 3 = 6$. Since $11 \mid (28 - 6)$, it follows that $11 \mid 8,241,739$.

Example 7.11

Express $n = 33,660$ as a product of primes.

Solution. There are several ways one might proceed to write $33,660$ as a product of primes but using some of the facts mentioned above, we find that $33,660 = 2^2 \cdot 3^2 \cdot 5 \cdot 11 \cdot 17$.   ♦

Definition 7.12   *An integer $n \geq 2$ that is not prime is called a* **composite number** (*or simply* **composite**).

The 15 smallest composite numbers are

$$4 \quad 6 \quad 8 \quad 9 \quad 10 \quad 12 \quad 14 \quad 15 \quad 16 \quad 18 \quad 20 \quad 21 \quad 22 \quad 24 \quad 25.$$

If n is a composite number, it then follows that there is an integer a with $1 < a < n$ such that $a \mid n$. Therefore, there exists an integer b such that $n = ab$. Necessarily, $1 < b < n$ as well. This is summarized in the next result.

Theorem 7.13   *An integer $n \geq 2$ is composite if and only if there exist integers a and b with $1 < a < n$ and $1 < b < n$ such that $n = ab$.*

There is an interesting consequence of this theorem.

Corollary to Prove: If n is a composite number, then n has a prime factor p such that $p \leq \sqrt{n}$.

Proof Strategy. If we use a direct proof, then we begin with a composite number n. Our goal is to show that n has a prime factor p such that $p \leq \sqrt{n}$. Because n is a composite number, it follows from Theorem 7.13 that n can be expressed as the product ab of integers such that $1 < a < n$ and $1 < b < n$. By Theorem 7.9 each of a and b can be expressed as a product of primes and so there is a prime that divides a and a prime that divides b. Certainly, one of a and b is less than or equal to the other, say $a \leq b$. Also, let p be a prime that divides a. Then

$$n = ab \geq a \cdot a = a^2 \text{ and so } \sqrt{n} \geq a.$$

Since $p \mid a$, it follows that $p \mid n$ and that $p \leq a \leq \sqrt{n}$. This gives us what we want.   ♦

Corollary 7.14   *If n is a composite number, then n has a prime factor p such that $p \leq \sqrt{n}$.*

Proof. Let n be a composite number. By Theorem 7.13, there are integers a and b with $1 < a < n$ and $1 < b < n$ such that $n = ab$. Suppose that $a \leq b$. Then $a \cdot a = a^2 \leq ab = n$ and so $a \leq \sqrt{n}$. Since $a \geq 2$, it follows by Theorem 7.9 that there is a prime p such that $p \mid a$ and so $p \leq a \leq \sqrt{n}$. By Theorem 7.3, $p \mid ab$ and so $p \mid n$.   ■

Corollary 7.14 can be useful to determine whether a given integer is a prime.

Example 7.15

Show that 103 is a prime.

Solution. If 103 is not a prime, then 103 is composite and by Corollary 7.14 has a prime factor that is not greater than $\sqrt{103}$. Since $10 < \sqrt{103} < 11$, we need only determine whether any prime less than 11 is a factor of 103. However, since none of 2, 3, 5, 7 is a factor of 103, it follows that 103 is a prime. ◆

There Are Infinitely Many Primes

It probably comes as no surprise that there are infinitely many primes.

Theorem to Prove: There are infinitely many primes.

Proof Strategy. Since "infinite" means "not finite," it is natural to attempt a proof of this theorem by contradiction. So we assume that there are not infinitely many primes. This means that there is only a finite number of primes, say $p_1, p_2, \ldots, p_k$ for some positive integer k. Our goal is to produce a contradiction. We need to contradict some definition, some assumption or some theorem. According to Theorem 7.9, if we have any integer $a \geq 2$, then some prime must divide a. However, from our assumption, the only primes are $p_1, p_2, \ldots, p_k$. So at least one of these primes must divide a. If we can construct an integer $n \geq 2$ such that none of $p_1, p_2, \ldots, p_k$ divides n, then a contradiction has been produced. One idea is to consider $n = p_1 p_2 \cdots p_k + 1$. ◆

Theorem 7.16 *There are infinitely many primes.*

Proof. Assume, to the contrary, that there are finitely many primes, say $p_1, p_2, \ldots, p_k$. Let

$$n = p_1 p_2 \cdots p_k + 1.$$

Since n is greater than each prime, n must be composite. By the Fundamental Theorem of Arithmetic, at least one of the primes $p_1, p_2, \ldots, p_k$ must divide n, say $p_j \mid n$. Thus $n = p_j r$ for some integer r. Hence

$$p_j r = n = p_1 p_2 \cdots p_k + 1 = p_1 p_2 \cdots p_{j-1} p_j p_{j+1} \cdots p_k + 1.$$

Therefore,

$$1 = p_j (r - p_1 p_2 \cdots p_{j-1} p_{j+1} \cdots p_k).$$

Since $r - p_1 p_2 \cdots p_{j-1} p_{j+1} \cdots p_k$ is an integer, $p_j \mid 1$. However, $p_j \geq 2$ and this is impossible. ∎

Actually it's been known for a long time (a *very* long time) that there are infinitely many primes. Over 2300 years ago, the famous mathematician Euclid, well known for his contributions to geometry, proved that there are infinitely many primes. Primes have fascinated people, mathematicians and non-mathematicians alike, for centuries. Many have not been satisfied to know *only* that there are infinitely many primes, however.

For a given positive integer n, how many primes are there less than n? If n is relatively small, then it is not too difficult to answer this question. But as anyone would probably guess, this question becomes increasingly difficult to answer the larger that n becomes. It is common to let $\pi(n)$ denote the number of primes that are less than n. So $\pi(2) = 0$, $\pi(3) = 1$, $\pi(5) = 2$, $\pi(10) = 4$ and $\pi(100) = 25$. In fact, Figure 7.1 gives the value of $\pi(n)$ for a few positive integers n.

n	$\pi(n)$
10^3	168
10^6	78,498
10^9	50,847,534
10^{12}	37,607,912,018
10^{21}	21,127,269,486,018,731,928

Figure 7.1: The number of primes that are less than n

Looking at the values of $\pi(n)$ for the positive integers n in Figure 7.1 certainly makes one think that there is no formula for $\pi(n)$ in terms of n – at least no simple formula, at least no familiar formula. But could it be that $\pi(n)$ is "close" to something familiar? This is a subject that intrigued many famous mathematicians and led to a theorem known as the **Prime Number Theorem**.

Theorem 7.17 (The Prime Number Theorem) *The number $\pi(n)$ is approximately equal to* $n/\ln n$. *More precisely,*

$$\lim_{n\to\infty} \frac{\pi(n)}{n/\ln n} = 1.$$

In 1798 the famous French mathematician Adrien-Marie Legendre made the first conjecture (guess) of this theorem. Carl Friedrich Gauss also was thinking along the right lines in the mid-1800s. The Russian mathematician Pafnuty Lvovich Chebyshev made the first progress towards a proof. The first proof occurred however in 1896 when Jaques Salomon Hadamard and Charles de la Vallée Poussin independently proved the theorem. Their proofs used the calculus of complex numbers. The first proof to avoid this approach was given in 1949 independently by Atle Selberg and Paul Erdős.

Since there are infinitely many primes, there is surely no largest prime. But what is an example of a "large" prime? A more interesting question probably is: What is the largest known prime? This is a question whose answer changes often. The number $2^{43,112,609} - 1$ was shown to be prime in September 2008 by a team at UCLA. Writing out this number requires nearly 13 million digits. Since this was the first prime discovered with at least 10 million digits, this team was awarded $100,000 by the Electronic Frontier Foundation. (There is a prize of $150,000 for the first prime with 100 million digits and a prize of $250,000 for the first prime with one billion digits.)

Unsolved Problems Involving Primes

There is a host of unanswered questions concerning prime numbers. We mention just a few of these.

(1) Two positive integers p and $p+2$ are called **twin primes** if they are both primes. For example, the following are twin primes:

$$\{3, 5\}, \{5, 7\}, \{11, 13\}, \{17, 19\}, \{29, 31\}.$$

In addition,

$$p = 16,869,987,339,975 \cdot 2^{171,960} - 1 \text{ and } p+2 = 16,869,987,339,975 \cdot 2^{171,960} + 1$$

are twin primes.

The Twin Primes Conjecture There are infinitely many twin primes.

(2) Observe that $4 = 2+2$, $6 = 3+3$, $8 = 3+5$, $10 = 3+7 = 5+5$, $12 = 5+7$ and $14 = 3+11 = 7+7$. The following conjecture appeared in a letter written to Leonhard Euler by Christian Goldbach in 1742.

Goldbach's Conjecture Every even integer that is 4 or more can be expressed as the sum of two primes.

(3) Observe that the following Fibonacci numbers are primes: 2, 3, 5, 13. Are there infinitely many prime Fibonacci numbers?

(4) For every integer n with $1 \le n \le 40$, the polynomial $n^2 - n + 41$ produces a prime. For $n = 41$, however, $n^2 - n + 41 = 41^2 - 41 + 41 = 41^2$, which, of course, is composite. It is known, though, that no polynomial expression with integer coefficients produces only primes.

In 2008, Eric Rowland discovered a recursively defined sequence, every value of which is either 1 or a prime. Define the sequence $a_1, a_2, a_3, \ldots$ by $a_1 = 7$ and $a_n = a_{n-1} + \gcd(n, a_{n-1})$. The first 13 terms of this sequence are

$$7, 8, 9, 10, 15, 18, 19, 20, 21, 22, 33, 36, 37.$$

Now define the sequence $p_1, p_2, p_3, \ldots$ by

$$p_n = a_{n+1} - a_n \text{ for } n \geq 1.$$

The first 12 terms of this sequence are

$$1, 1, 1, 5, 3, 1, 1, 1, 1, 11, 3, 1.$$

Therefore, the sequence $p_1, p_2, p_3, \ldots$ can repeat a prime. Does every odd prime appear in this sequence?

Exercises for Section 7.2

1. Express each of the following integers as a product of primes.
   (a) 250.   (b) 297.   (c) 2662.   (d) 1225.   (e) 891.

2. Determine which of the following are primes.
   (a) 127.   (b) 129.   (c) 131.   (d) 133.

3. Give an example, with explanation, of a prime between 1000 and 2000.

4. Show that there exists a positive integer n such that there is no prime between $n + 1000$ and $n + 2000$.

5. Show that there are infinitely many primes greater than $1,000,000$.

6. Show that there are infinitely many composite numbers.

7. Of course, 11 is a prime.
   (a) Show that 111 is not a prime.   (b) Show that 1111 is not a prime.
   (c) Show that $111,111$ is not a prime.   (d) Is $11,111$ a prime?

8. Prove that every prime except one has the form $a^2 - b^2$ for some positive integers a and b.

9. Let a and b be any two digits with $0 \leq b < a \leq 9$. Show that at most one of the integers ab, $abab$, $ababab$, $\cdots$ is a prime.

10. Show that only one prime can be expressed as $n^3 - 1$ for some positive integer n.

11. Show that only one prime can be expressed as $n^3 + 1$ for some positive integer n.

12. If the Twin Primes Conjecture is true, then there are infinitely many primes that can be expressed as the sum of two distinct primes. Show that there is a prime that can be expressed as the sum of
    (a) three distinct primes.   (b) four distinct primes.   (c) five distinct primes.

13. Goldbach's Conjecture states that every even integer $n \geq 4$ can be expressed as the sum of two primes. Goldbach also conjectured that every integer $n \geq 3$ can be written as the sum of three integers, each of which is either 1 or a prime. Prove that if Goldbach's conjecture is true, then this conjecture is also true.

14. Two distinct odd primes are called consecutive if there is no odd prime between p and q. Investigate the conjecture: If p and q are consecutive odd primes and $5 \nmid (pq + 2)$, then $pq + 2$ is prime.

15. Investigate the following conjecture: For every positive even integer n, there exists a positive even integer k such that $n^k + 1$ is prime.

16. It was mentioned that the number $2^{43,112,609} - 1$ was the largest prime as of September 2008. Show that $43,112,609$ is also a prime.

17. All of the largest primes are of the form $2^k - 1$ for some positive integer k. Show that there is no prime of the form $3^k - 1$ or $4^k - 1$ for any integer $k \geq 2$.

7.3   The Division Algorithm

Although the word *divides* has been introduced as a technical term, we are far more familiar with the phrase *divided by*. For example, when 16 is divided by 5, we obtain a quotient of 3 and a remainder of 1. This is described in a theorem called the Division Algorithm.

Theorem 7.18   (The Division Algorithm) *For every two integers m and $n > 0$, there exist unique integers q and r such that*

$$m = nq + r, \text{ where } 0 \leq r < n.$$

The integer q in Theorem 7.18 represents the **quotient** obtained when m is divided by n and r represents the **remainder**, which is one of the n integers $0, 1, \ldots, n-1$. Hence if $m = 16$ and $n = 5$, then $q = 3$ and $r = 1$ and $16 = 5 \cdot 3 + 1$. When m is not positive, however, the resulting quotient and remainder may not be what one initially expects.

Example 7.19

For each of the following pairs m, n of integers, find the quotient q and remainder r when m is divided by n. Then write $m = nq + r$.

(a) $m = 58$, $n = 7$. (b) $m = 0$, $n = 7$. (c) $m = -58$, $n = 7$. (d) $m = 21$, $n = 7$.

Solution.

(a) $q = 8$, $r = 2$, $58 = 7 \cdot 8 + 2$.

(b) $q = 0$, $r = 0$, $0 = 7 \cdot 0 + 0$.

(c) $q = -9$, $r = 5$, $-58 = 7 \cdot (-9) + 5$.

(d) $q = 3$, $r = 0$, $21 = 7 \cdot 3 + 0$.   ◆

In Examples 7.19(b) and 7.19(d), the remainder $r = 0$. In particular, $0 = 7 \cdot 0$ and $21 = 7 \cdot 3$ and so $7 \mid 0$ and $7 \mid 21$. That is, for two integers m and n with $n > 0$, writing $m = nq + 0$ is the same as saying that $n \mid m$. Therefore, we have the following:

If $m = nq + r$ and $1 \leq r \leq n - 1$, then $n \nmid m$.

In general, the quotient q, obtained when m is divided by n, is $\lfloor m/n \rfloor$ while the remainder r is $m - n \lfloor m/n \rfloor$; that is:

If $m = nq + r$, where $0 \leq r < n$, then

$$q = \left\lfloor \frac{m}{n} \right\rfloor \quad \text{and} \quad r = m - n \left\lfloor \frac{m}{n} \right\rfloor.$$

Example 7.20

For each of the following pairs m, n of integers, determine $\lfloor m/n \rfloor$ and $m - n\lfloor m/n \rfloor$.

(a) $m = 18$, $n = 7$. (b) $m = 0$, $n = 7$. (c) $m = -18$, $n = 7$.

Solution.

(a) $\lfloor m/n \rfloor = \lfloor 18/7 \rfloor = 2$ and $m - n\lfloor m/n \rfloor = 18 - 7 \cdot 2 = 4$.

(b) $\lfloor m/n \rfloor = \lfloor 0/7 \rfloor = 0$ and $m - n\lfloor m/n \rfloor = 0 - 7 \cdot 0 = 0$.

(c) $\lfloor m/n \rfloor = \lfloor -18/7 \rfloor = -3$ and $m - n\lfloor m/n \rfloor = -18 - 7 \cdot (-3) = 3$. ◆

Of course, what is being computed for integers m and $n \geq 1$ in Example 7.20 are the quotient $q = \lfloor m/n \rfloor$ and remainder $r = m - n\lfloor m/n \rfloor$ when m is divided by n. Some calculators contain built-in integer functions that provide us with q and r for given integers m and $n > 0$. At times, these are indicated by **div** and **mod**, respectively. That is, if $m = nq + r$, where $0 \leq r < n$, then

$$m \text{ } \textbf{div} \text{ } n = q \quad \text{and} \quad m \text{ } \textbf{mod} \text{ } n = r.$$

The integer function **mod** is especially useful.

Example 7.21

Determine m **div** n and m **mod** n for each of the following pairs m, n of integers.

(a) $m = 75$, $n = 13$. (b) $m = 84$, $n = 14$. (c) $m = -37$, $n = 5$.

Solution.

(a) $75 \text{ } \textbf{div} \text{ } 13 = 5$ and $75 \text{ } \textbf{mod} \text{ } 13 = 10$.

(b) $84 \text{ } \textbf{div} \text{ } 14 = 6$ and $84 \text{ } \textbf{mod} \text{ } 14 = 0$.

(c) $-37 \text{ } \textbf{div} \text{ } 5 = -8$ and $-37 \text{ } \textbf{mod} \text{ } 5 = 3$. ◆

If we take $n = 2$, then Theorem 7.18 tells us that every integer m can be written as $2q + r$ for some integers q and r, where $0 \leq r \leq 1$, that is, either $m = 2q$ (and m is even) or $m = 2q + 1$ (and m is odd). For $n = 3$, we see that every integer m can be written as $3q$, $3q + 1$ or $3q + 2$ for some integer q.

Just as certain theorems concerning an integer n can be conveniently proved by dividing the proof into two cases, namely n is even or n is odd, a proof by cases may be appropriate according to the remainder obtained when n is divided by some integer. We consider such an example.

Theorem to Prove: Let n be an integer. Then $3 \mid n^2$ if and only if $3 \mid n$.

Proof Strategy. Because the statement is a bicondictional, there are two implications to verify, namely

(1) $3 \mid n^2$ if $3 \mid n$ (or if $3 \mid n$, then $3 \mid n^2$) and

(2) $3 \mid n^2$ only if $3 \mid n$ (or if $3 \mid n^2$, then $3 \mid n$).

The first of these is straighforward and can be proved with a direct proof. It therefore remains to prove that $3 \mid n^2$ only if $3 \mid n$. This implication can also be expressed as: if $3 \mid n^2$, then $3 \mid n$. It is natural to give a proof by contrapositive of this implication, namely, to show that if $3 \nmid n$, then $3 \nmid n^2$.

Thus we assume that $3 \nmid n$. Our goal is to show that $3 \nmid n^2$. By the Division Algorithm, we can write $n = 3q + r$, where $0 \leq r \leq 2$. Because $3 \nmid n$, it follows that $r \neq 0$. Hence $n = 3q + 1$ or $n = 3q + 2$. This suggests a proof by cases.

If $n = 3q + 1$, then

$$n^2 = (3q + 1)^2 = 9q^2 + 6q + 1 = 3(3q^2 + 2q) + 1.$$

Since $3q^2 + 2q$ is an integer, $3q^2 + 2q$ is the quotient and 1 is the remainder when n^2 is divided by 3. So $3 \nmid n^2$. The situation is similar when $n = 3q + 2$.   ♦

Theorem 7.22 *Let n be an integer. Then $3 \mid n^2$ if and only if $3 \mid n$.*

Proof. First, we prove that if $3 \mid n$, then $3 \mid n^2$. Assume that $3 \mid n$. Then $n = 3k$ for some integer k and so

$$n^2 = (3k)^2 = 9k^2 = 3(3k^2).$$

Since $3k^2$ is an integer, $3 \mid n^2$.

We now prove that if $3 \mid n^2$, then $3 \mid n$. Assume that $3 \nmid n$. Then n has a remainder of 1 or 2 when divided by 3. We consider these two cases.

Case 1. $n = 3q + 1$ for some integer q. Therefore,

$$n^2 = (3q + 1)^2 = 9q^2 + 6q + 1 = 3(3q^2 + 2q) + 1.$$

Since $3q^2 + 2q$ is an integer, n^2 has a remainder 1 when divided by 3 and so $3 \nmid n^2$.

Case 2. $n = 3q + 2$ for some integer q. Therefore,

$$n^2 = (3q + 2)^2 = 9q^2 + 12q + 4 = 3(3q^2 + 4q + 1) + 1.$$

Since $3q^2 + 4q + 1$ is an integer, n^2 has a remainder 1 when divided by 3 and so $3 \nmid n^2$.   ■

Exercises for Section 7.3

1. For each of the following pairs m, n of integers, find the quotient q and remainder r when m is divided by n. Then write $m = nq + r$.

   (a) $m = 48$, $n = 11$. (b) $m = 0$, $n = 11$. (c) $m = -48$, $n = 11$. (d) $m = 9$, $n = 11$.

2. For each of the following pairs m, n of integers, determine $\lfloor m/n \rfloor$ and $m - n\lfloor m/n \rfloor$.

   (a) $m = 38$, $n = 11$. (b) $m = 22$, $n = 11$. (c) $m = -38$, $n = 11$.

3. Determine m **div** n and m **mod** n for the following pairs m, n of integers.

   (a) $m = 47$, $n = 14$.   (b) $m = 81$, $n = 22$.   (c) $m = 180$, $n = 45$.

   (d) $m = 24$, $n = 25$.   (e) $m = 0$, $n = 15$.   (f) $m = -55$, $n = 27$.

4. Show that for each integer $n \geq 4$, at most two of the three integers n, $n + 2$ and $n + 4$ are primes.

5. Show that if n is an odd integer, then n^2 has a remainder of 1 when divided by 4.

6. Give an example of two integers a and b satisfying all three of the conditions below:

   (1) a **mod** $4 \neq b$ **mod** 4, (2) a^2 **mod** $4 \neq a$ **mod** 4 and (3) b^2 **mod** $4 \neq b$ **mod** 4.

7. Do there exist two distinct positive integers a and b such that $a \bmod b = b \bmod a$?

8. Show that if $3 \nmid n$, then n^2 has a remainder of 1 when divided by 3.

9. For each integer n, prove that

   (a) 3 divides one of the integers n, $n + 1$ and $2n + 1$.
   (b) 3 divides one of the integers n, $2n - 1$ and $2n + 1$.

10. According to the Division Algorithm, what are the possible ways that an integer n can be expressed when n is divided by 5?

11. Let a, b, c, d and e be consecutive integers. Use the Division Algorithm to prove that 5 divides one of these five integers.

12. Prove that 4 always divides the product of every four consecutive integers but never divides the sum of these integers.

13. Let p be a prime. Prove that

   (a) p divides the product of every p consecutive integers and
   (b) with exactly one exception for p, the prime p divides the sum of every p consecutive integers.

   (Recall that $1 + 2 + \cdots + n = \frac{n(n+1)}{2}$ for every positive integer n.)

14. Let n be an integer. Prove that if $3 \nmid n$, then $3 \mid (2n^2 - 5)$.

15. Let $a, b \in \mathbb{Z}$. Prove that $4 \mid (a^2 - b^2)$ if and only if a and b are of the same parity.

16. Let $a \in \mathbb{Z}$. Prove that if $3 \nmid a^2$, then $3 \mid (a^2 - 1)$.

17. Let $a, b \in \mathbb{Z}$. Prove that if $3 \nmid a$ and $3 \nmid b$, then $3 \mid (a^2 - b^2)$.

18. Let $n \in \mathbb{Z}$. Prove that $3 \mid (2n^2 + 1)$ if and only if $3 \nmid n$.

19. Show that, except for 2 and 5, every prime can be expressed as $10k + 1$, $10k + 3$, $10k + 7$ or $10k + 9$ for some integer k.

20. Let $p \geq 3$ be a prime. Show that $p = 4k + 1$ or $p = 4k + 3$ for some integer k.

21. Let $p \geq 5$ be a prime. Show that $p = 6k + 1$ or $p = 6k + 5$ for some integer k.

22. Let $n \in \mathbb{Z}$. Show that if $n = 6k + 5$ for some integer k, then $n = 3\ell + 2$ for some integer ℓ.

23. Let $n \in \mathbb{Z}$. Prove that if $3 \mid (n^2 + 1)$, then $3 \nmid (2n^2 + 1)$.

7.4 Congruence

We have often expressed an interest in two integers a and b that are of the same parity. Recall that this means that a and b are both even or both odd. If a and b are both even, then both have a remainder of 0 when divided by 2; while if a and b are both odd, then both have a remainder of 1 when divided by 2. More generality, for an integer $n \geq 2$, we will often be interested in integers that have the same remainder when divided by n. This is the topic of the current section. We begin with a definition that leads us to this discussion.

Definition 7.23 *For integers a, b and $n \geq 2$, the integer a is **congruent to** b **modulo** n if $n \mid (a - b)$.*

To indicate that a is congruent to b modulo n, we write $a \equiv b \pmod{n}$. If a is not congruent to b modulo n, then we write $a \not\equiv b \pmod{n}$. We have now seen that the notation "mod" has two meanings in mathematics. When **mod** is expressed in bold such as in $m \bmod n$, it indicates the remainder when m is divided by n. We now see some examples of integers a, b and $n \geq 2$, where $a \equiv b \pmod{n}$ or $a \not\equiv b \pmod{n}$.

Example 7.24

(a) $24 \equiv 14 \pmod{5}$ since $5 \mid (24 - 14)$.

(b) $-7 \equiv 9 \pmod{8}$ since $8 \mid ((-7) - 9)$.

(c) $12 \equiv 12 \pmod{9}$ since $9 \mid (12 - 12)$.

(d) $34 \not\equiv 16 \pmod{12}$ since $12 \nmid (34 - 16)$.   ◆

The following theorem provides a convenient way to determine whether an integer a is congruent to b modulo n.

Theorem 7.25 *Let a, b and $n \geq 2$ be integers. Then $a \equiv b \pmod{n}$ if and only if $a = b + kn$ for some integer k.*

Proof. First, we prove that if $a \equiv b \pmod{n}$, then $a = b + kn$ for some integer k. Assume that $a \equiv b \pmod{n}$. Hence $n \mid (a - b)$. Thus there exists an integer k such that $a - b = nk$ and so $a = b + kn$.

We now turn to the converse and prove that if $a = b + kn$ for some integer k, then $a \equiv b \pmod{n}$. Assume that $a = b + kn$ for some integer k. Then $a - b = kn$. Since k is an integer, $n \mid (a - b)$ and so $a \equiv b \pmod{n}$.   ∎

The next theorem gives us yet another useful way to determine whether an integer a is congruent to b modulo n.

Theorem 7.26 *Let a, b and $n \geq 2$ be integers. Then $a \equiv b \pmod{n}$ if and only if a and b have the same remainder when divided by n.*

Proof. First, we prove that if a and b have the same remainder when divided by n, then $a \equiv b \pmod{n}$. Assume that both a and b have the remainder r when divided by n. Then $a = nq_1 + r$ and $b = nq_2 + r$ for integers q_1 and q_2. Therefore,

$$a - b = (nq_1 + r) - (nq_2 + r) = n(q_1 - q_2).$$

Since $q_1 - q_2$ is an integer, $n \mid (a - b)$ and so $a \equiv b \pmod{n}$.

Next, we show that if $a \equiv b \pmod{n}$, then a and b have the same remainder when divided by n. Assume that $a \equiv b \pmod{n}$. Then $a = b + kn$ for some integer k by Theorem 7.25. Suppose that $a = nq_1 + r_1$ and $b = nq_2 + r_2$ for integers q_1, r_1, q_2, r_2, where $0 \leq r_1 \leq n - 1$ and $0 \leq r_2 \leq n - 1$. We may assume that $r_1 \geq r_2$. Then

$$kn = a - b = (nq_1 + r_1) - (nq_2 + r_2) = n(q_1 - q_2) + (r_1 - r_2).$$

Therefore,

$$r_1 - r_2 = kn - n(q_1 - q_2) = n(k - q_1 + q_2).$$

Since $k - q_1 + q_2$ is an integer, it follows that $n \mid (r_1 - r_2)$. However,

$$0 \leq r_1 - r_2 \leq n - 1.$$

Because $n \mid (r_1 - r_2)$, we must have $r_1 - r_2 = 0$ and so $r_1 = r_2$.   ∎

Another way to state Theorem 7.26 is as follows:

Corollary 7.27   *Let a, b and $n \geq 2$ be integers. Then $a \equiv b \pmod{n}$ if and only if*

$$a \bmod n = b \bmod n.$$

Example 7.28

For each of the following pairs a, b of integers and an integer $n \geq 2$, compute $a \bmod n$ and $b \bmod n$ and use this to determine whether $a \equiv b \pmod{n}$.

$$\text{(a) } a = 17,\ b = -11,\ n = 7. \quad \text{(b) } a = 30,\ b = 10,\ n = 15.$$

Solution.

(a) Since $a \bmod n = 17 \bmod 7 = 3$ and $b \bmod n = -11 \bmod 7 = 3$, it follows by Corollary 7.27 that $17 \equiv -11 \pmod{7}$.

(b) Since $a \bmod n = 30 \bmod 15 = 0$ and $b \bmod n = 10 \bmod 15 = 10$, it follows by Corollary 7.27 that $30 \not\equiv 10 \pmod{15}$.   ♦

Exercises for Section 7.4

1. For the following integers a, b and n, determine whether a is congruent to b modulo n. If so, write $a \equiv b \pmod{n}$; if not, write $a \not\equiv b \pmod{n}$.

   (a) $a = 47,\ b = 23,\ n = 8$.   (b) $a = 18,\ b = 38,\ n = 5$.   (c) $a = 20,\ b = 10,\ n = 3$.

   (d) $a = 12,\ b = 12,\ n = 13$.   (e) $a = 37,\ b = 35,\ n = 2$.

2. Classify each of the following statements as true or false.

   (a) $24 \equiv 3 \pmod{7}$. (b) $-17 \equiv 9 \pmod{8}$.   (c) $-5 \equiv -5 \pmod{4}$. (d) $24 \equiv -3 \pmod{3}$.

3. Give an example of four integers that are congruent to 4 modulo 13.

4. Determine whether each of the following integers is congruent to 5 modulo 7.

   (a) 40.   (b) 108.   (c) -29.   (d) -122.

5. Recall Theorem 7.25: *Let a, b and $n \geq 2$ be integers. Then $a \equiv b \pmod{n}$ if and only if $a = b + kn$ for some integer k.* Prove the following:

   Let a, b and $n \geq 2$ be integers. Then $a \equiv b \pmod{n}$ if and only if $b = a + \ell n$ for some integer ℓ.

6. Let a, b, c and $n \geq 2$ be integers. Prove that if $a \equiv b \pmod{n}$ and $b \equiv c \pmod{n}$, then $a \equiv c \pmod{n}$.

7. Let a, b, m and n be integers with $m, n \geq 2$ and $m \mid n$. Show that if $a \equiv b \pmod{n}$, then $a \equiv b \pmod{m}$.

8. Let a, b, c, d and $n \geq 2$ be integers. Show that if $a \equiv b \pmod{n}$ and $c \equiv d \pmod{n}$, then $a + c \equiv b + d \pmod{n}$.

9. Let a, b, c, d and $n \geq 2$ be integers. Show that if $a \equiv b \pmod{n}$ and $c \equiv d \pmod{n}$, then $a - c \equiv b - d \pmod{n}$.

10. Let a, b, c, d and $n \geq 2$ be integers. Show that if $a \equiv b \pmod{n}$ and $c \equiv d \pmod{n}$, then $ac \equiv bd \pmod{n}$.

11. Let a, b and $n \geq 2$ be integers. Show that if $a \equiv b \pmod{n}$, then $a^r \equiv b^r \pmod{n}$ for every positive integer r.

12. Let a, b and $n \geq 2$ be integers.

    (a) Prove that if $a \equiv b \pmod{n}$, then $ak \equiv bk \pmod{nk}$ for every positive integer k.

    (b) Disprove: If $ak \equiv bk \pmod{n}$ for some integer k, then $a \equiv b \pmod{n}$.

13. Let $a, b \in \mathbb{Z}$. Prove that if $a \equiv 0 \pmod{5}$ and $b \equiv 2 \pmod{5}$, then $a^2 + b^2 \equiv 4 \pmod{5}$.

14. Let $a \in \mathbb{Z}$. Prove that if $a^2 \not\equiv a \pmod{3}$, then $a \not\equiv 0 \pmod{3}$ and $a \not\equiv 1 \pmod{3}$.

15. Let $a, b \in \mathbb{Z}$. Prove that if $a^2 + 2b^2 \equiv 0 \pmod{3}$, then either both a and b are congruent to 0 modulo 3 or neither a nor b is congruent to 0 modulo 3.

16. Let $a, b, c \in \mathbb{Z}$. Prove that if $abc \equiv 1 \pmod{3}$, then an odd number of a, b and c are congruent to 1 modulo 3.

17. For each of the following pairs a, b of integers and an integer $n \geq 2$, compute $a \bmod n$ and $b \bmod n$ and use this to determine whether $a \equiv b \pmod{n}$.

    (a) $a = 38$, $b = 14$, $n = 8$. (b) $a = 31$, $b = 43$, $n = 6$. (c) $a = 27$, $b = -15$, $n = 7$.

    (d) $a = 35$, $b = -11$, $n = 12$. (e) $a = 0$, $b = -2$, $n = 2$. (f) $a = 43$, $b = 29$, $n = 9$.

18. Prove or disprove: There exist distinct integers $a, b, c \geq 2$ such that $a \equiv b \pmod{c}$, $b \equiv c \pmod{a}$ and $c \equiv a \pmod{b}$.

7.5   Introduction to Cryptography

A modern day use of integers and congruences occurs in the area of cryptography. The word cryptography comes from the two Greek words *Kryptos* (meaning "hidden") and *graphein* (meaning "to write"). (Perhaps you recall that the imaginary element kryptonite, against which Superman is defenseless.) Formally, **cryptography** is the study of sending and receiving secret messages. The related term **cryptology** is the study of systems, called **cryptosystems**, for secure communications. Cryptography is also used to express the act of writing in codes and cryptology is the science of analyzing and deciphering codes.

In a cryptosystem, the sender of a message (written in **plaintext**) first transforms the message (into **ciphertext**) before transmitting it so that only individuals authorized to do so will be able to reconstruct the original message. **Encryption** is the process of transforming the original message into a disguised message, while **decryption** is the process of transforming the disguised message (in ciphertext) back to the original message (in plaintext).

The goal is to have a secure cryptosystem so that even if unauthorized individuals read a disguised message, they are unable to discover the decryption technique that will allow them to obtain the original message. Cryptosystems are essential for certain organizations such as governments, the military and internet-based businesses. In particular, if a business transaction were to take place over the internet during which a credit card number is sent, then it is critical that this information can be read only by those for whom it is intended. The same concern exists if financial or other private information were to be transmitted over the internet.

Cryptography is used when an individual wants to send a secret message to someone else, where there is a possibility that some third individual might intervene and read the message. It has been used by generals to send orders to their armies. Perhaps the best known example of this was the Enigma machine, an encryption machine invented and used during World War II to send military messages. Several books and motion pictures have dealt with this subject.

In one of the oldest and simplest cryptosystems, the sender and receiver possess a "key" that provides a substitute character for each character that could be contained in any message to be transmitted. Such a key is called a **private key**. A simple example of a cryptosystem deals with identifying each letter of the English alphabet by an integer from 0 to 25 as shown in Figure 7.2.

$A \leftrightarrow 0$	$B \leftrightarrow 1$	$C \leftrightarrow 2$	$D \leftrightarrow 3$	$E \leftrightarrow 4$	$F \leftrightarrow 5$	$G \leftrightarrow 6$
$H \leftrightarrow 7$	$I \leftrightarrow 8$	$J \leftrightarrow 9$	$K \leftrightarrow 10$	$L \leftrightarrow 11$	$M \leftrightarrow 12$	$N \leftrightarrow 13$
$O \leftrightarrow 14$	$P \leftrightarrow 15$	$Q \leftrightarrow 16$	$R \leftrightarrow 17$	$S \leftrightarrow 18$	$T \leftrightarrow 19$	$U \leftrightarrow 20$
$V \leftrightarrow 21$	$W \leftrightarrow 22$	$X \leftrightarrow 23$	$Y \leftrightarrow 24$	$Z \leftrightarrow 25$		

Figure 7.2: An example of a cryptosystem

Julius Caesar was known to have used this method to send secret messages to his generals. What he did was to replace each letter in a message, corresponding to the integer x, say, where $0 \leq x \leq 25$, by the letter corresponding to the integer $f(x)$, where

$$f(x) = (x + 3) \bmod 26.$$

That is, each integer was replaced by rotating the letter 3 units to the right. For example, the message

SEE   YOU   IN   ROME.

would be transformed to

VHH   BRX   LQ   URPH.

Of course, the intended individual receiving the message would have to transform (decrypt) the message back to its original plaintext in which a letter in the disguised message corresponding to an integer x is then replaced by the letter corresponding to

$$f^{-1}(x) = (x - 3) \bmod 26.$$

In the **Caesar shift** decryption technique, each letter in the plaintext is replaced by a letter obtained by rotating the letter a fixed number of positions. So a rotation of 3 units to the right is the actual Caesar shift used by Julius Caesar himself. In the Caesar shift employed by Caesar, it is very difficult to find English words in plaintext that are transformed into English words in ciphertext. Thus it would be clear quite quickly that a secret message had been sent by this method. The Caesar shift obtained by rotating the letter one unit to the left could be described mathematically by the function

$$g(x) = (x - 1) \bmod 26.$$

In this case,

IBM would be transformed into HAL.

(This might remind you of the motion picture 2001: *A Space Odyssey.*)

While Caesar's use of this scheme appears to be the first recorded use, other substitution codes have been known to be used earlier. It is not known how successful Caesar's scheme was, as this code does not appear difficult to break.

Sending and decoding secret messages has been popular among youngsters for decades. So-called "secret decoder" toys have been enclosed in boxes of breakfast cereals and snack foods as far back as the 1930s. The best known decoders were badges given to children by manufacturers of Ovaltine (a chocolatey drink whose name is an anagram for "Vital One"), which sponsored the *Little Orphan Annie* radio show and the *Captain Midnight* radio and television shows. Decoder badges appeared in a scene from the popular holiday movie *A Christmas Story*.

The methods we described for encrypting messages are based on congruences. Words (strings of characters) are transformed into numbers. The numbers are then transformed into other numbers which are then transformed back into strings of characters. These methods are examples of what we call private key cryptosystems. Knowledge of the encryption key allows one to find the (inverse) decryption key rather easily. For example, when a Caesar shift is used with encryption key k, an integer x representing a letter is sent to the letter represented by the integer

$$y = (x + k) \bmod 26.$$

Decryption is then accomplished by a shift of $-k$, that is,

$$x = (y - k) \bmod 26.$$

Example 7.29

A secret message in ciphertext of a plaintext message using the Caesar shift defined by $f(x) = (x - 6) \bmod 26$ is

$$\text{KOCT} \quad \text{NIXUS}.$$

What is the original message in plaintext?

Solution. In this case, decryption is defined according to

$$f^{-1}(x) = (x + 6) \bmod 26.$$

Identifying each letter of the English alphabet with an integer x, where $0 \le x \le 25$, as in Figure 7.2, we have

$$
\begin{array}{ccccccc}
K & \leftrightarrow & 10 & \rightarrow & 16 & \leftrightarrow & Q \\
O & \leftrightarrow & 14 & \rightarrow & 20 & \leftrightarrow & U \\
C & \leftrightarrow & 2 & \rightarrow & 8 & \leftrightarrow & I \\
T & \leftrightarrow & 19 & \rightarrow & 25 & \leftrightarrow & Z \\
N & \leftrightarrow & 13 & \rightarrow & 19 & \leftrightarrow & T \\
I & \leftrightarrow & 8 & \rightarrow & 14 & \leftrightarrow & O \\
X & \leftrightarrow & 23 & \rightarrow & 3 & \leftrightarrow & D \\
U & \leftrightarrow & 20 & \rightarrow & 0 & \leftrightarrow & A \\
S & \leftrightarrow & 18 & \rightarrow & 24 & \leftrightarrow & Y
\end{array}
$$

So the original message in plaintex is

$$\text{QUIZ} \quad \text{TODAY.} \blacklozenge$$

The following example of a cryptosystem also identifies each letter of the English alphabet with an integer x, where $0 \le x \le 25$, as in Figure 7.2.

Example 7.30

A secret message in a cryptosystem where the integer x associated with a letter is transformed into $f(x) = 5x \bmod 26$ is:

$$\text{QUM, ANN.}$$

It turns out that decryption is defined in this case by $f^{-1}(x) = 21x \bmod 26$. What is the original message?

Solution. In this case,

$$
\begin{array}{ccccccccccc}
Q & \leftrightarrow & 16 & \rightarrow & 21 \cdot 16 & \bmod & 26 & = & 24 & \rightarrow & Y \\
U & \leftrightarrow & 20 & \rightarrow & 21 \cdot 20 & \bmod & 26 & = & 4 & \rightarrow & E \\
M & \leftrightarrow & 12 & \rightarrow & 21 \cdot 12 & \bmod & 26 & = & 18 & \rightarrow & S \\
A & \leftrightarrow & 0 & \rightarrow & 21 \cdot 0 & \bmod & 26 & = & 0 & \rightarrow & A \\
N & \leftrightarrow & 13 & \rightarrow & 21 \cdot 13 & \bmod & 26 & = & 13 & \rightarrow & N \\
N & \leftrightarrow & 13 & \rightarrow & 21 \cdot 13 & \bmod & 26 & = & 13 & \rightarrow & N
\end{array}
$$

Therefore, the original message in plaintex is

YES, ANN.

(Curiously, ANN is transformed into ANN in this case.) ◆

When a **private key cryptosystem** is used, two individuals who wish to communicate in secret have the same private key. This key can both encrypt and decrypt messages. In the mid-1970s the idea of a **public key cryptosystem** was introduced. In this case, there are two keys – a public encryption key and private decryption key. While any one with an encryption key can send a disguised message to some person, having such a key should not help you to decrypt a disguised message that has been sent to this person. The decryption keys are kept private and only the intended recipient knows how to decrypt the message.

In 1976 Richard Rivest, Adi Shamir and Leonard Adleman (all researchers at MIT during the time) introduced a public key cryptosystem, which has since become known as the **RSA System** (for the initials of the three inventors), based on the idea that it is very easy for computers to determine the product of two very large primes but very difficult, even for a computer, to factor an integer that is the product of two very large primes.

Exercises for Section 7.5

1. Encrypt the message "YOU ARE CORRECT SIR." by transforming letters into integers using the encryption function defined by $f(x) = (x + 5) \bmod 26$ for $x \in \mathbb{Z}$, $0 \le x \le 25$.

2. Decrypt the message "NJN MVYVM!" where encryption is defined by $f(x) = (x - 5) \bmod 26$.

3. Using a certain Caesar cipher, a message is transformed into the secret message "LIPT MW LIVI." What is the original message?

4. Decrypt the secret message "UVTU ABBA." where encryption is defined by $f(x) = (x + 13) \bmod 26$.

5. Consider the cryptosystem in which the integer x associated with a letter is transformed into $f(x) = 3x \bmod 26$. In this case, decryption is defined by $f^{-1}(x) = 9x \bmod 26$.

   (a) Into which secret word is the word GUM transformed?

   (b) Which word is transformed into the secret word FOYN?

   (c) Which word is transformed into the secret word JAN?

6. Consider the cryptosystem in which the integer x associated with a letter is transformed into $f(x) = 7x$.

   (a) Determine $f^{-1}(x)$.

   (b) Into which secret word is the word SIT transformed?

   (c) Which word is transformed into the secret word BUD?

7. It is decided to have a cryptosystem in which the integer x associated with a letter is transformed into $f(x) = 2x \bmod 26$. Why is this a bad idea?

8. Find a cryptosystem in which the word NO is transformed into the word ON. What is the word BIN transformed into in this cryptosystem?

9. Consider the cryptosystem in which the integer x associated with a letter is transformed into $f(x) = x + (-1)^x$ for $x \in \{0, 1, \ldots, 25\}$.

   (a) Which word is transformed into the secret word BAPUF? (b) Determine f^{-1}.

10. Consider the cryptosystem in which the integer $x \in \{0, 1, \ldots, 25\}$ associated with a letter is transformed into

$$f(x) = \begin{cases} 12 - x & \text{if } x \in \{0, 1, \ldots, 12\} \\ 38 - x & \text{if } x \in \{13, 14, \ldots, 25\}. \end{cases}$$

(a) Which word is transformed into USA? (b) Determine f^{-1}.

11. A secret message is being sent where encryption is defined by $f(x) = (ax + b) \bmod 26$ for some integers a and b. One of the words in the message is RATS, but this is actually the word HARM. Another word in the message is BEN but this actually the word FUN. The word WING appears in this message. What is the actual word most likely to be in this case?

7.6   Greatest Common Divisors

We now turn our attention from one integer dividing another to one integer dividing two others.

Definition 7.31 *Let a, b and d be integers, where a and b are not both 0 and $d \neq 0$. The integer d is a* **common divisor** *of a and b if $d \mid a$ and $d \mid b$.*

Example 7.32

The common divisors of 8 and 20 are -4, -2, -1, 1, 2 and 4.   ♦

The concept that has attracted the most attention in this connection is the greatest common divisor.

Definition 7.33 *For integers a and b not both 0, the* **greatest common divisor** *of a and b is the greatest positive integer that is a common divisor of a and b. This number is denoted by $\gcd(a, b)$.*

By Example 7.32, $\gcd(8, 20) = 4$. If two integers a and b are not very large, then finding $\gcd(a, b)$ can be accomplished by observation.

Example 7.34

Determine by observation the greatest common divisor of each of the following pairs a, b of integers.

(a) $a = 15$, $b = 25$.   (b) $a = 30$, $b = 42$.   (c) $a = 15$, $b = 17$.
(d) $a = -14$, $b = -18$. (e) $a = 0$, $b = 6$. (f) $a = 16$, $b = 80$.

Solution.

(a) $\gcd(15, 25) = 5$. (b) $\gcd(30, 42) = 6$. (c) $\gcd(15, 17) = 1$.
(d) $\gcd(-14, -18) = 2$. (e) $\gcd(0, 6) = 6$. (f) $\gcd(16, 80) = 16$.   ♦

Example 7.34 illustrates a number of facts concerning the greatest common divisor of two integers. First, the value of $\gcd(a, b)$ is not affected by the sign of a or b, that is, $\gcd(a, b) = \gcd(|a|, |b|)$. Consequently, we may assume that a and b are both nonnegative. Second, if one of a and b is 0, say $b = 0$, then $a > 0$ and $\gcd(a, b) = \gcd(a, 0) = a$. Hence when determining $\gcd(a, b)$, we may assume that a and b are both positive, where $0 < a \leq b$ say. Also, if $a \mid b$, then $\gcd(a, b) = a$. We summarize these facts about the greatest common divisor of two integers a and b:

(1) $\gcd(a, b) = \gcd(|a|, |b|)$,

(2) $\gcd(a, 0) = |a|$ and

(3) if $a, b \neq 0$ and $a \mid b$, then $\gcd(a, b) = |a|$.

The Euclidean Algorithm

As we mentioned, what makes $\gcd(a, b)$ easy to compute for the various pairs a, b of integers in Example 7.34 is that the numbers are relatively small. But what if they are not small? There are a couple of ways to proceed in this case. One of these methods makes use of a technique credited to Euclid. Before describing this method, we first present a useful theorem that involves the Division Algorithm.

Theorem to Prove: Let a and b be two positive integers. If $b = aq + r$ for some integers q and r, then

$$\gcd(a, b) = \gcd(r, a).$$

Proof Strategy. What this theorem states is that the greatest common divisor of a pair a, b of positive integers is the greatest common divisor of another pair r, a of integers if b can be expressed as $aq + r$ for some integer q. Since we are interested in showing $\gcd(a, b)$ and $\gcd(r, a)$ are equal, this suggests introducing separate symbols for these two numbers, say

$$d = \gcd(a, b) \text{ and } e = \gcd(r, a),$$

and then showing that $d = e$. Since $d = \gcd(a, b)$, we know that $d \mid a$ and $d \mid b$; while since $e = \gcd(r, a)$, we know that $e \mid r$ and $e \mid a$. Because $b = aq + r$, we can write $r = b - aq$. Since $d \mid a$ and $d \mid b$, it is not difficult to show that $d \mid r$. So now we know that $d \mid r$ and $d \mid a$, that is, d is a common divisor of r and a. However, e is the *greatest* common divisor of r and a, which says that $d \leq e$. Starting with $e = \gcd(r, a)$ and showing that $e \mid b$ as well will allow us to show that $e \leq d$, giving us the desired result. ♦

Theorem 7.35 *Let a and b be two positive integers. If $b = aq + r$ for some integers q and r, then*

$$\gcd(a, b) = \gcd(r, a).$$

Proof. Suppose that $\gcd(a, b) = d$ and $\gcd(r, a) = e$. We show that $d = e$. Since $d = \gcd(a, b)$, it follows that $d \mid a$ and $d \mid b$. So $a = ds$ and $b = dt$ for integers s and t. Therefore,

$$r = b - aq = dt - (ds)q = d(t - sq).$$

Since $t - sq$ is an integer, $d \mid r$. Thus $d \mid r$ and $d \mid a$. Because e is the largest integer that is a common divisor of r and a, it follows that $d \leq e$.

Also, because $e = \gcd(r, a)$ is a common divisor of r and a, it follows that $e \mid r$ and $e \mid a$ and so $r = ex$ and $a = ey$ for integers x and y. Hence

$$b = aq + r = (ey)q + ex = e(yq + x).$$

Since $yq + x$ is an integer, $e \mid b$. So $e \mid a$ and $e \mid b$. However, because d is the greatest common divisor of a and b, it follows that $e \leq d$.

Since $d \leq e$ and $e \leq d$, we conclude that $d = e$, that is, $\gcd(a, b) = \gcd(r, a)$. ∎

Let's see how Theorem 7.35 can be used to determine $\gcd(a, b)$ for two positive integers a and b. We may assume that $a < b$. Although Theorem 7.35 does not require that q and r be the quotient and the remainder, respectively, when b is divided by a, it does say that if this is the case, then

$$\gcd(a, b) = \gcd(r, a), \text{ where } 0 \leq r < a < b.$$

If $r = 0$, then $a \mid b$ and $\gcd(a, b) = \gcd(0, a) = a$. Suppose then that $r \neq 0$. In this case, we apply Theorem 7.35 again, where this time a is divided by r, obtaining the remainder r_1. Then

$$\gcd(r, a) = \gcd(r_1, r), \text{ where } 0 \leq r_1 < r < a < b.$$

Continuing in this manner, we eventually arrive at the last nonzero remainder r_k and so

$$\gcd(a, b) = \gcd(r, a) = \gcd(r_1, r) = \gcd(r_2, r_1) = \cdots = \gcd(r_k, r_{k-1}) = \gcd(0, r_k) = r_k.$$

That is, the greatest common divisor of a and b is the last nonzero remainder obtained when the sequence of divisions described above are performed. This method for determining $\gcd(a, b)$ is called the **Euclidean Algorithm**. Let's see how this works with a specific example.

Example 7.36

Use the Euclidean Algorithm to determine $\gcd(384, 477)$.

Solution. For given integers m and $n > 0$, we have seen that the remainder r, obtained when m is divided by n, is $m \bmod n$. Since

$$
\begin{aligned}
477 \bmod 384 &= 93 \\
384 \bmod 93 &= 12 \\
93 \bmod 12 &= 9 \\
12 \bmod 9 &= 3 \\
9 \bmod 3 &= 0,
\end{aligned}
$$

it follows that $\gcd(477, 384) = 3$. ♦

Actually, many calculators have an integer-valued function that determines $\gcd(a, b)$ directly, but Theorem 7.35 and the resulting Euclidean Algorithm give us an explanation as to why this method works.

We mentioned that there is more than one way to determine the greatest common divisor of two positive integers a and b. We have seen that one of these involves using the Euclidean Algorithm. We now describe a second method. We have seen by Theorem 7.9 that every integer $n \geq 2$ can be uniquely expressed as a product of primes. Suppose that we express each of a and b as a product of primes, say in terms of the same primes. If some prime p appears as a factor in only one of a and b, then we include $p^0 = 1$ in the factorization of the other. For example, for $a = 45$ and $b = 84$, we write $a = 2^0 \cdot 3^2 \cdot 5^1 \cdot 7^0$ and $b = 2^2 \cdot 3^1 \cdot 5^0 \cdot 7^1$. In general, if

$$a = p_1^{a_1} p_2^{a_2} \cdots p_k^{a_k} \text{ and } b = p_1^{b_1} p_2^{b_2} \cdots p_k^{b_k}, \tag{7.1}$$

where $p_1, p_2, \cdots p_k$ are distinct primes and $a_1, a_2, \ldots, a_k, b_1, b_2, \ldots, b_k$ are nonnegative integer exponents, then

$$\gcd(a, b) = p_1^{\min(a_1,\ b_1)} p_2^{\min(a_2,\ b_2)} \cdots p_k^{\min(a_k,\ b_k)}. \tag{7.2}$$

Therefore, in the case of $a = 45 = 2^0 \cdot 3^2 \cdot 5^1 \cdot 7^0$ and $b = 84 = 2^2 \cdot 3^1 \cdot 5^0 \cdot 7^1$,

$$\gcd(a, b) = 2^{\min(2,\ 0)} 3^{\min(1,\ 2)} 5^{\min(0,\ 1)} 7^{\min(1,\ 0)} = 2^0 \cdot 3^1 \cdot 5^0 \cdot 7^0 = 3.$$

Example 7.37

Express each of the following pairs a, b of positive integers as a product of the same primes and use this to determine $\gcd(a, b)$.

(a) $a = 200, b = 2750$.   (b) $a = 147, b = 2100$.

(c) $a = 143, b = 338$. (d) $a = 2320, b = 12,100$.

Solution.

(a) For $a = 200 = 2^3 \cdot 5^2 \cdot 11^0$ and $b = 2750 = 2^1 \cdot 5^3 \cdot 11^1$,
$$\gcd(a, b) = 2^1 \cdot 5^2 \cdot 11^0 = 50.$$

(b) For $a = 147 = 2^0 \cdot 3^1 \cdot 5^0 \cdot 7^2$ and $b = 2100 = 2^2 \cdot 3^1 \cdot 5^2 \cdot 7^1$,
$\gcd(a, b) = 2^0 \cdot 3^1 \cdot 5^0 \cdot 7^1 = 21$.

(c) For $a = 143 = 2^0 \cdot 11^1 \cdot 13^1$ and $b = 338 = 2^1 \cdot 11^0 \cdot 13^2$,
$\gcd(a, b) = 2^0 \cdot 11^0 \cdot 13^1 = 13$.

(d) For $a = 2320 = 2^4 \cdot 5^1 \cdot 11^0 \cdot 29^1$ and $b = 12,100 = 2^2 \cdot 5^2 \cdot 11^2 \cdot 29^0$,
$\gcd(a, b) = 2^2 \cdot 5^1 = 20$. $\blacklozenge$

Least Common Multiples

While it is often of interest to know the greatest common divisor of two positive integers a and b, there are times when we are also interested in those integers that are divisible by both a and b.

Definition 7.38 *For two positive integers a and b, an integer n is a **common multiple** of a and b if n is a multiple of a and b. The smallest positive integer that is a common multiple of a and b is the **least common multiple** of a and b. This number is denoted by $\mathrm{lcm}(a, b)$.*

While ab is always a common multiple of two positive integers a and b, the least common multiple of a and b is often smaller than ab.

Example 7.39

For each of the following pairs a, b of positive integers, determine by observation the least common multiple of a and b.

(a) $a = 6, b = 9$. (b) $a = 8, b = 12$. (c) $a = 10, b = 10$.
(d) $a = 5, b = 7$. (e) $a = 15, b = 30$.

Solution.

(a) $\mathrm{lcm}(6, 9) = 18$. (b) $\mathrm{lcm}(8, 12) = 24$. (c) $\mathrm{lcm}(10, 10) = 10$.
(d) $\mathrm{lcm}(5, 7) = 35$. (e) $\mathrm{lcm}(15, 30) = 30$. $\blacklozenge$

Example 7.39 illustrates the following facts about the least common multiple of two integers a and b with $1 \le a \le b$:

(1) $b \le \mathrm{lcm}(a, b) \le ab$ and

(2) if $a \mid b$, then $\mathrm{lcm}(a, b) = b$.

Again, if a and b are expressed in (7.1) as a product of powers of the same distinct primes, that is,

$$a = p_1^{a_1} p_2^{a_2} \cdots p_k^{a_k} \text{ and } b = p_1^{b_1} p_2^{b_2} \cdots p_k^{b_k},$$

then

$$\mathrm{lcm}(a, b) = p_1^{\max(a_1,\, b_1)} p_2^{\max(a_2,\, b_2)} \cdots p_k^{\max(a_k,\, b_k)}. \tag{7.3}$$

Example 7.40

(a) For $a = 200 = 2^3 \cdot 5^2 \cdot 11^0$ and $b = 2750 = 2^1 \cdot 5^3 \cdot 11^1$,
$\mathrm{lcm}(a, b) = 2^3 \cdot 5^3 \cdot 11^1 = 11,000$.

(b) For $a = 147 = 2^0 \cdot 3^1 \cdot 5^0 \cdot 7^2$ and $b = 2100 = 2^2 \cdot 3^1 \cdot 5^2 \cdot 7^1$,
$\mathrm{lcm}(a, b) = 2^2 \cdot 3^1 \cdot 5^2 \cdot 7^2 = 14,700$.

(c) For $a = 143 = 2^0 \cdot 11^1 \cdot 13^1$ and $b = 338 = 2^1 \cdot 11^0 \cdot 13^2$,

$$\text{lcm}(a, b) = 2^1 \cdot 11^1 \cdot 13^2 = 3718.$$

(d) For $a = 2320 = 2^4 \cdot 5^1 \cdot 11^0 \cdot 29^1$ and $b = 12,100 = 2^2 \cdot 5^2 \cdot 11^2 \cdot 29^0$,

$$\text{lcm}(a, b) = 2^4 \cdot 5^2 \cdot 11^2 \cdot 29^1 = 1,403,600. \qquad\blacklozenge$$

The following theorem is a consequence of equations (7.2) and (7.3), which gives a relationship between the greatest common divisor and least common multiple of two integers (see Exercise 8).

Theorem 7.41 *For every two positive integers a and b,*

$$ab = \gcd(a, b) \, \text{lcm}(a, b).$$

From what we have seen, if a and b are integers with $1 \le a \le b$, then

$$1 \le \gcd(a, b) \le a \le b \le \text{lcm}(a, b) \le ab.$$

Example 7.42

For $a = 12$ and $b = 54$, we have $\gcd(a, b) = 6$ and, by Theorem 7.41,

$$\text{lcm}(a, b) = \frac{ab}{\gcd(a, b)} = \frac{(12)(54)}{6} = 108. \quad\blacklozenge$$

Relatively Prime Integers

For integers a and b with $1 \le a \le b$, we saw that $\gcd(a, b) = a$ if and only if $a \mid b$. However, $\gcd(a, b) = 1$ provided there is no integer greater than 1 that is a common divisor of a and b. If $\gcd(a, b) = 1$, then this means that there is *no prime* that divides both a and b.

Definition 7.43 *Two integers a and b, not both 0, are* **relatively prime** *if* $\gcd(a, b) = 1$.

For example, 5 and 7 are relatively prime. If two integers a and b are relatively prime, then neither a nor b need be prime, however. For example, $\gcd(4, 9) = 1$. Also, every two consecutive positive integers are relatively prime. For example, $\gcd(20, 21) = 1$. We now verify the statement that every two consecutive positive integers are relatively prime.

Result to Prove: Every two consecutive positive integers are relatively prime.

Proof Strategy. First, it seems as if it would be helpful to represent the two consecutive positive integers by symbols. Let's denote these integers by n and $n + 1$. Our goal then becomes to show that

$$\gcd(n, n + 1) = 1.$$

There appear to be two options to show that $\gcd(n, n + 1) = 1$. One of these is to use the Euclidean Algorithm. So we could begin by dividing $n + 1$ by n. Actually, if $n = 1$, then $n + 1 = 2$ and it is obvious that $\gcd(n, n + 1) = \gcd(1, 2) = 1$. On the other hand, if $n \ge 2$, then when $n + 1$ is divided by n the quotient is 1 and the remainder is 1, that is, $n + 1 = n \cdot 1 + 1$ and so $\gcd(n, n + 1) = \gcd(1, n) = 1$.

Another strategy to show that $\gcd(n, n+1) = 1$ is to introduce a symbol, namely d, to represent $\gcd(n, n+1)$, say $d = \gcd(n, n+1)$. Our goal is then to show that $d = 1$. Since $d = \gcd(n, n+1)$, it follows that d is a common divisor of n and $n + 1$. Hence there are integers r and s such that $n = dr$ and $n + 1 = ds$. This probably suggests writing $n + 1$ in two ways, namely $n + 1 = ds$ and $n + 1 = dr + 1$. Thus $ds = dr + 1$ and so $ds - dr = 1$. Therefore, $d(s - r) = 1$. Since $s - r$ is an integer, $d \mid 1$, which implies that $d \le 1$. However, d is a positive integer, which gives us the desired result. We use the second approach in the proof we give of this result. $\blacklozenge$

Result 7.44 *Every two consecutive positive integers are relatively prime.*

Proof. Let n and $n + 1$ be consecutive positive integers and let $d = \gcd(n, n + 1)$. Therefore, $d \mid n$ and $d \mid (n + 1)$. Hence there exist integers r and s such that $n = dr$ and $n + 1 = ds$. Thus $ds = n + 1 = dr + 1$ and so

$$1 = ds - dr = d(s - r).$$

Since $s - r$ is an integer, $d \mid 1$, which implies that $d \leq 1$. Since $d \geq 1$, it follows that $d = 1 = \gcd(n, n + 1)$. ∎

Linear Combinations of Integers

When dealing with certain kinds of objects in mathematics, including integers, it is often useful to know, as we are about to see, if these objects can be expressed as "linear combinations" of other objects.

Definition 7.45 *Let a and b be two integers. An integer of the form $ax + by$, where x and y are integers, is a* **linear combination** *of a and b.*

Example 7.46

The following integers are linear combinations of $a = 12$ and $b = 20$:

$$
\begin{aligned}
32 &= 12 \cdot 1 + 20 \cdot 1 \\
8 &= 12 \cdot (-1) + 20 \cdot 1 \\
-8 &= 12 \cdot 1 + 20 \cdot (-1) \\
-32 &= 12 \cdot (-1) + 20 \cdot (-1) \\
0 &= 12 \cdot (-5) + 20 \cdot 3 \\
16 &= 12 \cdot (-2) + 20 \cdot 2 \\
4 &= 12 \cdot (-3) + 20 \cdot 2. \blacklozenge
\end{aligned}
$$

You might notice that $\gcd(12, 20) = 4$. Also, 4 is a linear combination of 12 and 20. Of course, many other integers are linear combinations of 12 and 20 as well. Suppose that the positive integer n is a linear combination of 12 and 20. Then $n = 12x + 20y$ for some integers x and y. However, $n = 12x + 20y = 4(3x + 5y)$. Since $3x + 5y$ is an integer, $4 \mid n$ and so $n \geq 4$. This implies that 4 is the smallest possible positive integer that is a linear combination of 12 and 20. This observation illustrates the following theorem. The proof of this theorem makes use of the **Well-Ordering Principle** (see Exercise 28 in Supplementary Exercises for Chapter 4) which states that every nonempty subset of $\mathbb{N}$ has a smallest element.

Theorem 7.47 *Let a and b be integers that are not both 0. Then $\gcd(a, b)$ is the smallest positive integer that is a linear combination of a and b.*

Proof. First let S denote the set of all positive integers that are linear combinations of a and b, that is,

$$S = \{n \in \mathbb{N} : \ n = ax + by, x, y \in \mathbb{Z}\}.$$

If we let $x = a$ and $y = b$, then we obtain $a^2 + b^2$. Since $a^2 + b^2 > 0$, it follows that $a^2 + b^2 \in S$ and so $S \neq \emptyset$. By the Well-Ordering Principle, S has a least element, which we denote by d. Therefore, there exist integers x_0 and y_0 such that $d = ax_0 + by_0$. We now show that $d = \gcd(a, b)$.

Dividing a by d, we obtain (by the Division Algorithm) a quotient q and a remainder r, where $0 \leq r < d$. Thus

$$a = dq + r.$$

Hence

$$r = a - dq = a - (ax_0 + by_0)q = a(1 - qx_0) + b(-qy_0).$$

Therefore, r is a linear combination of a and b. If $r > 0$, then necessarily $r \in S$. However, since $r < d$ and d is the smallest element in S, this produces a contradiction. Hence $r = 0$, implying that $d \mid a$. By a similar argument, it follows that $d \mid b$ and so d is a common divisor of a and b.

It remains to show that d is, in fact, the *greatest* common divisor of a and b. Let e be a common divisor of a and b. By Theorem 7.4, e divides every linear combination of a and b. In particular, $e \mid d$. Thus $e \leq d$ and therefore $d = \gcd(a, b)$.   ∎

Example 7.48

For each of the following pairs a, b of positive integers, express $d = \gcd(a, b)$ as a linear combination of a and b.

(a) $a = 10$, $b = 14$.   (b) $a = 8$, $b = 24$.   (c) $a = 20$, $b = 25$.
(d) $a = 12$, $b = 12$.   (e) $a = 18$, $b = 30$.   (f) $a = 25$, $b = 27$.

Solution.

(a) $\gcd(10, 14) = 2 = 10 \cdot 3 + 14 \cdot (-2)$.

(b) $\gcd(8, 24) = 8 = 8 \cdot 1 + 24 \cdot 0$.

(c) $\gcd(20, 25) = 5 = 20 \cdot (-1) + 25 \cdot 1$.

(d) $\gcd(12, 12) = 12 = 12 \cdot 1 + 12 \cdot 0$.

(e) $\gcd(18, 30) = 6 = 18 \cdot (-3) + 30 \cdot 2$.

(f) $\gcd(25, 27) = 1 = 25 \cdot 13 + 27 \cdot (-12)$.   ◆

Analysis. Although it may not be too difficult to find integers x and y so that $\gcd(a, b) = ax + by$ for the pairs a, b of integers given in Example 7.48(a)–(e), such may not be the case for Example 7.48(f). Using the Euclidean Algorithm, we show how a pair x, y of integers can be found so that

$$\gcd(25, 27) = 1 = 25x + 27y.$$

As a first step in the Euclidean Algorithm, we use the Division Algorithm and divide 27 by 25, obtaining

$$27 = 25 \cdot 1 + 2. \tag{7.4}$$

Next, we divide $a = 25$ by $r_1 = 2$, obtaining

$$25 = 2 \cdot 12 + 1. \tag{7.5}$$

Since we have obtained a remainder of $r_2 = 1$, this is clearly the last nonzero remainder and so $\gcd(25, 27) = 1$. Solving the equation (7.5) for $\gcd(25, 27) = 1$, we obtain

$$1 = 25 - 2 \cdot 12 \tag{7.6}$$

and solving for the remainder 2 in (7.4), we have

$$2 = 27 - 25 \cdot 1. \tag{7.7}$$

Substituting the expression for 2 in (7.7) into expression (7.6) gives us

$$\begin{aligned} 1 &= 25 - 2 \cdot 12 = 25 - (27 - 25 \cdot 1) \cdot 12 \\ &= 25 \cdot 1 + 27 \cdot (-12) + 25 \cdot 12 = 25 \cdot 13 + 27 \cdot (-12). \end{aligned}$$

This algebra can be expressed more concisely as indicated below:

$$27 = 25 \cdot 1 + 2 \qquad 2 = 27 - 25 \cdot 1$$
$$25 = 2 \cdot 12 + 1 \qquad 1 = 25 - 2 \cdot 12.$$

Thus

$$1 = 25 - 2 \cdot 12 = 25 - (27 - 25 \cdot 1) \cdot 12 = 25 \cdot 13 + 27 \cdot (-12). \; \blacklozenge$$

If $d = \gcd(a, b)$ and $d = ax + by$ for integers x and y, then x and y are not unique. Indeed, d can also be expressed as the linear combination

$$d = a(x - b) + b(y + a).$$

Thus in Example 7.48(f), we can also write

$$
\begin{aligned}
1 &= 25 \cdot 13 + 27 \cdot (-12) = 25 \cdot (13 - 27) + 27 \cdot (-12 + 25) \\
&= 25 \cdot (-14) + 27 \cdot 13.
\end{aligned}
$$

We now state two consequences of Theorem 7.47. We saw in Example 7.48(e) that $\gcd(18, 30) = 6$. Actually, 3 is also a common divisor of 18 and 30, as is 2. Of course, $3 < 6$ and $2 < 6$ since 6 is the *greatest* common divisor of 18 and 30. However, both $3 \mid 6$ and $2 \mid 6$. This always occurs.

Corollary 7.49 *Let a and b be integers that are not both 0 and let $d = \gcd(a, b)$. If n is an integer that is a common divisor of a and b, then $n \mid d$.*

Proof. By Theorem 7.47, there are integers x and y such that $d = ax + by$. Since $n \mid a$ and $n \mid b$, there are integers s and t such that $a = ns$ and $b = nt$. Thus

$$d = ax + by = (ns)x + (nt)y = n(sx + ty).$$

Since $sx + ty$ is an integer, $n \mid d$. $\blacksquare$

We saw in Example 7.48(f) that 1 is a linear combination of 25 and 27. Actually, 2 is a linear combination of 25 and 27 as well since $2 = 25 \cdot (-1) + 27 \cdot 1$. Of course, just because we're able to express 2 as a linear combination of 25 and 27 does not mean that 2 is the greatest common divisor of 25 and 27. Indeed, it is not. However, if we are able to express 1 as a linear combination of two integers a and b, then $\gcd(a, b)$ must be 1 since 1 is the smallest positive integer. The following result is then a consequence of Theorem 7.47.

Corollary 7.50 *Two integers a and b are relatively prime if and only if 1 is a linear combination of a and b; that is, $\gcd(a, b) = 1$ if and only if $ax + by = 1$ for some integers x and y.*

Example 7.51

Use Corollary 7.50 to show that the following pairs of integers are relatively prime:

(a) every two consecutive integers.

(b) every two odd integers that differ by 2.

Solution.

(a) Let n and $n + 1$ be two consecutive integers. Since

$$1 = n \cdot (-1) + (n + 1) \cdot 1,$$

it follows by Corollary 7.50 that $\gcd(n, n + 1) = 1$.

(b) Let n and $n+2$ be two odd integers. Since n is odd, $n = 2k+1$ for some integer k. Consequently, $n + 2 = 2k + 3$. Since

$$1 = (2k + 1)(k + 1) + (2k + 3)(-k),$$

it follows by Corollary 7.50 that $\gcd(n, n + 2) = 1$.   ♦

If a, b and c are integers such that $a \mid bc$, then there is no reason to believe that $a \mid b$ or $a \mid c$. For example, let $a = 6, b = 4$ and $c = 3$. Then $6 \mid 4 \cdot 3$ but $6 \nmid 4$ and $6 \nmid 3$. On the other hand, if a and b are relatively prime, then we can make a different conclusion.

Theorem to Prove: Let a, b and c be integers with $a \neq 0$. If $a \mid bc$ and $\gcd(a, b) = 1$, then $a \mid c$.

Proof Strategy. If we attempt a direct proof of this theorem, then we would begin by assuming $a \mid bc$ and $\gcd(a, b) = 1$. Our goal is show that $a \mid c$, that is, to show we can express c as the product of a and some other integer. Let's see what information is available to us. Because $a \mid bc$, we can write $bc = aq$ for some integer q. Because a and b are relatively prime, it follows by Corollary 7.50 that 1 can be expressed as a linear combination of a and b, that is, there are integers x and y such that $1 = ax + by$. The question is how to put these two pieces of information together to show that $a \mid c$. Observe that if we were to multiply both sides of the equation $1 = ax + by$ by c, we would obtain $c = acx + bcy$. However, $bc = aq$. This observation leads to a proof.   ♦

Theorem 7.52 *Let a, b and c be integers with $a \neq 0$. If $a \mid bc$ and $\gcd(a, b) = 1$, then $a \mid c$.*

Proof. Assume that $a \mid bc$. Then $bc = aq$ for some integer q. Since a and b are relatively prime, it follows by Corollary 7.50 that there exist integers x and y such that $1 = ax + by$. Thus

$$c = c \cdot 1 = c(ax + by) = a(cx) + (bc)y = a(cx) + (aq)y = a(cx + qy).$$

Since $cx + qy$ is an integer, $a \mid c$.   ∎

Theorem 7.52 is of special interest when the integer a is a prime. The following is a consequence of Theorem 7.52.

Corollary 7.53 *Let b and c be integers and let p be a prime. If $p \mid bc$, then either $p \mid b$ or $p \mid c$.*

Proof. If p divides b, then the corollary is proved. Suppose then that p does not divide b. Since the only positive integer divisors of p are 1 and p, it follows that $\gcd(p, b) = 1$. It then follows by Theorem 7.52 that $p \mid c$.   ∎

Corollary 7.53 can be extended to the case when a prime p divides a product of any finite number of integers.

Theorem to Prove: Let $a_1, a_2, \ldots, a_n$ be $n \geq 2$ integers and let p be a prime. If

$$p \mid a_1\, a_2\, \cdots\, a_n,$$

then $p \mid a_i$ for some integer i with $1 \leq i \leq n$.

Proof Strategy: If $n = 2$ in the statement of the theorem, then what the theorem says is that if $p \mid a_1\, a_2$, then $p \mid a_1$ or $p \mid a_2$. But this an immediate consequence of Corollary 7.53. More generally, what we wish to show is the following:

> If p is a prime and $a_1, a_2, \ldots, a_n$ are any $n \geq 2$ integers such that $p \mid a_1 a_2 \cdots a_n$, then p divides at least one of the integers $a_1, a_2, \ldots, a_n$.

From the way this statement is worded, it appears that a natural proof technique to try is induction. The fact that this statement is known to be true when $n = 2$ verifies the basis step of the induction. To verify the inductive step, we assume that whenever a prime divides the product of $k \geq 2$ integers, then this prime divides at least one of these k integers. We are then required to show that if some prime divides a product of $k+1$ integers, then this prime divides at least one of these $k+1$ integers. Let's return to what we just said.

Assume that if some prime divides the product of $k \geq 2$ integers, then this prime must divide at least one of these k integers. Now let p be a prime and let $a_1, a_2, \ldots, a_{k+1}$ be $k+1$ integers such that $p \mid a_1 a_2 \cdots a_{k+1}$. We show that p divides at least one of the integers $a_1, a_2, \ldots, a_{k+1}$. If we write $b = a_1 a_2 \cdots a_k$, then the fact that $p \mid a_1 a_2 \cdots a_{k+1}$ can be expressed as $p \mid ba_{k+1}$. However then, the prime p divides the product of the two integers b and a_{k+1}. By Corollary 7.53, $p \mid b$ or $p \mid a_{k+1}$. Should it occur that $p \mid a_{k+1}$, then this conclusion gives us a proof of the inductive step. On the other hand, if $p \mid b$, then $p \mid a_1 a_2 \cdots a_k$. Since, however, p divides the product of k integers in this case, it follows from the induction hypothesis that p must divide one of the integers $a_1, a_2, \ldots, a_k$ and the inductive step has been verified in this case as well. We then have a proof of the theorem. $\blacklozenge$

Theorem 7.54 *Let $a_1, a_2, \ldots, a_n$ be $n \geq 2$ integers and let p be a prime. If*

$$p \mid a_1 \; a_2 \; \cdots \; a_n,$$

then $p \mid a_i$ for some integer i with $1 \leq i \leq n$.

Proof. We proceed by induction. By Corollary 7.53, the statement is true when $n = 2$. Thus the basis step of the induction is verified. Assume that

> if a prime p divides the product of k integers, where $k \geq 2$, then p divides at least one of these k integers.

Now let $a_1, a_2, \ldots, a_{k+1}$ be $k+1$ integers such that $p \mid a_1 a_2 \cdots a_{k+1}$. We show that $p \mid a_i$ for some integer i ($1 \leq i \leq k+1$). Let

$$b = a_1 a_2 \cdots a_k.$$

So $p \mid ba_{k+1}$. By Corollary 7.53, either $p \mid b$ or $p \mid a_{k+1}$. If $p \mid a_{k+1}$, then the proof is complete. Otherwise, $p \mid b$, that is,

$$p \mid a_1 a_2 \cdots a_k.$$

However, by the induction hypothesis, $p \mid a_i$ for some integer i ($1 \leq i \leq k$). In any case, $p \mid a_i$ for some integer i ($1 \leq i \leq k+1$).

By the Principle of Mathematical Induction, the theorem is true. $\blacksquare$

We are now in a position to present a proof of the Fundamental Theorem of Arithmetic (Theorem 7.9), whose statement is repeated below. Although the proof is not easy, it is good if we make an effort to understand it.

The Fundamental Theorem of Arithmetic *Every integer $n \geq 2$ is either prime or can be expressed as a product of (not necessarily distinct) primes, that is,*

$$n = p_1 p_2 \cdots p_k,$$

where $p_1, p_2, \ldots, p_k$ are primes. This factorization is unique except possibly for the order in which the primes appear.

Proof. To show the existence of such a factorization, we employ the Strong Principle of Mathematical Induction. Since 2 is prime, the statement is true for $n = 2$.

For an arbitrary integer $k \geq 2$, assume that every integer i, where $2 \leq i \leq k$, is either prime or can be expressed as a product of primes. We show that $k + 1$ is either prime or can be expressed as a product of primes. Of course, if $k + 1$ is prime, then there is nothing further to prove. We may assume, then, that $k + 1$ is composite. Then there exist integers a and b such that

$$k + 1 = ab, \text{ where } 2 \leq a \leq k \text{ and } 2 \leq b \leq k.$$

Therefore, by the induction hypothesis, each of a and b is prime or can be expressed as a product of primes. In any case, $k+1 = ab$ is a product of primes. By the Strong Principle of Mathematical Induction, every integer $n \geq 2$ is either prime or can be expressed as a product of primes.

To prove that such a factorization is unique, we proceed by contradiction. Assume, to the contrary, that there is an integer $n \geq 2$ that can be expressed as a product of primes in more than one way, say as

$$n = p_1 p_2 \cdots p_s \quad \text{and} \quad n = q_1 q_2 \cdots q_t,$$

where in each factorization, the primes are arranged in nondecreasing order, that is,

$$p_1 \leq p_2 \leq \cdots \leq p_s \text{ and } q_1 \leq q_2 \leq \cdots \leq q_t.$$

Since the factorizations are different, there must be a smallest positive integer r such that

$$p_r \neq q_r.$$

In other words, if $r \geq 2$, then $p_i = q_i$ for every i with $1 \leq i \leq r - 1$. After canceling, we have

$$p_r p_{r+1} \cdots p_s = q_r q_{r+1} \cdots q_t. \tag{7.8}$$

Consider the integer p_r. Either $s = r$ and the left side of (7.8) is exactly p_r or $s > r$ and $p_{r+1} p_{r+2} \cdots p_s$ is an integer that is the product of $s - r$ primes. In either case,

$$p_r \mid q_r q_{r+1} \cdots q_t.$$

By Theorem 7.54, $p_r \mid q_j$ for some integer j with $r \leq j \leq t$. Because q_j is prime, $p_r = q_j$. Since $q_r \leq q_j$, it follows that $q_r \leq p_r$. By considering the integer q_r (instead of p_r), we can use the same argument to show that $p_r \leq q_r$. Therefore, $p_r = q_r$. But this contradicts the fact that $p_r \neq q_r$. Hence, as claimed, every integer $n \geq 2$ has a unique factorization. $\blacksquare$

Exercises for Section 7.6

1. Determine, by observation, the greatest common divisor of each of the following pairs a, b of integers.

   (a) $a = 18$, $b = 26$. (b) $a = 11$, $b = 19$. (c) $a = 0$, $b = -10$.

   (d) $a = -8$, $b = 14$. (e) $a = 17$, $b = 51$.

2. Use the Euclidean Algorithm to determine $\gcd(a, b)$ for each of the following pairs a, b of integers. (a) $a = 558$, $b = 714$. (b) $a = 418$, $b = 648$.

3. For each of the following pairs a, b of positive integers, express $\gcd(a, b)$ as a linear combination of a and b.

   (a) $a = 4$, $b = 5$. (b) $a = 15$, $b = 35$. (c) $a = 12$, $b = 36$.

   (d) $a = 9$, $b = 12$. (e) $a = 30$, $b = 42$.

4. Let a and b be integers, not both 0. Show that there are infinitely many pairs x, y of integers such that $\gcd(a, b) = ax + by$.

5. Let a and b be integers, not both 0 and let $d = \gcd(a, b)$. Prove that n is a linear combination of a and b if and only if $d \mid n$.

6. Let a and b be integers, not both 0 and let $d = \gcd(a, b)$. Prove that if $a = da'$ and $b = db'$ for some integers a' and b', then $\gcd(a', b') = 1$.

7. Let a and b be integers, not both 0. Prove that $\gcd(ka, kb) = k \gcd(a, b)$ for every positive integer k.

8. Prove Theorem 7.41: *For every two positive integers a and b, $ab = \gcd(a,b)\operatorname{lcm}(a,b)$.*

9. Let $a \geq 2$ and $b \geq 2$ be two integers with the property that for every prime p the number p divides at most one of a and b. Determine $\operatorname{lcm}(a,b)$.

10. For each of the following pairs a, b of integers, determine $\gcd(a,b)$ and $\operatorname{lcm}(a,b)$.

   (a) $a = 0$, $b = 3$.   (b) $a = 5$, $b = 5^5$.   (c) $a = 2 \cdot 3 \cdot 5$, $b = 5 \cdot 7 \cdot 11 \cdot 13$.

   (d) $a = 2 \cdot 3 \cdot 5 \cdot 7 \cdot 11$, $b = 2^3 \cdot 5^2 \cdot 11 \cdot 13^2$.   (e) $a = 2^2 \cdot 3^3 \cdot 5^4 \cdot 7^5$, $b = 2^5 \cdot 3^4 \cdot 5^3 \cdot 7^2$.

11. If the product of two integers is $2^7 \cdot 3^8 \cdot 5^2 \cdot 7^{11}$ and their greatest common divisor is $2^3 \cdot 3^4 \cdot 7$, then what is their least common multiple?

12. If the product of two integers is $2^{11} \cdot 3^4 \cdot 5^7 \cdot 7^7$ and their least common multiple is $2^{10} \cdot 3^2 \cdot 5^3 \cdot 7^6$, then what is their greatest common divisor?

13. Find all pairs a, b of relatively prime integers with $0 < a < b$ such that $\operatorname{lcm}(a,b) = p^3$ for some prime p.

14. Find all pairs a, b of integers with $0 < a < b$ such that $\gcd(a,b) = 2^2$ and $\operatorname{lcm}(a,b) = 2^3 \cdot 3$.

15. Let $a, b \in \mathbb{Z}$. Prove that if $\gcd(a^n, b^n) = 1$, where $n \in \mathbb{N}$ and $n \geq 2$, then $\gcd(a,b) = 1$.

16. Which positive integers less than 11 are relatively prime to 11?

17. Which positive integers less than 18 are relatively prime to 18?

18. For which positive integers n are the integers 1, 3, 7, 9, 11, 13, 17, 19 the only positive integers less than and relatively prime to n?

19. Use Corollary 7.50 to show that every two integers that are not multiples of 3 but differ by 3 are relatively prime.

20. Let p be a prime and let a and b be relatively prime integers. Prove that if $p^2 \mid ab$, then $p^2 \mid a$ or $p^2 \mid b$.

21. Let $a, b, c \in \mathbb{Z}$, where a and b are relatively prime nonzero integers. Prove that if $a \mid c$ and $b \mid c$, then $ab \mid c$.

22. Let $p \geq 2$ be an integer. Prove that if for every pair a, b of integers $p \mid ab$ implies that $p \mid a$ or $p \mid b$, then p is a prime.

23. Suppose that a and b are relatively prime integers, where $1 \leq a < b$. Show that a and $b - a$ are also relatively prime.

24. Prove or disprove: The integers $2n + 7$ and $5n + 11$ are relatively prime for every positive integer n.

25. A sequence $a_1, a_2, a_3, \ldots$ is defined by $a_1 = 1$, $a_2 = 2$, $a_{2\ell+1} = a_\ell + a_{\ell+1}$ and $a_{2\ell+2} = a_{\ell+1}$ for $\ell \geq 1$. Prove that $\gcd(a_n, a_{n+1}) = 1$ for every positive integer n.

26. Show that $\log_2 3$ is irrational.

7.7   Integer Representations

When we write the integer 4587, each digit and its location have a meaning. In particular,

$$4587 = 4000 + 500 + 80 + 7 = 4 \cdot 10^3 + 5 \cdot 10^2 + 8 \cdot 10^1 + 7 \cdot 10^0,$$

that is, each digit represents a coefficient of 10^k for some nonnegative integer k. In fact, if we were working with decimal expansions of real numbers that are not integers, then each digit in the expansion represents a coefficient of 10^k for some (not necessarily nonnegative) integer k. For example,

$$23.506 = 2 \cdot 10^1 + 3 \cdot 10^0 + 5 \cdot 10^{-1} + 0 \cdot 10^{-2} + 6 \cdot 10^{-3}.$$

However, we will continue to restrict our discussion to integers. The representation of integers that we've just discussed is the decimal or base 10 representation of an integer. Representating integers is not restricted to base 10, however.

Theorem 7.55   *Let $b \geq 2$ be an integer. Then every positive integer n has a unique representation in base b as*

$$n = a_k b^k + a_{k-1} b^{k-1} + \cdots + a_1 b^1 + a_0 b^0,$$

where k is a nonnegative integer, $a_0, a_1, \ldots, a_k$ are nonnegative integers less than b and $a_k \neq 0$.

Proof. First, we divide n by b, obtaining (by the Division Algorithm) a quotient q_1 and a remainder a_0, where $0 \leq a_0 < b$. Then

$$n = bq_1 + a_0 = q_1 b + a_0.$$

Dividing q_1 by b, we obtain

$$q_1 = bq_2 + a_1 = q_2 b + a_1,$$

where $0 \leq a_1 < b$. Hence

$$
\begin{aligned}
n &= q_1 b + a_0 = (q_2 b + a_1)b + a_0 \\
  &= q_2 b^2 + a_1 b + a_0 \\
  &= q_2 b^2 + a_1 b^1 + a_0 b^0.
\end{aligned}
$$

Continuing in this manner, we arrive at the desired result.   ∎

The representation of n in base b given in Theorem 7.55 is called the **base b expansion** of n and is denoted by

$$n = (a_k a_{k-1} \cdots a_1 a_0)_b.$$

There are occasions when it is convenient to work with bases different than 10, especially when dealing with computers. In such instances, base 2 (**binary**), base 8 (**octal**) and base 16 (**hexadecimal**) are common and useful. When $b = 2$, the binary digits (bits) are 0 and 1; when $b = 8$, the octal digits are $0, 1, \ldots, 7$; and when $b = 16$, the hexadecimal digits are $0, 1, \ldots, 9, A, B, C, D, E, F$, where the letters $A, B, \ldots, F$ represent the numbers $10, 11, \ldots, 15$ (expressed decimally), respectively. Figure 7.3 gives the binary, octal and hexadecimal representations of each integer n with $0 \leq n \leq 15$.

We begin by converting an integer expressed in base 2 to its decimal representation.

Example 7.56

Determine the decimal expansions of the integers whose binary representations are given below.

(a) $(1101)_2$.   (b) $(100101)_2$.   (c) $(111000)_2$.

Decimal	0	1	2	3	4	5	6	7
Binary	0	1	10	11	100	101	110	111
Octal	0	1	2	3	4	5	6	7
Hexadecimal	0	1	2	3	4	5	6	7

Decimal	8	9	10	11	12	13	14	15
Binary	1000	1001	1010	1011	1100	1101	1110	1111
Octal	10	11	12	13	14	15	16	17
Hexadecimal	8	9	A	B	C	D	E	F

Figure 7.3: Binary, octal and hexadecimal representations of integers n for $0 \leq n \leq 15$

Solution.

(a) $(1101)_2 = 1 \cdot 2^3 + 1 \cdot 2^2 + 0 \cdot 2^1 + 1 \cdot 2^0 = 8 + 4 + 1 = 13$.

(b) $(100101)_2 = 1 \cdot 2^5 + 0 \cdot 2^4 + 0 \cdot 2^3 + 1 \cdot 2^2 + 0 \cdot 2^1 + 1 \cdot 2^0 = 32 + 4 + 1 = 37$.

(c) $(111000)_2 = 1 \cdot 2^5 + 1 \cdot 2^4 + 1 \cdot 2^3 + 0 \cdot 2^2 + 0 \cdot 2^1 + 0 \cdot 2^0 = 32 + 16 + 8 = 56$.   ◆

The next example concerns base 8 and base 16.

Example 7.57

Determine the decimal expansions of the integers whose octal and hexadecimal expansions are given below.

(a) $(53)_8$.   (b) $(172)_8$.   (c) $(1077)_8$.   (d) $(10)_{16}$.   (e) $(2A1)_{16}$. (f) $(BDF)_{16}$.

Solution.

(a) $(53)_8 = 5 \cdot 8^1 + 3 \cdot 8^0 = 40 + 3 = 43$.

(b) $(172)_8 = 1 \cdot 8^2 + 7 \cdot 8^1 + 2 \cdot 8^0 = 64 + 56 + 2 = 122$.

(c) $(1077)_8 = 1 \cdot 8^3 + 0 \cdot 8^2 + 7 \cdot 8^1 + 7 \cdot 8^0 = 512 + 56 + 7 = 575$.

(d) $(10)_{16} = 1 \cdot 16^1 + 0 \cdot 16^0 = 16$.

(e) $(2A1)_{16} = 2 \cdot 16^2 + 10 \cdot 16^1 + 1 \cdot 16^0 = 512 + 160 + 1 = 673$.

(f) $(BDF)_{16} = 11 \cdot 16^2 + 13 \cdot 16^1 + 15 \cdot 16^0 = 2816 + 208 + 15 = 3039$.   ◆

There is now a natural question: How can an integer represented in base 10 be converted to an expansion in base b, where $b \neq 10$? The following example suggests a general technique.

Example 7.58

What is the octal expansion of the number $(3947)_{10}$?

Solution. Dividing 3947 by 8, we obtain a quotient of 493 and a remainder of 3. Therefore,

$$3947 = 493 \cdot 8 + 3.$$

Dividing 493 by 8, we obtain a quotient of 61 and a remainder of 5. Thus

$$493 = 61 \cdot 8 + 5.$$

Hence

$$\begin{aligned} 3947 &= 493 \cdot 8 + 3 = (61 \cdot 8 + 5) \cdot 8 + 3 \\ &= 61 \cdot 8^2 + 5 \cdot 8 + 3. \end{aligned}$$

Dividing 61 by 8, we obtain

$$61 = 7 \cdot 8 + 5.$$

Thus

$$
\begin{aligned}
3947 &= 61 \cdot 8^2 + 5 \cdot 8 + 3 \\
&= (7 \cdot 8 + 5) \cdot 8^2 + 5 \cdot 8 + 3 \\
&= 7 \cdot 8^3 + 5 \cdot 8^2 + 5 \cdot 8 + 3 \cdot 8^0.
\end{aligned}
$$

Therefore, $(3947)_{10} = (7553)_8$.

In summary, we have found that

$$
\begin{aligned}
3947 \textbf{ div } 8 = 493 & \text{ and } 3947 \textbf{ mod } 8 = 3 \\
493 \textbf{ div } 8 = 61 & \text{ and } 493 \textbf{ mod } 8 = 5 \\
61 \textbf{ div } 8 = 7 & \text{ and } 61 \textbf{ mod } 8 = 5 \\
7 \textbf{ div } 8 = 0 & \text{ and } 7 \textbf{ mod } 8 = 7.
\end{aligned}
$$

Thus $(3947)_{10} = (7553)_8$.   ◆

We now consider such examples in base 2 and base 16.

Example 7.59

What is the binary expansion of the number $(43)_{10}$?

Solution. Since

$$
\begin{aligned}
43 \textbf{ div } 2 = 21 & \text{ and } 43 \textbf{ mod } 2 = 1 \\
21 \textbf{ div } 2 = 10 & \text{ and } 21 \textbf{ mod } 2 = 1 \\
10 \textbf{ div } 2 = 5 & \text{ and } 10 \textbf{ mod } 2 = 0 \\
5 \textbf{ div } 2 = 2 & \text{ and } 5 \textbf{ mod } 2 = 1 \\
2 \textbf{ div } 2 = 1 & \text{ and } 2 \textbf{ mod } 2 = 0 \\
1 \textbf{ div } 2 = 0 & \text{ and } 1 \textbf{ mod } 2 = 1,
\end{aligned}
$$

it follows that $(43)_{10} = (101011)_2$.   ◆

Example 7.60

What is the hexadecimal expansion of the number $(3162)_{10}$?

Solution. Since

$$
\begin{aligned}
3162 \textbf{ div } 16 = 197 & \text{ and } 3162 \textbf{ mod } 16 = 10 \\
197 \textbf{ div } 16 = 12 & \text{ and } 197 \textbf{ mod } 16 = 5 \\
12 \textbf{ div } 16 = 0 & \text{ and } 12 \textbf{ mod } 16 = 12,
\end{aligned}
$$

it follows that $(3162)_{10} = (C5A)_{16}$.   ◆

Converting the expansions of positive integers among the bases 2, 8 and 16 is, in general, a much easier process. The table shown in Figure 7.3 will be useful for the next three examples.

Example 7.61

Determine the octal and hexadecimal expansions of the number $(111001101)_2$.

Solution. The number $(111001101)_2$ represents

$$
\begin{aligned}
& 1 \cdot 2^8 + 1 \cdot 2^7 + 1 \cdot 2^6 + 0 \cdot 2^5 + 0 \cdot 2^4 + 1 \cdot 2^3 + 1 \cdot 2^2 + 0 \cdot 2^1 + 1 \cdot 2^0 \\
=\ & (1 \cdot 2^8 + 1 \cdot 2^7 + 1 \cdot 2^6) + (0 \cdot 2^5 + 0 \cdot 2^4 + 1 \cdot 2^3) + (1 \cdot 2^2 + 0 \cdot 2^1 + 1 \cdot 2^0) \\
=\ & (4 \cdot 2^6 + 2 \cdot 2^6 + 1 \cdot 2^6) + (0 \cdot 2^3 + 0 \cdot 2^3 + 1 \cdot 2^3) + (4 \cdot 2^0 + 0 \cdot 2^0 + 1 \cdot 2^0) \\
=\ & 7 \cdot 2^6 + 1 \cdot 2^3 + 5 \cdot 2^0 = 7 \cdot (2^3)^2 + 1 \cdot (2^3)^1 + 5 \cdot (2^3)^0 \\
=\ & 7 \cdot 8^2 + 1 \cdot 8^1 + 5 \cdot 8^0
\end{aligned}
$$

and so $(111001101)_2 = (715)_8$.

In short, what this tells us to do is to place the bits in the expansion $(111001101)_2$ into groups of three, namely, 111, 001, 101. Since $(111)_2 = (7)_8$, $(001)_2 = (1)_8$ and $(101)_2 = (5)_8$, we have the result that $(111001101)_2 = (715)_8$.

To convert $(111001101)_2$ to a hexadecimal expansion, we place the bits this time into groups of four, namely, 1, 1100, 1101. Since $(1)_2 = (1)_{16}$, $(1100)_2 = (C)_{16}$ and $(1101)_2 = (D)_{16}$, it follows that $(111001101)_2 = (1CD)_{16}$. ♦

Example 7.62

Express the octal expansion $(233)_8$ as an integer in hexadecimal expansion.

Solution. Since $(233)_8$ denotes $2 \cdot 8^2 + 3 \cdot 8^1 + 3 \cdot 8^0$ and in base 2, the coefficient 2 is expressed as 10 and the coefficient 3 is expressed as 011, we can express $(233)_8$ in base 2 as

$$10 \quad 011 \quad 011.$$

Regrouping the 1s and 0s in groups of four bits, we have

$$1001 \quad 1011.$$

However, $(1001)_2 = (9)_{16}$ and $(1011)_2 = (B)_{16}$. Thus $(233)_8 = (10011011)_2 = (9B)_{16}$. ♦

Example 7.63

Express the hexadecimal expansion $(6D)_{16}$ as an integer in octal expansion.

Solution. Since $(6D)_{16}$ denotes $6 \cdot 16^1 + D \cdot 16^0$ and in base 2, the coefficient 6 is expressed as 110, while in base 2 the coefficient D is expressed as 1101, we can express $(6D)_{16}$ in base 2 as

$$1101101.$$

Regrouping the 1s and 0s in groups of three bits, we have

$$1 \quad 101 \quad 101.$$

However, $(1)_2 = (1)_8$ and $(101)_2 = (5)_8$ and so $(6D)_{16} = (1101101)_2 = (155)_8$. ♦

Exercises for Section 7.7

1. Give the decimal expansions of the following integers:
   (a) $(101011)_2$. (b) $(1110111)_2$. (c) $(100010)_2$.

2. Give the decimal expansions of the following integers:
   (a) $(57)_8$. (b) $(262)_8$. (c) $(1717)_8$.

3. Give the decimal expansions of the following integers:
   (a) $(D)_{16}$. (b) $(99)_{16}$. (c) $(ABC)_{16}$.

4. What are the binary expansions of the following integers (expressed in base 10)?
   (a) 54. (b) 101. (c) 255.

5. What are the octal expansions of the following integers (expressed in base 10)?
   (a) 91. (b) 211. (c) 410.

6. What are the hexadecimal expansions of the following integers (expressed in base 10)?
   (a) 14. (b) 250. (c) 4090.

7. What are the decimal expansions of the following integers expressed in base 4?

    (a) $(2013)_4$.   (b) $(3001)_4$.   (c) $(2033)_4$.

8. What are the base 4 expansions of the following decimal integers?

    (a) 1492.   (b) 1776.   (c) 2007.

9. (a) Express $(79B)_{16}$ in base 2.

    (b) Express $(100110111)_2$ in base 8.

    (c) Express $(4731)_8$ in base 16.

10. Let a and b be integers such that $1 \leq a, b \leq 9$. If $(ab)_{10} = (ba)_{16}$, then what are a and b?

11. Suppose that the binary representation of an integer a uses ten bits. The 0s and 1s of this representation are interchanged resulting in the integer b. Express b in terms of a.

12. Perform the following operations in base 2.

    (a) Add $(11110)_2$ and $(10101)_2$. (b) Subtract $(1101)_2$ from $(11011)_2$.

    (c) Multiply $(1011)_2$ by $(110)_2$.

    (d) Divide $(10101)_2$ by $(110)_2$ and determine the resulting quotient and reminder.

13. Use base 2 arithmetic to show that $\sqrt{(1111001)_2} = (1011)_2$.

14. Let $a_k a_{k-1} \cdots a_1 a_0$ be the decimal expansion of a positive integer n, where $0 \leq a_i \leq 9$ for $0 \leq i \leq k - 1$ and $1 \leq a_k \leq 9$. Define the **reflection** $ref(n)$ of n as that positive integer whose decimal expansion is $a_0 a_1 \ldots a_k$. For example, if $n = 4307$, then $ref(n) = 7034$; while if $n = 19226$, then $ref(n) = 62291$.

    (a) Show that if n is a 6-digit integer, then $9 \mid (n - ref(n))$.

    (b) Show that if n is a 5-digit integer, then $99 \mid (n - ref(n))$.

Chapter 7   Highlights

Key Concepts

base b expansion, $n = (a_k a_{k-1} \cdots a_1 a_0)_b$: the expansion of a positive integer n in base b, where then $b \geq 2$ and $n = a_k b^k + a_{k-1} b^{k-1} + \cdots + a_1 b + a_0 b^0$.

binary expansion: the expansion of an integer in base 2.

Caesar shift: a decryption technique in which each letter in the plaintext is replaced by a letter obtained by rotating the letter a fixed number of positions.

common divisor (of a and b): an integer that divides both a and b.

common multiple (of a and b): an integer that is multiple of a and b.

composite: an integer greater than 1 that is not prime.

congruence (a is **congruent** to b modulo n), $a \equiv b \pmod{n}$: $n \mid (a - b)$.

cryptography: the study of sending and receiving secret messages.

cryptology: the study of systems for secure communications.

cryptosystem: system for secure communications.

decimal expansion: the expansion of an integer in base 10.

decryption: a process of transforming a disguised message back to the original message.

div (m **div** n): the quotient when the integer m is divided by the positive integer n.

divides (a divides b, where $a \neq 0$), $a \mid b$: there is an integer c such that $b = ac$.

divisor (of an integer b): an integer that divises b.

encryption: a process of transforming a message into a disguised message.

factor (of an integer b): an integer that divises b.

greatest common divisor (of a and b), $\gcd(a, b)$: the greatest positive integer that is a common divisor of a and b.

hexadecimal expansion: the expansion of an integer in base 16.

least common multiple (of integers a and b), $\mathrm{lcm}(a, b)$: the smallest positive integer that is a common multiple of a and b.

linear combination (of a and b): an integer of the form $ax + by$, where $x, y \in \mathbb{Z}$.

mod (m **mod** n): the remainder when the integer m is divided by the positive integer n.

multiple (of an integer a): an integer that a divises.
octal expansion: the expansion of an integer in base 8.

prime: an integer greater than 1 whose only positive integer divisors are 1 and itself.

private key: a substitute character for each character that could be contained in any message to be transmitted.

private key cryptosystem: a cryptosystem in which two individuals who wish to communicate in secret have a private key.

public key cryptosystem: a cryptosystem in which two keys are used, a public encryption key and private decryption key.

relatively prime (a and b relatively prime): $\gcd(a, b) = 1$.

RSA System: a public key cryptosystem introduced by Richard Rivest, Adi Shamir and Leonard Adleman.

Key Results

- Let a, b and c be integers with $a \neq 0$. If $a \mid b$ and $a \mid c$, then $a \mid (bx + cy)$ for every two integers x and y.

- Let a, b and c be integers with $a \neq 0$ and $b \neq 0$. If $a \mid b$ and $b \mid c$, then $a \mid c$.

- **The Fundamental Theorem of Arithmetic**: Every integer $n \geq 2$ is either prime or can be expressed as $n = p_1 p_2 \cdots p_k$, where $p_1, p_2, \ldots, p_k$ are primes. This factorization is unique except possibly for the order in which the primes appear.

- An integer $n \geq 2$ is composite if and only if there exist integers a and b with $1 < a < n$ and $1 < b < n$ such that $n = ab$.

- If n is a composite number, then n has a prime factor p such that $p \leq \sqrt{n}$.

- There are infinitely many primes.

- **The Prime Number Theorem**: The number $\pi(n)$ is approximately equal to $n / \ln n$. More precisely, $\lim_{n \to \infty} \frac{\pi(n)}{n / \ln n} = 1$.

- **The Division Algorithm**: For every two integers m and $n > 0$, there exist unique integers q and r such that $m = nq + r$, where $0 \leq r < n$.

- Let a, b and $n \geq 2$ be integers. Then $a \equiv b \pmod{n}$ if and only if $a = b + kn$ for some integer k.

- Let a, b and $n \geq 2$ be integers. Then $a \equiv b \pmod{n}$ if and only if a and b have the same remainder when divided by n.

- Let a, b and $n \geq 2$ be integers. Then $a \equiv b \pmod{n}$ if and only if $a \bmod n = b \bmod n$.

- Let a and b be two positive integers. If $b = aq + r$ for some integers q and r, then $\gcd(a, b) = \gcd(r, a)$.

- **Euclidean Algorithm:** An algorithm that determines $\gcd(a, b)$.

- For every two positive integers a and b, $ab = \gcd(a, b) \operatorname{lcm}(a, b)$.

- Every two consecutive integers are relatively prime.

- Let a and b be integers that are not both 0. Then $\gcd(a, b)$ is the smallest positive integer that is a linear combination of a and b.

- Two integers a and b are relatively prime if and only if 1 is a linear combination of a and b.

- Let a, b and c be integers with $a \neq 0$. If $a \mid bc$ and $\gcd(a, b) = 1$, then $a \mid c$.

- Let b and c be integers and p a prime. If $p \mid bc$, then either $p \mid b$ or $p \mid c$.

- Let $a_1, a_2, \ldots, a_n$ be $n \geq 2$ integers and let p be a prime. If $p \mid a_1 a_2 \cdots a_n$, then $p \mid a_i$ for some integer i with $1 \leq i \leq n$.

- Let $b \geq 2$ be an integer. Then every positive integer n has a unique representation in base b as $n = a_k b^k + a_{k-1} b^{k-1} + \cdots + a_1 b^1 + a_0 b^0$, where k is a nonnegative integer, the digits $a_0, a_1, \ldots, a_k$ are nonnegative integers less than b and $a_k \neq 0$.

Supplementary Exercises for Chapter 7

1. Let $A = \{n \in \mathbb{Z} : n \geq 2\}$. For $a \in A$, define $f(a)$ to be the largest positive integer k such that $k < a$ and $k \mid a$. Then f is a function from A to $\mathbb{N}$.
   (a) What is $f(12)$?   (b) What is $f(27)$?   (c) What is $f(5)$?
   (d) Is f one-to-one?   (e) Is f onto?

2. Let $A = \{n \in \mathbb{Z} : n \geq 2\}$. For $a \in A$, define $f(a)$ to be the the largest nonnegative integer k such that $2^k \mid a$. Then f is a function from A to $\mathbb{N} \cup \{0\}$.
   (a) What is $f(10)$?   (b) What is $f(24)$?   (c) What is $f(32)$?
   (d) What is $f(33)$?   (e) Is f one-to-one?   (f) Is f onto?

3. Let $a, b, c, d \in \mathbb{Z}$ such that $a \neq 0$. Prove that if $a \mid (b + c + d)$ and a divides any two of b, c and d, then a divides the third integer.

4. Let $a, b, c, d \in \mathbb{Z}$ such that $a \neq 0$ and $b \neq 0$. Prove that if $a \mid c$ and $b \mid d$, then $ab \mid cd$.

5. Let a and b be integers such that $a \neq 0$. Prove that if $a \mid b$, then $a^n \mid b^n$ for every positive integer n.

6. Let a and b be integers. Prove that if $3 \nmid ab$, then $3 \nmid a$ and $3 \nmid b$.

7. Prove that $3 \mid (n^3 + 2n)$ for every positive integer n.

8. Prove that $12 \mid (13^n - 1)$ for every positive integer n.

9. Express 234 as a product of primes.

10. For the following pairs m, n of integers, illustrate the Division Algorithm by finding integers q and r such that $m = nq + r$, where $0 \le r < n$. (a) $m = 78$, $n = 15$. (b) $m = -78$, $n = 15$.

11. Is $37 \equiv -19 \pmod{4}$?

12. Let x and y be integers such that $x + y \equiv 0 \pmod{3}$. Prove that if $a, b \in \mathbb{Z}$ such that $a \equiv b \pmod{3}$, then $ax + by \equiv 0 \pmod{3}$.

13. Prove or disprove:

    (a) There exists an integer a such that $ab \equiv 0 \pmod{5}$ for every integer b.

    (b) If $a \in \mathbb{Z}$, then $ab \equiv 0 \pmod{5}$ for every $b \in \mathbb{Z}$.

    (c) For every integer a, there exists an integer b such that $ab \equiv 0 \pmod{5}$.

14. (a) Use the Euclidean algorithm to determine $\gcd(70, 42)$.

    (b) Express $\gcd(70, 42)$ as a linear combination of 70 and 42.

15. Prove or disprove: Let a and b be two integers. If there is a linear combination of a and b that equals 2, then $\gcd(a, b) = 2$.

16. Let a and b be integers not both 0. Prove that if $\gcd(a, 4) = \gcd(b, 4) = 2$, then $\gcd(a+b, 4) = 4$.

17. Let a and b be integers not both 0. Prove that $\gcd(a, b) = \gcd(a, b + ak)$ for every integer k.

18. Find all integers a and b with $0 < a < b$ such that $\operatorname{lcm}(a, b) = p^4$ for some prime p.

19. Find all integers a and b with $0 < a < b$ such that $\gcd(a, b) = 2^3$ and $\operatorname{lcm}(a, b) = 2^6$.

20. Find all integers a and b with $0 < a < b$ such that $\gcd(a, b) = 2^3 \cdot 3$ and $\operatorname{lcm}(a, b) = 2^4 \cdot 3^2$.

21. (a) Let a, b and c be integers with $a \ne 0$. Prove that $\gcd(a, b) = \gcd(a, c) = 1$ if and only if $\gcd(a, bc) = 1$.

    (b) Let a and b be integers not both 0. Use (a) to prove that if a and b are relatively prime, then a^2 and b^2 are relatively prime.

22. Suppose that the greatest positive integer that divides three integers a, b and c is d. Prove that $d = 1$ if and only if there exist integers x, y and z such that $ax + by + cz = 1$.

23. Give the decimal expansion of $(123)_8$.

24. What is the hexadecimal expansion of 707?

25. Let $a_1, a_2, a_3, \ldots$ be a sequence of integers for which $a_1 \equiv 2 \pmod{3}$ and $a_2 \equiv 2 \pmod{3}$. For $n \ge 3$, $a_n = a_{n-1} + a_{n-2}$. Use induction to prove that $a_n \equiv 0 \pmod{3}$ if and only if $n \equiv 0 \pmod{4}$.

26. Let p be a prime and $n \ge 2$ an integer. Prove that $p^{1/n}$ is irrational.

27. Let k be a positive integer. Prove that if $2^k - 1$ is prime, then k is a prime.

28. Prove that if $a_1, a_2, \ldots, a_n$ are $n \ge 1$ integers such that $a_i \equiv 1 \pmod{3}$ for $1 \le i \le n$, then $a_1 a_2 \cdots a_n \equiv 1 \pmod{3}$.

29. Prove that for the nth Fibonacci number F_n, $2 \mid F_n$ if and only if $3 \mid n$.

30. For a sequence $s : a_1, a_2, \ldots, a_n$ of n integers, write an algorithm that determines how many integers in the sequence are even and how many are odd.

31. Write an algorithm that determines in a sequence of integers the number of times that the terms in the sequence change from even to odd or from odd to even.

32. Show that if p and q are primes with $p < q$, then $\log_p q$ is irrational.

33. Show that 20 and 33 are relatively prime by showing that 1 is a linear combination of 20 and 33.

34. Show that $3k + 11$ and $5k + 18$ are relatively prime for every positive integer k.

35. Show that the least common multiple of two positive integers a and b is ab if and only if a and b are relatively prime.

36. Let p, q, r, s, m and n be positive integers such that $p \equiv q \pmod{m}$, $r \equiv s \pmod{n}$ and $d = \gcd(m, n)$. Show that $p + r \equiv q + s \pmod{d}$.

37. Suppose that a, b and c are nonzero integers such that $a + b + c = 0$. Show that $\gcd(a, b) = \gcd(b, c) = \gcd(a, c)$.

38. A class has 30 students. Each student is assigned an integer x with $0 \le x \le 29$ according to where on the (alphabetical) class list the student's name appears. So the first name on the class list is assigned 0; while the last name on the class list is assigned 29. Near the end of the semester, each student is required to present a short report on his or her class project. The order in which the presentations are to be made is given by the function f, defined by $f(x) = (7x + 3) \bmod 30$, where $0 \le x \le 29$.

    (a) When does the last person on the class list speak?

    (b) Which student gives the first presentation?

    (c) What would happen if f is defined by $f(x) = (3x + 7) \bmod 30$, where $0 \le x \le 29$?

39. Suppose that 444 is the representation of an integer in base 5. What is its representation in base 4?

40. Suppose that 212 is the representation of an integer in base 9. What is its representation in base 3?

41. (a) For which values of a, where $1 \le a \le 9$, does there exist an integer x such that $ax \equiv 1 \pmod{10}$?

    (b) Can you guess what property a must have for such an integer x to exist?

42. An integer n has a remainder of 2 when divided by 3 and a remainder of 1 when divided by 2. What is the remainder of n when n is divided by 6?

43. Give an example of two bases a and b such that $(111)_a = (421)_b$.

44. Let $a_0, a_1, a_2, \ldots$ be a sequence of positive integers for which (1) $a_0 = 1$, (2) $a_{2n+1} = a_n$ for $n \ge 0$ and (3) $a_{2n+2} = a_n + a_{n+1}$ for $n \ge 0$. Prove that a_n and a_{n+1} are relatively prime for every nonnegative integer n.

Chapter 8

Introduction to Counting

Discrete mathematics consists of many areas of mathematics. One of these areas is combinatorics. **Combinatorics** is the branch of mathematics that deals with the study of configurations or arrangements of objects. Combinatorics evidently came into prominence during 1915-1916 only after the publication of the classic two-volume treatise *Combinatory Analysis* by Percy Alexander MacMahon. Quite possibly combinatorics did not become a respected branch of mathematics until the 1960s when the eminent combinatorialist Gian-Carlo Rota wrote a series of ten papers on "Foundations of Combinatorics." He was a professor at MIT during most of his career and, with Frank Harary, founded the *Journal of Combinatorial Theory*.

Another mathematician who played a major role in the development and popularization of combinatorics was Paul Erdős. Erdős was a Hungarian mathematician who traveled constantly, visiting mathematicians in many parts of the world to do research in combinatorics, number theory, set theory and probability theory (a subject we'll visit in Chapter 10). He was one of the most prolific mathematicians of all time, having authored or co-authored more than 1500 research articles. A humorous concept associated with Erdős is the **Erdős number**. Only Paul Erdős has Erdős number 0, while any person who co-authored a paper with Erdős has Erdős number 1. An individual who does not have Erdős number 1 but who has co-authored a paper with someone who has Erdős number 1 has Erdős number 2 and so on. Over 500 individuals have Erdős number 1, while over 6500 people have Erdős number 2.

As we mentioned, combinatorics deals with configurations or arrangements of objects. Many of the questions in this area concern one or more of the following topics:

(1) Existence: Is such a configuration or arrangement possible?
(2) Enumeration: How many such configurations are there?
(3) Optimization: Is some arrangement of a certain type more desirable in some way?

In this chapter and the next, we will be particularly concerned with questions in combinatorics of the type (2). This area of combinatorics is generally called **enumerative combinatorics** – the subject of counting. In this chapter, some of the fundamental principles of counting will be introduced. These will allow us to answer many types of questions concerning enumeration.

8.1   The Multiplication and Addition Principles

There are a number of instances when a certain procedure consists of a sequence of two tasks in which the second task is performed only after the first task has been performed. We say that the procedure itself is performed when the two tasks are performed. If we know the number of ways in which each of the two tasks can be performed, then the multiplication principle tells us the number of ways that the procedure can be performed.

The Multiplication Principle

The Multiplication Principle *A procedure consists of a sequence of two tasks. To perform this procedure, one performs the first task followed by performing the second task. If there are n_1 ways to perform the first task and n_2 ways to perform the second task after the first task has been performed, then there are $n_1 n_2$ ways to perform the procedure.*

Let's look at two examples where the Multiplication Principle can be applied.

Example 8.1

An employee of a software company in Washington, DC has been asked to attend a 2-day meeting in Atlanta followed by a 2-day meeting in Orlando. When checking airline web sites, he learns that there are four different flights from Washington to Atlanta that are convenient for him (namely at 1:30, 3:00, 4:00 and 5:30 pm) and three different flights from Atlanta to Orlando that meet his requirements (at 7 am, 8 am and 10 am). How many different satisfactory flight schedules from Washington to Orlando through Atlanta are there for him?

Solution. Since there are four satisfactory flights from Washington to Atlanta and three acceptable flights from Atlanta to Orlando, it follows by the Multiplication Principle that the number of flight schedules for him from Washington to Orlando through Atlanta is $4 \cdot 3 = 12$.   ♦

Example 8.2

At a certain university, a student is required to take a 2-course sequence during her senior year. The first course can be any of the three courses CS 410, CS 420 or CS 430. To complete a sequence, the student has two choices after taking any of these three courses; namely, she can take CS 411 or CS 412 after taking CS 410, she can take CS 421 or CS 422 after taking CS 420 or she can take CS 431 or CS 432 after taking CS 430. How many choices does she have for a required 2-course sequence?

Solution. Since there are three possible courses for the first course in a 2-course sequence and once the first course has been taken, there are two possible courses for the second course in the sequence, it follows by the Multiplication Principle that the total number of possibilities for a 2-course sequence is $3 \cdot 2 = 6$.   ♦

The six possible 2-course sequences discussed in Example 8.2 can be described by means of a diagram, called a **tree diagram**. The tree diagram for Example 8.2 is shown in Figure 8.1. Each sequence can be obtained by starting at the top of the diagram and proceeding downward through the diagram, resulting in the six ordered pairs (2-course sequences), where we omit "CS" for simplicity:

$$(410, 411), \ (410, 412), \ (420, 421), \ (420, 422), \ (430, 431), \ (430, 432).$$

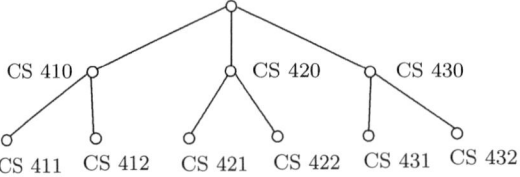

Figure 8.1: A tree diagram for Example 8.2

The Multiplication Principle, as stated, tells us how to compute the number of ways of performing a procedure that consists of performing two tasks in succession. Procedures need not be limited to two tasks, however, for suppose that a certain procedure A consists of performing three tasks, say T_1, T_2, T_3 in that order. That is, performing procedure A requires us to first perform task T_1, then perform T_2 and finally T_3. Let us assume that there are n_1 ways to perform T_1, n_2 ways to perform T_2 after T_1 has been performed and n_3 ways to perform T_3 after T_2 has been performed (so there are n_i ways to perform T_i for $i = 1, 2, 3$). In how many ways can procedure A be performed?

What we can do in this case is to think of linking the two tasks T_2 and T_3 together as a procedure, say procedure B. In other words, to perform procedure A, we perform task T_1 followed by performing procedure B. We can think of procedure B as a task itself, albeit a somewhat more complex task. By the Multiplication Principle then, the number of ways to perform procedure A is the number of ways of performing task T_1 (which we know is n_1) times the number of ways of performing procedure B. However, performing procedure B requires us to perform task T_2 and then task T_3. So, by the Multiplication Principle, the number of ways of performing procedure B is $n_2 n_3$ and so the total number of ways of performing procedure A is $n_1(n_2 n_3) = n_1 n_2 n_3$.

This suggests the following more general principle. In fact, the discussion just given for determining the number of procedures consisting of three tasks indicates how this general principle can be proved (by mathematical induction).

The (General) Multiplication Principle *Performing a certain procedure consists of performing a sequence of $m \geq 2$ tasks T_1, T_2, ..., T_m. If there are n_i ways of performing T_i after any preceding tasks have been performed for $i = 1, 2, \ldots, m$, then the total number of ways of performing the procedure is $n_1 n_2 \cdots n_m$.*

Let's now consider several examples of this more general principle. Whether we use the Multiplication Principle (in which two tasks are involved) or the General Multiplication Principle, we will henceforth refer to either simply as the Multiplication Principle.

Example 8.3

Several years ago an automobile license plate number in a certain state consisted of two letters followed by four digits. How many different license plates were available under that scheme?

Solution. We can think of a license plate looking like the following:

$$\underline{\quad} \ \underline{\quad} \qquad \underline{\quad} \ \underline{\quad} \ \underline{\quad} \ \underline{\quad}$$

The first two slots are to be filled in by letters and the last four by digits. Since there are 26 letters and 10 digits, it follows by the Multiplication Principle that there are

$$26 \cdot 26 \cdot 10 \cdot 10 \cdot 10 \cdot 10 = 26^2 \cdot 10^4 = 6{,}760{,}000$$

different license plates. ♦

Example 8.4

In a certain computer science course, there is a weekly quiz. The quiz for today consists of ten true-false questions. How many different sequences of answers are possible for this quiz?

Solution. Each of the ten questions can be answered in one of two ways (true or false); so the number of different sequences of responses for this quiz is

$$2 \cdot 2 \cdot 2 \cdot 2 \cdot 2 \cdot 2 \cdot 2 \cdot 2 \cdot 2 \cdot 2 = 2^{10} = 1024. \ ♦$$

The preceding example can be interpreted in a different way. Recall that a bit is a 0 or a 1 and a bit string of length n (or an n-bit string) is a sequence of n bits. Since each bit has two possible values, the number of different bit strings of length n is 2^n. In Example 8.4, we can consider sequences of responses to the 10-question true-false quiz as 10-bit sequences (where 0 corresponds to false, say, and 1 corresponds to true) and so the number of different sequences of responses is 2^{10}.

Counting the number of different 10-bit strings is a special case of a more general problem. Let $A = \{1, 2, \ldots, 10\}$, where the integer i ($1 \le i \le 10$) corresponds to question i on the quiz and let $B = \{0, 1\}$. Suppose that we were to count the number of different functions $f : A \to B$. Each such function f has the appearance

$$f = \{(1, \underline{\hspace{1em}}), (2, \underline{\hspace{1em}}), \ldots, (10, \underline{\hspace{1em}})\},$$

where each blank is to be filled in with 0 or 1. Thus, the number of functions from A to B is 2^{10}, which, of course, is the number of different 10-bit strings. This suggests a more general result.

Theorem 8.5 *If A and B are two finite nonempty sets with $|A| = m$ and $|B| = n$, then the number of different functions from A to B is $|B|^{|A|} = n^m$.*

Proof. Let $A = \{a_1, a_2, \ldots, a_m\}$. Any function $f : A \to B$ has the appearance

$$f = \{(a_1, \underline{\hspace{1em}}), (a_2, \underline{\hspace{1em}}), \ldots, (a_m, \underline{\hspace{1em}})\},$$

where each blank (the image of an element of A) is to be filled in with an element of the codomain B. Since there are n choices for each image, it then follows by the Multiplication Principle that the number of such functions is the product of m factors n, that is, $n \cdot n \cdot \cdots \cdot n = n^m$. ∎

Example 8.6

Determine the number of possible responses to a 20-question multiple choice exam, with five allowable choices for each question (assuming that every question is answered with exactly one of these five choices).

Solution. When such a multiple choice exam is taken, this produces a function. In particular, each response to a 20-question multiple choice exam, with five choices (say a, b, c, d, e) for each question, can be thought of as a function $f : A \to B$, where $A = \{1, 2, \cdots, 20\}$ and $B = \{a, b, c, d, e\}$. It then follows by Theorem 8.5 that there are 5^{20} possible responses to such a multiple choice exam. ◆

We now present a result that corresponds to Theorem 8.5 for one-to-one functions. We saw in Chapter 5 that in order for a function $f : A \to B$ to be one-to-one, we must have $|A| \le |B|$. Hence if $|A| > |B|$, then there are no one-to-one functions from A to B. Recall for a positive integer n that n factorial is $n! = n(n-1)\cdots 3 \cdot 2 \cdot 1$.

Theorem 8.7 *If A and B are two sets with $|A| = m$ and $|B| = n$, where $m \le n$, then the number of different one-to-one functions from A to B is $\dfrac{n!}{(n-m)!}$.*

Proof. Let's first consider the special case where $m = n$ so that $A = \{a_1, a_2, \ldots, a_n\}$. A one-to-one function $f : A \to B$ can be expressed as

$$f = \{(a_1, \underline{\hspace{1em}}), (a_2, \underline{\hspace{1em}}), \ldots, (a_n, \underline{\hspace{1em}})\}.$$

Since B has n elements, there are n possible images for a_1. Once an image for a_1 has been determined, this element of B cannot be the image of any other element of A since f is one-to-one. Hence there are $n-1$ possible images for a_2. Similarly, there are $n-2$ possible images for a_3, etc., until arriving at only one possible image for a_n. Consequently, the total number of one-to-one functions from A to B is

$$n(n-1)(n-2)\cdots 3 \cdot 2 \cdot 1 = n!.$$

If $m < n$, then each one-to-one function $f : A \to B$ can be expressed as

$$f = \{(a_1, \underline{\hspace{1em}}), (a_2, \underline{\hspace{1em}}), \ldots, (a_m, \underline{\hspace{1em}})\},$$

Proceeding as above, we see that the number of one-to-one functions from A to B is

$$n(n-1)(n-2)\cdots(n-m+1). \tag{8.1}$$

Multiplying the number in (8.1) by $(n-m)!/(n-m)!$, we see that the answer to this question can be expressed more simply as

$$n(n-1)(n-2)\cdots(n-m+1)\cdot\frac{(n-m)!}{(n-m)!}=\frac{n!}{(n-m)!}.$$

Using the standard convention that $0!=1$, we see that the number of one-to-one functions from A to B is given by $\dfrac{n!}{(n-m)!}$. ∎

Example 8.8

Determine the number of the one-to-one functions from A to B, where $|A|=4$ and $|B|=7$.

Solution. From Theorem 8.7, we see that the number of one-to-one functions from A to B is

$$\frac{7!}{(7-4)!}=\frac{7!}{3!}=7\cdot6\cdot5\cdot4=840. \blacklozenge$$

Suppose that a certain procedure consists of a sequence of $m\geq 2$ tasks T_1, T_2, ..., T_m. For $1\leq i\leq m$, let A_i be the set of all possible ways of performing the task T_i, where $|A_i|=n_i$. By the Multiplication Principle, the number of ways of performing the procedure is $n_1n_2\cdots n_m$. Each element of the Cartesian product $A_1\times A_2\times\cdots\times A_m$ is one way to perform this procedure and $A_1\times A_2\times\cdots\times A_m$ is the set of all possible ways to perform this procedure. Hence the number of ways to perform the procedure is

$$|A_1\times A_2\times\cdots\times A_m|=n_1n_2\cdots n_m.$$

Therefore, we have the following.

For $m\geq 2$ sets $A_1,A_2,\ldots,A_m$,

$$|A_1\times A_2\times\cdots\times A_m|=|A_1|\cdot|A_2|\cdot\ \cdots\ \cdot|A_m|.$$

In particular, for every two sets A and B,

$$|A\times B|=|A|\cdot|B|,$$

as we saw in Section 2.3.

The Addition Principle

We now turn to a second principle. There are instances when performing a certain procedure consists of performing one of two tasks that cannot be performed at the same time. In this case, we say that the procedure is performed if either task is performed. The Addition Principle informs us of the number of ways that such a procedure can be performed.

The Addition Principle *A procedure consists of two tasks that cannot be performed simultaneously. To perform this procedure, either of the two tasks is performed. If the first task can be performed in n_1 ways and the second can be performed in n_2 ways, then the number of ways of performing this procedure is n_1+n_2.*

Example 8.9

A student majoring in computer science needs only one additional approved course to complete a minor in either mathematics or physics. If there are three approved courses to complete a minor in mathematics and five approved courses to complete a minor in physics where no single course will complete both minors, then how many choices does the student have to complete a minor?

Solution. By the Addition Principle, the total number of ways to complete a minor in either mathematics or physics is $3 + 5 = 8$.   ♦

As with the Multiplication Principle, there is a more general Addition Principle when there are more than two tasks involved. Here too, we will use the term Addition Principle, regardless of the number of tasks involved.

The (General) Addition Principle *Performing a certain procedure consists of performing one of $m \geq 2$ tasks T_1, T_2, ..., T_m, no two of which can be performed at the same time. If the task T_i can be performed in n_i ways for $1 \leq i \leq m$, then the number of ways of performing this procedure is $n_1 + n_2 + \cdots + n_m$.*

Example 8.10

After a student graduates from college, he wants to work on a Master's degree in computer science. He is considering two universities in South Carolina, three universities in Georgia and five universities in Florida. By the Addition Principle, the total number of choices he has for his graduate studies is $2 + 3 + 5 = 10$.   ♦

Suppose that a certain procedure consists of $m \geq 2$ tasks T_1, T_2, ..., T_m, no two of which can be performed at the same time. For $1 \leq i \leq m$, let A_i be the set of all possible ways of performing the task T_i, where $|A_i| = n_i$. Thus $\{A_1, A_2, \ldots, A_m\}$ is a collection of *pairwise disjoint* finite sets. By the Addition Principle, the number of ways of performing the procedure is $n_1 + n_2 + \cdots + n_m$. Each element of the union $A_1 \cup A_2 \cup \ldots \cup A_m$ is one way to perform this procedure and $A_1 \cup A_2 \cup \ldots \cup A_m$ is the set of all possible ways to perform the procedure. Hence the number of ways to perform the procedure is

$$|A_1 \cup A_2 \cup \cdots \cup A_m| = n_1 + n_2 + \cdots + n_m.$$

This argument provides us with the following.

If $A_1, A_2, \cdots, A_m$ are $m \geq 2$ pairwise disjoint sets, then

$$|A_1 \cup A_2 \cup \cdots \cup A_m| = |A_1| + |A_2| + \cdots + |A_m|.$$

In the following example, both the Multiplication and Addition Principles are employed.

Example 8.11

A recent graduate has obtained a position with a communications company and needs to spend the first six months in training, which can be done in the Western United States or Eastern United States. She can spend the first three months in Seattle or Portland and the next three months in Sacramento, San Jose or San Francisco; or she can spend the first three months in New York City, Boston or Newark and the next three months in Philadelphia, Pittsburgh, New Haven or Buffalo. What is the number of choices this individual has for her training schedule?

Solution. By the Multiplication Principle, she has $2 \cdot 3 = 6$ choices for a training schedule if she decides to spend her training period in Western United States, while, again by the Multiplication Principle, she has $3 \cdot 4 = 12$ choices for a training schedule if she decides to spend her training period in Eastern United States. By the Addition Principle, the total number of possible training schedules for her is $6 + 12 = 18$. ◆

Let's now consider several examples involving the Multiplication and/or Addition Principles, beginning with two examples concerning bit sequences.

Example 8.12

How many 10-bit sequences begin with 001 *and* end with 11?

Solution. Such a sequence has the following appearance:

$$\underline{0}\ \underline{0}\ \underline{1}\ \_\ \_\ \_\ \_\ \_\ \underline{1}\ \underline{1}$$

There are five terms of the sequence to be determined and each term is either 0 or 1. Since there are two choices for each of the five terms to be determined, it follows by the Multiplication Principle that the total number of sequences with the desired property is $2 \cdot 2 \cdot 2 \cdot 2 \cdot 2 = 2^5 = 32$. ◆

Example 8.13

How many 8-bit sequences begin with 11011 *or* 0100?

Solution. We are considering 8-bit sequences of the type

$$\underline{1}\ \underline{1}\ \underline{0}\ \underline{1}\ \underline{1}\ \_\ \_\ \_ \qquad \text{or} \qquad \underline{0}\ \underline{1}\ \underline{0}\ \underline{0}\ \_\ \_\ \_\ \_$$

By the Multiplication Principle, the number of 8-bit sequences that begin with 11011 is $2 \cdot 2 \cdot 2 = 2^3 = 8$ and the number of 8-bit sequences that begin with 0100 is $2 \cdot 2 \cdot 2 \cdot 2 = 2^4 = 16$. Since an 8-bit sequence cannot begin with both 11011 and 0100, it follows by the Addition Principle that the number of 8-bit sequences with the desired property is $8 + 16 = 24$. ◆

We will see some variations of the question asked in Example 8.13 in the next section.

Example 8.14

In a school election, three students are to be elected to the student council for the next year. One of the students to be elected must be a freshman, one a sophomore and one a junior. Voting consists of listing three names:

$$1. \underline{\hspace{2cm}} \qquad 2. \underline{\hspace{2cm}} \qquad 3. \underline{\hspace{2cm}}$$

The first name listed must be a freshman, the second a sophomore and the third a junior.

(a) In how many different ways can a ballot be marked if there are 8 candidates for the freshman position, 3 for the sophomore position and 4 for the junior position?

(b) After the election, the president of the student council can choose any one of those not elected to the student council as an at-large member of the council. In how many ways can this be done?

Solution.

(a) By the Multiplication Principle, the total number of possible outcomes for a ballot is $8 \cdot 3 \cdot 4 = 96$.

(b) After the election there are 7 freshmen, 2 sophomores and 3 juniors who were not elected to the student council. By the Addition Principle, the number of choices for the at-large member is $7 + 2 + 3 = 12$. ◆

Exercises for Section 8.1

1. On a particular afternoon, a student plans to take her niece to a movie followed by dessert. The possibilities for the movie are an animated movie, a comedy and an adventure movie while the choices for dessert are hot fudge sundae, strawberry shortcake, apple pie and chocolate milkshake. How many possibilities are there for the afternoon events?

2. Draw a tree diagram for Example 8.1.

3. Each seat in an arena is labeled with a letter of the alphabet followed by a positive integer not exceeding 50. What is the largest number of seats that can be labeled differently?

4. A new automobile can be ordered with 8 different exterior colors and 6 different interior colors. How many possibilities for color combinations are there?

5. At a certain restaurant, a dinner starts with soup, salad or juice, followed by the main course (steak, chicken, shrimp or fish), followed by dessert (pie or cake). How many possibilities for dinner are there?

6. Suppose that a quiz in a discrete mathematics course consists of 15 questions. How many different sequences of answers are possible for this quiz if

    (a) each question is a true-false question?

    (b) each question is a multiple-choice question with 4 possible answers?

7. Let A and B be two sets with $|A| = 5$ and $|B| = 6$.

    (a) How many different functions from A to B are there?

    (b) How many different functions from B to A are there?

    (c) How many different one-to-one functions from A to B are there?

    (d) How many different one-to-one functions from B to A are there?

8. Each student at a certain university is given a 6-digit code (such as 123789 or 001122).

    (a) How many different codes are there?

    (b) How many codes read the same forward and backward?

    (c) How many codes contain only odd digits?

    (d) How many codes contain at least one even digit?

9. In how many orders can 4 married couples be seated in a row of 8 chairs if everyone must sit next to his or her spouse and no two people of the same sex can sit next to each other?

10. How many different 3-digit numbers can be formed using the digits 1, 2, 3, 4, 5 if

    (a) digits can be repeated?

    (b) digits cannot be repeated?

11. How many different 4-digit numbers can be formed from the digits 1, 2, 3, 4, 5, 6 if

    (a) digits can be repeated?

    (b) digits cannot be repeated?

12. How many 4-digit numbers can be formed from the digits $1, 2, \ldots, 7$ under the following conditions?

    (a) the numbers must be less than 5000 and no digits can be repeated.

(b) the numbers must be less than 5000 and digits can be repeated.

13. A student is preparing to go to class on a winter morning. He has three coats, four sweaters and two hats. How many possibilities does he have of what to wear if he must wear a coat?

14. A man leaves for work on a rainy morning. He has a choice of three raincoats, four umbrellas and two hats. Assuming that he must take a coat and an umbrella (but not necessarily a hat), how many possibilities for raingear does he have?

15. It is decided to have dinner at a Chinese, Italian or Mexican restaurant. If there are 7 possible Chinese restaurants, 5 possible Italian restaurants and 4 possible Mexican restaurants, what is the total number of possible choices for restaurants?

16. If radio station call letters consist of either 3 or 4 letters and begin with K or W, then how many different radio station call letters are there?

17. At a certain university, a telephone number begins with 355 or 357 and is then followed by four digits.

    (a) How many telephone numbers are possible?
    (b) How many telephone numbers have no digits repeated?

18. How many different 8-bit strings begin with 010 and end with 11?

19. How many different 10-bit strings begin and end with the same 5-bit string?

20. How many different 7-bit strings

    (a) begin with 1011 and end with 1100?
    (b) begin with 1100 and end with 1011?

21. How many different 7-bit strings begin with 11 or 0011?

22. How many different 10-bit strings begin with 1011 or 0110?

23. How many 10-bit strings begin with 1011 or 001100?

24. How many different 8-bit strings end with 011 or end with 01011?

25. A license plate consists of a sequence of three letters followed by three digits or three digits followed by three letters. How may different license plates are there?

26. A password on a computer system consists of four characters, each of which is either a digit or a letter of the alphabet. Suppose that each password must contain at least one digit and at least one letter. How many different such passwords are there?

27. A faculty committee has decided to choose one or more students to join the committee. A total of 5 juniors and 6 seniors have volunteered to serve on this committee. How many different choices are there if the committee decides to select

    (a) one junior and one senior?
    (b) exactly one student?

28. Give an example of a counting problem whose answer is $15 \cdot 4^4$. (Possible hint: Consider the number $8^4 - 4^4$.)

8.2   The Principle of Inclusion-Exclusion

According to the Addition Principle, if performing a procedure consists of performing one of two tasks, where the first task can be performed in n_1 ways and the second task can be performed in n_2 ways but the two tasks cannot be performed at the same time, then the number of ways of performing the procedure is $n_1 + n_2$. But what if it's possible that the two tasks can be performed at the same time? For example, suppose, on a United States geography exam, a student is asked to list a city that is either a State capital or is one of the 50 largest cities, according to the most recent United States census. A correct answer therefore consists either of listing a State capital or listing one of the 50 largest cities. This should be an easy question to answer, but here's a more challenging question: How many correct answers are there to this question? Since there are 50 State capitals and obviously 50 largest cities, it may appear that there are $50 + 50 = 100$ correct answers to this question. But this is not so. For example, Phoenix is not only the capital of Arizona, it is also one of the 50 largest cities in the United States. Consequently, the number of correct answers to the question above is less than 100.

Observe that if A and B are two disjoint finite sets, then $|A \cup B| = |A| + |B|$. But what does $|A \cup B|$ equal if A and B are not disjoint? Figure 8.2 suggests the answer to this question. Suppose that there are a elements in A that are not in B, there are b elements in B that are not in A and x elements that belong to both A and B. Therefore, $|A| = a + x$, $|B| = b + x$, $|A \cup B| = a + b + x$ and $|A \cap B| = x$. Since $a + b + x = (a + x) + (b + x) - x$, it follows that

$$|A \cup B| = |A| + |B| - |A \cap B|.$$

Therefore, to determine the number of elements in $A \cup B$, we count the number of elements belonging to A, count the number of elements belonging to B, count the number of elements belonging to both A and B and subtract the third number from the sum of the first two.

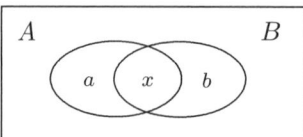

Figure 8.2: The Venn diagram for two sets A and B.

The Principle of Inclusion-Exclusion   *A procedure consists of two tasks. To perform the procedure, one performs either of the two tasks. If the first task can be performed in n_1 ways, the second task can be performed in n_2 ways and the two tasks can be simultaneously performed in n_{12} ways, then the total number of ways of performing the procedure is*

$$n_1 + n_2 - n_{12}.$$

The Addition Principle is therefore a special case of the Principle of Inclusion-Exclusion, as $n_{12} = 0$ in that case. The Principle of Inclusion-Exclusion can also be stated in terms of finite sets (as we indicated above).

The Principle of Inclusion-Exclusion (for two sets)   *For every two finite sets A and B,*

$$|A \cup B| = |A| + |B| - |A \cap B|.$$

In particular, if A and B are disjoint, then $|A \cup B| = |A| + |B|$.

The Principle of Inclusion-Exclusion is often attributed to the French mathematician Abraham de Moivre. Although famous for a number of accomplishments in mathematics, he was one of the first to use complex numbers in trigonometry. In fact, for the imaginary number $i = \sqrt{-1}$, de Moivre's formula states that

$$(\cos x + i \sin x)^n = \cos(nx) + i \sin(nx)$$

for every real number x and every integer n. A set-theoretic version of the Principle of Inclusion-Exclusion appeared in a 1718 textbook of de Moivre. The principle was used in 1713 by Pierre Rémond de Montmort to solve a problem he formulated in 1708. The Principle of Inclusion-Exclusion was known to have been used by Nicholas (also occasionally spelled Nicolaus) Bernoulli, a member of a large and distinguished family of Swiss mathematicians. The development of the principle is due in large degree to James Joseph Sylvester; however, many mathematicians were not aware of the principle until 1867 when it appeared in a book written by William Allen Whitworth.

We now consider some applications of the Principle of Inclusion-Exclusion.

Example 8.15

How many 8-bit sequences begin with 110 *or* end with 1100?

Solution. We are considering 8-bit sequences of one of the types

$$\underline{1}\ \underline{1}\ \underline{0}\ \_\ \_\ \_\ \_\ \_ \qquad \text{or} \qquad \_\ \_\ \_\ \_\ \underline{1}\ \underline{1}\ \underline{0}\ \underline{0}$$

By the Multiplication Principle, the number of 8-bit sequences that begin with 110 is $2 \cdot 2 \cdot 2 \cdot 2 \cdot 2 = 2^5 = 32$ and the number of 8-bit sequences that end with 1100 is $2 \cdot 2 \cdot 2 \cdot 2 = 2^4 = 16$. Observe that an 8-bit sequence could begin with 110 *and* end with 1100. Such a sequence has the following appearance:

$$\underline{1}\ \underline{1}\ \underline{0}\ \_\ \underline{1}\ \underline{1}\ \underline{0}\ \underline{0}$$

Therefore, the number of 8-bit sequences that begin with 110 and end with 1100 is 2. It then follows by the Principle of Inclusion-Exclusion that the number of 8-bit sequences with the desired property is $32 + 16 - 2 = 46$. ◆

Example 8.16

In a discrete mathematics course there are 17 students majoring in computer science, 11 students majoring in mathematics and 5 students majoring in both. How many students major in computer science or mathematics?

Solution. Let C be the set of all students majoring in computer science and let M be the set of all students majoring in mathematics. Then $|C| = 17$, $|M| = 11$ and $|C \cap M| = 5$. What we seek is $|C \cup M|$. By the Principle of Inclusion-Exclusion,

$$|C \cup M| = |C| + |M| - |C \cap M| = 17 + 11 - 5 = 23.$$

So there are 23 students majoring in computer science *or* mathematics. ◆

In the case of three pairwise disjoint sets A, B and C, the Addition Principle states that $|A \cup B \cup C| = |A| + |B| + |C|$. However, what if A, B and C are *not* pairwise disjoint? Figure 8.3 gives a Venn diagram for three arbitrary sets A, B and C, where a denotes the number of elements belonging to A but to neither B nor C (that is, there are a elements in $A - (B \cup C)$), r denotes the number of elements in both A and B but not C and so on.

Therefore, $|A \cup B \cup C| = a + b + c + r + s + t + x$. Since

$$|A| = a + r + s + x, \quad |B| = b + r + t + x \text{ and } |C| = c + s + t + x,$$

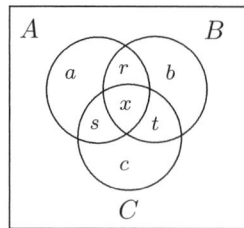

Figure 8.3: A Venn diagram for sets A, B and C

it follows that

$$|A| + |B| + |C| = a + b + c + 2r + 2s + 2t + 3x \neq |A \cup B \cup C|$$

if any of the integers r, s, t and x is not 0. Since $|A \cap B| = r + x$, $|A \cap C| = s + x$ and $|B \cap C| = t + x$, it follows that

$$|A| + |B| + |C| - |A \cap B| - |A \cap C| - |B \cap C| = a + b + c + r + s + t.$$

Because $|A \cap B \cap C| = x$, it follows that

$$|A| + |B| + |C| - |A \cap B| - |A \cap C| - |B \cap C| + |A \cap B \cap C| = a + b + c + r + s + t + x = |A \cup B \cup C|.$$

The Principle of Inclusion-Exclusion (for three sets) *For every three finite sets A, B and C,*

$$|A \cup B \cup C| = |A| + |B| + |C| - |A \cap B| - |A \cap C| - |B \cap C| + |A \cap B \cap C|.$$

Example 8.17

The director of a dormitory cafeteria interviews 60 students to determine their likes and dislikes among certain foods. Here is what she learned:

   31 students like beef.
   32 students like chicken.
   31 students like fish.
   15 students like beef and chicken.
   12 students like beef and fish.
   19 students like chicken and fish.
   8 students like all three.

How many like none of these?

Solution. Let B be the set of students who like beef, C the set of students who like chicken and F the set of students who like fish. By the Principle of Inclusion-Exclusion,

$$\begin{aligned} |B \cup C \cup F| &= |B| + |C| + |F| - |B \cap C| - |B \cap F| - |C \cap F| + |B \cap C \cap F| \\ &= 31 + 32 + 31 - 15 - 12 - 19 + 8 = 56. \end{aligned}$$

Since 56 of the 60 students like at least one of beef, chicken and fish, there are $60 - 56 = 4$ students who like none of these.   ♦

The next example describes a problem where the Principle of Inclusion-Exclusion does not appear to work directly and which perhaps requires some thought.

Example 8.18

A number of travelers are surveyed as to which of the three popular vacation cities Las Vegas, New York City and Orlando they would like to visit. Here are their responses:

> 26 travelers want to visit Las Vegas.
> 31 travelers want to visit New York City.
> 36 travelers want to visit Orlando.
> 12 travelers want to visit Las Vegas and New York City.
> 11 travelers want to visit Las Vegas and Orlando.
> 13 travelers want to visit New York City and Orlando.
> 5 travelers want to visit all three.

How many of these travelers want to visit at least two of Las Vegas, New York City and Orlando?

Solution. With the information given, it would be a simple application of the Principle of Inclusion-Exclusion to determine the number of travelers who want to visit at least one of Las Vegas (L), New York City (N) and Orlando (R). But this is not the question. What we can do, however, is look at the Venn diagram that represents this situation.

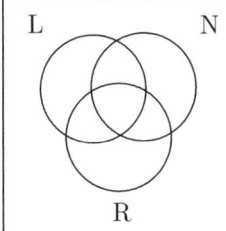 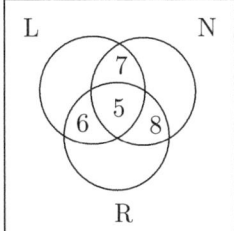

Figure 8.4: A Venn diagram for Example 8.18

Since 5 want to visit all three cities, there are 5 elements in the centermost region of the Venn diagram. We now work in reverse order from the information given. Since 13 want to visit New York City and Orlando but 5 want to visit all three cities, $13 - 5 = 8$ want to visit New York City and Orlando but not Las Vegas. We place 8 in the corresponding region. Continuing in this manner, we can fill in the number of elements in other regions. (In fact, we could find the numbers of elements in all eight regions of the Venn diagram if this were useful.) In any case, the answer to the question is $5 + 6 + 7 + 8 = 26$. We might also notice that the answer to this question is

$$|L \cap N| + |L \cap R| + |N \cap R| - 2|L \cap N \cap R| = 12 + 11 + 13 - 2 \cdot 5 = 26. \; \blacklozenge$$

Perhaps the connection between the cardinality of the union of a finite number of finite sets and the sum of the cardinalities of these sets is becoming clear. For example, in the case of four sets A, B, C and D, the cardinality $|A \cup B \cup C \cup D|$ equals the sum of the cardinalities of the four sets minus the sum of the cardinalities of the intersections of all pairs of the sets, plus the sum of cardinalities of the intersections of all triples of the sets minus the cardinality of the intersection of all four sets – thus the name: inclusion/exclusion (that is, we add, subtract, add, subtract and so on). In general, we have the following.

The Principle of Inclusion-Exclusion (for $n \geq 2$ sets) *If $A_1, A_2, \ldots, A_n$ are $n \geq 2$ finite sets, then*

$$|A_1 \cup A_2 \cup \cdots \cup A_n| = \sum_{1 \leq i \leq n} |A_i| - \sum_{1 \leq i < j \leq n} |A_i \cap A_j| + \sum_{1 \leq i < j < k \leq n} |A_i \cap A_j \cap A_k| - \cdots$$
$$+ \; (-1)^{n+1} |A_1 \cap A_2 \cap \cdots \cap A_n|$$

Exercises for Section 8.2

1. (a) How many different 8-bit strings begin with 010 or begin with 11?

    (b) How many different 8-bit strings begin with 010 or end with 11?

2. (a) How many different 9-bit strings begin with 101 or begin with 1010?

    (b) How many different 9-bit strings begin with 101 or end with 1010?

3. (a) How many different 10-bit strings begin with 1010 or begin with 0111?

    (b) How many different 10-bit strings begin with 1010 or end with 0111?

    (c) How many different 10-bit strings begin with 1010 and end with 0111?

4. Each entry of a string is an element of the set $S = \{0, 1, 2\}$. How many such strings of length 6 are there that begin with 022 or end with 01?

5. Let $n \in \mathbb{N}$ and let $S = \{1, 2, \ldots, n\}$. Let s_1 and s_2 be strings of length n, each of whose coordinates is an element of S. How many strings of length $3n$ are there beginning with s_1 and ending with s_2?

6. How many different 8-bit strings begin with 10, end with 01 or have 00 as its two middle bits?

7. How many different 8-bit strings have 00 as consecutive bits as either (a) its first and second bits, (b) its third and fourth bits, (c) its fifth and sixth bits or (d) its seventh and eighth bits?

8. The president of a company gave two talks, one on Monday and one on Tuesday. A total of 405 employees attended the Monday talk, 275 attended the Tuesday talk and 55 went to both. How many employees attended at least one of the talks?

9. There are 64 students who take at least one class during the morning, afternoon or evening. How many of these students take at least one morning class, at least one afternoon class *and* at least one evening class if

    29 students take morning classes,

    38 students take afternoon classes,

    35 students take evening classes,

    19 students take morning and afternoon classes,

    8 students take morning and evening classes and

    19 students take afternoon and evening classes?

10. A total of 36 students plan to take at least one of the courses Discrete Mathematics, Algebra and Calculus during the coming semester. Of these 36 students, it is known that

    23 students plan to take Discrete Mathematics,

    19 students plan to take Algebra,

    18 students plan to take Calculus,

    7 students plan to take Discrete Mathematics and Algebra,

    9 students plan to take Discrete Mathematics and Calculus and

    11 students plan to take Algebra and Calculus.

    (a) How many students plan to take all three courses?

    (b) How many students plan to take exactly one of the courses?

    (c) How many students plan to take exactly two of the courses?

11. A total of 60 students studying in a mathematics library are interviewed as to what courses they are taking this semester. Here are the responses:

    26 students are taking mathematics.

    35 students are taking computer science.

    28 students are taking statistics.

    15 students are taking mathematics and computer science.

    13 students are taking mathematics and statistics.

    18 students are taking computer science and statistics.

    7 students are taking all three.

    (a) How many students are taking none of these courses?
    (b) How many students are taking exactly one of the courses?
    (c) How many students are taking exactly two of the courses?

12. A total of 70 students who go to football, basketball or hockey games on a regular basis are surveyed as to which of these three events they attend. They responded:

    38 students go to football games.

    38 students go to basketball games.

    35 students go to hockey games.

    17 students go to both football and basketball games.

    15 students go to both football and hockey games.

    16 students go to both basketball and hockey games.

    How many go to all three?

13. As in Example 8.18, let L, N and R be the sets of all people who want to visit Las Vegas, New York City and Orlando, respectively. Using the numbers from this example, determine the number of travelers who want to visit *exactly two* of Las Vegas, New York City and Orlando.

14. From a group of students surveyed, it is determined that 28 students like Chinese food, 23 students like Italian food and 35 students like Mexican food. A total of 15 students like Chinese and Italian food, 20 students like Chinese and Mexican food and 12 students like Italian and Mexican food. A total of 9 students like all 3.

    (a) How many students like at least one of Chinese, Italian and Mexican food?
    (b) How many students like *exactly two* of Chinese, Italian and Mexican food?

15. A total of 100 clients are surveyed as to what kind of computer system is available to them. It is learned that 66 have a Windows-based system available to them, 35 have a Unix system available to them, 28 have a Macintosh system available to them, 20 have access to both Windows and Unix systems, 10 have access to both Unix and Macintosh systems and 25 have access to both Windows and Macintosh systems. How many have access to all three of these systems?

16. Each of the four sets A_1, A_2, A_3 and A_4 contains four elements. The intersection of every i of these four sets ($2 \leq i \leq 4$) consists of $5 - i$ elements. What is $|A_1 \cup A_2 \cup A_3 \cup A_4|$?

17. Give an example of a problem whose solution uses the Principle of Inclusion-Exclusion and whose answer is $3^7 - 3^4 + 1 \ (= 3 \cdot 3^6 - 3 \cdot 3^3 + 1)$.

8.3   The Pigeonhole Principle

It is probably only common sense that if in a particular Olympic Games some athlete won four Olympic medals (gold, silver or bronze), then at least two of these medals must be of the same kind.

In any collection of five integers, at least three are even or at least three are odd. In any set of 12 integers, at least 6 of these 12 integers are even or at least 6 are odd. In fact, in any set of 12 integers, at least 6 of these integers are even or at least 7 of these integers are odd. (If at least 6 of these integers are not even, then at most 5 are even, implying that at least 7 are odd.) These observations can be said more mathematically. In any finite set of integers, at least half of the integers are even or at least half are odd. Recall, for a real number x, that the ceiling $\lceil x \rceil$ of x is the smallest integer greater than or equal to x. Therefore, the observation we made above can be stated as follows: For any set of $n \geq 1$ integers, at least $\lceil n/2 \rceil$ of these integers are even or at least $\lceil n/2 \rceil$ are odd.

Let's look at another example that is similar to those we have considered.

Example 8.19

A student has been collecting stamps for a number of years from England, France and Germany. Regardless of how many stamps she has, at least one-third are from the same country. That is, if she has n stamps, then at least $\lceil n/3 \rceil$ stamps are from England, at least $\lceil n/3 \rceil$ stamps are from France or at least $\lceil n/3 \rceil$ stamps are from Germany.   ♦

All of these examples are special cases of what is commonly called the Pigeonhole Principle, one version of which is stated next.

The Pigeonhole Principle   *If a set S with n elements is divided into k pairwise disjoint subsets S_1, S_2, ..., S_k, then at least one of these subsets has at least $\lceil n/k \rceil$ elements.*

Before we give a proof of this principle, we note that $0 \leq \lceil x \rceil - x < 1$ and so

$$x \leq \lceil x \rceil < x + 1 \tag{8.2}$$

for every real number x. (See Exercise 22 in Section 4.4.)

Proof of the Pigeonhole Principle.   Assume, to the contrary, that none of the sets S_1, S_2, ..., S_k has at least $\lceil n/k \rceil$ elements. Since $\lceil n/k \rceil$ is an integer, every one of the sets S_1, S_2, ..., S_k has at most $\lceil n/k \rceil - 1$ elements. By (8.2), $\lceil n/k \rceil - 1 < n/k$, that is, each of the sets S_1, S_2, ..., S_k has less than n/k elements. Since S_1, S_2, ..., S_k are pairwise disjoint and $S = S_1 \cup S_2 \cup \cdots \cup S_k$, it follows that

$$n = |S| = |S_1 \cup S_2 \cup \cdots \cup S_k| = |S_1| + |S_2| + \cdots + |S_k| < k(n/k) = n,$$

and so $n < n$. This is clearly impossible.   ∎

In its simplest form, the Pigeonhole Principle states that if n pigeons are placed into m pigeonholes, where $m < n$, then at least one pigeonhole must contain at least two pigeons. Many decades ago, long before personal computers existed, some desks were built with small compartments, also called pigeonholes, which were used for filing papers, receipts, bills and the like.

Example 8.20

The desk shown in Figure 8.5 has 8 pigeonholes. By the Pigeonhole Principle, if 100 papers were to be filed in these pigeonholes, then at least one pigeonhole must contain at least $\lceil 100/8 \rceil = 13$ papers.   ♦

The first statement of this principle is believed to have been made in 1834 by Lejeune Dirichlet under the name Schubfachprinzip (drawer principle). Indeed, many mathematicians referred to this principle as Dirichlet's principle.

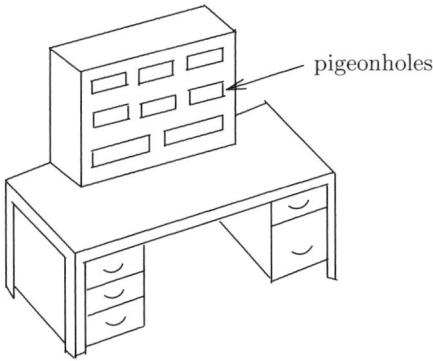

pigeonholes

Figure 8.5: A desk with 8 pigeonholes

Example 8.21

A teacher of a discrete mathematics course has 28 students.

(a) The teacher must assign each student one of the letter grades A, B, C, D, F. What is the largest number of students who *must* be assigned the same grade?

(b) It is known that the ages of the students in class range from 16 to 43. What is the maximum number of students in the class who *must* be the same age?

(c) In order to qualify for this discrete mathematics class, each student must have passed one of three prerequisite courses. What is the maximum number of students in the class who *must* have passed the same prerequisite course?

Solution.

(a) By the Pigeonhole Principle, at least $\lceil 28/5 \rceil = 6$ students must be assigned the same grade. Since there is no guarantee that 7 or more students will receive the same grade, it follows that 6 is the maximum.

(b) There are 28 integers in the range $16, 17, \ldots, 43$ (namely $43 - 16 + 1 = 28$). Therefore, the maximum number of students who must be the same age is $\lceil 28/28 \rceil = 1$. (Note: Of course, this can only occur if there is exactly one student of each of the ages $16, 17, \ldots, 43$ – probably not too likely.)

(c) The maximum number of students who must have passed the same prerequisite course is $\lceil 28/3 \rceil = 10$. (Note: Again, the key word here is *must*. That is, there *must* be 10 students who passed the same prerequisite course, but it is not necessary that there are 11 students who passed the same prerequisite course.) ◆

One question related to the Pigeonhole Principle is the following: For given positive integers r and k, what is the minimum cardinality N of a set S such that if S is divided into k subsets, then at least one of these subsets has at least r elements? According to the Pigeonhole Principle, $\lceil N/k \rceil = r$. Since $\lceil x \rceil < x + 1$ for every real number x, it follows that

$$r = \left\lceil \frac{N}{k} \right\rceil < \frac{N}{k} + 1$$

and so $\frac{N}{k} > r - 1$ or $N > k(r-1)$. Since N and $k(r-1)$ are integers, this implies that $N \geq k(r-1)+1$. Since N is the minimum integer with this property, $N = k(r-1)+1$. Therefore, we have the following:

Let r and k be positive integers. If N is the minimum cardinality of a set S such that however S is divided into k subsets, at least one of these subsets must contain at least r elements, then

$$N = k(r-1) + 1. \tag{8.3}$$

Example 8.22

(a) How many people are needed to be sure that at least two of them have a birthday during the same month?

(b) How many people are needed to be sure that at least three of them have a birthday during the same month?

Solution.

(a) Let N be the minimum number of people needed to guarantee that at least two of them have a birthday during the same month. Since there are 12 months, it follows that $\lceil N/12 \rceil = 2$. The smallest such integer N is 13 and so $N = 13$. Notice also that $N = 1 \cdot 12 + 1 = 13$ by (8.3).

(b) Let N be the minimum number of people needed to guarantee that at least three of them have a birthday during the same month. Thus $\lceil N/12 \rceil = 3$. The smallest such integer N is 25 and so $N = 25$. Notice also that $N = 2 \cdot 12 + 1 = 25$ by (8.3).   ♦

A problem related to those given in Example 8.22 will be discussed in Section 10.2. We now consider a more general example.

Example 8.23

An archaeologist collects ancient coins. He keeps a number of gold, silver and bronze coins in a desk drawer in his office. In fact, this drawer contains 15 gold coins, 20 silver coins and 12 bronze coins. He randomly takes 20 coins from the drawer to show his colleague. Show that he must have selected at least 8 gold coins, at least 9 silver coins or at least 5 bronze coins.

Solution. If the archaeologist did not select at least 8 gold coins, at least 9 silver coins or at least 5 bronze coins, then he would have selected at most 7 gold coins, at most 8 silver coins and at most 4 bronze coins. But this would mean that he selected at most $7 + 8 + 4 = 19$ coins, which is impossible.   ♦

Example 8.23 illustrates a more general version of the Pigeonhole Principle.

The (General) Pigeonhole Principle *A set S with n elements is partitioned into k pairwise disjoint subsets $S_1, S_2, \ldots, S_k$, where $|S_i| \geq n_i$ for a positive integer n_i for $i = 1, 2, \ldots, k$. Then each subset of S with at least*

$$1 + \sum_{i=1}^{k} (n_i - 1)$$

elements contains at least n_i elements of S_i for some integer i with $1 \leq i \leq k$.

Proof. We use a proof by contradiction. Suppose that there is some subset A of S such that

$$|A| \geq 1 + \sum_{i=1}^{k} (n_i - 1)$$

but A does not contain at least n_i elements of S_i for any i with $1 \leq i \leq k$. Necessarily then, the set A contains at most $n_i - 1$ elements of S_i for every integer i. Then

$$|A| \leq \sum_{i=1}^{k} (n_i - 1) \leq |A| - 1,$$

which is a contradiction. ∎

In the (General) Pigeonhole Principle, if N is the minimum cardinality of a subset A of the set S such that A contains at least n_i elements of S_i for some integer i with $1 \le i \le k$, then

$$N = 1 + \sum_{i=1}^{k}(n_i - 1).$$

Example 8.24

At a certain university, discrete mathematics is often taught as a large lecture class. Let's suppose that, on the average, 10% of the students receive A's, 25% receive B's, 40% receive C's, 20% receive D's and 5% receive E's. How many students would have to be in the class to be certain that the professor assigns either 10 A's, 25 B's, 40 C's, 20 D's or 5 E's?

Solution. According to the (General) Pigeonhole Principle, there would have to be at least

$$1 + (10 - 1) + (25 - 1) + (40 - 1) + (20 - 1) + (5 - 1) = 96$$

students in the class. ♦

Example 8.25

How many people are needed to be present at a party to be sure that at least 3 of them have a birthday during June, at least 3 have a birthday during July, at least 3 have a birthday during August or at least 4 of them have a birthday during the same month for one of the other months?

Solution. By the (General) Pigeonhole Principle, the smallest number of people needed to be present at the party is $1 + 2 \cdot 3 + 3 \cdot 9 = 34$. ♦

The Pigeonhole Principle can also be used to solve other types of problems.

Example 8.26

For a positive integer n, let $S = \{1, 2, \ldots, 2n\}$.

(a) Show that S contains a subset S' with $|S'| = n$ such that for every two distinct elements a and b in S', $a \nmid b$ and $b \nmid a$.

(b) Show that for each subset A of S, where $|A| = n + 1$, there exist distinct elements $a, b \in A$ such that $a \mid b$.

Solution.

(a) Let $S' = \{n+1, n+2, \ldots, 2n\}$. Then $|S'| = n$. Let a and b be two distinct elements of S', where, say, $n + 1 \le a < b \le 2n$. Then $b \nmid a$. Observe that $a \ge n + 1$ and so $b \le 2n < 2(n + 1) \le 2a$. Since $b < 2a$, it follows that $a \nmid b$ as well.

(b) **Proof.** Let $A \subseteq S$ with $|A| = n + 1$. For each $x \in A$, we write $x = x' \cdot 2^k$, where x' is an odd integer (with $x' \in S$) and k is a nonnegative integer. Hence x' is the largest odd integer divisor of x. Since there are exactly n odd integers in S (namely $1, 3, \ldots, 2n - 1$) and $|A| = n + 1$, it follows by the Pigeonhole Principle that there are two distinct integers a and b in A (with $a < b$ say) having the same largest odd integer divisor c. Thus

$$a = c \cdot 2^i \text{ and } b = c \cdot 2^j,$$

where $0 \le i < j$ and so

$$b = c \cdot 2^j = 2^{j-i}(c \cdot 2^i) = 2^{j-i}a.$$

Since 2^{j-i} is an integer, it follows that $a \mid b$.   ∎

We give one additional application of the Pigeonhole Principle.

Example 8.27

A graduate student serves as a mathematics tutor during a semester. Each day, including Saturday and Sunday, she has set aside some 1-hour periods for tutoring. She charges \$10 per hour. During the last 20 days of the semester, she tutored a total of 28 hours and at least one hour each day. Show that there was a certain period of consecutive days when she earned exactly \$100.

Solution. Let d_1 be the number of hours she tutored on the first day of the 20 days, d_2 the number of hours she tutored on the first two days and more generally, for $1 \le i \le 20$, let d_i denotes the number of hours she tutored during the first i days. Since she tutored at least one hour per day,

$$1 \le d_1 < d_2 < \cdots < d_{20} = 28,$$

that is, the 20 numbers $d_1, d_2, \ldots, d_{20}$ are distinct. Notice also that

$$d_1 + 10 < d_2 + 10 < \ldots < d_{20} + 10 = 38.$$

That is, the 20 numbers $d_1 + 10$, $d_2 + 10$, ..., $d_{20} + 10$ are also distinct. Since the 40 positive integers

$$d_1, d_2, \ldots, d_{20}, \ d_1 + 10, d_2 + 10, \ldots, d_{20} + 10$$

are all at most 38, it follows by the Pigeonhole Principle that at least two of these 40 numbers are equal. Since no two of the integers $d_1, d_2, \ldots, d_{20}$ are equal and no two of the integers $d_1 + 10, d_2 + 10, \ldots, d_{20} + 10$ are equal, some integer d_j must equal some integer $d_k + 10$, where $1 \le k < j \le 20$. That is, $d_j = d_k + 10$ and so $d_j - d_k = 10$. This says that during the $j - k$ days following day k, she tutored for exactly 10 hours and therefore earned exactly \$100.   ◆

Exercises for Section 8.3

1. Let n be a positive integer. Show that $n = \lceil n/2 \rceil + \lfloor n/2 \rfloor$.

2. Show, for any set of $n \ge 2$ integers, that at least $\lceil n/2 \rceil$ of these integers are even or at least $\lfloor n/2 \rfloor + 1$ of these integers are odd.

3. According to the Pigeonhole Principle, if 420 notebooks are distributed to 40 students, there must be at least one student who receives at least n notebooks for some positive integer n, but there is no guarantee that any student will receive $n + 1$ notebooks. What is n?

4. A bowl contains a large number of red and blue marbles. If 10 marbles are selected at random, show that at least 5 of the marbles are red or at least 6 are blue.

5. In a set of 27 English words, what is the largest number of words that must begin with the same letter?

6. In a discrete mathematics class of 32 students, what is the largest number of students who must receive the same grade if there are only 5 possible grades?

7. A bowl contains 9 marbles, each of which is red, blue or green. Show that at least 3 of these marbles are red, 3 are blue or 5 are green.

8. In a set of 7 integers, what is the largest number of integers that must have the same remainder when divided by 5?

9. There are 115 books to be placed on 9 bookshelves. What is the smallest possible number of books that can be placed on the shelf with the largest number of books?

10. At a certain high school, most students go to one of 7 universities. If 100 students go to these 7 universities some year, what is the smallest possible number of students who could attend the university where most students go?

11. A conference room contains 8 tables and 107 chairs. What is the smallest possible number of chairs at the table having the most chairs?

12. How many books must be chosen from 24 mathematics books, 25 computer science books, 21 literature books and 10 economics books in order to be certain there are at least 12 books on the same subject?

13. A bowl contains a large number of red, blue and yellow marbles. What is the fewest number of marbles that need to be randomly selected from the bowl to be guaranteed that 9 marbles of the same color are chosen?

14. A bag contains many marbles, each of which is colored red, white, blue, green or yellow. A number of marbles are selected from the bag without checking the colors first. What is the fewest number of marbles that need to be selected randomly to guarantee at least six marbles of the same color are selected?

15. A coin is flipped three times and the outcomes are recorded. For example, if heads comes up the first two times and tails the third time, then we write HHT. How many times must we flip a coin three times to be guaranteed that there are two identical outcomes?

16. A total of 100 students are taking a history exam. Each student is required to write an essay on one of six possible topics. What is the fewest possible number of students who could have chosen the topic selected by most students?

17. Two teams play baseball games until one team wins four games. What is the fewest number of games they must play to be guaranteed that one team will win four games?

18. A man runs a grocery store in a small town. He keeps a number of $1 bills, $5 bills, $10 bills and $20 bills in the cash register to make change. A good friend of the man often comes to the grocery store, asking the man to change a $100 bill into smaller bills all of the same denomination. The store owner never knows beforehand which denomination his friend will request. How many bills would the man have to keep in the cash register at all times to be certain that he can satisfy his friend's request?

19. Each piece of fruit in a fruit basket is either an apple, a banana, an orange, a pear or a peach. How many pieces of fruit must be in the basket to guarantee that there is at least one apple, at least two bananas, at least three oranges, at least four pears or at least five peaches?

20. How many people must be present to guarantee that

    (a) at least two have the same birthday?

    (b) at least two of their birthdays are in the same month?

    (c) at least three of their birthdays are in one of the months January, February, March, April or at least four of their birthdays are in one of the remaining months?

21. A young lady has eleven pairs of shoes that she wears on special occasions. Each pair is either black, brown or white. Show that she either has at least four pairs of black shoes, at least four pairs of brown shoes or at least five pairs of white shoes.

22. The solution of Example 8.26 actually shows, for the set $S = \{1, 2, \ldots, 2n\}$, where $n \in \mathbb{N}$, that (1) there is a subset $S' \subseteq S$ with $|S'| = n$ such that for every two distinct elements $a, b \in S'$ both $2a \nmid b$ and $2b \nmid a$ and (2) for each subset $A \subseteq S$ with $|A| = n + 1$, there exist distinct elements $a, b \in A$ such that $2a \mid b$. For a positive integer n, let $T = \{1, 2, \ldots, 3n\}$.

   (a) Show that T contains a subset T' with $|T'| = 2n$ such that for every two distinct elements a and b in T', $3a \nmid b$ and $3b \nmid a$.

   (b) Show that for each subset A of T, where $|A| = 2n + 1$, there exist distinct elements $a, b \in A$ such that $3a \mid b$.

23. Give an example of a problem whose solution uses the Pigeonhole Principle and whose answer is $1 + 6 \cdot 6^3 = 1 + 6^4 = 1297$.

24. Mary decides to save money this year so that she will have spending money for a vacation trip next year. During the 365 days she saves at least one dollar bill each day and a total of 540 dollar bills. Show that there is a period of consecutive days during the year when she saved exactly 175 dollar bills.

8.4 Permutations and Combinations

In the previous three sections, we introduced four counting principles, namely the Multiplication Principle, the Addition Principle, the Principle of Inclusion-Exclusion and the Pigeonhole Principle. These were used to answer several questions concerning the enumeration of objects possessing some specified properties. We now see how these principles can be used to solve more varied counting problems. Most of these problems deal with two fundamental concepts in discrete mathematics: permutations and combinations.

Many of the counting problems that we will encounter deal with the number of certain types of subsets of a given set S or the number of ways to order some or all of the elements of S. For example, suppose that there is a faculty position available in a department in a university and that 20 people apply for this position. The "Search Committee" for this position is required to select eight of the 20 applicants for what is called their "short list," from which a person will be selected for the position. In how many ways can eight of the 20 applicants be selected? The Dean then informs the Committee that they are permitted to invite five of these eight applicants for on-campus interviews. In how many ways can this be done? After the five on-campus interviews take place, the Dean then asks the Department to submit its preferred ranked list of three of these five applicants. One way for the Committee to determine the preferred ranking by the Department is to list all possible orderings of three of the five applicants and have each Department member vote for which of these he or she prefers. In how many ways can three of these five applicants be ordered? In order to describe how to answer questions such as these (see Exercise 23), we begin with a fundamental concept.

Permutations

Definition 8.28 *A **permutation** of a nonempty set S is an arrangement or ordered list of the elements of S.*

When dealing with a permutation of a set S, we sometimes refer to this as well as a permutation of *the elements of* S. The first appearance in print of the word *permutation* with its current mathematical meaning is due to Jacques Bernoulli, a member of a famous mathematical family.

Example 8.29

Consider the set $S = \{1, 2, 3, 4\}$. One permutation of S is 2, 1, 3, 4 (or 2134). That is, when expressing permutations, at times we separate the elements in a permutation by commas while at other times it may be more convenient not to do this. In fact, all of the permutations of the set S are listed below:

1234	2134	3124	4123
1243	2143	3142	4132
1324	2314	3214	4213
1342	2341	3241	4231
1423	2413	3412	4312
1432	2431	3421	4321

Therefore, there are 24 permutations of the integers 1, 2, 3, 4 (if you agree that we haven't missed or duplicated any permutation). We now explain why there are 24 permutations without actually listing them all. It is important that we learn how to compute the number of permutations of a set without having to list them since it is clearly not practical to list all permutations of a set having a large number of elements. To make an ordered list of the integers 1, 2, 3, 4, we need to fill in four positions, say

_____   _____   _____   _____

with distinct elements of $\{1, 2, 3, 4\}$. Let's see how many ways there are of doing this. There are four possible numbers for the first position. Once we've decided on the number for the first position, we have three choices remaining for the number in the second position. Continuing this, we see that there are two possible numbers for the third position and only one number for the last position. Therefore, by the Multiplication Principle, the number of permutations of the integers $1, 2, 3, 4$ is $4 \cdot 3 \cdot 2 \cdot 1 = 4! = 24$. A tree diagram illustrating how the permutations of $S = \{1, 2, 3, 4\}$ can be obtained is shown in Figure 8.6. ◆

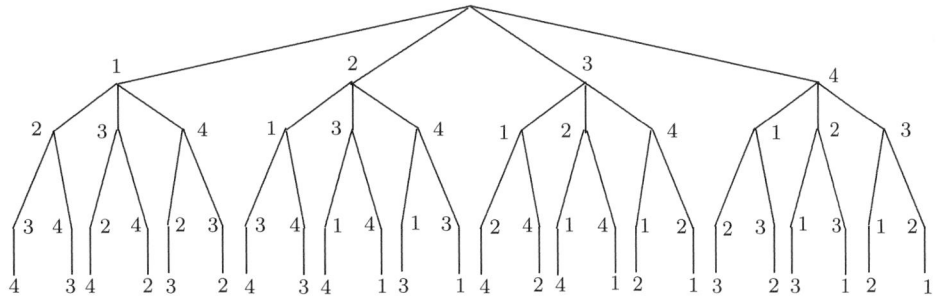

Figure 8.6: A tree diagram giving the 24
permutations of the set $\{1, 2, 3, 4\}$

More generally, for each positive integer n, the number of permutations of the integers $1, 2, \ldots, n$ (or of any n objects) is given by

$$n(n-1)(n-2) \cdots 3 \cdot 2 \cdot 1 = n! \ (n \text{ factorial}).$$

Example 8.30

The number of permutations of the integers $1, 2, 3, 4, 5$ is $5 \cdot 4 \cdot 3 \cdot 2 \cdot 1 = 5! = 120$. ◆

You might recall that we have encountered the term "permutation" earlier in another context in Section 4.2 and may wonder why we are using the same term to mean something else. Let us first remind ourselves where we saw and how we used this term before. For a (nonempty) set A, a function f from A to A that is bijective (that is, f is both one-to-one and onto) is called a **permutation** of A. Suppose that $A = \{1, 2, 3, 4\}$. An example of a permutation of A is the function

$$g = \{(1,3),(2,2),(3,4),(4,1)\}.$$

In particular, for this function the image of 1 is 3, the image of 2 is 2, the image of 3 is 4 and the image of 4 is 1. Consequently, we know how the function (permutation) g is defined once we know the images of the elements of A, that is, once we know that

$$g(1) = 3, \ g(2) = 2, \ g(3) = 4 \text{ and } g(4) = 1.$$

The function g is therefore completely determined by the sequence $3, 2, 4, 1$ of images of $1, 2, 3, 4$, respectively. That is, both interpretations of permutation have the same meaning.

As we saw in our opening discussion concerning ordering three applicants of the five applicants interviewed, there are occasions when we are not interested in ordered lists of *all* elements of a set S but only in ordered lists of a certain number of elements of S.

Example 8.31

Suppose that we wish to know the number of ordered lists of two elements of the set $S = \{1, 2, 3, 4\}$. There are 12 such ordered lists, namely

$$\begin{array}{cccc} 12 & 21 & 31 & 41 \\ 13 & 23 & 32 & 42 \\ 14 & 24 & 34 & 43 \end{array}$$

In this example, we were asked only for the *number* of such lists, not to display these lists. This number can be determined by another application of the Multiplication Principle. We are looking at sequences ___ ___ of two distinct numbers, chosen from 1, 2, 3, 4. For the first number in the sequence, there are four possibilities. Once this choice has been made, there are three possible numbers for the second number in the sequence. Hence the total number of sequences of two elements from a 4-element set is $4 \cdot 3 = 12$.   ♦

More generally, if we are seeking the number of ordered lists of r elements of an n-element set (where then $1 \leq r \leq n$), we are considering sequences of r elements:

$$\underline{\hspace{2cm}} \ \underline{\hspace{2cm}} \ \underline{\hspace{2cm}} \ \cdots \ \underline{\hspace{2cm}} \ \underline{\hspace{2cm}}$$
$$\text{position 1} \quad \text{position 2} \quad \text{position 3} \quad \cdots \quad \text{position } r-1 \quad \text{position } r$$

There are n choices for the element in the first position of the sequence. Once this element has been chosen, there are $n-1$ possibilities for the element in the second position and $n-2$ possibilities for the element in the third position. Continuing in this manner, we see that the number of possibilities for an element in position $r-1$ is $n - (r-2) = n - r + 2$ and finally, the number of possibilities for an element to be placed in position r is $n - (r-1) = n - r + 1$. Again, by the Multiplication Principle, the total number of sequences of length r whose elements are chosen from an n-element set is

$$n(n-1)(n-2)\cdots(n-r+2)(n-r+1). \tag{8.4}$$

This number can be expressed more simply and probably be remembered more easily by multiplying the expression in (8.4) by $(n-r)!/(n-r)!$, that is,

$$\begin{aligned} & n(n-1)(n-2)\cdots(n-r+2)(n-r+1) \\ = \ & n(n-1)(n-2)\cdots(n-r+2)(n-r+1)\frac{(n-r)!}{(n-r)!} \\ = \ & \frac{n!}{(n-r)!}. \end{aligned}$$

You may recognize this number, which is the number of ordered lists of r elements of an n-element set, as appearing in Theorem 8.7 and giving the number of different one-to-one functions from an r-element set to an n-element set.

Definition 8.32 *An ordered list of r elements of an n-element set S is called an r-**permutation** of the elements of S. The number of r-permutations of an n-element set is denoted by $P(n, r)$.*

Let's summarize what we've just discovered.

Theorem 8.33 *The number of r-permutations of an n-element set (where $1 \leq r \leq n$) is*

$$P(n, r) = n(n - 1)(n - 2) \cdots (n - r + 2)(n - r + 1) = \frac{n!}{(n - r)!}.$$

For the case when $r = n$, we are interested in the n-permutations or more simply the permutations, of an n-element set. The number of these permutations is

$$P(n, n) = \frac{n!}{(n - n)!} = \frac{n!}{0!}.$$

Since $0!$ is defined as 1, we have $P(n, n) = n!$, which is the desired answer. Consequently,

$$P(n, n) = \frac{n!}{0!} = n!.$$

Example 8.34

How many different arrangements of the five vowels a, e, i, o, u are there?

Solution. Since we are seeking the number of permutations of a 5-element set, this number is given as $P(5, 5) = 5! = 120$. $\blacklozenge$

Example 8.35

Let $S = \{a, b, c, d, e, f, g\}$.

(a) Give two examples of permutations of S.

(b) How many permutations of S are there?

(c) Give two examples of 4-permutations of S.

(d) How many 4-permutations of S are there?

Solution.

(a) Two permutations of S are $a\,b\,c\,d\,e\,f\,g$ and $g\,f\,e\,d\,c\,b\,a$.

(b) $P(7, 7) = 7! = 7 \cdot 6 \cdot 5 \cdot 4 \cdot 3 \cdot 2 \cdot 1 = 5040$.

(c) Two 4-permutations of S are $a\,b\,c\,d$ and $g\,e\,c\,a$.

(d) $P(7, 4) = \dfrac{7!}{(7 - 4)!} = \dfrac{7!}{3!} = \dfrac{7 \cdot 6 \cdot 5 \cdot 4 \cdot 3 \cdot 2 \cdot 1}{3 \cdot 2 \cdot 1} = 7 \cdot 6 \cdot 5 \cdot 4 = 42 \cdot 20 = 840$. $\blacklozenge$

Example 8.36

Seven students, namely 4 men and 3 women, are to present their solutions to seven different problems in class.

(a) In how many different orderings can this be done?

(b) In how many orderings can these presentations be made if the presentations are to alternate between men and women?

(c) In how many orderings can these presentations be made if the women are to present their problems consecutively and the men are to present their problems consecutively?

Solution.

(a) $P(7,7) = 7! = 5040$.

(b) Necessarily, the first and the last presentations must be made by men and appear as:

$$\text{man, woman, man, woman, man, woman, man.}$$

By the Multiplication Principle, the number of different orderings is $4 \cdot 3 \cdot 3 \cdot 2 \cdot 2 \cdot 1 \cdot 1 = 144$.

(c) If the 4 men present their solutions first, then, according to the Multiplication Principle, the number of orderings is $(4!)(3!) = 144$. If the 3 women present their solutions first, then, again by the Multiplication Principle, the number of orderings is $(3!)(4!) = 144$. By the Addition Principle, the total number of orderings satisfying the conditions is $144 + 144 = 288$.   ◆

Example 8.37

Seven graduating seniors majoring in computer science have applied for three positions with a well-known software company. One position is a permanent position with the company, one is a 2-year position and the other is a visiting research position. Assuming that three of the seven applicants are hired for these positions, how many outcomes are possible?

Solution. We can think of filling in the three slots

| _____ | _____ | _____ |
| permanent | 2-year | visiting |

There are 7 choices for the first slot (position), 6 choices for the second position once the first position has been filled and then 5 choices for the third position once the first two positions have been filled. By the Multiplication Principle, the number of possible outcomes is $7 \cdot 6 \cdot 5 = 210$. Notice also that the answer to this question is the number of 3-permutations of the set of 7 applicants. This number is therefore

$$P(7,3) = \frac{7!}{(7-3)!} = \frac{7!}{4!} = \frac{7 \cdot 6 \cdot 5 \cdot 4!}{4!} = 7 \cdot 6 \cdot 5 = 210.$$

Consequently, these two methods give the same answer (as they should).   ◆

Combinations

Whenever we talk about permutations or r-permutations of a set S, we are concerned with sequences, arrangements or ordered lists of r elements of S. That is, *order* is important – the order in which the elements are written matters. These are not the only kinds of questions in which we are interested, however.

Example 8.38

How many subsets of the set $A = \{1, 2, 3, 4\}$ contain exactly two elements?

Solution. The 2-element subsets of A are

$$\{1, 2\}, \{1, 3\}, \{1, 4\}, \{2, 3\}, \{2, 4\}, \{3, 4\}$$

and so the answer to this question is 6.   ◆

Example 8.39

How many 3-element subsets of the set $B = \{1, 2, 3, 4, 5\}$ are there?

Solution. Since the 3-element subsets of B are

$$\{1,2,3\}, \quad \{1,2,4\}, \quad \{1,2,5\}, \quad \{1,3,4\}, \quad \{1,3,5\},$$
$$\{1,4,5\}, \quad \{2,3,4\}, \quad \{2,3,5\}, \quad \{2,4,5\}, \quad \{3,4,5\},$$

the answer to this question is 10. ♦

While order is important for questions pertaining to permutations order is not relevant for the questions asked in Examples 8.38 and 8.39, as the order in which elements are written in subsets does not matter. Furthermore, the questions in Examples 8.38 and 8.39 asked for the *number* of subsets of a certain type. We were not asked to list all these subsets. Yet, this is the way we answered these questions. Since this will not always be practical and it is far too easy to make errors listing subsets, we need another approach to answer these questions.

For a set S with n elements, we denote the number of subsets of S with exactly r elements, where $0 \leq r \leq n$, by $C(n, r)$. From Examples 8.38 and 8.39, we see that $C(4, 2) = 6$ and $C(5, 3) = 10$.

Definition 8.40 *Let S be an n-element set. An r-element subset of S, where $0 \leq r \leq n$, is called an r-**combination** of S. Therefore, the number of r-combinations of an n-element set is $C(n, r)$. An r-combination is also referred to as an unordered list of r-elements or as an r-**selection**.*

The first known use of the word *combination* with the meaning we just described evidently occurred in a letter dated July 29, 1654 written by Blaise Pascal to another famous French mathematician, Pierre de Fermat.

Let $S = \{a_1, a_2, \ldots, a_n\}$ be an n-element set and let T be one of the $C(n, r)$ r-element subsets (r-combinations) of S, where $1 \leq r \leq n$. One possible choice for T is $\{a_1, a_2, \ldots, a_r\}$. The number of ways to order the r elements of T is $r!$. Consequently, the number of ways of ordering *all* r-element subsets of S is $r!C(n, r)$; however, this is the number of r-permutations of S, which we know is $P(n, r)$. Therefore,

$$r!C(n, r) = P(n, r) = \frac{n!}{(n-r)!}.$$

Solving for $C(n, r)$, we have

$$C(n, r) = \frac{n!}{r!(n-r)!}.$$

Notice that when $r = 0$ or $r = n$, we have

$$C(n, 0) = \frac{n!}{0!\, n!} = 1 \quad \text{and} \quad C(n, n) = \frac{n!}{n!\, 0!} = 1.$$

Since there is only one 0-element subset (namely the empty set) and only one n-element subset (namely the set S), this expression gives the expected answer when $r = 0$ and when $r = n$ as well.

In mathematics, a common symbol for the number of r-element subsets of an n-element set is $\binom{n}{r}$, which is read as "n **choose** r" to indicate that we are choosing or selecting r-element subsets of an n-element set. Consequently, $C(n, r)$ and $\binom{n}{r}$ represent the same number. Since both symbols are commonly used, we should be familiar with both and will use them interchangeably. Let's summarize what we've just discovered.

Theorem 8.41 *For integers r and n with $0 \leq r \leq n$, the number of r-element subsets of an n-element set (also called r-combinations or r-selections of an n-element set) is*

$$C(n, r) = \binom{n}{r} = \frac{n!}{r!(n-r)!}.$$

Thus,

$$C(4,2) = \binom{4}{2} = \frac{4!}{2!(4-2)!} = \frac{4!}{2!\,2!} = \frac{4 \cdot 3 \cdot 2 \cdot 1}{2 \cdot 1 \cdot 2 \cdot 1} = 6$$

and

$$C(5,3) = \binom{5}{3} = \frac{5!}{3!(5-3)!} = \frac{5!}{3!\,2!} = \frac{5 \cdot 4 \cdot 3 \cdot 2 \cdot 1}{3 \cdot 2 \cdot 1 \cdot 2 \cdot 1} = 10.$$

These two numbers agree, of course, with our answers to the questions in Examples 8.38 and 8.39.

Example 8.42

Determine the number of 4-element subsets of the set $A = \{1, 2, \ldots, 7\}$ and the number of 6-element subsets of the set $B = \{1, 2, \ldots, 8\}$.

Solution. The number of 4-element subsets of A is

$$C(7,4) = \binom{7}{4} = \frac{7!}{4!(7-4)!} = \frac{7 \cdot 6 \cdot 5 \cdot 4!}{4! \cdot 3 \cdot 2 \cdot 1} = 35,$$

while the number of 6-element subsets of B is

$$C(8,6) = \binom{8}{6} = \frac{8!}{6!\,2!} = \frac{8 \cdot 7 \cdot 6!}{6!\,2!} = 28. \qquad \blacklozenge$$

Example 8.43

A certain committee is required to meet 3 days during each February (not on weekends however). If the coming February does not occur during a leap year, how many different choices for meeting days are there?

Solution. Since the coming February does not occur during a leap year, there are exactly 4 weeks during the month, each week containing five weekdays. Therefore, there are exactly 20 weekdays. The total number of choices for meeting days (3-combinations) is

$$C(20,3) = \binom{20}{3} = \frac{20!}{3!\,17!} = \frac{20 \cdot 19 \cdot 18 \cdot 17!}{6 \cdot 17!} = 1140. \qquad \blacklozenge$$

Example 8.44

A student is trying to remember a classmate's 7-digit telephone number. He recalls that the first digit is 7 and then there are three 1s and three 3s in some order. How many telephone calls would he have to make to be certain that he finally calls his friend's telephone number?

Solution. Since he knows the first digit is 7, *only* the remaining six digits are in doubt. He knows that three of these six digits are 1 and the remaining three digits are 3. Once he has determined the locations of the 1s in the 7-digit telephone number, he knows the exact number. Since there are 6 possible locations for the three 1s, the maximum number of phone calls he would have to make is

$$C(6,3) = \binom{6}{3} = \frac{6!}{3!\,3!} = \frac{6 \cdot 5 \cdot 4 \cdot 3!}{6 \cdot 3!} = 20. \qquad \blacklozenge$$

There are many interesting properties concerning the numbers $\binom{n}{r}$. We now look at a few of these. First, observe that $\binom{4}{0} = 1$, $\binom{4}{1} = 4$, $\binom{4}{2} = 6$, $\binom{4}{3} = 4$, $\binom{4}{4} = 1$ and that

$$\binom{4}{0} + \binom{4}{1} + \binom{4}{2} + \binom{4}{3} + \binom{4}{4} = 1 + 4 + 6 + 4 + 1 = 16. \qquad (8.5)$$

Actually, equation (8.5) is not surprising, as we now see. We know that $\binom{4}{0}$ represents the number of 0-element subsets (the empty subset) of a 4-element set, $\binom{4}{1}$ is the number of 1-element subsets and so on. Therefore, $\binom{4}{0} + \binom{4}{1} + \binom{4}{2} + \binom{4}{3} + \binom{4}{4}$ is the number of *all* subsets of a 4-element set. However, this number is $2^4 = 16$ since, as we have seen in Sections 2.1 and 5.2, the number of subsets of an n-element set is 2^n. Therefore, more generally, we have the following.

Theorem 8.45 *For each integer $n \geq 0$,*

$$\binom{n}{0} + \binom{n}{1} + \binom{n}{2} + \cdots + \binom{n}{n-1} + \binom{n}{n} = 2^n.$$

As we have seen, for each integer $n \geq 0$, the number $\binom{n}{0} = 1$ since there is only one 0-element subset (the empty subset) of any set. Also, $\binom{n}{n} = 1$ since there is only one n-element subset of an n-element set, namely the set itself. Suppose that S is a set with $n \geq 1$ elements. The 1-element subsets of S are the sets $\{x\}$, where $x \in S$. Since there are n such subsets, $\binom{n}{1} = n$. Furthermore, $\binom{n}{n-1}$ also equals n for when we consider the subsets of S containing exactly $n-1$ elements, then we are leaving out exactly one element from the set S. Since there are $\binom{n}{1} = n$ choices for the element of S to leave out when constructing an $(n-1)$-element subset of S, it follows that $\binom{n}{n-1} = n$. More generally, we have the following result. We give two proofs of this result, the first of which is a "combinatorial" proof and the second an algebraic proof.

Theorem 8.46 *For every two integers r and n with $0 \leq r \leq n$,*

$$C(n, r) = C(n, n - r) \quad or \quad \binom{n}{r} = \binom{n}{n-r}.$$

Proof #1. Observe that $C(n, r)$ is the number of r-element subsets of an n-element set S. However, for each choice of r elements of S belonging to a subset, there are $n - r$ elements of S not belonging to the subset. Therefore, $C(n, r) = C(n, n - r)$. ∎

Proof #2. The number of subsets with $n - r$ elements in an n-element set is

$$
\begin{aligned}
C(n, n - r) &= \frac{n!}{(n-r)![n-(n-r)]!} = \frac{n!}{(n-r)!r!} \\
&= \frac{n!}{r!(n-r)!} = C(n, r),
\end{aligned}
$$

as desired. ∎

Example 8.47

How many subsets of $S = \{1, 2, \ldots, 8\}$ contain three or more elements?

Solution. To answer this question, we could compute the sum

$$\binom{8}{3} + \binom{8}{4} + \binom{8}{5} + \binom{8}{6} + \binom{8}{7} + \binom{8}{8},$$

which is $56 + 70 + 56 + 28 + 8 + 1 = 219$.

There is, however, a somewhat simpler way to answer this question. The total number of subsets of S is $2^8 = 256$. The number of subsets of S with 0, 1 and 2 elements is $\binom{8}{0} = 1$, $\binom{8}{1} = 8$ and $\binom{8}{2} = (8 \cdot 7)/2 = 28$, respectively. Therefore, the number of subsets of S with three or more elements is the total number of subsets of S except for those having 2 or fewer elements, that is,

$$2^8 - \binom{8}{0} - \binom{8}{1} - \binom{8}{2} = 256 - 1 - 8 - 28 = 219.$$

♦

In the previous example, we had the occasion to compute $\binom{8}{2}$, which is $(8 \cdot 7)/2 = 28$. Since we encounter the number $\binom{n}{2}$ so very often, it is good to know how to compute this number quickly. In particular,

$$C(n, 2) = \binom{n}{2} = \frac{n!}{2!(n-2)!} = \frac{n(n-1)(n-2)!}{2 \cdot (n-2)!} = \frac{n(n-1)}{2},$$

which gives us the following. For every integer $n \geq 2$,

$$C(n, 2) = \binom{n}{2} = \frac{n(n-1)}{2}. \tag{8.6}$$

Example 8.48

By (8.6) then,

$$C(5, 2) = \binom{5}{2} = \frac{5 \cdot 4}{2} = 10 \quad \text{and} \quad C(10, 2) = \binom{10}{2} = \frac{10 \cdot 9}{2} = 45.$$

By Theorem 8.46 and (8.6), we have

$$C(6, 4) = C(6, 2) = \binom{6}{2} = \frac{6 \cdot 5}{2} = 15 \quad \text{and} \quad C(12, 10) = C(12, 2) = \binom{12}{2} = \frac{12 \cdot 11}{2} = 66. \; \blacklozenge$$

Exercises for Section 8.4

1. Compute (a) $\frac{10!}{7! \, 3!}$.   (b) $P(8, 2)$.   (c) $\frac{6!}{0!}$.   (d) $\frac{P(7,3)}{P(7,4)}$.   (e) $\frac{n!}{(n-1)!}$.

2. Let $S = \{1, 2, 3, 4, 5, 6\}$.

   (a) How many 4-permutations of S are there?

   (b) Give an example of one of the 4-permutations in (a) that you are counting.

   (c) How many 6-permutations of S are there?

   (d) Give an example of one of the 6-permutations in (c) that you are counting.

3. Compute (a) $P(7, 2)$ and $C(7, 2)$. (b) $P(8, 3)$ and $C(8, 3)$. (c) $P(9, 4)$ and $C(9, 4)$. (d) $P(10, 10)$ and $C(10, 10)$.

4. There are 4 mathematics books, 3 computer science books and 2 engineering books to be placed on a book shelf.

   (a) In how many ways can this be done?

   (b) In how many ways can these books be placed on a book shelf if the books on the same subject must be grouped together?

5. How many different arrangements are there of the letters in the word "string"?

6. From a student organization with 10 members, a president, secretary and treasurer are to be chosen. In how many ways can this be done?

7. How many different ways are there of selecting 5 people from a group of 7 and seating them in a row of 5 chairs?

8. (a) In how many ways can 7 girls and 6 boys be seated in a row?

   (b) In how many ways can 7 girls and 6 boys be seated in a row so that no two people of the same sex sit next to each other?

9. In how many orders can 4 married couples be seated in a row of 8 chairs if every one must sit next to his or her spouse?

10. Compute (a) $C(n,0)$. (b) $C(n,1)$. (c) $C(n,n)$. (d) $C(n,n-1)$. (e) $C(n,n-2)$.

11. How many subsets of $\{a,b,c,d,e,f\}$ contain exactly three elements?

12. How many ways are there of selecting two students from a group of nine students to serve on a committee?

13. How many different 3-member committees can be formed from a group of 10 people?

14. If a student opens a checking account at a local bank, then he can receive any two items as gifts from a collection of 12 items. How many choices for two gifts does he have?

15. How many subsets of $\{1,2,3,4,5,6\}$ contain two or more elements?

16. Let $S = \{1,2,3,4,5,6,7\}$.

    (a) How many subsets of S are there?
    (b) How many nonempty subsets of S are there?
    (c) How many subsets of S contain exactly 3 elements?
    (d) How many subsets of S contain exactly 5 elements?
    (e) How many subsets of S contain at least 6 elements?

17. How many 3-element subsets of $\{1,2,\ldots,10\}$ contain only even integers?

18. How many subsets of $\{1, 2, \ldots, 9\}$ contain three elements, all of which are odd integers?

19. How many different 10-bit strings contain exactly 7 1s?

20. How many different 8-bit strings contain exactly 5 1s?

21. Let $S = \{x : x$ is an integer, $1 \le x \le 80, x \ne 75\}$

    (a) How many subsets of S have exactly one element?
    (b) How many subsets of S have at least one element?
    (c) How many subsets of S have at least two elements?

22. A pizza can be ordered with any combination of (or none of) pepperoni, ham, sausage, mushrooms and olives. In how many ways can a pizza be ordered?

23. A total of 20 people apply for a university position.

    (a) In how many ways can 8 of the 20 applicants be selected to form a "short list" of applicants?
    (b) In how many ways can 5 candidates out of 8 be selected on a "short list" to be interviewed?
    (c) In how many ways can 3 of the 5 candidates who are interviewed be ordered for ranking purposes?

24. How many 7-digit numbers contain exactly three 1s and four 2s?

25. How many 6-digit numbers contain exactly two 1s, two 2s and two 3s?

26. How many 8-digit numbers contain four 0s and four 5s? (Note: Since the number is an 8-digit number, the first digit cannot be 0.)

27. If $\binom{n}{2} = 45$ for some positive integer n, then what is $\binom{n}{3}$?

28. If $C(n, r) = 56$ for some n and r, then what is $C(n, n - r)$?

29. Prove that $rC(n, r) = nC(n - 1, r - 1)$ for $1 \le r \le n$.

30. What should $C(n, n - r)$ equal if $0 \le n < r$?

31. Give an example of a problem whose solution has $\binom{7}{1} + \binom{7}{3} + \binom{7}{5} + \binom{7}{7} = 7 + 35 + 21 + 1 = 64$ as its answer.

32. Show that $\binom{2n}{2} = 2\binom{n}{2} + n^2$.

33. Show that $\binom{3n}{3} = 3\binom{n}{3} + 6n\binom{n}{2} + n^3$.

34. Show that $\binom{n^2}{2} = n\binom{n}{2} + n^2\binom{n}{2} = (n + n^2)\binom{n}{2}$.

35. Find an expression for $\binom{n^2}{3}$ analogous to that in Exercise 34.

8.5   Applications of Permutations and Combinations

In order to gain a better understanding of how permutations and combinations can be used to solve certain counting problems, we consider a variety of examples.

Example 8.49

A committee of 5 is to consist of 3 faculty members and 2 students. The 3 faculty members are to be chosen from 6 eligible faculty members and the 2 students are to be chosen from 5 eligible students. How many different committees are possible?

Solution. Constructing such a committee can be considered as performing a procedure that consists of the task of choosing 3 faculty members for the committee followed by the task of choosing 2 students for the committee. Since the number of ways of choosing 3 faculty members from the 6 eligible faculty members is $C(6, 3) = \frac{6!}{3! \, 3!} = 20$ and the number of ways of choosing 2 students from the 5 eligible students is $C(5, 2) = \frac{5 \cdot 4}{2} = 10$, it follows by the Multiplication Principle that the number of ways of selecting the entire committee is $20 \cdot 10 = 200$.   ♦

Example 8.50

A total of 6 seniors and 5 juniors have been nominated for a 6-person committee, which is to consist of 3 seniors and 3 juniors. One senior will be chosen as president, one as vice-president, and one as secretary. The three juniors will not have any special title. How many such committees are possible?

Solution. There are 6 ways to select a senior to be president. Once this person has been chosen, there are 5 choices for the senior who will be vice-president. After these two selections have been made, there are 4 possible choices for the senior who will be secretary. By the Multiplication Principle, the total number of ways of selecting 3 seniors to serve in these positions for the committee is $6 \cdot 5 \cdot 4 = 120$. This number is also $P(6, 3)$. Moreover, there are $C(5, 3) = C(5, 2) = \frac{5 \cdot 4}{2} = 10$ ways of selecting 3 juniors for the committee. Again, by the Multiplication Principle, the total number of ways of selecting such a 6-person committee is $120 \cdot 10 = 1200$.   ♦

Example 8.51

A total of 6 freshmen, 5 sophomores and 4 juniors have volunteered to serve on a 4-person committee. How many such committees are possible if

(a) any student may serve on the committee?

(b) at least one freshman, at least one sophomore and at least one junior must serve on the committee?

(c) at least one freshman must serve on the committee?

Solution.

(a) Since any of the 15 students may serve on the committee, the number of possible 4-person committees is
$$\binom{15}{4} = \frac{15!}{4! \; 11!} = \frac{15 \cdot 14 \cdot 13 \cdot 12}{4 \cdot 3 \cdot 2} = 1365.$$

(b) In this case, the make-up of the committee must be one of the following types:

    (1) two freshmen, one sophomore and one junior.

    (2) one freshman, two sophomores and one junior.

    (3) one freshman, one sophomore and two juniors.

The number of 4-person committees

    of type (1) is $\binom{6}{2}\binom{5}{1}\binom{4}{1} = 15 \cdot 5 \cdot 4 = 300,$

    of type (2) is $\binom{6}{1}\binom{5}{2}\binom{4}{1} = 6 \cdot 10 \cdot 4 = 240$ and

    of type (3) is $\binom{6}{1}\binom{5}{1}\binom{4}{2} = 6 \cdot 5 \cdot 6 = 180.$

Thus the total number of such committees is $300 + 240 + 180 = 720.$

(c) In this case, we could compute the number of committees containing (1) exactly one freshman, (2) exactly two freshmen, (3) exactly three freshmen or (4) exactly four freshmen. These numbers are, respectively,
$$\binom{6}{1}\binom{9}{3} = 6 \cdot 84 = 504, \quad \binom{6}{2}\binom{9}{2} = 15 \cdot 36 = 540,$$
$$\binom{6}{3}\binom{9}{1} = 20 \cdot 9 = 180 \text{ and } \binom{6}{4}\binom{9}{0} = 15 \cdot 1 = 15.$$

Thus the total number of such committees is $504 + 540 + 180 + 15 = 1239.$

There is, however, another way to compute this number. The number of committees with no freshmen is
$$\binom{9}{4} = \frac{9!}{4! \; 5!} = 126.$$

Therefore, the number of committees with at least one freshman is

$$\binom{15}{4} - \binom{9}{4} = 1365 - 126 = 1239. \qquad\qquad \blacklozenge$$

Before looking at another example, a remark concerning the solution of question (b) in Example 8.51 might be useful. When determining the number of 4-person committees with at least one freshman, at least one sophomore and at least one junior, you may have thought of selecting one freshman, one sophomore, one junior and one from the remaining 12 students, giving the answer

$$\binom{6}{1}\binom{5}{1}\binom{4}{1}\binom{12}{1} = 6 \cdot 5 \cdot 4 \cdot 12 = 1440. \qquad\qquad (8.7)$$

This, of course, is not the same answer we obtained in our solution of (b); in fact, it is twice the number we obtained. The number in (8.7) does not give the correct answer to (b). In order to see why this is so, suppose that the 6 freshmen who volunteered for the committee are $F_1, F_2, F_3, F_4, F_5, F_6,$ the 5 sophomores are S_1, S_2, S_3, S_4, S_5 and the 4 juniors are $J_1, J_2, J_3, J_4.$ One of the committees we should be counting in the solution of (b) is $\{F_1, F_2, S_1, J_1\}$. This is counted in the solution of (b) we gave. However, in the "solution" described in the number in (8.7), one of the six freshmen is selected, say F_1, one sophomore is selected, say S_1, and one junior is selected, say J_1. Then one student is selected who has not already been selected, say F_2. So the committee $\{F_1, S_1, J_1, F_2\}$ is counted in (8.7). However, when one of the 6 freshmen is selected, we may have chosen F_2 instead, then selected one sophomore, namely S_1 and selected one junior, namely J_1. Then a student is selected who has not already be chosen. One possibility is F_1. This counts the committee $\{F_2, S_1, J_1, F_1\}$ *for a second time.* Indeed, every committee is counted twice in (8.7). The lesson here is that care must be taken in solving problems of this kind.

After studying Example 8.51, let's now look at a related example.

Example 8.52

A total of 5 seniors, 4 juniors and 3 sophomores have been nominated to serve on a 3-person committee. How many different committees are possible if

(a) there is no restriction on membership?

(b) the committee must consist of one senior, one junior and one sophomore?

(c) at least one senior must be on the committee?

Solution.

(a) Since there is no restriction on who can serve on the 3-person committee, the number of possible 3-person committees is $\binom{12}{3} = \frac{12!}{3! \; 9!} = \frac{12 \cdot 11 \cdot 10}{6} = 220$.

(b) There are 5 choices for the senior who would serve on the committee, 4 choices for the junior and 3 choices for the sophomore. By the Multiplication Principle, the number of possible 3-person committees is $5 \cdot 4 \cdot 3 = 60$. This can also be expressed as $\binom{5}{1}\binom{4}{1}\binom{3}{1} = 5 \cdot 4 \cdot 3 = 60$.

(c) To answer this question, we could compute the number of possible 3-person committees containing exactly one senior, the number of possible 3-person committees containing exactly two seniors and the number of possible 3-person committees consisting entirely of seniors and then add these three numbers. An alternative, however, is to compute the number of possible 3-person committees containing no seniors. This number is $\binom{7}{3} = \frac{7!}{3! \; 4!} = 35$. By (a) the number of possible 3-person committees with no restriction is $\binom{12}{3} = 220$. Hence the number of possible 3-person committees with at least one senior is $\binom{12}{3} - \binom{7}{3} = 220 - 35 = 185$. ♦

Example 8.53

A total of 3 freshmen, 5 sophomores, 3 juniors and 4 seniors are eligible for a \$3000 scholarship, a \$2000 scholarship and a \$1000 scholarship.

(a) How many distributions of scholarships are there if none go to a freshman?

(b) How many distributions of scholarships are there if at least two must go to seniors?

(c) How many distributions of scholarships are there if no two students from the same class can receive a scholarship?

Solution.

(a) There are 12 students who are not freshmen. So there are 12 choices for the recipient of the \$3000 scholarship. Once the recipient of this scholarship has been determined, there are 11 choices for the recipient of the \$2000 scholarship and then 10 choices for the recipient of the \$1000 scholarship. By the Multiplication Principle, the number of distributions of these scholarships is $12 \cdot 11 \cdot 10 = 1320$. This number is also $P(12, 3) = 1320$.

(b) If at least two scholarships must go to seniors, then either all three scholarships go to seniors or exactly two scholarships go to seniors. If all three scholarships go to seniors, then the number of distributions of scholarships is $P(4,3) = 4 \cdot 3 \cdot 2 = 24$. Next, suppose that exactly two scholarships go to seniors. If seniors are awarded the \$3000 and \$2000 scholarships (equivalently, one of the 11 non-seniors is awarded the \$1000 scholarship), then the number of distributions of scholarships is $4 \cdot 3 \cdot 11 = 132$. Since there are 3 possible scholarships that can be distributed to a non-senior, the number of distributions of scholarships that could be awarded to exactly two seniors is $3 \cdot 132 = 396$. By the Addition Principle, the total number of distributions of scholarships such that at least two go to seniors is $24 + 396 = 420$.

(c) There are four possibilities according to which class has no student receiving a scholarship. First, suppose that no senior receives a scholarship. If a freshmen receives the \$3000 scholarship, a sophomore receives the \$2000 scholarship and a junior receives the \$1000 scholarship, then the number of possibilities is $3 \cdot 5 \cdot 3 = 45$. Since there are $3! = 6$ ways of arranging (permuting) these three classes, the number of distributions of scholarships that no senior receives is $6 \cdot 45 = 270$. Similarly, the number of distributions of scholarships that no junior receives is $6 \cdot (3 \cdot 5 \cdot 4) = 360$; the number of distributions of scholarships when only a sophomore receives no scholarship is $6 \cdot (3 \cdot 3 \cdot 4) = 216$; and the number of distributions of scholarships when only a freshmen receives no scholarship is $6 \cdot (5 \cdot 3 \cdot 4) = 360$. By the Addition Principle, the total number of distributions of scholarships in this case is $270 + 360 + 216 + 360 = 1206$. ◆

Example 8.54

From a group of 5 men and 7 women in a company, 4 are to be chosen as branch managers in 4 different cities. How many different appointments can be made if

(a) every person is eligible?

(b) only the women are eligible?

(c) two men and two women are to be appointed?

(d) at least three women are to be appointed?

Solution.

(a) Since every person is eligible and the cities are different, the number of different appointments is $P(12,4) = 12 \cdot 11 \cdot 10 \cdot 9 = 11,880$.

(b) Since only the women are eligible, the number of different appointments is $P(7,4) = 7 \cdot 6 \cdot 5 \cdot 4 = 840$.

(c) Suppose that we denote the cities by A, B, C and D. First, suppose that men are chosen as branch managers in A and B and women are chosen as branch managers in C and D. Then the number of different appointments is $P(5,2) \cdot P(7,2) = 5 \cdot 4 \cdot 7 \cdot 6 = 840$. Since there are $C(4,2) = \binom{4}{2} = 6$ pairs of cities where men (or women) can be branch managers, the total number of different appointments is $6 \cdot 840 = 5040$ (by the Multiplication Principle).

(d) If at least three women are to be appointed as branch managers, then either all branch managers are women or exactly three branch managers will be women. If all four branch managers are women, then the number of appointments is $P(7,4) = 7 \cdot 6 \cdot 5 \cdot 4 = 840$ as we saw in (b). Suppose that exactly three women are branch managers of the four cities. If three women are branch managers in A, B and C (equivalently, a man is the branch manager only in D), then the number of appointments is $P(7,3) \cdot P(5,1) = 7 \cdot 6 \cdot 5 \cdot 5 = 1050$. Since there are $C(4,1) = 4$ possibilities for the city in which the man is the branch manager, the number of appointments is $4 \cdot (1050) = 4200$. By the Addition Principle, the total number of appointments is $840 + 4200 = 5040$. ◆

Example 8.55

A woman intends to invest $10,000 in 4 mutual funds from a list of 7 mutual funds recommended to her by her financial advisor. How many options does she have if

(a) $2500 will be invested in each of the 4 funds?

(b) amounts of $4000, $3000, $2000 and $1000 will be invested in the funds selected?

(c) amounts of $4000, $3000, $1500 and $1500, will be invested in the funds selected?

Solution.

(a) Since the woman will invest the same amount of money in 4 of the 7 funds, she has $C(7,4) = \binom{7}{4} = \frac{7!}{4!\,3!} = \frac{7 \cdot 6 \cdot 5}{3!} = 35$ options.

(b) Since the woman will invest different amounts of money in 4 of the 7 funds, she has $P(7,4) = 7 \cdot 6 \cdot 5 \cdot 4 = 840$ options.

(c) There are 7 possible funds in which to invest $4000 and, afterwards, 6 possible funds in which to invest $3000. Then there are $\binom{5}{2} = \frac{5 \cdot 4}{2} = 10$ choices for a pair of funds in which to invest $1500 each. By the Multiplication Principle, the number of options is $7 \cdot 6 \cdot 10 = 420$. ◆

Exercises for Section 8.5

1. A total of 10 prizes of equal value are to be given to 10 graduating seniors. A total of 12 seniors are eligible for these prizes. In how many ways can these prizes be distributed?

2. A student committee is to consist of 3 seniors and 4 juniors. A total of 6 seniors and 8 juniors have volunteered to serve on the committee. How many committees are possible?

3. An experiment uses three jars A, B and C. Jar A contains five balls numbered 1-5, Jar B contains ten balls numbered 6-15 and Jar C contains ten balls numbered 16-25. The experiment consists of selecting three balls from Jar A, two balls from Jar B and one from Jar C. How many outcomes are possible?

4. From a group of 6 juniors and 5 sophomores, a committee of 3 juniors and 3 sophomores is to be selected. One junior is to receive a 12-month term on the committee, one junior is to receive a 9-month term, one junior is to receive a 6-month term and all sophomores are to receive 4-month terms. How many different committees are possible?

5. An election will be held to fill three faculty seats and three student seats on a certain committee. The faculty member with the most votes gets a 3-year term, the one receiving the second highest total gets a 2-year term and the one receiving the third highest total gets a 1-year term. All three elected students receive a 1-year term. If there are 7 faculty members and 8 students on the ballot, how many different election outcomes are possible?

6. A group of three women and seven men have been nominated for a 3-person committee. One person serves a 1-year term, one a 2-year term and the third a 3-year term. In how many ways can three people be chosen for this committee if

    (a) every person is eligible?

    (b) only the men are eligible?

    (c) at least one woman must be selected?

    (d) at least one woman and at least one man are to be selected?

7. From a group of 6 sophomores and 5 freshmen, a committee of 3 sophomores and 2 freshmen is to be selected. One sophomore is to receive a 3-semester term, one sophomore is to receive a 2-semester term and one sophomore is to receive a 1-semester term. Both freshmen are to receive 1-semester terms. How many committees are possible?

8. A total of 2 freshmen, 3 sophomores, 4 juniors and 5 seniors have been nominated to serve on a committee. How many different committees are possible if:

   (a) the committee can consist of any 4 students?

   (b) the committee must consist of one student from each class?

   (c) the committee is to consist of 3 students from 3 different classes?

9. A total of 4 sophomores, 5 juniors and 6 seniors are eligible for 5 identical awards. In how many ways can the recipients be selected under the following conditions?

   (a) Any candidate can receive an award.

   (b) Only seniors can receive an award.

   (c) At least one senior must receive an award.

   (d) At least two juniors and at least two seniors must receive an award.

   (e) At least one person from each class must receive an award but no more than two members of the same class can receive an award.

10. A total of 5 seniors, 3 juniors and 4 sophomores have volunteered to serve on a 4-person committee. How many committees are possible if

    (a) there is no other restriction on membership for the committee?

    (b) at least one senior, one junior and one sophomore must serve on the committee?

    (c) at least 3 seniors must serve on the committee?

    (d) at least one senior must serve on the committee?

11. An investor intends to buy shares of stock in 3 companies, chosen from a list of 12 companies. How many different investment options are there if

    (a) equal amounts are invested in each company?

    (b) equal amounts are invested in exactly two companies?

    (c) different amounts are invested in each company?

12. Give an example of a problem involving students serving on a committee whose solution has the answer

$$[C(6,3)]^2 + [C(6,2)]^3 + [C(6,1)]^2 \cdot C(6,4) + C(6,6) = 400 + 3375 + 540 + 1 = 4316.$$

13. For a positive integer n, a total of $19n$ students have volunteered to serve on a committee. Of these, $3n$ are freshmen, $6n$ are sophomores and $10n$ are juniors. A total of $9n$ students are wanted for the committee, namely $2n$ freshmen, $3n$ sophomores and $4n$ juniors. How many committees are possible? Express your answer in terms of factorials.

Chapter 8   Highlights

Key Concepts

combination (*r*-**combination**): an unordered selection of r elements of a set.

$C(n, r) = \binom{n}{2}$: the number of r-element subsets of an n-element set.

permutation (of a set): an arrangement (an ordered list) of the elements of the set.

$P(n, r)$: the number of r-permutations of a set with n elements.

r-**permutation**: an arrangement (an ordered list) of r elements of a set.

Key Results

- **The Multiplication Principle:** A procedure consists of a sequence of two tasks. To perform this procedure, one performs the first task followed by performing the second task. If there are n_1 ways to perform the first task and n_2 ways to perform the second task after the first task has been performed, then there are $n_1 n_2$ ways to perform the procedure.

- **The (General) Multiplication Principle:** Performing a certain procedure consists of performing a sequence of $m \geq 2$ tasks $T_1, T_2, \ldots, T_m$. If there are n_i ways of performing T_i after any preceding tasks have been performed for $i = 1, 2, \ldots, m$, then the total number of ways of performing the procedure is $n_1 n_2 \cdots n_m$.

- **The Addition Principle:** A procedure consists of two tasks that cannot be performed simultaneously. To perform this procedure, either of the two tasks is performed. If the first task can be performed in n_1 ways and the second can be performed in n_2 ways, then the number of ways of performing this procedure is $n_1 + n_2$.

- **The (General) Addition Principle:** Performing a certain procedure consists of performing one of $m \geq 2$ tasks $T_1, T_2, \ldots, T_m$, no two of which can be performed at the same time. If the task T_i can be performed in n_i ways for $1 \leq i \leq m$, then the number of ways of performing this procedure is $n_1 + n_2 + \cdots + n_m$.

- **The Principle of Inclusion-Exclusion:** A procedure consists of two tasks. To perform the procedure, one performs either of the two tasks. If the first task can be performed in n_1 ways, the second task can be performed in n_2 ways and the two tasks can be simultaneously performed in n_{12} ways, then the total number of ways of performing the procedure is $n_1 + n_2 - n_{12}$.

- **The Principle of Inclusion-Exclusion (for two sets):** For every two finite sets A and B,
$$|A \cup B| = |A| + |B| - |A \cap B|.$$

  In particular, if A and B are disjoint, then $|A \cup B| = |A| + |B|$.

- **The Principle of Inclusion-Exclusion (for three sets):** For every three finite sets A, B and C,
$$|A \cup B \cup C| = |A| + |B| + |C| - |A \cap B| - |A \cap C| - |B \cap C| + |A \cap B \cap C|.$$

- **The Principle of Inclusion-Exclusion (for $n \geq 2$ sets):** *Let $A_1, A_2, \ldots, A_n$ be $n \geq 2$ finite sets, then*
$$
\begin{aligned}
|A_1 \cup A_2 \cup \cdots \cup A_n| &= \sum_{1 \leq i \leq n} |A_i| - \sum_{1 \leq i < j \leq n} |A_i \cap A_j| + \sum_{1 \leq i < j < k \leq n} |A_i \cap A_j \cap A_k| \\
&\quad - \cdots + (-1)^{n+1} |A_1 \cap A_2 \cap \cdots \cap A_n|
\end{aligned}
$$

- **The Pigeonhole Principle**: If a set S with n elements is divided into k pairwise disjoint subsets S_1, S_2, ..., S_k, then at least one of these subsets contains at least $\lceil n/k \rceil$ elements.

- Let r and k be positive integers. If N is the minimum cardinality of a set S such that however S is divided into k subsets, at least one of these subsets must contain at least r elements, then $N = k(r-1) + 1$.

- **The (General) Pigeonhole Principle**: A set S with n elements is divided into k pairwise disjoint subsets S_1, S_2, ..., S_k, where $|S_i| \geq n_i$ for a positive integer n_i for $i = 1, 2, \ldots, k$. Then each subset of S with at least $1 + \sum_{i=1}^{k}(n_i - 1)$ elements contains at least n_i elements of S_i for some integer i with $1 \leq i \leq k$

- $P(n,r) = \frac{n!}{(n-r)!}$.

- $C(n,r) = \binom{n}{r} = \frac{n!}{r!(n-r)!}$.

- **Identities:** $2^n = \binom{n}{0} + \binom{n}{1} + \binom{n}{2} + \cdots + \binom{n}{n-1} + \binom{n}{n}$.

  $\binom{n}{r} = \binom{n}{n-r}$ for $0 \leq r \leq n$.

Supplementary Exercises for Chapter 8

1. For the final exam in a discrete mathematics class, the professor gives her students a list of 20 topics to study. She states that she will choose 10 of these topics for the actual exam and the students will be required to write short essays on 6 of them. What number of topics will a student need to study to be certain that he/she will be prepared to do well on the final exam?

2. How many 3-digit numbers abc are there, where $4 \leq a \leq 6$, $3 \leq b \leq 7$ and $2 \leq c \leq 8$?

3. On the right side of an auditorium every seat is labeled with its row (A, B, $\cdots$, P) followed by a 2-digit even numbers from 02 to 36. On the left side of the auditorium, the situation is the same except the seat numbers in each row are odd (from 01 through 35). How many seats are there in the auditorium?

4. How many 4-digit even numbers with no repeating digits are there that are less than 6000?

5. In one portion of a city, a telephone number begins with 381, 383 or 385, followed by four digits.

   (a) How many such telephone numbers are possible?
   (b) How many such telephone numbers have no digits repeated?

6. How many different 12-bit strings begin with 0101, end with 1100 or have 0110 as the four middle bits?

7. A total of 100 people attending a movie were asked if they normally have popcorn, candy or a soft drink when they go to a movie. They answered:

   62 have popcorn, 20 have candy, 70 have a soft drink, 5 have popcorn and candy,

   51 have popcorn and a soft drink, 11 have candy and a soft drink and 3 have all 3.

   How many have none of these?

8. A movie theater complex contains several auditoriums. There are 8 movies that start at approximately the same time. If a total of 550 people go to one of these 8 movies, then what is the smallest possible number of people in the theater containing the largest number of people?

9. Let A and B be two sets and let $f : A \to B$ be a function. Use the Pigeonhole Principle to show that if $|A| > |B|$, then there exist $a_1, a_2 \in A$ such that $a_1 \neq a_2$ and $f(a_1) = f(a_2)$ (that is, f is not one-to-one).

10. Let S be a set of $k + 1 \geq 2$ integers. Use the Pigeonhole Principle to show that there are at least two integers in S that have the same remainder when divided by k.

11. How many integers must a set S contain to be certain that at least 35 integers in S have the same remainder when divided by 4?

12. A boy has a collection of 20 pennies, 8 nickels, 10 dimes and 15 quarters. How many coins must the boy take from his collection to be certain that he has

    (a) 12 coins of the same kind?

    (b) 10 pennies, 5 nickels, 6 dimes or 8 quarters?

    (c) 15 pennies, 10 nickels, 12 dimes or 8 quarters?

13. (a) How many 6-bit strings begin with 0101 and end with 111?

    (b) How many 6-bit strings begin with 111 and end with 0101?

    (c) How many 6-bit strings begin with 1011 or end with 1001?

    (d) How many 6-bit strings begin with 1010 or end with 1010?

14. Suppose that we are interested in determining the number of 4-bit strings containing two consecutive 0s. This could be determined by listing all of these 4-bit strings:

$$0000, 0001, 0010, 0011, 0100, 0101, 0110, 0111,$$
$$1000, 1001, 1010, 1011, 1100, 1101, 1110, 1111$$

    Thus 8 of the 16 4-bit strings contain two consecutive 0s.

    (a) Show that this question also be answered by performing the computation below.
    $$3 \cdot 2^2 - (2 + 1 + 2) + 1 = 8.$$

    (b) Determine the number of 5-bit strings containing two consecutive 0s without listing them.

15. How many 4-digit numbers less than 6000 are there with distinct digits that have a remainder of 0 when divided by 5?

16. Let B be a set with $|B| = k \in \mathbb{N}$. If, for every function from a set A to B, there must be at least k elements of A having the same image in B, then what is the minimum possible cardinality of A?

17. There is a computer game in which a sequence of three digits (each digit is one of $0, 1, \ldots, 9$) is randomly generated.

    (a) How many different sequences are possible?

    (b) How many different sequences contain exactly two equal digits?

18. Four sisters and their four husbands have been invited to a reception where these 8 people are seated at 8 chairs in a row. In how many ways can they be seated if each married couple must sit next to each other and either (1) no sister sets next to any other sister or (2) each sister must set next to one of her sisters?

19. A collector of coins has three valuable coins: one black, one gold and one silver. He decides to place these three coins in four drawers. For example, he might place all 3 coins in the second drawer or place the silver coin in the first drawer and the black and gold coins in the 4th drawer.

    (a) In how many ways can the 3 coins be placed in the 4 drawers?

    (b) In how many ways can the 3 coins be placed in the 4 drawers such that exactly two of the coins are in the same drawer?

20. How many 5-bit strings begin with 1, 11, 111, 1111 or 11111?

21. The first day on the job for a college graduate is January 2. He owns 4 sports jackets, 8 pairs of dress pants and 5 ties. Any jacket, pair of pants and tie can be worn together. If he wears a different combination to work each day (weekday), then what is the earliest month that he will need to buy new clothes to avoid the same combination of clothes that he wore earlier?

22. A bowl contains 18 blue marbles, 15 red marbles, 15 green marbles and 12 yellow marbles. What is the minimum number of marbles that must be removed from the bowl to be guaranteed that the number of marbles selected of some color is at least two more than the number of marbles of some other color?

23. Among the 6-bit strings in which there is a 1 in the 4th position, how many contain two consecutive 0s? Answer this question without listing all such strings.

24. For $n \in \mathbb{N}$, let $S = \{1, 2, \ldots, n\}$. Determine the number of sequences (x, Y, Z), where $x \in S$, $Y \subseteq S$ and $Z \subseteq S - \{x\}$?

25. How many 6-bit strings have 1010 as the first 4 bits, have 101 as bits 2, 3 and 4 or have 10 for bits 3 and 4?

26. For a positive integer n, let t_n be the number of different ways to divide $2n$ people into pairs to play n tennis matches.

    (a) Determine t_1 and t_2.

    (b) Give a recursive definition of t_n for $n \geq 1$.

    (c) Use the recursive definition in (b) to compute t_3 and t_4.

27. Let A and B be sets such that $|A| = 17$ and $|B| = 5$. Show that for every function $f : A \to B$, there exists an element $b \in B$ that is the image of at least 4 elements of A.

28. The integer $300 = 3 \cdot 4 \cdot 25$ is divisible by 3, 4 and 25. How many integers n with $1 \leq n \leq 300$ are divisible by none of 3, 4 and 25?

29. A bowl contains 75 balls. Of these, 10 are white, 11 are red, 12 are blue, 13 are green, 14 are yellow and 15 are black. What is the smallest number of balls that can be selected from the bowl to be certain that at least one ball of every color is chosen?

30. Let $S = \{1, 2, \ldots, 10\}$. Then S has $2^{10} - 1 = 1023$ nonempty subsets. For a nonempty subset S_k of S, let a_k be the sum of the numbers in S_k. Show that for each collection $A = \{S_1, S_2, \ldots, S_{10}\}$ of 10 nonempty subsets of S, there exists an integer i such that $a_i \equiv 0 \pmod{10}$ or there exist two integers i and j such that $a_i \equiv a_j \pmod{10}$.

31. Let $S = \{1, 2, \ldots, 100\}$.

    (a) Use the Principle of Inclusion-Exclusion to determine the number of integers $a \in S$ such that $3 \mid a$, $5 \mid a$ or $7 \mid a$.

(b) How many integers a belong to S such that $\gcd(a, 100) = 1$?

32. The value of each term of a sequence $a_1, a_2, \ldots, a_{2k}$ is a positive integer and $\sum_{i=1}^{2k} a_i = 3k$. Show that for each positive integer $m \leq k$, there exist integers r and s with $1 \leq r < s \leq 2k$ such that $\sum_{i=r+1}^{s} a_i = m$. [Hint: Consider the cases when $m < k$ and $m = k$ separately.]

33. How many subsets of a 10-element set contain 3 or more elements?

34. Each card in a deck of 12 cards is either red, blue, green or yellow. Show that the deck contains at least three red cards, at least three blue cards, at least three green cards or at least six yellow cards.

35. A young girl has a collection of 9 different stuffed animals: 3 bears, 3 dogs, 3 tigers. In how many ways can these stuffed animals be placed in a row so that all animals of the same type are grouped together?

36. From a group of 10 students, 6 are to be selected to serve on a committee, where one will chair the committee, one will be assistant to the chair, one will be secretary and the other 3 have no special title. How many such committees are possible?

37. A total of 5 seniors, 4 juniors and 3 sophomores have been nominated to serve on a 4-person committee. How many different committees are possible if

    (a) the committee must consist of students from exactly two of the three classes?

    (b) at least one student from each class must serve on the committee?

38. A group of 4 men and 5 women have been nominated to serve on a 4-person committee. One person is to serve a 1-month term, one a 2-month term, one a 3-month term and the other a 4-month term. How many committees are possible if

    (a) there is no restriction on who can serve on the committee?

    (b) only men can serve on the committee?

    (c) at least one woman must serve on the committee?

    (d) exactly two men and two women must serve on the committee?

39. There are 100 students in a school, assigned numbers 1 through 100. Each student has a locker with the same number. When the school day starts, all lockers are closed. Student #1 then opens all 100 lockers. Student #2 then closes every other locker (namely those numbered $2, 4, \ldots, 100$). Student #3 then reverses every third locker (that is, opens a closed locker or closes an open locker beginning with the locker numbered 3). This continues with all 100 students. How many lockers are open at the end? How many lockers would be open if there are 1000 students and 1000 lockers?

Chapter 9

Advanced Counting Methods

In Chapter 8 we introduced four counting principles, namely the Multiplication Principle, the Addition Principle, the Principle of Inclusion-Exclusion and the Pigeonhole Principle. These were used to answer several questions concerning the enumeration of objects possessing some specified properties. In Chapter 8, we also introduced the concepts of permutations and combinations, which are fundamental in solving certain kinds of counting problems. In this chapter, we consider even more varied counting problems.

9.1   The Pascal Triangle and the Binomial Theorem

Considering that he lived for only 39 years, it is quite remarkable how much the French mathematician and physicist Blaise Pascal accomplished. He was fascinated with geometry during his early teens and discovered several geometric facts on his own. Pascal's father was employed as a tax collector. In order to help his father, Pascal spent the 3-year period 1642-1645 attempting to invent a digital calculator, which he succeeded in doing. This device, called the Pascaline, became the first digital calculator. He is quite possibly the second person to invent a mechanical calculator. The German mathematician and astronomer Wilhelm Schickard had invented one in 1623. In 1968 a programming language (PASCAL) was named for him.

The Pascal Triangle

What most people in mathematics probably think of when they hear the name "Pascal" is the Pascal triangle. The Pascal triangle is a certain triangular array of positive integers. One way to construct this triangle is to begin with an infinite row of 1s. Under this row are the positive integers in their natural order $1, 2, 3, \ldots$. The next few rows are shown in Figure 9.1. There are patterns to each row. It may not be so easy to see what these patterns are, but we'll describe them now.

Once a certain number of rows of positive integers have been listed in the Pascal triangle, each number in the next row is obtained from the row immediately above it by adding those numbers directly above and to the left of it. Rotating Figure 9.1 clockwise 45° and adjusting the entries appropriately produces the **Pascal triangle**, as it is usually shown (in Figure 9.2).

Although the triangular array of integers given in Figure 9.2 is named for Pascal, he was certainly not the first mathematician to consider it. The Chinese mathematician Chu-Shih-Chien had previously considered this triangular diagram of numbers.

The elements of row 4 of the Pascal triangle in Figure 9.2 might look familiar. They are

$$\binom{4}{0} = 1 \quad \binom{4}{1} = 4 \quad \binom{4}{2} = 6 \quad \binom{4}{3} = 4 \quad \binom{4}{4} = 1.$$

1	1	1	1	1	1	......
1	2	3	4	5	6	......
1	3	6	10	15	21	......
1	4	10	20	35	56	......
1	5	15	35	70	126	......
1	6	21	56	126	252	......

Figure 9.1: Constructing the Pascal triangle

row 0   1

row 1   1   1

row 2   1   2   1

row 3   1   3   3   1

row 4   1   4   6   4   1

row 5   1   5   10   10   5   1

row 6   1   6   15   20   15   6   1

row 7   1   7   21   35   35   21   7   1

        ...   .

Figure 9.2: The Pascal triangle

That is, the elements of row 4 are the numbers of r-combinations $(0 \le r \le 4)$ of a 4-element set. In fact, *every* number in the Pascal triangle is of the form $\binom{n}{r}$ for some pair r, n of integers with $0 \le r \le n$. The Pascal triangle is shown again in Figure 9.3 written in this form.

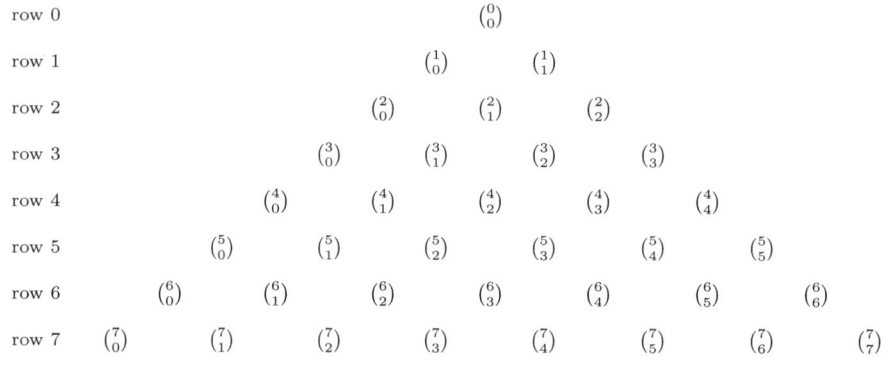

row 0 $\binom{0}{0}$

row 1 $\binom{1}{0}$ $\binom{1}{1}$

row 2 $\binom{2}{0}$ $\binom{2}{1}$ $\binom{2}{2}$

row 3 $\binom{3}{0}$ $\binom{3}{1}$ $\binom{3}{2}$ $\binom{3}{3}$

row 4 $\binom{4}{0}$ $\binom{4}{1}$ $\binom{4}{2}$ $\binom{4}{3}$ $\binom{4}{4}$

row 5 $\binom{5}{0}$ $\binom{5}{1}$ $\binom{5}{2}$ $\binom{5}{3}$ $\binom{5}{4}$ $\binom{5}{5}$

row 6 $\binom{6}{0}$ $\binom{6}{1}$ $\binom{6}{2}$ $\binom{6}{3}$ $\binom{6}{4}$ $\binom{6}{5}$ $\binom{6}{6}$

row 7 $\binom{7}{0}$ $\binom{7}{1}$ $\binom{7}{2}$ $\binom{7}{3}$ $\binom{7}{4}$ $\binom{7}{5}$ $\binom{7}{6}$ $\binom{7}{7}$

        ...   .

Figure 9.3: The Pascal triangle revisited

The numbers in the Pascal triangle have many interesting and unexpected properties. We have already mentioned in Theorem 8.46 that for every two integers r and n with $0 \le r \le n$,

$$\binom{n}{r} = \binom{n}{n-r}.$$

One interpretation of this is that each row of the Pascal triangle is symmetric about the middle number or middle two numbers in the row (see Figure 9.4). So once we know the left half (approximately) of a row in the Pascal triangle, we can easily complete the entire row.

If we look at any integer that is not 1 in the Pascal triangle in Figure 9.2, we see that it is the sum of the two numbers directly above and on either side of it. For example, the 3rd number in row

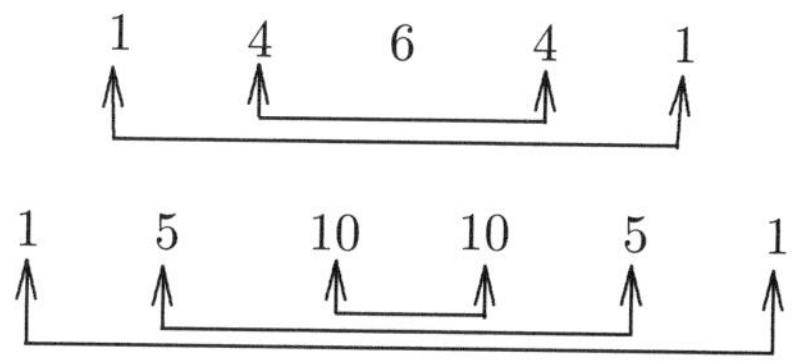

Figure 9.4: The symmetry of the rows of the Pascal triangle

6 is 15. It is the sum of the 2nd and 3rd numbers in row 5, that is, $15 = 5 + 10$. From Figure 9.3, we see that this says $\binom{6}{2} = \binom{5}{1} + \binom{5}{2}$. In general, we have the following.

Theorem 9.1 *For integers r and n with $1 \le r \le n-1$,*

$$\binom{n}{r} = \binom{n-1}{r-1} + \binom{n-1}{r}.$$

Proof. Let S be an n-element set, say $S = \{a_1, a_2, \ldots, a_n\}$. The number of r-element subsets of S is therefore $\binom{n}{r}$. Of course, some of these subsets contain the element a_n, while the others do not.

Let's determine the number of r-element subsets of S that contain a_n. Since one of the r elements is a_n, the remaining $r-1$ elements are to be chosen from the set $\{a_1, a_2, \ldots, a_{n-1}\}$ and the number of such subsets of S is $\binom{n-1}{r-1}$.

Next we determine the number of r-element subsets of S that do not contain a_n. Then the r elements must be selected from the set $\{a_1, a_2, \ldots, a_{n-1}\}$ and the number of such subsets of S is $\binom{n-1}{r}$.

By the Addition Principle, the number of r-element subsets of S is $\binom{n-1}{r-1} + \binom{n-1}{r}$ and so $\binom{n}{r} = \binom{n-1}{r-1} + \binom{n-1}{r}$. ∎

The Hockey Stick Theorem

From the way that the numbers in Figure 9.1 were described, we see that the 4th number in the 5th row, namely 35, is the sum of the 1st, 2nd, 3rd and 4th numbers in row 4, that is, $35 = 1 + 4 + 10 + 20$ (see Figure 9.5). Interpreting this in the Pascal triangle shown in Figure 9.3, we see that

$$\binom{7}{3} = \binom{3}{0} + \binom{4}{1} + \binom{5}{2} + \binom{6}{3}$$

or by Theorem 8.46,

$$\binom{7}{4} = \binom{3}{3} + \binom{4}{3} + \binom{5}{3} + \binom{6}{3}.$$

See Figure 9.5. This suggests another property of the Pascal triangle, which is sometimes called the **Hockey Stick Theorem** because the locations of these numbers in the Pascal triangle resemble the shape of a hockey stick.

Theorem 9.2 (**Hockey Stick Theorem**) *For every two integers r and n with $0 \le r \le n$,*

$$\binom{n+1}{r+1} = \binom{r}{r} + \binom{r+1}{r} + \binom{r+2}{r} + \cdots + \binom{n}{r}.$$

Proof. The number of bit strings (that is, sequences of 0s and 1s) of length $n+1$ containing exactly $r+1$ 1s is $\binom{n+1}{r+1}$. Consider a bit string of length $n+1$ as follows:

___	___	⋯	___	⋯	___	___
position 1	position 2	⋯	position $r-1$	⋯	position n	position $n+1$

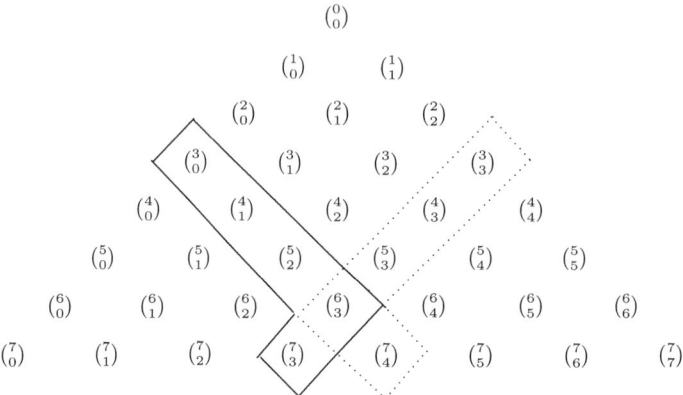

Figure 9.5: Illustrating the Hockey Stick Theorem in the Pascal triangle

Since each string contains $r+1$ 1s, the location of the last 1 in the sequence is either position $r+1$ or a position to the right of position $r+1$. If the final 1 is in position $r+1$, then the remaining r 1s must appear in positions 1 through r and the number of ways of selecting r possible positions for these r 1s is $\binom{r}{r}$. If the final 1 is in position $r+2$, then the remaining r 1s must occur in positions 1 through $r+1$ and the number of ways of selecting r of these $r+1$ positions for the remaining r 1s is $\binom{r+1}{r}$. We continue this until we arrive at the final possibility, where the last 1 occurs in position $n+1$. Then the remaining r 1s must be in positions 1 through n and the number of ways of selecting r of these n positions for the remaining r 1s is $\binom{n}{r}$. By the Addition Principle, we arrive at

$$\binom{n+1}{r+1} = \binom{r}{r} + \binom{r+1}{r} + \binom{r+2}{r} + \cdots + \binom{n}{r}. \qquad \blacksquare$$

Although the Hockey Stick Theorem may seem like a curious equation with little applicability, it is, in fact, quite useful, as we will see.

For $r=1$ and $r=2$, the Hockey Stick Theorem states that

$$1 + 2 + 3 + \cdots + n = \binom{n+1}{2} = \frac{n(n+1)}{2}, \qquad (9.1)$$

which we saw in Section 5.1 and

$$\binom{2}{2} + \binom{3}{2} + \binom{4}{2} + \cdots + \binom{n}{2} = \binom{n+1}{3} = \frac{(n+1)n(n-1)}{6} = \frac{n^3 - n}{6}.$$

The Binomial Theorem

The rows in the Pascal triangle may seem familiar for another reason. Observe that

$$(x + y)^0 = 1$$
$$(x + y)^1 = 1x + 1y$$
$$(x + y)^2 = 1x^2 + 2xy + 1y^2$$
$$(x + y)^3 = 1x^3 + 3x^2y + 3xy^2 + 1y^3$$
$$(x + y)^4 = 1x^4 + 4x^3y + 6x^2y^2 + 4xy^3 + 1y^4.$$

That is, the coefficients of the terms in the nth power ($n \geq 0$) of the binomial $x + y$ are precisely the numbers in row n of the Pascal triangle. This is a well-known theorem in discrete mathematics.

We are surely familiar with the expansion $(x + y)^2 = x^2 + 2xy + y^2$, which occurs in many forms in mathematics, such as

$$
\begin{aligned}
(a - b)^2 &= (a + (-b))^2 = a^2 + 2a(-b) + (-b)^2 = a^2 - 2ab + b^2 \\
(3x + 5)^2 &= (3x)^2 + 2(3x)(5) + 5^2 = 9x^2 + 30x + 25 \\
(x^2 + 1)^2 &= = (x^2)^2 + 2(x^2) \cdot 1 + 1^2 = x^4 + 2x^2 + 1.
\end{aligned}
$$

An interesting application of this fact concerns squaring positive integers whose last digit is 5, such as $(35)^2$ or $(75)^2$. Suppose that n is a positive integer whose last digit is 5. Then the decimal representation of n is

$$n = \underbrace{\underline{\quad}\ \underline{\quad} \cdots \underline{\quad}}_{a}\ \underline{\ 5\ }.$$

So if $n = 35$, then $a = 3$; while if $n = 125$, then $a = 12$. Hence $n = 10a + 5$; that is, $35 = 3 \cdot 10 + 5$ and $125 = 12 \cdot 10 + 5$. So

$$
\begin{aligned}
n^2 &= (10a + 5)^2 = (10a)^2 + 2 \cdot (10a) \cdot 5 + 5^2 \\
&= 100a^2 + 100a + 25 = 100(a^2 + a) + 25 = 100a(a + 1) + 25.
\end{aligned}
$$

Therefore, the decimal representation of n^2 is

$$n^2 = \underbrace{\underline{\quad}\ \underline{\quad} \cdots \underline{\quad}}_{a(a+1)}\ \underline{\ 2\ }\ \underline{\ 5\ }.$$

Example 9.3

To square 35, we therefore multiply 3 and $3 + 1$ (obtaining 12) and follow it by 25, that is, $(35)^2 = 1225$. Also, $(125)^2$ is $12 \cdot 13 = 156$ followed by 25 or $(125)^2 = 15,625$. ◆

To see why $(x + y)^3 = 1x^3 + 3x^2y + 3xy^2 + 1y^3$, first recall that $(x + y)^3$ represents the product of $x + y$ (3 times), that is, $(x + y)^3 = (x + y)(x + y)(x + y)$. If we multiply these binomials without collecting like terms, then we obtain

$$
\begin{aligned}
(x + y)^3 &= (x + y)(x + y)(x + y) = (x + y)(xx + xy + yx + yy) \\
&= xxx + xxy + xyx + xyy + yxx + yxy + yyx + yyy \\
&= x^3 + 3x^2y + 3xy^2 + y^3.
\end{aligned}
$$

So, for example, where does the term $3x^2y$ come from? It is obtained by adding the terms: $xxy + xyx + yxx$. To obtain one of these terms in the expansion of $(x + y)^3$, we need to select the term x from two of the three binomials $x + y$ and select the term y from one of the three binomials $x + y$ and then multiply them to obtain a term x^2y in the expansion of $(x + y)^3$. The number of ways of selecting the term y, say, is $\binom{3}{1} = 3$ and so the coefficient of x^2y in $(x + y)^3$ is $\binom{3}{1} = 3$.

Similarly, each term in the expansion $(x+y)^4 = (x+y)(x+y)(x+y)(x+y)$ is obtained by selecting either x or y in each binomial and then multiplying them. If we select y twice (and necessarily x twice), then the number of ways of doing this is $\binom{4}{2} = 6$. Hence the coefficient of x^2y^2 in $(x + y)^4$ is $\binom{4}{2} = 6$. In general, we have the famous Binomial Theorem.

Theorem 9.4 (**The Binomial Theorem**) *For every nonnegative integer n,*

$$(x+y)^n = \binom{n}{0}x^n + \binom{n}{1}x^{n-1}y + \binom{n}{2}x^{n-2}y^2 + \cdots + \binom{n}{r}x^{n-r}y^r + \cdots + \binom{n}{n}y^n$$

$$= \sum_{r=0}^{n}\binom{n}{r}x^{n-r}y^r.$$

Because the coefficient of $x^{n-r}y^r$ in the expansion of $(x+y)^n$ is $\binom{n}{r}$, the numbers $\binom{n}{r}$, where $r, n \in \mathbb{Z}$ and $0 \le r \le n$, are often referred to as **binomial coefficients**.

Mathematicians have been aware of the Binomial Theorem, at least for small values of n, for centuries. The great geometer Euclid knew that $(a+b)^2 = a^2 + 2ab + b^2$ by considering areas of squares and rectangles (see Figure 9.6).

Figure 9.6: $(a+b)^2 = a^2 + 2ab + b^2$

Although it seems as if it would have been relatively easy for Euclid and others to consider $(a+b)^n$ for $n \ge 3$, generalizing results was not a common way to think among the ancient Greeks. On the other hand, Diophantus, the father of Algebra to some, is believed to have been familiar with the expansion of $(a+b)^n$. Certainly, the Indian mathematician Aryabhata was aware of this. The Binomial Theorem stated in its standard form is due to Pascal. It appeared in a paper published in 1665 (after Pascal's death). There is a more general form of the Binomial Theorem, due to Sir Isaac Newton, which we will encounter in Section 9.3.

Example 9.5

According to the Binomial Theorem,

$$(x+y)^5 = \binom{5}{0}x^5 + \binom{5}{1}x^4y + \binom{5}{2}x^3y^2 + \binom{5}{3}x^2y^3 + \binom{5}{4}xy^4 + \binom{5}{5}y^5$$

$$= x^5 + 5x^4y + 10x^3y^2 + 10x^2y^3 + 5xy^4 + y^5$$

and

$$(x+y)^6 = \binom{6}{0}x^6 + \binom{6}{1}x^5y + \binom{6}{2}x^4y^2 + \binom{6}{3}x^3y^3 + \binom{6}{4}x^2y^4 + \binom{6}{5}xy^5 + \binom{6}{6}y^6$$

$$= x^6 + 6x^5y + 15x^4y^2 + 20x^3y^3 + 15x^2y^4 + 6xy^5 + y^6. \blacklozenge$$

Rows 0, 1, 2, 3 and 4 of the Pascal triangle are

$$
\begin{array}{ccccccccc}
 & & & & 1 & & & & \\
 & & & 1 & & 1 & & & \\
 & & 1 & & 2 & & 1 & & \\
 & 1 & & 3 & & 3 & & 1 & \\
1 & & 4 & & 6 & & 4 & & 1
\end{array}
$$

If we consider each of these rows as a single integer, then what we have is

$$
\begin{aligned}
(11)^0 &= 1 \\
(11)^1 &= 11 \\
(11)^2 &= 121 \\
(11)^3 &= 1331 \\
(11)^4 &= 14641.
\end{aligned}
$$

Actually, none of this is surprising. Consider $(x+y)^4$, for example. By the Binomial Theorem,

$$(x+y)^4 = 1x^4 + 4x^3y + 6x^2y^2 + 4xy^3 + 1y^4.$$

Letting $x = 10$ and $y = 1$, we obtain

$$(11)^4 = (10+1)^4 = 1 \cdot 10^4 + 4 \cdot 10^3 + 6 \cdot 10^2 + 4 \cdot 10 + 1 = (14641)_{10}.$$

For $n \geq 5$, we must make some adjustments to this line of reasoning, however. Again, by the Binomial Theorem,

$$(x+y)^5 = x^5 + 5x^4y + 10x^3y^2 + 10x^2y^3 + 5xy^4 + y^5.$$

Letting $x = 10$ and $y = 1$, we have

$$
\begin{aligned}
(11)^5 &= (10+1)^5 = 1 \cdot 10^5 + 5 \cdot 10^4 + 10 \cdot 10^3 + 10 \cdot 10^2 + 5 \cdot 10 + 1 \\
&= 1 \cdot 10^5 + 5 \cdot 10^4 + 1 \cdot 10^4 + 1 \cdot 10^3 + 5 \cdot 10 + 1 \\
&= 1 \cdot 10^5 + 6 \cdot 10^4 + 1 \cdot 10^3 + 5 \cdot 10 + 1 = (161051)_{10}.
\end{aligned}
$$

Therefore, the number 161051 can be obtained from the 5th row

$$1 \quad 5 \quad 10 \quad 10 \quad 5 \quad 1$$

of the Pascal triangle by first writing down the rightmost 1, then 5 to its immediate left, and then 0 (of the number 10). The digit 1 of 10 added to 10 on its left, resulting in 11. We write down 1 and add the leftmost 1 of 11 to 5 obtaining 6 and the finally write down the final 1. These steps are summarized below.

$$
\begin{array}{ccccccc}
1 & 5 & 10 & 10 & 5 & 1 \\
 & & \downarrow & & & \\
1 & 5 & 10+1 & 0 & 5 & 1 \\
 & & \downarrow & & & \\
1 & 5 & 11 & 0 & 5 & 1 \\
 & & \downarrow & & & \\
1 & 5+1 & 1 & 0 & 5 & 1 \\
 & & \downarrow & & & \\
1 & 6 & 1 & 0 & 5 & 1
\end{array}
$$

Thus $(11)^5 = 161,051$.

Even though the beginning (and ending) coefficient in the expansion of $(x+y)^n$ is 1, the second (and the second from the last) coefficient is n and we need only compute approximately half of the coefficients, it can still be a bit tedious to compute the binomial coefficients $\binom{n}{r}$ for $2 \leq r \leq n/2$. There is a shortcut for writing out the expansion of $(x+y)^n$, however (assuming that n is not terribly large).

Since the coefficient of $x^{n-r}y^r$ in the expansion of $(x+y)^n$ is $\binom{n}{r}$, one of the terms in this expansion is $\binom{n}{r}x^{n-r}y^r$. The term that follows $\binom{n}{r}x^{n-r}y^r$ in the expansion is $\binom{n}{r+1}x^{n-r-1}y^{r+1}$. Observe what we obtain if we were to multiply $\binom{n}{r}$ by $n-r$ and divide it by $r+1$.

Theorem 9.6   *For integers r and n with $0 \le r < n$,*

$$\binom{n}{r}\frac{n-r}{r+1} = \binom{n}{r+1}.$$

Proof. Observe that

$$\binom{n}{r}\frac{n-r}{r+1} = \frac{n!}{r!(n-r)!}\frac{n-r}{r+1} = \frac{n!}{(r+1)!(n-r-1)!} = \binom{n}{r+1},$$

producing the desired result.   ∎

Let's see what this says in an example.

Example 9.7

Suppose that we wish to expand $(x + y)^7$. Of course, we know that the expansion begins with $x^7 + 7x^6y$. To obtain the coefficient in the next term (that is, the coefficient of x^5y^2), we can multiply 7 by 6 and divide by 2. Said in other words, to find the term following $7x^6y$ in the expansion of $(x + y)^7$, Theorem 9.6 tells us to multiply the coefficient 7 by 6 (the exponent of x) and divide by 2 (because we're in the 2nd term). This gives us 21 and so we have

$$(x + y)^7 = x^7 + 7x^6y + 21x^5y^2 + \cdots.$$

The term immediately following $21x^5y^2$ is therefore

$$\left(\tfrac{21\cdot 5}{3}\right)x^4y^3 = 35x^4y^3.$$

This can be done in our head. Because of the symmetry of the coefficients in any binomial expansion,

$$(x + y)^7 = x^7 + 7x^6y + 21x^5y^2 + 35x^4y^3 + 35x^3y^4 + 21x^2y^5 + 7xy^6 + y^7.$$

Therefore, the entire expansion of $(x + y)^7$ can be written in a matter of seconds.   ♦

We mention one other curious property of some of the numbers in the Pascal triangle. First observe that $\binom{0}{0} = 1 = F_1$ and $\binom{1}{0} = 1 = F_2$. Also

$$
\begin{aligned}
\binom{2}{0} + \binom{1}{1} &= 2 = F_3 \\
\binom{3}{0} + \binom{2}{1} &= 3 = F_4 \\
\binom{4}{0} + \binom{3}{1} + \binom{2}{2} &= 5 = F_5 \\
\binom{5}{0} + \binom{4}{1} + \binom{3}{2} &= 8 = F_6.
\end{aligned}
\tag{9.2}
$$

Recall that the Fibonacci numbers $F_1, F_2, F_3, \ldots$ are defined recursively in Section 4.6 by

$$F_1 = 1, \ F_2 = 1 \text{ and } F_n = F_{n-2} + F_{n-1} \text{ for } n \ge 3.$$

Thus $F_3 = 2$, $F_4 = 3$, $F_5 = 5$ and $F_6 = 8$. The pattern indicated in (9.2) can be stated as follows. (See Exercise 18.)

Theorem 9.8   *For each positive integer n, the nth Fibonacci number is*

$$
F_n = \begin{cases}
\binom{n-1}{0} + \binom{n-2}{1} + \binom{n-3}{2} + \cdots + \binom{k}{k-1} & \text{if } n = 2k \\[2mm]
\binom{n-1}{0} + \binom{n-2}{1} + \binom{n-3}{2} + \cdots + \binom{k}{k} & \text{if } n = 2k+1.
\end{cases}
$$

Weighted Averages

The identities that we've described with connections to the Pascal triangle and the Binomial Theorem are referred to as combinatorical identities and applications of these can occur when we least expect them. We now describe an example of this.

Example 9.9

Suppose that during a 15-week semester, the instructor of a Discrete Mathematics course gives a quiz every week, each quiz valued at 100 points. Two students, Allyson and Brian, have the following quiz scores.

Week	1	2	3	4	5	6	7	8	9	10	11	12	13	14	15
Allyson	65	67	63	71	75	84	90	81	91	88	90	94	91	92	97
Brian	95	97	92	93	87	95	91	82	84	77	73	75	71	62	66

When the quiz averages are computed at the end of the semester, Brian's average is 83.1 while Allyson's is 82.6. Brian receives a grade of B^+ according to the instructor's grading scale and Allyson receives a grade of B.

If we were to look at Allyson's and Brian's quiz scores more closely, we see that Brian's scores have been decreasing overall, while Allyson's scores have been improving. Despite the fact that their averages are close, Brian receives a higher grade from the instructor. Are these grades appropriate? The problem we're facing is that all quizzes are being treated equally, regardless of when during the semester that they are taken. One might think that a student's quiz score should count more if the quiz is taken later in the semester.

For example, suppose that a student's quiz scores are $s_1, s_2, \ldots, s_{15}$, that is, the student's score on the quiz given during Week i is s_i. On the quiz given during the first week, the student's score is therefore s_1 and so the student's quiz average after Week 1 is also s_1. Let us denote the student's quiz average after the first week by $\overline{s_1}$, that is, $\overline{s_1} = s_1$. Now the student takes a quiz during the 2nd week and receives a score of s_2. Since this quiz is more recent, we decide that this score should count more and we decide to give that score double the "weight" of the score on the first quiz. So the student's quiz score average after Week 2, namely $\overline{s_2}$, is obtained by counting $s_1 = \overline{s_1}$ once and s_2 twice. It is as if the student has taken three quizzes. The weighted quiz average after Week 2 is therefore

$$\overline{s_2} = \frac{\overline{s_1} + 2s_2}{3} = \frac{s_1 + 2s_2}{3}.$$

Now the student takes a quiz during Week 3 and receives a score of s_3. Prior to taking that quiz, the student's 2-week quiz average was $\overline{s_2}$. However, because the quiz taken in Week 3 was taken more recently, we again decide to count that double. So the student's 3-week average is computed by counting $\overline{s_2}$ twice (because it's the average over a 2-week period) and s_3 twice (because it's a most recent quiz score), that is,

$$\begin{aligned}
\overline{s_3} &= \frac{2\overline{s_2} + 2s_3}{4} = \frac{2 \cdot \frac{\overline{s_1} + 4s_2}{3} + 2s_3}{4} \\
&= \frac{2s_1 + 4s_2 + 6s_3}{12} = \frac{s_1 + 2s_2 + 3s_3}{6}.
\end{aligned}$$

This is the same, therefore, as counting s_1 once, s_2 twice and s_3 three times. Continuing in this manner, we arrive at

$$\begin{aligned}
\overline{s_4} &= \frac{3\overline{s_3} + 2s_4}{5} = \frac{s_1 + 2s_2 + 3s_3 + 4s_4}{10} \\
&= \frac{s_1 + 2s_2 + 3s_3 + 4s_4}{\binom{5}{2}}
\end{aligned}$$

and more generally, for $1 \le k \le 15$,

$$\overline{s_k} = \frac{\sum_{i=1}^{k} i s_i}{\binom{k+1}{2}}$$

and so

$$\begin{aligned}
\overline{s_{15}} &= \frac{\sum_{i=1}^{15} i s_i}{\binom{16}{2}} = \frac{\binom{1}{1} s_1 + \binom{2}{1} s_2 + \cdots + \binom{15}{1} s_{15}}{\binom{16}{2}} \\
&= \frac{1 s_1 + 2 s_2 + \cdots + 15 s_{15}}{120}.
\end{aligned}$$

Note that

$$\binom{1}{1} + \binom{2}{1} + \cdots + \binom{15}{1} = \binom{16}{2} = 120,$$

which is a consequence of (9.1). When Allyson's weighted quiz average is computed, the result is 88.1 and now she receives an A^-. When Brian's weighted quiz average is computed, the result is 77.0 and Brian receives a C^+. ♦

The preceding example is one of many where it might seem more logical to compute a weighted average of a collection of numbers rather than compute the standard average. This is especially the case when the numbers to be averaged have occurred over a period of time and the more recent numbers should be given more emphasis (weight) than the earlier numbers.

In Example 9.9, we assigned weights of $1, 2, \ldots, 15$ to sets of 15 numbers to compute a weighted average and used this instead of the standard average (where the weights $1, 1, \ldots, 1$ are used). Since we are using the binomial coefficients $\binom{1}{1}$, $\binom{2}{1}$, $\ldots$, $\binom{15}{1}$ as weights, we might refer to this as a binomial weighted average. Of course, other weights can be used, such as $\binom{2}{2}$, $\binom{3}{2}$, $\binom{4}{2}$, $\ldots$, $\binom{16}{2}$.

Exercises for Section 9.1

1. Another property of the Pascal triangle concerns 4 numbers lying within certain rhombuses (parallellograms with equal sides). See Figure 9.7.

   (a) Compute the sum of these numbers. Based on this, make a guess and show that your guess is correct.

   (b) For each such rhombus, compute the product of the elements in the upper left and lower right positions minus the product of the remaining two elements. Based on this, make a guess and show that your guess is correct.

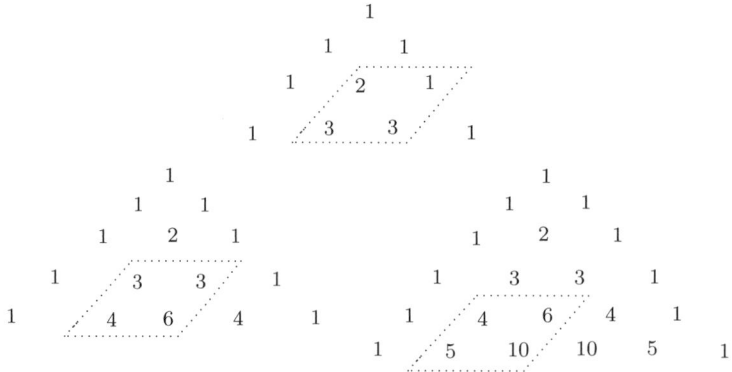

Figure 9.7: Rhombuses containing 4 elements in the Pascal triangle in Exercise 1

2. If $\binom{n}{r} = 56$ and $\binom{n-1}{r-1} = 35$ for some integers r and n with $1 \leq r \leq n-1$, what is $\binom{n-1}{r}$?

3. Is the number 22 one of the numbers in the Pascal triangle?

4. Define a function $f : \mathbb{N} \to \mathbb{N}$ by $f(a) = n$ if n is the smallest positive integer for which there exists an integer r with $0 \leq r \leq n$ such that $a = \binom{n}{r}$. What is $f(6)$?

5. Expand $(x - y)^6$, giving precise coefficients.

6. Expand $(x - y)^7$, giving precise coefficients.

7. Expand $(x + y)^8$, giving precise coefficients.

8. What is the exact coefficient of x in $\left(x^2 + \frac{1}{x}\right)^8$?

9. What is the exact term in $(x + y)^{10}$ containing x^4?

10. Expand $(2x - y)^5$, giving precise coefficients.

11. What is the exact term in $\left(2x^2 + \frac{y}{2}\right)^8$ containing x^6?

12. What is the exact term in $\left(x^3 - 2y\right)^7$ containing x^{12}?

13. What is the exact coefficient of $x^4 y^3$ in $(2x - 3y)^7$?

14. What is the exact coefficient of $a^7 b^3$ in $\left(2a - \frac{b}{2}\right)^{10}$?

15. What is the exact term in $\left(2x^2 - y\right)^8$ containing x^6?

16. What is the exact term in $\left(\sqrt{x} - \frac{1}{\sqrt{x}}\right)^{10}$ containing x?

17. One term in the expansion of $\left(2x - \frac{1}{2x^2}\right)^{10}$ is cx for some constant c. Determine c.

18. Prove Theorem 9.8: *For each positive integer n, the nth Fibonacci number is*

$$F_n = \begin{cases} \binom{n-1}{0} + \binom{n-2}{1} + \binom{n-3}{2} + \cdots + \binom{k}{k-1} & \text{if } n = 2k \\ \binom{n-1}{0} + \binom{n-2}{1} + \binom{n-3}{2} + \cdots + \binom{k}{k} & \text{if } n = 2k+1. \end{cases}$$

19. Without computing any of the numbers $\binom{n}{r}$, show that

$$1 \cdot \binom{4}{0} + 4 \cdot \binom{4}{1} + 6 \cdot \binom{4}{2} + 4 \cdot \binom{4}{3} + 1 \cdot \binom{4}{4} = 1 \cdot \binom{5}{1} + 3 \cdot \binom{5}{2} + 3 \cdot \binom{5}{3} + 1 \cdot \binom{5}{4}.$$

20. Do there exist constants a and b such that every coefficient in $(ax + by)^3$ is an even integer but no coefficient is divisble by 4?

21. During a particular semester, four exams are given in a Calculus course. The exam scores for three students in the course are (in order):

Carlos: 78, 78, 78, 78   Dina: 82, 80, 84, 66   Eric: 66, 82, 80, 84.

   (a) What is the (standard) exam average for all three students?

   (b) What is the weighted exam average for all three students, where Exam i is assigned weight i for $1 \leq i \leq 4$?

22. During a particular semester, ten 10-point quizzes are given in a statistics course. The quiz scores for one of the students in class are (in order): 1, 2, 3, 4, 5, 6, 7, 8, 9, 10.

(a) What is the (standard) quiz average for this student? [Recall that $1 + 2 + 3 + \cdots + n = \frac{n(n+1)}{2}$.]

(b) What is the weighted quiz average for this student, where Exam i is assigned weight i for $1 \leq i \leq 10$? [Recall by Result 5.5 that $1^2 + 2^2 + 3^2 + \cdots + n^2 = \frac{n(n+1)(2n+1)}{6}$.]

23. In Example 9.9 concerning weighted averages, what conclusion can be reached if the most recent quiz score in each case is given a weight of 3 rather than 2?

24. Give an example of a problem involving the Binomial Theorem whose solution results in the trigonometric identity $\sin^6 x + 3 \sin^2 x \cos^2 x + \cos^6 x = 1$.

25. Use the Binomial Theorem to verify the trigonometric identity
$$\sin^8 x + 4 \sin^2 x \cos^2 x - 2 \sin^4 x \cos^4 x + \cos^8 x = 1.$$

26. Use the Binomial Theorem to verify the trigonometric identity
$$\sec^6 x - 3 \sec^2 x \tan^2 x - \tan^6 x = 1.$$

27. Show that $\displaystyle\sum_{r=0}^{n} \binom{n}{r} 2^{2n-2r} = 5^n$.

28. Use the Binomial Theorem to simply the sum
$$2^{n^2} + \binom{n}{1} 2^{n(n-1)} + \binom{n}{2} 2^{n(n-2)} + \cdots + \binom{n}{n-1} 2^n + 1.$$

9.2   Permutations and Combinations with Repetition

In the permutations and combinations that we have seen thus far, the elements involved have been distinct. This, in fact, is the standard understanding for permutations and combinations. There are occasions, however, when this restriction is not appropriate. For example, if we wish to know how many 7-digit telephone numbers there are with the same area code, then we certainly must allow digits to be repeated. Indeed, this number can be computed by an application of the Multiplication Principle. Since there are 10 choices for each of the 7 digits, the number of different telephone numbers with the same area code is $10^7 = 10,000,000$. This illustrates a more general result.

Theorem 9.10   *The number of r-permutations with repetition of an n-element set is n^r.*

Permutations with Repetition

Next we turn to permutations with repetition where there are some additional restrictions.

Example 9.11

How many different permutations are there of the letters of the word PEPPER?

Solution. We have to be a bit careful here. Despite the fact that this is a 6-letter word, the answer to this question is *not* 6! = 720. For example, the permutation that interchanges the 3rd and 4th letters does not produce a new permutation. So the number of different permutations of the letters of the word PEPPER is surely less than 720. Suppose that the number of distinct permutations of the letters of PEPPER is N. One of these permutations is PPPEER. We can make the 3 Ps distinct by replacing them by P_1, P_2 and P_3. From the permutation PPPEER, we obtain 3! = 6 different permutations, namely

$$P_1 P_2 P_3 EER \quad P_2 P_1 P_3 EER \quad P_3 P_1 P_2 EER$$
$$P_1 P_3 P_2 EER \quad P_2 P_3 P_1 EER \quad P_3 P_2 P_1 EER.$$

Therefore, with the 3 letters P distinguishable, we now have $3!N$ permutations. Consider the permutation $P_1 P_2 P_3 EER$. We can make the 2 Es distinguishable by replacing one by E_1 and the other by E_2. For the permutation $P_1 P_2 P_3 EER$, we obtain $2! = 2$ different permutations, namely $P_1 P_2 P_3 E_1 E_2 R$ and $P_1 P_2 P_3 E_2 E_1 R$. If we make all 3 Ps, both Es and R distinguishable, we arrive at the 6! permutations of the 6 distinct elements. Thus $3!\, 2!\, 1!\, N = 6!$ and so

$$N = \frac{6!}{3!\, 2!\, 1!} = 60.$$
♦

Example 9.11 is a special case of a more general result.

Theorem 9.12 *Suppose that there are n objects, exactly k of which are distinguishable. If there are n_1 objects of type 1, n_2 objects of type 2 and so on, up to n_k objects of type k, where $n_1 + n_2 + \cdots + n_k = n$, then the number of different permutations of these n objects is*

$$\frac{n!}{n_1!\, n_2! \cdots n_k!}.$$

Proof. Of the n positions, there are $\binom{n}{n_1}$ possible locations for the n_1 objects of type 1. Once the location for these n_1 objects has been determined, there are $n - n_1$ positions remaining for the other objects. There are $\binom{n-n_1}{n_2}$ possible locations for the n_2 objects of type 2 and $\binom{n-n_1-n_2}{n_3}$ possible locations for the n_3 objects of type 3 and so on, until arriving at

$$\binom{n - n_1 - n_2 - \cdots - n_{k-1}}{n_k} = \binom{n_k}{n_k}$$

possible locations for the n_k objects of type k. By the Multiplication Principle, the number of different permutations of these n objects is

$$\binom{n}{n_1}\binom{n-n_1}{n_2}\binom{n-n_1-n_2}{n_3}\cdots\binom{n - n_1 - n_2 - \cdots - n_{k-1}}{n_k}$$

$$= \frac{n!}{n_1!\,(n-n_1)!} \cdot \frac{(n-n_1)!}{n_2!\,(n-n_1-n_2)!} \cdot \frac{(n-n_1-n_2)!}{n_3!\,(n-n_1-n_2-n_3)!} \cdots \frac{(n-n_1-n_2-\cdots-n_{k-1})!}{n_k!\, 0!}$$

$$= \frac{n!}{n_1!\, n_2! \cdots n_k!},$$

as claimed. ∎

Example 9.13

Determine the number of distinct permutations of the letters of each of the following words:

(a) HAWAIIAN. (b) DEED. (c) STREETS. (d) POSSESSES.

Solution. (a) $\frac{8!}{3!\, 2!\, 1!\, 1!\, 1!} = 3360$. (b) $\frac{4!}{2!\, 2!} = 6$. (c) $\frac{7!}{2!\, 2!\, 2!\, 1!} = 630$. (d) $\frac{9!}{5!\, 2!\, 1!\, 1!} = 1512$. ♦

Example 9.14

Suppose that we have 10 letters, 4 of which are x, 3 of which are y and 3 of which are z. What is the number of distinct arrangements of these letters?

Solution. $\frac{10!}{4!\, 3!\, 3!} = 4200$. ♦

Example 9.15

How many distinct 8-digit numbers can be formed from the digits in the number 23,532,353?

Solution. $\frac{8!}{2!\ 2!\ 4!} = 420.$ ◆

The type of argument we've been using to compute the number of distinct permutations with repetition can also be used to solve problems which may not appear to fit into this category.

Example 9.16

Jennie, who is about to be married, has purchased 13 different gifts for the 4 girls in her wedding party. Jennie has decided to give 4 gifts to her maid of honor and 3 gifts each to her other three friends. In how many ways can this be done?

Solution. Suppose that Ann (A) is the maid of honor and Jennie's other three friends are Betty (B), Connie (C) and Debbie (D). The number of ways of giving 4 of the 13 gifts to Ann is $\binom{13}{4}$. The number of ways of giving 3 of the remaining 9 gifts to Betty is $\binom{9}{3}$. Moreover, the number of ways of giving 3 of the remaining 6 gifts to Connie is $\binom{6}{3}$ and finally there are $\binom{3}{3}$ ways of giving the final 3 gifts to Debbie. By the Multiplication Principle, the number of ways of distributing all of these gifts is

$$\binom{13}{4}\binom{9}{3}\binom{6}{3}\binom{3}{3} = \frac{13!}{4!\ 9!} \cdot \frac{9!}{3!\ 6!} \cdot \frac{6!}{3!\ 3!} \cdot \frac{3!}{3!\ 0!} = \frac{13!}{4!\ 3!\ 3!\ 3!} = 1,201,200.$$

There is another way to look at this problem. Suppose that we consider 13 slots, numbered 1 through 13, corresponding to the 13 gifts:

$$\overline{}\ \overline{}\ \overline{}\ \overline{}\ \overline{}\ \overline{}\ \overline{}\ \overline{}\ \overline{}\ \overline{}\ \overline{}\ \overline{}\ \overline{}$$
$$\ \ 1\quad 2\quad 3\quad 4\quad 5\quad 6\quad 7\quad 8\quad 9\quad 10\quad 11\quad 12\quad 13$$

If we were to write AAAABBBCCCDDD, this would correspond to Ann receiving gifts 1 – 4, Betty receiving gifts 5 – 7, Connie receiving gifts 8 – 10 and Debbie receiving gifts 11 – 13. Therefore, the number of ways of distributing the 13 gifts equals the number of distinct permutations of the letters of AAAABBBCCCDDD. This number is

$$\frac{13!}{4!\ 3!\ 3!\ 3!} = 1,201,200. \ ◆$$

Combinations with Repetition

We now turn from problems involving permutations with repetition to problems involving combinations with repetition. We have already seen that the number of r-element subsets of an n-element set is

$$C(n,r) = \binom{n}{r} = \frac{n!}{r!\ (n-r)!}.$$

Of course, in this case, we were assuming that the n elements in the set are distinct. Let's consider an example where the elements in a set are not distinct.

Example 9.17

A young boy has a large collection of red (R), blue (B) and green (G) marbles that he keeps in a bowl. How many different outcomes are possible if he randomly selects 3 marbles from the bowl?

Solution. The possibilities are

RRR, BBB, GGG, RRB, RRG, BBR, BBG, GGR, GGB, RBG,

where the order in which the colors are listed is immaterial. Hence there are 10 different outcomes. ◆

The solution we gave in Example 9.17 consisted of listing the different possibilities. Since there weren't many different colors of the marbles (only 3) and the boy wasn't selecting many marbles (only 3), listing the possibilities was practical and not difficult. However, this would not have been the case if the number of marbles the boy selected was large and the number of colors of the marbles was large. Therefore, let's consider another solution to the problem in Example 9.17, one that does not depend on the small number of marbles selected and the small number of colors involved.

We can think of each marble in Example 9.17 as belonging to one of 3 categories depending on whether its color is red, blue or green. Therefore, there is a division of the marbles as follows:

Red | Blue | Green

We have already seen that there are 10 different outcomes if 3 marbles are selected. We repeat these 10 outcomes and show how these can be represented in the division of marbles listed above by the diagram shown below.

	Red	Blue	Green
RRR	o o o |	|	
BBB	|	o o o |	
GGG	|	|	o o o
RRB	o o |	o	|
RRG	o o |	|	o
BBR	o |	o o	|
BBG	|	o o |	o
GGR	o |	|	o o
GGB	|	o |	o o
RBG	o |	o |	o

Observe that for each of the 10 outcomes, the corresponding diagram consists of 5 symbols (3 os for the 3 marbles to be selected and 2 |s for the 2 separators needed to divide the marbles into 3 colors). The order in which these 5 symbols appear tell us how many marbles of a certain color are being selected. Therefore, we may think of 5 positions (or slots)

$$\underline{\quad}\ \underline{\quad}\ \underline{\quad}\ \underline{\quad}\ \underline{\quad}$$
$$\ \ 1 \quad 2 \quad 3 \quad 4 \quad 5$$

where in each slot, either the symbol o or the symbol | is to be placed (a total of 3 os and 2 |s in each case). The number of ways of selecting 3 of the 5 positions to place the symbol o is $C(5,3) = \binom{5}{3} = \binom{5}{2} = \frac{5 \cdot 4}{2} = 10$. This, of course, equals the number of ways of selecting 2 of the 5 positions to place the separator line symbol |, which is $C(5,2) = \binom{5}{2} = 10$.

Example 9.18

A lady enters Ben's Bagels for the purpose of purchasing 4 bagels. The restaurant currently has 5 kinds of bagels: plain, cherry, blueberry, crunchy and coconut. How many different assortments are possible?

Solution. We can think of dividing the bagels into the 5 kinds by 4 separating lines:

plain | cherry | blueberry | crunchy | coconut

Let the symbol ∗ indicate a bagel to be purchased. For example,

| ∗ | | ∗∗ | ∗

indicates that a cherry bagel, 2 crunchy bagels and a coconut bagel have been selected. Therefore, we can think of this problem involving 8 positions

$$\overline{}_{1} \quad \overline{}_{2} \quad \overline{}_{3} \quad \overline{}_{4} \quad \overline{}_{5} \quad \overline{}_{6} \quad \overline{}_{7} \quad \overline{}_{8}$$

where 4 of these 8 positions correspond to the 4 bagels being purchased and the other 4 positions correspond to the 4 separators that divide the bagels into 5 types. The number of different assortments that the lady can purchase is therefore the number of ways of selecting 4 of the 8 positions to place the symbol $*$, which is $\binom{8}{4} = \frac{8!}{4!\,4!} = 70$. ♦

The two examples we just presented illustrate the following theorem.

Theorem 9.19 *Let A be a set containing t different kinds of elements, where there are at least s elements of each kind. The number of different selections of s elements from A is*

$$\binom{s+t-1}{s}.$$

Proof. A selection of s elements from A corresponds to a sequence of $s+t-1$ elements

$$\overline{}_{1} \quad \overline{}_{2} \quad \overline{}_{3} \quad \overset{\cdots}{\cdots} \quad \overline{}_{s+t-2} \quad \overline{}_{s+t-1}$$

where s terms in the sequence correspond to the s elements selected and the remaining $t-1$ terms correspond to the $t-1$ separating lines (which separate the t different kinds of elements). Then the number of selections is the number of ways of choosing s of the $s+t-1$ terms in the sequence for the elements selected. This number is $\binom{s+t-1}{s}$. ∎

There is another interpretation of Theorem 9.19. When descussing a set S, with n elements say, we always assume that the n elements of S are distinct, that is, no two elements of S are the same. There are circumstances, however, when we consider collections of elements, some of which may be the same. A **multiset** is a collection of elements in which repetition of elements is permitted. Hence, considered as multisets, $\{a, a, b\}$ and $\{a, b, b\}$ are distinct 3-element multisets. In fact, the distinct 3-element multisets, every element of which is R, B or G, are:

$$\{R, R, R\}, \ \{B, B, B\}, \ \{G, G, G\}, \ \{R, R, B\}, \ \{R, R, G\},$$
$$\{B, B, R\}, \ \{B, B, G\}, \ \{G, G, R\}, \ \{G, G, B\}, \ \{R, B, G\}.$$

This, of course, corresponds to the solution given to Example 9.17. In fact, Theorem 9.19 can be restated in terms of multisets.

Theorem 9.20 *The number of s-element multisets whose elements belong to a t-element set is*

$$\binom{s+t-1}{s}.$$

We now return to our preceding discussion.

Example 9.21

A college student keeps his spending money in a drawer in no particular order. He has many $1, $5, $10 and $20 bills there. After just getting up one morning (and still half asleep), he takes 5 bills from the drawer. How many different assortments of bills could he have taken?

Solution. The student has made a selection of $s = 5$ bills and there are $t = 4$ kinds of bills, so the number of different assortments is

$$\binom{s+t-1}{s} = \binom{8}{5} = \frac{8!}{5!\,3!} = 56.$$

Once again this problem could be looked at as dividing the money in the drawer into 4 denominations by 3 separating lines:

$$\$1 \quad | \quad \$5 \quad | \quad \$10 \quad | \quad \$20$$

Since there are $s = 5$ bills to be selected and $t - 1 = 3$ separating lines, the number of possible outcomes is, as above, $\binom{s+t-1}{s} = \binom{8}{5} = 56.$ ♦

Example 9.22

Four students are finalists in a contest in which ten questions are asked of the four students. Only the first student to answer a particular question correctly receives credit for that question. We are interested in the number of questions answered by each student. Assuming that every question is answered correctly by some student, how many outcomes are possible?

Solution. Suppose that we denote the four students by S_1, S_2, S_3, S_4. For $1 \leq i \leq 4$, let a_i be the number of questions answered correctly by S_i. Then we can think of dividing the 10 answered questions among these four students as

$$a_1 \quad | \quad a_2 \quad | \quad a_3 \quad | \quad a_4$$

In this case, there are $s = 10$ answered questions and $t - 1 = 3$ separating lines; so the number of possible outcomes is

$$\binom{s+t-1}{s} = \binom{13}{10} = \frac{13!}{10! \, 3!} = 286.$$ ♦

Example 9.23

Three friends attend a young boy's birthday party. The mother has arranged 10 games where a prize is awarded for winning a game. The prizes are identical. If each of the 4 children receives at least one prize, how many distributions of prizes are possible?

Solution. Since every child will receive at least one prize, the only question concerns the distributions of the remaining 6 prizes among these 4 children. Since there are $s = 6$ remaining prizes and $t-1 = 3$ separating lines, the number of distributions of prizes is

$$\binom{s+t-1}{s} = \binom{9}{6} = \frac{9!}{6! \, 3!} = 84.$$ ♦

Example 9.24

A music program is to consist of 22 musical selections, namely 15 piano numbers and 7 violin numbers. How many programs are possible if no two violin selections are consecutive?

Solution. Necessarily, there is at least one piano selection between every two violin selections, so 13 of the 22 selections must appear in the program as

$$\text{V P V P V P V P V P V P V,} \tag{9.3}$$

where each of the remaining $15 - 6 = 9$ piano selections must be inserted somewhere in the sequence (9.3), including the possibility of inserting it at the beginning or at the end of the sequence. The only question is where in the schedule the remaining 9 piano selections will occur. Hence there are 8 possible locations for these 9 piano selections where each separating line corresponds to one of the 7 violin selections.

$$
\begin{array}{ccccccc}
V & V & V & V & V & V & V \\
| & | & | & | & | & | & |
\end{array}
$$

Thus there are $s = 9$ remaining piano selections and $t - 1 = 7$ separating lines (violin selections), and so the number of possible programs is

$$\binom{s+t-1}{s} = \binom{16}{7} = \frac{16!}{7!\,9!} = 11,440. \qquad \blacklozenge$$

Example 9.25

How many 3-digit numbers are there, the sum of whose digits is 9?

Solution. Any 3-digit number n can be expressed as $n_2 n_1 n_0$, where n_2, n_1 and n_0 are the digits of n. By assumption, $n_2 + n_1 + n_0 = 9$. Since n is a 3-digit number, $n_2 \geq 1$. The number n can be thought of being represented as follows:

$$n_2 \quad | \quad n_1 \quad | \quad n_0$$

For example, the integer 513 can be represented as:

$$* \; * \; * \; * \; * \quad | \quad * \quad | \quad * \; * \; *$$

Since $n_2 \geq 1$, one $*$ is located in the first location. The question remains of determining the number of ways of dividing the remaining $s = 8$ $*$s in the $t = 3$ locations. The answer to this is

$$\binom{s+t-1}{s} = \binom{10}{8} = \binom{10}{2} = \frac{10 \cdot 9}{2} = 45. \qquad \blacklozenge$$

Exercises for Section 9.2

1. Determine the number of distinct permutations (with repetition) of the letters of each of the following words.

   (a) LEVEL. (b) BABBLE. (c) TEETER. (d) REWIND. (e) HIAWATHA. (f) MISSISSIPPI.

2. Determine the number of distinct arrangements of 9 letters, 3 of which are A, 2 of which are B and 4 of which are C.

3. How many 7-digit numbers can be formed from the digits in the following numbers?

   (a) 4,116,461. (b) 8,555,858.

4. How many different 8-digit numbers can be obtained by permuting the digits in the number 70,440,704? (Since each number is an 8-digit number, the first digit cannot be 0.)

5. What is the number of 4-element multisets whose elements belong to the set $\{1, 2, 3, 4\}$?

6. What is the number of 5-element multisets whose elements belong to the set $\{1, 2, 3, 4, 5\}$ such that at least two elements in each multiset are distinct?

7. A 10-question multiple choice quiz is known to have 4 questions where the answer is (a), 3 questions where the answer is (b) and 3 questions where the answer is (c). How many possible assortments of answers are there?

8. Eight different computer science books have been purchased to give to 4 students who won a programming contest. If each student is to receive 2 of the 8 books, in how many ways can the 8 books be distributed to these students?

9. Seven young boys have moved into a new community and would like to participate in little league baseball. Three teams, namely the Falcons, the Jaguars and the Panthers, agree to take the 7 new players, with the Falcons taking 3 and the other teams taking 2 each. In how many ways can this be done?

10. For the final exam in a musical theater class, the three students John, Kim and Lisa are to present a program of 10 songs, with John singing 4 of them and Kim and Lisa singing 3 each. In how many different orders can the program be presented?

11. The Chair of a University Department is making committee assignments for the coming year. There are 8 faculty members who have not yet received their committee assignments. If there are 4 openings on the Undergraduate Committee and 2 openings each on the Graduate and Personnel Committees, then how many possible committee assignments are there for the 8 faculty members?

12. Before leaving with his girlfriend to visit her parents, a young man visits a fruit market to purchase a gift. He learns that for $20 he can purchase a fruit basket containing 10 pieces of fruit, chosen from apples, oranges, peaches, pears and plums. When asked what assortment of fruit he would like, he responds "I don't care." How many different assortments are possible?

13. At a buffet dinner, one of the people at a table (seating 6 people) volunteers to go to the dessert table to bring back a dessert for each person at the table. When he arrives at the dessert table, he learns that he has three choices: apple pie, chocolate cake and ice cream. How many different choices does he have for desserts to select?

14. A professor has 10 identical new pens that he no longer needs. In how many ways can these pens be given to 3 students if

    (a) there are no other conditions?
    (b) every student must receive at least one pen?
    (c) every student must receive at least two pens?
    (d) every student must receive at least three pens?

15. Ten people have been selected to receive gift certificates to a restaurant. Two people will receive $200 gift certificates, three will receive $100 gift certificates and five will receive $50 gift certificates. In how many ways can these gift certificates be distributed?

16. A total of 12 students have volunteered to participate in student government for the coming year. In an election, the top vote-getter becomes president of the student organization, the second highest vote-getter becomes vice-president and the next three highest vote-getters become members of the executive committee. How many outcomes are possible?

17. A total of 12 different computer science books are to be given to the top 3 participants in a programming contest. How many ways can the books be distributed if the person finishing first gets 6 books, the person finishing second gets 4 books and the person finishing third gets 2 books?

18. A bowl contains a large number of red, blue, green and yellow marbles (mixed together in the bowl). A young boy takes a handful of marbles and sees that he has selected 4 marbles. What is the number of possibilities for the 4 marbles he selected?

19. In how many ways can a man distribute 10 silver dollars to his three nephews if every nephew gets at least one silver dollar?

20. A coin collector has 10 identical silver coins and 6 identical gold coins. In how many ways can he place these 16 coins in a row if no two gold coins are placed next to each other?

21. In how many ways can 10 identical pens and 7 identical pocket calculators be distributed to 3 students?

22. How many 5-digit numbers are there, the sum of whose digits is 8?

23. How many 3-digit numbers are there, the sum of whose digits is 10?

24. How many 4-digit numbers are there, the sum of whose digits is 11?

25. A man enters a bakery just before it closes in order to buy donuts. There are only 4 kinds of donuts left, namely 5 plain donuts, 6 glazed donuts, 7 coconut donuts and 8 chocolate donuts. The man places no restrictions on the kinds of donuts he buys. How many different assortments are possible if the man decides to buy

    (a) 5 donuts?   (b) 6 donuts?   (c) 7 donuts?

26. A woman enters Doug's Donuts to purchase some donuts. Give an example of a question that has $\binom{10}{6} - 4 - 4^2 = 190$ as its answer.

9.3   Generating Functions

In this section, we consider the topic of generating functions, a concept that has a variety of applications, including (1) solving certain kinds of counting problems and (2) determining the general terms of some recursively-defined sequences of real numbers. Before introducing this concept, we consider a few counting problems. We begin with two examples, each of a type that we have encountered earlier.

Example 9.26

Suppose that we were to select two balls from a bowl consisting of one red ball (R), one blue ball (B), one green ball (G) and one yellow ball (Y). Then there are $\binom{4}{2} = 6$ possible outcomes, namely:

$$RB \quad RG \quad RY \quad BG \quad BY \quad GY. \qquad \blacklozenge$$

Example 9.27

Suppose that we were to select two balls from a bowl consisting of two red balls, two blue balls, two green balls and two yellow balls. Since we are selecting $s = 2$ objects of which there are at least two each of the $t = 4$ different kinds, it follows by Theorem 9.12 that the number of possible outcomes is

$$\binom{s+t-1}{s} = \binom{5}{2} = \frac{5 \cdot 4}{2} = 10.$$

These outcomes are

$$RR \quad BB \quad GG \quad YY \quad RB \quad RG \quad RY \quad BG \quad BY \quad GY. \qquad \blacklozenge$$

There are other ways to consider these two problems. Let's look at Example 9.26 again. We know by the Binomial Theorem that

$$
\begin{aligned}
(1+x)^4 &= \binom{4}{0} + \binom{4}{1}x + \binom{4}{2}x^2 + \binom{4}{3}x^3 + \binom{4}{4}x^4 \\
&= 1 + 4x + 6x^2 + 4x^3 + x^4.
\end{aligned}
$$

Of course, $(1 + x)^4$ is the product of four polynomials $1 + x$, that is,

$$(1 + x)^4 = (1 + x)(1 + x)(1 + x)(1 + x).$$

To obtain a term in the product

$$(1 + x)(1 + x)(1 + x)(1 + x),$$

we select either 1 or x in each of the four polynomials $1 + x$ and multiply them. In particular, to obtain the term x^2, we select x from two of the four polynomials $1 + x$ and select 1 from the remaining two polynomials $1 + x$. Since there are $\binom{4}{2} = 6$ ways to select x from two of the four polynomials, the coefficient of x^2 is 6. Of course, we may write $1 + x$ as $x^0 + x^1$ and think of each of the four polynomials $1 + x$ as representing one of the four balls (red, blue, green, yellow). That is, we are looking at these four polynomials representing the four colors:

$$\begin{array}{cccc} (1 + x) & (1 + x) & (1 + x) & (1 + x). \\ \text{red} & \text{blue} & \text{green} & \text{yellow} \end{array} \qquad (9.4)$$

Thus selecting two balls from a bowl of four balls (one of each of the four colors) is the same as selecting x in two of the four polynomials $1 + x$ and selecting 1 in the remaining two polynomials $1 + x$ in the product (9.4). For example, selecting $x = x^1$ in the first polynomial means that we are selecting a red ball (1 red ball); while selecting $1 = x^0$ in the first polynomial means that we are selecting 0 red balls. Consequently, if one of the two balls selected is blue and the other is green, then this is equivalent to selecting 1 in the first and fourth polynomials $1 + x$ in the product (9.4) and selecting x in the second and third polynomials $1 + x$ and then multiplying these (to obtain the term x^2).

We now turn to Example 9.27 and see how this problem can be looked at from a similar point of view. If we were to expand $\left(1 + x + x^2\right)^4$, we would obtain

$$\begin{aligned} \left(1 + x + x^2\right)^4 &= \left(1 + x + x^2\right)\left(1 + x + x^2\right)\left(1 + x + x^2\right)\left(1 + x + x^2\right) \\ &= 1 + 4x + 10x^2 + 16x^3 + 19x^4 + 16x^5 + 10x^6 + 4x^7 + x^8. \end{aligned}$$

To obtain the term x^2 in this product,

(i) we could select x from two of the four polynomials $1 + x + x^2$ and select 1 from the other two polynomials or

(ii) we could select x^2 from one of the four polynomials $1 + x + x^2$ and select 1 from the other three polynomials.

Since there are $\binom{4}{2} = 6$ ways to perform (i) and $\binom{4}{1} = 4$ ways to perform (ii), the total number of ways to perform (i) or (ii) is $6 + 4 = 10$ by the Addition Principle. Therefore, the coefficient of x^2 in $\left(1 + x + x^2\right)^4$ is 10. Each polynomial can be considered as representing balls of a certain color to be selected. That is, we have

$$\begin{array}{cccc} \left(1 + x + x^2\right) & \left(1 + x + x^2\right) & \left(1 + x + x^2\right) & \left(1 + x + x^2\right). \\ \text{red} & \text{blue} & \text{green} & \text{yellow} \end{array} \qquad (9.5)$$

Selecting two balls from a bowl of eight balls (two of each of the four colors) is the same as

(1) selecting x in two of the four polynomials $1 + x + x^2$ and 1 in the remaining two polynomials $1 + x + x^2$ in the product (9.5) or

(2) selecting x^2 from one of the four polynomials $1 + x + x^2$ and 1 in the remaining three polynomials $1 + x + x^2$.

For example, selecting $1 = x^0$ in the first, third and fourth polynomials $1 + x + x^2$ and x^2 in the second polynomial $1 + x + x^2$ means that we are selecting 0 red, 0 green and 0 yellow balls and 2 blue balls.

We also see that one of the other terms in the expansion of $\left(1 + x + x^2\right)^4$ is $16x^5$. This tells us that there are 16 different outcomes if we select 5 balls from a bowl of eight balls (two of each of the four colors). For example, we could select x from three of the four polynomials $1 + x + x^2$ and x^2 from the remaining polynomial. There are $\binom{4}{1} = 4$ ways to do this. Also, we could select 1 from one of the four polynomials, x from one of the other three polynomials and x^2 from the remaining two polynomials. There are $\frac{4!}{1!\,1!\,2!} = 12$ ways to do this and so the coefficient of x^5 in $\left(1 + x + x^2\right)^4$ is $4 + 12 = 16$, again by the Addition Principle.

The discussion above suggests a variety of additional problems.

Example 9.28

A bowl consists of 30 balls, namely 10 balls of each of the colors red, blue and green. In how many different ways can 10 balls be selected from the bowl such that (1) at least one and at most 7 balls are red, (2) at least 2 and at most 4 balls are blue and (3) at most 3 balls are green?

Solution. Let r, b and g denote the number of red, blue and green balls selected, respectively. Then $r + b + g = 10$, where $1 \le r \le 7$, $2 \le b \le 4$ and $g \le 3$. Since $b \le 4$ and $g \le 3$, it follows that $r = 10 - b - g \ge 10 - 4 - 3 = 3$. The selections with these restrictions are indicated below.

r	b	g
3	4	3
4	3	3
4	4	2
5	2	3
5	3	2
5	4	1
6	2	2
6	3	1
6	4	0
7	2	1
7	3	0

Consequently, the number of different ways of selecting 10 balls from the bowl (subject to the stated conditions) is 11.

There is another way to interpret the answer to our question. The number of red balls we can select is one of the exponents of x in the polynomial $x + x^2 + x^3 + x^4 + x^5 + x^6 + x^7$, the number of blue balls is an exponent of x in $x^2 + x^3 + x^4$, while the number of green balls is an exponent of x in $1 + x + x^2 + x^3$. Consider the product of these three polynomials:

$$(x + x^2 + x^3 + x^4 + x^5 + x^6 + x^7)(x^2 + x^3 + x^4)(1 + x + x^2 + x^3). \qquad (9.6)$$
$$\qquad\quad \text{red} \qquad\qquad\qquad\qquad \text{blue} \qquad\qquad \text{green}$$

When these polynomials are multiplied, a term x^r is chosen from the first polynomial (where $1 \le r \le 7$), a term x^b is chosen from the second polynomial (where $2 \le b \le 4$) and a term x^g is chosen from the third polynomial (where $0 \le g \le 3$). For example, three of these choices are $r = 3$, $b = 4$ and $g = 3$, which produces the term $x^3 x^4 x^3 = x^{3+4+3} = x^{10}$. Consequently, the answer to our question is the coefficient of x^{10} in the product (9.6), which, evidently, is 11. ◆

Example 9.29

A bowl consists of 30 balls, namely 10 balls of each of the colors red, blue and green. In how many different ways can 10 balls be selected from the bowl such that (1) at least 3 balls selected are red, (2) an even number of blue balls is selected and (3) an odd number of green balls is selected?

Solution. Again, let r, b and g denote the number of red, blue and green balls selected, respectively. Then $r + b + g = 10$, where $r \geq 3$, b is even and g is odd. Since $r + b + g$ is even, b is even and g is odd, it follows that r must be odd. The solutions with these restrictions are listed in the table below.

r	b	g
3	0	7
3	2	5
3	4	3
3	6	1
5	0	5
5	2	3
5	4	1
7	0	3
7	2	1
9	0	1

The number of different ways of selecting 10 balls, subject to the given conditions, is therefore 10. The answer to this question is also equal to the coefficient of x^{10} in the product

$$\underbrace{(x^3 + x^4 + x^5 + x^6 + x^7 + x^8 + x^9 + x^{10})}_{\text{red}}\underbrace{(1 + x^2 + x^4 + x^6 + x^8 + x^{10})}_{\text{blue}}\underbrace{(x + x^3 + x^5 + x^7 + x^9)}_{\text{green}}$$

of the three polynomials. Since certain terms in these polynomials cannot be used to produce a term x^{10}, we could consider the simpler product

$$(x^3 + x^5 + x^7 + x^9)(1 + x^2 + x^4 + x^6)(x + x^3 + x^5 + x^7).$$

From what we have seen, the coefficient of x^{10} in this product is 10. $\blacklozenge$

Example 9.30

A bowl consists of 30 balls, namely 10 balls of each of the colors red, blue and green. In how many different ways can 10 balls be selected from the bowl such that at least 2 but at most 5 balls of each color are selected?

Solution. Once again, let r, b and g denote the number of red, blue and green balls selected, respectively. Then $r + b + g = 10$, where $2 \leq r \leq 5$, $2 \leq b \leq 5$ and $2 \leq g \leq 5$. The solutions with these restrictions are listed below.

r	b	g
2	3	5
2	4	4
2	5	3
3	2	5
3	3	4
3	4	3
3	5	2
4	2	4
4	3	3
4	4	2
5	2	3
5	3	2

Thus the number of different ways of selecting 10 balls, subject to the given conditions, is 12. The answer to this questions is also the coefficient of x^{10} in the product

$$\underbrace{(x^2 + x^3 + x^4 + x^5)}_{\text{red}}\underbrace{(x^2 + x^3 + x^4 + x^5)}_{\text{blue}}\underbrace{(x^2 + x^3 + x^4 + x^5)}_{\text{green}} = (x^2 + x^3 + x^4 + x^5)^3$$

of the three (identical) polynomials. Consequently, the coefficient of x^{10} in this product is 12. $\blacklozenge$

Example 9.31

A man walks into Doug's Donuts to buy either 5 or 6 donuts (he hasn't decided which) and learns that there are only 2 chocolate donuts and 3 powdered donuts left but there is a large number of glazed donuts (Doug's specialty). How many different selections does the man have if he buys 5 or 6 donuts?

Solution. These questions can be answered by determining the coefficients of x^5 and x^6 in the product

$$\underset{\text{chocolate}}{\left(1 + x + x^2\right)} \underset{\text{powdered}}{\left(1 + x + x^2 + x^3\right)} \underset{\text{glazed}}{\left(1 + x + x^2 + x^3 + \cdots\right)}$$

of three expressions, the first two of which are polynomials and the third a power series. The product of these expressions is

$$
\begin{aligned}
&\left(1 + x + x^2\right) \left(1 + x + x^2 + x^3\right) \left(1 + x + x^2 + x^3 + \cdots\right) \\
= \; & 1 + 3x + 6x^2 + 9x^3 + 11x^4 + 10x^5 + 12x^6 + 12x^7 + 12x^8 + \cdots
\end{aligned}
$$

Consequently, there are 10 different ways to select 5 donuts and 12 different ways to select 6 donuts. ♦

We have seen therefore that to count the number of ways to select objects from a collection, where some of the objects may be the same and subject to some conditions, we can consider the product of certain polynomials or power series and compute an appropriate coefficient or coefficients. This leads us to the main concept of this chapter: generating functions. Before giving a formal definition of this concept, however, let's revisit the Binomial Theorem.

The Extended Binomial Theorem

According to the Binomial Theorem, for each nonnegative integer n,

$$(1 + x)^n = \sum_{r=0}^{n} \binom{n}{r} x^r,$$

where the binomial coefficient $\binom{n}{r} = 1$ when $r = 0$ and

$$\binom{n}{r} = \frac{n!}{r!(n-r)!} = \frac{n(n-1)\cdots(n-r+1)}{r!} \tag{9.7}$$

when $0 < r \leq n$. More generally, for a real number α and a nonnegative integer r, the **extended binomial coefficient** $\binom{\alpha}{r}$ is defined to be 1 if $r = 0$; otherwise,

$$\binom{\alpha}{r} = \frac{\alpha(\alpha-1)\cdots(\alpha-r+1)}{r!}. \tag{9.8}$$

Of course, if α and r are integers with $0 < r \leq \alpha$, then the definition of $\binom{\alpha}{r}$ in (9.8) agrees with that in (9.7). Moreover, if α is a nonnegative integer and r is an integer with $r > \alpha$, then one of the terms in the product $\alpha(\alpha-1)\cdots(\alpha-r+1)$ is 0 and so $\binom{\alpha}{r} = 0$.

The Binomial Theorem was extended by Sir Isaac Newton around 1676 to obtain an expression for $(1+x)^\alpha$ as an infinite series for any real number α, sometimes referred to as a **binomial series**.

Theorem 9.32   (The Extended Binomial Theorem) *Let $\alpha \in \mathbb{R}$. If $x \in \mathbb{R}$ with $|x| < 1$, then*

$$(1 + x)^\alpha = \sum_{r=0}^{\infty} \binom{\alpha}{r} x^r. \tag{9.9}$$

The condition that $|x| < 1$ tells us from calculus that if $-1 < x < 1$, then the infinite series in (9.9) converges and therefore has a numerical value, which in fact equals $(1 + x)^\alpha$.

In the particular case when α is a negative integer in $\binom{\alpha}{r}$, say $\alpha = -n$ where $n \in \mathbb{N}$, we have

$$
\begin{aligned}
\binom{\alpha}{r} = \binom{-n}{r} &= \frac{(-n)(-n-1)\cdots(-n-r+1)}{r!} \\
&= (-1)^r \frac{n(n+1)\cdots(n+r-1)}{r!} \\
&= (-1)^r \frac{(n+r-1)(n+r-2)\cdots n}{r!} \\
&= (-1)^r \binom{n+r-1}{r};
\end{aligned}
$$

that is,

$$
\binom{-n}{r} = (-1)^r \binom{n+r-1}{r} \quad \text{for } n \in \mathbb{N}. \tag{9.10}
$$

When $\alpha = -1$ in (9.9), we therefore have

$$
\begin{aligned}
(1+x)^{-1} &= \sum_{r=0}^{\infty} \binom{-1}{r} x^r = \sum_{r=0}^{\infty} (-1)^r \binom{r}{r} x^r \\
&= \sum_{r=0}^{\infty} (-1)^r x^r;
\end{aligned}
$$

that is,

$$
\frac{1}{1+x} = 1 - x + x^2 - x^3 + \cdots. \tag{9.11}
$$

The formula in (9.11) might seem familiar to you from the study of infinite series in calculus since the geometic series $1 - x + x^2 - x^3 + \cdots$ is known to converge to $1/(1+x)$ when $|x| < 1$.

Generating Functions

We now consider sequences whose domain is $\mathbb{N} \cup \{0\}$, that is, $\{a_k\}$ denotes the sequence a_0, a_1, a_2, ... With each sequence, there is associated a power series in a most natural way, which has applications to counting and to recursively-defined sequences.

Definition 9.33 *Let $a_0, a_1, a_2, \ldots$ be a sequence of real numbers. The* **generating function** *of this sequence $\{a_k\}$ is defined as the infinite series*

$$
f(x) = a_0 + a_1 x + a_2 x^2 + \cdots = \sum_{k=0}^{\infty} a_k x^k.
$$

The variable k in $\{a_k\}$ and in $\sum_{k=0}^{\infty} a_k x^k$ is not crucial to the meaning of the sequence or infinite series. Other variables could be used just as well. For example, $\{a_n\}$ also represents the sequence $a_0, a_1, a_2, \ldots$; while

$$
\sum_{n=0}^{\infty} a_n x^n = a_0 + a_1 x + a_2 x^2 + \cdots.
$$

The generating function of a sequence is what is called a *formal* power series. That is, unlike the situation in calculus, where we are often concerned with knowing those real numbers x for which a power series converges, our interest in generating functions is as a means of keeping track of the terms of the sequence; that is, the coefficient of x^k in a generating function acts as a "place holder" for the term a_k in the sequence $a_0, a_1, a_2, \ldots$.

If $\{a_k\}$ is a finite sequence, then there is some nonnegative integer n such that $a_k = 0$ for all $k > n$. Therefore, the generating function in this case is

$$f(x) = a_0 + a_1 x + a_2 x^2 + \cdots + a_n x^n = \sum_{k=0}^{n} a_k x^k,$$

which is a polynomial (of degree n if $a_n \neq 0$).

Example 9.34

The generating function of the sequence $1, 3, 5, 7, \ldots$ is the infinite series

$$f(x) = 1 + 3x + 5x^2 + 7x^3 + \cdots = \sum_{k=0}^{\infty} (2k+1)x^k.$$

Example 9.35

The generating function of the finite sequence $1, 3, 5, 7, 9$ is the polynomial

$$f(x) = 1 + 3x + 5x^2 + 7x^3 + 9x^4. \qquad \blacklozenge$$

Before seeing applications of generating functions to counting problems and later to recursively-defined sequences of real numbers, we first look at some specific generating functions.

Example 9.36

The generating function of the sequence $1, -1, 1, -1, \cdots$ is the infinite series

$$f(x) = 1 - x + x^2 - x^3 + \cdots = \sum_{k=0}^{\infty} (-1)^k x^k.$$

We have already seen in (9.11) that $f(x) = \dfrac{1}{1+x}$. $\qquad \blacklozenge$

Example 9.37

The generating function of the sequence $1, 1, 1, 1, \ldots$ is the infinite series

$$f(x) = 1 + x + x^2 + x^3 + \cdots = \sum_{k=0}^{\infty} x^k.$$

Since we saw in (9.11) that

$$\frac{1}{1+x} = 1 - x + x^2 - x^3 + \cdots,$$

it follows that

$$\begin{aligned}
\frac{1}{1-x} &= \frac{1}{1+(-x)} = 1 - (-x) + (-x)^2 - (-x)^3 + \cdots \\
&= 1 + x + x^2 + x^3 + \cdots
\end{aligned}$$

and so

$$f(x) = 1 + x + x^2 + x^3 + \cdots = \frac{1}{1-x} \qquad (9.12)$$

is the generating function of the sequence $1, 1, 1, 1, \ldots$. $\qquad \blacklozenge$

Example 9.38

Let $a \in \mathbb{R} - \{0\}$. Show that $f(x) = \dfrac{1}{1 - ax}$ is the generating function of some sequence.

Solution. By (9.12), we have

$$\frac{1}{1 - x} = 1 + x + x^2 + x^3 + \cdots.$$

Replacing x by ax, we obtain

$$
\begin{aligned}
\frac{1}{1 - ax} &= 1 + (ax) + (ax)^2 + (ax)^3 + \cdots \\
&= 1 + ax + a^2 x^2 + a^3 x^3 + \cdots \\
&= \sum_{k=0}^{\infty} a^k x^k.
\end{aligned}
$$

Therefore, $f(x) = \dfrac{1}{1 - ax}$ is the generating function of the sequence $1, a, a^2, a^3, \ldots$. ◆

Example 9.39

Show that

$$f(x) = \frac{1}{(1 - x)^2} \quad \text{and} \quad g(x) = \frac{x}{(1 - x)^2}$$

are generating functions for some sequences of real numbers.

Solution. By the Extended Binomial Theorem,

$$
\begin{aligned}
\frac{1}{(1 - x)^2} &= (1 + (-x))^{-2} = \sum_{r=0}^{\infty} \binom{-2}{r} (-x)^r \\
&= \sum_{r=0}^{\infty} (-1)^r \binom{r + 1}{r} (-x)^r \\
&= \sum_{r=0}^{\infty} (r + 1) x^r = 1 + 2x + 3x^2 + 4x^3 + \cdots.
\end{aligned}
$$

Therefore,

$$\frac{1}{(1 - x)^2} = 1 + 2x + 3x^2 + 4x^3 + \cdots \tag{9.13}$$

and so $f(x) = \dfrac{1}{(1 - x)^2}$ is the generating function of the sequence $1, 2, 3, 4, \ldots$.

Multiplying the expression in (9.13) by x, we obtain

$$g(x) = \frac{x}{(1 - x)^2} = x + 2x^2 + 3x^3 + 4x^4 + \cdots.$$

Thus $g(x) = \dfrac{x}{(1 - x)^2}$ is the generating function of the sequence $0, 1, 2, 3, \ldots$. ◆

The problem in Example 9.39 can be solved in another and simpler way if we recall some differentiation rules from calculus. We have already noted that

$$\frac{1}{1 - x} = 1 + x + x^2 + \cdots.$$

Since
$$\frac{1}{1-x} = (1-x)^{-1},$$

it follows by differentiating the equal expressions $(1-x)^{-1}$ and $1 + x + x^2 + \cdots$ that

$$-(1-x)^{-2}(-1) = (1-x)^{-2} = \frac{1}{(1-x)^2} = 1 + 2x + 3x^2 + 4x^3 + \cdots$$

and so $f(x) = \dfrac{1}{(1-x)^2}$ is the generating function of the sequence $1, 2, 3, 4, \ldots$.

Example 9.40

Let $n \in \mathbb{N}$. Find the sequence of which $f(x) = \dfrac{1}{(1-x)^n}$ is the generating function.

Solution. By the Extended Binomial Theorem and (9.10),

$$
\begin{aligned}
\frac{1}{(1-x)^n} &= (1 + (-x))^{-n} = \sum_{r=0}^{\infty} \binom{-n}{r} (-x)^r \\
&= \sum_{r=0}^{\infty} (-1)^r \binom{n+r-1}{r} (-x)^r = \sum_{r=0}^{\infty} \binom{n+r-1}{r} x^r \\
&= 1 + \binom{n}{1} x + \binom{n+1}{2} x^2 + \binom{n+2}{3} x^3 + \cdots.
\end{aligned}
$$

Therefore,

$$\frac{1}{(1-x)^n} = \sum_{r=0}^{\infty} \binom{n+r-1}{r} x^r = 1 + \binom{n}{1} x + \binom{n+1}{2} x^2 + \binom{n+2}{3} x^3 + \cdots. \tag{9.14}$$

Thus $f(x)$ is the generating function of the sequence

$$1, \binom{n}{1}, \binom{n+1}{2}, \binom{n+2}{3}, \cdots$$

of binomial coefficients.   ♦

   The problem in Example 9.40 can also be solved with the aid of differentiation from calculus. Now

$$\frac{1}{1-x} = 1 + x + x^2 + x^3 + \cdots = 1 + \binom{1}{1} x + \binom{2}{2} x^2 + \binom{3}{3} x^3 + \cdots$$

and

$$\frac{1}{(1-x)^2} = 1 + 2x + 3x^2 + 4x^3 + \cdots = 1 + \binom{2}{1} x + \binom{3}{2} x^2 + \binom{4}{3} x^3 + \cdots.$$

An induction proof (see Exercise 8) can then be used to establish formula (9.14), thereby showing that $\dfrac{1}{(1-x)^n}$ is the generating function of the sequence $1, \binom{n}{1}, \binom{n+1}{2}, \binom{n+2}{3}, \cdots$.

   Two generating functions can be added and (as we have already suggested) multiplied.

Definition 9.41 *Let $f(x) = a_0 + a_1 x + a_2 x^2 + \cdots$ and $g(x) = b_0 + b_1 x + b_2 x^2 + \cdots$ be two generating functions. Their* **sum** *$f(x) + g(x)$ and* **product** *$f(x)g(x)$ are defined as*

$$f(x) + g(x) = (a_0 + b_0) + (a_1 + b_1)x + (a_2 + b_2)x^2 + \cdots = \sum_{k=0}^{\infty} (a_k + b_k)x^k$$

and

$$f(x)g(x) = a_0b_0 + (a_0b_1 + a_1b_0)x + (a_0b_2 + a_1b_1 + a_2b_0)x^2 + \cdots = \sum_{k=0}^{\infty} c_k x^k,$$

where

$$c_k = a_0b_k + a_1b_{k-1} + a_2b_{k-2} + \cdots + a_kb_0 = \sum_{i=0}^{k} a_ib_{k-i}.$$

Generating Functions and Counting

Now we see how generating functions can be used to solve certain kinds of counting problems.

Example 9.42

A bowl contains a large number of marbles. Each marble is one of 8 colors. A total of r marbles ($r \geq 0$) is selected from the bowl. Use generating functions to determine the number of different outcomes. What is this number when $r = 5$?

Solution. With each of the 8 colors, there is associated a power series, namely,

$$1 + x + x^2 + \cdots.$$

Determining the answer to this question is the same as computing the coefficient of x^r in the product

$$(1 + x + x^2 + \cdots)(1 + x + x^2 + \cdots) \cdots (1 + x + x^2 + \cdots) = (1 + x + x^2 + \cdots)^8$$

of eight power series. Since

$$\frac{1}{1 - x} = 1 + x + x^2 + x^3 + \cdots,$$

as we saw in (9.12), it follows that

$$(1 + x + x^2 + \cdots)^8 = \left(\frac{1}{1 - x}\right)^8 = \frac{1}{(1 - x)^8} = \sum_{r=0}^{\infty} \binom{r + 7}{r} x^r,$$

which is a consequence of (9.14). Thus the number of different outcomes is $\binom{r+7}{r}$. When $r = 5$, the number of different ways to select 5 marbles from the bowl is $\binom{r+7}{r} = \binom{12}{5} = 792$. ♦

Example 9.43

In a certain room at a casino, there is a card game in which bets are made with chips. There are three different kinds of chips, one valued at \$3, one at \$5 and one at \$7. Determine the generating function of the sequence $\{a_k\}$, where a_k is the number of different ways that chips can be selected that have a total value of \$$k$ for $k \geq 0$.

Solution. The \$3 chips can be represented by the power series $1 + x^3 + x^6 + x^9 + \cdots$, the \$5 chips by $1 + x^5 + x^{10} + x^{15} + \cdots$ and the \$7 chips by $1 + x^7 + x^{14} + x^{21} + \cdots$. Hence the answer to the question is the coefficient of x^k in the product

$$\left(1 + x^3 + x^6 + x^9 + \cdots\right)\left(1 + x^5 + x^{10} + x^{15} + \cdots\right)\left(1 + x^7 + x^{14} + x^{21} + \cdots\right)$$
$$= 1 + x^3 + x^5 + x^6 + x^7 + x^8 + x^9 + 2x^{10} + x^{11} + 2x^{12} + 2x^{13} + 2x^{14} + 3x^{15} + \cdots.$$

For example, the term $3x^{15}$ is obtained by

(1) multiplying x^{15}, 1, 1 in the three power series,

(2) multiplying 1, x^{15}, 1,

(3) multiplying x^3, x^5, x^7

and then adding the results in (1)-(3). The product $x^{15} \cdot 1 \cdot 1 = x^{5(3)} \cdot x^0 \cdot x^0$ corresponds to \$15 worth of chips obtained from five \$3 chips.

Since

$$\frac{1}{1-x} = 1 + x + x^2 + x^3 + \cdots,$$

it follows that for $m \in \mathbb{N}$,

$$\frac{1}{1-x^m} = 1 + x^m + x^{2m} + x^{3m} + \cdots.$$

In particular,

$$
\begin{aligned}
\frac{1}{1-x^3} &= 1 + x^3 + x^6 + x^9 + \cdots \\
\frac{1}{1-x^5} &= 1 + x^5 + x^{10} + x^{15} + \cdots \\
\frac{1}{1-x^7} &= 1 + x^7 + x^{14} + x^{21} + \cdots
\end{aligned}
$$

and so the generating function under consideration is

$$f(x) = \left(\frac{1}{1-x^3}\right)\left(\frac{1}{1-x^5}\right)\left(\frac{1}{1-x^7}\right) = \frac{1}{(1-x^3)(1-x^5)(1-x^7)}. \qquad \blacklozenge$$

Recursively-Defined Sequences Revisited

In Section 4.3, we considered several examples of recursively-defined sequences of real numbers. In these examples, based on knowing the first few terms of a sequence, we attempted to conjecture the general term of the sequence and then prove that our guess was correct, often with the aid of the Strong Principle of Mathematical Induction. In many cases, however, it may not be clear what the general term appears to be. We will now see that generating functions can often be useful in finding the general term of a recursively-defined sequence, thereby taking the guesswork out of the problem. Before illustrating this application of generating functions with several examples, we emphasize once again that it is the coefficients of the various powers of x that provide the answers to questions. It is *not* the case that values are assigned to x.

Example 9.44

A sequence $\{a_n\}$ is defined recursively by

$$a_0 = 3, \ a_1 = -2 \text{ and } a_n = 2a_{n-1} - a_{n-2} \text{ for } n \geq 2.$$

Find a generating function for $\{a_n\}$ and use this to determine a_n for $n \geq 0$.

Solution. Let $f(x)$ be the generating function of the sequence $\{a_n\}$. Thus

$$
\begin{aligned}
f(x) &= a_0 + a_1 x + a_2 x^2 + \cdots = \sum_{n=0}^{\infty} a_n x^n = a_0 + a_1 x + \sum_{n=2}^{\infty} a_n x^n \\
&= a_0 + a_1 x + \sum_{n=2}^{\infty} (2a_{n-1} - a_{n-2}) x^n
\end{aligned}
$$

$$= a_0 + a_1x + 2\sum_{n=2}^{\infty} a_{n-1}x^n - \sum_{n=2}^{\infty} a_{n-2}x^n$$

$$= a_0 + a_1x + 2x\sum_{n=2}^{\infty} a_{n-1}x^{n-1} - x^2\sum_{n=2}^{\infty} a_{n-2}x^{n-2}$$

$$= a_0 + a_1x + 2x\sum_{n=1}^{\infty} a_nx^n - x^2\sum_{n=0}^{\infty} a_nx^n$$

$$= a_0 + a_1x + 2x(f(x) - a_0) - x^2 f(x).$$

Hence

$$f(x) = a_0 + (a_1 - 2a_0)x + 2xf(x) - x^2 f(x)$$

and so

$$(1 - 2x + x^2)f(x) = (1 - x)^2 f(x) = a_0 + (a_1 - 2a_0)x = 3 - 8x.$$

Solving for $f(x)$, we obtain

$$f(x) = \frac{3 - 8x}{(1 - x)^2}.$$

By (9.13),

$$\frac{1}{(1 - x)^2} = 1 + 2x + 3x^2 + \cdots + (n + 1)x^n + \cdots,$$

and so

$$f(x) = (3 - 8x)\left[1 + 2x + 3x^2 + \cdots + (n + 1)x^n + \cdots\right].$$

Hence the coefficient of x^n in $f(x)$ is

$$3(n + 1) - 8n = 3 - 5n.$$

Therefore, $a_n = 3 - 5n$ for $n \geq 0$.   ◆

Example 9.45

A sequence $\{a_n\}$ is defined recursively by

$$a_0 = 0, \ a_1 = 1 \text{ and } a_n = 2a_{n-1} - a_{n-2} + 2 \text{ for } n \geq 2.$$

Find a generating function for $\{a_n\}$ and determine a_n for $n \geq 0$.

Solution. Let $f(x)$ be the generating function of the sequence $\{a_n\}$. Then

$$f(x) = a_0 + a_1x + a_2x^2 + \cdots = \sum_{n=0}^{\infty} a_nx^n = a_0 + a_1x + \sum_{n=2}^{\infty} a_nx^n$$

$$= a_0 + a_1x + \sum_{n=2}^{\infty} (2a_{n-1} - a_{n-2} + 2)\, x^n$$

$$= a_0 + a_1x + 2\sum_{n=2}^{\infty} a_{n-1}x^n - \sum_{n=2}^{\infty} a_{n-2}x^n + 2\sum_{n=2}^{\infty} x^n.$$

By (9.12), $\sum_{n=0}^{\infty} x^n = \frac{1}{1-x}$. Therefore,

$$f(x) = a_0 + a_1x + 2x\sum_{n=2}^{\infty} a_{n-1}x^{n-1} - x^2\sum_{n=2}^{\infty} a_{n-2}x^{n-2} + 2\left(\frac{1}{1-x} - 1 - x\right)$$

$$= a_0 + a_1x + 2x\sum_{n=1}^{\infty} a_nx^n - x^2\sum_{n=0}^{\infty} a_nx^n + \frac{2}{1-x} - 2 - 2x$$

$$= a_0 + a_1x + 2x(f(x) - a_0) - x^2 f(x) + \frac{2}{1-x} - 2 - 2x.$$

Thus

$$f(x) = (a_0 - 2) + (a_1 - 2a_0 - 2)x + \frac{2}{1-x} + 2xf(x) - x^2 f(x)$$

and so

$$
\begin{aligned}
(1 - 2x + x^2)f(x) &= (1-x)^2 f(x) = (a_0 - 2) + (a_1 - 2a_0 - 2)x + \frac{2}{1-x} \\
&= -2 - x + \frac{2}{1-x} = \frac{x + x^2}{1-x}.
\end{aligned}
$$

Solving for $f(x)$, we obtain

$$f(x) = \frac{x + x^2}{(1-x)^3}.$$

By (9.14),

$$\frac{1}{(1-x)^3} = \binom{2}{2} + \binom{3}{2}x + \binom{4}{2}x^2 + \cdots + \binom{n+2}{2}x^n + \cdots$$

and so

$$f(x) = (x + x^2)\left(\binom{2}{2} + \binom{3}{2}x + \binom{4}{2}x^2 + \cdots + \binom{n+2}{2}x^n + \cdots \right).$$

Therefore, the coefficient of x^n is

$$\binom{n+1}{2} + \binom{n}{2} = n^2,$$

implying that $a_n = n^2$ for all $n \geq 0$. ♦

Generating Functions and Partial Fractions

For our final example, we employ the topic of partial fractions. In college algebra, we often encounter rational expressions (ratios of polynomials) that we wish to add. For example, to compute

$$\frac{3}{2x+1} + \frac{5}{x-3},$$

we obtain a common denominator, namely $(2x+1)(x-3)$ and then add. In this case, we have

$$
\begin{aligned}
\frac{3}{2x+1} + \frac{5}{x-3} &= \frac{3(x-3)}{(2x+1)(x-3)} + \frac{5(2x+1)}{(x-3)(2x+1)} = \frac{3(x-3) + 5(2x+1)}{(2x+1)(x-3)} \\
&= \frac{3x - 9 + 10x + 5}{(2x+1)(x-3)} = \frac{13x - 4}{2x^2 - 5x - 3}.
\end{aligned}
$$

However, there are occasions when it is useful to reverse this process. That is, beginning with

$$\frac{13x - 4}{2x^2 - 5x - 3}, \tag{9.15}$$

we wish to express (9.15) as the sum of two simpler fractions. The procedure one uses is called the method of **partial fractions**. (If you have had experience with integration in calculus, you might recall an integration technique called integration by partial fractions, where this algebraic method is used.)

We will describe only a simplified version of the method of partial fractions. For example, suppose that

$$f(x) = \frac{p(x)}{p_1(x)p_2(x)},$$

where $p_1(x)$ and $p_2(x)$ are two first degree (linear) polynomials and $p(x)$ is a polynomial of degree 0 (a constant) or of degree 1. In this case, it is possible to write

$$f(x) = \frac{p(x)}{p_1(x)p_2(x)} = \frac{A}{p_1(x)} + \frac{B}{p_2(x)}, \tag{9.16}$$

where A and B are constants. The next step is to determine the values of A and B. One way to do this is to first multiply (9.16) by $p_1(x)p_2(x)$, producing

$$p(x) = Ap_2(x) + Bp_1(x) \tag{9.17}$$

and then select two specific values x_1 and x_2 of x such that $p_1(x_1) = 0$ and $p_2(x_2) = 0$. We then substitute these values for x in (9.17) and solve for A and B. For example, let's suppose that

$$f(x) = \frac{x}{1 - 5x + 6x^2}.$$

Our first step is to factor $1 - 5x + 6x^2$. In this case,

$$1 - 5x + 6x^2 = (1 - 2x)(1 - 3x).$$

We then have

$$\frac{x}{1 - 5x + 6x^2} = \frac{x}{(1 - 2x)(1 - 3x)} = \frac{A}{1 - 2x} + \frac{B}{1 - 3x}.$$

Thus

$$x = A(1 - 3x) + B(1 - 2x).$$

Letting $x = \frac{1}{2}$, we have $\frac{1}{2} = -\frac{1}{2}A$ and so $A = -1$. Letting $x = \frac{1}{3}$, we have $\frac{1}{3} = \frac{1}{3}B$ and so $B = 1$. Therefore,

$$\frac{x}{1 - 5x + 6x^2} = \frac{x}{(1 - 2x)(1 - 3x)} = \frac{-1}{1 - 2x} + \frac{1}{1 - 3x}. \tag{9.18}$$

We now see how partial fractions play a role in the following example.

Example 9.46

A sequence $\{a_n\}$ is defined recursively by

$$a_0 = 0, \ a_1 = 1 \text{ and } a_n = 5a_{n-1} - 6a_{n-2} \text{ for } n \geq 2.$$

Find a generating function for $\{a_n\}$ and determine a_n for $n \geq 0$.

Solution. Let $f(x)$ be the generating function of the sequence $\{a_n\}$. Then

$$
\begin{aligned}
f(x) &= a_0 + a_1 x + a_2 x^2 + \cdots = \sum_{n=0}^{\infty} a_n x^n = a_0 + a_1 x + \sum_{n=2}^{\infty} a_n x^n \\
&= a_0 + a_1 x + \sum_{n=2}^{\infty} (5a_{n-1} - 6a_{n-2}) x^n \\
&= a_0 + a_1 x + 5 \sum_{n=2}^{\infty} a_{n-1} x^n - 6 \sum_{n=2}^{\infty} a_{n-2} x^n \\
&= a_0 + a_1 x + 5x \sum_{n=2}^{\infty} a_{n-1} x^{n-1} - 6x^2 \sum_{n=2}^{\infty} a_{n-2} x^{n-2} \\
&= a_0 + a_1 x + 5x \sum_{n=1}^{\infty} a_n x^n - 6x^2 \sum_{n=0}^{\infty} a_n x^n \\
&= a_0 + a_1 x + 5x(f(x) - a_0) - 6x^2 f(x).
\end{aligned}
$$

Thus
$$f(x) = a_0 + (a_1 - 5a_0)x + 5xf(x) - 6x^2 f(x)$$

and so
$$(1 - 5x + 6x^2)f(x) = a_0 + (a_1 - 5a_0)x = x.$$

Solving for $f(x)$, we obtain
$$f(x) = \frac{x}{1 - 5x + 6x^2}.$$

By (9.18), $f(x) = \frac{-1}{1-2x} + \frac{1}{1-3x}$. By Example 9.38,

$$\frac{-1}{1-2x} = -\sum_{n=0}^{\infty} 2^n x^n \quad \text{and} \quad \frac{1}{1-3x} = \sum_{n=0}^{\infty} 3^n x^n.$$

Therefore,

$$f(x) = -\sum_{n=0}^{\infty} 2^n x^n + \sum_{n=0}^{\infty} 3^n x^n = \sum_{n=0}^{\infty} (3^n - 2^n) x^n$$

and so $a_n = 3^n - 2^n$ for $n \geq 0$.   ♦

Some of the most common power series are shown in Figure 9.8.

Exercises for Section 9.3

1. A bowl contains 9 balls, 3 balls of each of the colors red, blue and green. Suppose that r balls $(1 \leq r \leq 9)$ are selected from the bowl.

    (a) Express the number of outcomes obtained as a product of polynomials. Determine the product of these polynomials.

    (b) What are the outcomes in (a) when $r = 3$?

    (c) What are the outcomes in (a) when $r = 5$?

2. A bowl consists of 10 balls, each colored red, blue, green or yellow. There is one red ball and there are two blue balls, three green balls and four yellow balls. Suppose that r balls $(1 \leq r \leq 10)$ are selected from the bowl.

    (a) Express the number of outcomes obtained as a product of polynomials. Determine the product of these polynomials.

    (b) What are the outcomes in (a) when $r = 3$?

    (c) What are the outcomes in (a) when $r = 4$?

3. A bowl consists of 24 balls, namely 8 balls of each of the colors red, blue and green. In how many different ways can 8 balls be selected from the bowl satisfying the three conditions: (1) at least 3 balls are red, (2) at most 2 balls are blue, (3) an even number of green balls is selected?

    (a) Answer the question above by listing all the possibilities in a table.

    (b) Answer the question above by computing a product of polynomials.

4. A man enters Ben's Bagels to buy a dozen bagels only to learn that even though Ben has a large supply of plain bagels and blueberry bagels, he only has two cinnamon bagels left and no other kinds.

    (a) Express the number of selections of r bagels $(r \geq 0)$ the man can make as a product of polynomials and/or power series.

$$(1+x)^n = \sum_{r=0}^{n} \binom{n}{r} x^r = 1 + \binom{n}{1}x + \binom{n}{2}x^2 + \binom{n}{3}x^3 + \cdots$$

$$(1+ax)^n = \sum_{r=0}^{n} \binom{n}{r} a^r x^r + 1 + \binom{n}{1}ax + \binom{n}{2}a^2x^2 + \binom{n}{3}a^3x^3 + \cdots$$

$$(1+x^k)^n = \sum_{r=0}^{n} \binom{n}{r} x^{kr} = 1 + \binom{n}{1}x^k + \binom{n}{2}x^{2k} + \binom{n}{3}x^{3k} + \cdots$$

$$\frac{1}{1-x} = \sum_{r=0}^{n} x^r = 1 + x + x^2 + x^3 + \cdots$$

$$\frac{1}{1+x} = \sum_{r=0}^{n} (-1)^r x^r = 1 - x + x^2 - x^3 + \cdots$$

$$\frac{1}{1-ax} = \sum_{r=0}^{\infty} a^r x^r = 1 + ax + a^2x^2 + a^3x^3 + \cdots$$

$$\frac{1}{1-x^k} = \sum_{r=0}^{\infty} x^{kr} = 1 + x^k + x^{2k} + x^{3k} + \cdots$$

$$\frac{1}{(1-x)^2} = \sum_{r=0}^{\infty} (r+1)x^r = 1 + 2x + 3x^2 + 4x^3 + \cdots$$

$$\frac{1}{(1-x)^n} = \sum_{r=0}^{\infty} \binom{n+r-1}{r} x^r = 1 + \binom{n}{1}x + \binom{n+1}{2}x^2 + \binom{n+2}{3}x^3 + \cdots$$

$$\frac{1}{(1+x)^n} = \sum_{r=0}^{\infty} (-1)^r \binom{n+r-1}{r} x^r = 1 - \binom{n}{1}x + \binom{n+1}{2}x^2 - \binom{n+2}{3}x^3 + \cdots$$

$$\frac{1}{(1-ax)^n} = \sum_{r=0}^{\infty} \binom{n+r-1}{r} a^r x^r = 1 + \binom{n}{1}ax + \binom{n+1}{2}a^2x^2 + \binom{n+2}{3}a^3x^3 + \cdots$$

$$(1+x)^\alpha = \sum_{r=0}^{\infty} \binom{\alpha}{r} x^r = 1 + \binom{\alpha}{1}x + \binom{\alpha}{2}x^2 + \binom{\alpha}{3}x^3 + \cdots$$

$$e^x = \sum_{r=0}^{\infty} \frac{x^r}{r!} = 1 + x + \frac{x^2}{2!} + \frac{x^3}{3!} + \cdots$$

Figure 9.8: Common power series

(b) Determine the coefficient of x^{12} of the product in (a). What information does this coefficient provide?

5. Compute the following binomial or extended binomial coefficients:

   (a) $\binom{\pi}{0}$.   (b) $\binom{e}{2}$.   (c) $\binom{\frac{1}{2}}{3}$.   (d) $\binom{4}{7}$.   (e) $\binom{7}{4}$.

   (f) $\binom{-3}{3}$.   (g) $\binom{0}{2}$.   (h) $\binom{2}{0}$.   (i) $\binom{-\frac{1}{2}}{2}$.   (j) $\binom{-1}{1}$.

6. Express $(1+x)^{1/2}$ as a power series.

7. Express $(1+x)^{-3}$ as a power series.

8. Use induction to prove that $\frac{1}{(1-x)^n}$ is the generating function of the sequence $1, \binom{n}{1}, \binom{n+1}{2}, \binom{n+2}{3}, \cdots$ for each positive integer n.

9. What is the generating function of the sequence $1, -2, 3, -4, 5, -6, \cdots$?

10. What is the generating function of the sequence $\{a_k\}$ where $a_k = \frac{1}{k!}$, $k \geq 0$?

11. Show that $f(x) = \frac{x^2-x+2}{(1+x)(1-x)^2}$ is the generating function of some sequence.

12. Suppose that we are interested in the number of nonnegative integer solutions a, b and c of the equation $a + 2b + 3c = r$, where $r \in \mathbb{N}$.

    (a) What is this number when $r \in \{1, 2, 3, 4\}$?

    (b) Use generating functions to describe this number for a general $r \in \mathbb{N}$.

13. Carla's Cookies sells five different kinds of cookies: sugar, oatmeal, peanut butter, coconut, chocolate chip.

    (a) Use generating functions to determine the number of ways that 12 cookies can be selected at Carla's Cookies.

    (b) What generating function has as its coefficient of x^r, $r \geq 0$, the number of different ways of selecting r cookies?

14. Use the method of partial fractions to express $(13x - 4)/(2x + 1)(x - 3)$ as the sum of two simpler fractions.

15. A sequence $\{a_n\}$ is defined recursively by $a_0 = 1$ and $a_n = 2a_{n-1}$ for $n \geq 1$. Use generating functions to determine a_n for $n \geq 0$.

16. A sequence $\{a_n\}$ is defined recursively by $a_0 = -2$, $a_1 = 1$ and $a_n = 2a_{n-1} - a_{n-2}$ for $n \geq 2$. Use generating functions to determine a_n for $n \geq 0$.

17. A sequence $\{a_n\}$ is defined recursively by $a_0 = -1$, $a_1 = 2$ and $a_n = 3a_{n-1} - 2a_{n-2} - 3$ for $n \geq 2$. Use generating functions to determine a_n for $n \geq 0$.

18. A sequence $\{a_n\}$ is defined recursively by $a_0 = 2$, $a_1 = 6$ and $a_n = 2a_{n-1} + 3a_{n-2}$ for $n \geq 2$. Use generating functions to determine a_n for $n \geq 0$.

19. A sequence $\{a_n\}$ is defined recursively by $a_0 = 1$, $a_1 = -2$ and $a_n = -a_{n-1} + 6a_{n-2}$ for $n \geq 2$. Use generating functions to determine a_n for $n \geq 0$.

Chapter 9   Highlights

Key Concepts

binomial coefficient $\binom{n}{r}$: the number of r-element subsets of an n-element set.

extended binomial coefficient $\binom{\alpha}{r}$: $\binom{\alpha}{0} = 1$ and $\binom{\alpha}{r} = \frac{\alpha(\alpha-1)\cdots(\alpha-r+1)}{r!}$ for $r \neq 0$.

generating function (of a sequence $\{a_k\}$): the series $\sum_{k=0}^{\infty} a_k x^k$.

multiset: a collection of elements in which repetition of elements is permitted.

Pascal triangle: a representation of the binomial coefficients such that the nth row of the triangle contains $C(n, r)$ for $0 \leq r \leq n$.

Key Results

- **Identities:**

  $\binom{n}{r} = \binom{n-1}{r-1} + \binom{n-1}{r}$ for $1 \leq r \leq n - 1$.

  $\binom{n}{r} \frac{n-r}{r+1} = \binom{n}{r+1}$ for $0 \leq r < n$.

- **Hockey Stick Theorem :**

  $\binom{n+1}{r+1} = \binom{r}{r} + \binom{r+1}{r} + \binom{r+2}{r} + \cdots + \binom{n}{r}$ for $0 \leq r \leq n$.

- **The Binomial Theorem** : $(x + y)^n = \sum_{r=0}^{n} \binom{n}{r} x^{n-r} y^r$.

- There are n^r r-permutations of an n-element set with repetition permitted.

- There are $\frac{n!}{n_1!\, n_2! \cdots n_k!}$ permutations of n objects, where there are n_i objects of type i for $1 \le i \le k$ and $n_1 + n_2 + \cdots + n_k = n$.

- There are $C(s + t - 1, s)$ s-combinations of a t-element set with repetition permitted.

- **Extended Binomial Theorem** : Let $\alpha \in \mathbb{R}$. If $x \in \mathbb{R}$ with $|x| < 1$, then
$$(1 + x)^\alpha = \sum_{r=0}^{\infty} \binom{\alpha}{r} x^r.$$

Supplementary Exercises for Chapter 9

1. Al has 9 different books. He decides to give 4 to Bob, 3 to Charles and 2 to Don. In how many ways can this be done?

2. In order to gain access to a private web site, each person must have his or her own 6-digit password, where some digit is used 3 times and the remaining 3 digits are distinct. How many such passwords are there?

3. A deck of 12 cards consists of 3 identical red cards, 3 identical blue cards, 3 identical green cards and 3 identical yellow cards. The cards are mixed up (shuffled). Three cards are selected and the result is recorded. The three cards are then returned to the deck and the process is repeated. How many times must this be done to be guaranteed that there are three identical outcomes?

4. How many 10-digit numbers contain five 1s, three 2s and two 3s?

5. What is the coefficient of x in $\left(2x - \frac{1}{2x}\right)^7$? Give the exact answer.

6. Expand $(a + b)^8$, giving precise coefficients.

7. What is the exact coefficient of $a^4 b^3 c^2$ in $(a + b + c)^9$?

8. (a) Let $a_k a_{k-1} \ldots a_1 a_0$ be the decimal representation of an integer n. Show that
$$n \bmod 9 = (a_k + a_{k-1} + \cdots + a_1 + a_0) \bmod 9.$$
   (b) Use (a) to determine the remainder when 8776 is divided by 9.

9. Determine the number of distinct permutations of the letters of the word ROOSTERS.

10. Ten students have decided to join a club. If 4 choose the computer science club and 3 each choose to join the mathematics club and statistics club, in how many ways can this be done?

11. What is the number of possible birthday months for three people?

12. How many 4-digit numbers are there that are divisible by 10 and the sum of whose digits is 10?

13. A boy has been saving for a vacation trip and keeps $1 bills, $5 bills, $10 bills and $20 bills in a drawer. If he selects 6 bills at random from the drawer, how many different assortments of bills are possible?

14. One evening each year, a baseball team has "two brothers" night, where two brothers are admitted to the baseball game for the price of one. A total of 75 pairs of brothers take advantage of this offer. All pairs of brothers fill out a form to be eligible for prizes to be awarded later. One piece of information requested is the birthday months of the two brothers. Is it necessary that two pairs of brothers have the same pair of birthday months?

15. A total of 10 different books are to be given to the top 3 mathematics students at an awards ceremony. If the top student gets 5 books, the second best gets 3 books and the 3rd best gets 2 books, in how many ways can the books be distributed?

16. A man goes into a donut shop to buy 7 donuts. At the time he is there, there are only 5 possible donuts: plain, cherry, jelly, lemon, peach. What is the total number of possibilities for his purchase?

17. A little boy is placing 20 of his marbles in a row. If 12 are red and 8 are blue and he decides not to place 2 blue marbles next to each other, how many arrangements of the marbles are possible?

18. How many 4-digit numbers are there, the sum of whose digits is 10?

19. A bowl contains 10 balls of each of the colors red, blue and green. In how many different ways can 10 balls be selected from the bowl such that there are at least two balls selected of each of the three colors?

    (a) Answer the question above by listing all the possibilities in a table.
    (b) Answer the question above by computing a product of polynomials.
    (c) Answer the question above by computing $\binom{s+t-1}{s}$ for an appropriate choice of s and t.

20. A man buys 6 donuts, each of which is a plain donut, a powdered donut or a glazed donut. How many possible selections are there if he buys at least one donut of each kind?

    (a) Answer the question above by listing all possible selections in a table.
    (b) Answer the question above by determining the coefficient of x^6 in a product of polynomials and/or power series.
    (c) Answer the question above by computing $\binom{s+t-1}{s}$ for an appropriate choice of s and t.

21. Compute the following binomial or extended binomial coefficients:

    (a) $\binom{\frac{1}{3}}{2}$.   (b) $\binom{5}{8}$.   (c) $\binom{8}{5}$.   (d) $\binom{-5}{3}$.   (e) $\binom{\sqrt{2}}{1}$.

22. Express $\left(1 + x^2\right)^{-1}$ as a power series.

23. Express $(1 - 3x)^{-2}$ as a power series.

24. A sequence $\{a_n\}$ is defined recursively by $a_0 = 1$, $a_1 = 2$ and $a_n = a_{n-1} + 2a_{n-2}$ for $n \geq 2$. Find a generating function for $\{a_n\}$ and determine a_n for $n \geq 0$.

25. A sequence $\{a_n\}$ is defined recursively by $a_0 = 0$, $a_1 = -1$ and $a_n = -2a_{n-1} - a_{n-2}$ for $n \geq 2$. Find a generating function for $\{a_n\}$ and determine a_n for $n \geq 0$.

26. A sequence $\{a_n\}$ is defined recursively by $a_0 = 0$, $a_1 = -1$ and $a_n = 2a_{n-1} + 15a_{n-2}$ for $n \geq 2$. Find a generating function for $\{a_n\}$ and determine a_n for $n \geq 0$.

27. At a bookstore, there is a special on children's books that have been in stock for a long time. The bookstore has several barrels each containing many books that are identically wrapped. For $20, a customer can select 10 books from each barrel, not knowing exactly which books he will be getting. He decides to do this with the barrel labeled "Arabian Nights." The sign on the barrel says that there are 5 different books (many of each), namely (1) "Scheherezade," (2) "Sinbad," (3) "Harun al-Rashid," (4) "Aladdin" and (5) "Ali Babba and the Forty Thieves." How many different assortments of books are possible?

Chapter 10

Discrete Probability

Although ideas from probability occurred thousands of years ago, probability as a mathematical theory came into existence only during the middle of the 17th century. The Frenchman Chevalier de Méré was known to gamble often. From his own experiences, he had observed that if a single die was rolled four times, then it was more likely than not that 6 would occur at least once. Consequently, he would bet with others that this would happen. Having grown tired of this game, de Méré decided to try a new game. He bet with others that if a pair of dice was rolled 24 times, a total of 12 (6 on each die) would occur at least once. He soon learned, however, that he was not winning as often with the second game as he did with the first game, so de Méré then asked his friend Blaise Pascal (after whom the Pascal triangle is named) why this was happening. Pascal determined that de Méré should expect to win the first game 51.8% of the time and the second game only 49.1% of the time.

The story goes that de Méré's problem initiated correspondence between Pascal and another famous French mathematician, Pierre de Fermat. It is believed that this correspondence went beyond de Méré's problem to other problems dealing with probability. Pascal and Fermat are often credited with founding probability theory. Since then, probability has become not only an important subject in mathematics and statistics courses but a topic that intersects many aspects of society, including genetics, medicine, industrial quality control and insurance.

In this chapter, we will be introduced to discrete probability, the probability of discrete mathematics.

10.1 The Probability of an Event

The communication between Pascal and Fermat led them to define the probability of winning certain kinds of games. Specifically, if a game has n equally likely outcomes, m of which are winning outcomes, then the probability of winning the game was defined to be m/n. While this definition requires the outcomes of a game to be equally likely, this is not always the case. Indeed, it's not always clear when the outcomes of a game are equally likely.

Pascal and Fermat also defined probability when a game is repeated a large number of times under the same conditions. In 1933 the Russian mathematician Andrey Kolmogorov developed the first rigorous approach to probability. We now consider some of the fundamental concepts associated with probability.

Suppose that a certain procedure has a number of possible outcomes. Each implementation of the procedure is referred to as an experiment. More formally, we have the following.

Definition 10.1 *An* **experiment** *is a procedure that results in one of a number of possible outcomes. The set of all possible outcomes of an experiment is called the* **sample space** *for the experiment. Each subset of a sample space is called an* **event**.

The sample space of an experiment is usually denoted by S and an event by E. Thus $E \subseteq S$. If S is a sample space and $s \in S$, then s is an **outcome**. If some outcome in an event occurs when an experiment is performed, then the event is said to occur.

Let's look at a number of examples of sample spaces.

Example 10.2

Flipping a coin once is an experiment. If we're interested in whether heads or tails occurs, then the two outcomes of this experiment are heads (H) and tails (T). The sample space here is therefore $S = \{H, T\}$ so that the possible events in this case are $\emptyset$, $\{H\}$, $\{T\}$ and $\{H, T\}$.   ◆

Example 10.3

A coin is flipped n times and we're interested in the possible sequences of heads and tails that can occur. What is the number of outcomes in the sample space S of this experiment? For $n = 3$, what is S?

Solution. Since there are two possible results with each flip of a coin, it follows by the Multiplication Principle that $|S| = 2^n$. If $n = 3$, then

$$S = \{HHH, HHT, HTH, HTT, THH, THT, TTH, TTT\}$$

and so $|S| = 2^3 = 8$. Therefore, there are $2^8 = 256$ possible events in this case, including $\emptyset$, $\{HHH\}$, $\{HTH, TTH\}$ and S.   ◆

Example 10.4

Selecting one ball at random from a bowl containing three red balls, two blue balls and one green ball is an experiment that yields one of three possible outcomes, namely a red ball (R), a blue ball (B) or a green ball (G). In this case, the sample space is $S = \{R, B, G\}$.   ◆

Example 10.5

You probably know that a die is a cube where each of its six faces has a number of spots, ranging from 1 to 6 (see Figure 10.1). When a die is tossed, rolled or thrown (on a flat surface), we typically consider the number of spots on the upper face as the outcome of this experiment. Therefore, the experiment of rolling a single die has six possible outcomes: 1, 2, 3, 4, 5, 6. The sample space here is $S = \{1, 2, 3, 4, 5, 6\}$.   ◆

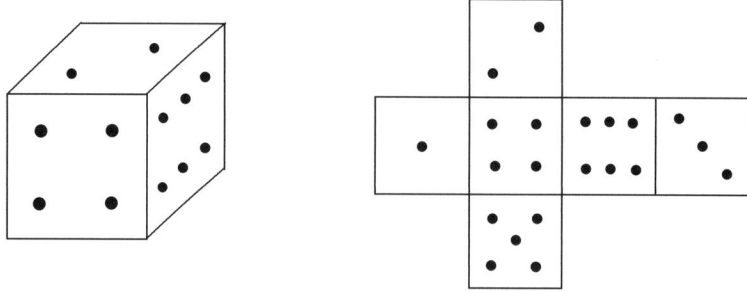

Figure 10.1: A die

Many games concerning dice (the plural of die), however, involve a pair of dice rather than a single die. When we toss a pair of dice, obtaining (a, b), where a is the number obtained for the first die and b is the number resulting from the second die, then we are typically interested in the number $a + b$, the sum of the numbers of the two dice. Since there are 6 possible numbers for the first die and 6 for the second die, the number of possible ordered pairs (a, b) from tossing two dice is $6 \cdot 6 = 36$ by the Multiplication Principle. However, there are only 11 distinct sums, namely, $2, 3, \ldots, 12$. These sums are shown in Figure 10.2.

Value of second die

	1	2	3	4	5	6
1	2	3	4	5	6	7
2	3	4	5	6	7	8
3	4	5	6	7	8	9
4	5	6	7	8	9	10
5	6	7	8	9	10	11
6	7	8	9	10	11	12

Value of first die

Figure 10.2: The sum of the values of two dice

It may seem that the sample space in this case should consist of the 11 outcomes $2, 3, \ldots, 12$ rather than the 36 outcomes mentioned above, but the choice of the sample space depends on which questions interest us. Suppose that we are only interested in the sum of the numbers of two dice when the dice are rolled. Then the appropriate sample space may very well be

$$S_1 = \{2, 3, \ldots, 12\}.$$

On the other hand, if we are interested in whether the number obtained on at least one of the two dice is 3, then this event E can be described more clearly by writing

$$E = \{(3, 1), (3, 2), (3, 3), (3, 4), (3, 5), (3, 6), (1, 3), (2, 3), (4, 3), (5, 3), (6, 3)\}.$$

An appropriate sample space in this case would most likely be

$$S_2 = \{(1, 1), (1, 2), (1, 3), \ldots, (6, 5), (6, 6)\},$$

which consists of all 36 possible ordered pairs (a, b) from tossing two dice. Moreover, if the question that interests us is how many times we need to roll a pair of dice before we obtain 7 (the sum of the numbers on the two dice), then the sample space in this case is most likely $S_3 = \mathbb{N}$.

Although sample spaces can be any kinds of sets, we will only consider discrete sample spaces, that is, sets containing countably many elements. Hence a sample space can be a finite nonempty set or a denumerable (countably infinite) set. For the most part, we will only be concerned with finite sample spaces. We now consider two experiments to describe appropriate sample spaces in each case.

Example 10.6

Suppose that a casino sells a die, where instead of having a face with a single spot, the logo of the casino is on that face and the remaining five faces have spots as with a standard die. If such a die is rolled and what interests us is whether the logo occurs, then one outcome is the logo and the other outcome is no logo, that is, the sample space here is $S = \{\text{logo, no logo}\}$. ♦

Example 10.7

Suppose that a die is white and that each spot on a face is colored either red or blue. If a face has an even number of spots, then there is an equal number of red and blue spots. On the other hand, if a face has an odd number of spots, then there is one more red spot than blue spot. If we are interested only in the *number* of spots, then the sample space is the set $S = \{1, 2, 3, 4, 5, 6\}$. If, instead, we are interested in the number of *blue* spots when the die is tossed, then one face of the die has 3 blue spots, two faces have 2 blue spots, two faces have 1 blue spot and the remaining face has no blue spots. The sample space in this case would be $S' = \{0, 1, 2, 3\}$.   ♦

We are now prepared to present the main concept of this chapter.

Definition 10.8   *For a set S, a function $p : S \to [0, 1]$ is a* **probability function** *if*

$$\sum_{s \in S} p(s) = 1.$$

For an element $s \in S$, the number $p(s)$ is the **probability** *of s.*

In the definition above of a probability function, the sample spaces S that we encounter will typically be finite and this is our primary interest; however, the set S can be denumerable (see Section 5.5) and so, in general, S can be a countable set. In any case,

$$0 \le p(s) \le 1$$

for each element $s \in S$ and the sum of the probabilities of the elements of S is 1.

Definition 10.9   *The* **probability** *$p(E)$ of an event E in a sample space S is the sum of the probabilities of the outcomes in that event, that is,*

$$p(E) = \sum_{s \in E} p(s).$$

Consequently,

$$0 \le p(E) \le 1$$

for every event E. In particular,

$$p(\emptyset) = 0 \text{ and } p(S) = 1.$$

If S is the sample space of an experiment, then the way that the probability $p(s)$ of an outcome $s \in S$ is defined can vary widely. It usually depends, however, on our experience of what we think a reasonable definition should be. We now look at two experiments we visited earlier and describe reasonable ways of defining the probability of an outcome in the resulting sample space.

Example 10.10

In Example 10.4, we considered selecting one ball at random from a bowl containing three red balls, two blue balls and one green ball and stated that the sample space is $S = \{R, B, G\}$. In this case, it is reasonable to assume that selecting any one of the six balls is equally likely. It makes sense, therefore, to define a probability function p on S by

$$p(R) = \tfrac{3}{6} = \tfrac{1}{2}, \quad p(B) = \tfrac{2}{6} = \tfrac{1}{3} \text{ and } p(G) = \tfrac{1}{6}.$$ ♦

Example 10.11

In Example 10.6, we considered a die that contains the logo of a casino on one face of the die and no logo on the other five faces. The sample space in this case is $S = \{\text{logo, no logo}\}$. Here too, it is reasonable to assume that obtaining the logo or any of the other faces is equally likely. Since one of the six faces has the logo and five of the six faces have no logo, it is logical to define a probability function p on S by

$$p(\text{logo}) = \tfrac{1}{6} \quad \text{and} \quad p(\text{no logo}) = \tfrac{5}{6}.$$ ♦

In general then, if a sample space S is finite, say $|S| = n \geq 1$ and each outcome in S is equally likely to occur, then it is logical to define

$$p(s) = \frac{1}{|S|} = \frac{1}{n}$$

for each outcome $s \in S$. In this case, p is called a **uniform probability function**. If a (finite) sample space S is assigned a uniform probability function and E is an event in S, then in this case,

$$p(E) = \frac{|E|}{|S|}.$$

This is the definition of probability given by Pascal and Fermat.

Example 10.12

A coin is flipped twice. If heads and tails are equally likely to occur, then what is the probability that heads and tails occur once each (in either order)?

Solution. The sample space in this case is

$$S = \{\text{HH, HT, TH, TT}\}.$$

The event that heads and tails occur once each is $E = \{\text{HT, TH}\}$. Since the probability is uniform in this case, the probability of E is

$$p(E) = \frac{|E|}{|S|} = \frac{2}{4} = \frac{1}{2}.$$ ♦

Example 10.13

A die is tossed once and each of the six possible outcomes is equally likely to occur. What is the probability that the number obtained is odd?

Solution. The sample space is $S = \{1, 2, 3, 4, 5, 6\}$ and the event of obtaining an odd number on this toss of the die is $E = \{1, 3, 5\}$. Since this probability function is also uniform, the probability of E is

$$p(E) = \frac{|E|}{|S|} = \frac{3}{6} = \frac{1}{2}.$$ ♦

Since an event in a sample space S is a subset of S, the union and intersection of events in S are also events in S. There are some basic observations concerning probabilities of events that are subsets of the same finite sample space. All of these are consequences of facts involving cardinalities of sets. Two events E and F in a sample space S are **mutually exclusive** if the sets E and F are disjoint, that is, if E and F cannot occur simultaneously. In particular, if E and F are mutually exclusive events, then

$$p(E \cup F) = \frac{|E \cup F|}{|S|} = \frac{|E| + |F|}{|S|} = \frac{|E|}{|S|} + \frac{|F|}{|S|} = p(E) + p(F).$$

If E and F are events that are not necessarily mutually exclusive, then it follows from the Principle of Inclusion-Exclusion (see Section 8.3) that

$$
\begin{aligned}
p(E \cup F) &= \frac{|E \cup F|}{|S|} = \frac{|E| + |F| - |E \cap F|}{|S|} \\
&= \frac{|E|}{|S|} + \frac{|F|}{|S|} - \frac{|E \cap F|}{|S|} = p(E) + p(F) - p(E \cap F).
\end{aligned}
$$

Similarly, $n \geq 2$ events $E_1, E_2, \ldots, E_n$ are **mutually exclusive** if $E_1, E_2, \ldots, E_n$ are pairwise disjoint. More generally then, if $E_1, E_2, \ldots, E_n$ are mutually exclusive events, then

$$
p(E_1 \cup E_2 \cup \cdots \cup E_n) = p(E_1) + p(E_2) + \cdots + p(E_n).
$$

For an event E in a sample space S, the **complementary event** $\overline{E}$ of E consists of those outcomes in S not belonging to E, that is, $\overline{E} = S - E$. Therefore,

$$
p(\overline{E}) = \frac{|\overline{E}|}{|S|} = \frac{|S - E|}{|S|} = \frac{|S| - |E|}{|S|} = \frac{|S|}{|S|} - \frac{|E|}{|S|} = 1 - p(E).
$$

The observations above are summarized in the theorem below.

Theorem 10.14   *Let E and F be two events in a sample space.*

(a) *Then*
$$
p(E \cup F) = p(E) + p(F) - p(E \cap F).
$$

(b) *If E and F are mutually exclusive, then*
$$
p(E \cup F) = p(E) + p(F).
$$

(c) *If $\overline{E}$ is the complementary event of E, then*
$$
p(\overline{E}) = 1 - p(E).
$$

We now turn to some probability questions involving tossing a pair of dice. There is every reason to believe that each ordered pair (a, b) is equally likely to occur when a pair of dice is tossed, where again, a is the number obtained on the first die and b is the number obtained on the second die. Since there are 36 ordered pairs and the probability function is uniform here,

$$
p((1,1)) = p((1,2)) = \cdots = p((6,5)) = p((6,6)) = \frac{1}{36}.
$$

Example 10.15

A pair of dice is tossed.

(a) What is the probability of obtaining a sum of 7?

(b) What is the probability of obtaining a sum of 7 or a sum of 10?

(c) What is the probability of obtaining 3 on at least one of the two dice?

(d) What is the probability of obtaining a sum that is different from 7?

Solution.

(a) The sample space $S_1 = \{2, 3, \ldots, 12\}$ seems like a good choice here and we wish to determine $p(7)$. This will not be a uniform probability function in this case as surely $p(7) \neq p(2)$, for example. Looking at the table in Figure 10.2, we see that 7 occurs in 6 of the 36 entries in the table, each of which is equally likely to occur. Thus $p(7) = 6/36 = 1/6$.

(b) Looking at Figure 10.2, we see that 10 occurs in 3 of the 36 entries in the table. Since obtaining a sum of 7 and obtaining a sum of 10 are mutually exclusive events, it follows by Theorem 10.14(b) that

$$p(\{7, 10\}) = p(7) + p(10) = \frac{1}{6} + \frac{3}{36} = \frac{1}{6} + \frac{1}{12} = \frac{1}{4}.$$

(c) As we saw earlier, the sample space

$$S_2 = \{(1,1), (1,2), (1,3), \ldots, (6,5), (6,6)\}$$

seems more appropriate here, where each of the 36 outcomes (ordered pairs) is equally likely to occur. The event that interests us is

$$E = \{(3,1), (3,2), (3,3), (3,4), (3,5), (3,6), (1,3), (2,3), (4,3), (5,3), (6,3)\}.$$

Since $|E| = 11$ and the probability function here is uniform,

$$p(E) = \frac{|E|}{|S|} = \frac{11}{36}.$$

We may look at the event E a bit differently. Let E_1 be the event that 3 is the number obtained on the first die and E_2 the event that 3 is the number obtained on the second die. Then the event under consideration is $E = E_1 \cup E_2$. By Theorem 10.14(a),

$$p(E) = p(E_1) + p(E_2) - p(E_1 \cap E_2).$$

Since $|E_1| = |E_2| = 6$ and $E_1 \cap E_2 = \{(3,3)\}$, it follows that $p(E_1) = p(E_2) = 1/6$ and $p(E_1 \cap E_2) = 1/36$. Therefore,

$$p(E) = \frac{1}{6} + \frac{1}{6} - \frac{1}{36} = \frac{11}{36}.$$

(d) Since the event $\overline{E}$ of not obtaining a sum of 7 is the complementary event of the event E of obtaining a sum of 7, it follows by (a) and Theorem 10.14(c) that

$$p(\overline{E}) = 1 - p(E) = 1 - \frac{1}{6} = \frac{5}{6}. \qquad \blacklozenge$$

Example 10.16

Two dice are tossed. What is the probability that the sum of the numbers on the dice is

(a) even?

(b) divisible by 3?

(c) divisible by 4?

Solution. Here too we choose the sample space to be

$$S = \{(1,1), (1,2), (1,3), \ldots, (6,5), (6,6)\}.$$

(a) According to Figure 10.2, there are 18 outcomes that produce an even sum. Therefore, the probability that the sum is even is $18/36 = 1/2$.

(b) There are 12 outcomes that produce a sum that is divisible by 3. So the probability in this case is $12/36 = 1/3$.

(c) There are 9 outcomes that produce a sum that is divisible by 4. Hence the probability that the sum is divisible by 4 is $9/36 = 1/4$. ♦

Example 10.17

Two dice are tossed.

(a) What is the probability that the numbers on the two dice are consecutive integers?

(b) What is the probability that the number on neither die is even?

Solution.

(a) The event in this case is

$$E = \{(1,2), (2,3), (3,4), (4,5), (5,6), (2,1), (3,2), (4,3), (5,4), (6,5)\}.$$

Since $|E| = 10$, it follows that $p(E) = |E|/|S| = 10/36 = 5/18$.

(b) If the number on neither die is even, then the numbers on both dice must be odd. That is, we are looking at outcomes (a,b), where a and b are both odd. Since there are 3 possibilities for a (namely, 1, 3, 5) and 3 possibilities for b, the number of outcomes is $3 \cdot 3 = 9$ by the Multiplication Principle. The probability that the number on neither die is even is therefore $9/36 = 1/4$. ♦

We now consider some examples that do not involve dice.

Example 10.18

A large bowl contains 15 balls that are identical except for their colors: 6 are colored red, 5 are colored blue and 4 are colored green.

(a) What is the probability that two balls selected at random from the bowl have the same color?

(b) What is the probability that two balls selected at random from the bowl are colored the same if the first ball is returned to the bowl before the second ball is selected?

Solution.

(a) There are $\binom{15}{2} = \frac{15 \cdot 14}{2} = 105$ ways to select two balls from the bowl and so the sample space S in this case has 105 elements. Let E be the event that two balls selected at random from the bowl have the same color. There are $\binom{6}{2} = \frac{6 \cdot 5}{2} = 15$ ways to select two red balls, $\binom{5}{2} = \frac{5 \cdot 4}{2} = 10$ ways to select two blue balls and $\binom{4}{2} = \frac{4 \cdot 3}{2} = 6$ ways to select two green balls. It follows by the Addition Principle that the number of ways to select two balls of the same color is therefore $|E| = 15 + 10 + 6 = 31$. Hence the probability of selecting two balls of the same color is $p(E) = |E|/|S| = 31/105 \approx 0.295$.

(b) Since there are $15 \cdot 15 = 225$ ways to select two balls from the bowl in this manner, the sample space has 225 elements. There are $6 \cdot 6 = 36$ ways to select two red balls, $5 \cdot 5 = 25$ ways to select two blue balls and $4 \cdot 4 = 16$ ways to select two green balls. Hence the number of ways to select two balls of the same color is $36 + 25 + 16 = 77$. So the probability of selecting two balls of the same color in this case is $77/225 \approx 0.342$. Therefore, it is a bit more likely to select two balls of the same color if the first ball is returned to the bowl before the second ball is selected. ♦

A standard deck of playing cards consists of 52 cards. These 52 cards are divided into 4 types of 13 cards each. The types are referred to as **suits**, called hearts ($\heartsuit$), diamonds ($\diamondsuit$), clubs ($\clubsuit$) and spades ($\spadesuit$). In each suit, there is one card of each of the following kinds: two, three, four, five, six, seven, eight, nine, ten, jack (J), queen (Q), king (K), ace (A). In the table below, each row corresponds to one of the four suits and each column corresponds to one of the 13 kinds. Hence one of the cards is the three of diamonds and another is the jack of clubs.

	2	3	4	5	6	7	8	9	10	J	Q	K	A
$\heartsuit$													
$\diamondsuit$		3$\diamondsuit$											
$\clubsuit$										J$\clubsuit$			
$\spadesuit$													

There are many games that use such a deck of cards. One of the best known games is poker, where each player in the game is dealt (supplied with) five of the 52 cards, called a **hand**. This game has achieved increased popularity in recent years as poker competitions have become the subject of television programs such as the *World Series of Poker*. In the example to follow, it is useful to know that the number of different poker hands is

$$\binom{52}{5} = \frac{52!}{5! \ 47!} = 2,598,960.$$

Example 10.19

What is the probability of being dealt a poker hand that consists of the three of hearts, the five of clubs, the six of spades, the ten of clubs and the queen of diamonds?

Solution. Since this hand consists of five specific cards, there is only one such hand and the probability of being dealt this hand is

$$\frac{1}{\binom{52}{5}} = \frac{1}{2,598,960} \approx 0.000000385.$$

It is highly unlikely therefore that a poker player has been dealt this particular hand and perhaps even more unlikely that the player would remember (or want to remember) being dealt this hand even if it did occur. On the other hand, it is just as likely to be dealt this hand as any other hand. $\blacklozenge$

A hand that consists of five cards of the same suit is called a **flush**.

Example 10.20

What is the probability of being dealt a flush?

Solution. The number of ways of being dealt a flush where the cards are of a particular suit is $\binom{13}{5} = 1287$. Since there are four different suits, the number of ways of being dealt a flush is $4 \cdot 1287 = 5148$. Therefore, the probability of being dealt a flush is $5148/2,598,960 \approx 0.00198$. $\blacklozenge$

By a **pair** in poker, we mean two cards of the same kind such as two twos, two eights, two queens and so on. A hand has **two pairs** if it has a pair of one kind, a second pair of another kind and one card of a third kind. For example, a hand consisting of two fives, two kings and an ace has two pairs. By **three of a kind** in poker is meant a hand consisting of three cards of the same kind, where the other two cards are of different kinds. For example, a hand consisting of three fives, a ten and a queen is three of a kind. Also, **four of a kind** is a hand in which four cards are of the same kind while the fifth card is obviously of a different kind.

Example 10.21

What is the probability of being dealt a hand of two pairs in poker?

Solution. Since there are 13 different kinds, there are $\binom{13}{2} = 78$ different choices for the kinds of the two pairs. Since there are 4 cards of each kind, there are $\binom{4}{2} = 6$ possibilities for the two cards of the same kind in each pair. There are $52 - 2 \cdot 4 = 44$ possibilities for the card of the third kind. Hence the number of hands consisting of two pairs is $78 \cdot 6 \cdot 6 \cdot 44 = 123,552$ by the Multiplication Principle and the probability of being dealt such a hand is $123,552/2,598,960 \approx 0.048$.   ♦

  Probably one of the best known games of chance involves playing the lottery. Actually, it is inappropriate to say *the* lottery as there are many lotteries. The example below deals with one of them.

Example 10.22

In a certain lottery, called a power ball lottery, 5 balls are selected at random from a collection of 53 white balls numbered 1 to 53 and one ball (called the power ball) is selected at random from a collection of 42 red balls numbered 1 to 42. To win the grand prize in this lottery, a lottery ticket must be purchased that lists the numbers of all 5 white balls (in any order) as well as the number of the power ball. What is the probability of winning the grand prize in this power ball lottery?

Solution. There are $\binom{53}{5}$ ways to select 5 white balls from the collection of 53 white balls and 42 ways to select a red ball from the collection of 42 red balls. Thus the number of different lottery tickets is $\binom{53}{5} \cdot 42 = 120,526,770$. Therefore, the probability of winning the grand prize in this power ball lottery is $1/120,526,770 \approx 0.0000000083$.   ♦

Example 10.23

A total of 20 students participate in a drawing in which one name is chosen at random. There are 6 seniors (2 men and 4 women), 5 juniors (3 men and 2 women) and 9 freshmen (5 men and 4 women). What is the probability that the name that is drawn belongs to

  (a) a woman?

  (b) a freshman?

  (c) a woman who is a freshman?

  (d) either a woman or a freshman?

  (e) either a junior or a freshman?

Solution. The sample space S has 20 names and so $|S| = 20$. Let W be the event that the name selected belongs to a woman, let J be the event that the selected name belongs to a junior, and let F be the event that the selected name belongs to a freshman.

  (a) Since $|W| = 4 + 2 + 4 = 10$, it follows that $p(W) = |W|/|S| = 10/20 = 1/2$.

  (b) Since $|F| = 9$, it follows that $p(F) = |F|/|S| = 9/20$.

  (c) Observe that $|W \cap F| = 4$. Thus the probability that the name that is drawn belongs to a woman who is a freshman is

$$p(W \cap F) = \frac{|W \cap F|}{|S|} = \frac{4}{20} = \frac{1}{5}.$$

(d) By Theorem 10.14(a), the probability that the selected name belongs to either a woman or a freshman is

$$p(W \cup F) = p(W) + p(F) - p(W \cap F) = \frac{1}{2} + \frac{9}{20} - \frac{1}{5} = \frac{3}{4}.$$

Also, observe that $|W \cup F| = 15$ and so

$$p(W \cup F) = \frac{|W \cup F|}{|S|} = \frac{15}{20} = \frac{3}{4}.$$

(e) Observe that J and F are mutually exclusive. Since $p(J) = |J|/|S| = 5/20$ and $p(F) = 9/20$ by (b), it follows by Theorem 10.14(b) that the probability that the selected name belongs to either a junior or a freshman is

$$p(J \cup F) = p(J) + p(F) = \frac{5}{20} + \frac{9}{20} = \frac{14}{20} = \frac{7}{10}.$$

On the other hand, observe that $|J \cup F| = 14$ and so $p(J \cup F) = 14/20 = 7/10$. ◆

There are situations when we may want to perform an experiment over and over again, or perhaps perform two experiments, one after the other. The following theorem is a consequence of the Multiplication Principle (described in Chapter 8).

Theorem 10.24 *Let $S_1, S_2, \ldots, S_n$ be $n \geq 2$ sample spaces with probability functions $p_1, p_2, \ldots, p_n$, respectively and let $S = S_1 \times S_2 \times \cdots \times S_n$. For $(s_1, s_2, \ldots, s_n) \in S$, define*

$$p((s_1, s_2, \ldots, s_n)) = p(s_1) \cdot p(s_2) \cdot \cdots \cdot p(s_n).$$

Then p is a probability function on S. In particular, if an event E of S consists of an event E_1 in S_1, followed by an event E_2 in S_2 and so on, then

$$p(E) = p(E_1) \cdot p(E_2) \cdot \cdots \cdot p(E_n).$$

We now consider three examples that illustrate this theorem.

Example 10.25

In baseball, a player attempts to get a "hit" each time he is "at bat." Suppose that a certain baseball player has a "batting average" of .300. This means that he gets a hit (H) 30% of the time and does not get a hit (N) 70% of the time. Suppose that in a certain game, this player is at bat 5 times, that is, he has 5 opportunities to get a hit.

(a) What is the probability that he will get exactly two hits in the game?

(b) What is the probability that he will get at least one hit in the game?

Solution.

(a) The probability that the baseball player gets hits in his first two times at bat and does not get a hit in his last three times at bat (HHNNN) is

$$(0.3)(0.3)(0.7)(0.7)(0.7) = 0.03087.$$

Since there are $\binom{5}{2} = 10$ ways of getting two hits in his five times at bat, the probability of getting exactly two hits in five times at bat is $10(0.03087) = 0.3087$.

(b) The probability that the baseball player gets no hits in his five times at bat (NNNNN) is $(0.7)^5 = 0.16807$. Therefore, the probability of getting at least one hit in the game is $1 - 0.16807 = 0.83193$. ◆

Recall from calculus that if a and r are real numbers such that $-1 < r < 1$, then

$$a + ar + ar^2 + ar^3 + \cdots = \frac{a}{1-r}. \tag{10.1}$$

This is the sum of the terms in a convergent geometric series. It is likely that you encountered this in a calculus course in which the subject of infinite series was studied (and perhaps likely that you don't remember this).

We return to an example involving the tossing of dice.

Example 10.26

A pair of dice is tossed until a sum of 7 is obtained for the two dice.

(a) What is the probability that 7 is obtained on the first toss of the dice?

(b) What is the probability that 7 is obtained for the first time on the second toss of the dice?

(c) What is the probability that 7 is obtained for the first time on the third toss of the dice?

(d) What is the probability that 7 is obtained for the first time on some toss of the dice?

(e) Use (d) to determine the probability of never getting a 7 on successive tosses of the dice.

Solution. Here we can let the sample space $S = \mathbb{N} = \{1, 2, 3, \ldots\}$, where k is the outcome that successive tosses of a pair of dice results in the sum of 7 for the first time on the kth toss. So the sample space here is countably infinite. For $i \in \mathbb{N}$, let $E_i = \{i\}$ be the event that 7 occurs for the first time on the ith toss.

(a) We have seen in Example 10.15(a) that the probability of getting a 7 on the toss of two dice is $1/6$. So $p(E_1) = 1/6$.

(b) In this case, the first toss of the dice results in a number other than 7, while the second toss results in 7. The probability of this occurring is $p(E_2) = \frac{5}{6} \cdot \frac{1}{6} = \frac{5}{36}$.

(c) The probability of this occurring is $p(E_3) = \frac{5}{6} \cdot \frac{5}{6} \cdot \frac{1}{6} = \left(\frac{5}{6}\right)^2 \cdot \frac{1}{6} = \frac{25}{216}$.

(d) Since $p(E_i) = \left(\frac{5}{6}\right)^{i-1} \cdot \frac{1}{6}$, the event that 7 is obtained for the first time on some toss of the dice is

$$E = E_1 \cup E_2 \cup E_3 \cup \cdots.$$

Since $E_i \cap E_j = \emptyset$ for all positive integers i, j with $i \neq j$, it follows that

$$p(E) = \sum_{i=1}^{\infty} p(E_i) = \sum_{i=1}^{\infty} \left(\frac{5}{6}\right)^{i-1} \frac{1}{6}.$$

Using (10.1) with $a = 1/6$ and $r = 5/6$, we see that the probability of getting a 7 for the first time on some toss of the dice is

$$p(E) = \frac{1}{6} + \frac{5}{6} \cdot \frac{1}{6} + \left(\frac{5}{6}\right)^2 \cdot \frac{1}{6} + \left(\frac{5}{6}\right)^3 \cdot \frac{1}{6} + \cdots = \frac{\frac{1}{6}}{1 - \frac{5}{6}} = \frac{\frac{1}{6}}{\frac{1}{6}} = 1.$$

(e) The event of never getting a 7 on successive tosses of the dice is the complement $\overline{E}$ of E. Thus the probability of never getting a 7 on successive tosses of the dice, according to (d) and Theorem 10.14(c), is $p(\overline{E}) = 1 - p(E) = 1 - 1 = 0$.

Example 10.27

A coin is flipped four times.

(a) What is the probability that heads occurs every time?

(b) What is the probability that heads occurs exactly twice?

(c) What is the probability that heads occurs at least once?

Solution. The sample space in this case consists of $2^4 = 16$ outcomes (see Example 10.3). The probability of getting heads (H) is $1/2$ and the probability of getting tails (T) is also $1/2$.

(a) The probability of getting heads every time is $\left(\frac{1}{2}\right)^4 = \frac{1}{16}$.

(b) The probability of getting heads exactly twice is $\binom{4}{2}\left(\frac{1}{2}\right)^4 = \binom{4}{2}\frac{1}{16} = \frac{6}{16} = \frac{3}{8}$.

(c) If we let E be the event that heads occurs at least once, then $\overline{E}$ is the event that heads never occurs on any of the four flips of the coins. Necessarily then, $\overline{E}$ is the event that tails occurs each time. Thus TTTT is the only one of the 16 outcomes in $\overline{E}$, that is, $\overline{E} = \{\text{TTTT}\}$. Therefore, $p(\overline{E}) = 1/16$ and so $p(E) = 1 - 1/16 = 15/16$.   ♦

Exercises for Section 10.1

1. A white die has each of its spots colored red or blue. A face with an even number of spots has an equal number of red and blue spots. A face with an odd number of spots has one more red spot than blue. What is the probability of getting exactly one blue spot when the die is tossed?

2. Before doing this problem, review Example 10.16. Two dice are tossed.

   (a) What is the probability that the sum of the numbers of the two dice is divisible by 6?

   (b) What is the probability that the sum of the numbers of the two dice is divisible by 5?

   (c) Is your answer to (b) surprising?

3. Two dice are tossed. What is the probability that the numbers of the two dice differ by at least 2?

4. Two dice are tossed and we are interested in the product of the numbers of the two dice. Which product has the highest probability and what is this probability?

5. Three dice are tossed. We are interested in the sum of the numbers of the dice.

   (a) What is the largest sum and what is the probability of getting that sum?

   (b) What is the probability of getting the sum that is most likely?

6. A coin is flipped 6 times. What is the probability that heads and tails occur an equal number of times?

7. A coin is flipped 8 times. What is the probability that heads occurs at least twice?

8. A bowl contains one red ball, two blue balls and three green balls. Three balls are selected at random from the bowl. What is the probability that the three balls selected are of different colors?

9. A bowl contains one red ball, two blue balls and three green balls. Three balls are selected at random from the bowl, but each time a ball is selected it is returned to the bowl before the next ball is selected. What is the probability that the three balls selected are of different colors?

10. A bowl contains 15 balls. There are 5 balls on which the number 1 is written, 5 balls on which the number 2 is written and 5 balls on which 3 appears. Three balls are selected at random. What is the probability that the sum of the numbers on the three balls selected is 6?

11. What is the probability of being dealt a poker hand where no two cards in the hand are of the same kind?

12. The best poker hand is called a **royal flush**, which consists of a ten, jack, queen, king and ace, all of the same suit. What is the probability of being dealt a royal flush?

13. A wealthy gambler plays the power ball lottery described in Example 10.22 every week. He is well aware that the number of different lottery tickets is $\binom{53}{5} \cdot 42 = 120,526,770$. It costs him one dollar to purchase a lottery ticket. If, as time goes by, the grand prize is not won, then the grand prize increases in value. It is announced that the grand prize will be worth approximately \$150,000,000 this coming Saturday. The gambler decides to buy all possible lottery tickets to be assured that he will win. What do you think of this strategy?

14. A total of 20 students participate in a drawing in which one name is chosen. There are 7 seniors (4 men and 3 women), 8 juniors (2 men and 6 women) and 5 freshmen (2 men and 3 women). Determine the probability of drawing the name of

    (a) a woman. (b) a man. (c) a woman who is a senior.

    (d) a man who is a junior. (e) either a man or a junior.

    (f) either a senior or a junior. (g) either a junior or a freshmen.

15. There are 24 finalists for a 1-year scholarship. It is decided that all 24 finalists are equally qualified, so a winner will be chosen at random. Of the 24 finalists,

    (1) 6 are freshmen: 2 mathematics majors (ages: 18, 19), 2 statistics majors (ages: 18, 21), 2 computer science majors (ages: 18, 20);

    (2) 8 are sophomores: 3 mathematics majors (ages: 19, 20, 22), 1 statistics major (age: 19), 4 computer science majors (ages: 18, 19, 21, 21);

    (3) 10 are juniors: 3 mathematics majors (ages: 19, 19, 21), 2 statistics majors (ages: 21, 22), 5 computer science majors (ages: 19, 20, 21, 21, 22).

    What is the probability that the person selected for the scholarship is a sophomore, a mathematics major or is 21 years old or older?

16. After two light bulbs burned out, they were mistakenly placed in a drawer with 8 good light bulbs. If two light bulbs are randomly selected from the drawer (now containing 10 light bulbs), what is the probability that both light bulbs are good?

17. After missing several classes, a student comes to class one day only to learn that there is a 10-question true-or-false quiz to be given that day. Since the student is totally unprepared for the quiz, he guesses the answer to each question. He needs to get at least 5 right to pass the quiz. What is the probability that he will pass the quiz?

18. During the last class period of the semester, each student in a graduate computer science class with 10 students is required to give a brief report on his or her class project. The professor randomly selects the order in which the reports are to be given. Two students have been working on similar projects and would like to give their reports consecutively. What is the probability that this will happen?

19. During the coming commencement, 500 students will graduate. Of these 500, a total of 270 are women and 90 of these 270 women are among the 150 students who will be graduating from the College of Arts and Sciences. One graduating student is always selected at random to speak at the commencement. What is the probability this year that the speaker will either be a woman or a student from the College of Arts and Sciences?

20. Is it possible to have a uniform probability function on a denumerable (countably infinite) sample space?

21. After the beginning of the school year, a local bank runs a promotion where each student who opens a checking account and who agrees to participate in a computer survey will receive a \$25 gift certificate to either McDonald's, Burger King or Wendy's restaurant. Each of these occurring is equally likely. During the first morning of the promotion, 10 students open checking accounts and complete the computer survey.

    (a) What is the probability that everyone received a gift certificate to Burger King?

    (b) What is the probability that no one received a gift certificate to Wendy's?

    (c) What is the probability that at least three gift certificates are awarded for each restaurant?

22. A coin is flipped until heads occurs for the first time.

    (a) What is the probability that heads occurs on the first flip of the coin?

    (b) What is the probability that heads occurs for the first time on the second flip of the coin?

    (c) What is the probability that heads occurs for the first time on the third flip of the coin?

    (d) What is the probability that heads occurs for the first time on some flip of the coin?

    (e) What is the probability that heads never occurs on any flip of the coin?

23. A coin is flipped 10 times. What is the probability that heads has occurred an odd number of times after each even number n of flips of the coin ($n \in \{2, 4, 6, 8, 10\}$)?

24. A man is playing a game with a pair of dice and has just rolled a 6 (that is, the sum of the numbers of the two dice is 6). To win the game, he needs to roll another 6 before obtaining 7. Is he likely to win the game?

25. A man is playing a game with a pair of dice and has just rolled x, where $x \neq 7$ (that is, the sum of the numbers of the two dice is x). To win the game, he must roll another x before rolling $14 - x$. What is the probability that he will win the game?

10.2 Conditional Probability and Independent Events

During the first week of a semester, the 200 students taking Calculus II at a certain university are required to pass two of three quizzes that cover material which the students are expected to know from Calculus I. A student who doesn't pass two of these quizzes is required either to drop the course or to spend ten hours in the tutor lab. A student who passes the first two quizzes is excused from taking the third quiz.

The first two quizzes are then given. A total of 120 students pass the first quiz and 90 students pass both quizzes. The probability that a student is excused from taking the third quiz is therefore $\frac{90}{200} = 0.45$. But what is the probability that a student is excused from taking the third quiz *given that the student passed the first quiz*?

In order for a student to be excused from taking the third quiz given that the student passed the first quiz, the student (1) must be one of those 120 students who passed the first quiz and (2) among these 120 students, he or she must be one of the 90 students who also passed the second quiz. The probability that this occurs is $\frac{90}{120} = 0.75$.

Conditional Probability

The probability that we have just described is called conditional probability. This is the probability that an event occurs given that some other event has already occurred. This is the main topic of the current section. Before giving a formal definition of conditional probability, we present one additional example.

Example 10.28

We have seen that the probability of rolling a 7 with a pair of dice is $1/6 \approx 0.167$. Suppose that a woman is watching a man roll a pair of dice. It is important that the man rolls a 7 in the game he is playing. Because the woman is not wearing her glasses, she is unable to see the outcome of the throw except that she can tell there is more than one spot on each die. With this information, is it more or less likely that the man rolled a 7?

Solution. Since 1 does not occur on either die, there are now only five possibilities for each die and thus $5 \cdot 5 = 25$ possibilities for the pair of dice. That is, the sample space in this case now consists of 25 ordered pairs (rather than 36 ordered pairs), namely all those ordered pairs (a, b) for which $a \neq 1$ and $b \neq 1$. The only ways to roll a 7 without a 1 on either die are the following: (2, 5), (3, 4), (4, 3), (5, 2). Hence the probability that the man has rolled a 7 is now $4/25 = 0.16$ and it is less likely that he rolled a 7.   ◆

What we have just observed is that the probability of an event E occurring may be different in a more restricted sample space F. This is referred to as the conditional probability of E given F. We write $p(E \mid F)$ for the conditional probability of the event E given that event F has already occurred. Consequently, from what we have seen, if E is the event of rolling a 7 on a pair of dice and F is the event of not rolling a 1 on either of the two dice, then $p(E \mid F) = 0.16$.

More generally, suppose that E and F are events in a sample space S and $F \neq \emptyset$. If we wish to compute $p(E \mid F)$, then we are only concerned with the outcomes in E that belong to F. That is, we are interested in the probability of the event $E \cap F$ in the new sample space F. Therefore,

$$p(E \mid F) = \frac{|E \cap F|}{|F|}.$$

This can also be written as

$$p(E \mid F) = \frac{|E \cap F|/|S|}{|F|/|S|} = \frac{p(E \cap F)}{p(F)}.$$

We then have the following definition.

Definition 10.29 *Let E and F be events in a sample space with $p(F) > 0$. Then the* **conditional probability** *of the event E given that event F has occurred is*

$$p(E \mid F) = \frac{p(E \cap F)}{p(F)}.$$

We now look at some examples of conditional probability.

Example 10.30

When a coin is flipped 4 times, the sample space S consists of 16 outcomes, namely,

$$S = \{\text{HHHH, HHHT, HHTH, HHTT, HTHH, HTHT, HTTH, HTTT,}$$
$$\text{THHH, THHT, THTH, THTT, TTHH, TTHT, TTTH, TTTT}\}.$$

The probability of getting an equal number of heads and tails when a coin is flipped 4 times is $6/16 = 3/8$. Suppose that a coin is flipped 4 times but we don't get the same result on the first and last flips. Does this information affect the probability of getting an equal number of heads and tails?

Solution. If F is the event that a different result is obtained on the first and last flips of a coin, then

$$F = \{\text{HHHT, HHTT, HTHT, HTTT, THHH, THTH, TTHH, TTTH}\}.$$

If we let E be the event that there is an equal number of heads and tails, then

$$E = \{\text{HHTT, HTHT, HTTH, THHT, THTH, TTHH}\}.$$

Thus $E \cap F = \{\text{HHTT, HTHT, THTH, TTHH}\}$ and so

$$p(E \mid F) = \frac{|E \cap F|}{|F|} = \frac{4}{8} = \frac{1}{2}.$$

Thus under the condition that the results of the first and last flips are different on four flips of a coin, we now see that it is *more* likely that there will be an equal number of heads and tails. ◆

We now look at a problem concerning conditional probability that results when colored balls are selected from a bowl.

Example 10.31

A bowl contains two red balls and three blue balls. Four balls are selected from the bowl, one at a time and when a ball is selected, it is returned to the bowl before the next ball is selected.

(a) What is the probability that two of the four balls selected are red and the other two are blue?

(b) What is the probability that two of the four balls selected are red and the other two are blue given that the first two balls selected are of different colors?

Solution. Let E be the event that two of the four balls selected are red and the remaining two balls are blue.

(a) Since there are 5 choices for each ball selected, the number of possible outcomes is $5\cdot5\cdot5\cdot5 = 625$ and the sample space S has 625 elements. One way of selecting two red balls and two blue balls is to select them in the order RRBB. There are two ways to select each red ball and three ways to select each blue ball, so the number of ways to select two red balls and two blue balls in this order is $2 \cdot 2 \cdot 3 \cdot 3 = 36$. However, the number of orders in which two red balls can be selected is $\binom{4}{2} = 6$. By the Multiplication Principle, the number of ways to select two red balls and two blue balls is $6 \cdot 36 = 216$ and so $|E| = 216$. Therefore, the probability of selecting two red balls and two blue balls is

$$p(E) = \frac{|E|}{|S|} = \frac{216}{625} \approx 0.346.$$

(b) Let F be the set of outcomes in which the first two balls selected have different colors. The number of ways of selecting 4 balls where the first ball is red and the second ball is blue is $2 \cdot 3 \cdot 5 \cdot 5 = 150$. There are also 150 ways of selecting 4 balls where the first ball is blue and the second ball is red. Therefore, there are 300 ways of selecting 4 balls where the first two balls selected are of different colors and so $|F| = 300$. Since there are 12 ways of selecting the first two balls if they are of different colors and 12 ways of selecting the last two balls if they are of different colors, there are $12 \cdot 12 = 144$ ways to select two red balls and two blue balls where the first two balls selected are of different colors. Hence $|E \cap F| = 144$ and the conditional probability of E given F is

$$p(E \mid F) = \frac{|E \cap F|}{|F|} = \frac{144}{300} = 0.48.$$

Consequently, it is more likely that two of the four balls selected are red and two are blue given that the first two balls selected are of different colors. ◆

Example 10.32

When a coin is flipped four times, we have seen in Example 10.27(b) that the probability that heads comes up exactly twice is $6/16 = 3/8$. What is the probability that heads comes up exactly twice given that heads comes up on the first flip?

Solution. Let E be the event that heads comes up exactly twice when a coin is flipped four times. Therefore,

$$E = \{\text{HHTT, HTHT, HTTH, THHT, THTH, TTHH}\}.$$

Let F be the event that heads comes up on the first flip. Then

$$F = \{\text{HHHH, HHHT, HHTH, HHTT, HTHH, HTHT, HTTH, HTTT}\}.$$

Thus,

$$E \cap F = \{\text{HHTT, HTHT, HTTH}\}.$$

Hence the probability of E given F is

$$p(E \mid F) = \frac{|E \cap F|}{|F|} = \frac{3}{8}.$$

Therefore, $p(E \mid F) = p(E)$ by Example 10.27(b). ◆

Independent Events

In the preceding example, we saw that the probability that heads comes up exactly twice when a coin is flipped four times is the same as the probability that heads comes up exactly twice given that the first flip of the coin is heads. In other words, to have heads occur exactly twice is not affected by the knowledge that the first flip of the coin is heads. (See Exercise 1.) This leads us to the next concept.

Definition 10.33 *Two events E and F in the same sample space are* **independent** *if*

$$p(E \cap F) = p(E) \cdot p(F).$$

If E and F are independent events in the same sample space, then

$$p(E \mid F) = \frac{p(E \cap F)}{p(F)} = \frac{p(E) \cdot p(F)}{p(F)} = p(E)$$

and

$$p(F \mid E) = \frac{p(F \cap E)}{p(E)} = \frac{p(F) \cdot p(E)}{p(E)} = p(F).$$

That is, the probability of either event E or F occurring does not depend on the knowledge that the other event has already occurred. We now consider three examples in which we are asked to determine whether two given events are independent.

Example 10.34

Two dice are rolled. Let E be the event that (the sum) 7 is rolled on the two dice and let F be the event that the first die is 1. Are E and F independent events?

Solution. Here $E \cap F$ is the event that 1 is rolled on the first die and that 7 is the outcome of the two dice. Thus $E \cap F$ consists of the single outcome in which 1 occurs on the first die and 6 occurs on the second die. So $p(E \cap F) = 1/36$. Since $p(E) = p(F) = 1/6$, it follows that

$$p(E \cap F) = \frac{1}{36} = \frac{1}{6} \cdot \frac{1}{6} = p(E)p(F)$$

and so E and F are independent. ◆

Example 10.35

A coin is flipped 6 times. Let E be the event that heads and tails come up an equal number of times. Let F be the event that heads and tails come up once each during the first two flips of the coin. Are E and F independent events?

Solution. The sample space S in this case has $2^6 = 64$ elements. One of the outcomes in E is HHHTTT. Since there are $\binom{6}{3} = 20$ possible locations for the three heads, $|E| = 20$. Since there are two possibilities for the first two flips for each outcome in F and $2^4 = 16$ possibilities for the remaining four flips, $|F| = 2 \cdot 16 = 32$. Since there are two possibilities for the first two flips in $E \cap F$ and $\binom{4}{2} = 6$ possible outcomes for the two heads occurring during the last four flips of the coin, $|E \cap F| = 2 \cdot 6 = 12$. Therefore,

$$p(E \cap F) = \frac{|E \cap F|}{|S|} = \frac{12}{64} = \frac{3}{16}$$

and

$$p(E)p(F) = \frac{|E|}{|S|} \cdot \frac{|F|}{|S|} = \frac{20}{64} \cdot \frac{32}{64} = \frac{5}{32}.$$

Since

$$p(E \cap F) \neq p(E)p(F),$$

it follows that E and F are not independent. ◆

Example 10.36

If a coin is flipped three times, then it is unlikely that heads comes up all three times or that tails comes up all three times. Let E be the event that when a coin is flipped three times, we don't get all heads or all tails. Let F be the event that when a coin is flipped three times, heads comes up at most once. Are E and F independent events?

Solution. Observe that

$$E = \{\text{HHT, HTH, HTT, THH, THT, TTH}\} \text{ and } F = \{\text{HTT, THT, TTH, TTT}\}.$$

Since the sample space S has $2^3 = 8$ elements,

$$p(E) = \frac{|E|}{|S|} = \frac{6}{8} = \frac{3}{4} \text{ and } p(F) = \frac{|F|}{|S|} = \frac{4}{8} = \frac{1}{2}.$$

Now $E \cap F = \{\text{HTT, THT, TTH}\}$ and so

$$p(E \cap F) = \frac{|E \cap F|}{|S|} = \frac{3}{8}.$$

Since

$$p(E \cap F) = \frac{3}{8} = \frac{3}{4} \cdot \frac{1}{2} = p(E)p(F),$$

it follows that E and F are independent events.   ♦

We now use the fact that two events are independent to answer a few questions.

Example 10.37

Determine the probability that two flips of a coin result in two heads.

Solution. Since the outcome on the second flip of a coin is not influenced by the first flip of the coin, it follows that the probability of getting heads both times is the product of probabilities of getting heads on the first flip and getting heads on the second flip, which is $\frac{1}{2} \cdot \frac{1}{2} = \frac{1}{4}$.   ♦

Example 10.38

Determine the probability that heads occurs for the second time on the 4th flip of a coin.

Solution. Let E be the event that heads comes up on the 4th flip of a coin and let F be the event that heads comes up exactly once during the first three flips of the coin. Then we wish to determine $p(E \cap F)$. Certainly, whether heads comes up on the 4th flip of a coin is not affected by what happens during the first three flips of the coin, that is, E and F are independent events and so $p(E \cap F) = p(E)p(F)$. Now $p(E) = 1/2$. Since there are exactly three outcomes of the eight that result in heads coming up exactly one, it follows that $p(F) = 3/8$. Therefore,

$$p(E \cap F) = p(E)p(F) = \frac{1}{2} \cdot \frac{3}{8} = \frac{3}{16}.$$ ♦

Example 10.39

It's Friday and the TV weather forecaster states that there is a 50% chance of rain on Saturday and a 50% chance of rain on Sunday. If we were to assume that having rain on Saturday and having rain on Sunday are independent events, then what is the probability of having no rain over the weekend?

Solution. Observing that the probability of having no rain on Saturday is $1/2$ and no rain on Sunday is $1/2$, we see that the probability of having no rain over the weekend is $\frac{1}{2} \cdot \frac{1}{2} = \frac{1}{4}$.   ♦

Example 10.40

Three jars, denoted by J_1, J_2 and J_3, contain three balls each, which are identical except for their colors. The jar J_1 contains a red ball (R), a blue ball (B) and a green ball (G); the jar J_2 contains two red balls and a blue ball; and the jar J_3 contains three red balls. The jars are not labeled and the balls cannot be seen from the outside. A jar is selected at random and a ball is selected at random from that jar. If the jar J_2 was selected and the blue ball, say, was selected from J_2, then this outcome is denoted by (J_2, B). What is the probability that J_1 was the jar selected given that a red ball was selected from a jar?

Solution. Let E be the event that jar J_1 was selected and let F be the event that a red ball was selected from a jar. Hence we seek the conditional probability $p(E \mid F)$, which equals $p(E \cap F)/p(F)$. Thus $E \cap F = \{(J_1, R)\}$ and $F = \{(J_1, R), (J_2, R), (J_3, R)\}$. To help us determine $p(E \cap F)$ and $p(F)$, we make use of the tree diagram shown in Figure 10.3. Therefore, $p(E \cap F) = p(\{(J_1, R)\}) = \frac{1}{3} \cdot \frac{1}{3} = \frac{1}{9}$ and

$$
\begin{aligned}
p(F) &= p(\{(J_1, R), (J_2, R), (J_3, R)\}) = \frac{1}{3} \cdot \frac{1}{3} + \frac{1}{3} \cdot \frac{2}{3} + \frac{1}{3} \cdot 1 \\
&= \frac{1}{9} + \frac{2}{9} + \frac{1}{3} = \frac{2}{3}.
\end{aligned}
$$

Therefore, $p(E \mid F) = \frac{p(E \cap F)}{p(F)} = \frac{1/9}{2/3} = \frac{1}{6}.$ ◆

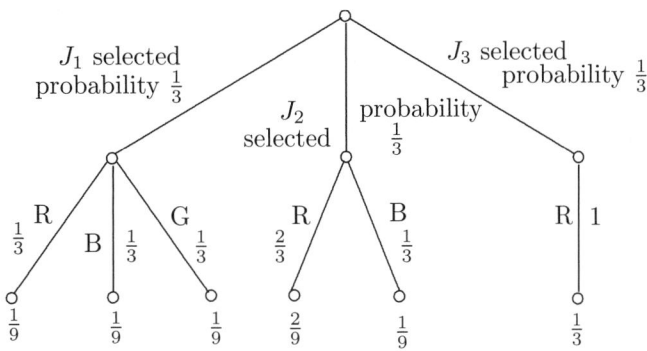

Figure 10.3: A tree diagram for Example 10.40

By the Pigeonhole Principle, if there are 13 people in a room, then there must be (at least) two people whose birthdays occur during the same month. On the other hand, if there are only 12 people in the room, then there is a possibility that these 12 people have their birthdays during different months and therefore no two of them have birthdays during the same month. It seems likely, though, that some of the 12 people would have their birthdays during the same month. That is, it would seem that the probability that some two people in a group of 12 having their birthdays during the same month is greater than 0.5. This brings up a question. What is the smallest number of people for which it is more likely than not that two people have their birthdays during the same month? We are assuming here of course that there are no connections among any of these people that might influence when their birthday months are (such as three of the people being triplets). That is, we are assuming that the birthday months of the people are independent.

Example 10.41

What is the minimum number of people needed for which it is more likely than not that two of them have birthdays during the same month?

Solution. The minimum number is 5. First, let the five people be denoted by A_0, A_1, A_2, A_3, A_4. We compute the probability that no two of these individuals have their birthdays during the same month. It doesn't matter when the birthday of A_0 is. The probability that the birthday of A_1 is during a month different from that of A_0 is 11/12. The probability that A_2 has his/her birthday during a month different from both A_0 and A_1 is 10/12. Hence the probability that all five people have their birthdays during different months is

$$\frac{11}{12} \cdot \frac{10}{12} \cdot \frac{9}{12} \cdot \frac{8}{12} = \frac{7920}{20,736} \approx 0.382.$$

Thus the probability that some pair of people have their birthdays during the same month is approximately $1 - 0.382 = 0.618$. Since the probability that some pair of people in a group of 4 have their birthdays during different months is approximately 0.427, the minimum number of people is 5. ◆

Exercises for Section 10.2

1. (a) What is the probability of the event that heads occurs exactly twice when a coin is flipped 4 times given that the first flip is tails?

   (b) Based on (a) and Example 10.32, what can you say about the probability that heads occurs exactly twice when a coin is flipped 4 times?

2. A coin is flipped three times.

   (a) What is the probability of getting three heads?

   (b) What is the probability of getting three heads given that the first flip came up heads?

   (c) What is the probability of getting three heads given that the first two flips resulted in two heads?

   (d) What is the probability of getting three heads given that the first three flips resulted in all heads?

   (e) What is the probability of getting three heads given that at least one of the first two flips resulted in heads?

   (f) What is the probability of getting three heads given that at most one of the first two flips resulted in heads?

3. A coin is flipped twice.

   (a) What is the probability of getting heads and tails once each?

   (b) What is the probability of getting heads and tails once each given that the first flip results in tails?

4. Suppose that a coin is flipped 4 times. Let E be the event that heads occurs an odd number of times and let F be the event that heads comes up on the first flip. Are E and F independent?

5. A coin is flipped three times. Let E be the event that heads and tails occur at least once each and let F be the event that heads occurs at least twice. Are E and F independent events?

6. Determine the probability that tails occurs for the 3rd time on the 5th flip of a coin.

7. A red bowl contains five red balls numbered 1 to 5 while a blue bowl contains five blue balls also numbered 1 to 5. A ball is selected from the red bowl and then a ball is selected from the blue bowl.

   (a) What is the probability that the sum of the numbers on the balls selected is even?

   (b) What is the probability that the sum of the numbers on the balls selected is even given that the number on the red ball is 1?

(c) What is the probability that the sum of the numbers on the balls selected is even given that the number on the red ball is 2?

8. Two dice are rolled. Let E be the event that the sum of 5 is rolled and F the event that 2 is rolled on the first die. Are E and F are independent?

9. Two dice are rolled. Let E be the event that 8 is rolled on the two dice and let F be the event that the first die is 4. Are E and F independent?

10. A coin is flipped and then a die is tossed. What is the probability that heads comes up on the coin and an odd number occurs on the die?

11. (a) How many times does a die have to be tossed for it to be possible that the same number comes up twice?

    (b) How many times does a die have to be tossed to be guaranteed that the same number comes up twice?

    (c) How many times does a die have to be tossed for it to be more likely than not that the same number comes up twice?

12. Your discrete mathematics class meets on Monday, Wednesday and Friday each week. Your instructor announces that during each class period next week, there is a 50% chance of having a quiz. What is the probability of having at least one quiz next week?

13. The probability of a student earning less than C on an exam in a certain course is 0.2, while the probability of a student earning less than C on an exam and being absent the day after the exam is given is 0.15. What is the probability that a student is absent the day after an exam given that the student earned less than C on the exam?

14. Al and Laura have been taking a 2-course sequence in computer science. A quiz has been given every week in both courses. Al has received a B or better 80% of the time, while Laura has received a B or better 90% of the time. Assuming that this trend continues and that the scores of these two students are independent, what is the probability that

    (a) both get a B or better on the next quiz?

    (b) at least one gets a B or better on the next quiz?

    (c) neither gets a B or better on the next quiz?

15. Suppose that the probability of a person being overweight is 0.3, the probability of a person having high blood pressure is 0.2 and that the probability of a person having high blood pressure given that he/she is overweight is 0.4. What is the probability that a person is overweight given that he/she has high blood pressure?

16. A graduate class in computer science consists of six students, three men and three women. During the last two days of the semester, presentations are made by the students, three presentations on each day. The professor randomly selects the students to make presentations. If, on the first of the two days, a man is selected to give the first presentation, what is the probability that two women and a man will give presentations on the second day?

17. You have two coins: a fair coin C_1 and a fake coin C_2. Therefore, the probability of getting heads when C_1 is flipped is 1/2 and the probability of getting tails when C_1 is flipped is also 1/2. The fake coin C_2 has heads on both sides. A friend of yours chooses one of these two coins at random and then the coin is flipped. The outcome that C_1 is chosen and heads is obtained when C_1 is flipped is denoted by (C_1, H).

    (a) What is the sample space S of this experiment?

(b) After your friend selects a coin and flips it, heads is obtained. What is the probability that he selected the fair coin?

18. A man has a pair of electronic dice. Initially, the spots on each die cannot be seen until the die is tossed in which case the spot(s) on the top face (only) light up. One die D_1 is fair and the six faces of D_1 have 1, 2, 3, 4, 5 and 6 spots. The other die D_2, however, is a fake die and has only one spot on each face. One of these two dice is selected randomly and is tossed. If the die D_1 is thrown and the number 3, say, is obtained, then this outcome is denoted by $(D_1, 3)$.

   (a) What is the sample space S of this experiment?

   (b) Given that a die was selected and 1 was obtained after it was tossed, what is the probability that the fair die D_1 was selected?

19. There is a dangerous disease that has infected 5% of the population of a large area and those with the disease show no initial symptoms. A scientist has discovered a test for this. However, 10% of the time the test says you don't have the disease when you actually do (a "false negative") and 20% of the time the test says you have the disease when you don't (a "false positive"). The test is performed on a randomly selected person. If the test is positive, what is the probability that the person has the disease?

20. At a class reunion, you meet an old friend whom you haven't seen in years. You learn that she has two children and one of her children is a girl. What is the probability that this girl has a sister?

21. The jar J_1 contains one red ball (R) and one blue ball (B) while the jar J_2 contains a red ball and two blue balls. A coin is flipped. If the coin comes up heads, a ball is taken from J_1; while if the coin comes up the tails, a ball is taken from J_2. What is the probability that a ball was selected from J_1 given that the color of the ball selected is red?

22. Three dice are tossed. What is the probability that 1 was obtained on two of the dice given that the sum of the numbers on the three dice is 7?

23. A magician has five coins in his pocket. Four of the coins are fair but the other is a fake coin and has two heads. He pulls a coin at random from his pocket and flips it five times. It comes up heads all five times. What is the probability that he selected a fair coin from his pocket?

24. A student has an exam in a discrete mathematics class tomorrow. The exam is divided into two parts. The first part consists only of true-false questions and the second part consists only of multiple choice questions. She estimates that the probability of getting B or better on Part 1 is 0.8 and the probability of getting B or better on Part 2 is 0.6. She also estimates that the probability of getting B or better on both parts is 0.5.

   (a) What is the probability of getting B or better on Part 1 if she got B or better on Part 2?

   (b) Are the events "getting B or better on Part 1" and "getting B or better on Part 2" independent?

25. Suppose that E and F are independent events.

   (a) Show that E and $\overline{F}$ are independent.

   (b) Show that $\overline{E}$ and $\overline{F}$ are independent.

10.3 Random Variables and Expected Values

A computer science professor routinely gives a 20-point quiz each week in every class she teaches. During the current semester, she is teaching a graduate class with four students and an undergraduate class with 40 students. During the most recent quiz in the graduate class, the quiz scores were 14, 15, 17 and 18. What is the average quiz score? This question can be answered quite easily. The average quiz score is 16, although, of course, not even one student had a quiz score of 16. During the most recent quiz in the undergraduate class, it turned out that every student also received one of the quiz scores 14, 15, 17 and 18. What is the average quiz score for the undergraduate class? This question is impossible to answer from the information that was given. We need to know how many of the 40 students received each of the scores 14, 15, 17, 18 or to put in another way, we need to know the probability of a student getting each of the scores 14, 15, 17, 18. Having this information will enable us to compute a "weighted average" or "expected value" for a quiz score (see Exercise 10). This is one of the major concepts of this section. To prepare ourselves to study this idea, we consider ways that a number can be associated with each outcome in the sample space of an experiment. This is a function from the sample space to some set of numbers and brings up the concept of a random variable, which is the second major topic of this section.

Random Variables

Definition 10.42 *Let S be a sample space on which is defined a probability function. A **random variable** on S is a function X from S to the set **R** of real numbers. The **range** $X(S)$ of X is the set of functional values of X.*

The choice of the term "random variable" for this concept, although standard, is not a particularly good choice since a random variable is neither random nor a variable. Let's consider some examples of random variables.

Example 10.43

A bowl contains two red balls and three blue balls. Three balls are selected at random from the bowl. For each outcome t of this experiment, let $X(t)$ denote the number of red balls selected in the outcome t. Then X is a random variable on the sample space of this experiment. Since each outcome contains at most two red balls, the range of X is $\{0, 1, 2\}$. ◆

Example 10.44

A coin is flipped three times, resulting in the sample space

$$S = \{\text{HHH, HHT, HTH, HTT, THH, THT, TTH, TTT}\}.$$

One random variable X defined on the sample space of this experiment is

$$X(TTT) = 0, \quad X(HHH) = X(HHT) = X(HTH) = X(HTT) = 1,$$
$$X(THH) = X(THT) = 2, \quad X(TTH) = 3.$$

We can also present this information in the following table:

t	HHH	HHT	HTH	HTT	THH	THT	TTH	TTT
$X(t)$	1	1	1	1	2	2	3	0

That is, for an element t in the sample space S, $X(t) = 0$ if $t = TTT$; otherwise $X(t) = k$ ($1 \le k \le 3$) if heads occurs for the first time on the kth flip. The range of X in this case is $\{0, 1, 2, 3\}$. ◆

For a sample space S of an experiment and a random variable X defined on S, the set of outcomes in S having a given value (image) is an event of special interest.

Definition 10.45   *Let* $X : S \to \mathbf{R}$ *be a random variable defined on the sample space* S *of some experiment. For each* $r \in \mathbf{R}$*, the set*

$$\{s \in S :\ X(s) = r\}$$

is therefore an event, which we denote by $X = r$*. We write* $p(X = r)$ *for the probability of the event in which each outcome has the value* r*.*

Example 10.46

For the experiment of rolling a pair of dice, suppose that the sample space is

$$S = \{(1,1), (1,2), (1,3), \ldots, (6,5), (6,6)\}.$$

Define a random variable X on S by

$$X((a,b)) = \max\{a, b\}$$

for each outcome $(a,b) \in S$. For $1 \le n \le 6$, let E_n denote the event

$$E_n = \{s \in S :\ X(s) = n\}.$$

Determine the probability $p(E_5)$ of E_5.

Solution. The event E_5 consists of those ordered pairs $(a,b) \in S$ where the maximum of a and b is 5. Therefore,
$$E_5 = \{(1,5), (2,5), (3,5), (4,5), (5,5), (5,1), (5,2), (5,3), (5,4)\}.$$
Since $|E_5| = 9$ and $|S| = 36$, it follows that

$$p(E_5) = \frac{|E_5|}{|S|} = \frac{9}{36} = \frac{1}{4}.$$

Thus $p(X = 5) = 1/4$.   ♦

We saw in Example 10.46 that $|E_5| = 9$. Indeed, for $1 \le n \le 6$, $|E_n| = 2n - 1$. Because $\{E_1, E_2, \ldots, E_6\}$ is a partition of S, it follows that

$$|E_1| + |E_2| + |E_3| + |E_4| + |E_5| + |E_6| = |S|$$

and so

$$1 + 3 + 5 + \cdots + (2n - 1) = n^2$$

for $n = 6$. This illustrates a formula we saw (and proved) in Result 4.4 in Section 4.1.

Definition 10.47   *For a random variable* X *on a sample space* S*, the set of pairs*

$$(r,\ p(X = r))$$

for each $r \in X(S)$ *is called the* **distribution** *of* X *on* S*. The distribution of a random variable* X *is typically expressed by writing* $p(X = r)$ *for each* $r \in X(S)$*.*

Therefore, for a random variable X on a sample space S,

$$\sum_{r \in X(S)} p(X = r) = 1. \tag{10.2}$$

Example 10.48

Determine the distribution of the random variable given in Example 10.43.

Solution. In Example 10.43, three balls are selected at random from a bowl containing two red balls and three blue balls. Since there are $\binom{5}{3} = 10$ ways to choose three balls from the bowl, the sample space S consists of 10 elements. For each outcome t of this experiment, $X(t)$ is the number of red balls selected in the outcome t. Since each outcome contains at most 2 red balls, we saw in Example 10.43 that the range of X is $X(S) = \{0, 1, 2\}$. The distribution of the random variable X is

$$p(X = 0) \;=\; \frac{\binom{3}{3}}{10} = \frac{1}{10} = 0.1$$

$$p(X = 1) \;=\; \frac{\binom{2}{1}\binom{3}{2}}{10} = \frac{2 \cdot 3}{10} = 0.6$$

$$p(X = 2) \;=\; \frac{\binom{2}{2}\binom{3}{1}}{10} = \frac{1 \cdot 3}{10} = 0.3.$$

Thus $p(X = 0) + p(X = 1) + p(X = 2) = 1$, as anticipated in (10.2). $\blacklozenge$

In Example 10.48 the distribution of the random variable X can also be presented in the following table.

r	0	1	2
$p(X = r)$	0.1	0.6	0.3

Expected Value

Our primary interest in random variables lies with the associated concept of expected value.

Definition 10.49 *The **expected value** $E(X)$ of a random variable X on a sample space S is the sum over all elements s of S of the product of the probability of s and the value $X(s)$, that is,*

$$E(X) = \sum_{s \in S} p(s)X(s). \tag{10.3}$$

In particular, if $S = \{s_1, s_2, \ldots, s_n\}$, then

$$E(X) = \sum_{i=1}^{n} p(s_i)X(s_i).$$

Therefore, the expected value of a random variable X on a sample space S can be considered as a "weighted average" of the values $X(s)$ for $s \in S$. When the sample space S of an experiment is relatively small, we can compute the expected value of a random variable directly from the definition in (10.3). On the other hand, when S consists of a large number of outcomes, it is convenient to compute the expected value of a random variable by grouping all outcomes in S assigned the same value by the random variable. This suggests expressing the expected value of a random variable X on a sample space S in terms of the distribution of the random variable X on S.

Theorem 10.50 *For a random variable X on a sample space S, the expected value of X on S is*

$$E(X) = \sum_{r \in X(S)} p(X = r) \cdot r.$$

For the random variable X described in Example 10.48, the expected value is

$$E(X) = \sum_{r \in X(S)} p(X = r) \cdot r = (0.1) \cdot 0 + (0.6) \cdot 1 + (0.3) \cdot 2 = 1.2.$$

In this case, the outcome we might expect to get is not the expected value. Next we consider two problems where the opposite is true.

Example 10.51

A fair coin is flipped four times. How many heads would one expect to get?

Solution. Let S be the sample space of the $2^4 = 16$ outcomes and let X be the random variable that assigns to each outcome the number of heads that occurs. Since there is 1 outcome with no heads, 4 outcomes with one head, 6 outcomes with two heads, 4 outcomes with three heads and 1 outcome with four heads, it follows that

$$p(X = 0) = \tfrac{1}{16}, \quad p(X = 1) = \tfrac{4}{16}, \quad p(X = 2) = \tfrac{6}{16},$$
$$p(X = 3) = \tfrac{4}{16}, \quad p(X = 4) = \tfrac{1}{16}.$$

This information is also presented in the following table:

r	0	1	2	3	4
$p(X = r)$	$\frac{1}{16}$	$\frac{4}{16}$	$\frac{6}{16}$	$\frac{4}{16}$	$\frac{1}{16}$

By Theorem 10.50,

$$E(X) = 0 \cdot \frac{1}{16} + 1 \cdot \frac{4}{16} + 2 \cdot \frac{6}{16} + 3 \cdot \frac{4}{16} + 4 \cdot \frac{1}{16} = \frac{32}{16} = 2,$$

as expected.   ◆

Example 10.52

Two fair dice are tossed. What is the expected sum of the numbers of the two dice?

Solution. Let S be the sample space consisting of the 36 outcomes (see Figure 10.2) and let X be the random variable that assigns to each outcome the sum of the values of the two dice. The distribution of the random variable X is given in the following table:

r	2	3	4	5	6	7	8	9	10	11	12
$p(X = r)$	$\frac{1}{36}$	$\frac{2}{36}$	$\frac{3}{36}$	$\frac{4}{36}$	$\frac{5}{36}$	$\frac{6}{36}$	$\frac{5}{36}$	$\frac{4}{36}$	$\frac{3}{36}$	$\frac{2}{36}$	$\frac{1}{36}$

By Theorem 10.50, the expected value of X is

$$\begin{aligned}
E(X) &= 2 \cdot \frac{1}{36} + 3 \cdot \frac{2}{36} + 4 \cdot \frac{3}{36} + 5 \cdot \frac{4}{36} + 6 \cdot \frac{5}{36} + 7 \cdot \frac{6}{36} + \\
&\quad 8 \cdot \frac{5}{36} + 9 \cdot \frac{4}{36} + 10 \cdot \frac{3}{36} + 11 \cdot \frac{2}{36} + 12 \cdot \frac{1}{36} \\
&= \frac{252}{36} = 7,
\end{aligned}$$

again as you probably expected.   ◆

In terms of gambling, expected values give us an idea of the likelihood of our winning games of chance. We consider two games for the purpose of determining the wisdom of participating in them.

Example 10.53

At a carnival it costs $10 to play a game, where three balls are selected from a bowl consisting of three red balls, three blue balls and three green balls. If all three balls selected have the same color, then you win $63, while if all balls selected have different colors, you win $21. If you were to play this game 10 times, how much would you expect to win or lose?

Solution. First, there are $\binom{9}{3} = 84$ different outcomes and so the sample space has 84 elements. Define a random variable X on the sample space by assigning 63 to an outcome in which all three balls selected have the same color, 21 to an outcome in which all balls selected have different colors and 0 to all other outcomes. Only one of these outcomes consists of three red balls (or three blue balls or three green balls). Therefore, $p(X = 63) = 3/84 = 1/28$. Since there are three balls of each color, there are $3 \cdot 3 \cdot 3 = 27$ outcomes where all the balls have different colors. Hence $p(X = 21) = 27/84 = 9/28$. Hence $p(X = 0) = 1 - 1/28 - 9/28 = 18/28 = 9/14$. The distribution of the random variable X is given in the following table:

r	63	21	0
$p(X = r)$	$\frac{1}{28}$	$\frac{9}{28}$	$\frac{9}{14}$

By Theorem 10.50, the expected value of X is

$$E(X) = 63 \cdot \frac{1}{28} + 21 \cdot \frac{9}{28} + 0 \cdot \frac{9}{14} = \frac{252}{28} = 9.$$

Consequently, each time the game is played, you would expect to get $9. However, you pay $10 to play the game, so you would expect to lose $1. If the game is played 10 times, then you would expect to lose $10. ◆

In Example 10.53, a game is described at a carnival. Although you would expect to lose money by playing it, you would not expect to lose much. Let's look at another game that might take place at a carnival.

Example 10.54

It costs $10 to play a game that consists of selecting a total of three balls from three bowls, where one ball is selected from each bowl. Each bowl contains three balls, one of each of the colors red, blue and green. If all three balls selected have the same color, then you win $63; while if all balls selected have different colors, you win $21. If you were to play this game 10 times, how much would you expect to win or lose?

Solution. There are $3 \cdot 3 \cdot 3 = 27$ different outcomes and, consequently, the sample space consists of 27 elements. Define a random variable X on the sample space by assigning 63 to an outcome in which the three balls selected have the same color, 21 to an outcome in which the three balls selected have different colors and 0 to all other outcomes. Since there are 3 outcomes where the three balls have the same color and 6 outcomes where the three balls have different colors, there are 18 outcomes that are of neither type. The distribution of the random variable X is given in the following table:

r	63	21	0
$p(X = r)$	$\frac{1}{9}$	$\frac{2}{9}$	$\frac{2}{3}$

By Theorem 10.50, the expected value of X is

$$E(X) = 63 \cdot \frac{1}{9} + 21 \cdot \frac{2}{9} + 0 \cdot \frac{2}{3} \approx 11.67.$$

Therefore, each time you play the game, you would expect to win about $1.67. If the game is played 10 times, then you would expect to win about $16.67. (For this reason, you probably won't see this game played at a carnival.) ◆

Exercises for Section 10.3

1. A bowl contains two red balls, one blue ball and one green ball. A ball is selected at random from the bowl and then a second ball is selected at random from the remaining three balls in the bowl.

   (a) What is the probability of each outcome in the experiment?

   (b) Let X be the random variable defined on the sample space S such that $X(t)$ is the number of green balls selected for each $t \in S$. Determine $X(t)$ for each $t \in S$.

   (c) Determine the range $X(S)$ of the random variable X defined in (b).

   (d) Determine the distribution of the random variable X defined in (b).

2. A bowl contains two red balls and three blue balls. Two balls are selected at random from the bowl and the number of the red balls selected is recorded. This is done a total of ten times and all recorded numbers are added. What is the expected total number of red balls selected?

3. A bowl contains four red balls and two blue balls. Three balls are selected at random from the bowl. How many red balls would one expect to select?

4. You pay \$4.50 to play the following game. A pair of dice is rolled and you receive n dollars if the largest number on one of the dice is n. How much would you expect to win or lose if you play this game?

5. A game is played on nine squares, numbered 1 through 9. It costs \$1 to play the game. A computer program randomly selects 4 of the 9 squares to be colored red, 3 of the squares to be colored blue and the remaining 2 squares to be colored white. Before the program is run, the player selects 4 squares to color red, 3 squares to color blue and 2 squares to color white. If the color of each square chosen by the player matches that selected by the computer program, then the player wins \$1000.

1	2	3
4	5	6
7	8	9

   (a) What is the expected gain or loss if this game is played 100 times?

   (b) Why is knowing this expected value of little use in this situation?

6. A coin is flipped three times.

   (a) What is the sample space S of this experiment? What is $|S|$?

   (b) What is the probability of each outcome in this experiment?

   (c) Let X be the random variable defined on S such that $X(t)$ is the largest number of consecutive heads for each $t \in S$. Determine $X(t)$ for each $t \in S$.

   (d) Determine the range of the random variable X defined in (c).

   (e) Determine the distribution of the random variable X defined in (c).

   (f) Determine the expected value $E(X)$ of the random variable X defined in (c).

7. A pair of dice is thrown. What is the expected value of the absolute value of the difference of the numbers on the two dice?

8. A pair of dice is thrown. What is the expected value of the product of the numbers on the two dice?

9. It costs $10 to play the following game. On Table # 1 is a coin, on Table #2 is a die and on Table #3 is a bowl containing three red balls and two blue balls. The game is played by flipping the coin, throwing the die and selecting (at random) two balls from the bowl. If heads is flipped, you win $5. If you win $5 and then throw a 6 on the die, you win an extra $20. If you win $25 and then select two balls of the same color, you win an extra $100. How much would you expect to win or lose if you play this game?

10. Suppose that 40 students take a 20-point quiz. A total of 5 of these students obtain a score 14 and 15 students obtain a score of 18. If 10 students obtain a score of 15 and 10 students obtain a score of 17, what is the expected value (average) of the quiz scores?

11. On a particular die, three of the faces have no spots, two have one spot and one face has two spots. When a pair of dice is thrown, what is the expected value of

   (a) the sum of the number of spots on the two dice?

   (b) the product of the number of spots on the two dice?

12. Early each fall, a department store manager purchases a large number of winter sweaters. He pays $60 for each sweater. Any sweater that isn't sold by Christmas will be sold for a $10 loss. Experience says that he can sell 40% of them by Christmas if he prices the sweaters at $100 each, he can sell 60% of them if each is priced at $90 and he can sell 70% of them if they are priced at $80 each. How should the manager price the sweaters?

13. One year, a political party is likely to select one of three people (A, B or C) as its candidate for president. The probability of selecting A is 4/9, the probability of selecting B is 1/3 and the probability of selecting C is 2/9. In addition, it is likely that this party will select one of three other people (D, E or F) as its candidate for vice-president. The probability of selecting D is 1/2, the probability of selecting E is 1/3 and the probability of selecting F is 1/6. The ages of these six individuals are as follows:

   A – 68, B – 53, C – 47, D – 41, E – 57, F – 50.

   What is the expected average age of the two people selected as candidates for president and vice-president?

14. You pay 1 dollar to play the following electronic game at a casino. There are two buttons. First you push button #1 and then button #2. When each button is pushed, two randomly generated distinct integers between 1 and 10 are produced and displayed. If the two pairs of integers selected are the same, you win k dollars (for a net gain of $k - 1$ dollars); otherwise you lose the dollar. What should k equal so that the game is fair, this is, you expect to break even?

15. An insurance agent insures student apartments against burglaries in a large off-campus apartment development. Statistics show that 2% the apartments will be burglarized in a given year and that a typical payout by the insurance company for each such robbery is $5000. How much should the insurance agent charge to insure each apartment in order for him to receive 10% over the break-even amount to cover his expenses and to make a profit?

Chapter 10 Highlights

Key Concepts

complementary event (of an event E), $\overline{E}$: the set of outcomes in the sample space not belonging to E.

conditional probability of E **given** F, $p(E \mid F)$: the probability of an event E given that the event F has already occurred.

distribution (of a random variable X on the sample S of an experiment): the set of pairs $(r, p(X = r))$ for each $r \in X(S)$, where $p(X = r)$ denotes the probability that X has the value r.

event: a subset of a sample space.

expected value (of a random variable X on a sample space S): $E(X) = \sum_{s \in S} p(s)X(s)$.

experiment: a procedure that results in one of a number of possible outcomes.

independent events: the occurrence of either of two events does not influence the likelihood of the occurrence of the other.

mutually exclusive events E **and** F: $E \cap F = \emptyset$.

outcome: an element in a sample space.

probability (of an event): the sum of the probabilities of the outcomes in the event.

probability (of an outcome in a sample space): the value assigned to the outcome by a probability function.

probability function (on a sample space S): a function $p : S \to [0, 1]$ for which $\sum_{s \in S} p(s) = 1$.
random variable: a function assigning a real number to each outcome in an experiment.

sample space: the set of all possible outcomes of an experiment.

uniform probability function: a probability function p defined by $p(s) = \frac{1}{|S|}$, where S is a finite sample space and each outcome in S is equally likely to occur.

Key Results

- For an event E in a sample space, $0 \le p(E) \le 1$.

- For any two events E and F in a sample space, $p(E \cup F) = p(E) + p(F) - p(E \cap F)$.

- If E and F are mutually exclusive events, then $p(E \cup F) = p(E) + p(F)$.

- If $E_1, E_2, \ldots, E_n$ are $n \ge 2$ mutually exclusive events, then
  $$p(E_1 \cup E_2 \cup \cdots \cup E_n) = p(E_1) + p(E_2) + \cdots + p(E_n).$$

- If E and F are independent events, then $p(E \cap F) = p(E)p(F)$.

- For every event E in a sample space, $p(\overline{E}) = 1 - p(E)$.

- For a random variable X on a sample space S, the expected value of X on S is
  $$E(X) = \sum_{r \in X(S)} p(X = r) \cdot r.$$

Supplementary Exercises for Chapter 10

1. Suppose that there is a talent competition called University Idol and there are 12 students in the competition. Each week the students vote for their favorite person in the competition and the person with the smallest number of votes is eliminated from future competitions. So after the first week, only 11 competitors remain. If each person in the competition is equally likely to win (or lose), what is the probability that a particular student is in the competition after the first two weeks but has been eliminated after the third week?

2. A high school puts on a musical play each year. In order to raise money for this, the students have been visiting businesses throughout the community asking if they would contribute money for this purpose. A family restaurant in town agrees to contribute three $50 gift certificates to dine at the restaurant. The school has decided to hold a raffle for these certificates and sell raffle tickets for $1 each. A drawing of three tickets will be held from those sold to determine who wins the gift certificates. One of the parents has purchased a raffle ticket. Just before the raffle is held, he learns that 250 tickets have been sold. How would his chances of winning a gift certificate be affected if he were to purchase a second raffle ticket just before the drawing?

3. A bowl contains 50 balls numbered from 1 to 50.

   (a) A ball is selected at random from the bowl. What is the probability that the number on this ball is an odd integer?

   (b) A ball is selected at random from the bowl. What is the probability that the number on this ball is greater than 10?

   (c) A ball is selected at random from the bowl. What is the probability that the number on this ball is prime or ends in 9?

   (d) Two balls are selected at random from the bowl. What is the probability that the numbers on both balls are prime?

   (e) Two balls are selected at random from the bowl. What is the probability that the number on one ball is prime and the number on the other ball is a composite number?

4. Suppose that a day is selected at random from the days of a year that is not a leap year.

   (a) What is the probability that the day selected is a day in May?

   (b) What is the probability that the day selected is the 29th day of some month?

   (c) What is the probability that the day selected is the 31st day of some month?

5. What is the probability that at least two people in a group of 10 people have the same birthday (assuming that a year has 365 days)?

6. Let $S = \{i : 1 \leq i \leq 30\}$. In a certain lottery, a subset L of S consisting of six numbers is selected at random. These are the numbers on a winning lottery ticket.

   (a) What is the probability of winning this lottery by purchasing a lottery ticket that contains the same six integers that belong to L?

   (b) What is the probability that none of the six integers on your lottery ticket belong to L?

   (c) Determine the probability that exactly one of the six integers on your lottery ticket belongs to L.

7. A letter is selected at random from the English alphabet and then a second letter is chosen randomly from the remaining 25 letters. Given that the six letters a, e, i, o, u, y are vowels, what is the probability that

   (a) at least one of the two letters selected is a vowel?

   (b) both letters selected are vowels?

   (c) the second letter selected is a vowel given that the first letter is not a vowel?

   (d) both letters selected are vowels given that the first letter is a vowel?

8. In a drug study of a group of 500 patients, 150 patients responded positively to drug #1, 200 patients responded positively to drug #2 and 90 patients responded positively to both drug #1 and drug #2.

(a) What is the probability that a patient responds positively to either drug #1 or drug #2?

(b) What is the probability that a patient responds positively to drug #1 given that this patient responded positively to drug #2?

9. Let E be the event of generating at random a 3-bit string that contains an odd number of 1s and let F be the event of generating at random a 3-bit string that starts with 1. Are E and F independent?

10. A coin is flipped k times. Let E be the event that when a coin is flipped k times, we don't get all heads or all tails. Let F be the event that when a coin is flipped k times, heads comes up at most once. Determine whether E and F are independent if

    (a) $k = 2$,   (b) $k = 4$. (See Example 10.36.)

11. Suppose that the probability that a child is a girl is 0.49 and the sexes of children born into a family are independent. Suppose that a family has five children. Determine the probability of each of the following events:

    (a) exactly three children are girls.

    (b) exactly three children are boys.

    (c) at least one child is a girl.

    (d) at least one child is a boy.

    (e) all five children are boys or all five are girls.

12. Consider a set C of three cards, one of which is red on both sides, one of which is blue on both sides and one of which is red on one side and blue on the other side. Suppose that a card is selected from C and is placed on the table such that the side on top is red. What is the probability that the bottom side of this card is also red?

13. A final exam in a discrete mathematics course consists of 20 true-false questions, each worth 3 points, and 10 multiple-choice questions, each worth 4 points. Suppose that the probability that Mary gives a correct answer to a true-false question is 0.9 and the probability that Mary gives a correct answer to a multiple-choice question is 0.8. What is Mary's expected score on the final exam?

14. When John leaves the university each afternoon to return to his apartment, he encounters three traffic lights. It ordinarily takes him 4 minutes to drive to the first traffic light, 4 minutes of driving after the 3rd traffic light and 4 minutes of driving between two consecutive traffic lights. When he encounters a traffic light, there is a probability of 1/3 that the traffic light is green and a probability of 2/3 that the traffic light is red and that he will have to wait 2 minutes for it to turn green. What is the expected amount of time it takes for John to arrive home after leaving the university?

15. Two students are asked to select an integer from 1 to 10 and each records his or her number so that it is out of view of the other student. What is the probability that the numbers selected by the two students are the same?

16. A bowl contains 40 balls, numbered from 1 to 40.

    (a) Two balls are selected from the bowl at random. What is the probability that the sum of the numbers on the two balls selected is 40?

    (b) Two balls are selected from the bowl at random, where the first ball is returned to the bowl before the second ball is selected. What is the probability that the sum of the numbers on the two balls selected is 40?

    (c) What is a possible explanation why the answers in (a) and (b) are different?

17. A die is tossed and the number obtained is recorded on a piece of paper. Then a coin is flipped twice. If heads comes up both times, then the die is tossed a second time and the number obtained on the die is added to the number on the paper. What is the probability that the final number on the paper is 7?

18. There is a mathematics competition that is given in four parts, numbered 1, 2, 3 and 4. A student who passes part i is permitted to take part $i + 1$ for $1 \leq i \leq 3$. A total of 50 students sign up for the competition, 25 pass the first part, 25 pass the second part, 15 pass the third part but no one passes all four parts. For $1 \leq i \leq 4$, let A_i be the set of students who pass part i. Determine, with explanation, whether the statements (a), (b) and (c) are true or false.

    (a) If $a \in A_1$, then $a \in A_2$. (b) If $a \in A_3$, then $a \in A_4$. (c) If $a \in A_2$, then $a \in A_3$.

    (d) Does it make sense to assign a probability to any of the statements in (a), (b) and (c)?

19. Your history professor has listed four possible topics for an essay for a forthcoming exam. At the beginning of the exam, the professor will select two of these topics on which to write. If you only have time to prepare for two of these topics, what is the probability that the professor will select at least one of the two topics you studied?

20. A bowl contains 3 red balls, 2 white balls and 1 blue ball.

    (a) What is the expected number of white balls obtained if three balls are selected at random from the bowl?

    (b) What is the expected number of white balls obtained if three balls are selected at random from the bowl, one at a time, where a ball is returned to the bowl after it is selected?

21. An athlete is one of eight finalists for a running event at an Olympics games. The probability of him finishing 1st is 0.05, 2nd is 0.15, 3rd is 0.25, 4th is 0.20, 5th is 0.15, 6th is 0.10, 7th is 0.05 and 8th is 0.05. In what position would this athlete be expected to finish the race?

Chapter 11

Partially Ordered Sets and Boolean Algebras

In Chapter 7 we introduced the concept of a relation and discussed the most studied type of relation: an equivalence relation. In this chapter, we introduce a class of relations called partial orders. This brings us to the concept of partially ordered sets, which then leads us to a study of the discrete structures of lattices and Boolean algebras.

11.1   Partially Ordered Sets

For a relation to be an equivalence relation, it must possess the reflexive, symmetric and transitive properties. Recall that a relation R on a set S is symmetric if whenever $a \ R \ b$ for $a, b \in S$, it must also be the case that $b \ R \ a$. Therefore, for relations R with this property, it can never occur that exactly one of $a \ R \ b$ and $b \ R \ a$ is true for distinct elements a and b in S. There are relations R on a set S for which $a \ R \ b$ and $b \ R \ a$ can occur simultaneously for $a, b \in S$ only when $a = b$. When this property replaces the symmetric property in an equivalence relation, we have the major concept of this chapter: a partial order relation.

Definition 11.1   *A relation R on a set S is* **antisymmetric** *if for all $a, b \in S$,*

$$\text{whenever } a \ R \ b \text{ and } b \ R \ a, \text{ then } a = b.$$

We now consider two examples of relations on the set $\{x, y, z\}$ for the purpose of determining whether these relations are antisymmetric.

Example 11.2

Let $S = \{x, y, z\}$. Consider the relation

$$R = \{(x, x), (x, y), (y, z), (z, z)\}$$

defined on S. The only pairs a, b of elements of S such that both $a \ R \ b$ and $b \ R \ a$ are those for which (1) $a = x$ and $b = x$ or (2) $a = z$ and $b = z$. In each case, $a = b$ and so R is antisymmetric. ◆

Example 11.3

Let $S = \{x, y, z\}$. Then

$$R = \{(x, x), (x, y), (y, x), (y, z)\}$$

is a relation defined on S. The only pairs a, b of elements of S such that both $a\,R\,b$ and $b\,R\,a$ are those for which (1) $a = x$ and $b = x$ or (2) $a = x$ and $b = y$ (or $a = y$ and $b = x$). In (2), however, $a \neq b$ and so R is not antisymmetric.   ♦

We have just seen that the relation R described in Example 11.3 is not antisymmetric. In that example, $y\,R\,z$ but $z\,\not{R}\,y$. Thus R is not symmetric either. Therefore, *not symmetric* does not mean *antisymmetric*. Indeed, the relation

$$R' = \{(x,x), (y,y)\}$$

on the set $S = \{x, y, z\}$ is *both* symmetric and antisymmetric, while the relation

$$R'' = \{(x,y), (y,x), (y,z), (z,y)\}$$

on S is symmetric but not antisymmetric. Thus a relation on a set may

- be both symmetric and antisymmetric,
- be neither symmetric nor antisymmetric or
- possess exactly one of these two properties.

In this section, we will be especially interested in relations having all three of the properties reflexive, antisymmetric and transitive.

Definition 11.4   *A relation R on a nonempty set S is called a* **partial order relation** *or simply a* **partial order** *on S if R is reflexive, antisymmetric and transitive.*

That is, a partial order R on a nonempty set S has the following properties:

(1) **reflexive:** $a\,R\,a$ for all $a \in S$;

(2) **antisymmetric:**   if $a\,R\,b$ and $b\,R\,a$, then $a = b$ for all $a, b \in S$;

(3) **transitive:**   if $a\,R\,b$ and $b\,R\,c$, then $a\,R\,c$ for all $a, b, c \in S$.

Definition 11.5 *A nonempty set S together with a partial order R on S is called a* **partially ordered set** *or, more simply, a* **poset**. *Such a partially ordered set is commonly denoted by (S, R).*

The term *poset*, as an abbreviation for *partially ordered set*, was coined by Garrett Birkhoff in his 1948 book *Lattice Theory*. (We'll be introduced to the term *lattice* in Section 11.2.) We now consider one of the best known partially ordered sets.

Example 11.6

The relation "less than or equal to" on the set $\mathbb{Z}$ of integers is a partial order. That is $(\mathbb{Z}, \leq)$ is a poset. In particular, if a and b are two integers such that $a \leq b$ and $b \leq a$, then $a = b$, which implies that $\leq$ is antisymmetric. In fact, $(S, \leq)$ is a poset for every nonempty set S of integers or real numbers.   ♦

Just as the properties possessed by the equals relation $=$ on the set $\mathbb{Z}$ can be considered the inspiration for equivalence relation, the properties possessed by "less than or equal to" relation $\leq$ on $\mathbb{Z}$ can be considered the inspiration for the partial order relation. For this reason, the similar and suggestive symbol $\preceq$ is commonly used to represent a general partial order. Hence $(S, \preceq)$ typically represents a partially ordered set. In fact, it is common to refer to the symbol $\preceq$ by **less than or equal to**. Thus if a and b are elements of S and $a \preceq b$, then we say that a is **less than or equal to** b or b is **greater than or equal to** a. If $a \preceq b$ but $a \neq b$, then we write $a \prec b$ and say that a is **less than** b or b is **greater than** a.

Example 11.7

The divisibility relation $|$ on the set $\mathbb{N}$ of positive integers is a partial order. Since $a \mid a$ for every positive integer a, the relation $|$ is reflexive. Furthermore, if $a \mid b$ and $b \mid a$ for positive integers a and b, then $a = b$. Thus the relation $|$ is antisymmetric. In addition, if $a \mid b$ and $b \mid c$ for positive integers a, b and c, then $a \mid c$. Hence the relation $|$ is transitive. Thus $(\mathbb{N}, |)$ is a partially ordered set. ♦

Hasse Diagrams

Finite partially ordered sets can often be visualized with the aid of a diagram called a Hasse diagram. Such diagrams are named for Helmut Hasse, an important German mathematician of the 20th century. Specifically, a **Hasse diagram** of a poset $(S, \preceq)$ consists of a collection of points in the plane (called **vertices**), one vertex for each element of S, such that

(1) if $a \prec b$, then the vertex b is placed higher than a in the diagram and

(2) if $a \prec b$ and there is no element $c \in S$ such that $a \prec c \prec b$, then a line is drawn joining a and b.

Therefore, if a, b and c are elements of a poset $(S, \preceq)$ such that $a \prec c \prec b$, then a, b and c are placed successively higher in a Hasse diagram of $(S, \preceq)$ and there is no line joining a and b. Nevertheless, $a \preceq b$ (in fact, $a \prec b$) is inferred by the transitivity of the partial order $\preceq$.

Example 11.8

For $S = \{2, 3, 6, 12, 18\}$, Hasse diagrams are drawn in Figure 11.1 both for $(S, \leq)$ and $(S, |)$. ♦

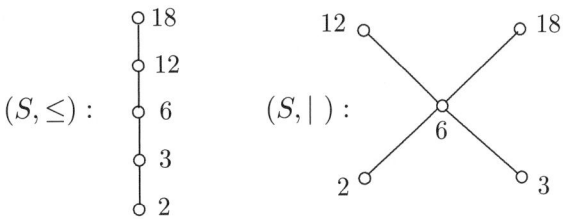

Figure 11.1: Hasse diagrams for $(S, \leq)$ and $(S, |)$, where $S = \{2, 3, 6, 12, 18\}$

Example 11.9

Let X be a set and S a collection of subsets of X, that is, $S \subseteq \mathcal{P}(X)$. Then $(S, \subseteq)$ is a partially ordered set. To see this, first observe that $A \subseteq A$ for every subset A of X in S. Thus $\subseteq$ is reflexive. If A and B are elements of S such that $A \subseteq B$ and $B \subseteq A$, then $A = B$. Hence $\subseteq$ is antisymmetric. Finally, for elements A, B and C of S such that $A \subseteq B$ and $B \subseteq C$, it follows that $A \subseteq C$. Therefore, $\subseteq$ is transitive. In particular, for the power set $\mathcal{P}(X)$ itself, it follows that $(\mathcal{P}(X), \subseteq)$ is a poset. ♦

Example 11.10

Let $X = \{a, b, c\}$ and let $S = \mathcal{P}(X)$. By Example 11.9, $(S \subseteq)$ is a poset. In this case,

$$S = \mathcal{P}(X) = \{\emptyset, \{a\}, \{b\}, \{c\}, \{a, b\}, \{a, c\}, \{b, c\}, X\}.$$

A Hasse diagram of $(S, \subseteq)$ is shown in Figure 11.2. ♦

Despite the great similarity of the partially ordered sets $(\mathbb{Z}, \leq)$, $(\mathbb{N}, |)$ and $(\mathcal{P}(X), \subseteq)$ for a set X, there are also some important differences.

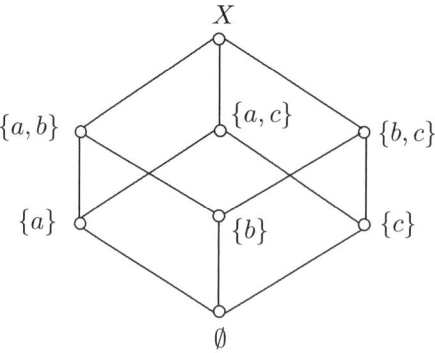

Figure 11.2: A Hasse diagram for $(\mathcal{P}(X), \subseteq)$, where $X = \{a, b, c\}$

Definition 11.11   *Two elements a and b in a partially ordered set $(S, \preceq)$ are called* **comparable** *if either $a \preceq b$ or $b \preceq a$. If neither $a \preceq b$ nor $b \preceq a$, then a and b are* **incomparable**.

Example 11.12

In the partially ordered set $(\mathbb{Z}, \leq)$ every two integers a and b are comparable, that is, $a \leq b$ or $b \leq a$. This is not the case for the partially ordered set $(\mathbb{N}, |\,)$, however. For example, neither $2 \mid 3$ nor $3 \mid 2$; that is, 2 and 3 are incomparable in $(\mathbb{N}, |\,)$. If $X = \{1, 2\}$, then $\{1\}, \{2\} \in \mathcal{P}(X)$. However, $\{1\} \not\subseteq \{2\}$ and $\{2\} \not\subseteq \{1\}$ and so $\{1\}$ and $\{2\}$ are incomparable elements in the poset $(\mathcal{P}(X), \subseteq)$. ♦

If $(S, \preceq)$ is a partially ordered set, then the adjective "ordered" in "partially ordered" is meant to suggest that the elements of S are ordered in some way. If $a, b \in S$ and $a \prec b$, then one can think of a as *preceding* b in some manner. On the other hand, the adverb "partially" in "partially ordered" is meant to suggest that not all pairs of elements of S may be ordered (related), that is, some pairs of elements of S may be incomparable.

Partially ordered sets occur naturally in many common situations. We often encounter projects that need to be performed. If a project is sufficiently complex, it may be useful, if not necessary, to divide the project into a number of smaller steps or activities. Ordinarily, these activities are interrelated where a certain activity in the project cannot be started until other activities have been completed. We consider two examples of this.

Example 11.13

Suppose that constructing a house consists of performing the following activities. (Each of these activities could very well be considered a project itself and then divided into smaller activities.)

     Clear Land (C)
     Build Foundation (F)
     Build Upper Structure (U)
     Put on Roof (R)
     Do Electrical Work (E)
     Do Plumbing Work (P)
     Place Siding on House (S)
     Complete Interior Work on House (I)
     Do Landscaping (L)

Let $A = \{C, F, U, R, E, P, S, I, L\}$. Then $(A, \preceq)$ is a poset, where for $X, Y \in A$, we have $X \prec Y$ if activity X must be completed before activity Y can be started. A possible Hasse diagram of $(A, \preceq)$ is shown in Figure 11.3. This Hasse diagram provides a pictorial blueprint for this project. ♦

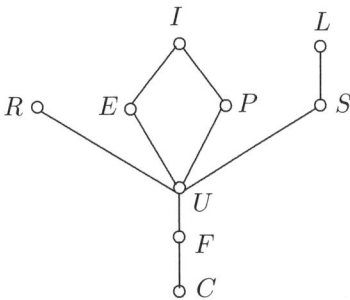

Figure 11.3: A Hasse diagram for the poset $(A, \preceq)$ in Example 11.13

Example 11.14

A married couple has decided to convert the lower level of their house into a recreation room. To complete this project, they have decided to divide the project into a number of activities.

Clean the Floor (F)
Place Padding on the Floor (P)
Place Carpeting on the Padding (C)
Assemble Pool Table (A)
Place Pool Equipment on Pool Table (E)

Let $S = \{F, P, C, A, E\}$. Then $(S, \preceq)$ is a poset, where for $X, Y \in S$, $X \prec Y$ if activity X must be completed before activity Y can be started. A possible Hasse diagram of $(S, \preceq)$ is shown in Figure 11.4. ♦

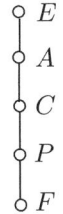

Figure 11.4: A Hasse diagram for the poset $(S, \preceq)$ in Example 11.14

Totally Ordered Sets

Notice that in the poset $(S, \preceq)$ of Example 11.14 (involving converting the lower level of a house into a recreation room) every two elements (activities) of S are comparable. This is an example of a special kind of poset.

Definition 11.15 *A partially ordered set $(S, \preceq)$ in which every two elements of S are comparable is called a* **totally ordered set** (*or* **linearly ordered set**) *and the partial order $\preceq$ is called a* **total order** (*or* **linear order**) *on S.*

Therefore, the poset $(S, \preceq)$ of Example 11.14 is totally ordered; while the poset $(A, \preceq)$ of Example 11.13 is not totally ordered. Since the poset $(S, \preceq)$ of Example 11.14 is totally ordered, there is a unique order for the steps to be performed when preparing the lower level of a house for a recreation room.

Definition 11.16 *Every totally ordered subset of a poset $(S, \preceq)$ is called a **chain** in S.*

Therefore, the poset $(S, \preceq)$ of Example 11.14 is itself a chain, while $\{C, F, U, E, I\}$ is a chain in the poset $(A, \preceq)$ of Example 11.13 since $C \prec F \prec U \prec E \prec I$.

Example 11.17

The poset $(\mathbb{Z}, \leq)$ in Example 11.12 is a totally ordered set, while the poset $(\mathbb{N}, |\ )$ is not totally ordered. The elements of the subset $S = \{1, 2, 2^2, 2^3, \ldots\}$ produce a chain in $(\mathbb{N}, |\ )$. Furthermore, if $X = \{1, 2\}$, then $(\mathcal{P}(X), \subseteq)$ is not a totally ordered set either. On the other hand, $\emptyset \subset \{1\} \subset \{1, 2\}$ is a chain in $(\mathcal{P}(X), \subseteq)$.   ♦

A well-known partial order that is obtained from two or more partial orders is the lexicographic order. Let $(A, \preceq_A)$ and $(B, \preceq_B)$ be two partially ordered sets. Therefore, $\preceq_A$ and $\preceq_B$ denote partial orders on A and B, respectively. The **lexicographic order** $\preceq$ is defined on the Cartesian product $A \times B$ of A and B by

$$(a, b) \preceq (c, d)$$

if

$$(1) \quad a \prec_A c \quad \text{or} \quad (2) \quad a = c \text{ and } b \preceq_B d. \tag{11.1}$$

The lexicographic order is a partial order on $A \times B$.

Example 11.18

For $A = \{1, 2, 3\}$ and $B = \{2, 3, 6\}$, both $(A, |\ )$ and $(B, |\ )$ are partially ordered sets. Thus $(A \times B, \preceq)$ is a partially ordered set for the lexicographic order $\preceq$ on $A \times B$. In particular, $(1, 3) \preceq (2, 3)$ since $1 \mid 2$ and $(2, 3) \preceq (2, 6)$ since $2 = 2$ and $3 \mid 6$, while $(2, 3)$ and $(3, 6)$ are incomparable elements of $A \times B$.   ♦

If $(A, \preceq_A)$ and $(B, \preceq_B)$ are two totally ordered sets, then even more can be said about $(A \times B, \preceq)$.

Theorem 11.19 *Let $(A, \preceq_A)$ and $(B, \preceq_B)$ be totally ordered sets. Then the lexicographic order $\preceq$ defined on $A \times B$ in (11.1) is a total order.*

Proof. Let (a, b) and (c, d) be two elements of $A \times B$. We show that either $(a, b) \preceq (c, d)$ or $(c, d) \preceq (a, b)$. Since $\preceq_A$ is a total order on A, either $a = c$, $a \prec_A c$ or $c \prec_A a$. If $a \prec_A c$, then $(a, b) \preceq (c, d)$. If $c \prec_A a$, then $(c, d) \preceq (a, b)$. Hence we may assume that $a = c$. If $b \preceq_B d$, then $(a, b) \preceq (c, d)$; otherwise, $d \prec_B b$ and $(c, d) \preceq (a, b)$.   ∎

A lexicographic order can also be defined on the Cartesian product of more than two sets. Let $(A_1, \preceq_1), (A_2, \preceq_2), \ldots, (A_n, \preceq_n)$ be $n \geq 2$ partially ordered sets. The **lexicographic order** $\preceq$ is defined on $A_1 \times A_2 \times \cdots \times A_n$ by

$$(a_1, a_2, \ldots, a_n) \preceq (b_1, b_2, \ldots, b_n)$$

if

$$(1) \quad a_1 \prec_1 b_1 \quad \text{or} \quad (2) \quad a_i = b_i \text{ for all } i \text{ with } 1 \leq i \leq k \tag{11.2}$$
$$\text{for some } k \leq n \text{ and } a_{k+1} \prec_{k+1} b_{k+1} \text{ if } k < n.$$

If $(A_1, \preceq_1), (A_2, \preceq_2), \ldots, (A_n, \preceq_n)$ are totally ordered sets, the lexicographic order $\preceq$ defined on $A_1 \times A_2 \times \cdots \times A_n$ as in (11.2) is also a total order (see Exercsie 29). The best known example of this concerns ordering English words in a dictionary.

Let A denote the set consisting of the 26 letters of the English alphabet, that is, $A = \{\alpha_1, \alpha_2, \ldots, \alpha_{26}\}$, where α_i indicates the ith letter of the alphabet. In other words, $\alpha_1 = a$, $\alpha_2 = b$, $\ldots$, $\alpha_{26} = z$. We define $\alpha_i \leq \alpha_j$ if $i \leq j$. Then $(A, \leq)$ is a totally ordered set. The Cartesian product of

$n \geq 1$ sets A is denoted by A^n. Then $(A^n, \preceq)$ is a totally ordered set, where $\preceq$ is the lexicographic order on A^n. In this context, it is customary to write $a_1 a_2 \cdots a_n$ rather than $(a_1, a_2, \ldots, a_n)$ for an element of A^n, which we refer to as a **word of length** n. Let W denote the set of all words (of any length), that is,

$$W = \{w \in A^n : \ n \in \mathbb{N}\}.$$

Let $a_1 a_2 \cdots a_m$ be a word of length m, let $b_1 b_2 \cdots b_n$ be a word of length n and let r denote the minimum of m and n, that is, $r = \min(m, n)$. We define the relation R on W by

$$a_1 a_2 \cdots a_m \ R \ b_1 b_2 \cdots b_n$$

if

(1) $a_1 a_2 \cdots a_r \prec b_1 b_2 \cdots b_r$ in $(A^r, \preceq)$ or

(2) $a_1 a_2 \cdots a_m = b_1 b_2 \cdots b_m$ and $m \leq n$.

This total lexicographic order is also called the **dictionary order** on the poset (W, R). The words below would therefore occur in a dictionary in the order indicated:

cane cat catch map mat matter mop move movers moving.

Maximal and Minimal Elements

An element M in a poset $(S, \preceq)$ is called a **maximal element** of S if M is not less than any other element of S; that is, M is a maximal element of S if there is no element $a \in S$ such that $M \prec a$. Similarly, an element m in S is a **minimal element** of S if there is no element of S that is less than m; that is, m is a minimal element of S if there is no element $a \in S$ such that $a \prec m$.

Example 11.20

The element 18 in the poset $(S, \leq)$ of Figure 11.1 is the only maximal element, while 2 is the only minimal element. In the poset $(S, |\ )$ of Figure 11.1, the integers 12 and 18 are both maximal elements and 2 and 3 are both minimal elements. In the poset $(\mathcal{P}(X), \subseteq)$ of Figure 11.2, X is the only maximal element and $\emptyset$ is the only minimal element. ◆

Every finite nonempty poset has both a minimal and a maximal element. We prove the first of these facts. The proof of the second of these is left as an exercise (see Exercise 21).

Theorem 11.21 *Every finite nonempty poset has a minimal element.*

Proof. Let $(S, \preceq)$ be a finite nonempty poset. Among the chains in S, let

$$x_1 \prec x_2 \prec \cdots \prec x_k$$

be one containing a maximum number of elements. There can be no element $x_0 \in S$ such that $x_0 \prec x_1$, for otherwise, $x_0 \prec x_1 \prec x_2 \prec \cdots \prec x_k$ would be a chain with more than k elements. Hence x_1 is a minimal element of S. ∎

While Theorem 11.21 guarantees that every finite poset $(S, \preceq)$ has at least one minimal element, it does not provide us with a method for finding such an element. Of course, if we are given a Hasse diagram of $(S, \preceq)$, then it is quite easy to locate a minimal element as any element at the lowest level in the diagram is a minimal element. On the other hand, if $(S, \preceq)$ is presented as a pair of sets, namely the set S and the set $\preceq$ of ordered pairs, then there is a variety of methods that can be used to locate a minimal element in $(S, \preceq)$. As an illustration, we present one such algorithm. Although it is a minor point, we can assume that only the ordered pairs (a, b) of $\preceq$ for which $a \neq b$ are given.

Algorithm 11.22 *Find a minimal element in a finite poset.*

This algorithm finds a minimal element in a finite poset $(S, \preceq)$, where $s : a_1, a_2, \ldots, a_n$ is a list of the n elements of S and $s' : e_1, e_2, \ldots, e_m$ is a list of the m elements (a, b) of $\preceq$ for which $a \neq b$.

> **Input**: Two positive integers n and m, a list $s : a_1, a_2, \ldots, a_n$ of the n elements of S and a list $s' : e_1, e_2, \ldots, e_m$ of the m elements (a, b) of $\preceq$ for which $a \neq b$.
>
> **Output**: A minimal element x of $(S, \preceq)$.

1. $x := a_1$ [x is initially assigned the element a_1]

2. **for** $i := 2$ **to** n **do** [i is assigned the integers from 2 to n]

3.   **for** $j := 1$ **to** m **do** [j is assigned the integers from 1 to m]

4.   **if** $e_j = (a_i, x)$ **then** $x := a_i$ and $j := m + 1$. [If e_j is the ordered pair (a_i, x), then x is reassigned the element a_i and j is assigned the value $m + 1$.]

5. **output** x

That a minimal element of $(S, \preceq)$ is indeed produced by Algorithm 11.22 is verified next.

Theorem 11.23 *Algorithm 11.22 produces a minimal element in a finite poset $(S, \preceq)$.*

Proof. We use the notation introduced in Algorithm 11.22. Suppose that the element $x = a_r$ $(1 \leq r \leq n)$ is output by the algorithm. If $r = 1$, then no element of S distinct from a_1 is less than a_1 and so a_1 is certainly a minimal element of $(S, \preceq)$. Hence we may assume that $r \neq 1$. Suppose that the distinct elements of S that are assigned to x as we proceed through Algorithm 11.22 are, in order, $a_1, a_t, \ldots, a_s, a_r$. Therefore, $1 < t < \cdots < s < r$ and

$$C : \; a_r \prec a_s \prec \cdots \prec a_t \prec a_1$$

is a chain. By the transitive property, a_r is less than every element that follows a_r in the chain C. We claim that a_r is a minimal element of $(S, \preceq)$. Assume, to the contrary, that $a_k \prec a_r$ for some $a_k \in S$. By Algorithm 11.22, we cannot have $k > r$. Thus $k < r$. Since $a_k \prec a_r$ and $a_r \prec a_1$, it follows by the transitive property that $a_k \prec a_1$. If a_k is in the chain C, then $a_r \prec a_k$. This, however, is impossible since $\preceq$ is antisymmetric. Thus a_k is not in the chain C. Therefore, there are positive integers p and q such that $p < k < q$, where a_p and a_q are in C and $a_q \prec a_p$. Since $a_k \prec a_r$ and $a_r \prec a_p$, it follows that $a_k \prec a_p$. Since $k < q$, it follows by Algorithm 11.22 that a_q cannot be the element of S that immediately precedes a_p in the chain C. This is a contradiction. ∎

We now give an illustration of Algorithm 11.22.

Example 11.24

Illustrate Algorithm 11.22 for the poset $(S, \preceq)$ whose Hasse diagram is shown in Figure 11.5.
Solution. Suppose that the elements of S listed in the expected order $s : a_1, a_2, a_3, a_4, a_5$. Furthermore, suppose that the ordered pairs (a, b) of $\preceq$ for which $a \neq b$ are listed in the sequence

$$s' : \; (a_3, a_1), \, (a_2, a_1), \, (a_4, a_1), \, (a_4, a_2), \, (a_4, a_3), \, (a_5, a_1), \, (a_5, a_2).$$

In this case, $n = 5$, $m = 7$ and s and s' are input. In Step 1, x is initialized as a_1. In the for loop at Step 2, i is initially assigned the value 2. In the for loop at Step 3, j is initially assigned the value 1. In Step 4, it is first determined whether e_1 is (a_2, a_1). It is not. The value of j is increased to 2. It is then determined whether e_2 is (a_2, a_1). It is. By Step 4, x is reassigned the element a_2 and j is assigned the value $m + 1 = 8$. This completes Step 3 and so i is increased to 3. In the for loop at Step 3, j runs through the values from 1 to $m = 7$. In Step 4, it is determined whether any ordered

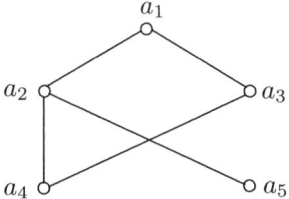

Figure 11.5: A Hasse diagram of the poset in Example 11.24

pair e_j is (a_3, a_2). This does not occur. So Step 3 has been completed, i is increased to 4 and we return to the for loop at Step 3. Starting with $j = 1$, it is determined whether e_j is (a_4, a_2). This occurs when $j = 4$. So the conclusion of Step 4 is performed, reassigning x to a_4 and assigning j the value $m + 1 = 8$. This completes the for loop at Step 3. The integer i is then increased to 5 in the for loop at Step 3. In Step 4, it is determined whether any e_j is (a_5, a_4). This does not occur and completes the for loop at Step 3 and, in turn, the for loop at Step 2. We therefore move on to Step 5, where $x = a_4$ is output as a minimal elememt in $(S, \preceq)$. ◆

Topological Sorting

A recipe for preparing a food dish also gives rise to a partially ordered set.

Example 11.25

One of the recipes for making fresh cherry pie involves performing a number of steps:

D: Prepare pie dough
L: Line pie pan with pie dough
C: Wash, drain and pit 4 cups of fresh cherries
M: Mix cherries with 3 tablespoons of tapioca and $1\frac{1}{2}$ cups of sugar
S: Let fruit mixture stand for 15 minutes
P: Preheat oven to 450°
F: Pour fruit mixture into pie shell
B: Bake pie for 10 minutes at 450° and another 30 minutes at 350°
      or until golden brown.

Some of these steps can only be performed after other steps have been completed. A partial order on the set A of all steps is defined by step X $\prec$ step Y if step Y can be performed only after step X has been completed. This produces a poset $(A, \preceq)$, a Hasse diagram of which is shown in Figure 11.6. ◆

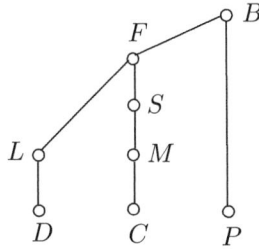

Figure 11.6: A Hasse diagram for the poset $(A, \preceq)$ in Example 11.25

If a cook has decided to follow the recipe in Example 11.25 to make fresh cherry pie, then a natural question to ask is: In what order should steps be taken to make the pie (since the cook cannot do more than one step at a time)? One possible order is

$$D \ L \ C \ M \ S \ P \ F \ B.$$

This is an acceptable order since

$$D \prec L \prec C \prec M \prec S \prec P \prec F \prec B.$$

This total order then provides us with an order in which all steps in the recipe can be performed such that when any step is ready to be performed, all predecessors of this step have already been performed. That is, the steps in this recipe for making a fresh cherry pie can be listed in a cookbook in the order given above. This gives rise to a new concept.

Definition 11.26 *Let* (S, R) *be a poset. A total order* $\preceq$ *on* S *is said to be* **compatible** *with the partial order* R *if* $a \preceq b$ *whenever* $a \ R \ b$ *for all* $a, b \in S$. *That is, a total order* $\preceq$ *on* S *is compatible with* R *if the set* $\preceq$ *of ordered pairs contains* R. *In this case,* $\preceq$ *is a* **topological sorting** *of* (S, R).

This concept is illustrated with a simple example.

Example 11.27

Let $S = \{x, y, z\}$. A Hasse diagram of a poset (S, R) is shown in Figure 11.7. A Hasse diagram of a totally ordered set $(S, \preceq)$ is also shown in Figure 11.7. Observe that (y, x) and (z, x) are elements of R, that is, $y \ R \ x$ and $z \ R \ x$. Since $y \preceq x$ and $z \preceq x$, it follows that $\preceq$ is a topological sorting of (S, R). The fact that $z \preceq y$ and that y and z are incomparable in (S, R) is irrelevant. If we think of the elements of S in the poset (S, R) as three tasks that are to be performed and that x cannot be performed until y and z are first performed, then the topological sorting of $(S, \preceq)$ of (S, R) tells us that we could perform the three tasks of S in the order z, y, x. ◆

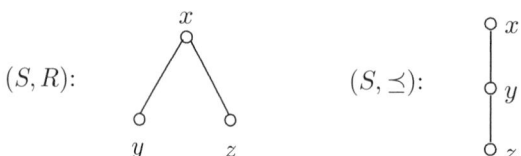

Figure 11.7: A topological sorting of the poset in Example 11.27

We now present an algorithm that describes a method of constructing a topological sorting of any finite poset. Recall (from Chapter 6) that in a sequence $s : a_1, a_2, \ldots, a_n$, where $1 \le i < j \le n$, the insertion command **insert**$(a_j \to a_i)$ places the term a_j into the position where a_i is located, while the elements $a_i, a_{i+1}, \ldots, a_{j-1}$ are moved one position to the right and then the terms of s are relabeled, denoting the resulting sequence by $a_1, a_2, \ldots, a_n$ again. This is reviewed Figure 11.8.

$$
\begin{array}{cccccccccccc}
a_1 & a_2 & \cdots & a_{i-1} & \mathbf{a_i} & a_{i+1} & \cdots & a_{j-1} & \mathbf{a_j} & a_{j+1} & \cdots & a_n \\
\end{array}
$$

$$
\begin{array}{cccccccccccc}
a_1 & a_2 & \cdots & a_{i-1} & \mathbf{a_j} & \mathbf{a_i} & \cdots & a_{j-2} & a_{j-1} & a_{j+1} & \cdots & a_n \\
\end{array}
$$

$$
\begin{array}{cccccccccccc}
a_1 & a_2 & \cdots & a_{i-1} & a_i & a_{i+1} & \cdots & a_{j-1} & a_j & a_{j+1} & \cdots & a_n \\
\end{array}
$$

Figure 11.8: Illustrating **insert**$(a_j \to a_i)$

The algorithm for finding a topological sorting of a finite poset that we are about to present uses Algorithm 11.22, which locates a minimal element in a finite poset. If a Hasse diagram of the poset is available, then a minimal element can be found by inspection.

Algorithm 11.28 *Find a topological sorting of a finite poset.*

This algorithm finds a topological sorting of a finite poset $(S, \preceq)$ of cardinality n whose elements are listed in a sequence $s : a_1, a_2, \ldots, a_n$.

> **Input**: A positive integer n and a list $s : a_1, a_2, \ldots, a_n$ of the n elements of S
> **Output**: A topological sorting of the elements of $(S, \preceq)$.

1. **for** $i := 1$ **to** $n - 1$ **do**

> **begin**

2.   Algorithm 11.22 is used to find a minimal element x of S

3. $\text{insert}(x \to a_i)$

4. $S := S - \{a_i\}$

> **end**

5. **output** s

If Algorithm 11.28 employs Algorithm 11.22 to find a minimal element in $(S, \preceq)$ and in the subsequent posets produced in Step 4 of Algorithm 11.28, then we also need to input a list s' : $e_1, e_2, \ldots, e_m$ of all m elements (a, b) of the partial order relation $\preceq$ for which $a \neq b$ as well as the positive integer m. We illustrate Algorithm 11.28 by finding a topological sorting of a finite poset.

Example 11.29

Construct a topological sorting of the poset $(S, \preceq)$, a Hasse diagram of which is given in Figure 11.9 and where $s : u, v, w, x, y, z$ is a listing of the elements of S. So $a_1 = u$, $a_2 = v$, $a_3 = w$, $a_4 = x$, $a_5 = y$, $a_6 = z$.

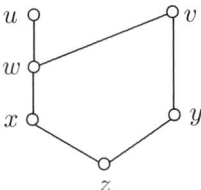

Figure 11.9: A Hasse diagram for the poset in Example 11.29

Solution. In the for loop at Step 1 in Algorithm 11.28, i is initially assigned the value 1. In Step 2, Algorithm 11.22 is applied to find a minimal element, namely z. In Step 3, the sequence s is reordered, so now $s : z, u, v, w, x, y$, the set S is redefined as $S = \{u, v, w, x, y\}$ and j is reassigned the value $n + 1 = 7$ in Step 4. This completes Step 2 for $i = 1$ and i is increased to 2. Algorithm 11.22 is applied again to the new set S, arriving at x. The sequence s is reordered once again and now $s : z, x, u, v, w, y$. The value of j is changed to $n + 1 = 7$, completing Step 2 for $i = 2$. Continuing in this manner, we eventually arrive at a final ordering $s : z, x, w, u, y, v$, which is then a topological sorting of $(S, \preceq)$. ♦

Since the poset $(S, \preceq)$ in Example 11.29 is sufficiently simple and we have a Hasse diagram for it in Figure 11.9, we can construct a topological sorting without applying Algorithm 11.22. First, we choose a minimal element of $(S, \preceq)$, namely z, which is the only minimal element of $(S, \preceq)$. There are two minimal elements in $(S - \{z\}, \preceq)$, namely x and y. Either can be chosen next, say we choose x. In the set $S - \{z, x\}$, there are two minimal elements, namely y and w. We choose w, say. In the set $S - \{z, x, w\}$, y and u are the minimal elements. We choose u. Since y is the only minimal element in $S - \{z, x, w, u\}$, we choose y and v is the only element remaining. This produces the topological sorting

$$z \prec x \prec w \prec u \prec y \prec v.$$

This procedure is shown in Figure 11.10.

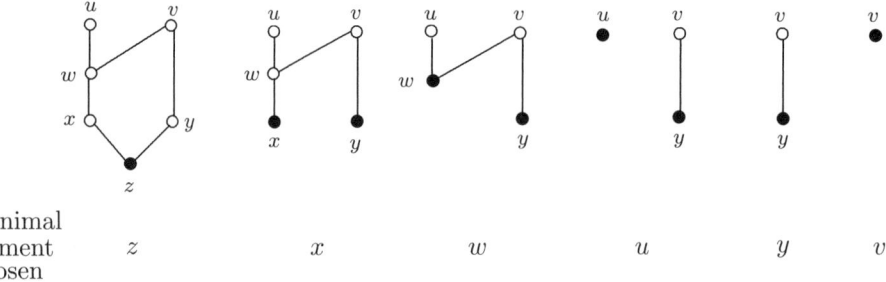

Minimal
element z x w u y v
chosen

Figure 11.10: Constructing a topological sorting in a poset

Example 11.30

A certain program at a university requires a student to take the following mathematics and statistics courses (depending on the mathematics courses the student took in high school):

PC:	Precalculus
C_1:	Calculus I
C_2:	Calculus II
PB:	Probability
S:	Statistics
D:	Discrete Mathematics

In order to take some of these courses, it is required that a student has successfully completed other courses (in other words, some courses have required prerequisites). A partial order R is defined on the set A of these courses by $X \, R \, Y$ if course X is a prerequisite for course Y. This produces a poset (A, R), a Hasse diagram of which is shown in Figure 11.11.

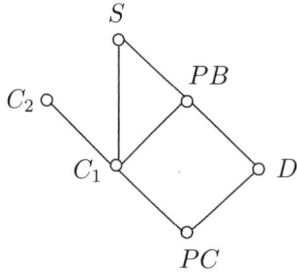

Figure 11.11: A Hasse diagram for the poset (A, R) in Example 11.30

One topological sorting of the courses in the poset (A, R) is

$$\text{PC} \prec \text{D} \prec \text{C}_1 \prec \text{C}_2 \prec \text{PB} \prec \text{S},$$

which implies that if exactly one of these courses is taken by a student during each of six consecutive semesters, then the courses could be taken in the order

1. Precalculus	2. Discrete Mathematics	3. Calculus I
4. Calculus II	5. Probability	6. Statistics.

When a course is taken, the student can be assured that he or she has taken all prerequisite courses. Consequently, $(A, \preceq)$ is a totally ordered set. ◆

Greatest and Least Elements

An element M in a poset $(S, \preceq)$ is a **greatest element** of S if $a \preceq M$ for every $a \in S$; while an element m of S is a **least element** of S if $m \preceq a$ for all $a \in S$. Therefore, if M is a greatest element of a poset, then M is greater than any other element in the poset; while if m is a least element of a poset, then m is less than any other element in the poset.

Example 11.31

In the poset $(S, \leq)$ in Figure 11.1, the integer 18 is the greatest element of S and 2 is the least element. The poset $(S, | \,)$ in Figure 11.1 has neither a greatest nor a least element. In the poset $(\mathcal{P}(X), \subseteq)$ of Figure 11.2, the element X is the greatest element and $\emptyset$ is the least element. ◆

Certainly every greatest element in a poset is a maximal element and every least element in a poset is a minimal element. Neither converse is true however. While a poset can have several maximal elements or minimal elements, this is not the case for greatest elements or least elements.

Theorem 11.32 *Every nonempty poset contains at most one greatest element and at most one least element.*

Proof. Let $(S, \preceq)$ be a poset. We show only that S has at most one greatest element. A proof showing that S has at most one least element is similar and is left as an exercise (see Exercise 34).

The theorem is certainly true if S contains no greatest element. So we can assume that S contains a greatest element. We show that such an element is unique. Suppose that M_1 and M_2 are both greatest elements in S. Since M_1 is a greatest element, it follows that $M_2 \preceq M_1$. On the other hand, M_2 is also a greatest element, so $M_1 \preceq M_2$. However, $\preceq$ is antisymmetric, which implies that $M_1 = M_2$. Consequently, S has only one greatest element. ∎

Let $(S, \preceq)$ be a poset. For a nonempty subset A of S, there may exist some element $u \in S$ such that $a \preceq u$ for all $a \in A$. If this happens, then u is called an **upper bound** for A. If there is some element $\ell \in S$ such that $\ell \preceq a$ for all $a \in A$, then ℓ is a **lower bound** for A. If S itself has an upper bound, then this upper bound is the greatest element of S. If S has a lower bound, then this lower bound is the least element of S.

Example 11.33

In the poset $(\mathcal{P}(X), \subseteq)$ of Figure 11.2, where $X = \{a, b, c\}$, let $A_1 = \{\emptyset, \{a\}, \{b\}\}$. Then both $\{a, b\}$ and X are upper bounds for A_1, while $\emptyset$ is the only lower bound for A_1. In the poset $(S, \leq)$ of Figure 11.1, let $A_2 = \{2, 6\}$. Then $6, 12$ and 18 are all upper bounds of A_2 and 2 is the only lower bound for A_2. On the other hand, in the poset $(S, | \,)$ of Figure 11.1, let $A_3 = \{2, 3\}$. The elements $6, 12$ and 18 are all upper bounds of A_3, while there is no lower bound for A_3. ◆

Exercises for Section 11.1

1. Let $S = \{a, b, c, d\}$. Give an example of a relation R on S that is

   (a) antisymmetric but not symmetric.

   (b) symmetric but not antisymmetric.

   (c) antisymmetric and symmetric.

   (d) neither antisymmetric nor symmetric.

2. Let $S = \{2, 21, 25\}$. A relation R is defined on S by $a\ R\ b$ if $a \mid b$. Determine whether R is (a) antisymmetric, (b) symmetric.

3. Let $S = \{-25, -21, -2, 2, 21, 25\}$. A relation R defined on S by $a\ R\ b$ if $a \mid b$. Determine whether R is (a) antisymmetric, (b) symmetric.

4. Let $S = \{3, 5, 7\}$. Define a relation R on S by $a\ R\ b$ if $ab \leq 12$. Determine whether R is (a) antisymmetric, (b) symmetric.

5. Let $S = \{3, 5, 7\}$. Define a relation R on S by $a\ R\ b$ if $ab \leq 16$. Determine whether R is (a) reflexive, (b) antisymmetric, (c) symmetric, (d) transitive.

6. Let $S = \{w, x, y, z\}$. Give an example of a relation R on S that

    (a) is reflexive, antisymmetric and transitive.

    (b) is reflexive but neither antisymmetric nor transitive.

    (c) is antisymmetric but neither reflexive nor transitive.

    (d) is transitive but neither reflexive nor antisymmetric.

    (e) is reflexive and antisymmetric but not transitive.

    (f) is reflexive and transitive but not antisymmetric.

    (g) is antisymmetric and transitive but not reflexive.

    (h) has none of the properties reflexive, antisymmetric and transitive.

7. Let $S = \{1, 2, 3, 4\}$ and consider the relation $R = \{(1, 1), (3, 3), (4, 4)\}$ defined on S. Which of the properties reflexive, antisymmetric, transitive does R possess?

8. Let $S = \{3, 7, 15, 31\}$. A relation R is defined on S by $a\ R\ b$ if $a \leq 2b$. Which of the properties reflexive, antisymmetric, transitive does R possess?

9. Give an example of a relation R on $\mathbb{Z}$ such that R is both a partial order and an equivalence relation.

10. Which of the following relations are partial orders on the set $\mathbb{N}$ of positive integers?

    (a) $a\ R_1\ b$ if $a \geq b$. (b) $a\ R_2\ b$ if $a \neq b$. (c) $a\ R_3\ b$ if $a \nmid b$.

    (d) $a\ R_4\ b$ if $a + b$ is odd. (e) $a\ R_5\ b$ if $a + ab + b$ is even.

11. Let $S = \{a, b, c\}$ and let R be the set of all elements of $S \times S$ except (a, b), (b, c) and (c, a). Is (S, R) is a poset?

12. Let $S = \{2, 3, 4, 5\}$. Does there exist a partial order R on S such that a Hasse diagram of (S, R) appears as shown in Figure 11.12?

5

3

4

2

Figure 11.12: A Hasse diagram for Exercise 12

13. Give an example of a poset $(S, \preceq)$ so that the Hasse diagram (appropriately labeled) in Figure 11.13 results.

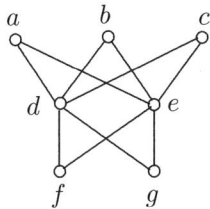

Figure 11.13: The diagram for Exercise 13

14. Let S be a nonempty set and let $\mathcal{P}$ be a partition of S. Define a relation R on $\mathcal{P}$ by $A \ R \ B$ if $A \cap B = \emptyset$. Determine whether R is a partial order.

15. Give an example of a poset of cardinality 2.

16. Let $(S, \preceq)$ be a poset and A is a nonempty subset of S. Prove that $(A, \preceq)$ is also a poset.

17. Suppose that R is a partial order on a nonempty set S. Define the **inverse relation** R^{-1} of R on S defined by $a \ R^{-1} \ b$ if $b \ R \ a$. Show that R^{-1} is also a partial order on S. The poset (S, R^{-1}) is called the **dual** of (S, R).

18. Two coins are placed on two of the four squares of a 2×2 checkerboard. There are therefore $\binom{4}{2} = 6$ possible configurations $s_1, s_2, \ldots, s_6$ that can occur (see Figure 11.21). Let $S = \{s_1, s_2, \ldots, s_6\}$. We write $s_i \prec s_j$ if one of the coins in configuration s_i can be moved horizontally or moved up (to a blank square) to obtain configuration s_j. Is $\preceq$ a partial order on S?

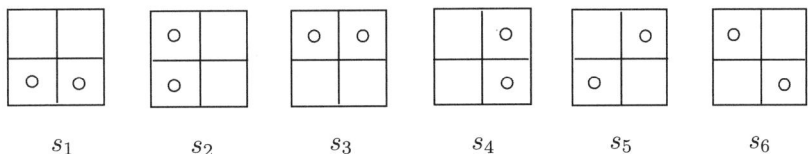

Figure 11.14: Six configurations $s_1, s_2, \ldots, s_6$ in Exercise 18

19. Which of the following pairs of elements are comparable in the poset $(\mathcal{P}(N), \subseteq)$?
    (a) $\{1, 2\}$ and $\emptyset$.   (b) $\{1, 2\}$ and $\{2, 3\}$.
    (c) $\{2, 4, 6, \ldots\}$ and $\{1, 3, 5, \ldots\}$.   (d) $\emptyset$ and $\mathbb{N}$.

20. Consider the poset $(\{2, 4, 6, 9\}, |\ )$.

    (a) Draw a Hasse diagram for this poset.

    (b) Find two comparable elements and two incomparable elements in this poset.

21. Prove that every finite nonempty poset has a maximal element.

22. (a) Give an example of a poset having exactly five maximal elements and exactly three minimal elements.

    (b) The example in (a) should suggest a question to you. Ask and answer this question.

23. Prove that if a finite poset $(S, \preceq)$ has a unique minimal element m, then m is the least element of $(S, \preceq)$.

24. Let $S = \{1, 2, 3, 5, 6, 10, 15, 30\}$.

(a) Draw a Hasse diagram for the poset $(S, |\ )$.

(b) Determine the maximal and minimal elements of $(S, |\ )$.

(c) Does $(S, |\ )$ have a greatest and least element? If so, what are they?

25. Let $S = \{2, 3, 5, 6, 7, 10, 14, 15, 21, 35\}$.

(a) Draw a Hasse diagram for the poset $(S, |\ )$.

(b) Determine the maximal and minimal elements of $(S, |\ )$.

(c) Does $(S, |\ )$ have a greatest and least element? If so, what are they?

26. (a) Use Algorithm 11.22 to find a minimal element in the finite poset $(S, \preceq)$, where s : $a_1, a_2, a_3, a_4, a_5, a_6$ is a given list of the elements of S and s' : (a_1, a_2), (a_3, a_2), (a_4, a_1), (a_4, a_2), (a_4, a_3), (a_5, a_2), (a_6, a_2), (a_6, a_3), (a_6, a_5) is a given list of the elements (a, b) of $\preceq$ for which $a \neq b$.

(b) Draw a Hasse diagram for the poset $(S, \preceq)$.

27. Write an algorithm that finds a maximal element in a finite poset.

28. Let $S = \{n \in \mathbb{N} : n \leq 6\}$. Let $S_1 = S_2 = S$, let $\preceq_1$ denote $\leq$ and let $\preceq_2$ denote $|$. A partial order $\preceq$ is defined on $S_1 \times S_2$ by $(a, b) \preceq (c, d)$ if (1) $a \prec_1 b$ or (2) $a = b$ and $c \preceq_2 d$.

(a) Give an example of two comparable elements of $(S_1 \times S_2, \preceq)$.

(b) Give an example of two incomparable elements of $(S_1 \times S_2, \preceq)$.

(c) Give an example of a chain in $(S_1 \times S_2, \preceq)$ containing a maximum number of distinct elements.

29. Prove that if $(A_1, \preceq_1)$, $(A_2, \preceq_2)$, ..., $(A_n, \preceq_n)$ are $n \geq 2$ totally ordered sets, then the lexicographic order $\preceq$ defined on $A_1 \times A_2 \times \cdots \times A_n$ as in (11.2) is also a total order.

30. Two roommates have invited their boyfriends over for dinner. In order to prepare the dinner table, they have decided to perform the following activities:

T: Place tablecloth on table.

N: Fold napkins and place them on the table.

D: Set dishes and silverware on the table.

G: Place water glasses and wine glasses on the table.

I: Pour ice water into water glasses.

W: Pour wine into wine glasses.

F: Place food on table.

A partial order is defined on the set A of all activities by activity X $\prec$ activity Y if activity Y can only be performed after activity X has been completed. This produces a poset.

(a) Draw a Hasse diagram for this poset $(A, \preceq)$.

(b) Find a topological sorting of $(A, \preceq)$.

31. Use Algorithm 11.28 to find a topological sorting of the finite poset $(S, \preceq)$, a Hasse diagram of which is given in Figure 11.15 and where $s : a_1, a_2, \ldots, a_8$ is a listing of the elements of S.

32. Let $X = \{1, 2, 3\}$. Find two different topological sortings of the poset $(\mathcal{P}(X), \subseteq)$.

33. Let $S = \{2, 4, 8, 24, 40, 72, 200, 1800\}$.

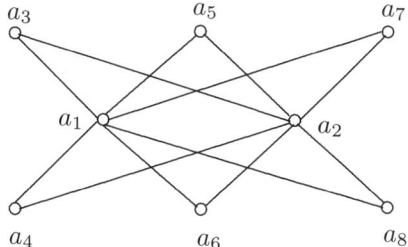

Figure 11.15: A Hasse diagram for the poset in Exercise 31

(a) Draw a Hasse diagram for the poset $(S, | \,)$.

(b) Find two different topological sortings of $(S, | \,)$.

34. Prove that every nonempty poset contains at most one least element.

35. An element c in a poset $(S, \preceq)$ is called an *intermediate element* if there exist elements $a, b \in S$ such that $a \prec c \prec b$. Prove that for every rational number $q \in (0, 1)$, there exists a poset $(S, \preceq)$ for which its set I of intermediate elements satisfies $|I|/|S| = q$.

11.2 Lattices

One of the best known discrete structures is the lattice. Not only can lattices be described as partially ordered sets having certain additional requirements, they are also algebraic structures possessing certain properties.

Let $(S, \preceq)$ be a poset and let A be a nonempty subset of S. An element $x \in S$ is a **least upper bound** of A if x is an upper bound of A and $x \preceq u$ for every upper bound u of A. An element $y \in S$ is a **greatest lower bound** of A if y is a lower bound of A and $\ell \preceq y$ for every lower bound ℓ of A. If a set A has a least upper bound, then this element is necessarily unique. The same is true of greatest lower bounds. (See Exercise 1.) For this reason, we speak of *the* least upper bound of A and *the* greatest lower bound of A. The least upper bound and the greatest lower bound of a subset A in a poset $(S, \preceq)$, should they exist, are denoted by $\mathrm{lub}(A)$ and $\mathrm{glb}(A)$, respectively. If the set A consists of two elements of S, say $A = \{a, b\}$, then $\mathrm{lub}(A)$ is denoted by $a \vee b$ and $\mathrm{glb}(A)$ is denoted by $a \wedge b$. The element $a \vee b$, if it exists, is called the **join** of a and b; while $a \wedge b$, if it exists, is called the **meet** of a and b.

Example 11.34

Figure 11.16 shows a Hasse diagram of a poset $(S, \preceq)$, where $S = \{a, b, c, d, e, f, g\}$. This poset has a as its greatest element and g as its least element. Let $A = \{b, c, d\}$. Then a is the only upper bound of A and so necessarily a is the least upper bound of A. All elements e, f and g are lower bounds of A. However, A has no greatest lower bound. Also, $b \vee d = a$, while $b \wedge d$ doesn't exist. Furthermore, $e \vee f$ doesn't exist, while $e \wedge f = g$. $\blacklozenge$

Definition 11.35 *A **lattice** is a poset in which every two elements have both a least upper bound and a greatest lower bound.*

The poset $(S, \preceq)$ described in Figure 11.16 is not a lattice. As we saw in Example 11.34, the elements b and d, for example, do not have a greatest lower bound. We now consider some examples of posets that are lattices.

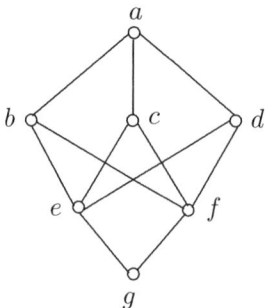

Figure 11.16: A Hasse diagram for the poset in Example 11.34

Example 11.36

Every totally ordered poset $(S, \preceq)$ is necessarily a lattice. In particular, if a and b are two distinct elements of S, then either $a \prec b$ or $b \prec a$. If $a \prec b$, then $a \vee b = b$ and $a \wedge b = a$. As a consequence of this, the poset $(\mathbb{N}, \leq)$ is a lattice. In this case, for positive integers a and b, $a \vee b$ is the larger of a and b and $a \wedge b$ is the smaller of a and b. ♦

Example 11.37

The poset $(\mathbb{N}, \mid)$ is a lattice. For positive integers a and b, the join $a \vee b$ of a and b is the least common multiple of a and b, while the meet $a \wedge b$ of a and b is the greatest common divisor of a and b. ♦

Example 11.38

For a nonempty set X, the poset $(\mathcal{P}(X), \subseteq)$ is a lattice. For subsets A and B of X, $A \vee B = A \cup B$ and $A \wedge B = A \cap B$. ♦

If $(L, \preceq)$ is a lattice and $a, b \in L$, then both $a \vee b$ and $a \wedge b$ are unique elements of L. This says that $\vee$ and $\wedge$ associate with each element of the Cartesian product $L \times L$ a unique element of L. For this reason, $\vee$ and $\wedge$ are **binary operations** on the set L. The most common binary operations are addition $+$ and multiplication $\cdot$ on the set $\mathbb{Z}$ of integers or on the set $\mathbb{R}$ of real numbers. For a set X, both union $\cup$ and intersection $\cap$ are binary operations on $\mathcal{P}(X)$. The binary operations $\vee$ and $\wedge$ are illustrated for the three lattices described above.

Example 11.39

For the lattice $(\mathbb{N}, \leq)$,

$$6 \vee 15 = 15 \text{ and } 6 \wedge 15 = 6.$$

For the lattice $(\mathbb{N}, \mid)$,

$$6 \vee 15 = 30 \text{ and } 6 \wedge 15 = 3.$$

For the lattice $(\mathcal{P}(X), \subseteq)$, where $X = \{2, 3, 5\}$,

$$\{2, 3\} \vee \{3, 5\} = \{2, 3, 5\} \text{ and } \{2, 3\} \wedge \{3, 5\} = \{3\}. ♦$$

There are some familiar properties that many common binary operations satisfy. For example, addition and multiplication of integers are commutative, namely

$$a + b = b + a \text{ and } ab = ba \text{ for } a, b \in \mathbb{Z},$$

and are associative, namely

$$(a + b) + c = a + (b + c) \text{ and } (ab)c = a(bc) \text{ for } a, b, c \in \mathbb{Z}.$$

For 2×2 matrices, addition is commutative but multiplication is not; for example,

$$\begin{bmatrix} 1 & 2 \\ 2 & 3 \end{bmatrix} \begin{bmatrix} 1 & 1 \\ 1 & 0 \end{bmatrix} = \begin{bmatrix} 3 & 1 \\ 5 & 2 \end{bmatrix} \neq \begin{bmatrix} 3 & 5 \\ 1 & 2 \end{bmatrix} = \begin{bmatrix} 1 & 1 \\ 1 & 0 \end{bmatrix} \begin{bmatrix} 1 & 2 \\ 2 & 3 \end{bmatrix}.$$

While the union and intersection of sets are both commutative and associative, there are two added properties, namely, for any two sets A and B,

$$A \cap (A \cup B) = A \text{ and } A \cup (A \cap B) = A.$$

These are the absorption properties. Sets also satisfy the idempotent properties, namely for every set A,

$$A \cup A = A \text{ and } A \cap A = A.$$

Neither addition or multiplication of real numbers nor addition or multiplication of 2×2 matrices satisfy the absorption properties and idempotent properties. However, every lattice satisfies all four properties mentioned above.

Theorem 11.40 *Let $(L, \preceq)$ be a lattice. Then the following hold.*

- **Commutative Laws**

  $a \vee b = b \vee a \quad and \quad a \wedge b = b \wedge a \text{ for all } a, b \in L.$

- **Associative Laws**

  $(a \vee b) \vee c = a \vee (b \vee c) \text{ and } (a \wedge b) \wedge c = a \wedge (b \wedge c) \text{ for all } a, b, c \in L.$

- **Absorption Laws**

  $a \wedge (a \vee b) = a \text{ and } a \vee (a \wedge b) = a \text{ for all } a, b \in L.$

- **Idempotent Laws**

  $a \vee a = a \text{ and } a \wedge a = a \text{ for each } a \in L.$

Proof.   The element $a \vee b$ is the least upper bound of the set $\{a, b\}$. Since $\{a, b\} = \{b, a\}$, it follows that $a \vee b$ is also the least upper bound of $\{b, a\}$, which is $b \vee a$. That is, $a \vee b = b \vee a$. Similarly, $a \wedge b = b \wedge a$.

Since $(a \vee b) \vee c$ is the least upper bound of $a \vee b$ and c, it follows that $a \vee b \preceq (a \vee b) \vee c$ and $c \preceq (a \vee b) \vee c$. Furthermore, $a \preceq a \vee b$ and $b \preceq a \vee b$. Since the relation $\preceq$ is transitive, $a \preceq (a \vee b) \vee c$ and $b \preceq (a \vee b) \vee c$. Thus $(a \vee b) \vee c$ is an upper bound for b and c. Therefore,

$$b \vee c \leq (a \vee b) \vee c.$$

Now, $(a \vee b) \vee c$ is an upper bound for a and $b \vee c$, so

$$a \vee (b \vee c) \preceq (a \vee b) \vee c.$$

Similarly, $(a \vee b) \vee c \preceq a \vee (b \vee c)$. Since $\preceq$ is antisymmetric, $(a \vee b) \vee c = a \vee (b \vee c)$. A similar argument shows that $(a \wedge b) \wedge c = a \wedge (b \wedge c)$.

Since $a \vee b$ is the least upper bound of a and b, it follows that $a \preceq a \vee b$ and $b \preceq a \vee b$. Since $\preceq$ is reflexive, $a \preceq a$. Thus a is a lower bound of both a and $a \vee b$. Therefore, $a \preceq a \wedge (a \vee b)$. Because $a \wedge (a \vee b)$ is a lower bound of both a and $a \vee b$, it follows that $a \wedge (a \vee b) \preceq a$. Since $\preceq$ is antisymmetric, $a \wedge (a \vee b) = a$. A similar argument shows that $a \vee (a \wedge b) = a$.

Since $a \vee a$ is the least upper bound of a and a, it follows that $a \preceq a \vee a$. On the other hand, since a is an upper bound for both a and a, it follows that $a \vee a \preceq a$. Thus $a \vee a = a$. Similarly, $a \wedge a = a$. ∎

In mathematics, an **algebraic structure** typically consists of a nonempty set on which is defined one or more binary operations. A lattice can also be defined independent of partial orders as an algebraic structure with two binary operations. The common symbols for these operations are $\vee$ and $\wedge$ but these are not the join and meet as we do not even have a partial order in this case.

Definition 11.41   *A **lattice** is a nonempty set L on which are defined two binary operations $\vee$ and $\wedge$ satisfying the following three properties:*

- **Commutative Laws**

  $a \vee b = b \vee a$ *and* $a \wedge b = b \wedge a$ *for all $a, b \in L$.*

- **Associative Laws**

  $(a \vee b) \vee c = a \vee (b \vee c)$ *and* $(a \wedge b) \wedge c = a \wedge (b \wedge c)$ *for all $a, b, c \in L$.*

- **Absorption Laws**

  $a \wedge (a \vee b) = a$ *and* $a \vee (a \wedge b) = a$ *for all $a \in L$.*

It was the research of and 1940 book *Lattice Theory* by the American mathematician Garrett Birkhoff that turned lattice theory into a major branch of modern algebra.

We saw in Theorem 11.40 that a lattice, defined in terms of posets, satisfies a fourth property, namely the idempotent laws. Actually, any algebraic structure on which are defined two binary operations satisfying the commutative laws, associative laws and absorption laws must also satisfy the idempotent laws as well.

Theorem 11.42   *A lattice L defined as an algebraic structure in Definition 11.41 satisfies the idempotent laws.*

Proof.   Let $a \in L$. We show that $a \vee a = a$ and $a \wedge a = a$. By the absorption laws,

$$a \wedge (a \vee b) = a \text{ and } a \vee (a \wedge b) = a$$

for every $a \in L$. Therefore, $a \wedge (a \vee a) = a$ and $a \vee (a \wedge a) = a$ and so

$$
\begin{aligned}
a \vee a &= a \vee (a \wedge (a \vee a)) = a \\
a \wedge a &= a \wedge (a \vee (a \wedge a)) = a,
\end{aligned}
$$

and so the idempotent laws hold. ∎

Bounded Lattices

Thus, when a lattice is defined in terms of posets, as in Definition 11.35, we have a nonempty set L and a partial order $\preceq$ defined on L such that every two elements of L have a least upper bound and a greatest lower bound. The least upper bound of a and b is denoted by $a \vee b$ and the greatest lower bound of a and b is denoted by $a \wedge b$, thereby giving rise to two binary operations $\vee$ and $\wedge$ satisfying several properties.

However, when a lattice is defined as an algebraic structure in Definition 11.41, we have a nonempty set L on which are defined two binary operations $\vee$ and $\wedge$ satisfying the commutative laws, associative laws and absorption laws. For $a, b \in L$, suppose that we define a relation $\preceq$ on L by

$$a \preceq b \text{ if } a \vee b = b \text{ (or } a \wedge b = a).$$

Then $\preceq$ is a partial order and the poset $(L, \preceq)$ is a lattice. Hence lattices can be looked at from two points of view. There are other properties of interest that a lattice may possess.

Definition 11.43 *A lattice $(L, \preceq)$ is* **bounded** *if L has both a greatest element and a least element.*

Example 11.44

For a fixed integer $n \geq 2$, the lattice $(S, \leq)$ of all positive integers less than or equal to n is bounded. In fact, the greatest element in S is n and the least element in S is 1. $\blacklozenge$

Example 11.45

For a fixed integer $n \geq 2$, the lattice $(S, |\,)$ of all positive integer divisors of n is bounded. The greatest element in S is n and the least element is 1. $\blacklozenge$

Example 11.46

For a nonempty set X with $S = \mathcal{P}(X)$, the lattice $(S, \subseteq)$ is bounded. The greatest element in this lattice is X and the least element is $\emptyset$. $\blacklozenge$

In a general bounded lattice $(L, \preceq)$, it is customary to denote the greatest element by 1 (typically called the **unity**) and the least element by 0 (called the **zero**). Thus for every element $a \in L$, we have $0 \preceq a$ and $a \preceq 1$. In fact,

$$0 \wedge a = 0 \text{ and } 1 \vee a = 1$$

for all $a \in L$, while

$$0 \vee a = 1 \wedge a = a.$$

Two common properties satisfied by set unions and intersections are the distributive properties, namely for sets A, B and C,

$$A \cap (B \cup C) = (A \cap B) \cup (A \cap C) \text{ and } A \cup (B \cap C) = (A \cup B) \cap (A \cup C).$$

Some lattices possess a related property.

Distributive Lattices

Definition 11.47 *A lattice $(L, \preceq)$ is* **distributive** *if the following hold.*

- **Distributive Laws**

  (D1) $a \wedge (b \vee c) = (a \wedge b) \vee (a \wedge c)$ *for all $a, b, c \in L$,*
  (D2) $a \vee (b \wedge c) = (a \vee b) \wedge (a \vee c)$ *for all $a, b, c \in L$.*

Strictly speaking, a lattice need only satisfy one of the two properties stated in Definition 11.47 to be distributive.

Theorem 11.48 *A lattice $(L, \preceq)$ satisfies property (D1) if and only if $(L, \preceq)$ satisfies property (D2).*

Proof. Suppose first that $(L, \preceq)$ satisfies (D1). For $a, b, c \in L$,

$$
\begin{aligned}
(a \vee b) \wedge (a \vee c) &= ((a \vee b) \wedge a) \vee ((a \vee b) \wedge c) \quad \text{(distributive law (D1))} \\
&= (a \wedge (a \vee b)) \vee ((a \vee b) \wedge c) \quad \text{(commutative law)} \\
&= a \vee ((a \vee b) \wedge c) \quad \text{(absorption law)} \\
&= a \vee (c \wedge (a \vee b)) \quad \text{(commutative law)} \\
&= a \vee ((c \wedge a) \vee (c \wedge b)) \quad \text{(distributive law (D1))} \\
&= (a \vee (c \wedge a)) \vee (c \wedge b) \quad \text{(associative law)} \\
&= a \vee (c \wedge b) \quad \text{(absorption law)} \\
&= a \vee (b \wedge c) \quad \text{(commutative law)}.
\end{aligned}
$$

Thus $(L, \preceq)$ satisfies distributive law $(D2)$.

For the converse, assume that $(L, \preceq)$ satisfies distributive law $(D2)$. Then for all $a, b, c \in L$,

$$
\begin{aligned}
(a \wedge b) \vee (a \wedge c) &= ((a \wedge b) \vee a) \wedge ((a \wedge b) \vee c) \quad \text{(distributive law } (D2)) \\
&= (a \vee (a \wedge b)) \wedge ((a \wedge b) \vee c) \quad \text{(commutative law)} \\
&= a \wedge ((a \wedge b) \vee c) \quad \text{(absorption law)} \\
&= a \wedge (c \vee (a \wedge b)) \quad \text{(commutative law)} \\
&= a \wedge ((c \vee a) \wedge (c \vee b)) \quad \text{(distributive law } (D2)) \\
&= (a \wedge (c \vee a)) \wedge (c \vee b) \quad \text{(associative law)} \\
&= a \wedge (c \vee b) \quad \text{(absorption law)} \\
&= a \wedge (b \vee c) \quad \text{(commutative law)}.
\end{aligned}
$$

Thus $(L, \preceq)$ satisfies distributive law $(D1)$.   ∎

For a nonempty set X with $S = \mathcal{P}(X)$, the lattice $(S, \subseteq)$ of subsets of X is therefore distributive. The lattice $(S, |\,)$ of all positive integer divisors of an integer $n \geq 2$ is also distributive.

Example 11.49

The lattice $(L, \preceq)$ whose Hasse diagram is shown in Figure 11.17 is not distributive.

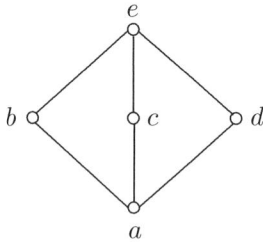

Figure 11.17: A Hasse diagram for the lattice in Example 11.49

Solution.   First observe that
$$b \wedge (c \vee d) = b \wedge e = b.$$
Since
$$(b \wedge c) \vee (b \wedge d) = a \vee a = a,$$
it follows that $b \wedge (c \vee d) \neq (b \wedge c) \vee (b \wedge d)$ and so $(L, \preceq)$ is not distributive.   ◆

Complemented Lattices

There are occasions in a bounded lattice when for a given element a of the lattice, there may be another element b such that the least upper bound and the greatest lower bound of $\{a, b\}$ are the unity and zero element, respectively.

Definition 11.50 *Let $(L, \preceq)$ be a bounded lattice with unity 1 and zero element 0. For $a \in L$, an element $\overline{a} \in L$ is called a* **complement** *of a if both*

$$a \vee \overline{a} = 1 \ \text{ and } \ a \wedge \overline{a} = 0.$$

The lattice $(L, \preceq)$ is **complemented** *if every element of L has a complement.*

There is an example of a complemented lattice that we have encountered earlier and often.

Example 11.51

For a nonempty set X, the lattice $(\mathcal{P}(X), \subseteq)$ is bounded and distributive. Show that $(\mathcal{P}(X), \subseteq)$ is also complemented.

Solution. We have seen that the unity in this lattice is X and the zero element is $\emptyset$. Let $A \in S$. Consider the set complement $\overline{A} = X - A$. Then $A \vee \overline{A} = A \cup \overline{A} = X$ and $A \wedge \overline{A} = A \cap \overline{A} = \emptyset$. Hence $\overline{A}$ is a complement of A in the lattice $(\mathcal{P}(X), \subseteq)$. ♦

Another lattice we have seen concerns the positive integer divisors of an integer $n \geq 2$.

Example 11.52

Let S denote the set of all positive integer divisors of 12. Thus $S = \{1, 2, 3, 4, 6, 12\}$. Then $(S, |\,)$ is a bounded distributive lattice. Is S complemented?

Solution. In this lattice, the unity is 12 and the zero element is 1. Furthermore, for $a, b \in S$, $a \vee b = \operatorname{lcm}(a, b)$ and $a \wedge b = \gcd(a, b)$. First, let $a = 4$. Since $4 \vee 3 = \operatorname{lcm}(3, 4) = 12$ and $4 \wedge 3 = \gcd(3, 4) = 1$, it follows that $\overline{a} = 3$ is a complement of 4. Next, let $a = 6$. The only elements b for which $a \vee b = 12$ are 4 and 12. However, $6 \wedge 4 = 2$ and $6 \wedge 12 = 6$. Therefore, the integer $a = 6$ does not have a complement in this lattice. Consequently, $(S, \preceq)$ is not complemented. ♦

Example 11.53

The lattice $(L, \preceq)$ whose Hasse diagram is shown in Figure 11.18, although bounded, is not distributive. Determine whether the element $a \in L$ has a complement.

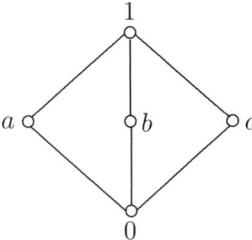

Figure 11.18: A Hasse diagram for the lattice in Example 11.53

Solution. Observe that $a \vee b = 1$ and $a \wedge b = 0$. Thus b is a complement of a. In fact, $a \vee c = 1$ and $a \wedge c = 0$; that is, c is also a complement of a. Therefore, the element a in this lattice has two complements. ♦

The element a (and the elements b and c as well) in the lattice discussed in Example 11.53 has two complements and this lattice is not distributive. Such a situation cannot occur in a bounded distributive lattice.

Theorem 11.54 *Let $(L, \preceq)$ be a bounded distributive lattice. If an element of L has a complement, then it is unique.*

Proof. Suppose that there is an element $a \in L$ that has two complements, say b and c. Then

$$a \vee b = a \vee c = 1 \text{ and } a \wedge b = a \wedge c = 0.$$

Therefore,

$$\begin{aligned} b &= b \vee 0 = b \vee (a \wedge c) = (b \vee a) \wedge (b \vee c) \\ &= 1 \wedge (b \vee c) = (a \vee c) \wedge (b \vee c) \\ &= (c \vee a) \wedge (c \vee b) = c \vee (a \wedge b) = c \vee 0 = c \end{aligned}$$

and so $b = c$.   ∎

Since every element in a bounded distributive lattice that has a complement has only one complement, we denote this unique complement of a by $\bar{a}$.

Exercises for Section 11.2

1. Let $(S, \preceq)$ be a poset. Prove that every nonempty subset of S has at most one least upper bound.

2. Give an example of a poset $(S, \preceq)$ and a nonempty subset of S that has exactly three upper bounds and exactly two lower bounds.

3. Figure 11.19 shows a Hasse diagram of a poset $(A, \preceq)$, where $A = \{a_1, a_2, \ldots, a_{11}\}$. Is $(A, \preceq)$ a lattice?

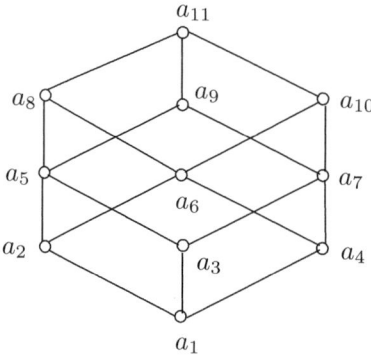

Figure 11.19: A Hasse diagram for the poset in Exercise 3

4. Figure 11.20 shows a Hasse diagram of the poset $(S, \preceq)$. Is $(S, \preceq)$ a lattice?

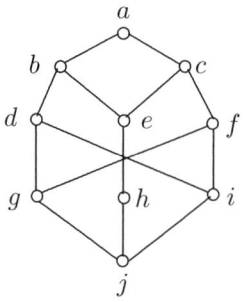

Figure 11.20: A Hasse diagram for the poset in Exercise 4

5. Let $(L, \preceq)$ be a lattice, where $a, b, c \in L$. Prove that if $a \preceq b$, then
   (a) $a \vee c \preceq b \vee c$ and (b) $a \wedge c \preceq b \wedge c$.

6. Let $(L, \preceq)$ be a lattice, where $a, b, c \in L$. Prove that $a \preceq c$ and $b \preceq c$ if and only if $a \vee b \preceq c$.

7. Let $(L, \preceq)$ be a lattice, where $a, b, c \in L$. Prove that $c \preceq a$ and $c \preceq b$ if and only if $c \preceq a \wedge b$.

8. Let $(L, \preceq)$ be a lattice, where $a, b, c, d \in L$. Prove that if $a \preceq b$ and $c \preceq d$, then
   (a) $a \vee c \preceq b \vee d$ and (b) $a \wedge c \preceq b \wedge d$.

9. Let $(L, \preceq)$ be a lattice, where $a, b \in L$. Prove the following.

   (a) $a \vee b = b$ if and only if $a \preceq b$. (b) $a \wedge b = a$ if and only if $a \preceq b$.

   (c) $a \wedge b = a$ if and only if $a \vee b = b$.

10. Let $(L, \preceq)$ be a bounded lattice and $a, b, c \in L$.

    (a) Prove that $a \wedge (b \vee c) = (a \wedge b) \vee (a \wedge c)$ and $a \vee (b \wedge c) = (a \vee b) \wedge (a \vee c)$ if (1) either one of a, b and c is 1 or one of a, b and c is 0 or (2) two of a, b and c are equal.

    (b) What does the result in (a) tell us?

11. Let $a = 6$, $b = 20$ and $c = 75$.

    (a) Compute $\gcd(a, \operatorname{lcm}(b, c))$ and $\operatorname{lcm}(\gcd(a, b), \gcd(a, c))$.

    (b) Compute $\operatorname{lcm}(a, \gcd(b, c))$ and $\gcd(\operatorname{lcm}(a, b), \operatorname{lcm}(a, c))$.

    (c) What do (a) and (b) illustrate?

12. Prove or disprove: The lattice $(\mathbb{N}, \leq)$ is distributive.

13. For an integer $n \geq 2$, let $S = \{k \in \mathbb{N} : k \leq n\}$. Is $(S, \leq)$ complemented?

14. Let $(L, \preceq)$ be a bounded distributive lattice. Prove that if $a, b \in L$ where a has the (unique) complement $\overline{a}$, then the following hold:

    (a) $a \vee (\overline{a} \wedge b) = a \vee b$ and (b) $a \wedge (\overline{a} \vee b) = a \wedge b$.

15. Let $(L, \preceq)$ be a distributive lattice where $a, b, c \in L$. Prove that if $a \wedge b = a \wedge c$ and $a \vee b = a \vee c$, then $b = c$.

16. Show that none of the lattices whose Hasse diagrams are given in Figures 11.21 (a), (b), (c) are distributive.

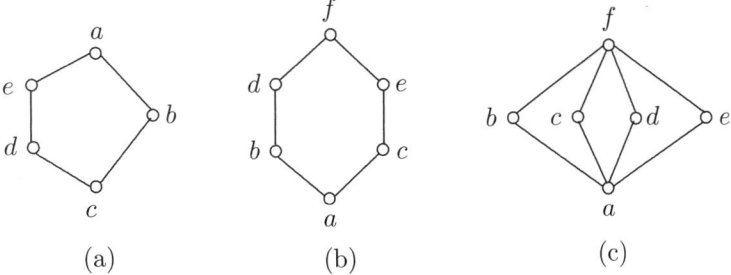

Figure 11.21: Hasse diagrams for the lattices in Exercise 16

17. A lattice $(L, \preceq)$ is **modular** if for all $a, b, c \in L$, $a \preceq c$ implies that $a \vee (b \wedge c) = (a \vee b) \wedge c$.

    (a) Prove that every distributive lattice is a modular lattice.

    (b) Neither of the lattices described in Exercise 16(a) and 16(b) is distributive. It turns out that the lattice in Exercise 16(a) is modular. Show that the lattice in Exercise 16(b) is not modular.

18. Show that every finite complemented lattice with at least two elements has an even number of elements.

19. Give an alternative proof of Theorem 11.54:

> Let $(L, \preceq)$ be a bounded distributive lattice. If an element of L has a complement, then it is unique.

by first observing that if an element a has two complements b and c, then $b = b \wedge 1$.

20. Let $n = pqr$, where p, q and r are distinct primes and let S be the set of all positive integer divisors of n. Then $(S, \preceq)$ is a lattice where $a \preceq b$ if $a \mid b$. Find a complement of each element of S that has a complement.

21. Let $n = p^3 q^2 r$, where p, q and r are distinct primes and let S be the set of all positive integer divisors of n. Then $(S, \preceq)$ is a lattice where $a \preceq b$ if $a \mid b$. Find a complement of each element of S that has a complement.

22. For a fixed integer $n \geq 2$, let S_n be the set of positive integer divisors of n. Determine all n such that $(S_n, \mid)$ is a complemented lattice.

11.3   Boolean Algebras

It is often the case in mathematics that while studying certain structures of interest, it is observed that these structures have some properties in common. This suggests introducing a new concept, defined as a structure possessing these properties. If it can be proved that each such newly defined structure possesses other properties, then the original structures being studied must satisfy these additional properties as well. That is, we can learn information about all of the given structures by studying properties possessed by this more general structure. This brings us to the main subject of this section.

In the preceding section, we saw that some lattices are bounded and therefore contain a unity 1 and zero element 0. Furthermore, some bounded lattices are distributive and some are complemented (each element has a complement). Bounded lattices with both of these properties are of special interest.

Definition 11.55 *A **Boolean algebra** is a bounded lattice that is both distributive and complemented.*

It is common to require 0 and 1 to be distinct elements in a Boolean algebra and for this reason we assume that every Boolean algebra contains at least two elements.

One might very well wonder why such lattices are referred to as a type of algebra. After all, the term *algebra* probably suggests algebraic expressions constructed by adding and multiplying variables. The properties possessed by a Boolean algebra are reminiscent of logical equivalences we've seen involving the disjunction, conjunction and negations of statements as well as properties involving the union, intersection and complements of subsets of a given set. This gives rise to an alternative interpretation of a Boolean algebra, namely one whose definition is independent of lattices. When defined in this new manner, it is common practice to use + (addition) as one binary operation, rather than $\vee$ or $\cup$ and to use $\cdot$ (multiplication) as the other binary operation, rather than $\wedge$ or $\cap$. From this point of view, we are now dealing with an algebraic structure.

Definition 11.56 *A **Boolean algebra** is a set S with two or more elements on which are defined two binary operations called addition $+$ and multiplication $\cdot$ (and so $a + b$ and $a \cdot b = ab$ are unique elements of S for $a, b \in S$) satisfying the following properties:*

- **Commutative Laws**

  $a + b = b + a$ *and* $a \cdot b = b \cdot a$ *for all* $a, b \in S$.

- **Associative Laws**

  $(a + b) + c = a + (b + c)$ *and* $(a \cdot b) \cdot c = a \cdot (b \cdot c)$ *for all* $a, b, c \in S$.

- **Distributive Laws**

  $a \cdot (b + c) = (a \cdot b) + (a \cdot c)$ *and* $a + (b \cdot c) = (a + b) \cdot (a + c)$ *for all* $a, b, c \in S$.

- **Existence of Zero and Unity**

  There exist distinct elements $0, 1 \in S$ *such that* $a + 0 = a$ *and* $a \cdot 1 = a$ *for each* $a \in S$.

- **Existence of Complements**

  For every $a \in S$, *there exists* $\overline{a} \in S$ *such that* $a + \overline{a} = 1$ *and* $a \cdot \overline{a} = 0$.

On occasion, a Boolean algebra defined in this way is denoted by $(S, +, \cdot)$. If the two binary operations are understood, then this Boolean algebra may be denoted simply by S. It should be emphasized that although the two binary operations mentioned in Definition 11.56 of a Boolean algebra are denoted by $+$ and $\cdot$ and are called addition and multiplication, these are not the addition and multiplication with which we are familiar when dealing with integers and real numbers. Furthermore, the elements 0 and 1 of S are not integers in general. Each is an element of S possessing a very special property.

From what we saw in the preceding section, it therefore follows that the power set S of a finite nonempty set X, where addition and multiplication are defined on S by

$$A + B = A \cup B \text{ and } A \cdot B = A \cap B$$

for $A, B \in S$, is a Boolean algebra. In this case, the unity 1 in the definition of a Boolean algebra is the set X and the zero element 0 is the empty set $\emptyset$.

Example 11.57

The set S of positive integer divisors of 30, where addition and multiplication are defined on S by

$$a + b = \text{lcm}(a, b) \text{ and } a \cdot b = \gcd(a, b)$$

for $a, b \in S$, is a Boolean algebra. Thus $S = \{1, 2, 3, 5, 6, 10, 15, 30\}$. In this case, the unity 1 is the integer 30 and the zero element 0 is the integer 1. Furthermore, the complement of the elements of S are $\overline{1} = 30, \overline{2} = 15, \overline{3} = 10, \overline{5} = 6, \overline{6} = 5, \overline{10} = 3, \overline{15} = 2$ and $\overline{30} = 1$. ◆

All Boolean algebras satisfy a number of other familiar properties, including the following.

If $(S, +, \cdot)$ *is a Boolean algebra, then*

- **Absorption Laws**

  $a \cdot (a + b) = a$ *and* $a + (a \cdot b) = a$ *for all* $a, b \in S$.

- **Idempotent Laws**

  $a + a = a$ *and* $a \cdot a = a$ *for every* $a \in S$.

- **Domination Laws**

  $a + 1 = 1$ *and* $a \cdot 0 = 0$ *for every* $a \in S$.

We now verify that every Boolean algebra satisfies these three laws. In the proof of the first two of these laws, the distributive laws play a key role.

Proposition 11.58 *Every Boolean algebra satisfies the absorption laws.*

Proof.   Let a and b be elements in a Boolean algebra. Then

$$
\begin{aligned}
a \cdot (a + b) &= (a + 0) \cdot (a + b) \quad \text{(property of zero)} \\
&= a + (0 \cdot b) \quad \text{(distributive law)} \\
&= a + (b \cdot 0) \quad \text{(commutative law)} \\
&= a + 0 \quad \text{(domination law)} \\
&= a \quad \text{(property of zero)}
\end{aligned}
$$

and

$$
\begin{aligned}
a + (a \cdot b) &= (a \cdot 1) + (a \cdot b) \quad \text{(property of unity)} \\
&= a \cdot (1 + b) \quad \text{(distributive law)} \\
&= a \cdot (b + 1) \quad \text{(commutative law)} \\
&= a \cdot 1 \quad \text{(domination law)} \\
&= a \quad \text{(property of unity)}.
\end{aligned}
$$

Therefore, the absorption laws hold.   ∎

Proposition 11.59   *Every Boolean algebra satisfies the idempotent laws.*

Proof.   Let a be an element in a Boolean algebra. Then

$$
\begin{aligned}
a + a &= (a + a) \cdot 1 \quad \text{(property of unity)} \\
&= (a + a) \cdot (a + \bar{a}) \quad \text{(property of complement)} \\
&= a + (a \cdot \bar{a}) \quad \text{(distributive law)} \\
&= a + 0 \quad \text{(property of complement)} \\
&= a \quad \text{(property of zero)}
\end{aligned}
$$

and

$$
\begin{aligned}
a \cdot a &= (a \cdot a) + 0 \quad \text{(property of zero)} \\
&= (a \cdot a) + (a \cdot \bar{a}) \quad \text{(property of complement)} \\
&= a \cdot (a + \bar{a}) \quad \text{(distributive law)} \\
&= a \cdot 1 \quad \text{(property of complement)} \\
&= a \quad \text{(property of unity)}.
\end{aligned}
$$

Therefore, the idempotent laws hold.   ∎

Proposition 11.60   *Every Boolean algebra satisfies the domination laws.*

Proof.   Let a be an element in a Boolean algebra. Then

$$
\begin{aligned}
a + 1 &= a + (a + \bar{a}) \quad \text{(property of complement)} \\
&= (a + a) + \bar{a} \quad \text{(associative law)} \\
&= a + \bar{a} \quad \text{(idempotent law)} \\
&= 1 \quad \text{(property of complement)}
\end{aligned}
$$

and

$$
\begin{aligned}
a \cdot 0 &= a \cdot (a \cdot \bar{a}) \quad \text{(property of complement)} \\
&= (a \cdot a) \cdot \bar{a} \quad \text{(associative law)} \\
&= a \cdot \bar{a} \quad \text{(idempotent law)} \\
&= 0 \quad \text{(property of complement)}.
\end{aligned}
$$

Therefore, the domination laws hold. ∎

Since the complement of an element in a distributive lattice is unique, the following result should not be surprising.

Theorem 11.61 *In a Boolean algebra, every element has a unique complement.*

Proof. Let S be a Boolean algebra and let $a \in S$. Then there is an element $\bar{a} \in S$ such that $a + \bar{a} = 1$ and $a \cdot \bar{a} = 0$. Suppose that b is an element of S with the properties $a + b = 1$ and $a \cdot b = 0$. We show that $b = \bar{a}$. Observe that

$$
\begin{aligned}
b &= b + 0 \quad \text{(property of zero)} \\
&= b + (a \cdot \bar{a}) \quad \text{(property of complement)} \\
&= (b + a) \cdot (b + \bar{a}) \quad \text{(distributive law)} \\
&= (a + b) \cdot (\bar{a} + b) \quad \text{(commutative law, twice)} \\
&= 1 \cdot (\bar{a} + b) \quad \text{(property of element } b) \\
&= (a + \bar{a}) \cdot (\bar{a} + b) \quad \text{(property of complement)} \\
&= (\bar{a} + a) \cdot (\bar{a} + b) \quad \text{(commutative law)} \\
&= \bar{a} + (a \cdot b) \quad \text{(distributive law)} \\
&= \bar{a} + 0 \quad \text{(property of element } b) \\
&= \bar{a} \quad \text{(property of zero).}
\end{aligned}
$$

Therefore, the complement of a is unique. ∎

Corollary 11.62 *Let S be a Boolean algebra. For every element a in S, $\bar{\bar{a}} = a$.*

Proof. Let $a \in S$. Therefore, $\bar{\bar{a}}$ represents the complement of the element $\bar{a} \in S$. By definition, $a + \bar{a} = \bar{a} + a = 1$ and $a \cdot \bar{a} = \bar{a} \cdot a = 0$. By Theorem 11.61, a is the (unique) complement of $\bar{a}$ and so $\bar{\bar{a}} = a$. ∎

All Boolean algebras also satisfy De Morgan's Laws.

Theorem 11.63 *Every Boolean algebra S satisfies De Morgan's Laws, namely*

$$\overline{a + b} = \bar{a} \cdot \bar{b} \text{ and } \overline{a \cdot b} = \bar{a} + \bar{b} \text{ for all } a, b \in S.$$

Proof. We prove only the first of these two laws. The second law can be proved in a similar manner (see Exercise 1). Let $a, b \in S$. To see that $\overline{a + b} = \bar{a} \cdot \bar{b}$, we show that $\bar{a} \cdot \bar{b}$ is the complement of $a + b$. Hence we must verify that $(a + b) + (\bar{a} \cdot \bar{b}) = 1$ and $(a + b) \cdot (\bar{a} \cdot \bar{b}) = 0$. First, observe that

$$
\begin{aligned}
(a + b) + (\bar{a} \cdot \bar{b}) &= ((a + b) + \bar{a}) \cdot ((a + b) + \bar{b}) \quad \text{(distributive law)} \\
&= ((b + a) + \bar{a}) \cdot ((a + b) + \bar{b}) \quad \text{(commutative law)} \\
&= (b + (a + \bar{a})) \cdot (a + (b + \bar{b})) \quad \text{(associative law)} \\
&= (b + 1) \cdot (a + 1) \quad \text{(property of complement)} \\
&= 1 \cdot 1 \quad \text{(domination law)} \\
&= 1 \quad \text{(property of unity).}
\end{aligned}
$$

Next,

$$
\begin{aligned}
(a + b) \cdot (\bar{a} \cdot \bar{b}) &= (\bar{a} \cdot \bar{b}) \cdot (a + b) \quad \text{(commutative law)} \\
&= ((\bar{a} \cdot \bar{b}) \cdot a) + ((\bar{a} \cdot \bar{b}) \cdot b) \quad \text{(distributive law)} \\
&= ((\bar{b} \cdot \bar{a}) \cdot a) + ((\bar{a} \cdot \bar{b}) \cdot b) \quad \text{(commutative law)}
\end{aligned}
$$

$$
\begin{aligned}
&= \; (\bar{b} \cdot (\bar{a} \cdot a)) + (\bar{a} \cdot (\bar{b} \cdot b)) && \text{(associative law)} \\
&= \; (\bar{b} \cdot (a \cdot \bar{a})) + (\bar{a} \cdot (b \cdot \bar{b})) && \text{(commutative law)} \\
&= \; (\bar{b} \cdot 0) + (\bar{a} \cdot 0) && \text{(property of complement)} \\
&= \; 0 + 0 && \text{(domination law)} \\
&= \; 0 && \text{(property of zero).}
\end{aligned}
$$

Therefore, $\overline{a + b} = \bar{a} \cdot \bar{b}$. ∎

The Duality Principle

You may have observed that many of the results concerning Boolean algebras occur in pairs. Furthermore, if the proofs of the two results in each pair are studied carefully, then one might also observe that these two proofs are closely related. Indeed, often the statement of one result in such a pair can be obtained from the other by interchanging $+$ and $\cdot$ and interchanging 0 and 1. From this, it follows that for each property of a Boolean algebra, there is a dual property. In fact, this principle is referred to as the **duality principle**.

The Duality Principle for Boolean Algebras *For each theorem concerning a Boolean algebra S, there is a dual theorem concerning S obtained by interchanging $+$ and $\cdot$ and by interchanging 0 and 1.*

For example, each of De Morgan's two laws in Theorem 11.63 can be obtained from the other by the duality principle. The following examples also illustrate the duality principle.

Theorem 11.64 *Let S be a Boolean algebra and let $a, b \in S$. Then*

$$
a \cdot \bar{b} + a \cdot b = a.
$$

Proof. Observe that $a \cdot \bar{b} + a \cdot b = a \cdot (\bar{b} + b) = a \cdot 1 = a$. ∎

We have an immediate corollary of Theorem 11.64 by the duality principle.

Corollary 11.65 *Let S be a Boolean algebra and let $a, b \in S$. Then*

$$
(a + \bar{b}) \cdot (a + b) = a.
$$

Theorem 11.66 *For elements a and b in a Boolean algebra, $a \cdot b = a$ if and only if $a \cdot \bar{b} = 0$.*

Proof. Assume first that $a \cdot b = a$. Then

$$
a \cdot \bar{b} = (a \cdot b) \cdot \bar{b} = a \cdot (b \cdot \bar{b}) = a \cdot 0 = 0.
$$

For the converse, assume that $a \cdot \bar{b} = 0$. Then

$$
a \cdot b = a \cdot b + 0 = a \cdot b + a \cdot \bar{b} = a \cdot (b + \bar{b}) = a \cdot 1 = a,
$$

giving the desired result. ∎

Again, we have a corollary of Theorem 11.66 by the duality principle.

Corollary 11.67 *For elements a and b in a Boolean algebra, $a + b = a$ if and only if $a + \bar{b} = 1$.*

The best known and most useful Boolean algebra is also the simplest, consisting only of two elements. Before describing this Boolean algebra, recall that when we encountered the operations of disjunction, conjunction and negation of statements P and Q in Chapter 1, we obtained the truth tables shown in Figure 11.22.

P	Q	$P \vee Q$
T	T	T
T	F	T
F	T	T
F	F	F

P	Q	$P \wedge Q$
T	T	T
T	F	F
F	T	F
F	F	F

P	$\sim P$
T	F
F	T

Figure 11.22: Truth tables for statements

Interpreting $\vee$ and $\wedge$ as binary operations on the set $\{T, F\}$, we have the following two tables for these binary operations.

$\vee$	T	F
T	T	T
F	T	F

$\wedge$	T	F
T	T	F
F	F	F

This gives rise to a Boolean algebra, namely one with two elements 1 and 0, where 1 denotes T and 0 denotes F.

Example 11.68

Let $B = \{0, 1\}$, where addition, multiplication and complementation are defined as shown below.

$+$	1	0
1	1	1
0	1	0

$\cdot$	1	0
1	1	0
0	0	0

$$\overline{1} = 0$$
$$\overline{0} = 1$$

Then B is a Boolean algebra. If we were to interpret 0 and 1 as integers, then we see that addition, multiplication and complementation can also be defined by

$$\begin{aligned} a + b &= \max(a, b) \text{ for } a, b \in B \\ a \cdot b &= \min(a, b) \text{ for } a, b \in B \\ \overline{a} &= 1 - a \text{ for } a \in B. \end{aligned}$$

From the symmetry of the addition and multiplication tables above, it is clear that the commutative laws hold. Since both $(a + b) + c$ and $a + (b + c)$ equal $\max(a, b, c)$, while both $(a \cdot b) \cdot c$ and $a \cdot (b \cdot c)$ equal $\min(a, b, c)$, the associative laws hold. The distributive laws are more troublesome to verify. Each of these properties requires eight cases to give a complete proof. For example, for $a = 1$, $b = 0$ and $c = 1$,

$$a \cdot (b + c) = 1 \cdot (0 + 1) = 1 \cdot 1 = 1 \text{ and } a \cdot b + a \cdot c = 1 \cdot 0 + 1 \cdot 1 = 0 + 1 = 1;$$

while

$$a + (b \cdot c) = 1 + 0 \cdot 1 = 1 + 0 = 1 \text{ and } (a + b) \cdot (a + c) = (1 + 0) \cdot (1 + 1) = 1 \cdot 1 = 1.$$

The remaining two properties of this Boolean algebra follow immediately. ♦

Electronic switches (which can either be in the on position or off position) and optical switches (which can either be lit or unlit) can be studied with the aid of the Boolean algebra B described in Example 11.68. Because of the importance of the Boolean algebra B, the remainder of this section is devoted exclusively to this Boolean algebra. Addition in B will be referred to as **Boolean addition** and multiplication in B as **Boolean multiplication**.

In the following example, the two Boolean operations and complementation in B are used to simplify two expressions.

Example 11.69

In the Boolean algebra B, compute and simplify the expressions $(1 + 0) + (\overline{0 \cdot 1})$ and $\overline{0 \cdot (1 + 0)}$.

Solution. Observe that

$$(1 + 0) + (\overline{0 \cdot 1}) = 1 + \overline{0} = 1 + 1 = 1$$

and

$$\overline{0 \cdot (1 + 0)} = \overline{0 \cdot 1} = \overline{0} + \overline{1} = 1 + 0 = 1. \; \blacklozenge$$

Boolean Functions

A variable x is called a **Boolean variable** if it takes on values from B, that is, x can either have the value 1 or the value 0. By a **Boolean function of degree** n we mean a function $f : B^n \to B$, where B^n is the Cartesian product of n sets B, that is, $B^n = B \times B \times \cdots \times B$. A Boolean function can be represented by a **Boolean expression**, constructed from Boolean variables and Boolean operations.

Example 11.70

For Boolean variables x, y and z, the function f defined by $f(x, y, z) = x \cdot \overline{y} + z$ is a Boolean function. (The product $x \cdot \overline{y}$ of x and $\overline{y}$ can also be expressed by $x\overline{y}$.) Since $f : B^3 \to B$, the function f has degree 3. The values of this function are displayed in the table below. $\blacklozenge$

x	y	z	$\overline{y}$	$x \cdot \overline{y}$	$f(x, y, z) = x \cdot \overline{y} + z$
1	1	1	0	0	1
1	1	0	0	0	0
1	0	1	1	1	1
1	0	0	1	1	1
0	1	1	0	0	1
0	1	0	0	0	0
0	0	1	1	0	1
0	0	0	1	0	0

Example 11.71

For Boolean variables x, y and z, determine the values of the function f of degree 3 defined by $f(x, y, z) = (\overline{x} + \overline{y}) \cdot z$.

Solution. The values of this function are displayed in the table below. $\blacklozenge$

x	y	z	$\overline{x}$	$\overline{y}$	$\overline{x} + \overline{y}$	$f(x, y, z) = (\overline{x} + \overline{y}) \cdot z$
1	1	1	0	0	0	0
1	1	0	0	0	0	0
1	0	1	0	1	1	1
1	0	0	0	1	1	0
0	1	1	1	0	1	1
0	1	0	1	0	1	0
0	0	1	1	1	1	1
0	0	0	1	1	1	0

The function f described in Example 11.71 can be expressed in several other equivalent ways. For example, by one of the commutative laws,

$$f(x, y, z) = (\overline{x} + \overline{y}) \cdot z = z \cdot (\overline{x} + \overline{y}).$$

By a distributive law,

$$f(x, y, z) = z \cdot (\overline{x} + \overline{y}) = z \cdot \overline{x} + z \cdot \overline{y}.$$

By one of De Morgan's Laws,

$$f(x, y, z) = (\overline{x} + \overline{y}) \cdot z = (\overline{x \cdot y}) \cdot z.$$

We now consider the reverse question of being given the values of a function f, of degree 3 say and this time asking for some Boolean expression that represents the function.

Example 11.72

The values of a Boolean function f of degree 3 are given in the table below. Find a Boolean expression that represents the function f.

x	y	z	f
1	1	1	1
1	1	0	0
1	0	1	1
1	0	0	1
0	1	1	0
0	1	0	1
0	0	1	0
0	0	0	0

Solution. First observe that $f(x, y, z) = 1$ when $x = y = z = 1$. Notice that the product xyz has the value 1 only when $x = y = z = 1$. Also, $f(1, 0, 1) = 1$. The product $x\overline{y}z$ has the value 1 only when $x = z = 1$ and $y = 0$. Similarly, the fact that $f(1, 0, 0) = f(0, 1, 0) = 1$ gives rise to the products $x\overline{y}\ \overline{z}$ and $\overline{x}\ y\overline{z}$. Hence if f is defined by $f(x, y, z) = xyz + x\overline{y}z + x\ \overline{y}\ \overline{z} + \overline{x}\ y\ \overline{z}$, then the values of this function are exactly those given in the table. ♦

The function f in Example 11.72 given by

$$f(x, y, z) = xyz + x\overline{y}z + x\ \overline{y}\ \overline{z} + \overline{x}y\overline{z}$$

is called the **sum-of-products expansion** or **disjunctive normal form** of the Boolean function f. From this example, we see that every Boolean function has a sum-of-products expansion.

Example 11.73

Find the sum-of-products expansion of the Boolean function f of degree 3 defined by

$$f(x, y, z) = (x + \overline{y})z.$$

Solution. There is more than one way to solve this problem. First, observe that

$$
\begin{aligned}
f(x, y, z) &= (x + \overline{y})z = xz + \overline{y}z \\
&= x \cdot 1 \cdot z + 1 \cdot \overline{y}z \\
&= x(y + \overline{y})z + (x + \overline{x})\overline{y}z \\
&= xyz + x\overline{y}z + x\overline{y}z + \overline{x}\ \overline{y}\ z \\
&= xyz + x\overline{y}z + \overline{x}\ \overline{y}\ z,
\end{aligned}
$$

where the last expression is obtained from the preceding expression since $x\overline{y}z + x\overline{y}z = x\overline{y}z$ by an idempotent law.

Another way to solve this problem is to first construct a table of functional values from the Boolean expression that represents this function. This table is given below.

x	y	z	$\overline{y}$	$x + \overline{y}$	$f(x,y,z) = (x + \overline{y}) \cdot z$
1	1	1	0	1	1
1	1	0	0	1	0
1	0	1	1	1	1
1	0	0	1	1	0
0	1	1	0	0	0
0	1	0	0	0	0
0	0	1	1	1	1
0	0	0	1	1	0

From this table, we obtain $f(x,y,z) = xyz + x\overline{y}z + \overline{x}\;\overline{y}\;z$, which, of course, agrees with our earlier answer.   ◆

There is another common form of a Boolean expression that represents a Boolean function. We illustrate this in the following example.

Example 11.74

The values of a Boolean function f of degree 3 are given in the table below. Find two Boolean expressions that represent f.

x	y	z	f
1	1	1	0
1	1	0	1
1	0	1	1
1	0	0	1
0	1	1	0
0	1	0	1
0	0	1	1
0	0	0	0

Solution. First we can consider the sum-of-products expansion of f, which can be obtained by looking at all triples (x,y,z) for which $f(x,y,z) = 1$. This gives us

$$f(x,y,z) = xy\overline{z} + x\overline{y}z + x\;\overline{y}\;\overline{z} + \overline{x}yz + \overline{x}\;\overline{y}\;z.$$

A second expression can be constructed from those triples (x,y,z) for which $f(x,y,z) = 0$. One of these triples is $(1,1,1)$. Thus $\overline{x} + \overline{y} + \overline{z} = 0$ only when $x = y = z = 1$. Furthermore, $x + \overline{y} + \overline{z} = 0$ only when $x = 0$ and $y = z = 1$, while $x + y + z = 0$ only when $x = y = z = 0$. Therefore,

$$(\overline{x} + \overline{y} + \overline{z})(x + \overline{y} + \overline{z})(x + y + z) = 1$$

for all triples (x,y,z) of 0s and 1s, except for $(1,1,1)$, $(0,1,1)$ and $(0,0,0)$. Hence

$$f(x,y,z) = (\overline{x} + \overline{y} + \overline{z})(x + \overline{y} + \overline{z})(x + y + z)$$

is another Boolean expression for f.   ∎

The second Boolean expression

$$f(x,y,z) = (\overline{x} + \overline{y} + \overline{z})(x + \overline{y} + \overline{z})(x + y + z)$$

for the function f in Example 11.74 is called the **product-of-sums expansion** or **conjunctive normal form** of the Boolean function f.

Combinatorial Circuits

Boolean algebra is named for George Boole, a British mathematician and philosopher of the 19th century. The Boolean algebra B forms the basis of modern computer arithmetic and Boole is regarded as one of the founders of computer science. Although Boole never regarded logic as a branch of mathematics, he described a close connection between the symbols of algebra and those that appear in logic. He considered formal logic as mathematics restricted to two symbols, namely 0 and 1. Boole's work was extended and refined by other mathematicians, including Augustus De Morgan. Except for mathematicians who worked with logic, Boole's work was rather obscure for a long time as his work was thought to have no practical use. In 1930s, some 70 years after Boole's death, Claude Shannon was introduced to Boole's work in a philosophy class he took at the University of Michigan. In his master's thesis at MIT, Shannon showed how Boolean algebra could be used to optimize the design of systems of electromechanical relays that were used in telephone routing switches. He proved that circuits with relays could solve Boolean algebra problems.

The Boolean algebra B is used to model the circuitry of electronic devices. Each input and each output can be thought of as an element of the set $\{0, 1\}$. Most computer processors are constructed from a few simple circuits that have been designed in a way to make it easy for a designer to connect the outputs of one circuit to the inputs of another circuit. In this manner, complex circuits can be constructed from simple components.

Definition 11.75 *A **combinatorial circuit** converts a combination of inputs, each 0 or 1, into a uniquely defined output, which is also either 0 or 1.*

Every combinatorial circuit can be designed using the rules of Boolean algebra. The basic elements of such circuits are called **gates**, each type of which implements a Boolean operation. An **OR gate** converts two inputs x and y into the output $x + y$. An OR gate is drawn as shown in Figure 11.23(a). An **AND gate** converts two inputs x and y into the output $x \cdot y = xy$. An AND gate is drawn as shown in Figure 11.23(b). An **inverter** or **NOT gate** converts an input x into the output $\bar{x}$. A NOT gate is drawn as shown in Figure 11.23(c).

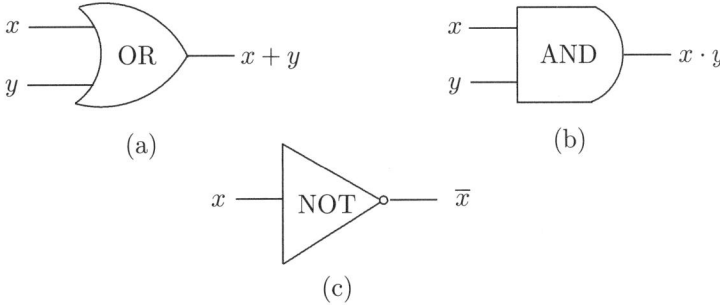

Figure 11.23: OR, AND and NOT gates

When combinations of circuits are formed, some gates may share inputs. There are two methods of showing this. One method uses **branchings**, which indicate all gates using a particular input. The other method indicates this input separately for each gate. This is illustrated in the circuit shown in Figure 11.24(a) that produces the output $xy + \bar{x}y$ for the two inputs x and y. The circuit in Figure 11.24(b) uses a branching.

Example 11.76

Find a combinatorial circuit that produces the output $xyz + xy\bar{z} + x\,\bar{y}\,\bar{z}$.

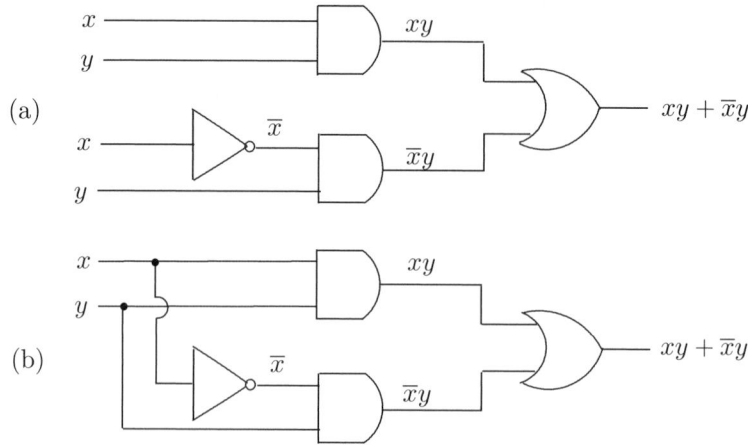

Figure 11.24: Circuits producing the output $xy + \overline{x}y$

Solution. Although there is a circuit with nine gates that produces the desired output, we can simplify the expression to obtain a simpler circuit with an equivalent output. Observe that

$$
\begin{aligned}
xyz + xy\overline{z} + x\,\overline{y}\,\overline{z} &= xy(z + \overline{z}) + x\,\overline{y}\,\overline{z} \\
&= xy \cdot 1 + x\,\overline{y}\,\overline{z} = xy \cdot (1 + \overline{z}) + x\,\overline{y}\,\overline{z} \\
&= (xy + xy\overline{z}) + x\,\overline{y}\,\overline{z} \\
&= xy + x(y + \overline{y})\overline{z} = xy + x \cdot 1 \cdot \overline{z} \\
&= xy + x\overline{z} = x(y + \overline{z}).
\end{aligned}
$$

The circuit with three gates shown in Figure 11.25 produces an output that is equivalent to $xyz + xy\overline{z} + x\,\overline{y}\,\overline{z}$. ◆

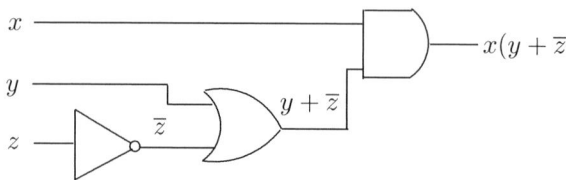

Figure 11.25: A circuit producing the output $x(y + \overline{z})$ in Example 11.76

Example 11.77

Find a combinatorial circuit that produces the output $x \cdot (\overline{y} + z) + y$.

Solution. The circuit in Figure 11.26 has the desired output. ◆

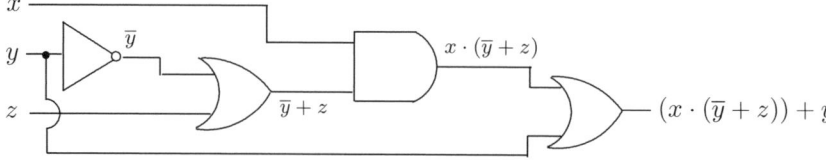

Figure 11.26: A circuit producing the output $x \cdot (\overline{y} + z) + y$ in Example 11.77

Exercises for Section 11.3

1. It was proved in Theorem 11.63 that for every two elements a and b in a Boolean algebra, $\overline{a + b} = \overline{a} \cdot \overline{b}$. Without using the duality principle, prove that $\overline{a \cdot b} = \overline{a} + \overline{b}$ for all $a, b \in S$.

2. What property is dual to each of the following properties?

   (a) $x + 1 = 1$.   (b) $x + \overline{x} \cdot y = x + y$.

3. In the Boolean algebra B, compute the following:

   (a) $\overline{1 \cdot 0} + (1 + 0)$. (b) $(\overline{1} + 0) \cdot (1 + \overline{0})$. (c) $\overline{1 \cdot 0 \cdot 1} + \overline{0} \cdot (1 + 1)$.

4. Let x and y be Boolean variables. Prove that $x \cdot (\overline{x} + y) = x \cdot y$ (a) by using a truth table and (b) algebraically (using laws of Boolean algebras).

5. Simplify each of the following Boolean expressions.

   (a) $(w + x) \cdot \overline{(y + z)} + (w + x) \cdot (y + z)$.   (b) $x \cdot y + y \cdot (\overline{x} + x \cdot z)$.

   (c) $\overline{(w + x)} \cdot \overline{(y + z)} + \overline{(w + x)}$.

6. In (a)-(e), find the values of the Boolean expressions for $x = 1$, $y = 0$, $z = 1$ and $w = 1$.

   (a) $\overline{x + z}$.   (b) $x \cdot (y \cdot \overline{z})$.   (c) $(\overline{x} + y) \cdot (x + \overline{z})$.

   (d) $(x \cdot (\overline{y} + \overline{w})) + (\overline{y} \cdot (\overline{z} + w))$.   (e) $((\overline{x \cdot y}) + z) \cdot \overline{(\overline{y} \cdot (z + w))}$.

7. Construct a table that displays the values of each of the following Boolean functions.

   (a) $f(x, y) = (x + \overline{y}) \cdot y + \overline{x}$. (b) $f(x, y, z) = (x + \overline{z}) \cdot (\overline{y} + z) + \overline{x}y$.

   (c) $f(x, y, z) = x\overline{y} + y\overline{z} + \overline{x}z$.

8. The values of three Boolean functions f_1, f_2 and f_3 of degree 3 are given in the table below. Find Boolean expressions that represent these three functions.

x	y	z	f_1	f_2	f_3
1	1	1	1	0	1
1	1	0	0	1	0
1	0	1	0	1	0
1	0	0	1	0	0
0	1	1	1	0	1
0	1	0	0	1	0
0	0	1	0	0	1
0	0	0	0	0	1

9. Find the sum-of-products expansion of the Boolean function f of degree 3 defined by $f(x, y, z) = x(y + \overline{z}) + \overline{x}$.

10. The values of two Boolean functions f and g of degree 3 are given in the table below. Express each function in their product-of-sums expansions.

x	y	z	f	g
1	1	1	0	1
1	1	0	1	0
1	0	1	1	1
1	0	0	0	0
0	1	1	0	1
0	1	0	0	0
0	0	1	1	0
0	0	0	0	1

11. The Boolean function $f(x, y, z) = (x + y + \overline{z})(\overline{x} + y + \overline{z})$ is expressed in the product-of-sums expansion. Express f in the sum-of-products expansion.

12. The Boolean function $f(x, y, z) = xy\overline{z} + \overline{x}yz + \overline{x}y\overline{z}$ is expressed in the sum-of-products expansion. Express f in the product-of-sums expansion.

13. Write the Boolean expression that represents the combinatorial circuit in Figure 11.27 and write the output of each gate symbolically.

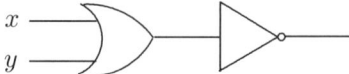

Figure 11.27: The combinatorial circuit in Exercise 13

14. Write the Boolean expression that represents the combinatorial circuit in Figure 11.28 and write the output of each gate symbolically.

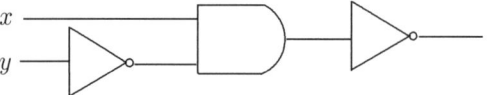

Figure 11.28: The combinatorial circuit in Exercise 14

15. Write the Boolean expression that represents the combinatorial circuit in Figure 11.29 and write the output of each gate symbolically.

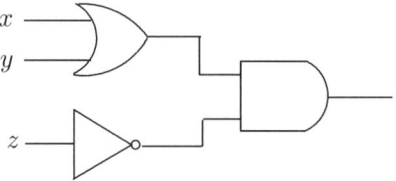

Figure 11.29: The combinatorial circuit in Exercise 15

16. The **exclusive-OR**, denoted by XOR, of two Boolean variables x and y is denoted by $x \oplus y$ and defined by $x \oplus y = x\overline{y} + \overline{x}y$.

    (a) Draw a combinatorial circuit that has the output $x \oplus y$.
    (b) Show that $x \oplus y = (x + y)\overline{xy}$.
    (c) Use (b) to draw a combinatorial circuit with fewer gates than that in (a).

Chapter 11 Highlights

Key Concepts

algebraic structure: a nonempty set on which is defined one or more binary operations.

AND gate: a gate in a combinatorial circuit that converts two inputs x and y into the output xy.

antisymmetric: a relation R on a set S is antisymmetric if whenever $a \, R \, b$ and $b \, R \, a$, then $a = b$ for all $a, b \in S$.

binary operation (on a lattice L): an operation that associates with each element of the Cartesian product $L \times L$ a unique element of a lattice L.

Boolean addition ($+$): a Boolean operation in the Boolean algebra $B = \{0, 1\}$ defined by $0 + 0 = 0$ and $0 + 1 = 1 + 0 = 1 + 1 = 1$.

Boolean algebra: a bounded lattice that is both distributive and complemented or a nonempty set S on which are defined two binary operations called addition $+$ and multiplication $\cdot$ satisfying commutative laws, associative laws and distributive laws. Also, S contains zero and unity and every element in S has a complement.

Boolean expression: an expression of a Boolean function constructed from Boolean variables and Boolean operations.

Boolean function of degree n: a function $f : B^n \rightarrow B$, where $B = \{0, 1\}$.

Boolean multiplication ($\cdot$): a Boolean operation in the Boolean algebra $B = \{0, 1\}$ defined by $0 \cdot 0 = 0 \cdot 1 = 1 \cdot 0 = 0$ and $1 \cdot 1 = 1$.

Boolean variable: a variable that takes on values from the set $\{0, 1\}$.

bounded lattice: a lattice that has both a greatest element and a least element.

branching: a method of combining gates to share an input.

chain: a totally ordered subset of a poset.

combinatorial circuit: a circuit that converts a combination of inputs (0 or 1) into a unique output (0 or 1).

comparable: elements a and b are comparable in a poset $(S, \preceq)$ if $a \preceq b$ or $b \preceq a$.

compatible: a total order $\preceq$ on a poset (S, R) is compatible with R if $a \preceq b$ whenever $a \, R \, b$ for all $a, b \in S$.

complement $\bar{a}$ (of an element a): an element satisfying $a \vee \bar{a} = 1$ and $a \wedge \bar{a} = 0$.

complemented lattice: every element in the lattice has a complement.

conjunctive normal form: a Boolean expression that is a product of sums of each variable or the complement of a variable.

disjunctive normal form: a Boolean expression that is a sum of products of each variable or the complement of a variable.

distributive lattice: a lattice L for which $a \wedge (b \vee c) = (a \wedge b) \vee (a \wedge c)$ and $a \vee (b \wedge c) = (a \vee b) \wedge (a \vee c)$ for all $a, b, c \in L$.

duality principle: for each statement concerning a Boolean algebra, there is a dual statement obtained by interchanging $+$ and $\cdot$ and interchanging 0 and 1.

gate: a basic element of a combinatorial circuit.

greatest element: an element in a poset that is greater than or equal to all elements of the poset.

greatest lower bound (of a set A), glb(A): a lower bound of A that is greater than all other lower bounds of A.

Hasse diagram (of a poset $(S, \preceq)$): a graphical representation of the poset that consists of a collection of vertices in the plane, one vertex for each element of S, such that (1) if $a \prec b$, then the vertex b is placed higher than a in the diagram and (2) if $a \prec b$ and there is no element $c \in S$ such that $a \prec c \prec b$, then a line is drawn joining a and b.

incomparable: elements in a poset that are not comparable.

inverter (or NOT gate): a gate in a combinatorial circuit that converts an input x into the output $\bar{x}$.

join (of a and b), $a \vee b$: least upper bound of $\{a, b\}$.

lattice: a poset in which every two elements have both a least upper bound and a greatest lower bound or a nonempty set on which are defined two binary operations $\vee$ and $\wedge$ satisfying commutative laws, associative laws and absorption laws.

least element: an element in a poset that is less than or equal to all elements of the poset.

least upper bound (of a set A), $\text{lub}(A)$: an upper bound of A that is less than all other upper bounds of A.

lexicographic order: a partial order $\preceq$ on the Cartesian product $A \times B$ of two posets $(A, \preceq_A)$ and $(B, \preceq_B)$ defined by $(a, b) \preceq (c, d)$ if (1) $a \prec_A c$ or (2) $a = c$ and $b \preceq_B d$.

linear order (or **total order**): a partial order for which every two elements are comparable.

linearly ordered poset (or **totally ordered poset**): a poset with a linear order.

lower bound (of a set): an element in a poset that is less than or equal to all other elements of the set.

maximal element: an element in a poset that is not less than any other element of the poset.

meet (of a and b), $a \wedge b$: greatest lower bound of $\{a, b\}$.

minimal element: an element in a poset that is not greater than any other element of the poset.

NOT gate (or inverter): a gate in a combinatorial circuit that converts an input x into the output $\bar{x}$.

OR gate: a gate in a combinatorial circuit that converts two inputs x and y into the output $x + y$.

partial order: a reflexive, antisymmetric and transitive relation.

partially ordered set (or **poset**): a nonempty set together with a partial order on it.

poset (or **partially ordered set**): a nonempty set together with a partial order on it.

product-of-sums expansion: a Boolean expression that is a product of sums of each variable or the complement of a variable.

sum-of-products expansion: a Boolean expression that is a sum of products of each variable or the complement of a variable.

topological sorting: a total order that is compatible with a given partial order.

total order (or **linear order**): a partial order for which every two elements are comparable.

totally ordered poset (or **linearly ordered poset**): a poset with a total order.

unity (of a bounded lattice L): the greatest element 1 in L.

upper bound (of a set): an element in a poset that is greater than or equal to all other elements of the set.

zero (of a bounded lattice L): the least element 0 in L.

Key Results

- Every finite nonempty poset has a maximal element and a minimal element.

- Every nonempty poset contains at most one greatest element and at most one least element.

- In a lattice $(L, \preceq)$, the following hold.

Commutative Laws $a \vee b = b \vee a$ and $a \wedge b = b \wedge a$ for all $a, b \in L$.

Associative Laws $(a \vee b) \vee c = a \vee (b \vee c)$ and $(a \wedge b) \wedge c = a \wedge (b \wedge c)$ for all $a, b, c \in L$.

Absorption Laws $a \wedge (a \vee b) = a$ and $a \vee (a \wedge b) = a$ for all $a, b \in L$.

Idempotent Laws $a \vee a = a$ and $a \wedge a = a$ for all $a \in L$.

- A lattice $(L, \preceq)$ satisfies one distributive law if and only if $(L, \preceq)$ satisfies the other distributive law

- If an element of a bounded distributive lattice has a complement, then it is unique.

- If $(S, +, \cdot)$ is a Boolean algebra, then

  Idempotent Laws $a + a = a$ and $a \cdot a = a$ for every $a \in S$.

  Domination Laws $a + 1 = 1$ and $a \cdot 0 = 0$ for every $a \in S$.

  Absorption Laws $a \cdot (a + b) = a$ and $a + (a \cdot b) = a$ for all $a, b \in S$.

- In a Boolean algebra, every element has a unique complement.

- Let S be a Boolean algebra. For every $a \in S$, $\overline{\overline{a}} = a$.

- Every Boolean algebra S satisfies De Morgan's Laws: $\overline{a + b} = \overline{a} \cdot \overline{b}$ and $\overline{a \cdot b} = \overline{a} + \overline{b}$ for all $a, b \in S$.

- **The Duality Principle for Boolean Algebras** For each statement concerning a Boolean algebra S, there is a dual statement concerning S obtained by interchanging $+$ and $\cdot$ and interchanging 0 and 1.

- Let S be a Boolean algebra and let $a, b \in S$. Then $a \cdot \overline{b} + a \cdot b = a$.

- For elements a and b in a Boolean algebra, $a \cdot b = a$ if and only if $a \cdot \overline{b} = 0$.

- For elements a and b in a Boolean algebra, $a + b = a$ if and only if $a + \overline{b} = 1$.

Supplementary Exercises for Chapter 11

1. Let $S = \{1, 2, 3\}$. A relation R is defined on $S \times S$ by $(a, b) \; R \; (c, d)$ if $a + b \le c + d$. Determine whether R is antisymmetric.

2. Let $S = \{1, 2, 3, 4\}$. Define a relation R on $\mathcal{P}(S)$ by $A \; R \; B$ if $|A| \le |B|$. Which of the properties reflexive, antisymmetric and transitive does R possess?

3. Give an example of a set $S = \{a, b, c, d\}$ of positive integers such that a Hasse diagram of $(S, |)$ is that shown in Figure 11.30.

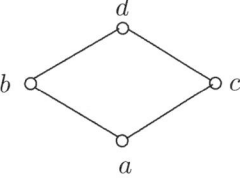

Figure 11.30: A Hasse diagram for the poset in Exercise 3

4. Give an example of a set S and a subset A of $\mathcal{P}(S)$ such that a Hasse diagram of $(A, \subseteq)$ is shown in Figure 11.31.

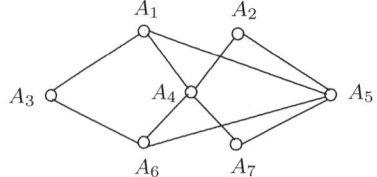

Figure 11.31: A Hasse diagram for the poset in Exercise 4

5. Let $S = \{1, 2, 3\}$. A relation R is defined on $S \times S$ by $(a, b)\ R\ (c, d)$ if $a \leq c$ and $b \leq d$.

   (a) Prove that (S, R) is a poset. (b) Draw a Hasse diagram of (S, R).

   (c) Give a topological sorting of (S, R). (d) Is (S, R) a lattice?

6. Let $S = \{1, 2, 3\}$.

   (a) For the poset $(S, |\ )$, draw a Hasse diagram of the poset $(S \times S, \preceq)$, where $\preceq$ is the lexicographic order on $S \times S$.

   (b) Is $(S \times S, \preceq)$ totally ordered?   (c) Give a topological sorting of the poset $(S \times S, \preceq)$.

   (d) Is $(S \times S, \preceq)$ a lattice?

7. Given in Figure 11.32 is a Hasse diagram of a poset $(S, \preceq)$, where $S = \{a, b, c, d, e, f, g, h, i\}$.

   (a) What are the maximal and minimal elements of $(S, \preceq)$?

   (b) For $A_1 = \{d, f\}$, determine $\mathrm{lub}(A_1)$ and $\mathrm{glb}(A_1)$.

   (c) For $A_2 = \{c, e\}$, determine $\mathrm{lub}(A_2)$ and $\mathrm{glb}(A_2)$.

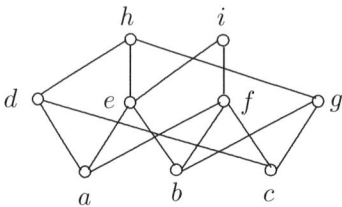

Figure 11.32: A Hasse diagram for the poset in Exercise 7

8. On a board with three rows and two columns of squares, a coin is placed on each of the two lowest squares. This is called position 1 and is denoted by P_1 (see Figure 11.33). A legal move on the board consists of moving one of the coins to the next higher square. This results in nine possible positions for the coins on the squares.

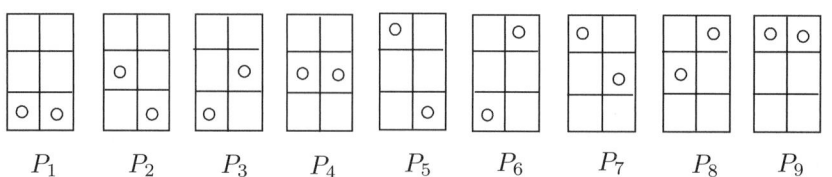

Figure 11.33: Nine possible positions in Exercise 8

Let $S = \{P_1, P_2, \ldots, P_9\}$. Define a relation $\preceq$ on S by $P_i \prec P_j$ if the coins must be in position P_i before they can be in position P_j. Then $(S, \preceq)$ is a poset.

(a) Draw a Hasse diagram of $(S, \preceq)$. (b) Is $(S, \preceq)$ a lattice?

9. A relation R on a nonempty set S is called **sequential** if for every sequence a, b, c of elements of S (distinct or not) at least one of the ordered pairs (a, b) and (b, c) belongs to R. If R is a symmetric, sequential relation on the set $S = \{x, y, z\}$, which of the following is true?

   (a) R is an equivalence relation.

   (b) It is impossible to determine whether R is an equivalence relation.

10. Let S be a finite set with $|S| = n \geq 2$ and let R be a relation on S. For each element $x \in S$, let $n_x = |\{y \in S : (x, y) \in R\}|$. Let $f : S \to \{1, 2, \ldots, n\}$ be a function defined by $f(x) = n_x$ for each $x \in S$. Prove that if R is an equivalence relation, then f is not bijective.

11. Let S be a finite set with $n \geq 2$ elements and let $\preceq$ be a partial order on S. For $x \in S$, let $S_x = \{y \in S : x \preceq y\}$. Let $f : S \to \{1, 2, \ldots, n\}$ be a function defined by $f(x) = |S_x|$ for each $x \in S$. Prove that f is bijective if and only if $(S, \preceq)$ is a totally ordered set.

12. In Figure 11.34 are shown Hasse diagrams of posets $(S, \preceq)$. Which posets are lattices?

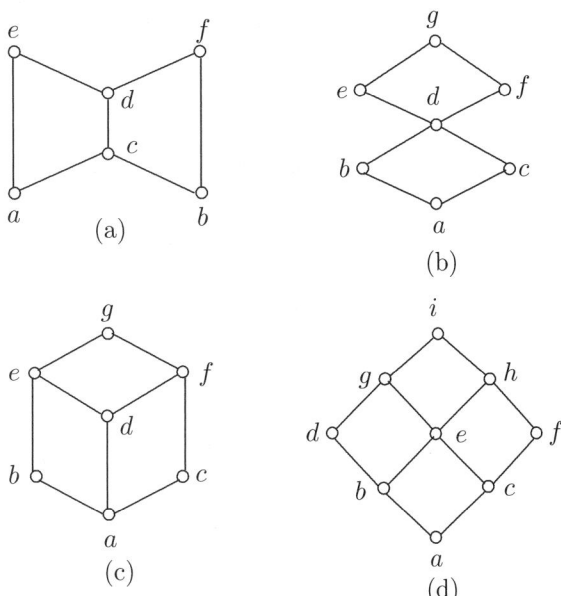

Figure 11.34: Hasse diagrams for posets in Exercise 12

13. Write the Boolean expression that represents the combinatorial circuit in Figure 11.35 and write the output of each gate symbolically.

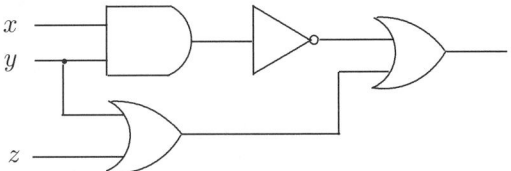

Figure 11.35: The combinatorial circuit in Exercise 13

14. Write the Boolean expression that represents the combinatorial circuit in Figure 11.36 and write the output of each gate symbolically.

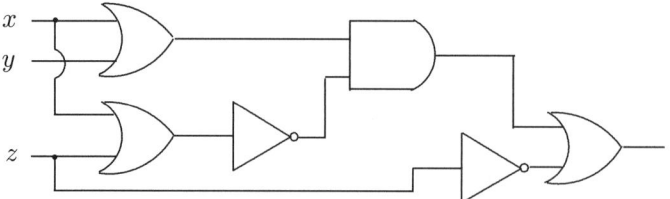

Figure 11.36: The combinatorial circuit in Exercise 14

15. Find the output of the circuit shown in Figure 11.37.

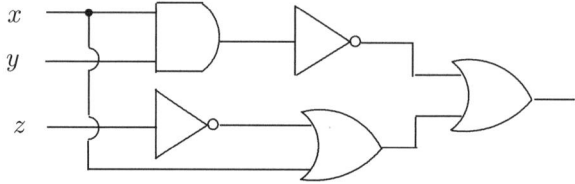

Figure 11.37: The combinatorial circuit in Exercise 15

Chapter 12

Introduction to Graphs

Figure 12.1 shows an art building, where there are paintings from an exhibit hung along each walkway in the building. Is it possible to enter the building, walk along each walkway exactly once and then exit the building?

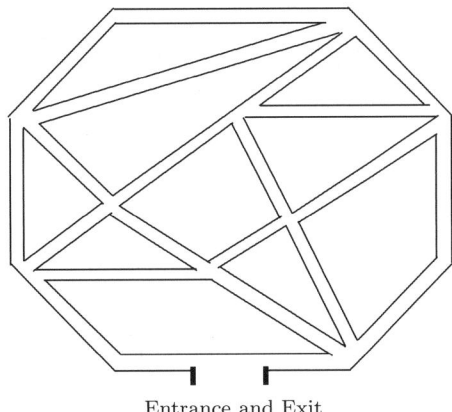

Entrance and Exit

Figure 12.1: The walkways in an art building

Three homes are under construction and each home must be provided with connections to each of three utilities, namely, water, electricity and natural gas. Each utility provider needs a direct line from the utility terminal to each house without passing through another provider's terminal or another house along the way. Furthermore, all three utility providers need to bury their lines at exactly the same depth underground without any lines crossing. Can this be done?

A total of eleven chemicals need to be shipped to a chemistry department at a university. Certain pairs of chemicals can react with each other and so should not be placed in the same container when shipped. Suppose that the chemicals are denoted by $c_1, c_2, \ldots, c_{11}$ and c_1 cannot be shipped in the same container with any of c_2, c_5, c_7 or c_{10}. The chemicals that cannot be shipped in the same container are listed below:

$$c_1 : \ c_2, c_5, c_7, c_{10} \qquad c_2 : \ c_3, c_6, c_8 \ \text{(as well as } c_1) \qquad c_3 : \ c_4, c_7, c_9$$
$$c_4 : \ c_5, c_8, c_{10} \qquad c_5 : \ c_6, c_9 \qquad c_{11} : \ c_6, c_7, c_8, c_9, c_{10}.$$

What is the smallest number of containers that can be shipped safely to the university?

These three questions and many others can be answered with the aid of a discrete structure called a *graph*, which serves as a useful mathematical model in a variety of situations. In this first of four chapters dealing with graphs, we introduce many of the fundamental concepts of graph theory and the basic notation used in this subject, as well as concepts related to paths and cycles in graphs.

12.1   Fundamental Concepts of Graph Theory

A **graph** G consists of a finite nonempty set V of objects called **vertices** (the singular is **vertex**) and a set E of 2-element subsets of V. Each element of E is called an **edge**. (Vertices are sometimes referred to as **nodes** or **points**, while edges are sometimes called **links** or **lines**.) The set V is called the **vertex set** of G, while E is the **edge set** of G. The fact that a graph G has vertex set V and edge set E is indicated by writing $G = (V, E)$. The vertex set and edge set of G are often denoted by $V(G)$ and $E(G)$, respectively. It is common and convenient to denote an edge $\{x, y\}$ of a graph by xy or yx. If $xy \in E$, then x and y are called **adjacent** vertices; otherwise, x and y are **nonadjacent** vertices. If x and y are adjacent vertices, then the edge xy is said to **join** x and y. Any vertex adjacent to a vertex x is called a **neighbor** of x and the set of neighbors of x is the **neighborhood** of x, denoted by $N(x)$. The edge xy is said to be **incident** to (or with) x and y. If xy and yz are distinct edges of G, then xy and yz are **adjacent** edges. Thus wx and yz are nonadjacent edges if and only if $\{w, x\} \cap \{y, z\} = \emptyset$.

The two sets $V = \{v_1, v_2, v_3, v_4, v_5\}$ and $E = \{v_1v_2, v_2v_3, v_2v_4, v_3v_4\}$ define a graph G. In G, no edge is incident to v_5. It is common to represent a graph by a diagram, which is also referred to as the graph itself. In the diagram, each vertex is represented by a point or a small circle and an edge is represented by drawing a line or curve joining the corresponding points. A diagram of this graph G is shown in Figure 12.2. Graphs can often be drawn in many ways. Sometimes different drawings of the same graph might suggest that we have two different graphs but such is not the case.

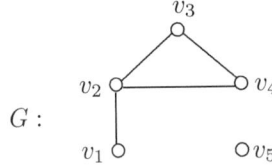

Figure 12.2: A graph

We now consider some situations that can be modeled by graphs.

Example 12.1

A representative of a company finds it convenient to travel to eight cities, denoted by $c_1, c_2, \ldots, c_8$. Her favorite airline has direct routes between various pairs of these cities. In particular, there are direct routes from c_1 to c_3, c_4, c_7 and back again, which we denote by $c_1 : c_3, c_4, c_7$. The complete set of direct routes is given below:

$$c_1 : c_3, c_4, c_7 \qquad c_2 : c_4, c_8 \qquad c_3 : c_6, c_8 \text{ (as well as } c_1) \qquad c_4 : c_5, c_7 \qquad c_5 : c_6, c_7$$

Model this situation by a graph.

Solution. We construct a graph G whose vertex set consists of these eight cities. Two vertices (cities) of G are adjacent if there is a direct flight between these two cities. This graph G is shown in Figure 12.3.   ♦

Example 12.2

Five graduating seniors (Avery, Benton, Casey, Donna and Edwin) have learned of six positions available at an appliance company, a bank, a chemical company, a defense company, an electronics company and a financial institution. The students decide to apply for these positions as listed below.

> Avery: bank, defense, financial;   Benton: appliance, electronics;
> Casey: bank, financial;   Donna: chemical, electronics;
> Edwin: everything.

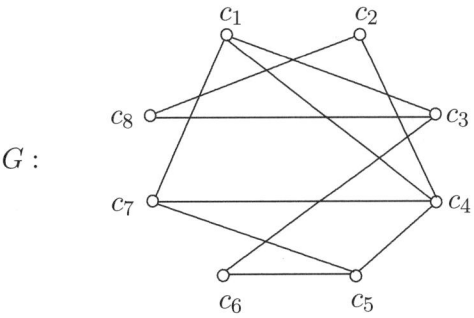

Figure 12.3: The graph in Example 12.1

Model this situation by means of a graph, where the students are A, B, C, D, E and the jobs are a, b, c, d, e, f.

Solution. We construct a graph G whose vertex set is the union of the set of students and the set of jobs. Two vertices of G are adjacent if one vertex is a student and the other is a position for which this student has applied. This graph G is shown in Figure 12.4. ♦

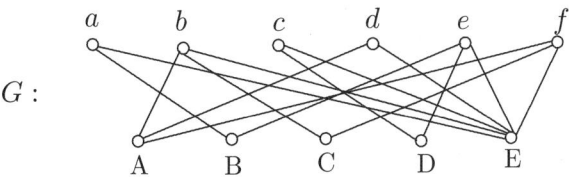

Figure 12.4: The graph in Example 12.2

Example 12.3

A student is studying divisibility of integers from the set $V = \{2, 3, 4, 6, 8, 9, 12, 18, 27\}$ and is interested in those pairs of distinct integers for which one of the integers divides the other. Model this situation by a graph.

Solution. We construct a graph G with vertex set V such that two vertices are adjacent if one divides the other. The graph G is shown in Figure 12.5. ♦

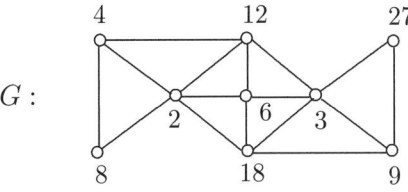

Figure 12.5: The graph in Example 12.3

The number of vertices in a graph G is referred to as the **order** of G, while the number of edges is the **size** of G. Therefore, the order of the graph G of Figure 12.2 is 5 and its size is 4. We commonly use n and m to denote the order and size, respectively, of a graph, that is, $n = |V(G)|$ and $m = |E(G)|$. If G is a graph of order n and size m, then $n \geq 1$ and $0 \leq m \leq \binom{n}{2}$. If $n = 1$, then G contains exactly one vertex and is called a **trivial graph**. A **nontrivial graph** then has at least two vertices. If G has no edges, then G is called an **empty graph**. A **nonempty graph** therefore has at least one edge.

Although graphs can be described in terms of sets (the definition) or in terms of diagrams, there are other ways in which graphs can be represented. One of these is by means of an **adjacency list**, which associates with each vertex a list of its adjacent vertices. In an adjacency list of a graph G each vertex v of G is listed in a column and every vertex of G adjacent to v is listed in a row to the right of v. This is illustrated in Figure 12.6 for the graph G of Figure 12.2.

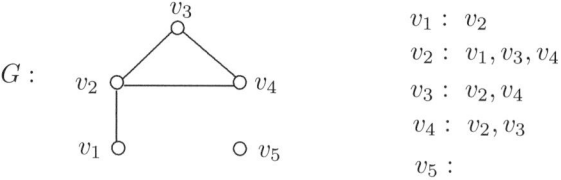

$$v_1 : \ v_2$$
$$v_2 : \ v_1, v_3, v_4$$
$$v_3 : \ v_2, v_4$$
$$v_4 : \ v_2, v_3$$
$$v_5 :$$

Figure 12.6: Adjacency lists of a graph

Graphs and Matrices

Another common way to represent a graph is by means of a matrix. Before describing this representation, we present (or perhaps review) some fundamental facts concerning matrices, beginning with the definition.

Definition 12.4 *A* **matrix** *is a rectangular array of numbers.*

A **matrix** $\mathbf{A}$ having m rows and n columns is an $m \times n$ **matrix**. The numbers m and n are referred to as the **size** of $\mathbf{A}$. If $m = n$, then $\mathbf{A}$ is a **square matrix**. Typically, the element in row i, column j of $\mathbf{A}$ is called the (i, j)-**entry** of $\mathbf{A}$ and is denoted by a_{ij}. The matrix $\mathbf{A}$ itself is then written as $\mathbf{A} = [a_{ij}]$. For example,

$$\mathbf{A} = \begin{bmatrix} 1 & 0 & 7 \\ 8 & 4 & 0 \end{bmatrix} \text{ is a } 2 \times 3 \text{ matrix.}$$

If every entry of a matrix is 0 or 1, then the matrix is called a **zero-one matrix**. For example,

$$\mathbf{A} = \begin{bmatrix} 0 & 1 & 1 \\ 1 & 0 & 0 \\ 1 & 0 & 0 \end{bmatrix} \text{ is a zero-one matrix.}$$

Two matrices $\mathbf{A}$ and $\mathbf{B}$ can be added or multiplied depending on their sizes. In order for $\mathbf{A}$ and $\mathbf{B}$ to be added, they must have the same size, that is, the same number of rows and columns.

Definition 12.5 *Let* $\mathbf{A} = [a_{ij}]$ *and* $\mathbf{B} = [b_{ij}]$ *be two* $m \times n$ *matrices. The* **sum** $\mathbf{A} + \mathbf{B}$ *of* $\mathbf{A}$ *and* $\mathbf{B}$ *is the* $m \times n$ *matrix*

$$\mathbf{A} + \mathbf{B} = [a_{ij} + b_{ij}].$$

For example,

$$\begin{bmatrix} -2 & 1 & 0 \\ 3 & -4 & 2 \end{bmatrix} + \begin{bmatrix} 5 & -4 & 3 \\ -1 & 2 & 6 \end{bmatrix} = \begin{bmatrix} (-2)+5 & 1+(-4) & 0+3 \\ 3+(-1) & (-4)+2 & 2+6 \end{bmatrix} = \begin{bmatrix} 3 & -3 & 3 \\ 2 & -2 & 8 \end{bmatrix}.$$

A **row matrix** has a single row and a **column matrix** has a single column.

Definition 12.6 *Let* $\mathbf{A}$ *be a row matrix and* $\mathbf{B}$ *a column matrix, each with n entries. Hence* $\mathbf{A}$ *is a* $1 \times n$ *matrix and* $\mathbf{B}$ *is an* $n \times 1$ *matrix. The* **inner product** *or* **dot product** $\mathbf{A} \cdot \mathbf{B}$ *of* $\mathbf{A}$ *and* $\mathbf{B}$ *is obtained by adding the products of the corresponding elements of* $\mathbf{A}$ *and* $\mathbf{B}$*. That is, if*

$$\mathbf{A} = \begin{bmatrix} a_1 & a_2 & \cdots & a_n \end{bmatrix} \text{ and } \mathbf{B} = \begin{bmatrix} b_1 \\ b_2 \\ \vdots \\ b_n \end{bmatrix},$$

then

$$\mathbf{A} \cdot \mathbf{B} = a_1 b_1 + a_2 b_2 + \cdots + a_n b_n = \sum_{i=1}^{n} a_i b_i.$$

The product $\mathbf{AB}$ of two matrices $\mathbf{A}$ and $\mathbf{B}$ is defined if the number of columns of $\mathbf{A}$ equals the number of rows of $\mathbf{B}$.

Definition 12.7 *Let $\mathbf{A}$ be an $m \times n$ matrix and $\mathbf{B}$ an $n \times p$ matrix, where $\mathbf{A} = [a_{ij}]$ with $1 \leq i \leq m$ and $1 \leq j \leq n$, and $\mathbf{B} = [b_{jk}]$, with $1 \leq j \leq n$ and $1 \leq k \leq p$. Then the* **product** $\mathbf{C} = \mathbf{AB}$ *of $\mathbf{A}$ and $\mathbf{B}$ is the $m \times p$ matrix, where $\mathbf{C} = [c_{ik}]$ with $1 \leq i \leq m$ and $1 \leq k \leq p$ such that*

$$c_{ik} = a_{i1} b_{1k} + a_{i2} b_{2k} + \cdots + a_{in} b_{nk} = \sum_{j=1}^{n} a_{ij} b_{jk}.$$

That is, to calculate the (i,k)-entry c_{ik} of $\mathbf{C}$, we compute the inner product of row i of $\mathbf{A}$ (which has n columns) and column k of $\mathbf{B}$ (which has n rows).

Example 12.8

Determine the inner product of row 1 of the matrix $\mathbf{A}$ below and column 2 of the matrix $\mathbf{B}$ and then compute the product $\mathbf{C} = \mathbf{AB}$.

$$\mathbf{A} = \begin{bmatrix} 1 & 2 & -2 \\ 0 & 4 & 3 \end{bmatrix} \text{ and } \mathbf{B} = \begin{bmatrix} 5 & 6 & 1 & -1 \\ 2 & 0 & 3 & 1 \\ 0 & 4 & 2 & 3 \end{bmatrix}$$

Solution. First, the inner product of row 1 of $\mathbf{A}$ and column 2 of $\mathbf{B}$ is

$$c_{12} = \begin{bmatrix} 1 & 2 & -2 \end{bmatrix} \cdot \begin{bmatrix} 6 \\ 0 \\ 4 \end{bmatrix} = 1 \cdot 6 + 2 \cdot 0 + (-2) \cdot 4 = -2.$$

In general,

$$\mathbf{C} = \mathbf{AB} = \begin{bmatrix} 1 & 2 & -2 \\ 0 & 4 & 3 \end{bmatrix} \begin{bmatrix} 5 & 6 & 1 & -1 \\ 2 & 0 & 3 & 1 \\ 0 & 4 & 2 & 3 \end{bmatrix} = \begin{bmatrix} 9 & -2 & 3 & -5 \\ 8 & 12 & 18 & 13 \end{bmatrix}. \qquad \blacklozenge$$

Definition 12.9 *For an $n \times n$ square matrix $\mathbf{A}$, the* **powers** *of $\mathbf{A}$ are defined by*

$$\mathbf{A}^1 = \mathbf{A} \quad \text{and} \quad \mathbf{A}^k = \underbrace{\mathbf{AA} \cdots \mathbf{A}}_{k \text{ factors}} \text{ for each integer } k \geq 2.$$

For example, if $\mathbf{A} = \begin{bmatrix} 1 & 2 \\ 1 & 0 \end{bmatrix}$, then

$$\mathbf{A}^1 = \mathbf{A} = \begin{bmatrix} 1 & 2 \\ 1 & 0 \end{bmatrix}, \mathbf{A}^2 = \mathbf{AA} = \begin{bmatrix} 3 & 2 \\ 1 & 2 \end{bmatrix} \text{ and } \mathbf{A}^3 = \mathbf{AAA} = \mathbf{AA}^2 = \begin{bmatrix} 5 & 6 \\ 3 & 2 \end{bmatrix}.$$

We can now describe a matrix representation of a graph. Suppose that G is a graph of order n, where $V(G) = \{v_1, v_2, \ldots, v_n\}$. The **adjacency matrix** of G is the $n \times n$ zero-one matrix $\mathbf{A} = [a_{ij}]$, where

$$a_{ij} = \begin{cases} 1 & \text{if } v_i v_j \in E(G) \\ 0 & \text{if } v_i v_j \notin E(G). \end{cases}$$

Figure 12.7 shows the adjacency matrix of the graph G of Figure 12.2.

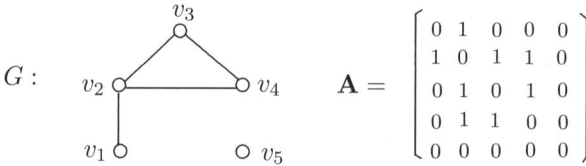

Figure 12.7: A graph and its adjacency matrix

There are several observations that can be made about the adjacency matrix $\mathbf{A}$ of a graph G of order n. First, all entries along the main diagonal of $\mathbf{A}$ are 0 since no vertex of G is adjacent to itself. Second, $\mathbf{A}$ is a symmetric matrix, that is, row i of $\mathbf{A}$ is identical to column i of $\mathbf{A}$ for every integer i with $1 \le i \le n$. Also, if we were to add the entries in row i (or in column i), then we obtain the number of vertices adjacent to v_i, which leads us to an important concept in graph theory.

The Degree of a Vertex

The number of neighbors of a vertex x in a graph G is the **degree** of x and is denoted by $\deg_G(x)$ or simply $\deg(x)$ or $\deg x$ if the graph G is clear. Thus the degree of a vertex x is also the number of edges incident with x, the number of vertices adjacent to x and $|N(x)|$. For the graph G of Figure 12.2,

$$\deg v_1 = 1, \deg v_2 = 3, \deg v_3 = \deg v_4 = 2, \deg v_5 = 0.$$

A vertex of degree 0 is called an **isolated vertex** and a vertex of degree 1 is an **end-vertex**, a **leaf** or a **pendant vertex**. In the graph G of Figure 12.2, v_5 is an isolated vertex and v_1 is an end-vertex. The **maximum degree** of a graph G is the maximum of the degrees of the vertices of G and is denoted by $\Delta(G)$, while the **minimum degree** of G is denoted by $\delta(G)$. Thus

$$0 \le \delta(G) \le \deg v \le \Delta(G) \le n - 1$$

for every vertex v in a graph G of order n. For the graph G of order 5 in Figure 12.2, $\Delta(G) = 3$ and $\delta(G) = 0$.

The sum of the degrees of the vertices of the graph G of Figure 12.2 is 8, which is twice the size of G. In fact, the sum of the degrees of the vertices of every graph is twice its size. This fact is often called *the First Theorem of Graph Theory*. This is actually an appropriate name for this observation as it was noticed by the famous Swiss mathematician Leonhard Euler, who used it in 1736 to prove the first major theorem of graph theory (which we will visit in Section 12.3) despite the fact that graphs did not appear in Euler's work and hadn't even been invented yet.

Theorem 12.10 (The First Theorem of Graph Theory) *If G is a graph of order n and size m with $V(G) = \{v_1, v_2, \ldots, v_n\}$, then*

$$\sum_{i=1}^{n} \deg v_i = 2m.$$

Proof. When the degrees of the vertices of G are summed, each edge of G is counted twice, once for each of its two incident vertices. ∎

Example 12.11 *A graph G of size 29 has three vertices of each of the degrees $3, 5$ and 6. The remaining vertices of G have degree 4. What is the order of G?*

Solution. Suppose that G has x vertices of degree 4. By Theorem 12.10, the sum of the degrees of the vertices of G is $3 \cdot 3 + 3 \cdot 5 + 3 \cdot 6 + 4x = 2 \cdot 29$ and so $x = 4$. Thus the order of G is therefore $9 + x = 13$. ◆

A vertex in a graph is **even** if its degree is even and is **odd** if its degree is odd. The graph G of Figure 12.2 has three even vertices, namely v_3, v_4 and v_5 and the two odd vertices v_1 and v_2. Every graph has an even number of odd vertices, another consequence of an observation by Euler in 1736.

Corollary 12.12 *Every graph contains an even number of odd vertices.*

Proof. Let G be a graph of size m. Since the sum of the degrees of the vertices of G is even, namely $2m$, and the sum of the degrees of the even vertices of G is even, it follows that the sum of the degrees of the odd vertices of G is even. Therefore, G has an even number of odd vertices. ∎

A graph G is **regular** if every vertex of G has the same degree and is r-**regular** if this degree is r. A 3-regular graph is also referred to as a **cubic graph**. Figure 12.8 shows r-regular graphs of order 5 for $r = 0, 2, 4$. By Corollary 12.12, there is no 1-regular graph or 3-regular graph of order 5.

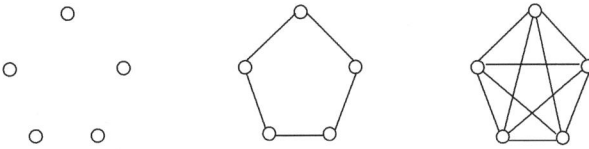

Figure 12.8: Regular graphs of order 5

A graph H is a **subgraph** of a graph G if $V(H) \subseteq V(G)$ and $E(H) \subseteq E(G)$. If H is a subgraph of a graph G and $H \neq G$, then H is a **proper subgraph** of G. A subgraph H of a graph G is an **induced subgraph** of G if whenever two vertices of H are adjacent in G, they are also adjacent in H. A subgraph H of a graph G is a **spanning subgraph** of G if $V(H) = V(G)$. Figure 12.9 shows a graph G and three subgraphs G_1, G_2, and G_3 of G. While G_3 is an induced subgraph of G, neither G_1 nor G_2 are induced subgraphs of G. The subgraph G_1 is a spanning subgraph of G but neither G_2 nor G_3 are spanning subgraphs of G.

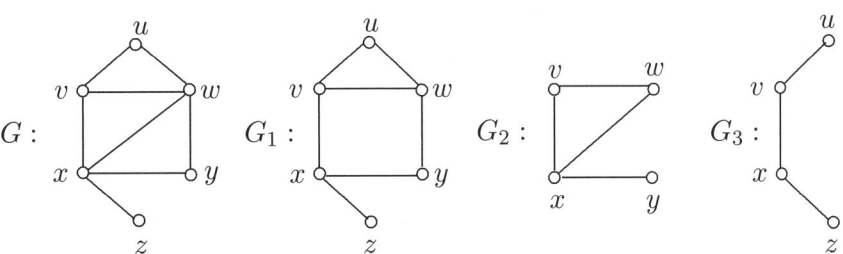

Figure 12.9: A graph and three subgraphs

A graph is **complete** if every two distinct vertices are adjacent. A complete graph of order n is denoted by K_n. The graph K_n is $(n-1)$-regular and its size is $\binom{n}{2}$. The five smallest complete graphs are shown in Figure 12.10.

$$K_1: \quad K_2: \quad K_3: \quad K_4: \quad K_5:$$

Figure 12.10: The five smallest complete graphs

The **complement** of a graph G is that graph $\overline{G}$ with $V(\overline{G}) = V(G)$ such that two distinct vertices u and v of G are adjacent in $\overline{G}$ if and only if u and v are not adjacent in G. Thus if G is a graph of order n and size m, then the order of $\overline{G}$ is n and the size of $\overline{G}$ is $\binom{n}{2} - m$. The empty graph of order n is therefore $\overline{K}_n$. A graph G and its complement $\overline{G}$ are shown in Figure 12.11.

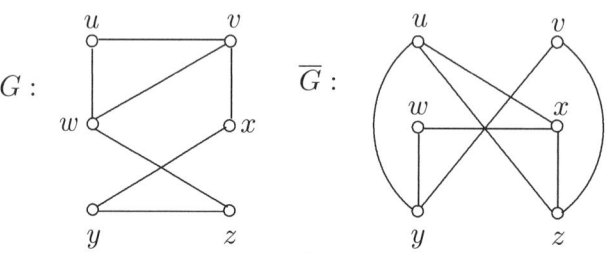

Figure 12.11: A graph and its complement

A graph G of order n is a **path** if the vertices of G can be labeled so that $V(G) = \{v_1, v_2, \ldots, v_n\}$ and $E(G) = \{v_i v_{i+1} : 1 \leq i < n\}$. A graph G of order $n \geq 3$ is a **cycle** if the vertices of G can be labeled so that $V(G) = \{v_1, v_2, \ldots, v_n\}$ and $E(G) = \{v_i v_{i+1} : 1 \leq i < n\} \cup \{v_1 v_n\}$. A path of of order n is denoted by P_n and a cycle of order n by C_n. The cycle C_3 is also called a **triangle**. The **length** of a path or cycle is the number of edges it contains. Hence the length of P_n is $n - 1$ and the length of C_n is n. Figure 12.12 shows several small paths and cycles. The three regular graphs of order 5 in Figure 12.8 are then $\overline{K}_5$, C_5 and K_5.

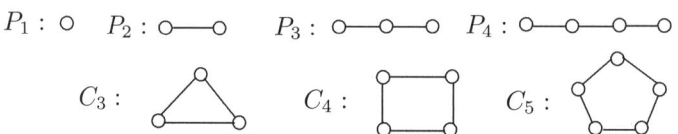

$$C_3: \quad C_4: \quad C_5:$$

Figure 12.12: Paths and cycles

Isomorphic Graphs

There is only one graph of order 1, two of order 2, four of order 3 and eleven of order 4. These are all shown in Figure 12.13. By this is meant that there are eleven *different* graphs of order 4. This brings up the question as to what it means for two graphs to be different or, for that matter, what it means for two graphs to be the same. Two graphs G and H are considered to be the same graph if they have *the same structure*. The technical term for this is *isomorphic*. In particular, whether two graphs are the same or different does not depend on how (or whether) the graphs are labeled or how they may be drawn. Informally, two graphs G and H are the same (isomorphic) if either can be drawn so that it looks like the other. Formally, we have the following.

Definition 12.13 *A graph G is **isomorphic** to a graph H (written $G \cong H$) if there exists a bijective function $\phi : V(G) \to V(H)$ such that two vertices u and v are adjacent in G if and only if*

$\phi(u)$ and $\phi(v)$ are adjacent in H. The graphs G and H are then called **isomorphic graphs** and the function ϕ an **isomorphism**.

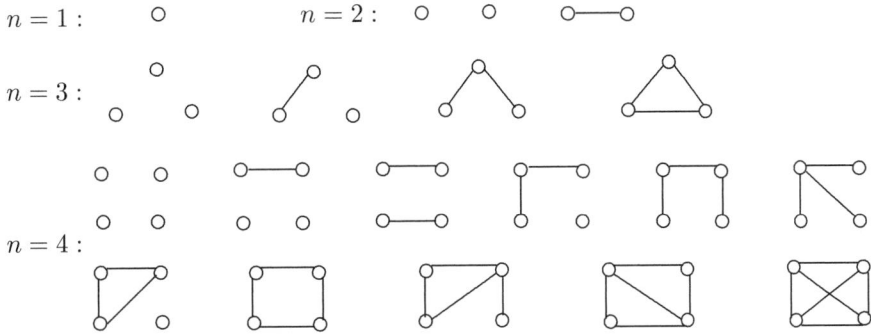

Figure 12.13: The graphs of order n for $1 \leq n \leq 4$

Example 12.14

The graphs G and H of Figure 12.14 are isomorphic and the function $\phi : V(G) \to V(H)$ defined by

$$\phi(u_1) = y_2, \ \phi(v_1) = x_2, \ \phi(w_1) = z_2, \ \phi(x_1) = u_2, \ \phi(y_1) = v_2, \ \phi(z_1) = w_2$$

is an isomorphism.   ◆

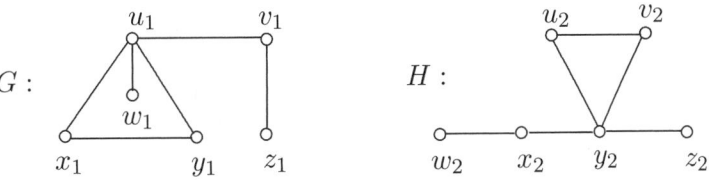

Figure 12.14: Isomorphic graphs

If G and H are isomorphic graphs, then there exists an isomorphism $\phi : V(G) \to V(H)$. Since ϕ is one-to-one and onto, $|V(G)| = |V(H)|$, that is, G and H have the same order. Furthermore, since uv is an edge of G if and only if $\phi(u)\phi(v)$ an edge of H, it follows that G and H have the same size. These two facts as well as one other are summarized next.

Theorem 12.15 *If G and H are isomorphic graphs, then*

(a) *G and H have the same order,*

(b) *G and H have the same size and*

(c) *the degrees of the vertices of G are the same as the degrees of the vertices of H.*

Proof. The truth of (a) and (b) have already been noted. To verify (c), let $\phi : V(G) \to V(H)$ be an isomorphism and let v be a vertex of G such that $\deg_G v = k$, where $N(v) = \{v_1, v_2, \ldots, v_k\}$. Let $\{u_1, u_2, \ldots, u_\ell\}$ be the set of vertices of G distinct from v that are not adjacent to v. Then $\phi(v)$ is adjacent to all of $\phi(v_1), \phi(v_2), \ldots, \phi(v_k)$ and to none of $\phi(u_1), \phi(u_2), \ldots, \phi(u_\ell)$. Thus $\deg_H \phi(v) = k$. ∎

The proof of Theorem 12.15(c) also shows that if G and H are isomorphic graphs and $\phi : V(G) \to V(H)$ is an isomorphism, then $\deg_G v = \deg_H \phi(v)$ for every vertex v of G.

Theorem 12.15 therefore states that every two isomorphic graphs must have the same order and same size and that the degrees of the vertices of one graph are the same as the degrees of the vertices of the other graph. Consequently, any two graphs having different orders, different sizes or whose vertices have different degrees are not isomorphic. The converse of the statement above is not true however; that is, if two graphs have the same order, the same size and whose vertices have the same degrees, then the graphs need not be isomorphic.

Example 12.16

The two graphs G and H in Figure 12.15 have order 6 and size 9 and the degree of every vertex of each graph is 3; yet these two graphs are not isomorphic. Let's see why. Assume, to the contrary, that G and H are isomorphic. Then there is an isomorphism $\phi : V(G) \to V(H)$. Since the three vertices u_1, v_1, w_1 are mutually adjacent in G, the vertices $\phi(u_1), \phi(v_1), \phi(w_1)$ are mutually adjacent in H. However, H does not contain three mutually adjacent vertices, which contradicts the assumption that ϕ is an isomorphism. Thus G and H are not isomorphic. In other words, the graph G of Figure 12.15 contains a triangle, while the graph H of Figure 12.15 does not.   ♦

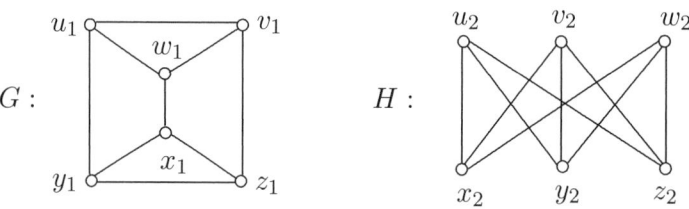

Figure 12.15: Two non-isomorphic graphs

A graph G can be used to represent a network of highways connecting cities in a certain region, where the vertices of G are the cities and an edge joins two vertices if there is a highway joining the two cities that does not pass through an intermediate city. If two cities are joined by more than one highway, then a graph does not accurately capture the structure of this network. A structure consisting of a nonempty set of vertices in which two distinct vertices are permitted to be joined by more than one edge is a **multigraph**. If two vertices are joined by two or more edges, then these edges are called **parallel edges** (or **multiple edges**). An edge that joins a vertex to itself is a **loop**. A structure that allows both parallel edges and loops (including multiple loops) is often called a **pseudograph**. Two multigraphs and a pseudograph are shown in Figure 12.16. (Some authors permit two vertices to be joined by more than one edge in a graph and use **simple graph** to refer to graphs in which two vertices are joined by at most one edge. In our case, however, a graph does not allow two vertices to be joined by two or more edges and does not allow loops.)

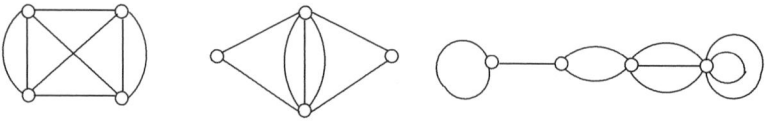

Figure 12.16: Two multigraphs and a pseudograph

Exercises for Section 12.1

1. In Example 12.1, suppose that the company representative wanted to fly from city c_2 to the city c_6. Can this be done?

2. On the first day of a semester, five students (Al, Bob, Charles, Dean and Evan) arrive early for a class. Al, Dean and Evan know one another; while Al and Bob know each other, as do

Charles and Dean. Model this situation by two graphs whose vertices are these five students, where in the first graph an edge is determined by two students who are acquainted and in the second graph an edge is determined by two students who are not acquainted.

3. Let $V = \{1, 2, 3, 4, 5\}$. Draw a graph G with vertex set V, where two vertices a and b are adjacent if $a + b$ is a prime.

4. Let $V = \{0, 1, 2, 3, 4, 5\}$. Draw a graph G with vertex set V, where two vertices a and b are adjacent if $a + b \in V$.

5. Let $S = \{2, 3, 9, 35, 55, 77\}$. The vertex set of a graph G is S and two vertices i and j are adjacent if i and j are relatively prime (the greatest common divisor of i and j is 1). Draw the graph G.

6. (a) Let $S = \{1, 2, 3, 4, 5\}$. The vertex set of a graph G is the set of 2-element subsets of S. Two vertices $\{w, x\}$ and $\{y, z\}$ are adjacent if $\{w, x\} \cap \{y, z\} = \emptyset$. Draw the graph G.

   (b) Use the idea expressed in (a) to define and draw another graph G'.

7. The vertex set of a graph G is a finite set S of positive integers, where two vertices a and b of G are adjacent if $\gcd(a, b)$ is a prime. For $S = \{4, 6, 9, 10, 12, 15\}$, draw the graph G.

8. Draw the graph G with $V(G) = \{1, 2, \ldots, 9\}$, where $ab \in E(G)$ if $a + b = \binom{n}{2}$ for some $n \geq 2$.

9. Draw the graph G whose vertex set is the power set of the set $S = \{a, b, c\}$, where two vertices A and B are adjacent in G provided either $A \subset B$ or $B \subset A$ and $||A| - |B|| = 1$.

10. What is the largest number of edges that a graph G of order 6 can have such that no two vertices of G have a common neighbor?

11. Determine the adjacency list and adjacency matrix for the graph in Figure 12.17.

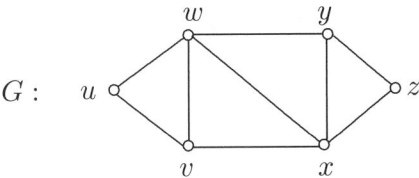

Figure 12.17: The graph in Exercise 11

12. Draw the graph G whose adjacency matrix is $\mathbf{A} = \begin{bmatrix} 0 & 1 & 1 & 0 & 0 \\ 1 & 0 & 1 & 1 & 0 \\ 1 & 1 & 0 & 0 & 1 \\ 0 & 1 & 0 & 0 & 1 \\ 0 & 0 & 1 & 1 & 0 \end{bmatrix}$.

13. Prove or disprove: There exists a graph whose adjacency matrix has an odd number of 1s.

14. Give an example of a graph G that contains some isolated vertices, some end-vertices and some vertices that are neither isolated vertices nor end-vertices such that every vertex that is not an end-vertex is adjacent to an end-vertex.

15. A certain graph G of order n and size $m = 17$ has one vertex of degree 1, two vertices of degree 2, one vertex of degree 3 and two vertices of degree 5. The remaining vertices of G have degree 4. What is n?

16. The degree of each vertex in a certain graph G of order 25 and size 62 is either 3, 4, 5 or 6. There are two vertices of degree 4 and 11 vertices of degree 6. How many vertices of G have degree 3?

17. Let G be a graph of order 21 and size 50. There are three vertices of degree 2, six vertices of degree 5 and four vertices of degree 8. The remaining vertices of G have the same degree r. What is r?

18. Let G be a graph of order 25 and size 63. There are five vertices of degree 4, eight vertices of degree 5 and six vertices of degree 7. Two-thirds of the remaining vertices of G have degree r, while the remaining one-third have degree $2r$. What is r?

19. Give an example of a graph G such that for some positive integer k, the graph G contains i vertices of degree i for every integer i with $1 \leq i \leq k$. What is the order and size of your graph G?

20. Give an example of three graphs G_1, G_2 and G_3 of order 7 such that every vertex of each graph has degree 2, 3 or 4, where there is at least one vertex of each of these degrees and

    (a) there are at least three vertices of degree 2 in G_1.

    (b) there are at least four vertices of degree 3 in G_2.

    (c) there are at least three vertices of degree 4 in G_3.

    Label the vertices of G_1, G_2 and G_3 in (a), (b) and (c) with their degrees.

21. Let G be a graph of order $6k + 1$ ($k \geq 1$) such that every vertex of G has degree $2k$, $2k + 1$ or $2k + 2$ and there is at least one vertex of each of these degrees. Prove that G contains at least $2k + 1$ vertices of degree $2k$, at least $2k + 2$ vertices of degree $2k + 1$ or at least $2k + 1$ vertices of degree $2k + 2$.

22. Eight students in a class are asked how many of the other seven students they know. One student knows all other students but only two students know the same number of students. Show that these two students know each other.

23. A graph G is called **self-complementary** if $G \cong \overline{G}$.

    (a) Give an example of a self-complementary graph of order 4

    (b) Give an example of a self-complementary graph of order 5.

    (c) Show that there is no self-complementary graph of order 6.

24. The size of an r-regular graph of order 10 is 30. What is r?

25. Prove or disprove: If G is a graph of order n such that $\Delta(G) = 5$, then there exists an r-regular graph H of order $2n + 1$ containing G as an induced subgraph, where $r = \Delta(G)$.

26. A nontrivial graph G is **irregular** if every two vertices of G have distinct degrees. Show that no nontrivial graph is irregular.

27. Let G be a graph with $V(G) = \{u, v, w, x, y\}$ and $E(G) = \{uv, ux, uy, vw, vx, xy\}$. Also, let $V(H) = \{u, v, x, y\}$ and $E(H) = \{ux, vx, vy\}$. Is H a subgraph of G?

28. Consider the graph G of Figure 12.18.

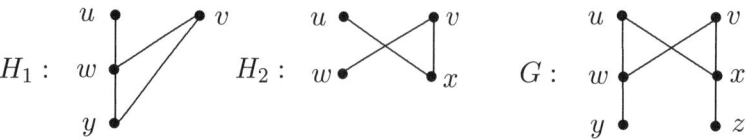

Figure 12.18: The graphs H_1, H_2 and G in Exercise 28

   (a) Determine whether H_1 and H_2 are induced subgraphs of G.

   (b) Find a subset S of $V(G)$ of maximum cardinality such that the induced subgraph $G[S]$ is empty.

   (c) For the graph G of Figure 12.18, draw the induced subgraphs $G[S_1]$, $G[S_2]$, $G[S_3]$ and $G[S_4]$, where $S_1 = \{u, x\}$, $S_2 = \{u, v, w, x\}$, $S_3 = \{w\}$ and $S_4 = V(G)$.

29. Prove or disprove: If G is a 4-regular graph of order 9, then $\overline{G} \cong G$.

30. Prove that a graph G is regular if and only if $\overline{G}$ is regular.

31. Prove that if G and $\overline{G}$ are both r-regular for some nonnegative integer r, then G has odd order.

32. Give an example of three different (non-isomorphic) graphs of order 6 and size 6.

33. Give an example of three graphs of the same order and same size, the degrees of whose vertices are the same, such that no two of these graphs are isomorphic.

34. For the graphs H_1, H_2, H_3, H_4, H_5 shown in Figure 12.19, determine whether H_i and H_{i+1} are isomorphic for $i = 1, 2, 3, 4$. If $H_i \cong H_{i+1}$, give an isomorphism; if $H_i \not\cong H_{i+1}$, then explain why.

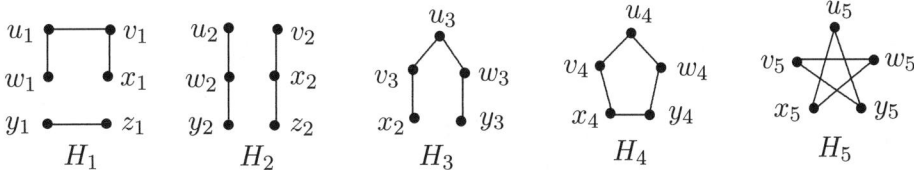

Figure 12.19: The graphs H_1, H_2, H_3, H_4, H_5 in Exercise 34

35. For each of the pairs G_1, G_2 and H_1, H_2 of graphs of Figure 12.20, show that the graphs are isomorphic and find an isomorphism in each case.

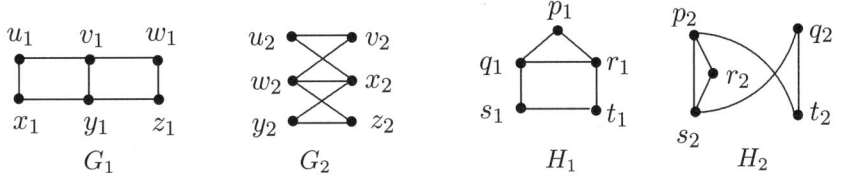

Figure 12.20: The graphs G_1, G_2, H_1, H_2 in Exercise 35

36. Suppose that G and H are isomorphic graphs. Answer each of the following questions with an explanation in each case.

   (a) If G contains two adjacent vertices of degree 4, must H contain two adjacent vertices of degree 4?

   (b) If G contains two triangles that have an edge in common, and H contains two triangles T_1 and T_2, must T_1 and T_2 have an edge in common?

   (c) If G is isomorphic to a graph F, must H be isomorphic to F?

37. Let G_1, G_2 and G_3 be three graphs of order n and size m having adjacency matrices $\mathbf{A}_1$, $\mathbf{A}_2$ and $\mathbf{A}_3$, respectively.

   (a) Prove or disprove: If $\mathbf{A}_1 = \mathbf{A}_2$, then $G_1 \cong G_2$.

(b) Prove or disprove: If $\mathbf{A}_2 \neq \mathbf{A}_3$, then $G_2 \ncong G_3$.

38. Let $S = \{G_1, G_2, \ldots, G_6\}$, where the graphs G_i $(1 \leq i \leq 6)$ are shown in Figure 12.21. Define a graph G with $V(G) = S$ and G_i and G_j $(i \neq j)$ are adjacent if one of these two graphs is isomorphic to an induced subgraph of the other. Draw the graph G.

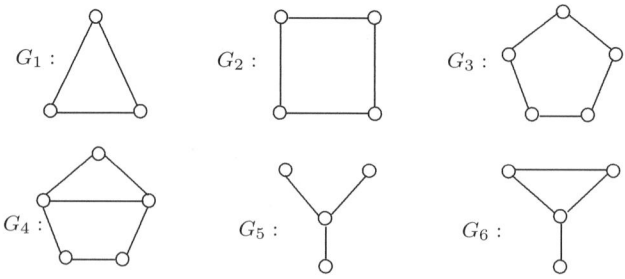

Figure 12.21: Graphs G_i $(1 \leq i \leq 6)$ in Exercise 38

39. Let G be a graph whose vertex set is a collection S of distinct points in the xy-plane such that P and Q are adjacent if P and Q lie on a line with positive slope. The graph G is called the **positive slope graph** of S. A graph G is **a positive slope graph** is there is a set S of distinct points in the xy-plane such that G is the positive slope graph of S.

   (a) Let $A_1 = (0, 2)$, $A_2 = (2, 1)$, $A_3 = (3, 4)$, $A_4 = (4, 3)$ and $A_5 = (5, 1)$ and let $S = \{A_1, A_2, \ldots, A_5\}$. Draw the positive slope graph of S.

   (b) Show that the path P_n is a positive slope graph for every $n \in \mathbb{N}$.

   (c) Is C_5 a positive slope graph?

12.2   Connected Graphs

Suppose that from a vertex u in a graph G, we proceed to a neighbor v of u, then to a neighbor of v and so on, until we finally conclude this procedure at some vertex w of G. (It may occur that some vertices are encountered more than once and in fact that $w = u$.) What we have just described is a *walk* from u to w in G.

Walks, Trails and Paths

Definition 12.17 *A* **walk** *in a graph G is a sequence $u_0, u_1, \ldots, u_k$ of vertices of G such that u_i and u_{i+1} are adjacent for $0 \leq i \leq k - 1$. We denote this sequence by*

$$W = (u_0, u_1, \ldots, u_k). \tag{12.1}$$

Then W is a $u_0 - u_k$ walk, where u_0 is the **initial vertex** *of W and u_k is the* **terminal vertex** *of W.*

Since the walk W in (12.1) encounters k edges (counting multiplicities), namely, u_0u_1, u_1u_2, ..., $u_{k-1}u_k$, the walk W is said to have **length** k. The walk W is also said to contain these edges. Therefore, a walk of length k contains at most k distinct edges. A walk of length 0 is a **trivial walk** and contains no edges.

In the graph G of Figure 12.22, (x, z, w, v, w, y) is an $x - y$ walk of length 5. In this walk, both the vertex w and edge vw are encountered twice. This is perfectly acceptable. Indeed, a vertex of G can appear in a walk any (finite) number of times and an edge can be encountered any (finite)

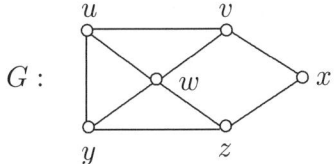

Figure 12.22: Walks in graphs

number of times. Each time an edge is encountered in a walk, however, it contributes 1 to the length of the walk.

The following are also walks in the graph G of Figure 12.22:

$$W_1 = (u, y, w, z), W_2 = (z, x, z, x, z), W_3 = (x), W_4 = (w, v, u, w, z, y). \qquad (12.2)$$

The walk W_1 is a $u - z$ walk of length 3, W_2 is a $z - z$ walk of length 4, W_3 is the trivial $x - x$ walk of length 0 and W_4 is a $w - y$ walk of length 5.

Definition 12.18 *A trail is a walk in which no edge is encountered more than once.*

For example,

$$T = (v, w, y, z, w, u) \qquad (12.3)$$

is a $v - u$ trail of length 5 in the graph G of Figure 12.22. Observe that w occurs twice in the trail T. Indeed, a vertex may occur more than once in a trail even though no edge can be encountered more than once.

Definition 12.19 *A path is a walk in which no vertex is repeated.*

Therefore, the trail T in (12.3) is not a path. On the other hand,

$$P = (x, z, w, v, u) \qquad (12.4)$$

is an $x - u$ path of length 4 in the graph G of Figure 12.22. While every path is a trail and every trail is a walk, the respective converses are not true.

A walk whose initial and terminal vertices are distinct is called an **open walk**. A **closed walk** is a walk whose initial and terminal vertices are the same. The walks W_1 and W_4 in (12.2) are open, while W_2 and W_3 are closed.

Definition 12.20 *A circuit is a closed trail of length 3 or more.*

For example, in the graph G of Figure 12.22,

$$C = (u, w, y, z, w, v, u) \qquad (12.5)$$

is a circuit of length 6 (whose initial and terminal vertex is u). As this circuit C illustrates, a vertex in a circuit may be repeated (in addition to the initial and terminal vertex).

Definition 12.21 *A cycle is a circuit in which no vertex is repeated except that the initial vertex is the terminal vertex. A cycle of length k is called a k-cycle and a 3-cycle is also referred to as a **triangle**.*

Hence the circuit C in (12.5) is not a cycle. The circuit

$$C^* = (w, u, v, w, y, z, w)$$

in the graph G of Figure 12.22 is also not a cycle. On the other hand,

$$C' = (u, v, w, z, y, u) \tag{12.6}$$

is a 5-cycle in the graph G of Figure 12.22. The cycle

$$C'' = (w, y, z, w)$$

is a triangle in this graph. A cycle of even length is called an **even cycle**, while a cycle of odd length is called an **odd cycle**. Triangles are therefore odd cycles.

The terms "path" and "cycle" therefore have two interpretations in a graph. They not only refer to types of graphs and subgraphs, they also refer to types of routes that can be used to proceed about in a graph.

Recall that the adjacency matrix of a graph G of order n with $V(G) = \{v_1, v_2, \ldots, v_n\}$ is the $n \times n$ zero-one matrix $\mathbf{A} = [a_{ij}]$ for which

$$a_{ij} \quad = \quad \begin{cases} 1 & \text{if } v_i v_j \in E(G) \\ 0 & \text{otherwise.} \end{cases}$$

Whenever $a_{ij} = 1$, this means that G contains the edge $v_i v_j$ and therefore a $v_i - v_j$ path of length 1 and, of course, a $v_i - v_j$ walk of length 1 as well. Not only can the adjacency matrix of G be used to identify whether G contains a $v_i - v_j$ walk of length 1, it can be used to determine whether G contains a $v_i - v_j$ walk of length k for every positive integer k and, in fact, the number of such walks. Before stating a theorem that provides us with this information, we need to know when two $u - v$ walks in a graph G are considered the same.

Two $u - v$ walks $W = (u = u_0, u_1, \ldots, u_k = v)$ and $W' = (u = v_0, v_1, \ldots, v_\ell = v)$ in a graph are **equal** if $k = \ell$ and $u_i = v_i$ for all i with $0 \leq i \leq k$.

Theorem 12.22 *Let G be a graph with vertex set $V(G) = \{v_1, v_2, \ldots, v_n\}$ and adjacency matrix $\mathbf{A}$. For each positive integer k, the number of different $v_i - v_j$ walks of length k in G is the (i, j)-entry in the matrix $\mathbf{A}^k$.*

Proof. Let $a_{ij}^{(k)}$ denote the (i, j)-entry in the matrix $\mathbf{A}^k$ for a positive integer k. Since $\mathbf{A}^1 = \mathbf{A}$, it follows that $a_{ij}^{(1)} = a_{ij}$. We proceed by induction. For vertices v_i and v_j of G, there can be only one $v_i - v_j$ walk of length 1 or no $v_i - v_j$ walks of length 1 and this occurs if $a_{ij} = 1$ or $a_{ij} = 0$, respectively. Therefore, the (i, j)-entry of the matrix $\mathbf{A}$ is the number of $v_i - v_j$ walks of length 1 in G. Thus the basis step of the induction is established.

We now verify the inductive step. Assume, for a positive integer k, that $a_{ij}^{(k)}$ is the number of different $v_i - v_j$ walks of length k in G. We show that the (i, j)-entry $a_{ij}^{(k+1)}$ in $\mathbf{A}^{k+1}$ gives the number of different $v_i - v_j$ walks of length $k+1$ in G. First, observe that every $v_i - v_j$ walk of length $k + 1$ in G is obtained from a $v_i - v_t$ walk of length k for some vertex v_t in G that is adjacent to v_j.

Since $\mathbf{A}^{k+1} = \mathbf{A}^k \cdot \mathbf{A}$, it follows that the (i, j)-entry $a_{ij}^{(k+1)}$ in $\mathbf{A}^{k+1}$ can be obtained by taking the inner product of row i of $\mathbf{A}^k$ and column j of $\mathbf{A}$. That is,

$$a_{ij}^{(k+1)} = a_{i1}^{(k)} a_{1j} + a_{i2}^{(k)} a_{2j} + \ldots + a_{in}^{(k)} a_{nj} = \sum_{t=1}^{n} a_{it}^{(k)} a_{tj}. \tag{12.7}$$

By the induction hypothesis, $a_{it}^{(k)}$ is the number of different $v_i - v_t$ walks of length k in G. If $a_{tj} = 1$, then v_t is adjacent to v_j and so there are $a_{it}^{(k)}$ different $v_i - v_j$ walks of length $k + 1$ in G whose next-to-last vertex is v_t. On the other hand, if $a_{tj} = 0$, then v_t is not adjacent to v_j and there are no $v_i - v_j$ walks of length $k + 1$ in G whose next-to-last vertex is v_t. In any case, $a_{it}^{(k)} \cdot a_{tj}$ gives the number of different $v_i - v_j$ walks of length $k + 1$ in G whose next-to-last vertex is v_t. Consequently, the total number of different $v_i - v_j$ walks of length $k + 1$ in G is the sum in (12.7), which is $a_{ij}^{(k+1)}$.

By the Principle of Mathematical Induction, $a_{ij}^{(k)}$ is the number of different $v_i - v_j$ walks of length k in G for every positive integer k. ∎

Before giving an example to illustrate Theorem 12.22, let's make some observations. Let G be a graph of order n with $V(G) = \{v_1, v_2, \ldots, v_n\}$ and $\mathbf{A}^k = \left[a_{ij}^{(k)} \right]$, where $\mathbf{A}$ is the adjacency matrix of G. By Theorem 12.22, $a_{ii}^{(2)}$ gives the number of different $v_i - v_i$ walks of length 2 in G. Since a $v_i - v_i$ walk of length 2 is (v_i, v_t, v_i) for some vertex v_t adjacent to v_i, it follows that $a_{ii}^{(2)} = \deg v_i$ for every vertex v_i of G. For $i \neq j$, $a_{ij}^{(2)}$ is the number of different $v_i - v_j$ *paths* of length 2 in G. Again, by Theorem 12.22, $a_{ii}^{(3)}$ gives the number of different $v_i - v_i$ walks of length 3 in G. Since a $v_i - v_i$ walk of length 3 is (v_i, v_s, v_t, v_i) for adjacent vertices v_s and v_t, each of which is adjacent to v_i, it follows that v_i must belong to a triangle. Not only is (v_i, v_s, v_t, v_i) a $v_i - v_i$ walk of length 3, so too is (v_i, v_t, v_s, v_i) a (different) $v_i - v_i$ walk of length 3 in G. Therefore, $a_{ii}^{(3)}$ is twice the number of triangles in G that contain v_i.

Example 12.23

For the graph G of Figure 12.23, use Theorem 12.22 to determine the number of different $v_i - v_j$ walks of length 2 and 3 for some pairs i, j of integers with $1 \leq i, j \leq 4$.

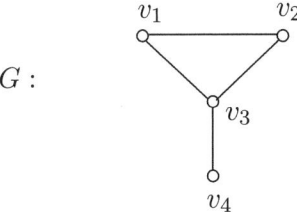

Figure 12.23: The graph G in Example 12.23

Solution. The adjacency matrix of the graph G of Figure 12.23 is

$$\mathbf{A} = \begin{bmatrix} 0 & 1 & 1 & 0 \\ 1 & 0 & 1 & 0 \\ 1 & 1 & 0 & 1 \\ 0 & 0 & 1 & 0 \end{bmatrix}.$$

The matrices $\mathbf{A}^2$ and $\mathbf{A}^3$ are therefore

$$\mathbf{A}^2 = \begin{bmatrix} 2 & 1 & 1 & 1 \\ 1 & 2 & 1 & 1 \\ 1 & 1 & 3 & 0 \\ 1 & 1 & 0 & 1 \end{bmatrix} \text{ and } \mathbf{A}^3 = \begin{bmatrix} 2 & 3 & 4 & 1 \\ 3 & 2 & 4 & 1 \\ 4 & 4 & 2 & 3 \\ 1 & 1 & 3 & 0 \end{bmatrix}.$$

Since $a_{11}^{(2)} = a_{22}^{(2)} = 2$, it follows that $\deg v_1 = \deg v_2 = 2$. Also, since $a_{22}^{(3)} = 2$, the vertex v_2 belongs to one triangle in G. Because $a_{23}^{(3)} = 4$, there are four $v_2 - v_3$ walks of length 3 in G, namely:

(1) (v_2, v_1, v_2, v_3), (2) (v_2, v_3, v_1, v_3), (3) (v_2, v_3, v_2, v_3), (4) (v_2, v_3, v_4, v_3). ◆

One of the most important properties that a graph can possess is that of being connected. Loosely speaking, a graph is connected if it consists of one piece. This highly informal description of a connected graph may provide us with a mental picture of what a connected graph is but is of little use when it is necessary to verify that a graph is connected.

Definition 12.24 *A graph G is* **connected** *if for every two vertices u and v, G contains a $u - v$ path. If G contains a $u - v$ path, then we say that u and v are* **connected** *in G. That is, G is a connected graph if every two vertices of G are connected.*

If a graph H contains two vertices u and v for which there is no $u - v$ path in H, then H is not connected. In this case, we say that H is **disconnected**.

Example 12.25

In Figure 12.24 the graph G_1 is connected, while G_2 is disconnected. In G_2, there is no $v_1 - v_4$ path, for example; so v_1 and v_4 are not connected in G_2.   ◆

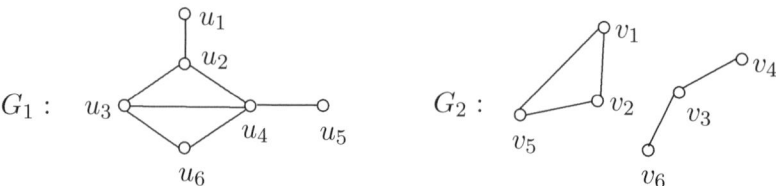

Figure 12.24: Connected and disconnected graphs in Example 12.25

The following theorem tells us that to determine whether a graph G is connected, we need only show that G contains a $u - v$ walk for every pair u, v of vertices of G.

Theorem 12.26 *If a graph G contains a $u - v$ walk for vertices u and v in G, then G also contains a $u - v$ path.*

Proof. This is certainly true if $u = v$, so we assume that $u \neq v$. Among all $u - v$ walks in G, let W be one of smallest length. We claim that W is a $u - v$ path. Suppose that it is not true. Then some vertex of G must be repeated in W. Let $W = (u = u_0, u_1, \ldots, u_k = v)$ and suppose that $u_i = u_j$, where $0 \leq i < j \leq k$. Then

$$W' = (u = u_0, u_1, \ldots, u_{i-1}, u_i = u_j, u_{j+1}, \ldots, u_k = v)$$

is also a $u - v$ walk in G but one whose length is less than that of W. This is impossible.   ■

Definition 12.27 *A subgraph G' of a graph G is a **component** of G if G' is connected but G' is not a proper subgraph of any connected subgraph of G.*

Again, loosely speaking, the components of a graph G are the connected pieces of G. A graph has only one component if and only if it is connected. The graph G_1 of Figure 12.24 has only one component, namely G_1 itself. The graph G_2 of Figure 12.24 is disconnected and has two components, namely a triangle and a path of length 2. The number of components of a graph G is denoted by $k(G)$. So, for the graphs G_1 and G_2 of Figure 12.24, $k(G_1) = 1$ and $k(G_2) = 2$.

Example 12.28

The graph F of Figure 12.25 has four components, namely, F_1, F_2, F_3 and F_4 and so $k(F) = 4$. ◆

An important observation to make about the components of a disconnected graph is the following.

If u and v are two vertices that belong to different components of a graph G, then u and v are not adjacent in G.

Distance in Graphs

In order for a graph G to be connected, we have seen that G must contain a $u - v$ path for every pair u, v of vertices of G. However, it is often the case that a connected graph G may contain several $u - v$ paths for two vertices u and v of G. For example, all of the following are $t - y$ paths in the graph H of Figure 12.26:

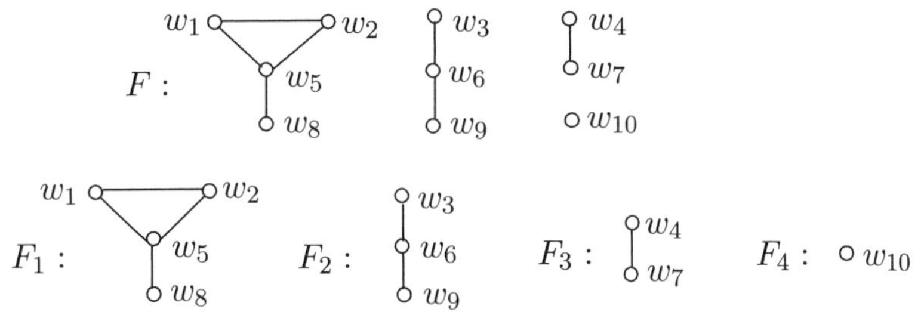

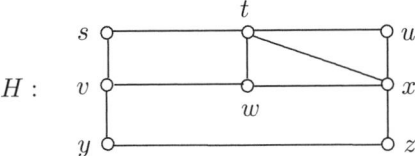

Figure 12.25: The components of the disconnected graph in Example 12.28

$$P' = (t, w, x, z, y), \ P'' = (t, s, v, w, x, z, y), \ P''' = (t, w, v, y).$$

The path P' has length 4, P'' has length 6 and P''' has length 3. There is no $t-y$ path in H having length less than 3 or more than 6.

Figure 12.26: Paths in a graph

Definition 12.29 *Let u and v be two vertices in a connected graph G. The **distance** $d(u, v)$ from u to v is the length of a shortest $u - v$ path in G.*

Therefore, $d(t, y) = 3$ for the vertices t and y in the graph H of Figure 12.26. By Theorem 12.22, if G is a connected graph of order n with $V(G) = \{v_1, v_2, \ldots, v_n\}$ and adjacency matrix $\mathbf{A}$, then the distance $d(v_i, v_j)$ from v_i to v_j ($i \neq j$) is the smallest positive integer k for which the (i, j)-entry of the matrix $\mathbf{A}^k$ is nonzero.

Bipartite Graphs

The connected graph G in Figure 12.27 has the property that its vertex set can be partitioned into two subsets $U = \{u_1, u_2, u_3\}$ and $W = \{w_1, w_2, w_3, w_4\}$ in such a way that every path connecting two vertices of U (or two vertices of W) has even length and every path connecting a vertex of U and a vertex of W has odd length. In particular, the distance between two vertices of U (or of W) is even and the distance between a vertex of U and a vertex of W is odd. Graphs with these properties are of special interest.

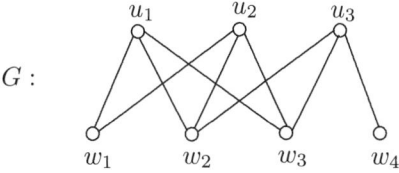

Figure 12.27: A bipartite graph

Definition 12.30 *A graph G is a **bipartite graph** if it is possible to partition its vertex set into two subsets U and W, called **partite sets**, so that every edge of G joins a vertex of U and a vertex of W.*

Both the path P_5 and the cycle C_6 are bipartite, as shown in Figure 12.28. Since there is no partition of the vertex set of C_3 that satisfies the requirement of the definition above, C_3 is not bipartite. The graphs K_4 and C_5 are also not bipartite (see Exercises 23 and 24).

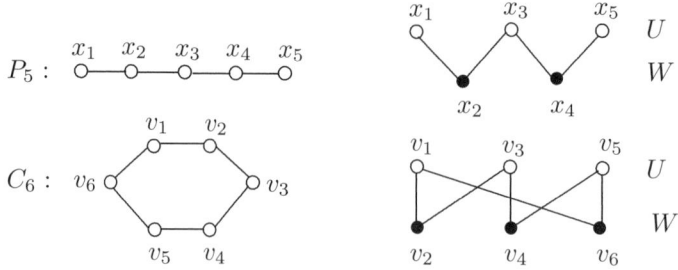

Figure 12.28: Bipartite graphs

The fact that none of C_3, C_5 and K_4 is bipartite but each contains odd cycles is no coincidence.

Theorem 12.31 *A nontrivial graph G is bipartite if and only if G does not contain an odd cycle.*

Proof. First, assume that G is a bipartite graph and that $C = (u_1, u_2, \ldots, u_k, u_1)$ is a cycle in G. Then there is a partition of the vertex set of G into two subsets U and W such that every edge of G joins a vertex of U and a vertex of W. We may assume that $u_1 \in U$. Then $u_2 \in W$, $u_3 \in U$ and so on. Since $u_1 \in U$ and $u_1 u_k \in E(G)$, it follows that $u_k \in W$ and that k is even. Thus C is an even cycle.

To prove the converse, let G be a nontrivial graph having no odd cycles. We show that G is bipartite. Since a graph is bipartite if and only if each of its components is bipartite, we may assume that G is connected. Let u be any vertex of G, let U consist of all vertices of G whose distance from u is even and let W consist of all vertices whose distance from u is odd. Thus $\{U, W\}$ is a partition of $V(G)$. Since $d(u, u) = 0$, it follows that $u \in U$. We claim that every edge of G joins a vertex of U and a vertex of W.

Assume, to the contrary, that there exist two adjacent vertices in U or two adjacent vertices in W. Since these two situations are similar, we will assume that there are vertices v and w in W such that $vw \in E(G)$. Since $d(u, v)$ and $d(u, w)$ are both odd, $d(u, v) = 2s + 1$ and $d(u, w) = 2t + 1$ for nonnegative integers s and t. Let $P' = (u = v_0, v_1, \ldots, v_{2s+1} = v)$ be a $u - v$ path of length $d(u, v)$ and let $P'' = (u = w_0, w_1, \ldots, w_{2t+1} = w)$ be a $u - w$ path of length $d(u, w)$ in G. Certainly, P' and P'' have their initial vertex u in common but they may have other vertices in common as well. Among the vertices P' and P'' have in common, let x be the last vertex. Perhaps $x = u$. In any case, $x = v_i$ for some integer $i \geq 0$. Thus $d(u, v_i) = i$. Since x is on P'' and w_i is the only vertex of P'' whose distance from u is i, it follows that $x = w_i$. So $x = v_i = w_i$. However then, $C = (v_i, v_{i+1}, \ldots, v_{2s+1}, w_{2t+1}, w_{2t}, \ldots, w_i = v_i)$ is a cycle of length

$$[(2s + 1) - i] + [(2t + 1) - i] + 1 = 2s + 2t - 2i + 3 = 2(s + t - i + 1) + 1$$

and so C is an odd cycle, which is a contradiction.   ∎

The proof of Theorem 12.31 above shows that if G is a connected bipartite graph with partite sets U and W, then every $x - y$ path in G where $x, y \in U$ or $x, y \in W$ is even and every $x - y$ path in G where $x \in U$ and $y \in W$ (or $x \in W$ and $y \in U$) is odd, as we noted earlier.

A graph G is called a **complete bipartite graph** if it is possible to partition $V(G)$ into two subsets U and W, again called **partite sets**, such that two vertices of G are adjacent if and only if they belong to different partite sets. If $|U| = s$ and $|W| = t$, then this graph is denoted by $K_{s,t}$ or $K_{t,s}$. If either $s = 1$ or $t = 1$, then this graph is called a **star**. The vertex in the star $K_{1,t}$ ($t \geq 2$) having degree t is called the **center** of the star. Some examples of complete bipartite graphs are

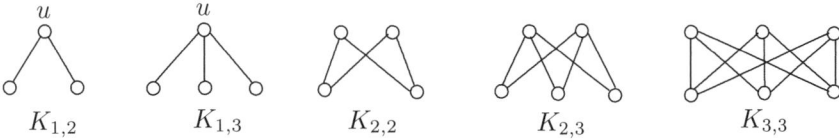

Figure 12.29: Complete bipartite graphs

shown in Figure 12.28. The graphs $K_{1,2}$ and $K_{1,3}$ are therefore stars (where u is the center in each case). Observe that $K_{1,2} = P_3$ and $K_{2,2} = C_4$.

Bipartite graphs belong to a more general class of graphs. A graph G is a k-**partite graph** if $V(G)$ can be partitioned into k subsets $V_1, V_2, \ldots, V_k$ (once again called **partite sets**) such that uv is an edge of G if u and v belong to different partite sets. If, in addition, every two vertices in different partite sets are joined by an edge, then G is a **complete k-partite graph**. If $|V_i| = n_i$ for $1 \le i \le k$, then we denote this complete k-partite graph by $K_{n_1,n_2,\ldots,n_k}$. The complete k-partite graphs are also referred to as **complete multipartite graphs**. If $n_i = 1$ for every i $(1 \le i \le k)$, then $K_{n_1,n_2,\ldots,n_k}$ is the complete graph K_k. Complete 2-partite graphs are thus complete bipartite graphs. Several complete multipartite graphs are shown in Figure 12.30.

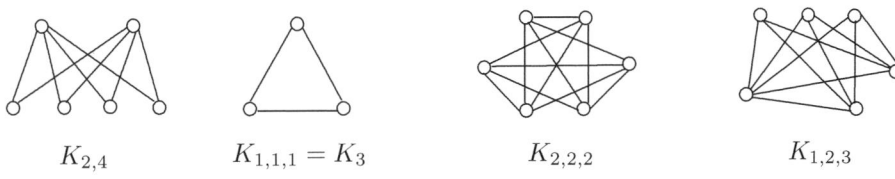

Figure 12.30: Complete multipartite graphs

Figure 12.31 shows two graphs G_1 and G_2 together with their complements $\overline{G}_1$ and $\overline{G}_2$. Observe that G_1 and $\overline{G}_1$ are both connected, while G_2 is connected and $\overline{G}_2$ is disconnected. There is no example, however, of a graph G for which both G and $\overline{G}$ are disconnected.

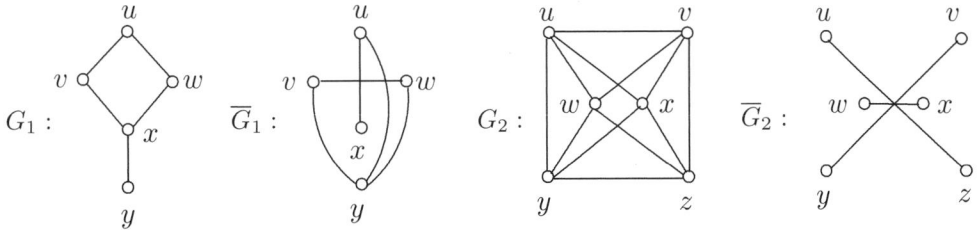

Figure 12.31: Connected and disconnected complements

Theorem 12.32 *If G is a disconnected graph, then $\overline{G}$ is connected.*

Proof. Let G be a disconnected graph. Then G contains two or more components. To verify that $\overline{G}$ is connected, we show that for every two vertices u and v of $\overline{G}$, there is a $u - v$ path in $\overline{G}$. We now consider two cases, according to whether u and v belong to different components of G or to the same component of G.

Case 1. u and v belong to different components of G. Then u and v are not adjacent in G and so u and v are adjacent in $\overline{G}$. Then (u, v) is a $u - v$ path of length 1 in $\overline{G}$.

Case 2. u and v belong to the same component of G. Let w be a vertex of G that belongs to a component different from the one containing u and v. Therefore, w is not adjacent to either u or v

in G, which implies that w is adjacent to both u and v in $\overline{G}$. Consequently, (u, w, v) is a $u - v$ path of length 2 in $\overline{G}$.

In either case, $\overline{G}$ contains a $u - v$ path and so $\overline{G}$ is connected. ∎

The complete graph K_n of order n is certainly connected. In fact, if u and v are any two vertices of K_n, then u and v are adjacent and so (u, v) is a path of length 1. Since each vertex of K_n has degree $n - 1$, we can conclude that if the degrees of the vertices of a graph G are large enough in terms of its order, then G *must* be connected. But how large is "large enough"? Figure 12.32 shows two disconnected graphs F and H. The order of F is 6 and the degree of every vertex of F is 2; while the order of H is 7 and the degree of every vertex of H is 2 or 3. Therefore, for graphs of order 6 or 7, knowing that the degree of every vertex is 2 or more is not sufficient to guarantee that the graphs must be connected. If we knew that the degrees of their vertices were just a bit larger, however, then we could reach a different conclusion.

Figure 12.32: Two disconnected graphs

Theorem 12.33 *If G is a graph of order n such that*

$$\deg v \geq \frac{n - 1}{2}$$

for every vertex v of G, then G is connected.

Proof. Let G be a graph of order n such that $\deg v \geq (n - 1)/2$ for every vertex v of G. To verify that G is connected, we show that there is a path between every two vertices of G. Let x and y be two vertices of G. We show that G contains an $x - y$ path. Of course, if x is adjacent to y, then xy is an edge of G and G contains the $x - y$ path (x, y). So we can assume that x and y are not adjacent. Since $\deg x \geq (n - 1)/2$ and $\deg y \geq (n - 1)/2$, each of x and y has at least $(n - 1)/2$ neighbors. Furthermore, neither x nor y is a neighbor of the other. If there is no vertex of G that is a neighbor of both x and y, then $N(x) \cup \{x\}$ and $N(y) \cup \{y\}$ are disjoint sets, each of which has cardinality at least $(n + 1)/2$. This implies that the order of G is at least $n + 1$; however, this is impossible since G has only n vertices. Therefore, there must be a vertex z that is a neighbor of both x and y. This says that (x, z, y) is an $x - y$ path in G. Hence x and y are connected and so G is connected. ∎

There is a special kind of edge and vertex in a graph whose deletion from the graph results in an increased number of components. For an edge e of G, the subgraph $G - e$ is the spanning subgraph of G whose edge set consists of all edges of G except e. For a vertex v of a nontrivial graph $G = (V, E)$, the subgraph $G - v$ has vertex set $V - \{v\}$ and consists of all edges of G except those incident with v. While $G - e$ is not an induced subgraph of G, the subgraph $G - v$ is necessarily an induced subgraph of G; in fact, $G - v = G[V - \{v\}]$. For a nonempty subset X of E and a nonempty subset U of V, the graphs $G - X$ and $G - U$ are similarly defined.

Definition 12.34 *An edge e in a graph G is called a **bridge** of G if $G - e$ has more components than G. A vertex v in a nontrivial graph G is a **cut-vertex** of G if $G - v$ has more components than G.*

In particular, for a connected graph G, an edge e of G is a bridge of G if $G - e$ is disconnected; while a vertex v of G is a cut-vertex of G if $G - v$ is disconnected.

Example 12.35

The vertex r in the graph G of Figure 12.33 is a cut-vertex of G, as are the vertices t and u in G. The graph G has three bridges, namely pr, qr and tu. The graph $G - r$ has three components and the graph $G - tu$ has two components, as shown in Figure 12.33. ♦

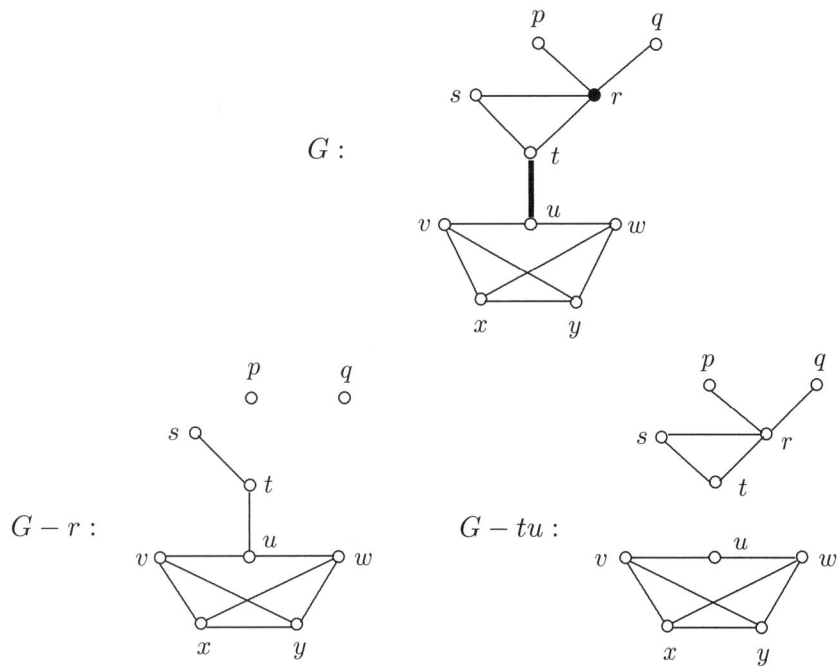

Figure 12.33: Bridges and cut-vertices in a connected graph in Example 12.35

In fact, if $e = uv$ is a bridge in a connected graph G, then $G - e$ always has exactly two components, one component containing u and the other containing v. The following two theorems provide characterizations of bridges and cut-vertices in a graph. (See Exercises 32 and 35.)

Theorem 12.36 *An edge e of a graph G is a bridge of G if and only if e lies on no cycle of G.*

Theorem 12.37 *A vertex v of a connected graph G is a cut-vertex of G if and only if there exist vertices u and w distinct from v such that v lies on every $u - w$ path of G.*

Exercises for Section 12.2

1. Give an example of a graph G and a walk W in G such that W contains at least distinct three edges, no two of which are encountered the same number of times in W.

2. Show that if a graph G contains a $u - v$ path and a $v - w$ path, then G contains a $u - w$ path.

3. For the graph G of Figure 12.34, give an example of each of the following kinds of walks:

   (a) an open trail that is not a path.   (b) a path.

   (c) a circuit that is not a cycle.   (d) a cycle.

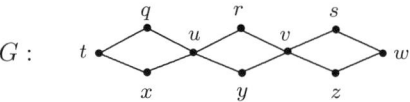

Figure 12.34: The graph in Exercise 3

4. For the graph G of Figure 12.35,

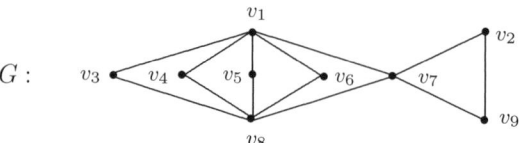

Figure 12.35: The graph in Exercise 4

   (a) give an example of a path of length k for as many nonnegative integers k as possible.

   (b) give an example of a k-cycle for as many positive integers k as possible.

5. For the graph G of Figure 12.36, give an example (if one exists) of a subgraph that is

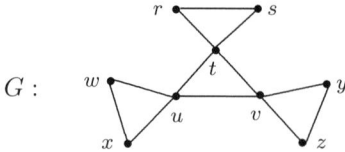

Figure 12.36: The graph in Exercise 5

   (a) a path of size 4. (b) a trail of size 5. (c) a triangle.

   (d) a circuit whose size is greater than 6.

6. For the graph G of Figure 12.36, determine the largest size of a subgraph that is

   (a) a cycle.   (b) a path.   (c) an open trail. (d) a circuit.

7. (a) Prove that if u and v are two vertices in a connected bipartite graph G such that there is a $u - v$ path of length 4 in G, then $uv \notin E(G)$.

   (b) State and prove a result that generalizes the statement in (a).

8. (a) Use Theorem 12.22 to compute $\mathbf{A}^4$ if $\mathbf{A} = \begin{bmatrix} 0 & 1 \\ 1 & 0 \end{bmatrix}$ without multiplying matrices.

   (b) Show that $\mathbf{A}^4$ could be computed more easily by first computing $\mathbf{A}^2$ using Theorem 12.22.

9. Use Theorem 12.22 to compute $\mathbf{A}^2$ and $\mathbf{A}^3$ (without multiplying matrices) where $\mathbf{A}$ is the adjacency matrix of the graph G of Figure 12.37.

10. Use Theorem 12.22 to compute $\mathbf{A}^4$ (without multiplying matrices), where $\mathbf{A}$ is the adjacency matrix of the graph G of Figure 12.38.

11. For an $n \times n$ matrix $\mathbf{A}$, it is common to define $\mathbf{A}^0$ as $\mathbf{I}_n$ (the $n \times n$ identity matrix). If this is done, how is this related to Theorem 12.22?

12. Give an example of two graphs G and H having the same order, the same size and whose vertices have the same degrees, where G is connected and H is disconnected.

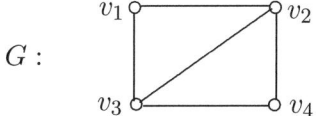

Figure 12.37: The graph G in Exercise 9

Figure 12.38: The graph G in Exercise 10

13. Give an example of a regular disconnected graph containing two non-isomorphic components.

14. If G and H are isomorphic graphs and G is connected, must H be connected?

15. Let G be a graph of order 12 with $k(G) = 3$ such that G_1, G_2 and G_3 are the three components of G. Show that either the order of G_1 is at least 4, the order of G_2 is at least 4 or the order of G_3 is at least 6.

16. Show that if G is a graph of order 11 such that every vertex of G has degree 3 or more, then G has at most two components.

17. Show that if a graph G of order $n \geq 3$ contains a vertex v such that $\deg v = n - 1$, then G is connected.

18. Let G be a graph of order $n \geq 4$ containing two adjacent vertices u and v, both of degree $n - 2$. Is it true that G must be connected? What if u and v are not adjacent?

19. Does there exist a connected graph G of order $n \geq 3$ and a vertex u of G such that $d(u, x) \neq d(u, y)$ for every two distinct vertices x and y?

20. Figure 12.39 shows a graph that models the street system of a community, where each edge represents a street and each vertex represents an intersection. What are the possible intersections where a security station can be constructed so that the number of blocks needed to drive from the station to an intersection farthest from the station will be as small as possible?

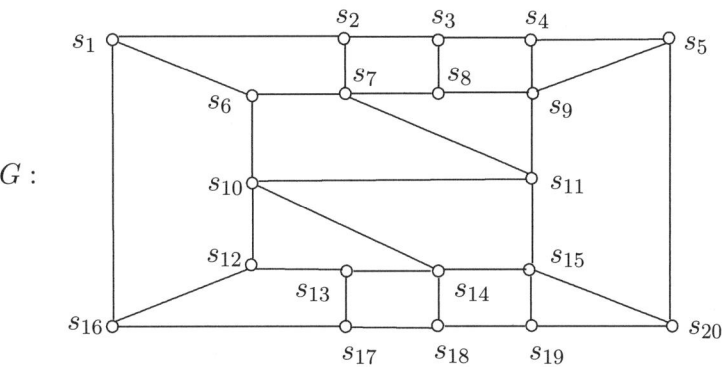

Figure 12.39: A graph representing a street system in Exercise 20

21. Let G be a nontrivial connected graph. For two distinct vertices u and v of G, a $u - v$ path of length $d(u, v)$ is called a $u - v$ **geodesic** in G. A graph H is defined by $V(H) = V(G)$ and $uv \in E(H)$ if G contains only one $u - v$ geodesic in G. The graph H is called the **unique geodesic graph** of G and is denoted by $ug(G)$.

(a) Draw $ug(G_i)$ for for the graphs G_i $(i = 1, 2, 3)$ in Figure 12.40.

(b) Give an example of a graph G of order 5 or more such that $ug(G) \cong G$.

(c) Give an example of a graph G such that $ug(G) \not\cong G$.

(d) Is there an example of a graph G such that $ug(ug(G)) \not\cong ug(G)$?

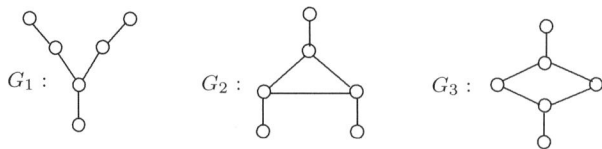

Figure 12.40: Graphs G_i $(i = 1, 2, 3)$ in Exercise 21(a)

22. Let G be a connected graph. The **odd distance graph** $OD(G)$ of G has the same vertex set as G and two vertices u and v are adjacent in $OD(G)$ if $d_G(u, v)$ is odd.

(a) Show that $OD(OD(G)) = OD(G)$.   (b) What is the complement of $OD(G)$?

(c) Does there exist a connected graph G of order 3 or more such that $OD(G)$ is complete?

23. Without using Theorem 12.31, show that for $n \geq 3$, the complete graph K_n is not bipartite.

24. Without using Theorem 12.31, show that the cycle C_5 is not bipartite.

25. What is the size of $\overline{C}_n$ for $n \geq 3$?

26. Determine the size of the complete bipartite graph $K_{s,t}$.

27. Let G be a graph of order 5 or more. Prove that at most one of G and $\overline{G}$ is bipartite.

28. Prove that if G is an r-regular bipartite graph with $r \geq 1$, then the two partite sets of G have the same cardinality.

29. (a) Let G be a 3-regular graph. Show that G contains a set X of edges such that $G - X$ is 2-regular if and only if G contains a set Y of edges such that $G - Y$ is 1-regular.

(b) If G is a 3-regular graph, can a set X of edges of G always be found such that $G - X$ is 2-regular?

30. Let G be a nontrivial connected graph in which every vertex of G is even. Show that it is possible for G to contain a cut-vertex but it is impossible for G to contain a bridge.

31. Show that if a graph G has a cut-vertex v, then v is not a cut-vertex of $\overline{G}$.

32. Prove Theorem 12.36: *An edge e of a graph G is a bridge of G if and only if e lies on no cycle of G.*

33. Prove that if uv is a bridge in a graph G, then there is a unique $u - v$ path in G.

34. Let G be a connected graph and let e_1 and e_2 be two edges of G. Prove that $G - e_1 - e_2$ has three components if and only if both e_1 and e_2 are bridges in G.

35. Prove Theorem 12.37: *A vertex v of a connected graph G is a cut-vertex of G if and only if there exist vertices u and w distinct from v such that v lies on every $u - w$ path of G.*

36. Prove that a 3-regular graph G has a cut-vertex if and only if G has a bridge.

37. (a) Let G be a graph of order 3 or more. Prove that G is connected if and only if G contains two distinct vertices u and v such that $G - u$ and $G - v$ are connected.

(b) Prove or disprove: Every connected graph G of order 4 or more contains three distinct vertices u, v and w such that $G - u$, $G - v$ and $G - w$ are connected.

38. Let G be a connected graph containing a vertex w such that $\deg w \not\equiv 0 \pmod 3$ and such that $\deg u + \deg v \equiv 0 \pmod 3$ for every two adjacent vertices u and v of G. Prove that G is bipartite and contains no vertex x such that $\deg x \equiv 0 \pmod 3$.

12.3   Eulerian Graphs

While it is unclear just how and when many areas of mathematics began, such is not the case with graph theory. It is generally accepted that graph theory was born in 1736. The story starts before then, however, in the city of Königsberg, which was located in Prussia during that period and is now the city of Kaliningrad in Russia. Königsberg (see Figure 12.41) was divided into four land regions (denoted by A, B, C and D) by the River Pregel. There were seven bridges that crossed the river at various locations allowing people to walk between the land regions. Evidently, the citizens of Königsberg would often take strolls about the city. A question arose that intrigued many people in Königsberg: Is it possible to take a stroll about Königsberg and cross each of its seven bridges exactly once? This problem became known as the **Königsberg Bridge Problem**. It is perhaps surprising that this question was not answered more quickly but eventually this problem came to the attention of the the great Swiss mathematician Leonhard Euler, who solved the problem in a very mathematical manner.

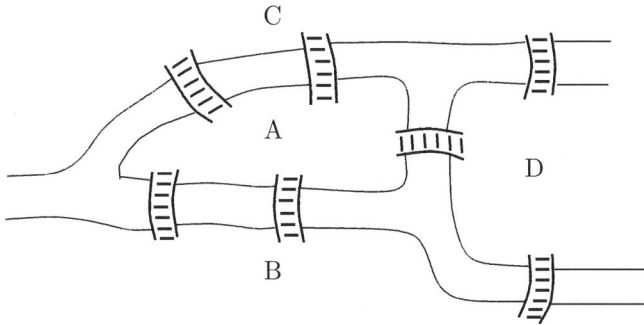

Figure 12.41: Königsberg in the early 18th century

Leonhard Euler (1707–1783), considered by many as the outstanding mathematician of the 18th century, was known to have an amazing memory. His list of publications (nearly 900 in all) is still exceeded by very few mathematicians. It is especially remarkable that many of his accomplishments occurred during the last few years of his life, while he was totally blind.

Let's see how Euler solved the Königsberg Bridge Problem. If there was a stroll through Königsberg that crossed each bridge exactly once, then such a stroll could be represented by a sequence

$$x_1, x_2, x_3, x_4, x_5, x_6, x_7, x_8 \tag{12.8}$$

where each term of the sequence is one of the letters A, B, C and D, indicating the land region to which the stroll had progressed. Since there are seven bridges, (12.8) is a sequence of eight letters.

Let's consider land region B for example. Each occurrence of B in the sequence (12.8) indicates that land region B had been entered or exited at that point of the stroll. Since each occurrence of B accounts for at most two bridges (one to enter B and one to exit B) and there are three bridges that enter (and exit) B, the symbol B must occur twice in the sequence. Similarly, C and D must occur twice each as well. By the same reasoning, A must occur three times. Therefore, A, B, C and D must occur 9 times in the sequence, which is impossible. This says that no stroll through Königsberg that crosses each bridge exactly once is possible.

The Königsberg Bridge Problem can be described as a problem involving graphs (actually multigraphs). Let G be the multigraph whose vertices are A, B, C and D and where two vertices are joined by a number of edges equal to the number of bridges joining the corresponding land regions. The multigraph so constructed is shown in Figure 12.42.

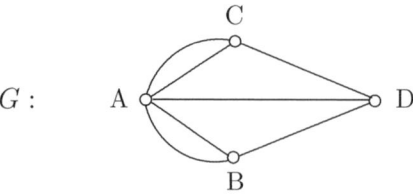

Figure 12.42: The Königsberg multigraph

The Königsberg Bridge Problem can now be stated as a graph theory problem: Does the multigraph G of Figure 12.42 have a trail containing every edge of G? Of course, we know from Euler's work that the answer to this question is no. However, this question suggested a considerably more general problem to Euler. Euler's 1736 paper, in which he solved the Königsberg Bridge Problem, was written in Latin. In an English translation of this paper, Euler wrote:

> ... I have formulated the general problem: whatever be the arrangement and division of the river into branches and however many bridges there be, can one find out whether or not it is possible to cross each bridge exactly once?

For the purpose of presenting this more general problem, we introduce some new terminology.

Definition 12.38 *Let G be a connected graph or connected multigraph. An open trail that contains every edge of G is called an* **Eulerian trail**, *while a circuit that contains every edge of G is an* **Eulerian circuit**. *A graph or multigraph containing an Eulerian circuit is itself called* **Eulerian**.

Example 12.39

The graph G_1 of Figure 12.43 is an Eulerian graph since it contains Eulerian circuits. One Eulerian circuit in G_1 is

$$C = (s_1, t_1, v_1, x_1, y_1, z_1, x_1, w_1, u_1, v_1, w_1, t_1, u_1, s_1).$$

The multigraph G_2 of Figure 12.43 contains Eulerian trails. Since G_2 contains parallel edges, it is useful (indeed essential) to label some of the edges so that an Eulerian trail can be accurately described. One Eulerian trail in G_2 is

$$T = (w_2, y_2, e_3, z_2, e_4, y_2, e_5, z_2, x_2, e_1, w_2, e_2, x_2). \blacklozenge$$

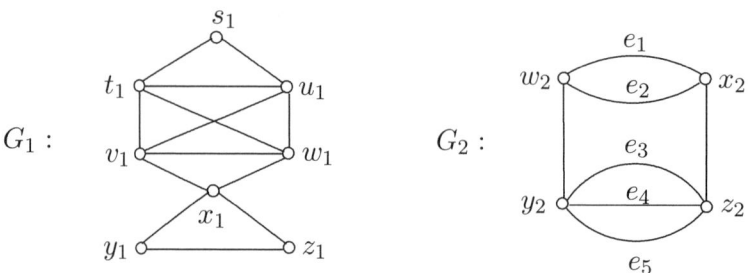

Figure 12.43: An Eulerian graph and a multigraph with an Eulerian trail

With the aid of this terminology, we know that the multigraph of Figure 12.42 contains neither an Eulerian circuit nor an Eulerian trail. Of course, Euler did not use this terminology. But the Königsberg Bridge Problem led Euler to ask (in current terminology):

Which connected graphs and multigraphs contain an Eulerian circuit or an Eulerian trail?

The **degree** of a vertex v in a multigraph G is the number of edges in G incident with v. Thus the degree of the vertex A in the Königsberg multigraph in Figure 12.42 is 5, while all other vertices have degree 3. Perhaps you have noticed that all four vertices of the Königsberg multigraph of Figure 12.42 have odd degree. Is this information relevant for determining which connected graphs or multigraphs contain neither an Eulerian circuit nor an Eulerian trail? It turns out that this information is not only relevant, it is critical. We begin with an important observation.

Theorem 12.40 *If G is an Eulerian graph, then every vertex of G is even.*

Proof. Let G be an Eulerian graph. Then G contains an Eulerian circuit $C = (x_0, x_1, \ldots, x_m = x_0)$. If v is a vertex of G that is not the initial vertex of C (and therefore not the terminal vertex of C either), then each time v is encountered on C, it is both entered and exited and therefore contributes 2 to the degree of v. Consequently, v is even. If v is the initial (and terminal) vertex of C, then C contributes 1 to the degree of v both as the initial vertex and the terminal vertex of C and contributes 2 to the degree of v for all other occurrences of v on C. Thus v is even here as well. ∎

Perhaps surprisingly, the converse of Theorem 12.40 is true as well.

Theorem 12.41 *Let G be a nontrivial connected graph. If every vertex of G is even, then G is Eulerian.*

Proof. Let G be a nontrivial connected graph in which every vertex of G is even. We show by means of an algorithm that G contains an Eulerian circuit.

Let u be a vertex of G. We begin a trail T with initial vertex u. Follow u by any vertex u_1 adjacent to u. Next, select any vertex u_2 adjacent to u_1 such that $u_1 u_2$ does not already appear on T. We continue this procedure until arriving at a vertex v, where every edge incident with v already belongs to T. Since v is even, it follows that $v = u$, that is, T is a $u - u$ circuit, which we now denote by C. If C contains every edge of G, then C is an Eulerian circuit and the proof is complete.

Suppose that C is not an Eulerian circuit of G. Then there are edges of G that do not belong to C. Since G is connected, there is some vertex v_1 on C that is incident with edges not on C. If we delete the edges of C from G, then every vertex of the resulting graph G_1 is even. In particular, every vertex in the component G'_1 in G_1 containing v_1 is even. As before, there is a circuit C' in G'_1 with initial vertex v_1. By inserting C' at v_1 in C, we obtain a circuit C'' of greater length than that of C. If C'' contains all edges of G_1, then C'' is an Eulerian circuit of G; otherwise, C'' contains a vertex v_2 that is incident with edges not on C''. We then continue this process as before until an Eulerian circuit of G is produced. ∎

The proof that Euler gave of Theorem 12.41 was incomplete and it wasn't until 1871 when the first satisfactory proof of this theorem (due to Carl Hierholzer) was published. Theorems 12.40 and 12.41 provide us with the following theorem of Euler.

Theorem 12.42 *Let G be a nontrivial connected graph. Then G is Eulerian if and only if every vertex of G is even.*

With the aid of Theorem 12.42, the following theorem can be established, telling us which connected graphs have Eulerian trails. (See Exercise 3.)

Theorem 12.43 *Let G be a connected graph. Then G has an Eulerian trail if and only if G has exactly two odd vertices. Furthermore, any Eulerian trail of G begins at one of the odd vertices and ends at the other.*

There are theorems for multigraphs whose statements and proofs are similar to Theorems 12.42 and 12.43.

Theorem 12.44 *Let G be a connected multigraph with one or more edges. Then*

(1) *G is Eulerian if and only if every vertex of G is even and*

(2) *G has an Eulerian trail if and only if G has exactly two odd vertices.*

Example 12.45

Figure 12.44 shows a downtown district of a town with nine street intersections, denoted by I_1, I_2, ..., I_9. Is it possible for a mail carrier to deliver mail along the streets of this district and drive along each street exactly once?

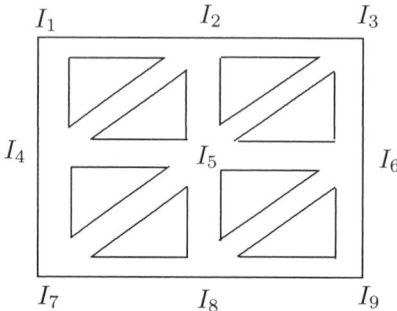

Figure 12.44: A downtown district of a town in Example 12.45

Solution. The graph G in Figure 12.45 represents this situation. Since G has exactly two odd vertices I_3 and I_7, such a mail delivery is possible if the mail route starts at either I_3 or I_7 and ends at the other. ◆

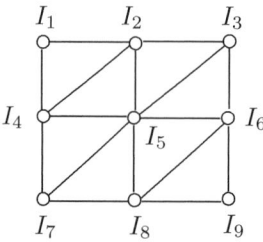

Figure 12.45: A graph representing the downtown district in Example 12.45

There is a related question whose answer may be clear. Suppose that G is a nontrivial graph (or multigraph). If the vertices of G have been drawn on a piece of paper, then under what circumstances can the edges of G be drawn in one continuous motion? The answer: G must have an Eulerian circuit or an Eulerian trail.

Example 12.46

Determine which graphs in Figure 12.46 contain an Eulerian circuit or Eulerian trail or neither.

Solution. Since G_1 and G_6 are connected graphs containing only even vertices, it follows by Theorem 12.42 that they are Eulerian and therefore contain Eulerian circuits. Since G_2, G_3, G_5 and G_8 are connected graphs containing exactly two odd vertices, they contain Eulerian trails (beginning at one odd vertex and ending at the other). Since G_4 and G_7 contain more than two odd vertices, neither graph contains an Eulerian circuit or an Eulerian trail. Since G_9 is disconnected, G_9 contains neither an Eulerian circuit nor an Eulerian trail even though every vertex of G_9 is even. ◆

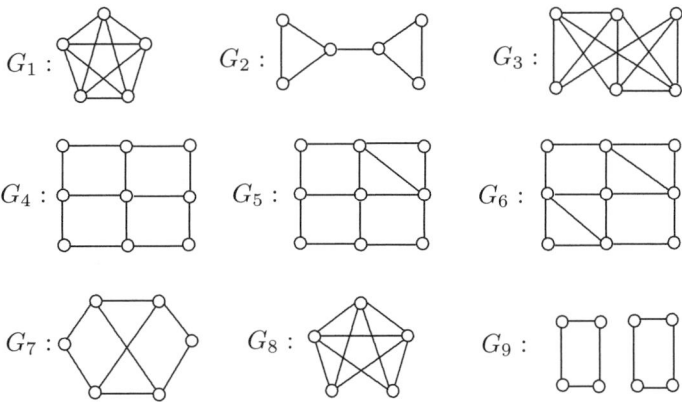

Figure 12.46: Graphs in Example 12.46

There are several operations on graphs that allow us to construct a new graph from two given graphs and permit us to give examples of graphs having some specified properties.

Definition 12.47 *The* **union** $G \cup H$ *of two graphs G and H with disjoint vertex sets is that graph whose vertex set is $V(G) \cup V(H)$ and whose edge set is $E(G) \cup E(H)$.*

The union $G \cup H$ of two graphs G and H is always a disconnected graph; in fact, the number of components of $G \cup H$ is $k(G \cup H) = k(G) + k(H)$.

Definition 12.48 *The* **join** $G + H$ *of two graphs G and H with disjoint vertex sets is that graph constructed from G and H by adding all edges joining the vertices of G and the vertices of H.*

The join of two two graphs, connected or not, is always a connected graph. Several examples of the joins of two graphs are shown in Figure 12.47. For an integer $n \geq 3$, the **wheel** W_n is the graph $C_n + K_1$. Thus the graph $C_5 + K_1$ of Figure 12.47 is the wheel W_5. For positive integers s and t, the graph $\overline{K}_s + \overline{K}_t$ is (isomorphic to) the complete bipartite graph $K_{s,t}$. Therefore, the graph $\overline{K}_2 + \overline{K}_3$ of Figure 12.47 is $K_{2,3}$.

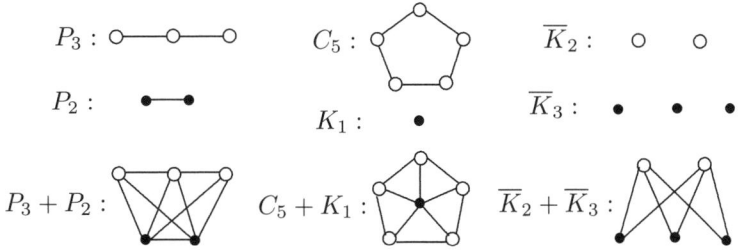

Figure 12.47: Joins of graphs

Example 12.49 *Let F and H be two connected non-Eulerian regular graphs and let $G = (F \cup H) + K_1$, that is, G is obtained from F and H by adding a new vertex v and joining v to each vertex in F and H. Prove that G is Eulerian.*

Proof. Suppose that F is an r_1-regular graph of order n_1 and H is an r_2-regular graph of order n_2. Then r_1 and r_2 are both odd and so n_1 and n_2 are both even. In the graph G, $\deg_G x = r_1 + 1$ if $x \in V(F)$, $\deg_G x = r_2 + 1$ if $x \in V(H)$ and $\deg_G v = n_1 + n_2$. Thus each vertex of G is even. Since G is connected, G is Eulerian. ∎

Definition 12.50 *The **Cartesian product** $G \times H$ of two graphs G and H (sometimes written as $G \,\square\, H$) is that graph having vertex set $V(G \times H) = V(G) \times V(H)$, that is, every vertex of $G \times H$ is an ordered pair (u, v), where $u \in V(G)$ and $v \in V(H)$, such that two distinct vertices (u, v) and (x, y) are adjacent in $G \times H$ if*

$$\text{either (1) } u = x \text{ and } vy \in E(H) \text{ or (2) } v = y \text{ and } ux \in E(G).$$

An important class of graphs defined in terms of Cartesian products are the n-**cubes** (or **hypercubes**). The 1-cube Q_1 is defined as K_2. The 2-cube Q_2 is the graph $K_2 \times K_2 = C_4$. The graph $C_4 \times K_2$ is the 3-cube Q_3. The n-cubes can then be defined recursively by $Q_n = Q_{n-1} \times K_2$ for $n \geq 2$. The n-cubes can also be described somewhat differently. For a positive integer n, the n-cube Q_n is that graph whose vertex set is the set of n-bit strings and such that two vertices are adjacent if their n-bit strings differ in exactly one coordinate. Using this definition, the n-cubes for $n = 1, 2, 3$ are shown in Figure 12.48. Thus Q_n is an n-regular graph of order 2^n.

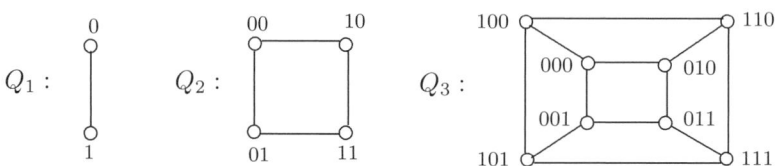

Figure 12.48: Three hypercubes

Example 12.51 *Find a necessary and sufficient condition for the Cartesian product $G \times H$ of two nontrivial connected graphs G and H to be Eulerian.*

Solution. We can think of $G \times H$ as being constructed by replacing each vertex v of G by a copy H_v of H. Let x be a vertex of $G \times H$. Then x belongs to H_v for some vertex v of G. The vertex x is adjacent to its neighbors in H_v as well as to one vertex in H_u for every neighbor u of v in G. Thus

$$\deg_{G \times H} x = \deg_{H_v} x + \deg_G v.$$

Hence $\deg_{G \times H} x$ is even if and only if $\deg_{H_v} x$ and $\deg_G v$ are both even or both odd (that is, they are of the same parity).

If $\deg_{H_v} x$ is even, then $\deg_G v$ is even for every vertex v of G; while if $\deg_{H_v} x$ is odd, then $\deg_G v$ is odd for every vertex v of G. Therefore, $G \times H$ is Eulerian if and only if both G and H are Eulerian or every vertex of G and H is odd. ◆

Exercises for Section 12.3

1. In the argument that Euler gave to solve the Königsberg Bridge Problem (see Figure 12.41), he indicated that any stroll could be represented as a sequence such as $x_1, x_3, x_3, x_4, x_5, x_6, x_7, x_8$ (see (12.8)), where each term is one of A, B, C, D, indicating the land region to which the stroll had progressed. Observe, for example, that at least one of B, C and D, say B, is neither x_1 nor x_8. Thus each occurrence of B in the sequence must be x_i where $2 \leq i \leq 7$.

   (a) Can B occur exactly once in the sequence?

   (b) Can B occur more than once in the sequence?

   (c) What information do (a) and (b) give?

2. Solve the problem dealing with walking through an art exhibit, stated at the beginning of this chapter.

3. Prove Theorem 12.43: *Let G be a connected graph. Then G has an Eulerian trail if and only if G has exactly two odd vertices. Furthermore, any Eulerian trail of G begins at one of the odd vertices and ends at the other.*

4. Classify each graph or multigraph in Figure 12.49 as containing (a) an Eulerian circuit, (b) an Eulerian trail or (c) neither.

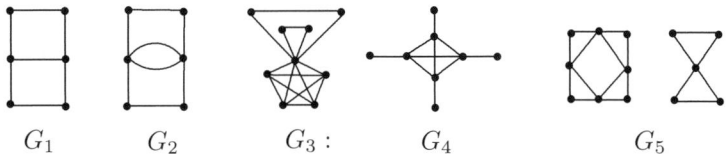

Figure 12.49: Graphs in Exercise 4

5. Give an example of a graph G of order 6, where

   (a) G is Eulerian and the vertices of G have two different degrees, half of the vertices having one degree and the other half having the other degree.

   (b) G has an Eulerian trail and the vertices of G have three different degrees, one third of each degree.

6. Prove or disprove the following.

   (a) Let G be a multigraph with an Eulerian trail. If an edge (possibly an additional edge) is added between two odd vertices, then the resulting multigraph is Eulerian.

   (b) Let G be a multigraph containing an Eulerian trail whose two odd vertices are adjacent. If the edge joining these two vertices is removed from G, then the resulting multigraph is Eulerian.

   (c) Let G be a graph containing an Eulerian trail. If a new vertex is added to G and joined to all odd vertices of G, then the resulting graph is Eulerian.

7. Let $n \geq 3$ be an integer. Prove that $\overline{C}_n$ is Eulerian if and only if $n \geq 5$ and n is odd.

8. Prove for $1 \leq s \leq t$ that $K_{s,t}$ is Eulerian if and only if both s and t are even.

9. Prove for $n \geq 2$, that Q_n is Eulerian if and only if n is even.

10. A graph G of order 8 has the property that $G - v$ is Eulerian for every vertex v of G. If $G = K_8$, then G has this property. Does any other graph of order 8 have this property?

11. It is known that every graph of order $n \geq 2$ contains at least two vertices of the same degree. Show that every Eulerian graph of order $n \geq 3$ contains at least three vertices of the same degree.

12. Suppose that G is an Eulerian graph of order $n \geq 3$ containing exactly three vertices of the same degree r and at most two vertices of any other degree.

   (a) Show that n is odd.

   (b) Show that $n = 4k - 1$ or $n = 4k + 1$ for some positive integer k.

   (c) Show that if $n = 4k - 1$ or $n = 4k + 1$, then $r = 2k$.

13. Let G be a connected graph of order $n \geq 6$ that has neither an Eulerian circuit nor an Eulerian trail. A graph H is constructed by adding a new vertex v to G and joining v to every odd vertex of G. Prove or disprove: H is Eulerian.

14. Draw $\overline{K}_{3,4}$, $\overline{W}_5$ and $\overline{P_4 + P_5}$.

15. Draw the following:

    (a) $P_4 + P_4$. $K_3 + K_3$. $K_{2,3} + K_2$. (b) $P_4 \times P_4$. $K_3 \times K_3$. $K_{2,3} \times K_2$.

16. Draw Q_4.

17. What is the order and the size of the n-cube Q_n?

18. For an integer $n \geq 2$, the vertex set of the graph R_n is the set of n-bit strings and two vertices of R_n are adjacent if they differ in exactly *two* coordinates.

    (a) Draw R_2 and R_3. (b) What is the size of R_4?

19. Let G_1, G_2 and G_3 be pairwise disjoint connected regular graphs and let $G = G_1 + (G_2 + G_3)$ be the graph obtained from G_1, G_2 and G_3 by adding edges between every two vertices belonging to two of G_1, G_2 and G_3. Prove that if G_1 and $\overline{G}_1$ are Eulerian but G_2 and G_3 are not Eulerian, then G is Eulerian.

20. Let G be an r-regular graph of odd order n and let $F \cong \overline{G}$, where F and G have disjoint vertex sets. A graph H is constructed from G and F by adding two new vertices u and v and joining u and v to each other as well as to every vertex of G and F. Which of the following is true?

    (a) H is Eulerian.

    (b) H has an Eulerian trail.

    (c) H has neither an Eulerian circuit nor an Eulerian trail.

21. Prove or disprove: There exist two connected graphs G and H both of order at least 3 and neither of which is Eulerian such that $G + H$ is Eulerian.

22. Figure 12.50 shows a map of city divided into six land regions, denoted by A, B, C, D, E, F and twelve bridges. Is it possible to take a stroll about this city where each bridge is crossed exactly once?

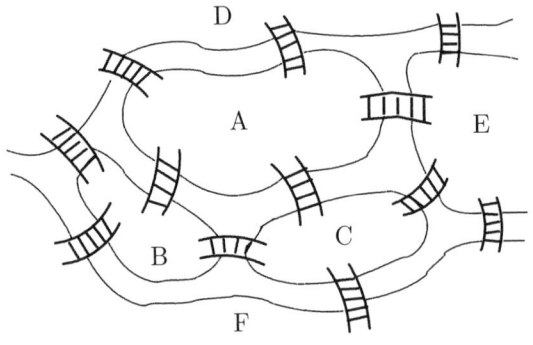

Figure 12.50: A map of a city

23. Inspector House is investigating the mysterious death of Count Evan Gilbertson III that has occurred at the estate of the Count (see Figure 12.51). The butler states that he saw a suspicious person enter the Computer Room (where the body was found) and then leave the room by the very same door. When the Inspector questions the Count's business partner Mr. Garfield Floyd, Floyd admits entering the estate by the front door and exiting by the rear door. However, Floyd says that he went through each doorway exactly once and so he could not have been the person the butler saw. Is someone lying?

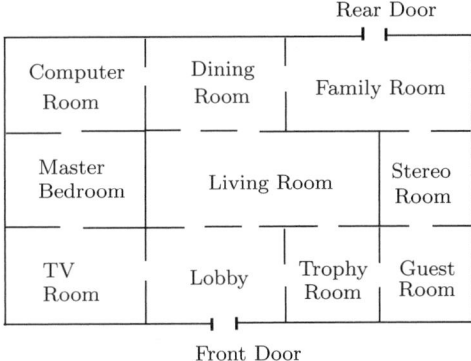

Figure 12.51: The estate of Count Evan Gilbertson III

12.4 Hamiltonian Graphs

For more than a century following Euler's solution of the Königsberg Bridge Problem, there was only minor progress in the development of the new field of graph theory. However, in the 1850s three events occurred that would have a major influence on graph theory. We will discuss one of these now and the remaining two in the next two chapters.

Born in Dublin, Ireland in 1805, William Rowan Hamilton was a child prodigy with a gift for languages and mathematics. Although he had mastered many languages before he was a teenager, his main interests would become mathematics and physics. Today Hamilton is recognized as a leading mathematician and physicist of the 19th century. It was Hamilton's work in physics, however, that led to him being knighted in 1835 and thus becoming *Sir* William Rowan Hamilton.

While Hamilton developed a great deal of mathematics, one of his final major ideas led to a subject he referred to as *icosian calculus*. In August 1856, Hamilton attended a meeting of the British Association and stayed at the home of his friend John Graves, who had an extensive library. It was during this time that Graves posed some puzzles to Hamilton. Whether it was Graves or the books Hamilton was reading at Graves' home, Hamilton began thinking about a connection between his new icosian calculus and traveling along the edges of a dodecahedron, visiting every vertex exactly once and returning to the starting point. A *dodecahedron* is a geometric solid with 20 vertices, 30 edges and 12 faces (see Figure 12.52). As it turned out, Hamilton's icosian calculus could be illustrated by a game with a game board containing the "graph of the dodecahedron" and holes at the vertices where numbered pegs could be placed to help keep track of a cycle that traveled about every vertex of the dodecahedron.

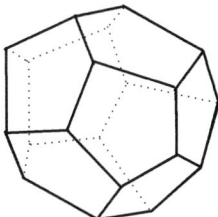

Figure 12.52: A dodecahedron

Graves put Hamilton in contact with John Jacques, a London toymaker, to whom Hamilton sold the rights for this game. Two versions of the game were manufactured, one for the parlor, played on a flat board and another for the "traveler," played on an actual dodecahedron. Hamilton called his game the Icosian Game (see Figure 12.53). The 20 vertices of the dodecahedron were labeled with the 20 consonants of the English alphabet.

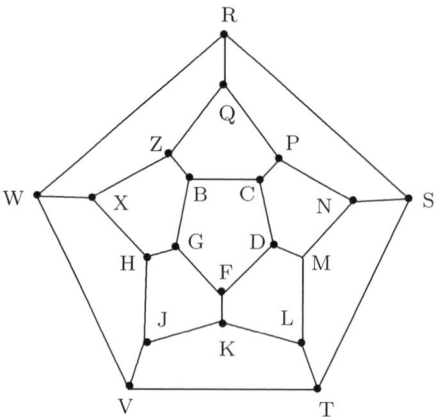

Figure 12.53: Hamilton's Icosian Game

The second version of the game was labeled as

NEW PUZZLE
TRAVELLER'S DODECAHEDRON
or
A VOYAGE ROUND THE WORLD

B. Brussels   N. Naples
C. Canton   P. Paris
D. Delhi   Q. Quebec
F. Frankfort   R. Rome
G. Geneva   S. Stockholm
H. Hanover   T. Toholsk
J. Jeddo   V. Vienna
K. Kashmere   W. Washington
L. London   X. Xenia
M. Moscow   Z. Zanzibar

The idea was thus to construct a round trip where each of the 20 cities is visited on the trip exactly once. Hamilton was also involved in marketing the game. Hamilton even wrote the preface to the instruction pamphlet, which began as follows:

In this new Game (invented by

Sir WILLIAM ROWAN HAMILTON, LL.D., & c., of Dublin,

and by him named Icosian from a Greek word signifying 'twenty') a player is to place the whole or part of a set of twenty numbered pieces or men upon the points or in the holes of a board, represented by the diagram above drawn, in such a manner as always to proceed along the lines of the figure and also to fulfill certain other conditions, which may in various ways be assigned by another player. Ingenuity and skill may thus be exercised in proposing as well as in resolving problems of the game. For example, the first of the two players may place the first five pieces in any five consecutive holes and then require the second player to place the remaining fifteen men consecutively in such a manner that the succession may be cyclical, that is, so that No. 20 may be adjacent to No. 1; and it is always possible to answer any question of this kind.

This game eventually led to a new concept in graph theory.

Definition 12.52 *A cycle in a graph G that contains every vertex of G is called a* **Hamiltonian cycle** *of G and a graph containing a Hamiltonian cycle is called a* **Hamiltonian graph**.

Example 12.53

The graph of the dodecahedron is Hamiltonian. A Hamiltonian cycle of this graph is shown (with solid edges) in Figure 12.54. This also provides a solution to the puzzle A Voyage Round the World. ♦

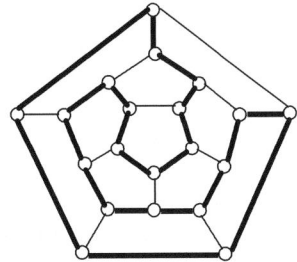

Figure 12.54: A Hamiltonian cycle in the graph of the dodecahedron

A simple but useful observation concerning Hamiltonian graphs is the following:

Let H be a spanning subgraph of a graph G. If H is Hamiltonian, then so is G.

Example 12.54

The cycle C_n ($n \geq 3$) is a Hamiltonian graph. In fact, C_n is itself a Hamiltonian cycle. Consequently, for $n \geq 3$, the complete graph K_n is Hamiltonian. ♦

Example 12.55

All of the graphs in Figure 12.55 are Hamiltonian (with a Hamiltonian cycle indicated with solid edges in each graph). ♦

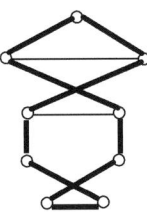 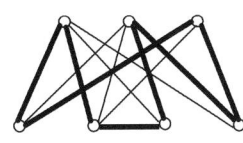 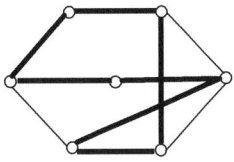

Figure 12.55: Hamiltonian graphs

Unlike the situation for Eulerian graphs, where there is a theorem (Theorem 12.42) that provides us with necessary and sufficient conditions for a graph G to be Eulerian (G is nontrivial, connected and every vertex is even), there is no such theorem for Hamiltonian graphs. That is, there is no known useful conditions that a graph G must satisfy to guarantee that G is Hamiltonian and such that if G does not satisfy these conditions then G is not Hamiltonian. In fact, the situation is even worse than this. Many mathematicians believe that no such conditions will ever be found. So we are faced with a rather unpleasant situation. It seems that to show a graph G is Hamiltonian, we must find a Hamiltonian cycle in G. And to show that G is not Hamiltonian – then what must we do? Surely, being unable to find a Hamiltonian cycle in G does not mean that G is not

Hamiltonian. After all, maybe G has a Hamiltonian cycle but we simply can't find one. Fortunately, the situation is not quite as bleak as this. Only connected graphs can be Hamiltonian. That is, if G is a disconnected graph, then G is definitely not Hamiltonian. Furthermore, if G is connected but G contains a cut-vertex, then G cannot be Hamiltonian. There is, in fact, a property that all Hamiltonian graphs must possess. (Recall that $k(H)$ denotes the number of components in a graph H.)

Theorem 12.56 *If G is a Hamiltonian graph, then*

$$k(G - S) \leq |S|$$

for every nonempty proper subset S of $V(G)$.

Proof. Let G be a Hamiltonian graph and let S be a nonempty proper subset of $V(G)$. Of course, if $G - S$ is connected, then $k(G - S) = 1$ and $k(G - S) \leq |S|$. So we may assume that $G - S$ is disconnected. Let $G_1, G_2, \ldots, G_k$, where $k = k(G - S) \geq 2$, be the components of $G - S$. Necessarily, no vertex of G_i is adjacent to a vertex of G_j $(i \neq j)$. Since G is Hamiltonian, G contains a Hamiltonian cycle C. We may assume that the initial (and terminal) vertex of C is v, where v belongs to G_1. For $1 \leq i \leq k$, let v_i be the last vertex of C that belongs to G_i and let x_i be the vertex that immediately follows v_i on C. Thus $x_i \in S$. Therefore, $|S| \geq k = k(G - S)$.   ∎

Theorem 12.56 is a *necessary* condition for a graph to be Hamiltonian, that is, every Hamiltonian graph satisfies this condition. Since necessary conditions are most useful in their contrapositive formulation, we also state a theorem that is equivalent to Theorem 12.56 and which provides a sufficient condition for a graph to be non-Hamiltonian.

Theorem 12.57 *If there exists a nonempty proper subset S of the vertex set of a graph G such that*

$$k(G - S) > |S|,$$

then G is not Hamiltonian.

For example, this says, as we noted earlier, that no graph with a cut-vertex is Hamiltonian.

Example 12.58

The graph $G = K_{2,3}$ of Figure 12.56 is not Hamiltonian. Letting $S = \{u, v\}$, we see that $k(G - S) = 3 > 2 = |S|$. By Theorem 12.57 then, G is not Hamiltonian.   ◆

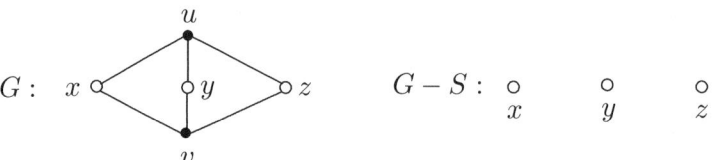

Figure 12.56: A non-Hamiltonian graph G

There are also some *sufficient* conditions for a graph to be Hamiltonian. That is, if a graph G satisfies one of these conditions, then G must be Hamiltonian. However, these conditions are strictly sufficient, that is, a graph G can be Hamiltonian without satisfying any of these. We look at one of the simplest and yet most useful of these conditions.

As we noted earlier, for $n \geq 3$, the complete graph K_n is Hamiltonian. Consequently, if every vertex of a graph G of order $n \geq 3$ has degree $n - 1$, then G is Hamiltonian. A Hamiltonian graph of order $n \geq 3$ must be connected. By Theorem 12.33, if every vertex of a graph G of order $n \geq 3$ has degree at least $(n - 1)/2$, then G must be connected. However, G need not be Hamiltonian. For

$G:$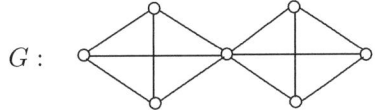

Figure 12.57: A non-Hamiltonian graph G of order n with $\delta(G) = (n-1)/2$

example, every vertex of the graph G of order $n = 7$ in Figure 12.57 has degree at least $(n-1)/2 = 3$ but G is not Hamiltonian.

This leads us to ask if every vertex in a graph G of order $n \geq 3$ has degree at least $n/2$, whether G must be Hamiltonian. This, in fact, turns out to be true and is an observation that was first made by the Danish mathematician Gabriel Andrew Dirac (the stepson of Paul Dirac, who was awarded the Nobel Prize in physics in 1933).

Theorem 12.59 *If G be a graph of order $n \geq 3$ such that*

$$\deg v \geq \frac{n}{2}$$

for each vertex v of G, then G is Hamiltonian.

Proof. Suppose that the theorem is false. Then there must exist some integer $n \geq 3$ and some graph G of order n such that $\deg v \geq n/2$ for each vertex v of G but yet G is not Hamiltonian. Among all such graphs G, let H be one with a maximum number of edges. Surely, $H \neq K_n$ since K_n is Hamiltonian. Therefore, H contains some pairs of nonadjacent vertices. Let u and w be two nonadjacent vertices in H. Because H is a graph of largest size with this property that is not Hamiltonian, $H + uw$ must be Hamiltonian and therefore contains Hamiltonian cycles. In fact, every Hamiltonian cycle of $H + uw$ must contain the edge uw.

Let C be a Hamiltonian cycle of $H + uw$. If we remove the edge uw from C, then we obtain a $u - w$ path

$$P = (u = u_1, u_2, \ldots, u_n = w)$$

that contains every vertex of H. Observe that if P contains a vertex u_i $(2 \leq i \leq n)$ such that uu_i and $u_{i-1}w$ are both edges of H (of course, this is impossible for $i = 2$ and $i = n$), then

$$C' = (u = u_1, u_i, u_{i+1}, \ldots, u_n = w, u_{i-1}, u_{i-2}, \ldots, u_1 = u)$$

is a Hamiltonian cycle in H, which cannot occur (see Figure 12.58). What this says is that for every vertex u_i adjacent to u, the vertex u_{i-1} cannot be adjacent to w. Since the degree of w is at most $n - 1$ in any case, it follows that

$$\deg w \leq n - 1 - \deg u \leq n - 1 - \frac{n}{2} = \frac{n}{2} - 1.$$

This, however, is impossible since $\deg w \geq n/2$. ∎

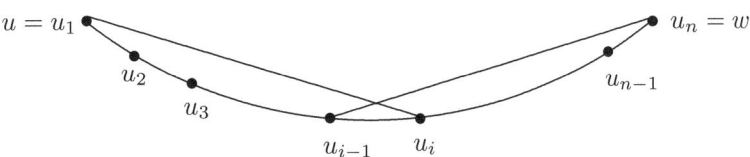

Figure 12.58: A step in the proof of Theorem 12.59

Example 12.60 *A 3-regular graph G of order 12 and a 4-regular graph H of order 20 need not be Hamiltonian. Show that $G + H$ must be Hamiltonian.*

Solution. The graph $G + H$ has order $n = 12 + 20 = 32$. Every vertex of G has degree $3 + 20 = 23$ in $G + H$, while every vertex of H has degree $4 + 12 = 16$ in $G + H$. Since every vertex of $G + H$ has degree at least $n/2$, it follows by Theorem 12.59 that $G + H$ is Hamiltonian.   ♦

Exercises for Section 12.4

1. Figure 12.59 shows a diagram of an art exhibit consisting of seven rooms labeled A, B, C, D, R, S, T. A visitor enters the art exhibit by going from the outside into room A.

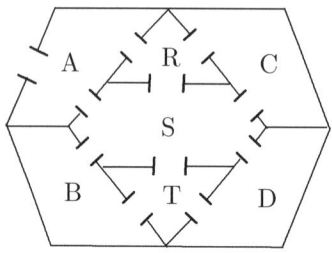

Figure 12.59: Strolling through an art exhibit

   (a) Is it possible to walk through the exhibit, entering each room exactly once, returning to room A and then exiting?

   (b) Once a visitor has entered room A, is it possible to stroll through the art exhibit, passing through each doorway exactly once, returning to room A and then exiting?

2. Determine which of the graphs in Figure 12.60 are Hamiltonian.

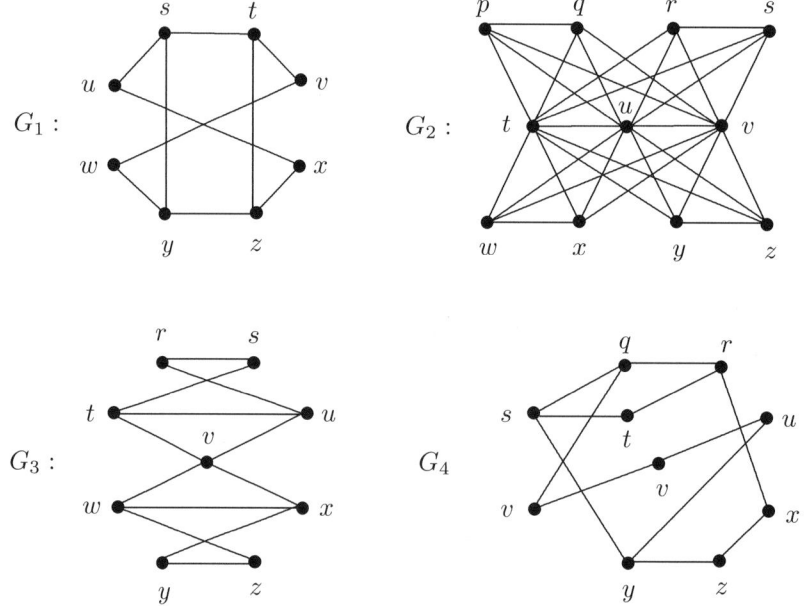

Figure 12.60: Graphs in Exercise 2

3. (a) What does Theorem 12.59 say about Q_3?

   (b) Is Q_3 Hamiltonian?

4. (a) Show that there exists no graph G such that both G and $\overline{G}$ satisfy the conditions in Theorem 12.59

   (b) Does there exist a graph G such that both G and $\overline{G}$ are Hamiltonian?

5. Show that if a connected graph G contains three vertices of degree 2, all of which are adjacent to two vertices u and v, then G is not Hamiltonian.

6. Suppose that G is a graph of order $n \geq 3$ such that $\deg v \geq (n-1)/2$ for every vertex v of G. Which of the following is true?

   (a) G is Hamiltonian. (b) G is not Hamiltonian.

   (c) It is impossible to determine whether G is Hamiltonian.

7. The vertices of a Hamiltonian graph G of order n have degrees $d_1, d_2, \ldots, d_n$ while the vertices of a graph H of order n have degrees $d'_1, d'_2, \ldots, d'_n$. If $d'_i \geq d_i$ for $i = 1, 2, \ldots, n$, is H Hamiltonian?

8. Determine which of the graphs $P_3 \times P_2$, $P_3 \times P_3$, $P_3 \times P_4$ and $P_3 \times P_5$ are Hamiltonian.

9. Determine whether the graph $C_3 \times C_3$ is Hamiltonian.

10. Does there exist a 4-regular graph G of order 6 containing two Hamiltonian cycles that have no edges in common?

11. Let $K_{s,t}$ be the complete bipartite graph where $2 \leq s \leq t$. Prove that $K_{s,t}$ is Hamiltonian if and only if $s = t$.

12. Prove or disprove: If u and v are two adjacent vertices in a graph G such that there is no $u-v$ path in G containing all of the vertices of G, then G is not Hamiltonian.

13. A Hamiltonian graph G of order n has the property that for every edge e of G, the graph $G-e$ is not Hamiltonian. Show that $G = C_n$.

14. Show that if every vertex of a graph G of order $n \geq 4$ has degree at least $(n+1)/2$, then $G-v$ is Hamiltonian for every vertex v of G.

15. Prove or disprove: If G is a graph of order $n \geq 3$ such that $\deg v < n/2$ for every vertex v of G, then G is not Hamiltonian.

16. Let $S = \{1, 2, 3, 4\}$ and let V be the set of 2-element and 3-element subsets of S. Let G be a graph with vertex set V such that two vertices A and B are adjacent if $|A \cap B| = 2$.

    (a) Which of the following is true?

       (i) G is Eulerian.
       (ii) G has an Eulerian trail.
       (iii) G has neither an Eulerian circuit nor an Eulerian trail.

    (b) Is G Hamiltonian?

12.5 Weighted Graphs

We have already mentioned that a network of highways connecting cities can be modeled by a graph (or perhaps a multigraph). Graphs can also be used to represent a collection of airline routes connecting pairs of the cities. What may interest us in both situations, however, is not the distance between the cities in the resulting graphs but, instead, the minimum driving distance or the minimum driving time or perhaps the minimum cost of flights between cities. What this says is that the edges in each such graph cannot be interpreted the same. In particular, with each edge in such a graph, there is associated a number (perhaps a distance or cost) called the **weight** of the edge.

Example 12.61 *A **weighted graph** is a graph G in which every edge e of G is assigned a real number weight (usually a positive number), which is denoted by $w(e)$ and called the weight of the edge e.*

Figure 12.61 shows two weighted graphs G and H.

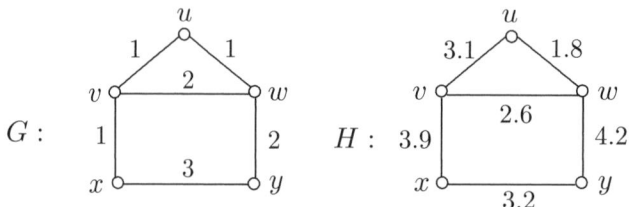

Figure 12.61: Two weighted graphs

Recall that the distance $d(u,v)$ between two vertices u and v in a connected graph G is the length of a shortest $u - v$ path in G. For example, in the graph G of Figure 12.62, $d(u,v) = 6$. The manner in which G is drawn suggests that G might model a street system in which u and v are street intersections. Considering distance in this manner, we see that the shortest drive between u and v involves traveling 6 blocks. Indeed, there are many routes from u to v that involve driving 6 blocks. Equivalently, there are many $u - v$ paths of length 6 in the graph G.

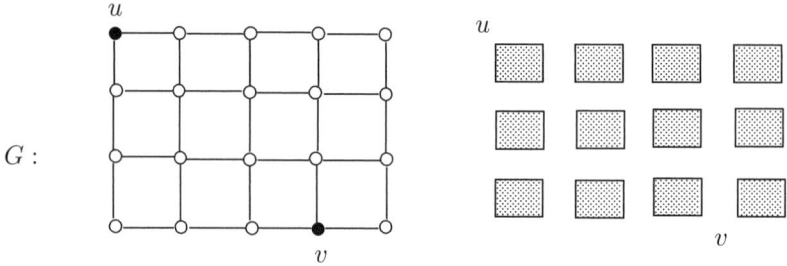

Figure 12.62: Distance between two vertices

In a particular situation, the problem that we have just described is probably too simplistic. Suppose that the blocks shown in Figure 12.62 are in a major metropolitan city. One would not expect all blocks to be equal. Typically, some blocks are longer than others and, depending on the time of day we are driving, the amount of time it takes to travel the length of a particular block can vary. If vertex u indicates the location of an emergency medical service and v indicates an intersection where a person is in need of help, then it is essential that a medical unit dispatched from u travel along a route to v that is likely to take the least amount of time. Indeed, it is conceivable that the route from u to v that takes the least time involves driving more than 6 blocks. In order to solve this problem and others like it, we consider distance in weighted graphs.

Definition 12.62 *Let u and v be two vertices in a weighted graph G. The **length** of a $u - v$ path P in G is the sum of the weights of the edges of P. Then the **distance** $d(u,v)$ between u and v is the minimum length of a $u - v$ path in G.*

Let's consider an example of this.

Example 12.63

A connected weighted graph G is shown in Figure 12.63. The graph G contains a number of $u - v$ paths. The $u - v$ path of minimum length, however, is

$$P = (u, r, s, p, q, t, w, y, v).$$

Since the length of this path is 120, it follows that $d(u, v) = 120$.

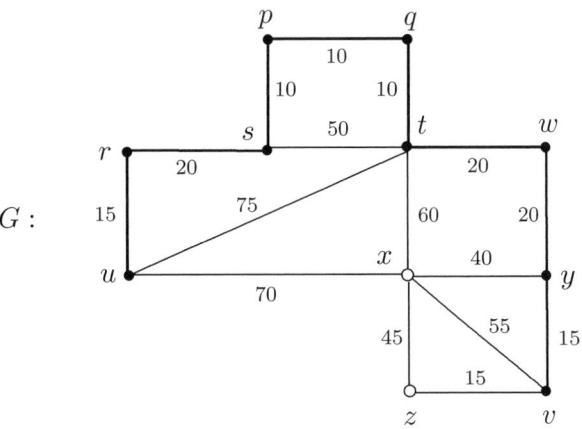

Figure 12.63: A shortest $u - v$ path in a connected
weighted graph G in Example 12.63

Dijkstra's Algorithm

The best known algorithm for determining the distance between a specific vertex r in a connected weighted graph G and the other vertices of G is one due to Edsger Wybe Dijkstra (1930–2002), a pioneer in the science and industry of computing. He believed that mathematical logic is and must be the basis for the construction of sensible computer science programs. He was well known for his love of writing, both professionally and through correspondence with his many friends. Despite his computer background, he greatly preferred writing longhand to using e-mail.

Dijkstra's algorithm is one of the most important algorithms in all of discrete mathematics. Before formally stating this algorithm, let us illustrate how this algorithm works for a particular vertex in a particular weighted graph, namely for the vertex r in the weighted graph G of Figure 12.64.

Example 12.64

Use Dijkstra's algorithm to find the distances from the vertex r to the other vertices in the connected weighted graph G of Figure 12.64.

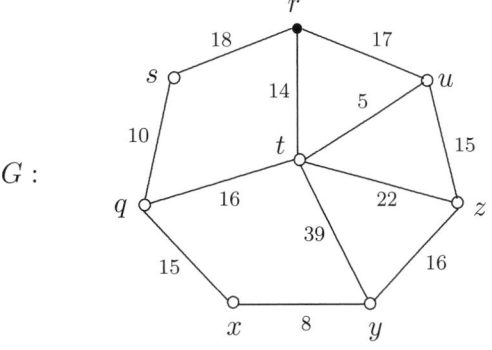

Figure 12.64: The weighted graph G in Example 12.64

Solution. We begin by assigning a label $L(v)$ to each vertex v of G. The vertex r is assigned the label $L(r) = 0$, while $L(v) = \infty$ is assigned to all vertices v distinct from r. (Rather than letting $L(v) = \infty$, we could let $L(v) = k(n-1)$, for example, where k is the largest weight of any edge in G and n is the order of G.) Since we know that $d(r, r) = 0$, we initially define a set S by letting $S = \{r\}$, which denotes the set of vertices to which the distance from r is now known. As we proceed through Dijkstra's algorithm, each vertex v distinct from r will eventually be assigned an ordered pair (a, b), possibly more than once, where a is the current value of $L(v)$ and b is a vertex adjacent to v that lies on a current $r - v$ path under consideration.

We begin with the vertices adjacent to r. Since s is adjacent to r and $w(rs) = 18$, we let $L(s) = 18$ and assign s the ordered pair $(18, r)$. This indicates that the length of the path (r, s) is 18. Similarly, t is assigned the ordered pair $(14, r)$ and u is is assigned the ordered pair $(17, r)$. Presently, the labels of the remaining vertices are not changed. Since 14 is the smallest number among the numbers 18, 14, 17, we claim that $d(r, t) = 14$. The path (r, t) is a shortest $r - t$ path, for otherwise any other $r - t$ path would have to begin with either rs or ru, each of which has weight exceeding 14. The final ordered pair assigned to t is then $(14, r)$. We now update S so that $S = \{r, t\}$. Thus, we have now found shortest paths from r to the vertices of S and the lengths of these paths.

Next, we look at the vertices adjacent to t that do not belong to S. Since q is adjacent to t and $w(qt) = 16$, we let $L(q) = w(qt) + L(t) = 16 + 14 = 30$. This indicates that there is an $r - q$ path of length 30, namely (r, t, q). Similarly, $L(y) = w(yt) + L(t) = 39 + 14 = 53$ and $L(z) = w(zt) + L(t) = 22 + 14 = 36$. However, $w(tu) = 5$ and $w(tu) + L(t) = 5 + 14 = 19$. This says that there is an $r - u$ path of length 19, namely (r, t, u). Since $L(u) = 17 < 19$, we are already aware of an $r - u$ path of smaller length, namely (r, u). Consequently, we do not change the value of $L(u)$ and so we still have $L(u) = 17$. Since 17 is the minimum of $18, 17, 30, 53, 36$, it follows that there is no $r - u$ path whose length is less than 17 and so $d(r, u) = 17$. Thus u is assigned the ordered pair $(17, r)$. What this says is that (r, u) is a shortest $r - u$ path in G. We then add u to S and so S is now $\{r, t, u\}$. Continuing in this manner, we obtain the table in Figure 12.65 below which gives the distances from r to the other vertices of G and provides the information that gives a shortest path.

r	s	t	u	q	x	y	z	S
0	∞	∞	∞	∞	∞	∞	∞	r
	$(18, r)$	$(14, r)$	$(17, r)$	∞	∞	∞	∞	t
	$(18, r)$		$(17, r)$	$(30, t)$	∞	$(53, t)$	$(36, t)$	u
	$(18, r)$			$(30, t)$	∞	$(53, t)$	$(32, u)$	s
				$(28, s)$	∞	$(53, t)$	$(32, u)$	q
					$(43, q)$	$(53, t)$	$(32, u)$	z
					$(43, q)$	$(48, z)$		x
						$(48, z)$		y

$$
\begin{aligned}
d(r, s) &= 18 & (r, s) \\
d(r, t) &= 14 & (r, t) \\
d(r, u) &= 17 & (r, u) \\
d(r, q) &= 28 & (r, s, q) \\
d(r, x) &= 43 & (r, s, q, x) \\
d(r, y) &= 48 & (r, u, z, y) \\
d(r, z) &= 32 & (r, u, z)
\end{aligned}
$$

Figure 12.65: Applying Dijkstra's algorithm to find the distances from the vertex r to the other vertices in the connected weighted graph G in Example 12.64

For example, the last entry in the column headed by x is $(43, q)$. This says that $d(r, x) = 43$ and that in some shortest $r - x$ path, the vertex adjacent to x is q. However, in some shortest $r - q$ path, the vertex adjacent to q is s. A shortest $r - s$ path is (r, s). Thus a shortest $r - x$ path is (r, s, q, x).   ◆

We now present a formal statement of Dijkstra's Algorithm.

Algorithm 12.65 (**Dijkstra's Algorithm**) *This algorithm determines the distances from a fixed vertex v_1 in a connected weighted graph G with $V(G) = \{v_1, v_2, \ldots, v_n\}$ to all other vertices of G and for each vertex v_i $(2 \leq i \leq n)$ gives the vertex adjacent to v_i that lies on some shortest $v_1 - v_i$ path.*

> **Input**: A positive integer n and a nontrivial connected weighted graph G with $V(G) = \{v_1, v_2, \ldots, v_n\}$.

> **Output**: $d(v_1, v_i)$ for each vertex v_i of G $(2 \leq i \leq n)$ and a vertex adjacent to v_i that lies on some shortest $v_1 - v_i$ path.

1. $L(v_1) := 0$
2. **for** $i := 2$ **to** n **do**
   $\qquad L(v_i) := \infty$
3. $S := \{v_1\}$
4. $u := v_1$
5. **while** $V(G) - S \neq \emptyset$ **do**
   $\qquad$ **begin**
6. $\qquad$ **for** $i := 2$ **to** n **do**
   $\qquad\qquad$ **begin**
7. $\qquad\qquad$ **while** $v_i \in V(G) - S$ and $uv_i \in E(G)$ **do**
8. $\qquad\qquad$ **if** $w(uv_i) + L(u) < L(v_i)$ **then**
   $\qquad\qquad\qquad L(v_i) := w(uv_i) + L(u)$ and $P(v_i) := (L(v_i), u)$
   $\qquad\qquad$ **end**
9. $\qquad a := \min\{L(v_i) : v_i \in V(G) - S\}$
10. $\qquad$ **for** $i := 2$ **to** n **do**
    $\qquad\qquad$ **begin**
11. $\qquad\qquad$ **while** $v_i \in V(G) - S$ **do**
12. $\qquad\qquad$ **if** $L(v_i) = a$ **then** $S := S \cup \{v_i\}$, $u := v_i$ and $a := a - 1$
    $\qquad\qquad$ **end**
    $\qquad$ **end**
13. **for** $i := 2$ **to** n **do**
14. $\qquad$ output $P(v_i)$

Now that we have given a formal statement of Dijkstra's algorithm, we use it in another example.

Example 12.66

Use Dijkstra's algorithm to find the distances from the vertex v_1 in the connected weighted graph G of Figure 12.66 to all other vertices in G.

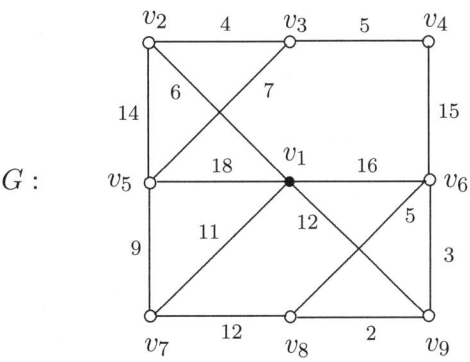

Figure 12.66: The weighted graph G in Example 12.66

Solution. Initially, v_1 is assigned the label $L(v_1) = 0$ and all other vertices of G are assigned the label ∞. The set S is initially defined to be $\{v_1\}$ and u is defined to be v_1. Step 5 is a while loop which is executed as long as $V(G) - S \neq \emptyset$. There is a for loop at Step 6. For each vertex $v_i \notin S$ ($2 \leq i \leq n$) that is adjacent to u, the new label $w(uv_i) + L(u)$ will be assigned to v_i if this is less than its previous label and each such vertex is assigned an ordered pair $P(v_i) = (L(v_i), u)$. (Certainly, this is the case whenever $L(v_i) = \infty$ is the current label.)

After completing the while loop at Step 7 and necessarily Step 8 as well, we have

$$L(v_2) = 6, \ L(v_5) = 18, \ L(v_6) = 16, \ L(v_7) = 11 \text{ and } L(v_9) = 12.$$

Since $u = v_1$, we now have

$$P(v_2) = (6, v_1), \ P(v_5) = (18, v_1), \ P(v_6) = (16, v_1), \ P(v_7) = (11, v_1), \ P(v_9) = (12, v_1).$$

The minimum value a of the labels of the vertices not in S is then computed in Step 9, where it is determined that $a = 6$. Proceeding through the for loop in Step 10, we see that the first vertex attaining the value $a = 6$ is v_2, which is added to S (that is, $S = \{v_1, v_2\}$ now) and becomes the vertex u in Step 12. In the same step, the number a is then reduced to $a - 1 = 5$ (which means that only one vertex will have the initial value $a = 6$).

Since $V(G) - S \neq \emptyset$, the while loop at Step 5 is repeated. We run through the for loop at Step 6, each time considering those vertices $v_i \in V(G) - S$ that are adjacent to $u = v_2$. There are two such vertices, namely v_3 and v_5. Since $w(uv_3) + L(u) = 10$, which is less than ∞, it follows that $L(v_3)$ is redefined as 10 and $P(v_3) = (10, v_2)$. However, since $w(uv_5) + L(u) = 20 > 18$, it follows that $L(v_5)$ remains at 18 and $P(v_5)$ remains at $(18, v_1)$. The minimum value a of the labels of the vertices not in S is computed in Step 9 and we find that $a = 10$. We run through the for loop in Step 10 and see that the first (actually *only*) vertex whose label is 10 is v_3. In Step 12, the vertex v_3 is now added to S and so currently $S = \{v_1, v_2, v_3\}$. Also, in Step 12, u is changed to v_3 and a is changed to $a - 1 = 10 - 1 = 9$. No vertex in $V(G) - S$ has the label 9, which completes all steps in the while loop at Step 5 and since $V(G) - S \neq \emptyset$, we repeat it beginning with Step 6. The table in Figure 12.67 gives all labels $L(v_i)$ and ordered pairs $P(v_i)$ for $2 \leq i \leq n$. When we reach $S = V(G)$, we now turn to the for loop in Step 13 and all ordered pairs $P(v_i)$ for $2 \leq i \leq n$ are output.

Below the table, the distances $d(v_1, v_i)$ are given for $2 \leq i \leq 9$ as well as the shortest $v_1 - v_i$ path determined by the entries in the table.   ♦

v_1	v_2	v_3	v_4	v_5	v_6	v_7	v_8	v_9	S
0	∞	∞	∞	∞	∞	∞	∞	∞	v_1
	$(6, v_1)$	∞	∞	$(18, v_1)$	$(16, v_1)$	$(11, v_1)$	∞	$(12, v_1)$	v_2
		$(10, v_2)$	∞	$(18, v_1)$	$(16, v_1)$	$(11, v_1)$	∞	$(12, v_1)$	v_3
			$(15, v_3)$	$(17, v_3)$	$(16, v_1)$	$(11, v_1)$	∞	$(12, v_1)$	v_7
			$(15, v_3)$	$(17, v_3)$	$(16, v_1)$		$(23, v_7)$	$(12, v_1)$	v_9
			$(15, v_3)$	$(17, v_3)$	$(15, v_9)$		$(14, v_9)$		v_8
			$(15, v_3)$	$(17, v_3)$	$(15, v_9)$				v_4
				$(17, v_3)$	$(15, v_9)$				v_6
				$(17, v_3)$					v_5

$$d(v_1, v_2) = 6 \qquad (v_1, v_2)$$
$$d(v_1, v_3) = 10 \qquad (v_1, v_2, v_3)$$
$$d(v_1, v_4) = 15 \qquad (v_1, v_2, v_3, v_4)$$
$$d(v_1, v_5) = 17 \qquad (v_1, v_2, v_3, v_5)$$
$$d(v_1, v_6) = 15 \qquad (v_1, v_9, v_6)$$
$$d(v_1, v_7) = 11 \qquad (v_1, v_7)$$
$$d(v_1, v_8) = 14 \qquad (v_1, v_9, v_8)$$
$$d(v_1, v_9) = 12 \qquad (v_1, v_9)$$

Figure 12.67: Applying Dijkstra's algorithm to find the distances from the vertex v_1
to the other vertices in the connected weighted graph G in Example 12.66

The Traveling Salesman Problem

There is another problem concerning connected weighted graphs, which, although similar to that of finding distances between pairs of vertices, is extraordinarily more difficult.

The Traveling Salesman Problem: A salesman wishes to make a round trip that visits a certain number of cities. He knows the distance between all pairs (or at least many pairs) of cities or perhaps knows instead the cost of traveling between certain pairs of cities. If he is to visit each city exactly once, then what is the minimum distance (or minimum cost) of such a round trip?

This problem can be modeled quite naturally by a weighted graph G whose vertex set is the set of cities and where two vertices u and v are joined by an edge having weight r if the distance between u and v (or the cost of traveling between u and v) is known and this distance (cost) is r. To solve this Traveling Salesman Problem, we need to determine the weight of a Hamiltonian cycle in G of minimum weight. Certainly, G must contain a Hamiltonian cycle for this problem to have a solution. However, if G is complete (that is, if we know the distance or the cost of traveling between every pair of cities), then there are many Hamiltonian cycles in G if its order n is large. Since every city must lie on every Hamiltonian cycle of G, we can think of a Hamiltonian cycle starting (and ending) with a city c. Then the remaining $n-1$ cities can follow c on the cycle in any of $(n-1)!$ orders. Indeed, if we have one of the $(n-1)!$ orderings of the $n-1$ cities, then we need to add the distances (costs) between consecutive cities in the sequence as well as the distances between c and the first city in the sequence and the last city in the sequence and c. We then need to compute the minimum of these $(n-1)!$ numbers. Actually, we need *only* find the minimum of $(n-1)!/2$ numbers since we would get the same sum if a sequence was traversed in reverse order.

Solving the problem in this manner has factorial time complexity and is certainly not efficient. On the other hand, there is no known efficient algorithm to solve the general problem.

Example 12.67

Solve the Traveling Salesman Problem for the complete weighted graph G in Figure 12.68.

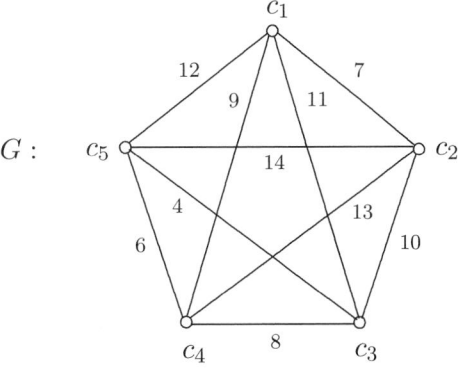

Figure 12.68: The graph in Example 12.67

Solution. Since the order of G is 5, there are $(5-1)!/2 = 12$ Hamiltonian cycles in G. These are listed below together with their weights.

Hamiltonian cycles	Weight of cycle
$s_1 = (c_1, c_2, c_3, c_4, c_5, c_1)$	$7 + 10 + 8 + 6 + 12 = 43$
$s_2 = (c_1, c_2, c_3, c_5, c_4, c_1)$	$7 + 10 + 4 + 6 + 9 = 36$
$s_3 = (c_1, c_2, c_4, c_3, c_5, c_1)$	$7 + 13 + 8 + 4 + 12 = 44$
$s_4 = (c_1, c_2, c_4, c_5, c_3, c_1)$	$7 + 13 + 6 + 4 + 11 = 41$
$s_5 = (c_1, c_2, c_5, c_3, c_4, c_1)$	$7 + 14 + 4 + 8 + 9 = 42$
$s_6 = (c_1, c_2, c_5, c_4, c_3, c_1)$	$7 + 14 + 6 + 8 + 11 = 46$
$s_7 = (c_1, c_3, c_2, c_4, c_5, c_1)$	$11 + 10 + 13 + 6 + 12 = 52$
$s_8 = (c_1, c_3, c_2, c_5, c_4, c_1)$	$11 + 10 + 14 + 6 + 9 = 50$
$s_9 = (c_1, c_3, c_4, c_2, c_5, c_1)$	$11 + 8 + 13 + 14 + 12 = 58$
$s_{10} = (c_1, c_3, c_5, c_2, c_4, c_1)$	$11 + 4 + 14 + 13 + 9 = 51$
$s_{11} = (c_1, c_4, c_2, c_3, c_5, c_1)$	$9 + 13 + 10 + 4 + 12 = 48$
$s_{12} = (c_1, c_4, c_3, c_2, c_5, c_1)$	$9 + 8 + 10 + 14 + 12 = 53$

We see therefore that the minimum weight of a Hamiltonian cycle is 36. To obtain this weight, the vertices of G should be visited in the order listed in the sequence s_2, that is, c_1, c_2, c_3, c_5, c_4, c_1.   ♦

The importance of the Traveling Salesman Problem is due to a number of related but useful applications, including, but definitely not restricted to, the following:

(1) Each morning a school bus leaves school to pick up students at a number of bus stops. It would be useful to know a route that would use the least time (to get the students to school quickly and to minimize the cost of gasoline for the bus).

(2) Late afternoon each day, a van leaves a restaurant to deliver "meals on wheels" to customers who prefer to have meals delivered to them.

(3) Each day a mail truck leaves the post office to pick up mail that has been left in mailboxes.

Indeed, there are numerous applications that have nothing to do with "traveling" but involve performing activities in some cyclic sequence that is least costly or most time efficient.

The Traveling Salesman Problem belongs to a class of problems called combinatorial optimization problems, where some quantity is being minimized or maximized. A number of these problems, including the Traveling Salesman Problem, are known to be NP-complete problems. At present, no efficient algorithm has even been found that solves any of these problems. However, if an efficient algorithm is ever found to solve any one of them, then an efficient algorithm must exist to solve every one of them. Despite this, there have been instances when a Traveling Salesman Problem has been solved for a large number of cities. For example, in 1998 David Applegate, Robert Bixby, Vasek Chvátal and William Cook solved a Traveling Salesman Problem for the 13,509 largest cities in the United States (those whose population exceeded 500 at that time). They also solved a Traveling Salesman Problem for 15,113 German cities in 2001 and for 24,978 Swedish cities in 2004. The ultimate goal is to solve the Traveling Salesman Problem for every registered city or town in the world plus a few research bases in Antartica (1,904,711 locations in all). Applegate, Bixby, Chvátal and Cook wrote a 2007 book titled *The Traveling Salesman Problem: A Computational Study*, in which they describe the history of the Traveling Salesman Problem as well as the method they used to solve a range of large-scale problems.

Exercises for Section 12.5

1. For the vertices u and v in the graph G of Figure 12.62, it was noticed that $d(u, v) = 6$. Therefore, G contains a $u - v$ path of length 6 but no $u - v$ path of length less than 6.

   (a) Give an example of a $u - v$ path of length greater than 6 in G.

   (b) Show that G contains no $u - v$ path of length 7.

2. Use Dijkstra's algorithm to compute $d(v_1, v_i)$ for $2 \leq i \leq 9$ for the connected weighted graph G of Figure 12.69 and find a shortest $v_1 - v_i$ path for each vertex v_i.

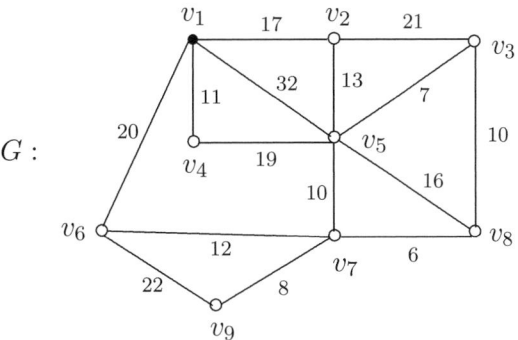

Figure 12.69: The weighted graph G in Exercise 2

3. Use Dijkstra's algorithm to compute $d(v_1, v_i)$ for $2 \leq i \leq 6$ for the connected weighted graph G of Figure 12.70 and find a shortest $v_1 - v_i$ path for each vertex v_i.

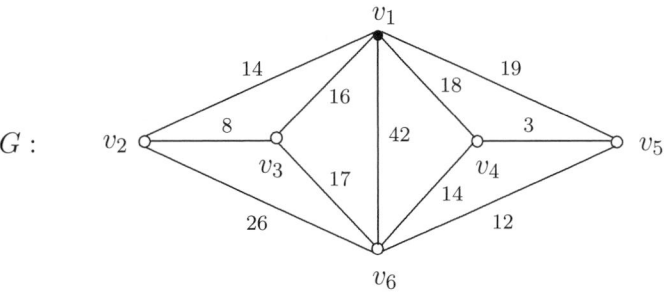

Figure 12.70: The weighted graph G in Exercise 3

4. It costs \$5 to play a game on each of the two weighted graphs G_1 and G_2 in Figure 12.71. The player finds two Hamiltonian cycles C and C' in each of the two graphs and is paid $|w(C) - w(C')|$ dollars. What is the maximum amount that anyone could win playing the game on each graph?

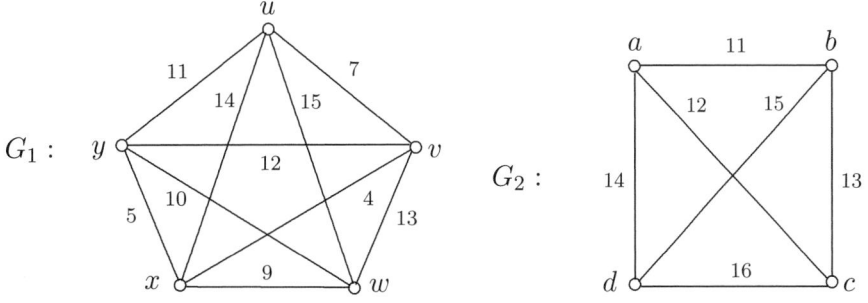

Figure 12.71: The weighted graphs G_1 and G_2 in Exercise 4

5. Solve the Traveling Salesman Problem for the weighted graph G in Figure 12.72.

6. Suppose that G_n is a weighted complete graph of order $n \geq 4$ such that $V(G) = \{v_1, v_2, \ldots, v_n\}$, where $w(v_i v_j) = i + j$ for $1 \leq i < j \leq n$.

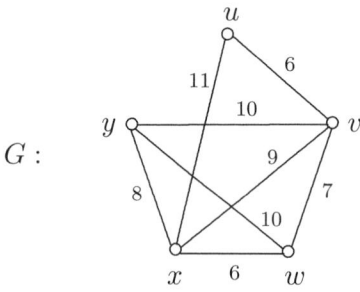

G :

Figure 12.72: The weighted graph G in Exercise 5

(a) Solve the Traveling Salesman Problem for G_4.

(b) Solve the Traveling Salesman Problem for G_5.

(c) Based on your answers for (a) and (b), what would be a reasonable conjecture for the minimum weight of a Hamiltonian cycle in G_n for $n \geq 4$?

7. A businessman in New York City learns that he will need to visit four branch offices (in Miami, Houston, Minneapolis and Los Angeles) of his company for one week each next year. When he checks into the costs of constructing trips between these cities, he learns that it is less expensive to purchase a round trip ticket between any two cities than to purchase a one-way ticket in either direction. The costs of theses trips are given below

NYC-MIA: 232   MIA-MIN: 279   NYC-HOU: 333   MIA-LOS: 322

NYC-MIN: 325   HOU-MIN: 292   NYC-LOS: 315   HOU-LOS: 552

MIN-HOU: 285   MIN-LOS: 260.

(These were the actual prices at one time.)

(a) In what order should he visit the cities and then return to New York City to minimize the cost?

(b) Later, it's determined that it will not be necessary for him to visit Minneapolis. How will this affect this trip and the cost?

(c) What interesting observation is there after comparing the 12 possible round trips in (a) with the 3 possible round trips in (b)?

Chapter 12 Highlights

Key Concepts

adjacency list (of a graph G): for each vertex of G, a list of all vertices adjacent to it is given.

adjacency matrix (of a graph G with $V(G) = \{v_1, v_2, \ldots, v_n\}$): the $n \times n$ matrix $\mathbf{A} = [a_{ij}]$, where

$$a_{ij} = \begin{cases} 1 & \text{if } v_i v_j \in E(G) \\ 0 & \text{if } v_i v_j \notin E(G). \end{cases}$$

bipartite graph: a graph whose vertex set can be partitioned into two subsets U and W (called **partite sets**) such that each edge joins a vertex in U and a vertex in W.

bridge (of a graph G): an edge e in G such that $G - e$ has more components than G.

Cartesian product of two graphs G and H, $G \times H$: a graph constructed by replacing each vertex of G by a copy of H and by joining two corresponding vertices of H by an edge if the vertices of G are adjacent.

center (of the star $K_{1,t}$, $t \geq 2$): the vertex of degree t in $K_{1,t}$.

circuit: a closed walk of length 3 or more in which no edge is repeated.

closed walk: a walk whose initial and terminal vertices are the same.

column matrix: a matrix with a single column.

complement $\overline{G}$ (of a graph G): a graph with vertex $V(G)$ such that two vertices u and v are adjacent in $\overline{G}$ if and only if u and v are not adjacent in G.

complete bipartite graph, $K_{s,t}$: a bipartite graph with partite sets U and W, where $|U| = s$ and $|W| = t$, such that two vertices are adjacent if and only if they belong to different partite sets.

complete graph (of order n), K_n: the graph of order n in which every two vertices are adjacent.

component (of a graph G): a connected subgraph of a graph G that is not a proper subgraph of any connected subgraph of G.

connected graph: a graph containing a $u - v$ path for every two vertices u and v in the graph.

connected vertices: two vertices u and v are connected in a graph G if G contains a $u - v$ path.

cubic graph: a 3-regular graph.

cut-vertex (of a graph G): a vertex v in G such that $G - v$ has more components than G.

cycle (of order $n \geq 3$), C_n: a graph whose vertices can be labeled as $v_1, v_2, \ldots, v_n$ and whose edges are $v_1v_2, v_2v_3, \ldots, v_{n-1}v_n$ and v_1v_n.

degree (of a vertex v), $\deg v$: the number of edges incident with v.

disconnected graph: a graph that is not connected.

distance (between vertices u and v in a graph), $d(u, v)$: the length of a shortest $u - v$ path.

distance (between two vertices u and v in a weighted graph), $d(u, v)$: the minimum length of a $u - v$ path in the weighted graph.

empty graph: a graph containing no edges.

end-vertex (or a **leaf**): a vertex of degree 1.

entry (of **a matrix**): a number in a matrix.

Eulerian circuit (of a graph G): a circuit that contains every edge of G.

Eulerian graph: a graph containing an Eulerian circuit.

Eulerian trail (of a graph G): an open trail that contains every edge of G.

even cycle: a cycle of even length.

even vertex: a vertex of even degree.

graph: a finite nonempty set V of **vertices** and a set E of **edges**, each a 2-element subset of V.

$G - e$ (for an edge e in a graph G): the spanning subgraph of G whose edge set consists of all edges of G except e.

$G - X$ (for a set X of edges of a graph G): the spanning subgraph of G with edge set $E(G) - X$.

$G - v$ (for a vertex v in a graph G): the subgraph of G induced by $V(G) - \{v\}$.

$G-U$ (for a proper subset U of the vertex set of a graph G): the subgraph of G induced by $V(G)-U$.

Hamiltonian cycle (of a graph G): a cycle in G that contains every vertex of G.

Hamiltonian graph: a graph containing a Hamiltonian cycle.

hypercube, n-**cube**, Q_n: $Q_1 = K_2$ and $Q_n = Q_{n-1} \times K_2$ for $n \geq 2$.

induced subgraph (of a graph G): a subgraph H such that for every two vertices u and v of H, whenever uv is an edge of G, uv is an edge of H.

inner product or dot product (of a row matrix $\mathbf{A}$ and a column matrix $\mathbf{B}$) $\mathbf{A} \cdot \mathbf{B}$: sum of the products of the corresponding elements of $\mathbf{A}$ and $\mathbf{B}$, each with the same number of entries.

isolated vertex: a vertex of degree 0.

isomorphic (two graphs G and H are **isomorphic**), $G \cong H$: there exists a bijective function (**isomorphism**) $f : V(G) \to V(H)$ such that two vertices u and v are adjacent in G if and only if $f(u)$ and $f(v)$ are adjacent in H.

join (of two graphs G and H), $G + H$: a graph constructed from G and H by adding all edges that join a vertex of G and a vertex of H.

Königsberg Bridge Problem: The problem of determining whether its possible to take a stroll through the city of Königsberg and cross each of its seven bridges exactly once.

leaf: a vertex of degree 1.

length (of a path in a weighted graph): the sum of the weights of the edges in the path.

length (of a walk in a graph): the number of occurrences of edges traversed in the walk.

loop: an edge that joins a vertex to itself.

matrix: a rectangular array of numbers. An $m \times n$ matrix is a matrix with m rows and n columns.

maximum degree (of a graph G), $\Delta(G)$: the maximum degree among the vertices of G.

minimum degree (of a graph G), $\delta(G)$: the minimum degree among the vertices of G.

multigraph: a finite nonempty set V (of vertices) such that every two distinct vertices are joined by a finite number (zero, one or more) of edges.

neighbor (of a vertex v): a vertex adjacent to v.

neighborhood (of a vertex v), $N(v)$: the set of vertices adjacent to v.

nonempty graph: a graph with one or more edges.

nontrivial graph: a graph with two or more vertices.

odd cycle: a cycle of odd length.

odd vertex: a vertex of odd degree.

open walk: a walk whose initial and terminal vertices are distinct.

order (of a graph): the number of vertices in the graph.

parallel edges: edges joining the same pair of distinct vertices in a multigraph.

path: a walk in which no vertex is repeated.

path of order n, P_n: a graph whose vertices can be labeled as $v_1, v_2, \ldots, v_n$ and whose edges are $v_1 v_2, v_2 v_3, \ldots, v_{n-1} v_n$.

pendant vertex: a vertex of degree 1.

power (of a matrix $\mathbf{A}$), $\mathbf{A}^k$: the product $\mathbf{AA} \cdots \mathbf{A}$ of k factors of $\mathbf{A}$.

product (of an $m \times n$ matrix $\mathbf{A} = [a_{ij}]$ and an $n \times p$ matrix $\mathbf{B} = [b_{jk}]$), $\mathbf{AB}$: an $m \times p$ matrix whose (i, k)-entry is $a_{i1}b_{1k} + a_{i2}b_{2k} + \cdots + a_{in}b_{nk}$ for $1 \leq i \leq m$ and $1 \leq k \leq p$.

proper subgraph (of a graph G): a subgraph H of G such that $H \neq G$.

pseudograph: a finite nonempty set V (of vertices) allowing both parallel edges and loops (including multiple loops).

regular graph (or r-**regular graph**): a graph each of whose vertices has degree r.

row matrix: a matrix with a single row.

size (of a graph): the number of edges in the graph.

size (of **a matrix**): the number of rows and columns in the matrix.

spanning subgraph (of a graph G): a subgraph H of G with $V(H) = V(G)$.

star: a complete bipartite graph, one of whose partite sets consists of one vertex.

subgraph (of a graph G): a graph H with $V(H) \subseteq V(G)$ and $E(H) \subseteq E(G)$.

sum (of two matrices $\mathbf{A} = [a_{ij}]$ and $\mathbf{B} = [b_{ij}]$ of the same size), $\mathbf{A} + \mathbf{B} = [a_{ij} + b_{ij}]$.

trail: a walk in which no edge is repeated.

Traveling Salesman Problem: the problem of finding a Hamiltonian cycle of minimum length in a weighted graph.

triangle: a 3-cycle.

trivial graph: a graph containing only one vertex.

walk (in a graph G): a sequence $(u_0, u_1, \ldots, u_k)$ of vertices of G such that u_i and u_{i+1} are adjacent for $0 \leq i \leq k - 1$.

weighted graph: a graph each of whose edges is assigned a **weight** (a real number).

wheel, W_n: the join of C_n and K_1.

zero-one matrix: a matrix each of whose entries is either 0 or 1.

Key Results

- **The First Theorem of Graph Theory**: If G is a graph or multigraph of size m, then $\sum_{v \in V(G)} \deg v = 2m$.

- Every graph contains an even number of odd vertices.

- If G and H are isomorphic graphs, then

  (a) G and H have the same order,

  (b) G and H have the same size and

  (c) the degrees of the vertices of G are the same as the degrees of the vertices of H.

- Let G be a graph with vertex set $V(G) = \{v_1, v_2, \ldots, v_n\}$ and adjacency matrix $\mathbf{A}$. For each positive integer k, the number of different $v_i - v_j$ walks of length k in G is the (i, j)-entry in the matrix $\mathbf{A}^k$.

- If a graph G contains a $u - v$ walk for vertices u and v in G, then G also contains a $u - v$ path.

- A nontrivial graph G is bipartite if and only if G does not contain an odd cycle.

- If G is a disconnected graph, then $\overline{G}$ is connected.

- If G is a graph of order n such that $\deg v \geq \frac{n-1}{2}$ for every vertex v of G, then G is connected.

- An edge e of a graph G is a bridge of G if and only if e lies on no cycle of G.

- A vertex v of a connected graph G is a cut-vertex of G if and only if there exist vertices u and w distinct from v such that v lies on every $u - w$ path of G.

- Let G be a nontrivial connected graph. Then G is Eulerian if and only if every vertex of G is even.

- Let G be a connected graph. Then G has an Eulerian trail if and only if G has exactly two odd vertices. Furthermore, any Eulerian trail of G begins at one of the odd vertices and ends at the other.

- Let G be a connected multigraph with one or more edges. Then

    (1) G is Eulerian if and only if every vertex of G is even and

    (2) G has an Eulerian trail if and only if G has exactly two odd vertices.

- If G is a Hamiltonian graph, then $k(G-S) \leq |S|$ for every nonempty proper subset S of $V(G)$.

- If there exists a nonempty proper subset S of the vertex set of a graph G such that $k(G-S) > |S|$, then G is not Hamiltonian.

- If G be a graph of order $n \geq 3$ such that $\deg v \geq \frac{n}{2}$ for each vertex v of G, then G is Hamiltonian.

- **Dijkstra's Algorithm**: an algorithm that determines the distances from a fixed vertex v_1 in a connected weighted graph G with $V(G) = \{v_1, v_2, \ldots, v_n\}$ to all other vertices of G and provides a shortest $v_1 - v_i$ path for each i ($2 \leq i \leq n$).

Supplementary Exercises for Chapter 12

1. Let $A = \{a, b, c\}$. Draw the graph G with $V(G) = \mathcal{P}(A)$ (the power set of A) such that two vertices X and Y of G are adjacent if either $|X \cup Y| = 1$ or $|X \cap Y| = 1$.

2. Let $A = \{3, 4, 5, 6, 7, 8\}$. Draw the graph G with $V(G) = A$ such that two vertices a and b of G are adjacent if there exists $d \in A$ for which $d \mid (a + b)$.

3. Give an example of a graph G of order 8 having the following two properties.

    (1) For every vertex u of G, there is exactly one vertex v of G such that $u \neq v$ and $\deg u = \deg v$.

    (2) If x and y are any two vertices of G, then the degree of x in G is different from the degree of y in $\overline{G}$.

4. Does there exist a graph G of order 6 such that

    (a) all vertices of G are odd and no more than three vertices of G have the same degree?

    (b) all vertices of G are even and no more than three vertices of G have the same degree?

    (c) all vertices of G are odd and no more than two vertices of G have the same degree?

    (d) all vertices of G are even and no more than two vertices of G have the same degree?

5. Give an example of a graph G containing a vertex of degree 4 such that for every vertex x in G no two vertices in the neighborhood $N(x)$ of x have the same degree in G.

6. Let u and v be two vertices of degree 3 in a graph G of order 6. If the degrees of the vertices of the graph $G - \{u, v\}$ are 2, 1, 1, 0, then what is the size of G?

7. Let v be a vertex in an r-regular graph G of order n. If $G - v$ is a regular graph, then what is r?

8. A graph G with $V(G) = \{v_1, v_2, \ldots, v_7\}$ is defined by the adjacency lists below. Is G bipartite?

$$v_1 : v_3, v_4, v_6; \quad v_2 : v_6, v_7; \quad v_3 : v_1, v_5;$$
$$v_4 : v_1, v_6; \quad v_5 : v_3, v_7; \quad v_6 : v_1, v_2, v_4; \quad v_7 : v_2, v_5.$$

9. A self-complementary graph has order n and size $m = 18$. What is n?

10. Let G be the graph shown in Figure 12.73. A graph H models all those subgraphs of order 3 in G containing exactly two adjacent edges. Two vertices of H are adjacent if these two subgraphs have exactly one edge in common. Draw the graph H. [Hint: If a subgraph has two adjacent edges numbered 1 and 2, then denote this vertex in H by 12.]

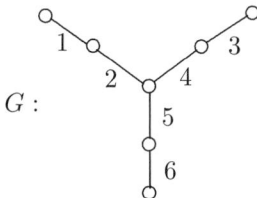

G:

Figure 12.73: The graphs in Exercise 10

11. Let G be a graph of order 6 and size 7. What is the order and size of $G + P_4$?

12. Let G be a graph of order 10 and size m. If the size of $G \times K_2$ is $2.5m$, what is m?

13. Show that there exists no graph G of order $2k + 1$ for some positive integer k such that k vertices of G have degree k and the remaining $k + 1$ vertices have degree $k + 1$.

14. Let G be a self-complementary graph of some order n. Let H be the graph of order $n + 4$ obtained from G and P_4 by joining every vertex of G to the two end-vertices of P_4. What can be said of $\overline{H}$?

15. A graph G of order n and size 17 has three end-vertices, four vertices of degree 2 and two vertices of degree 4. All other vertices have degree 3. How many vertices of degree 3 does G have?

16. For $k \geq 3$, a graph G_k has $k - i$ vertices of degree i for $1 \leq i \leq k - 1$,

    (a) Give an example of such a graph G_k for $k = 3, 4, 5$.
    (b) Show that there exists no such graph G_k for $k = 6$.

17. Explain why there are no example of the following.
    (a) A graph of order 10 and size 46. (b) A 7-regular graph of order 27.
    (c) A bipartite graph containing C_7 as a subgraph.
    (d) Two isomorphic graphs, one of size 14 and the other of size 15.

18. Determine whether the graphs G and H in Figure 12.74 are isomorphic.

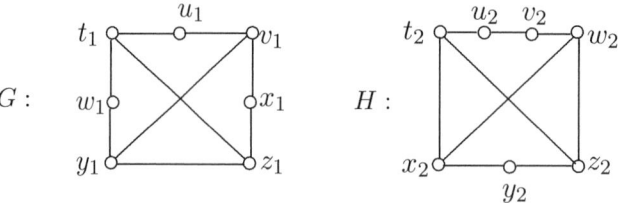

Figure 12.74: The graphs in Exercise 18

19. Determine whether the graphs G and H in Figure 12.75 are isomorphic.

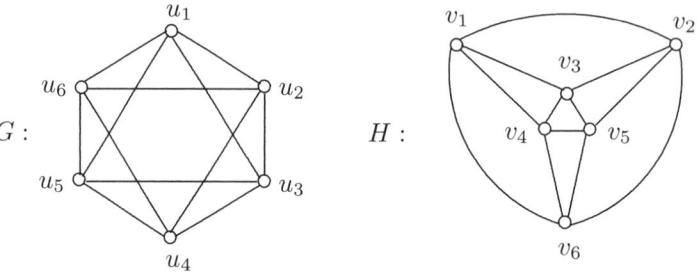

Figure 12.75: The graphs in Exercise 19

20. Determine whether the graphs G and H in Figure 12.76 are isomorphic.

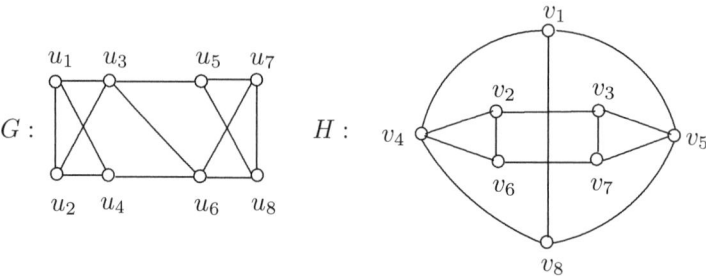

Figure 12.76: The graphs in Exercise 20

21. Draw a graph G with vertex set $V(G) = \{v_1, v_2, v_3, v_4, v_5\}$ whose adjacency matrix is

$$\mathbf{A} = \begin{bmatrix} 0 & 1 & 1 & 1 & 1 \\ 1 & 0 & 0 & 1 & 0 \\ 1 & 0 & 0 & 0 & 1 \\ 1 & 1 & 0 & 0 & 1 \\ 1 & 0 & 1 & 1 & 0 \end{bmatrix}.$$

22. Use Theorem 12.22 to compute $\mathbf{A}^2$ and $\mathbf{A}^3$ (without multiplying matrices) where $\mathbf{A}$ is the adjacency matrix of the graph G of Figure 12.77.

23. A graph G of size 44 has $r - 2$ vertices of degree r, 4 vertices of degree $r + 2$ and no other vertices. What is r?

24. Show that there is no bipartite self-complementary graph whose order exceeds 4.

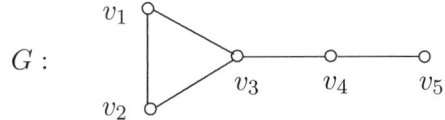

Figure 12.77: The graph G in Exercise 22

25. A vertex v in a graph G of order n is called a **double even vertex** if v has even degree in both G and $\overline{G}$. A vertex v is a **double odd vertex** if v has odd degree in both G and $\overline{G}$. Otherwise, v is a **mixed vertex**. Show that there exists no graph that contains a double even vertex, a double odd vertex and a mixed vertex.

26. Let G be a graph of order n and size m such that $n \equiv 3 \pmod 4$. Suppose that the complement $\overline{G}$ of G has size $\overline{m}$. Show that either $m > \frac{1}{2}\binom{n}{2}$ or $\overline{m} > \frac{1}{2}\binom{n}{2}$.

27. Let $H = C_5 \times K_2$. Show that it is possible to color each edge of H red, blue or green, resulting in three spanning subgraphs H_R (the red subgraph of H whose edges are red), H_B (the blue subgraph of H whose edges are blue) and H_G (the green subgraph of H whose edges are green) such that for every two distinct vertices u and v in H, the degrees of u and v are different in at least one of these three subgraphs H_R, H_B and H_G. After you color each edge of H, then label each vertex v of H by a triple (a, b, c), where a is the degree of v in H_R, b is the degree of v in H_B and c is the degree of v in H_G. (If you color the edges of H correctly, then no two vertices of H will have the same triple.)

28. In a nontrivial connected graph of order n, suppose that the number of vertices of degree 1 equals the number of vertices of degree $n - 1$. If this number is k, then what is k?

29. Show that if G is a graph of order $n \geq 2$ such that $\deg v \geq (n - 2)/3$ for every vertex v of G, then G has at most two components.

30. Let G be a connected graph of order $n \geq 3$. Suppose that each vertex of G is colored with one of the colors red, blue and green such that for each color, there exists at least one vertex of G assigned that color.

    (a) Show that G contains two adjacent vertices that are colored differently.

    (b) Show that, regardless of how large n may be, G may not contain two adjacent vertices that are colored the same.

    (c) Show that G has a path containing at least one vertex of each of these three colors.

    (d) The question in (c) should suggest another question to you. Ask and answer such a question.

31. It is known that if G is a graph of order $n \geq 2$, then it is impossible for all of the degrees of G to be distinct. However, it is possible for a connected graph G_n of order $n \geq 3$ to contain two distinct vertices u and v such that $\deg u = \deg v$ and such that the degree of every other vertex is different from that of all other $n - 1$ vertices. (In fact, there is only one such graph G_n for $n \geq 3$.)

    (a) What is G_n for $n = 3, 4, 5$?

    (b) For a graph G_n, where $n \geq 3$, let H_n be the graph obtained from $\overline{G}_n$ by deleting all isolated vertices if any. Show that $H_n = G_{n-1}$.

32. Let G be a disconnected graph of order $n \geq 6$. Prove that if $\Delta(\overline{G}) < \frac{2n}{3}$, then G has exactly two components.

33. Let G be a graph with $\delta(G) = \delta$. Prove the following:

(a) The graph G contains a path of length δ.

(b) If $\delta \geq 2$, then G contains a cycle of length at least $\delta + 1$.

[Hint: Let P be a longest path in G. Suppose that P is a $u - v$ path. Now consider which vertices of G can be adjacent to u (or v).]

34. Let G be a connected graph with the property that the smallest number of edges needed to be removed from G to produce a disconnected graph is 10.

    (a) Let X be a set of edges of G such that $|X| = 10$ and $G - X$ is disconnected. Suppose that $e \in X$. Show that e is a bridge of the graph $G - (X - \{e\})$.

    (b) Let Y be any set of 8 edges of G. Show that $G - Y$ contains no bridges.

35. Suppose that P is a longest path in a connected graph G. Let P be a $u - v$ path. Show that u is not a cut-vertex of G.

36. A graph G of order 5 has a cut-vertex but no bridge. Which of the following is true?

    (a) G is connected.   (b) G is disconnected.

    (c) There is no way to determine whether G is connected or not.

37. Let G be a nontrivial connected graph. Suppose that there exists a positive constant k such that $d(u, v) = k$ for every pair u, v of distinct vertices of G. Determine k and G.

38. A nontrivial connected graph G has the property that there exists a constant k such that for every two distinct vertices u and v of G, there are exactly k distinct $u - v$ paths of length $d(u, v)$.

    (a) Give an example of a graph satisfying this condition. What is k in this case?

    (b) Show that for every graph G satisfying this condition, only one value of k is possible.

39. Let G and H be two graphs such that G contains two vertices u and v of degree 3 with the property that there is a $u - v$ path of length 3 in G. On the other hand, H contains two vertices x and y of degree 3 such that there is no $x - y$ path of length 3 in H. Is it correct to conclude that $G \not\cong H$?

40. For a connected graph G and a vertex z of G, let $\ell(z)$ be a vertex that is farthest from z. For $k \geq 2$, let $\ell^k(z) = \ell(\ell^{k-1}(z))$. For the vertex u_1 of the graph G of Figure 12.78, find the smallest positive integer k such that $\ell^k(u_1) = \ell^j(u_1)$ for some j with $1 \leq j < k$.

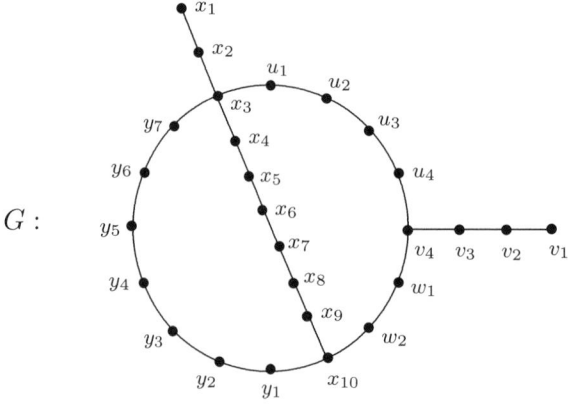

Figure 12.78: The graph G in Exercise 40

41. For a connected graph G and a positive integer k, define the *k-step graph* $G^{(k)}$ of G as that graph with $V(G^{(k)}) = V(G)$ and $uv \in E(G^{(k)})$ if $d_G(u, v) = k$. Define $G^{2(2)} = \left(G^{(2)}\right)^{(2)}$.

  (a) Find $H^{(2)}$, $F^{(2)}$ and $F^{2(2)}$ for the graphs H and F Figure 12.79.

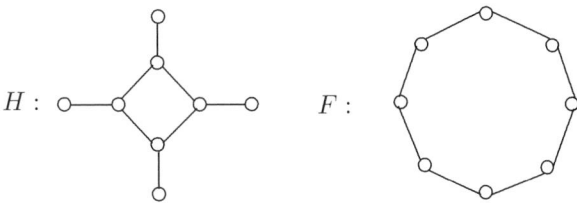

Figure 12.79: The graphs in Exercise 41

  (b) Give an example of a graph G such that $G^{(2)} \cong G$.

  (c) Show, for every positive integer k, that there exists a connected graph G such that $G \cong G^{(i)}$ for every i with $1 \le i \le k$.

42. For a connected graph G, let each vertex of G be colored with one of the three colors red, blue and green. For each vertex v of G, define the **color code** of v to be the ordered triple (a, b, c), where a is the distance from v to the closest red vertex in G, b is the distance from v to the closest blue vertex and c is the distance from v to the closest green vertex.

  (a) Show that it is possible to color each vertex of C_9 with one of these three colors so that no two vertices of C_9 have the same color code.

  (b) The question in (a) should suggest another question to you. Ask and answer such a question.

43. Let u be a cut-vertex in a connected graph G and let v be a vertex of G such that $d(u, v) = k \ge 1$. Show that G contains a vertex w such that $d(v, w) > k$.

44. A graph F is 3-regular of order 10 and a graph H is 4-regular of order 13. Let $G = F + H$.

  (a) Is G Eulerian? (b) Is G Hamiltonian?

45. Let F and H be two non-Eulerian regular graphs and let G be obtained from F and H by adding a new vertex v and joining v to each vertex in F and H. Prove or disprove: G is Eulerian.

46. Determine whether the graph G of Figure 12.80

  (a) is Eulerian, (b) has an Eulerian trail,

  (c) has neither an Eulerian circuit nor an Eulerian trail.

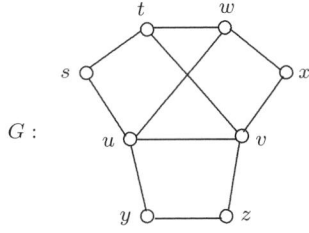

Figure 12.80: The graphs in Exercise 46

47. A graph G of order $n \geq 6$ has an Eulerian $u - v$ trail and $\deg u - \deg v \geq n - 2$.

    (a) Show that n must be even.   (b) Give an example of such a graph.

48. Suppose that G is an r-regular graph of order n such that both G and its complement $\overline{G}$ are connected. Is it possible that neither G nor $\overline{G}$ is Eulerian?

49. (a) Prove that if G is an Eulerian graph, then for every vertex v of G either $\deg v \equiv 0 \pmod{4}$ or $\deg v \equiv 2 \pmod{4}$.

    (b) Use (a) to prove that if an Eulerian graph G has even size, then there is an even number of vertices v for which $\deg v \equiv 2 \pmod{4}$.

50. A graph G of order 20 has $V(G) = \{v_1, v_2, \ldots, v_{20}\}$. Each vertex v_i ($1 \leq i \leq 6$) has degree 2 and is adjacent to exactly two vertices in the set $\{v_i : 16 \leq i \leq 20\}$. All other vertices v_i ($7 \leq i \leq 15$) have degree 9 or more. Prove that G is not Hamiltonian.

51. The vertex set of a graph G consists of the 3-element subsets of the set $\{1, 2, \ldots, 7\}$. Two vertices S and S' are adjacent in G if $S \cap S' = \emptyset$.

    (a) What is the order of G?   (b) Is G Eulerian?

    (c) What is the greatest distance between two vertices of G?

    (d) What is the length of a smallest cycle in G?

Chapter 13

Trees

The graphs known as trees constitute one of the simplest yet most important classes of graphs. Trees appeared implicitly in the 1847 work of the German physicist Gustav Kirchhoff (1824–1887) in his study of currents in electrical networks, while Arthur Cayley (1821–1895) used trees in 1857 to count certain types of chemical compounds. Trees are important to the understanding of the structure of graphs and are used to systematically visit the vertices of a graph. Trees are also widely used in computer science as a means to organize and utilize data. This entire chapter will be devoted to a study of this single class of graphs.

13.1   Fundamental Properties of Trees

In a region of an underdeveloped country are eight settlements, denoted by s_1, s_2, ..., s_8 (see Figure 13.1). In addition to rebuilding the settlements, several friendly countries have offered to construct paved roads between certain pairs of settlements so that it is possible to travel by way of paved roads between any two of the eight settlements. Because of the effort and expense needed to construct these roads, the plan is to build as few roads as possible to accomplish the stated goal. One way to do this is shown in Figure 13.1.

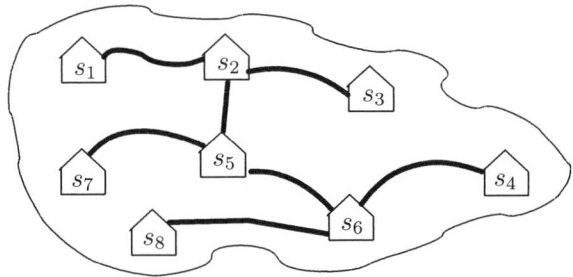

Figure 13.1: Connecting eight settlements by paved roads

Interpreting the settlements as vertices, we wish to construct a connected graph of order 8 that has as few edges as possible. Necessarily, this graph G can have no cycles, for if G has a cycle C, then for each edge e of C, the graph $G - e$ is connected but has fewer edges than G. Thus any graph with the desired properties is connected and contains no cycles. This is the class of graphs in which we are interested.

Definition 13.1 *A **tree** is a connected graph having no cycles.*

491

Thus the graph constructed in Figure 13.1 is a tree. It is easy to give examples of trees. All three graphs in Figure 13.2 are trees.

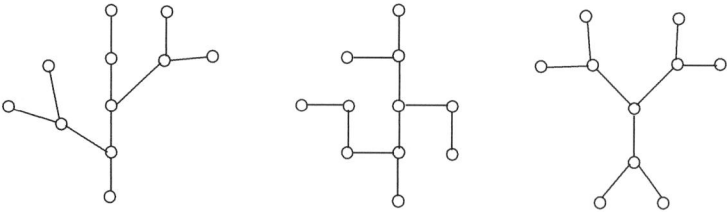

Figure 13.2: Three trees

As expected, there are relatively few different (non-isomorphic) trees of small orders. There is only one tree of each of the orders 1, 2 and 3, there are two trees of order 4 and three trees of order 5. All eight of these trees are shown in Figure 13.3.

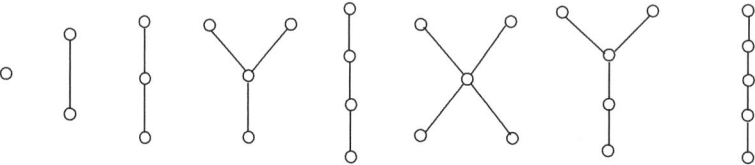

Figure 13.3: Trees of order 5 or less

Sometimes we are interested in graphs without cycles and not in whether they are connected.

Definition 13.2 *A graph containing no cycles is called a* **forest**.

Consequently, a connected forest is a tree. While every tree is a forest, the converse is not true. The graph F of Figure 13.4 is a forest with six components. Each component of the forest F is therefore a tree.

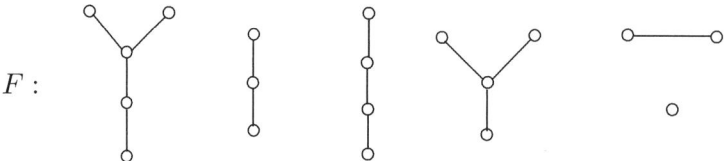

Figure 13.4: A forest with six components

Every graph contains subgraphs that are trees. In most instances, a graph contains many subgraphs that are trees. If the given graph is connected, then the subgraphs that are trees in which we will be most interested are spanning trees.

Definition 13.3 *A* **spanning tree** T *of a connected graph* G *is a spanning subgraph of* G *that is a tree, that is,* $V(T) = V(G)$.

If G is itself a tree, then G has only one spanning tree, namely itself.

Example 13.4

For the connected graph G of Figure 13.5, all of the trees T_1, T_2, T_3, T_4 are spanning trees of G. ♦

We now look at some properties possessed by all nontrivial trees. Since every tree is connected, no nontrivial tree can contain an isolated vertex. However, every nontrivial tree must contain some

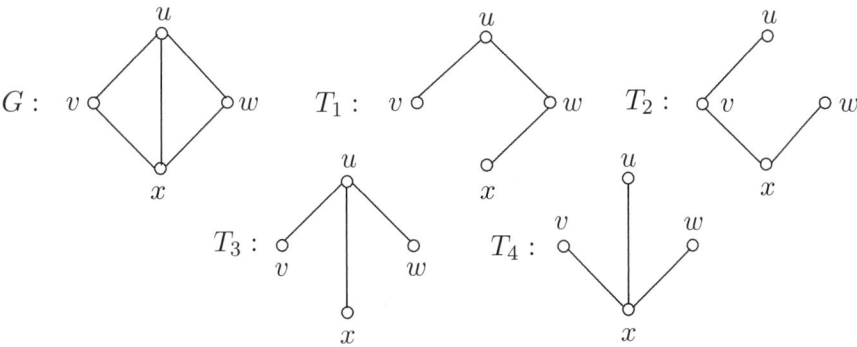

Figure 13.5: Spanning trees of a connected graph in Example 13.4

end-vertices. Especially when working with trees, it is common to refer to a vertex of degree 1 as a leaf. We now show that every tree with at least two vertices contains at least two leaves. In the proof, it is once again useful to consider a longest path.

Theorem 13.5 *Every nontrivial tree contains at least two leaves.*

Proof. Let $P = (v_1, v_2, \ldots, v_k)$, $k \geq 2$, be a longest path in a tree T. We claim that $\deg v_1 = \deg v_k = 1$. Since v_1 is adjacent to v_2, certainly $\deg v_1 \geq 1$. The vertex v_1 cannot be adjacent to a vertex v_0 that is not on P, for otherwise, $(v_0, v_1, v_2, \ldots, v_k)$ is a path whose length is greater than that of P. Also, v_1 cannot be adjacent to a vertex v_t on P with $3 \leq t \leq k$; for otherwise, $(v_1, v_2, \ldots, v_t, v_1)$ is a cycle in T, which is impossible since T has no cycles. Thus, as claimed, $\deg v_1 = 1$. By a similar argument, $\deg v_k = 1$. ∎

Let T be a nontrivial tree. By Theorem 13.5, T contains two or more leaves. Let v be one of them. Then $T - v$ is also a tree (see Exercise 8). This simple observation has useful consequences in certain induction proofs involving trees, which is illustrated in the proof of the following theorem.

Theorem 13.6 *Every tree of order n has size $n - 1$.*

Proof. The proof is by induction on the order n. Suppose first that $n = 1$. There is only one tree of order 1, namely the trivial tree K_1. Since the size of K_1 is $0 = 1 - 1$, the result is true for $n = 1$. This verifies the basis step. We next verify the inductive step. Assume that every tree of order k, where $k \geq 1$, has $k - 1$ edges and let T be a tree of order $k + 1$. Since T is nontrivial, T contains a leaf v. Then $T - v$ is a tree of order k. By the induction hypothesis, $T - v$ has $k - 1$ edges. Since $T - v$ has one less edge than T, it follows that T has $k = (k + 1) - 1$ edges, as desired.

By the Principle of Mathematical Induction, every tree of order n has size $n - 1$. ∎

By Theorem 13.6, if T is a tree of order n and size m, then $m = n - 1$. Another way to state this is:

> If T is a connected graph of order n and size m containing no cycles, then $m = n - 1$.

In fact, if G is a graph of order n and size $n - 1$ containing no cycles, then G must be a tree.

Theorem 13.7 *Every graph of order n and size $n - 1$ containing no cycles is a tree.*

Proof. Let G be a graph of order n and size m containing no cycles such that $m = n - 1$. We show that G is a tree. Since G has no cycles, it remains only to show that G is connected. Let the components of G be $G_1, G_2, \ldots, G_k$, where $k \geq 1$. Our goal is to show that $k = 1$. Suppose that G_i has order n_i and size m_i $(1 \leq i \leq k)$. Thus

$$m = \sum_{i=1}^{k} m_i \quad \text{and} \quad n = \sum_{i=1}^{k} n_i.$$

Since each component G_i is a tree, it follows by Theorem 13.6 that $m_i = n_i - 1$ for each integer i ($1 \le i \le k$). Therefore,

$$n - 1 = m = \sum_{i=1}^{k} m_i = \sum_{i=1}^{k} (n_i - 1) = \left(\sum_{i=1}^{k} n_i \right) - k = n - k.$$

This says that $k = 1$ and so G is connected. Therefore, G is a tree. ∎

From the proof of Theorem 13.7, we have the following corollary.

Corollary 13.8 *If F is a forest of order n and size m having k components, then*

$$m = n - k.$$

Furthermore, if G is a connected graph of order n and size m for which $m = n - 1$, then G too is a tree.

Theorem 13.9 *Every connected graph of order n and size $n - 1$ is a tree.*

Proof. Let G be a connected graph of order n and size $m = n - 1$. We show that G is a tree. Since G is connected, it remains only to show that G contains no cycles. Assume, to the contrary, that G contains cycles. If e_1 is an edge of G that lies on a cycle of G, then $G_1 = G - e_1$ is connected. If G_1 contains no cycles, then G_1 is a tree. Otherwise, G_1 contains cycles. In such a case, let e_2 be an edge of G_1 that lies on a cycle of G_1. Then $G_2 = G_1 - e_2$ is connected. We continue this until the edges $e_1, e_2 \ldots, e_k$ ($k \ge 1$) are removed, arriving at a connected graph G_k containing no cycles. Then G_k is a tree of order n and size $m - k \le m - 1 = n - 2$. This, however, contradicts Theorem 13.6. ∎

The following result is a consequence of Theorems 13.7 and 13.9 and the definition of a tree.

Theorem 13.10 *If G is a graph of order n and size m satisfying any two of the following three properties, then G is a tree:*

(1) *G is connected,*   (2) *G has no cycles,*   (3) *$m = n - 1$.*

There is also another property that only trees possess.

Theorem 13.11 *A graph G is a tree if and only if G has a unique $u - v$ path for every two vertices u and v in G.*

Proof. Let G be a tree and let u and v be two vertices of G. Since G is connected, G contains a $u - v$ path. If G contains two $u - v$ paths, then G contains a cycle, which is impossible since G is a tree. Thus G contains a unique $u - v$ path.

For the converse, suppose that G is a graph containing a unique $u - v$ path for every two vertices u and v of G. We claim that G is a tree. Certainly G is connected. Assume, to the contrary, that G contains a cycle C. Let x and y be two vertices on C. Then C and therefore G contains two $x - y$ paths, which is a contradiction. So, as claimed, G is a tree. ∎

Degrees of the Vertices of a Tree

We have mentioned that every nontrivial tree contains at least two leaves. A tree may contain exactly two leaves or exactly three leaves, as the trees T_1 and T_2 of Figure 13.6 show. The tree T_1 has exactly two leaves and no vertices of degree 3, while the tree T_2 has exactly three leaves and one vertex of degree 3. Each of the trees T_3 and T_4 of Figure 13.6 contains exactly four leaves. While the tree T_3 has two vertices of degree 3 and no vertices of degree more than 3, the tree T_4 has no

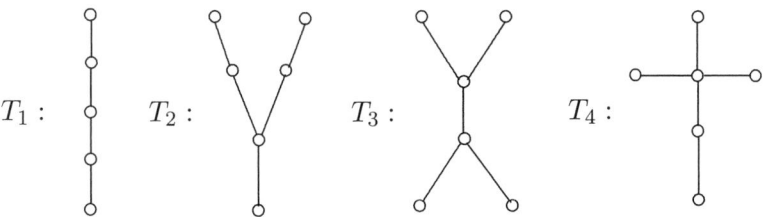

Figure 13.6: Some trees with a small number of leaves

vertices of degree 3 but does contain one vertex of degree 4. There is a formula for the number of leaves in a tree T in terms of the number of vertices of degree 3 or more that T contains.

Let T be a tree of order $n \geq 2$ and size m having maximum degree $\Delta(T) = k$. Suppose that T has n_i vertices of degree i for $i = 1, 2, \ldots, k$. The number n_1 is therefore the number of leaves in T. Then

$$\sum_{i=1}^{k} n_i = n_1 + n_2 + \cdots + n_k = n \quad \text{and}$$

$$\sum_{i=1}^{k} i n_i = 1 \cdot n_1 + 2 n_2 + \cdots + k n_k = 2m = 2(n-1) = 2n - 2.$$

Since the second sum adds the degrees of vertices of T, the result is twice the number of edges of T. With the aid of these two sums, we can now develop a formula for the number of leaves in a tree.

Theorem 13.12 *Let T be a tree of order $n \geq 2$ with maximum degree k having n_i vertices of degree i for $i = 1, 2, \ldots, k$. Then the number of leaves in T is*

$$n_1 = 2 + n_3 + 2n_4 + 3n_5 + \cdots + (k-2)n_k.$$

Proof. Since

$$\sum_{i=1}^{k} i n_i = 2n - 2 = 2 \left(\sum_{i=1}^{k} n_i \right) - 2,$$

it follows that

$$n_1 + 2n_2 + 3n_3 + \cdots + k n_k = 2n_1 + 2n_2 + 2n_3 + \cdots + 2n_k - 2$$

and so

$$n_1 = 2 + n_3 + 2n_4 + 3n_5 + \cdots + (k-2)n_k,$$

as desired. ∎

According to Theorem 13.12, if a tree T has exactly one vertex of degree 3 but no vertex of higher degree, then T has $2 + 1 = 3$ leaves. If T has exactly two vertices of degree 3 but none of any higher degree, then T has $2 + 1 \cdot 2 = 4$ leaves. If T has exactly one vertex of degree 4 but no other vertex of degree 3 or more, then T has $2 + 2 \cdot 1 = 4$ leaves.

Example 13.13

A certain tree T has five vertices of degree 2, six vertices of degree 3, two vertices of degree 4, four vertices of degree 5 and no vertices of degree 6 or more. How many leaves does T possess?

Solution. According to Theorem 13.12, the number of vertices of degree 2 has no effect on the number of leaves. However, the remaining information tells us that the number of leaves in T is

$$n_1 = 2 + n_3 + 2n_4 + 3n_5 = 2 + 6 + 2 \cdot 2 + 3 \cdot 4 = 24.$$

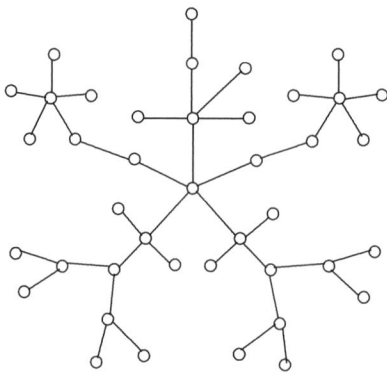

Figure 13.7: A tree with five vertices of degree 2, six vertices of degree 3,
two vertices of degree 4 and four vertices of degree 5

One such tree is given in Figure 13.7.   ◆

   The solution of Example 13.13 shows that if G is any tree containing exactly five vertices of degree 2, six vertices of degree 3, two vertices of degree 4, four vertices of degree 5 and no vertices of degree 6 or more, then G must have 24 leaves. However, suppose, instead of computing n, we simply draw a tree satisfying the conditions of the tree T specified in Example 13.13 (such as the tree in Figure 13.7) and then count the number of leaves of the tree we just drew. Of course, we would get 24. However, this is not a satisfactory solution of the problem, for how do we know that the number of leaves in the tree T described in Example 13.13 is the same as the number of leaves in the tree we drew? Certainly, these trees need not to be isomorphic. We don't know that these trees have the same number of leaves until we've shown it.

Exercises for Section 13.1

1. Draw all trees of order 6.

2. What is the smallest size of a graph G of order 5 such that G contains each of the three spanning trees of order 5?

3. Figure 13.5 shows four spanning trees of a connected graph G. There are others. Draw the remaining spanning trees of G.

4. Show that K_6 contains three spanning trees T_1, T_2 and T_3 such that $\{E(T_1), E(T_2), E(T_3)\}$ is a partition of $E(K_6)$ and the degrees of the vertices of each tree are 3, 3, 1, 1, 1, 1.

5. Give an example of two non-isomorphic trees of the same order whose vertices have the same degrees.

6. (a) By Theorem 13.5, the minimum number of leaves in a tree of order $n \geq 3$ is 2. What is the maximum number of leaves that a tree of order $n \geq 3$ can contain?

   (b) Show, for every two integers k and n with $2 \leq k \leq n - 1$, that there exists a tree of order n containing exactly k leaves.

7. How many trees are there whose complement is also a tree?

8. Show that if v is a leaf in a nontrivial tree T, then $T - v$ is also a tree.

9. If G is a graph of order n and size m with $m = n - 1$, is G a tree?

10. A tree T has 22 leaves, four vertices of degree 4 and three vertices of degree 6. The tree T also contains exactly two vertices of another degree x. What is x?

11. Show that if G is a connected graph of order n and size $m = n$, then G must contain cycles. How many cycles must G contain?

12. (a) Give an example of a graph G that is not a tree such that for every two vertices u and v, there is exactly one $u - v$ path of length $d(u, v)$ in G.

    (b) Give an example of a graph G such that for every two vertices u and v, there are exactly two $u - v$ paths in G.

13. A tree T has four vertices of degree 2, three vertices of degree 3, five vertices of degree 6 and no other vertices of degree 3 or more. How many leaves does T possess?

14. A tree T of order 100 has only vertices of degrees 1 and 3. How many vertices of each degree does T have?

15. A tree T has $7k$ end-vertices, $2k - 2$ vertices of degree 3 and $2k + 2$ vertices of degree 4 for some positive integer k, but no other vertices. What is the order of T?

16. A tree T has three vertices of degree 4 and five vertices of degree 6. The remaining vertices of T have degree 1 or 3. There are three times as many end-vertices as vertices of degree 3 in T. What is the order of T?

17. Show that no tree of order 100 can have only vertices of degrees 1 and 5.

18. What is the minimum order of a tree that contains at least one vertex of degree k for every k with $1 \le k \le 10$?

19. (a) Use Theorem 13.12 to show that if T is a tree with maximum degree $\Delta(T) \ge 3$, then T contains at least three leaves.

    (b) Use (a) to show that if a tree T has exactly two leaves, then T is a path.

20. What is the smallest positive integer k such that if F_1 and F_2 are two forests of order 15 each consisting of k non-isomorphic components, then F_1 must contain a component G_1 and F_2 must contain a component G_2 with $G_1 \cong G_2$?

21. Let G be a nontrivial connected graph of order n. Prove that G contains two vertices u and v with $d(u, v) = n - 1$ if and only if $G = P_n$.

22. Let T be a nontrivial tree of order n and let S be a set of vertices of T. By $T' = T(S)$, we mean a subtree of T of minimum size such that $S \subseteq V(T')$.

    (a) Prove that $\Delta(T') \le |S|$. Show that it is possible for $\Delta(T') < |S|$.

    (b) Determine the minimum cardinality of a set $S \subseteq V(T)$ such that $T(S) = T$.

23. Let T be a tree of order at least 3 and let G be a graph with $V(G) = V(T)$ such that $uv \in E(G)$ if and only if u and v are connected in $T - w$ for every vertex $w \in V(T) - \{u, v\}$. What is G?

13.2   Rooted and Spanning Trees

Many of the applications of trees concern the designation of a particular vertex r in a tree T and the distance from r to each vertex of T.

Definition 13.14   *A tree T is called a* **rooted tree** *if T contains a designated vertex r (called the* **root** *of T). The tree T is then said to be* **rooted at** r.

The tree T_0 in Figure 13.8(a) is a rooted tree as it is rooted at the vertex r. It is common to draw a rooted tree with the root at the top (and placed at what is referred to as level 0). The neighbors of r are placed at the same level below r (at level 1). Any vertex at distance 2 from r is placed one level below that (at level 2). Continuing in this manner, we say that a vertex s of T is at **level** i the distance from r to s is i. Drawing a rooted tree in this manner is illustrated with the tree T_0 of Figure 13.8(a), which is drawn again in Figure 13.8(b).

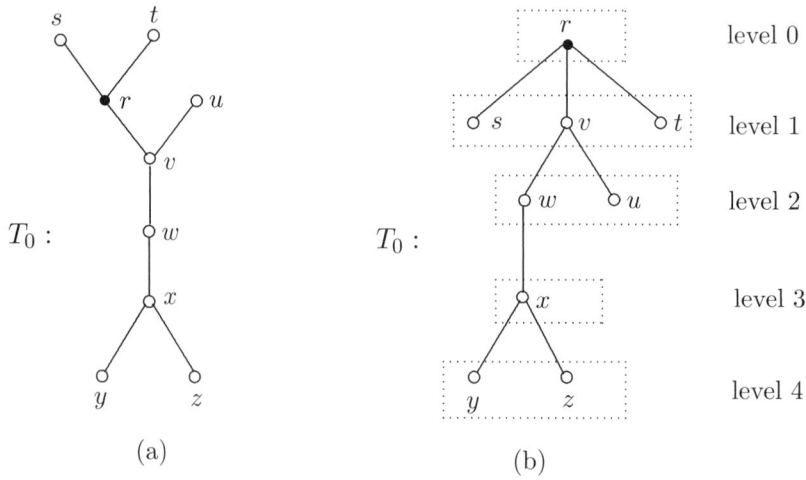

Figure 13.8: A rooted tree

The **height** of a rooted tree T with root r is the largest integer h for which there is a vertex v of T at level h. The height h of a tree T rooted at r is then the maximum distance from r to another vertex of T. The rooted tree T_0 of Figure 13.8 has height 4.

Rooted trees can be used to represent a variety of situations. One of these describes the hierarchical administrative structure that exists within a college or university.

Example 13.15

A possible portion of the administrative structure in a certain university is shown in Figure 13.9. In this case, the root is the president of the university.   ♦

Another common application of a rooted tree is its use as a family tree to show the descendants of some individual. Inspired by familiar terms in family trees, this terminology is often applied to all rooted trees. Let T be a tree rooted at the vertex r and let v be a vertex of T. Any vertex in T adjacent to v but belonging to the level one step below that of v is called a **child** of v. Any vertex w ($\neq v$) for which the $r - w$ path in T contains v is called a **descendant** of v. Any vertex u ($\neq v$) for which the $r - v$ path in T contains u is called an **ancestor** of v. The ancestor of v belonging to the level one step above v is the **parent** of v.

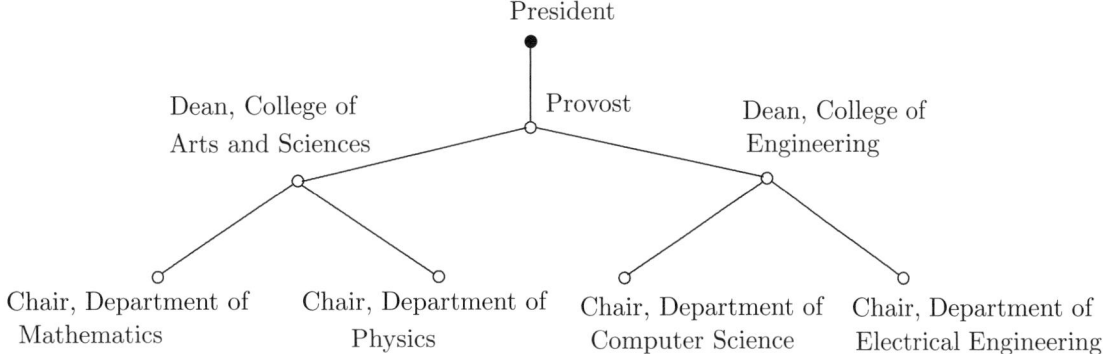

Figure 13.9: A rooted tree describing a portion of the hierarchical structure of a university administration

Example 13.16

Figure 13.10 shows a rooted tree that represents a possible family tree of someone's grandfather, who is the root in this instance. Grandfather Earl is a parent of Uncle Ed, while cousin Diane is a child of Uncle Ed. Grandfather Earl is an ancestor of Sister Teresa and Cousin Jim is a descendant of Grandfather Earl. ♦

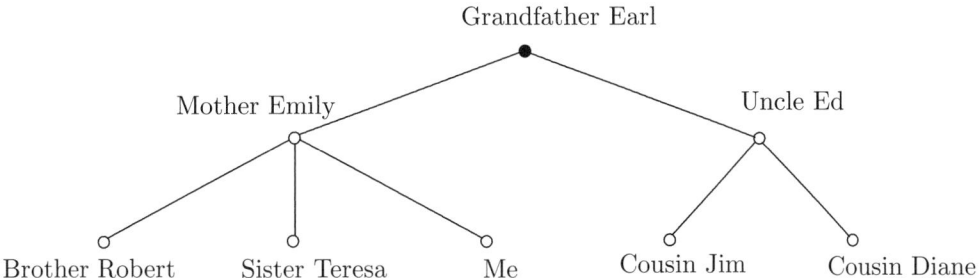

Figure 13.10: A rooted tree: a family tree in Example 13.16

Another familiar application of rooted trees concerns searching for a particular name in a list of names arranged alphabetically or perhaps searching for a particular number in a list of numbers given in increasing order. Actually, one approach to accomplish this is to use the Binary Search Algorithm (Algorithm 6.32). This gives rise to a rooted tree, called a binary search tree. A **binary tree** is a rooted tree in which every vertex has at most two children and in which each child is assigned a special designation, namely as a **left child** or a **right child**. If v is a vertex of a binary tree T with left child u and right child w, then the subtree of T induced by u and its descendants is the **left subtree** of v, while the subtree of T induced by w and its descendants is the **right subtree** of v. The rooted tree shown in Figure 13.11 is a binary tree rooted at r. The vertex u is the left child of r, while w is the right child of r. The vertex z is the left child of t but t has no right child. No leaves have any children. The right subtree T' of r (rooted at w) is also shown in Figure 13.11.

Binary Search Trees

For a list of names arranged alphabetically, a **binary search tree** T is a binary tree whose vertex set is the set of names on the list such that for every vertex v of T the vertices in the left subtree of v precede v alphabetically and those vertices in the right subtree of v follow v alphabetically. When

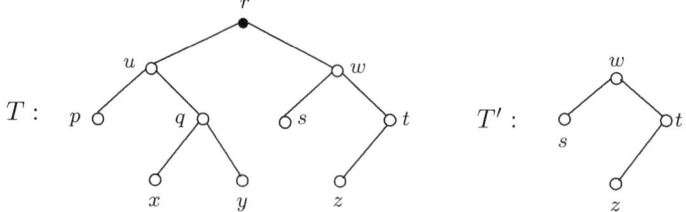

Figure 13.11: A binary tree and the right subtree of its root

using the Binary Search Algorithm to search for a name, a specific binary search tree is constructed. Namely, we select a name on the list that is in the middle of the list (or the first of two middle names alphabetically if there is no unique middle name). The left child of the root is the middle name of the sublist of names preceding the root alphabetically and the right child of the root is the middle name of the sublist of names following the root alphabetically. Continuing in this manner produces the binary search tree T that corresponds to the Binary Search Algorithm.

Example 13.17

Construct the binary search tree resulting from the following list of 15 names (arranged alphabetically):

Aaron, Abel, Allen, Benson, Bloom, Boston, Carter, Curry,
Daniels, Fletcher, Harper, Lumens, Parsons, Thomas, Waters.

Indicate in this binary search tree, how the name Harper is found using a binary search.

Solution. The binary search tree constructed from the given list of 15 names is shown in Figure 13.12. Searching for the name Harper results in the path (Curry, Lumens, Fletcher, Harper). ♦

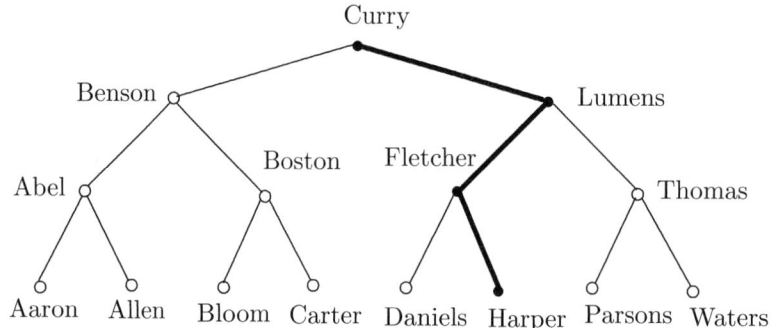

Figure 13.12: The binary search tree in Example 13.17

| **Decision Trees** |

We have seen illustrations of how rooted trees can be used to model a number of situations and, in the case of binary trees, can be used to conduct a search for a word in a alphabetized list of words (or for a number in a list of numbers arranged in increasing order). We now describe another problem, solutions of which can be found through a sequence of comparisons. Finding the solutions can be considered as an application of rooted trees, where each vertex of the tree corresponds to a decision that directs us to a subtree and eventually leads to a possible solution of the problem at a leaf of the tree. For this reason, such rooted trees are often called **decision trees**.

Suppose that we have a collection of $n \geq 2$ coins, exactly one of which is counterfeit and weighs (slightly) less than each of the $n - 1$ authentic coins. We have at our disposal a balance scale (see Figure 13.13). By placing coins on each of the two scale pans A and B, one can determine if the pans are balanced and so the coins on the two pans weigh the same, or if the pans are unbalanced and whether it is the coins on A or the coins on B that are heavier (or lighter). The problem of interest to us is to determine a sequence of weighings that results in an identification of the counterfeit coin. Moreover, we desire to accomplish this with a minimum number of weighings. We consider this question for $n = 4$.

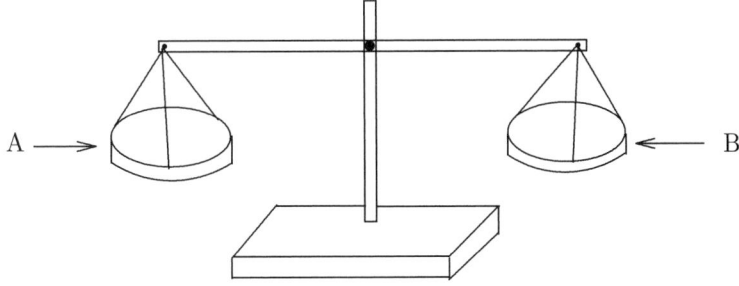

Figure 13.13: A balance scale

Example 13.18

Suppose that we have a collection of four coins numbered 1, 2, 3, 4, exactly one of which is counterfeit and weighs less than each of the three authentic coins. What is the minimum number k of weighings needed to be certain that we have identified the counterfeit coin? Describe how this problem can be solved by a sequence of k weighings.

Solution. It is probably obvious that there is no way to guarantee in a single weighing that we can discover the counterfeit coin. It is possible to be lucky, of course, where we place coin 1, say, on scale pan A and coin 2 on scale pan B. We indicate this by writing $\{1\}, \{2\}$. If the contents of scale pan A (namely coin 1) is lighter than the contents of scale pan B (which can be easily observed since A is higher than B), then coin 1 is the counterfeit coin. Of course, if the scale pans balance, then we know that neither coin 1 nor coin 2 is counterfeit. To determine which of coin 3 and coin 4 is counterfeit, another weighing is required.

There is more than one way to proceed to solve this problem. Two possible approaches are described by means of the two decision trees in Figure 13.14. In the decision tree in Figure 13.14(b), for example, coins 1 and 2 are placed on scale pan A, while coins 3 and 4 are placed on scale pan B. There is no possibility that these pans will balance. If the contents of scale pan A (namely coins 1 and 2) is lighter, then the counterfeit coin is on pan A. In this case, we conduct a second weighing, placing coin 1 on pan A and coin 2 on pan B. Whichever scale has the lighter contents identifies the the counterfeit coin. Otherwise, the contents of scale pan B is lighter and we proceed in a similar fashion to determine whether it is coin 3 or coin 4 that is counterfeit.   ◆

One question that might occur to you is the following. If, somehow, we knew that exactly one of four coins is counterfeit, then it would probably not be surprising to learn that the weight of this counterfeit coin is different from each of the three authentic coins. However, how likely is it that we would know beforehand how the weight of the counterfeit coin compares with the other three? Perhaps it is heavier. Perhaps it has a different weight than the three authentic coins but we don't know if the counterfeit coin is lighter or heavier.

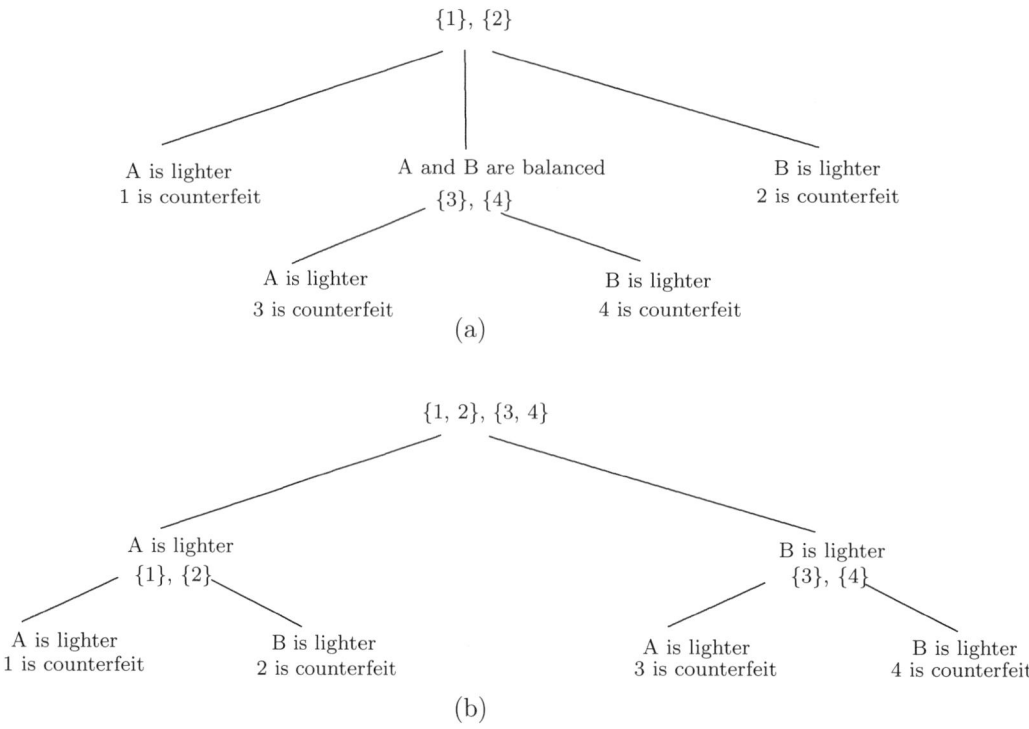

Figure 13.14: Identifying a counterfeit coin that is
lighter than three other coins in Example 13.18

Example 13.19

Suppose that we have a collection of four coins numbered 1, 2, 3, 4, exactly one of which is counterfeit and whose weight is different from each of the three authentic coins. What is the minimum number of weighings needed to be certain that have identified the counterfeit coin?

Solution. In Example 13.18, where we knew that the counterfeit coin is lighter than each of the three authentic coins, we saw that the minimum number of weighings needed to identify the counterfeit coin is 2. In the current situation, we have less information, however. Consequently, the number of weighings needed to solve this problem must be at least 2. But what is this minimum? The decision tree in Figure 13.15 shows us how we might proceed.

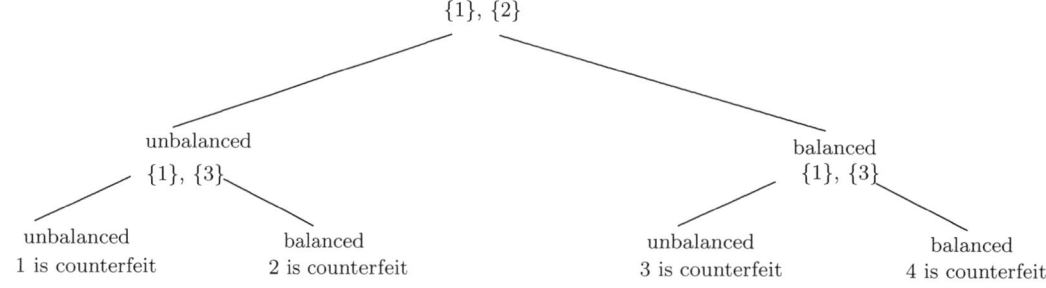

Figure 13.15: Identifying a counterfeit coin whose weight
is different from three other coins in Example 13.19

According to the decision tree in Figure 13.15, we begin by placing coin 1 on scale pan A and coin 2 on scale pan B. Either the scale pans balance or they don't. If pan A and pan B do not balance, then we know that either coin 1 or coin 2 is counterfeit and that both coin 3 and coin 4 are authentic. Next, we place coin 1 on scale pan A and (the authentic) coin 3 on scale pan B. If the pans do not balance, then coin 1 is counterfeit; while if they balance, then coin 2 is counterfeit. On the other hand, if pan A and pan B initially balance, then we have learned that coin 1 and coin 2 are authentic. We again place coin 1 on scale pan A and coin 3 on scale pan B. If the pans do not balance, then coin 3 is counterfeit; while if the pans balance, then coin 4 is counterfeit.

Therefore, even without knowing whether the counterfeit coin is lighter or heavier than the authentic coins, still only two weighings are required to discover which of the four coins is counterfeit. ♦

We now consider a larger number of coins, where we again assume that it is known that the counterfeit coin is lighter than each of the authentic coins.

Example 13.20

Suppose that we have a collection of nine coins numbered 1, 2, ..., 9, exactly one of which is counterfeit and weighs less than each of the eight authentic coins. What is the minimum number of weighings needed to be certain that we have identified the counterfeit coin? Describe a procedure for locating the counterfeit coin using this minimum number of weighings.

Solution. Perhaps surprisingly, the minimum number of weighings needed to solve this problem is 2 in this case as well. How the counterfeit coin can be discovered in two weighings is shown in the decision tree in Figure 13.16. ♦

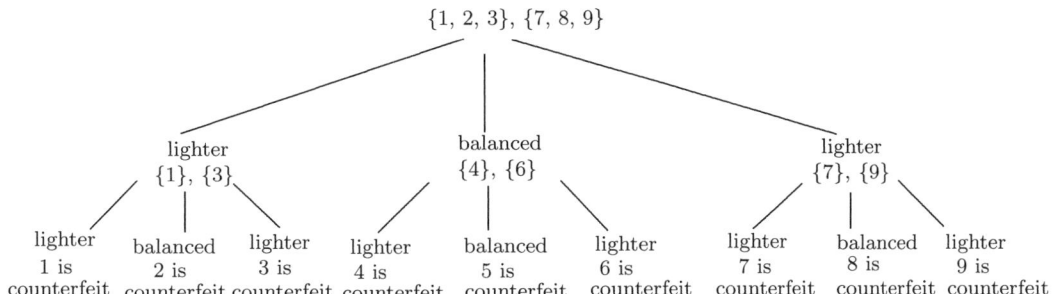

Figure 13.16: Identifying a counterfeit coin whose weight
is less than the eight other in Example 13.20

Notice that whenever we reach a leaf in a decision tree, we have reached a conclusion. In the decision tree in Figure 13.16, for example, there are nine leaves, each corresponding to the one of nine coins that turned out to be counterfeit. More generally, if we are seeking a counterfeit coin among n coins, then any decision tree used to identify which of the n coins is the fake coin must have at least n leaves. (Perhaps there is more than one sequence of comparisons that would show that a certain coin is counterfeit.)

A famous problem in this connection is the following.

The 12 Coins Problem: You are given 12 coins, one of which is counterfeit and weighs more or less than the 11 authentic coins. Show that in three weighings it can be determined which of the 12 coins is counterfeit and whether the counterfeit coin is heavier or lighter than each authentic coin. (See Exercise 10.)

Constructing Spanning Trees

We have already mentioned that every connected graph contains a spanning tree, perhaps several spanning trees. Let G be a connected graph of order n and size m. A spanning tree T of G can be constructed in a number of ways. Let's look at one such way. If G itself is a tree, then G is a spanning tree of itself. If G is not a tree, then G contains a cycle C. If e is an edge of C, then the graph $G - e$ is connected. If $G - e$ contains no cycles, then $G - e$ is a spanning tree of G. If $G - e$ contains a cycle C' and e' is an edge of C', then we consider $G - \{e, e'\}$ and proceed as above, until we arrive at a spanning tree T of G. Since T is a spanning tree of G, it follows that T has order n and size $n - 1$. Consequently, it is necessary to remove $m - (n - 1) = m - n + 1$ edges from G to produce T, although, as we have mentioned, each deleted edge must belong to a cycle in the graph currently being considered.

Let's look at another way to construct a spanning tree T of a connected graph G of order n. Suppose that $V(G) = \{v_1, v_2, \ldots, v_n\}$. Of course, $V(T) = V(G)$. It remains to determine the edge set of T. Select any two edges e_1 and e_2 of G. These are the first two edges of $E(T)$. Now select an edge e_3 of G not already selected and which does not produce a cycle with the edges already in $E(T)$ and place this edge in $E(T)$. Continuing in this manner produces a spanning tree of G.

We now illustrate these methods of constructing a spanning tree in a connected graph.

Example 13.21

Find a spanning tree in the connected graph G of Figure 13.17 by

(a) deleting appropriate edges from G and

(b) beginning with $V(G)$ and adding appropriate edges.

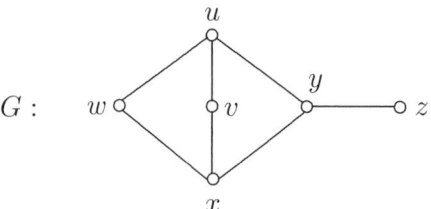

Figure 13.17: The graph G in Example 13.21

Solution.

(a) First, we delete from G any edge that belongs to a cycle. Every edge except yz has this property. We delete uw, producing G_1 (see Figure 13.18(b)). Then we delete an edge of G_1 belonging to the unique cycle of G_1, say we delete the edge uy. This produces the spanning tree T_1 of G shown in Figure 13.18(c).

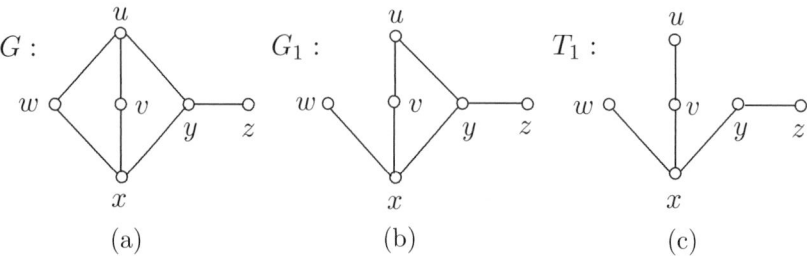

Figure 13.18: Constructing a spanning tree by deleting edges in Example 13.21(a)

(b) Beginning with the vertices of G, we successively add edges provided no cycles are created. For example, we could add the edges uw, xw, xv, uy, yz. This procedure and the resulting spanning tree T_2 of G are shown in Figure 13.19. ◆

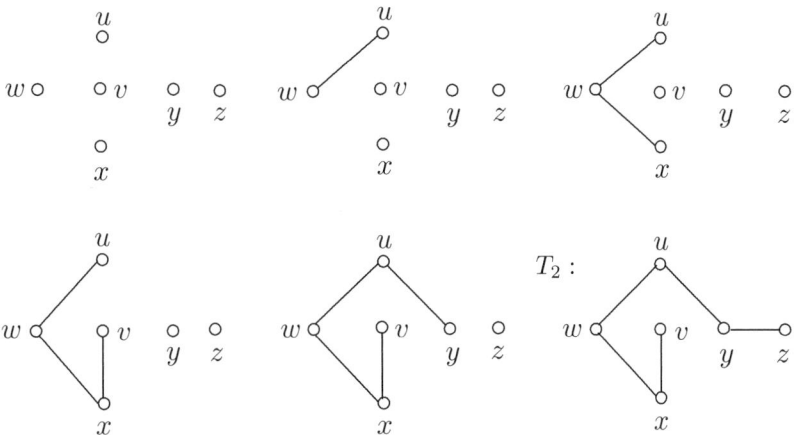

Figure 13.19: Constructing a spanning tree by adding edges in Example 13.21(b)

Since no disconnected graph can contain a spanning connected subgraph, we have the following observation.

Theorem 13.22 *A graph G contains a spanning tree if and only if G is connected.*

In 1847 Gustav Kirchhoff showed that the problem of finding a spanning tree in a connected graph G is related to the problem of resistance between the nodes of an electrical circuit.

Depth-First Search Trees

We now look at two well-known spanning trees of a connected graph G that are rooted trees. The first of these is called a **depth-first search tree** formed by conducting what is called a **depth-first search** in G. In a depth-first search, we begin at some vertex v (the root of the depth-first search tree). The idea is to systematically visit all vertices of G by proceeding deeply into G with a path P having v as its initial vertex. If P should contain all vertices of G, then P is a spanning tree and is a depth-first search tree. If P does not contain all vertices of G, then we backtrack along P until we reach a vertex v' that is adjacent to a vertex not on P and construct a new path P' with initial vertex v' that proceeds as deeply into G as possible. When P' terminates and if the tree T' constructed thus far contains all vertices of G, then this is a depth-first search tree. Otherwise, we backtrack along P' and possibly along P as well until a new vertex v'' in T' is found that is adjacent to a vertex not on T' and then proceed as before.

Before formally stating an algorithm for conducting a depth-first search in a connected graph G, we give an example. In order to indicate how a depth-first search tree can be constructed systematically, we assume that G is a connected graph with $V(G) = \{v_1, v_2, \ldots, v_n\}$ and that the adjacency lists of these vertices are given, whose subscripts are in increasing order. The resulting depth-first search tree T of G is rooted at v_1.

Example 13.23

Construct a depth-first search tree T rooted at v_1 for the graph G of Figure 13.20. The adjacency lists of these vertices of G are given below:

$$v_1 :\quad v_3, v_4, v_8, v_9, v_{11} \qquad v_2 :\quad v_3, v_8, v_{12} \qquad v_3 :\quad v_1, v_2, v_5, v_{11}$$
$$v_4 :\quad v_1, v_6, v_9, v_{10} \qquad v_5 :\quad v_3, v_{12} \qquad v_6 :\quad v_4, v_{10}$$
$$v_7 :\quad v_8, v_{12} \qquad v_8 :\quad v_1, v_2, v_7, v_{10} \qquad v_9 :\quad v_1, v_4, v_{11}$$
$$v_{10} :\quad v_4, v_6, v_8 \qquad v_{11} :\quad v_1, v_3, v_9 \qquad v_{12} :\quad v_2, v_5, v_7.$$

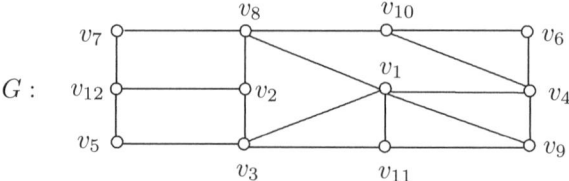

Figure 13.20: The graph G in Example 13.23

Solution. To construct a depth-first search tree T in this case, we begin at v_1. Since v_3 is the first vertex on the adjacency list of v_1, we start with the path (v_1, v_3). The first vertex on the adjacency list of v_3 not already used is v_2; so now we have the path (v_1, v_3, v_2). Continuing in this manner gives us the path

$$P = (v_1, v_3, v_2, v_8, v_7, v_{12}, v_5).$$

The two vertices on the adjacency list of v_5 (namely v_3 and v_{12}) already appear on P, however. Hence we backtrack along P to v_{12}. Since the three vertices on the adjacency list of v_{12} are on P, we backtrack along P again to v_7. The two vertices on the adjacency list of v_7 are already on P. We backtrack along P one step further, this time placing us at v_8. However, there is a vertex (namely v_{10}) on the adjacency list of v_8 that does not belong to P.

Hence we now start a new path P' with (v_8, v_{10}). Since the first vertex v_4 on the adjacency list of v_{10} has not been used, we extend P' so that it begins with (v_8, v_{10}, v_4). The first vertex on the adjacency list of v_4 that has not been used is v_6 and so P' is extended again to obtain

$$P' = (v_8, v_{10}, v_4, v_6).$$

The two vertices on the adjacency list of v_6 have been used. Thus P' terminates and the tree T' is produced. Since T' is not a spanning tree of G, we backtrack along P' to v_4. The first vertex on the adjacency list of v_4 that has not been used is v_9 and so a new path P'' is begun with (v_4, v_9). Since v_{11} is the final vertex of G and is on the adjacency list of v_9, the depth-first search tree T is completed by adding v_{11} to P'' and obtaining

$$P'' = (v_4, v_9, v_{11}).$$

(See Figure 13.21). ◆

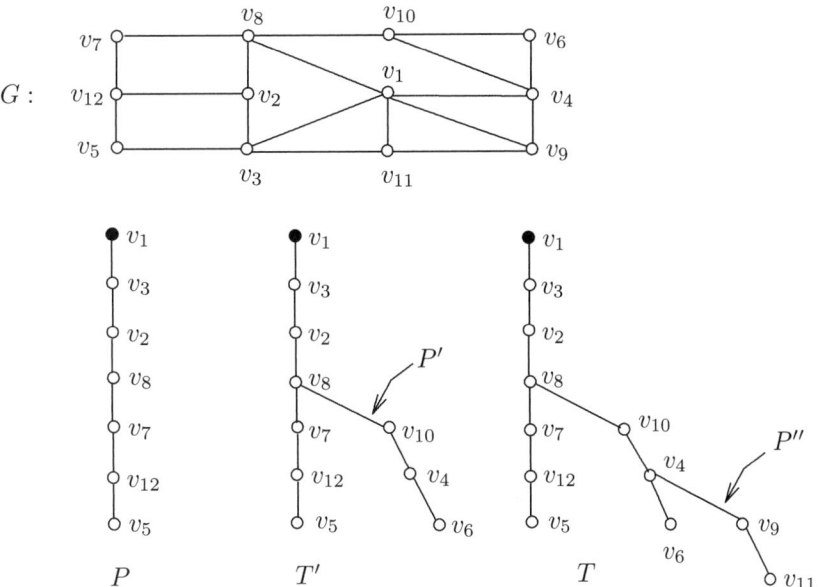

Figure 13.21: Constructing a depth-first search tree T
for the graph G in Example 13.23

Algorithm 13.24 (The Depth-First Search Algorithm) *This algorithm constructs a depth-first search tree in a connected graph.*

Input: A positive integer n and a connected graph G with vertex set $V(G) = \{v_1, v_2, \ldots, v_n\}$ described by means of the adjacency lists of its vertices, where the subscripts of the vertices in each adjacency list are listed in increasing order.

Output: A depth-first search tree T of G with vertex set $V' = V(G)$ and edge set E'.

1. $V' := \{v_1\}$ [v_1 is the root of T.]
      $E' := \emptyset$ [E' represents the edge set of the tree being considered.]
      $w := v_1$ [w is the vertex currently being considered.]
      $i := 0$ [i counts the number of edges in the tree being considered.]
2.   **while** $i < n - 1$ **do**
        begin
3.   **while** there is a vertex $v \notin V'$
         on the adjacency list of w such that
         no cycle is created when wv is added to E' **do**
           begin
             Choose the minimum k such that no cycle is created when wv_k
                       is added to E'
             $V' := V' \cup \{v_k\}$
             $E' := E' \cup \{wv_k\}$
             $w := v_k$
             $i := i + 1$
           end
         $w := $ parent of w
        end
4.   **output** E'

Breadth-First Search Trees

Another common spanning tree of a connected graph G with $V(G) = \{v_1, v_2, \ldots, v_n\}$ that is also a rooted tree is a **breadth-first search tree**. As in a depth-first search, we systematically visit all vertices here as well, but the vertices are visited in a completely different manner. We begin with a vertex v (the root of this tree) but rather than start with a path having initial vertex v, we start with the star T_1 having center v and containing all vertices adjacent to v. If T_1 should contain all vertices of G, then T_1 is a breadth-first search tree of G. If T_1 does not contain all vertices of G, then we go to the neighbor of v having the smallest subscript, say v'. If there are any neighbors of v' not in T_1, then we add a new star to T_1 at v' containing the remaining neighbors of v'. We then move to the neighbor v'' of v having the second smallest subscript. We continue this until either the tree constructed is a spanning tree of G or until we have run through all neighbors of v, in which case we start over again with the neighbors of v' and then the neighbors of v'' and so on. The resulting spanning tree is the breadth-first search tree of G. Thus, rather than conducting a search that proceeds deeply into G, as with a depth-first search, a **breadth-first search** of G conducts a broad search of G. While a depth-first search is more *vertical* in nature, a breadth-first search is more *horizontal*.

Before presenting a formal algorithm for conducting a breadth-first search in a connected graph, we illustrate this method with the graph G of Figure 13.21, also shown in Figure 13.22.

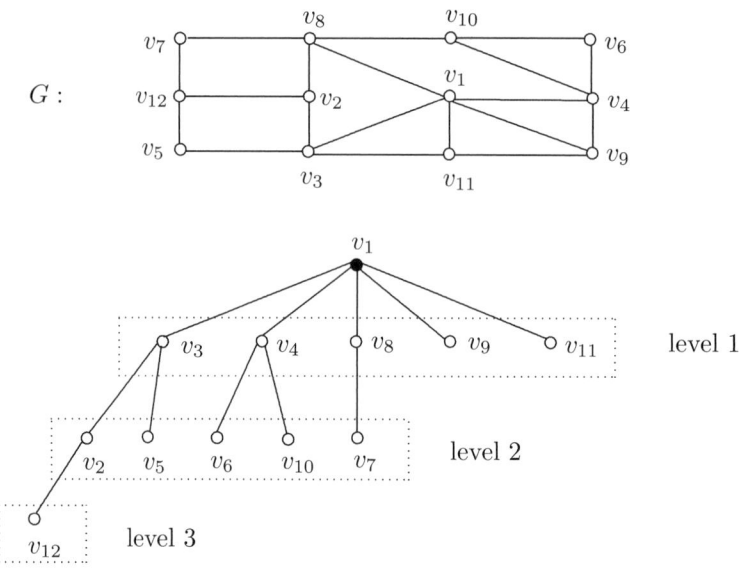

Figure 13.22: Constructing a breadth-first search tree T
for the graph G in Example 13.25

Example 13.25

Construct a breadth-first search tree T rooted at v_1 for the graph G of Figure 13.22. The adjacency lists of these vertices of G are given below:

$$
\begin{array}{lll}
v_1: \ v_3, v_4, v_8, v_9, v_{11} & v_2: \ v_3, v_8, v_{12} & v_3: \ v_1, v_2, v_5, v_{11} \\
v_4: \ v_1, v_6, v_9, v_{10} & v_5: \ v_3, v_{12} & v_6: \ v_4, v_{10} \\
v_7: \ v_8, v_{12} & v_8: \ v_1, v_2, v_7, v_{10} & v_9: \ v_1, v_4, v_{11} \\
v_{10}: \ v_4, v_6, v_8 & v_{11}: \ v_1, v_3, v_9 & v_{12}: \ v_2, v_5, v_7.
\end{array}
$$

Solution. To construct a breadth-first search tree T in this case, we begin at v_1 and select the star T_1 having all edges incident with v_1. The neighbor of v_1 in T_1 with the smallest subscript is v_3.

There are two vertices adjacent to v_3 (namely v_2 and v_5) not in T_1, so the vertices v_2 and v_5 and the edges v_3v_2 and v_3v_5 are added to the breadth-first search tree being constructed. We move to v_4. There are two vertices adjacent to v_4 (namely v_6 and v_{10}) that do not belong to the tree already constructed, so the vertices v_6 and v_{10} and the edges v_4v_6 and v_4v_{10} are added to form a new tree. (Since the tree is not a spanning tree of G, we continue.) There is one vertex adjacent to v_8 (namely v_7) that is not in this tree, so v_7 and v_7v_8 are added to this tree. There are no such neighbors of v_9 and v_{11}. Therefore, we now consider the vertex in level 2 with the smallest subscript and which is adjacent to v_3. This is v_2. The vertex v_2 is adjacent to the vertex v_{12}, which is not in the current tree; so the vertex v_{12} and the edge v_2v_{12} are added to this tree. Since all vertices of G have now been encountered, this is a breadth-first search tree T of G (see Figure 13.22). ♦

Algorithm 13.26 (The Breadth-First Search Algorithm) *This algorithm constructs a breadth-first search tree in a connected graph.*

> **Input**: A positive integer n and a connected graph G with vertex set $V(G) = \{v_1, v_2, \ldots, v_n\}$ described by means of the adjacency lists of its vertices, where the subscripts of the vertices in each adjacency list are listed in increasing order.
>
> **Output**: A breadth-first search tree T of G with vertex set $V' = V(G)$ and edge set E'.

1. $V' := \{v_1\}$ [v_1 is the root of T.]
      $E' := \emptyset$ [E' represents the edge set of the tree being considered.]
      $w := v_1$ [w is the vertex currently being considered.]
      $i := 1$ [i counts the number of vertices in the tree being considered.]
      $j := 1$ [u_j is the current value of w.]
      $u_1 := v_1$
2.   **while** $i < n$ **do**
        begin
3.   **while** there is a vertex $v \notin V'$ on the adjacency list of w **do**
            begin
              Choose the minimum k such that v_k is on the adjacency list
                       of w and $v_k \notin V'$.
              $V' := V' \cup \{v_k\}$
              $E' := E' \cup \{wv_k\}$
              $i := i + 1$
              $u_i := v_k$
            end
           $j := j + 1$
           $w := u_j$
        end
4.   **output** E'

For each labeling of the vertices of a connected graph G of order n by $v_1, v_2, \ldots, v_n$, there is a unique depth-first search tree and a unique breadth-first search tree when using the Depth-First Search Algorithm and Breadth-First Search Algorithm, respectively. Since the vertices of G can ordinarily be labeled in many ways, there are often many possible depth-first search trees and many possible breadth-first search trees of G.

Exercises for Section 13.2

1. Can a rooted tree have more than one root?

2. For each integer k with $1 \le k \le 6$, give an example of a rooted tree T_k of order 7 with height k.

3. Let T be a tree. Among all orientations of T that result in rooted trees, let T_1 be one whose height h_1 is minimum and let T_2 be one whose height h_2 is maximum.

(a) What familiar numbers are h_1 and h_2?

(b) If T_1 is rooted at r_1, what familiar property does r_1 have?

4. Construct the binary search tree resulting from the Binary Search Algorithm when applied to the following alphabetized list of words: range, roster, rust, sergeant, silver, trample, tulip. Indicate in this binary search tree how the word silver is found using a binary search.

5. Construct the binary search tree resulting from the Binary Search Algorithm when applied to the following alphabetized list of names: Aaron, Bonds, Griffey, Killebrew, Mays, McGwire, Robinson, Rodriguez, Ruth, Sosa, Thome. Indicate in this binary search tree how the name Killebrew is found using a binary search.

6. Suppose that we have a collection of six coins, exactly one of which is counterfeit and weighs less than each authentic coin. Use a decision tree to describe a method for identifying the counterfeit coin using two weighings of a balance scale.

7. Suppose that we have a collection of six coins, exactly one of which is counterfeit and does not weigh the same as an authentic coin. Use a decision tree to describe a method for identifying the counterfeit coin using at most three weighings of a balance scale.

8. Suppose that we have a collection of three coins, exactly two of which are counterfeit and weigh the same as each other but not the same as the authentic coin. What is the minimum number of weighings needed to identify the two counterfeit coins if

(a) each counterfeit coin is lighter than each authentic coin?

(b) it is only known that the authentic coin does not have the same weight as a counterfeit coin?

9. Suppose that we have a collection of four coins, two of which are counterfeit (and weigh the same as each other) and two of which are authentic (and weigh the same as each other). Describe a method to identify the counterfeit coins by successive weighings on a balance scale if

(a) each counterfeit coin is lighter than each authentic coin?

(b) it is only known that an authentic coin does not have the same weight as a counterfeit coin?

10. Solve **The 12 Coins Problem:** You are given 12 coins, one of which is counterfeit and weighs more or less than the 11 authentic coins. Show that in three weighings it can be determined which of the 12 coins is counterfeit and whether the counterfeit coin is heavier or lighter than the authentic coins. [Hint: First consider $\{1, 2, 3, 4\}$, $\{5, 6, 7, 8\}$ and then $\{1, 2, 3, 8\}$, $\{4, 9, 10, 11\}$.]

11. Use the Depth-First Search Algorithm (Algorithm 13.24) to construct the depth-first search tree having root v_1 for the graph G of Figure 13.23.

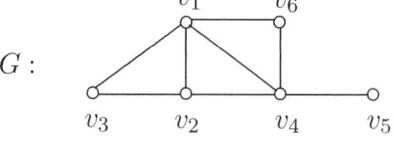

Figure 13.23: The graph in Exercise 11

12. Use the Depth-First Search Algorithm (Algorithm 13.24) to construct the depth-first search tree with root v_1 for the graph G of Figure 13.24.

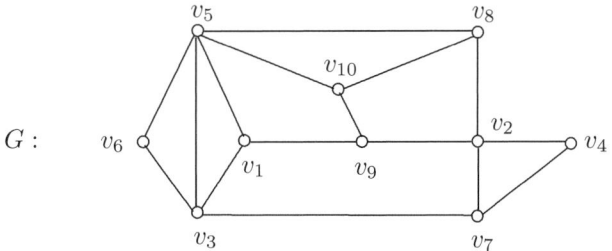

Figure 13.24: The graph in Exercise 12

13. Use the Depth-First Search Algorithm (Algorithm 13.24) to construct the depth-first search tree with root v_1 for the connected graph G with $V(G)=\{v_1, v_2, \ldots, v_{10}\}$, the adjacency lists of whose vertices are:

$v_1 : v_2, v_4, v_7, v_8, v_{10}$ $v_2 : v_1, v_4, v_6$ $v_3 : v_5, v_9$

$v_4 : v_1, v_2$ $v_5 : v_3, v_6, v_8$ $v_6 : v_2, v_5$

$v_7 : v_1, v_{10}$ $v_8 : v_1, v_5, v_9$ $v_9 : v_3, v_8, v_{10}$

$v_{10} : v_1, v_7, v_9.$

14. Use the Breadth-First Search Algorithm (Algorithm 13.26) to construct the breadth-first search tree with root v_1 for the graph G of Figure 13.25.

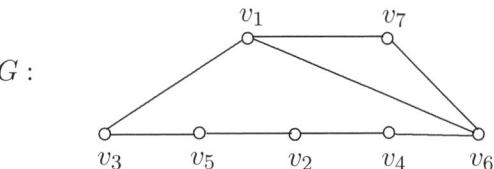

Figure 13.25: The graph in Exercise 14

15. Use the Breadth-First Search Algorithm (Algorithm 13.26) to construct the breadth-first search tree with root v_1 for the graph G with $V(G)=\{v_1, v_2, \ldots, v_8\}$ whose adjacency matrix $\mathbf{A}$ is given below:

$$\mathbf{A} = \begin{bmatrix} 0 & 0 & 0 & 0 & 1 & 0 & 1 & 0 \\ 0 & 0 & 1 & 1 & 1 & 0 & 0 & 0 \\ 0 & 1 & 0 & 0 & 1 & 0 & 0 & 0 \\ 0 & 1 & 0 & 0 & 0 & 0 & 0 & 1 \\ 1 & 1 & 1 & 0 & 0 & 0 & 0 & 0 \\ 0 & 0 & 0 & 0 & 0 & 0 & 1 & 1 \\ 1 & 0 & 0 & 0 & 0 & 1 & 0 & 1 \\ 0 & 0 & 0 & 1 & 0 & 1 & 1 & 0 \end{bmatrix}.$$

16. The adjacency list of a graph G with $V(G) = \{v_1, v_2, \ldots, v_{12}\}$ is given below:

$v_1 : v_2, v_7, v_9$ $v_2 : v_1, v_3, v_7$ $v_3 : v_2, v_4, v_5, v_8$

$v_4 : v_3, v_5, v_6$ $v_5 : v_3, v_6, v_8$ $v_6 : v_4, v_5$

$v_7 : v_1, v_2, v_8, v_9$ $v_8 : v_3, v_5, v_7, v_{10}$ $v_9 : v_1, v_7, v_{10}, v_{11}$

$v_{10} : v_8, v_9, v_{11}, v_{12}$ $v_{11} : v_9, v_{10}, v_{12}$ $v_{12} : v_{10}, v_{11}.$

Construct a depth-first search tree T_1 and a breadth-first search tree T_2 rooted at v_1.

13.3   The Minimum Spanning Tree Problem

In the example discussed in Figure 13.1, there are eight settlements $s_1, s_2, \ldots, s_8$. We mentioned that paved roads are to be constructed between some pairs of these settlements so that it is possible to travel between any two of these settlements on paved roads and that Figure 13.1 shows one way to do this. Since the idea is to construct paved roads as cheaply as possible, we noted that the resulting graph must be a tree. In an actual situation, however, we would know (or at least have an estimate of) the cost of constructing a paved road between each pair of settlements. If we were to construct paved roads between every pair of settlements, the possible road system might look something like that shown in Figure 13.26.

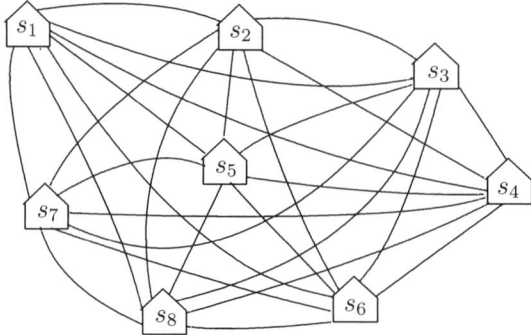

Figure 13.26: Constructing all possible paved roads between pairs of settlements

Suppose that we know the cost of constructing a paved road between every pair of settlements and this information is given in a cost matrix $\mathbf{C} = [c_{ij}]$, where c_{ij} represents the cost (in thousands of dollars) of constructing a paved road between s_i and s_j.

$$\mathbf{C} = \begin{bmatrix} 0 & 255 & 475 & 615 & 310 & 620 & 380 & 470 \\ 255 & 0 & 265 & 315 & 240 & 405 & 390 & 460 \\ 475 & 265 & 0 & 210 & 280 & 390 & 540 & 515 \\ 615 & 315 & 210 & 0 & 275 & 245 & 575 & 535 \\ 310 & 240 & 280 & 275 & 0 & 265 & 235 & 240 \\ 620 & 405 & 390 & 245 & 265 & 0 & 360 & 290 \\ 380 & 390 & 540 & 575 & 235 & 360 & 0 & 200 \\ 470 & 460 & 515 & 535 & 240 & 290 & 200 & 0 \end{bmatrix}.$$

One possible way to choose the paved roads to construct is to begin with the least expensive, which would be the road between s_7 and s_8, costing \$200,000. The next least expensive is the one between s_3 and s_4 (which would cost \$210,000). The next least expensive road would be the road between s_5 and s_7 costing \$235,000. The next least expensive road would cost \$240,000. There are two such roads, one between s_2 and s_5 and the other between s_5 and s_8. Since the road between s_5 and s_8 creates a cycle, we do not choose this one. We do select the road between s_2 and s_5, however. The next least expensive road is between s_4 and s_6, which is estimated to cost \$245,000. Since no cycle is created, we choose this road. Proceeding in this manner, we conclude by selecting the roads between s_1 and s_2 and between s_2 and s_3, which cost \$255,000 and \$265,000, respectively (see Figure 13.27). Thus the total estimated cost of constructing all these roads would be \$1,650,000. Of course, the crucial question here is the following: Is there some way of selecting roads to construct that would be less expensive? The answer is *no*. In fact, the method we used to choose the roads to pave cannot be improved upon.

The problem we have been describing is actually a special case of a well-known problem. Let G be a connected weighted graph. So each edge of G has a weight (or cost), which is a positive real number. The weight of an edge e of G is denoted by $w(e)$. The **weight** $w(H)$ of a subgraph H of G is the sum of the weights of the edges of H, that is,

$$w(H) = \sum_{e \in E(H)} w(e).$$

Definition 13.27 *A **minimum spanning tree** of a connected weighted graph G is a spanning tree of G whose weight is minimum among all spanning trees of G.*

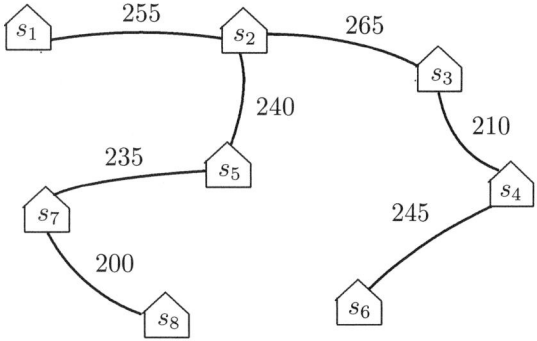

Figure 13.27: Constructing the least expensive roads

The major problem in this context is the following.

The Minimum Spanning Tree Problem *Find a minimum spanning tree in a connected weighted graph.*

In a paper written by Ronald L. Graham and Pavol Hell in 1985, they concluded that the Minimum Spanning Tree Problem appears to have its origin in 1926 when Otakar Borůvka was considering the rural electrification of Southern Moravia in the Czech Republic. There have been a number of algorithmic solutions of this problem. We present two of the best known of these, the first of which is an algorithm due to Joseph Kruskal.

Kruskal's Algorithm

Joseph Bernard Kruskal (1928-2010) was from a family of five children, three boys and two girls. All boys became mathematicians. Kruskal received his Ph.D. from Princeton University in 1954. Throughout his life he has been an active researcher, with a great deal of his work in mathematics and linguistics. He spent much of his life working at Bell Laboratories. However, it was only two years after completing his doctoral degree that the paper was published containing the algorithm that bears his name.

Before we formally state Kruskal's algorithm for constructing a minimum spanning tree T in a connected weighted graph G, let's see how this algorithm works. We start by selecting any edge of G of smallest weight for the first edge e_1 of T and then select any remaining edge of G with smallest weight for the second edge e_2 of T. For the third edge e_3 of T, we choose any remaining edge of G with smallest weight that does not produce a cycle with the previously selected edges. We continue in this manner until a spanning tree is produced. Therefore, the manner in which the roads were chosen to be paved in the example above uses Kruskal's algorithm. Let's consider another example.

Example 13.28

Figure 13.28 shows how a minimum spanning tree of a connected weighted graph G is constructed using Kruskal's algorithm. We begin by selecting the edges having weights 2 and 3. When selecting an edge with weight 4, only one can be selected since selecting both would produce a cycle. It doesn't matter, however, which edge of weight 4 is chosen. When choosing an edge having weight 5, only one edge of the two edges having weight 5 is possible, namely wx. This edge is chosen and a minimum spanning tree has been constructed. The weight of a minimum spanning tree of G is therefore $2 + 3 + 4 + 5 = 14$. ◆

We now present a formal statement of Kruskal's algorithm.

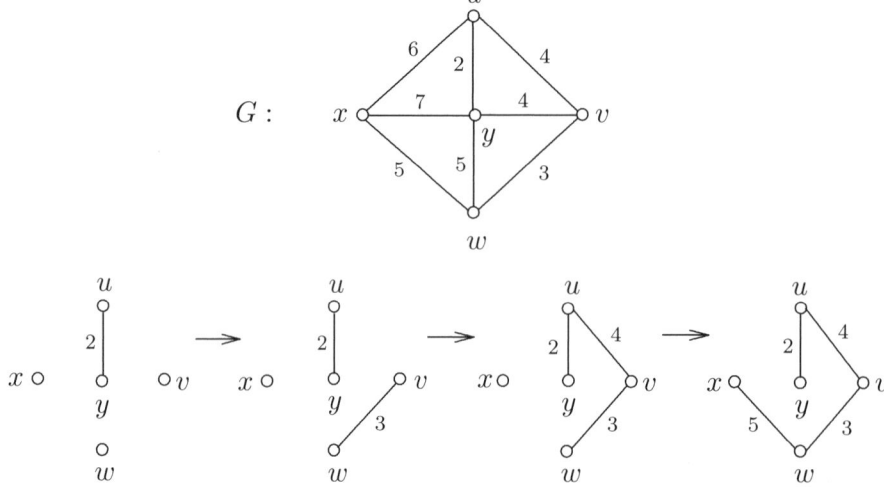

Figure 13.28: Constructing a minimum spanning tree by Kruskal's algorithm

Algorithm 13.29 (**Kruskal's Algorithm**) *This algorithm determines a minimum spanning tree in a connected weighted graph G of order n and size m.*

> **Input**: A positive integer n, a positive integer m and a connected weighted graph G expressed by its edge set $E(G) = \{e_1, e_2, \ldots, e_m\}$, where $w(e_i) \le w(e_{i+1})$ for $i = 1, 2, \ldots, m - 1$.

> **Output**: A minimum spanning tree of G with edge set E'.

> 1. $E' := \emptyset$ [E' represents the edge set of the tree being considered.]
> $i := 0$ [i counts the number of edges in the tree being considered.]
> 2.   **for** $j := 1$ **to** m **do** [e_j is the edge in G being considered.]
>   **begin**
> 3.   **while** $\langle E' \cup \{e_j\} \rangle$ contains no cycles **do**
>   **begin**
> $E' := E' \cup \{e_j\}$
> $i := i + 1$
>   **end**
>   **if** $i = n - 1$, then $j := m + 1$
>   **end**
> 4.   **output** E'

We illustrate Kruskal's algorithm one additional time by finding a minimum spanning tree in the connected weighted graph G in Figure 13.29.

Example 13.30

Use Kruskal's algorithm to find a minimum spanning tree in the connected weighted graph G shown in Figure 13.29.

Solution. Since 7 is the smallest weight of an edge of G, we select the edge tu as our first edge in the minimum spanning tree T being constructed. The next three edges having smallest weights are xy, uv and wz having weights 8, 10 and 12, respectively. Since no cycle is created from any of these four edges, the three edges xy, uv and wz are added to T. The next smallest weight of an edge is 14 and there are two such edges, namely tv and yz. Since adding tv to T would produce a cycle, this edge is not selected; however, yz is added to T. The next smallest weight of an edge is 16. Two edges

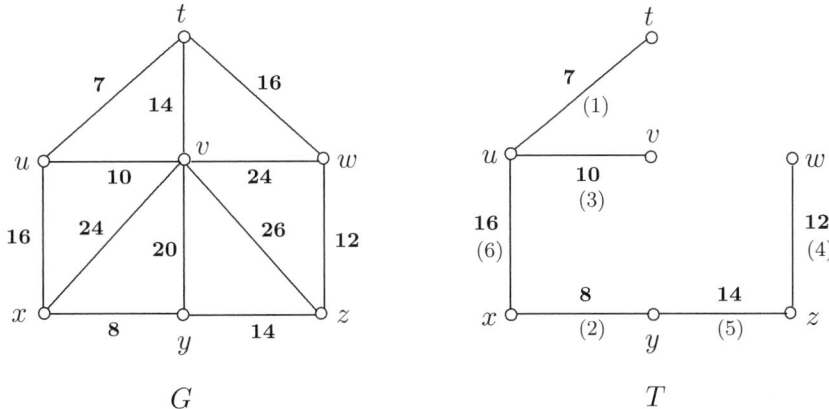

Figure 13.29: Constructing a minimum spanning tree
by Kruskal's algorithm in Example 13.30

have this weight, namely ux and tw. Either one of these would complete a minimum spanning tree, so ux is selected arbitrarily. The resulting minimum spanning tree is also shown in Figure 13.29, where (k), $1 \leq k \leq 6$, placed near an edge indicates that this is the kth edge chosen by Kruskal's algorithm. The weight of a minimum spanning tree of G is therefore $7+8+10+12+14+16 = 67$. ◆

We now verify that a spanning tree constructed by Kruskal's algorithm is, in fact, a *minimum* spanning tree.

Theorem 13.31 *Kruskal's algorithm produces a minimum spanning tree in every connected weighted graph.*

Proof. Suppose that the theorem is false. Then there exists a connected weighted graph G of order $n \geq 2$ such that a spanning tree T produced by Kruskal's algorithm is not a minimum spanning tree of G. Let $E(T) = \{e_1, e_2, \ldots, e_{n-1}\}$, where the edges of T have been selected in the order listed. Hence

$$w(e_1) \leq w(e_2) \leq \cdots \leq w(e_{n-1}).$$

The weight of T is therefore

$$w(T) = \sum_{i=1}^{n-1} w(e_i).$$

Since T is not a minimum spanning tree of G, there is a spanning tree of G whose weight is less than that of T. Among all minimum spanning trees of G, let T' be one having the greatest number of edges in common with T. Certainly, T contains one or more edges that do not belong to T'. Let k be the smallest integer $(1 \leq k \leq n-1)$ such that e_k is an edge of T that does not belong to T' and define $G' = T' + e_k$. Therefore, G' has exactly one cycle C. Since T has no cycles, C contains an edge e not belonging to T. Therefore,

$$T'' = G' - e = (T' + e_k) - e$$

is a spanning tree of G whose weight is

$$w(T'') = w(T') + w(e_k) - w(e).$$

Because T' is a minimum spanning tree of G, $w(T') \leq w(T'')$ and so $w(e) \leq w(e_k)$. If $k = 1$, then $w(e) = w(e_1)$ since e_1 has the smallest weight among all edges of G. For $k \geq 2$, by Kruskal's algorithm, e_k is an edge of minimum weight such that the subgraph induced by $\{e_1, e_2, \ldots, e_{k-1}\} \cup$

$\{e_k\}$ has no cycles. Since the subgraph induced by $\{e_1, e_2, \ldots, e_{k-1}\} \cup \{e\}$ is a subgraph of T', this subgraph has no cycles. By Kruskal's algorithm, it follows that $w(e_k) \leq w(e)$ and so $w(e_k) = w(e)$. This implies that $w(T'') = w(T')$ and so T'' is a minimum spanning tree of G. However, T'' has one more edge in common with T than T' does. This is a contradiction. $\blacksquare$

Prim's Algorithm

The second algorithm we present is named for Robert Clay Prim (born in 1921). Prim received his Ph.D. from Princeton University in 1949. He served as vice president of research at the Sandia Corporation. The paper containing the algorithm that bears his name was published in 1957, only one year after Kruskal's algorithm was published. This algorithm was originally discovered by the Czech mathematician Vojtěch Jarnik in 1930. Nevertheless, it is known as Prim's algorithm.

Prim's algorithm for constructing a minimum spanning tree T in a connected weighted graph G begins with selecting a vertex r of G and an edge of smallest weight incident with r as the first edge e_1 of T. For the second edge e_2 of T, we select any remaining edge of G of smallest weight incident with exactly one vertex already in T and add this edge to T. The resulting graph is therefore connected and contains no cycles, that is, the resulting graph is a tree. We continue this procedure until a spanning tree is produced. Before presenting a formal statement of Prim's algorithm for constructing a spanning tree in a connected weighted graph, let's illustrate how this method works.

Example 13.32

Figure 13.30 shows how a minimum spanning tree T of the connected weighted graph G is constructed using Prim's algorithm. Note that the graph G is the same graph shown in Figure 13.28. We choose to begin with the vertex v of G to construct a minimum spanning tree T of G. $\blacklozenge$

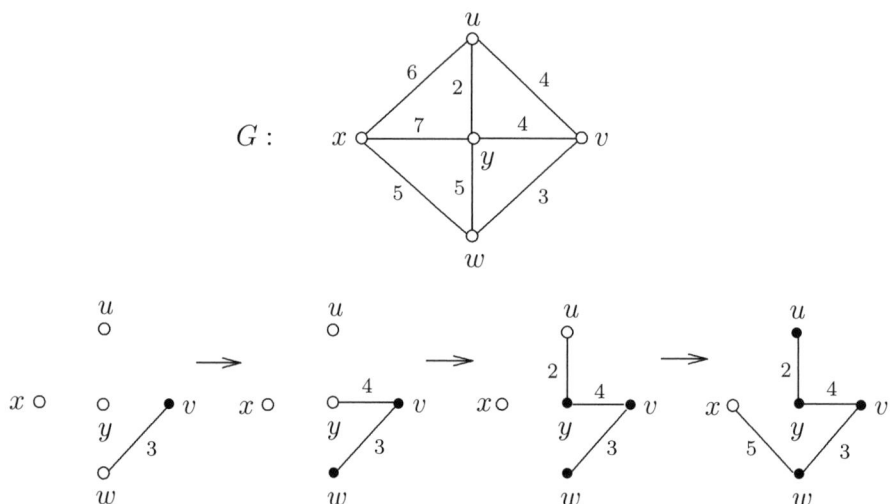

Figure 13.30: Constructing a spanning tree by Prim's algorithm

We now present a formal statement of Prim's algorithm. In this case, however, we do not include a proof that Prim's algorithm accomplishes what it promises.

Algorithm 13.33 (**Prim's Algorithm**) *This algorithm determines a minimum spanning tree in a connected weighted graph G of order n and size m.*

> **Input**: A positive integer n, a positive integer m, a connected weighted graph G expressed by its edge set $E(G) = \{e_1, e_2, \ldots, e_m\}$, where $w(e_i) \leq w(e_{i+1})$ for $i = 1, 2, \ldots, m-1$ and a vertex r of G.
>
> **Output**: The edge set E' of a minimum spanning tree of G rooted at the vertex r.
>
> 1. $E' := \{e\}$, where e is an edge of minimum weight incident with r.
>   [E' represents the edge set of the tree being considered.]
> $i := 1$ [i counts the number of edges in the tree being considered.]
> 2.   **while** $i < n - 1$ **do**
>   **begin**
> 3.   **for** $j := 1$ **to** m **do** [e_j is the edge of G being considered.]
> 4.   **while** $e_j \notin E'$ and the subgraph induced by $E' \cup \{e_j\}$ is connected
>   and contains no cycles **do**
>   **begin**
> $E' := E' \cup \{e_j\}$
> $i := i + 1$
> $j := m + 1$
>   **end**
>   **end**
> 5.   **output** E'

If all weights of the edges of a connected graph G are the same, say 1, then every spanning tree of G is a minimum spanning tree. Indeed, a depth-first search tree of G with root r and a breadth-first search tree of G with root r can both be considered as special cases of spanning trees of G with root r constructed from Prim's algorithm.

Prim's algorithm is illustrated again by finding a minimum spanning tree in a connected weighted graph.

Example 13.34

Use Prim's algorithm to find a minimum spanning tree in the connected weighted graph G shown in Figure 13.29, which is then redrawn in Figure 13.31.

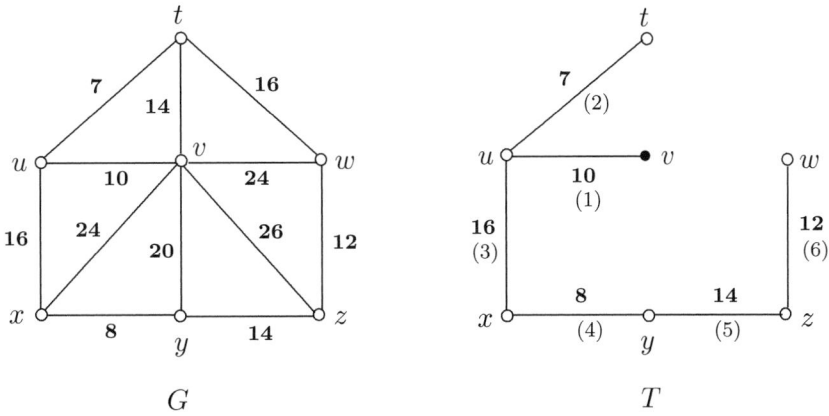

Figure 13.31: Constructing a minimum spanning tree
by Prim's algorithm in Example 13.34

Solution. We begin with a vertex of G, say v. Since 10 is the smallest weight of an edge of G that is incident with v, we select the edge uv as our first edge in the minimum spanning tree T being constructed. Among the remaining edges incident with u or v, the edge ut has the smallest weight, namely 7 and so we add ut to T as the second edge of T. The remaining edges that are incident with a vertex in T are tv, tw and ux. Although tv has the smallest weight, namely 14 (each of tw and ux has weight 16), this edge is not selected since adding tv to T would produce a cycle, namely the cycle (u, v, t, u). On the other hand, we can add either tw or ux to T, say we add ux to T as the third edge of T. Continuing in this manner, the next three edges we select for T are xy, yz and zw and this completes a minimum spanning tree T. The resulting minimum spanning tree is also shown in Figure 13.31. The order in which the six edges of T are selected is indicated by (1), (2), ..., (6).   ◆

Exercises for Section 13.3

1. Use Kruskal's algorithm to find a minimum spanning tree in the connected weighted graph G shown in Figure 13.32.

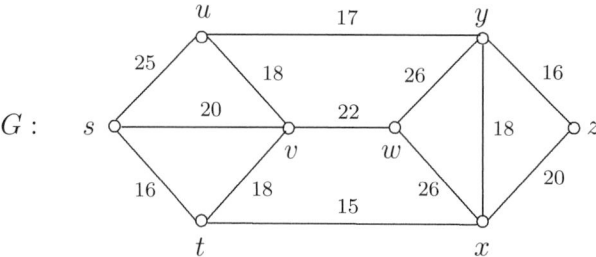

Figure 13.32: The graph G in Exercises 1 and 5

2. Use Kruskal's algorithm to find a minimum spanning tree in the connected weighted graph G shown in Figure 13.33.

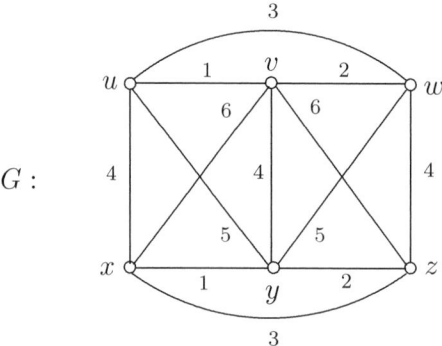

Figure 13.33: The graph G in Exercises 2 and 6

3. What does Kruskal's algorithm produce when no two weights in a connected weighted graph are equal?

4. What does Kruskal's algorithm produce if all weights in a connected weighted graph are equal?

5. Use Prim's algorithm to find a minimum spanning tree in the connected weighted graph shown in Figure 13.32.

6. Use Prim's algorithm to find a minimum spanning tree in the connected weighted graph shown in Figure 13.33.

7. Let G be a connected weighted graph. If every spanning tree of G is a minimum spanning tree, does every edge of G have the same weight?

8. Let G be a connected weighted graph and let $w_1 < w_2 < \cdots < w_k$ be the $k \geq 3$ weights of the edges of G. Thus w_1 is the minimum weight and w_k is a maximum weight of the edges of G. Prove or disprove:

   (a) Every minimum spanning tree of G must contain an edge with the minimum weight w_1.

   (b) Every minimum spanning tree of G must contain an edge with the weight w_1 and an edge with the weight w_2.

   (c) Every minimum spanning tree of G must contain an edge e_i with the weight w_i for each i with $1 \leq i \leq 3$.

   (d) No minimum spanning tree of G contains an an edge with the maximum weight w_k.

9. Let G be a connected weighted graph. Describe, with justification, a method for finding a maximum spanning tree of G.

10. The weights of the edges of a complete graph of order 5 are $1, 2, 2, 3, 3, 3, 4, 4, 4, 4$. Show that the weight of a minimum spanning tree in this graph cannot exceed 10.

11. For each of the weighted graphs G_1 and G_2 in Figure 13.34, determine the number of positive integers that can be the weight of some spanning tree of the graph.

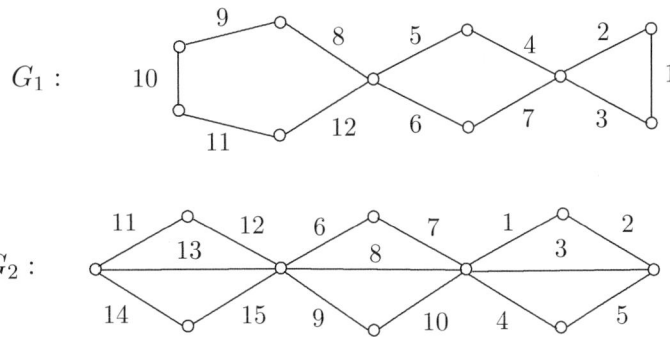

Figure 13.34: The weighted graphs in Exercise 11

12. Let G is a connected weighted graph of order $n \geq 4$ all of whose edges have the same weight. If every collection of $n - 1$ edges of G forms a minimum spanning tree, then what is G?

13. Let G be a connected weighted graph and T a minimum spanning tree of G. Prove that T is a unique minimum spanning tree of G if and only if the weight of each edge e of G that is not in T exceeds the weight of every other edge on the cycle in $T + e$.

14. Show, for each integer $k \geq 2$, that there exists a connected weighted graph containing exactly k minimum spanning trees.

Chapter 13 Highlights

Key Concepts

ancestor (of a vertex v in a tree rooted at r): a vertex u ($\neq v$) that lies on the $r - v$ path in the tree.

binary search tree: a tree produced by conducting the Binary Search Algorithm in a graph.

breadth-first search tree: a tree produced by conducting a **breadth-first search** in a graph.

child (of a vertex v in a rooted tree): a descendant of v and belonging to the level one step below that of v.

depth-first search tree: a tree produced by conducting a **depth-first search** in a graph.

descendant (of a vertex v in a tree rooted at r): a vertex w ($\neq v$) such that v lies on the $r - w$ path in the tree.

forest: a graph containing no cycles.

height (of a rooted tree T with root r): the largest integer h for which there is a vertex of T at level h.

level i: the set of vertices in a rooted tree T with root r whose distance from r is i.

minimum spanning tree (of a weighted graph G): a spanning tree of G whose weight is minimum among all spanning trees of G.

The Minimum Spanning Tree Problem: the problem of finding a minimum spanning tree in a connected weighted graph.

parent (of a vertex v in a rooted tree): the ancestor of v belonging to the level one step above v.

rooted tree: a tree T with a designated vertex r (called the **root** of T).

spanning tree (of a graph G): a spanning subgraph of G that is a tree.

tree: a connected graph having no cycles.

The 12 Coins Problem: You are given 12 coins, one of which is counterfeit and weighs more or less than the 11 authentic coins. This is the problem of showing that in three weighings it can be determined which of the 12 coins is counterfeit and whether the counterfeit coin is heavier or lighter than each authentic coin.

weight of a subgraph (of a weighted graph): the sum of the weights of the edges of the subgraph.

Key Results

- Every nontrivial tree contains at least two leaves.

- Every tree of order n has size $n - 1$.

- If T is a connected graph of order n and size m containing no cycles, then $m = n - 1$.

- Every graph of order n and size $n - 1$ containing no cycles is a tree.

- If F is a forest of order n and size m having k components, then $m = n - k$.

- Every connected graph of order n and size $n - 1$ is a tree.

- If G is a graph of order n and size m satisfying any two of the following three properties, then G is a tree: (1) G is connected, (2) G has no cycles and (3) $m = n - 1$.

- A graph G is a tree if and only if G has a unique $u - v$ path for every two vertices u and v in G.

- Let T be a tree of order $n \geq 2$ with maximum degree k having n_i vertices of degree i for $i = 1, 2, \ldots, k$. Then the number of leaves in T is $n_1 = 2 + n_3 + 2n_4 + 3n_5 + \cdots + (k - 2)n_k$.

- A graph G contains a spanning tree if and only if G is connected.

- **The Depth-First Search Algorithm**: This algorithm constructs a depth-first search tree in a connected graph.

- **The Breadth-First Search Algorithm**: This algorithm constructs a breadth-first search tree in a connected graph.

- **Kruskal's Algorithm**: an algorithm that produces a minimum spanning tree in a connected weighted graph by constructing forests of increasing size.

- **Prim's Algorithm**: an algorithm that produces a minimum spanning tree in a connected weighted graph by constructing trees of increasing size.

Supplementary Exercises for Chapter 13

1. Determine the number of forests of order 6.

2. Give an example of a connected graph G that is not a tree such that

   (a) every two spanning trees of G are isomorphic.
   (b) no two spanning trees of G are isomorphic.

3. It is known that if T is a nontrivial tree and v is a leaf of T, then $T - v$ is also a tree.

   (a) Give an example of a connected graph G with $\Delta(G) = 2$ such that $G - v$ is a tree for every vertex v of degree 2 in G.
   (b) Give an example of a connected graph G with $\Delta(G) = 3$ such that $G - v$ is a tree for every vertex v of degree 3 in G.
   (c) Give an example of a connected graph G with $\Delta(G) = 4$ containing at least one vertex of degree 3 such that $G - v$ is a tree for every vertex v of degree 3 in G.

4. Prove that an edge e of a connected graph G is a bridge if and only if e belongs to every spanning tree of G.

5. Let G be a connected graph that is not a tree containing two distinct vertices u and v such that $G - u$ and $G - v$ are both trees. Show that $\deg u = \deg v$.

6. Let G be a connected graph of order $n \geq 4$ and size m such that $m = (3n - 2)/2$.

   (a) Prove or disprove: G contains at least two vertices of degree 2.
   (b) The question in (a) might suggest another question to you. Ask and answer another such question.

7. A rooted tree T has maximum degree $\Delta(T) = 4$ and height 5. What is the maximum possible order of T?

8. What is the largest number of vertices of degree 5 that a tree with 1001 leaves can have?

9. What is the maximum number of spanning trees that a connected graph of order $n \geq 4$ can have, no two of which have an edge in common?

10. Construct the binary search tree resulting from the following list of 15 terms (arranged alphabetically): bridge, connected, cut-vertex, cycle, degree, forest, graph, induced, leaf, multigraph, order, root, size, subgraph, tree. Indicate in this binary search tree, how the term *size* is found using a binary search.

11. Give an example of a connected graph G of order 8 or more that is not a tree with the property that for every two adjacent vertices, exactly one is a cut-vertex of G.

12. A tree T has three vertices of degree 4 and five vertices of degree 6. The remaining vertices of T have degree 1 or 3. There are three times as many end-vertices as vertices of degree 3 in T. What is the order of T?

13. The degrees of the vertices of a connected graph G are 5, 5, 4, 4, 4, 3, 3, 3, 3, 2. How many edges must be removed from G to produce a spanning tree T of G?

14. How many edges must be deleted from a 4-regular connected graph G of order n to obtain a spanning tree of G?

15. A certain tree T has three vertices of degree 10 and two vertices of degree 20. All other vertices of T are leaves. How many leaves does T have?

16. A tree T of order 52 has only vertices of degree 1, 3 and 5. There are twice as many vertices of degree 5 as there are vertices of degree 3. How many end-vertices does T have?

17. A tree T of order n is known to satisfy the following properties: (1) $95 < n < 100$, (2) the degree of every vertex of T is either 1, 3 or 5 and (3) T has twice as many vertices of degree 3 as that of degree 5. What is n?

18. Suppose that we have a collection of four coins, two authentic and two counterfeit. Neither counterfeit coin weights the same as an authentic coin but the sum of the weights of the counterfeit coins equals the sum of the weights of the authentic coins. Use a decision tree to describe a method for identifying the two counterfeit coins in at most three weighings.

19. Use Kruskal's algorithm to find a minimum spanning tree T in the connected weighted graph G shown in Figure 13.35. Indicate the order in which the edges are added to T.

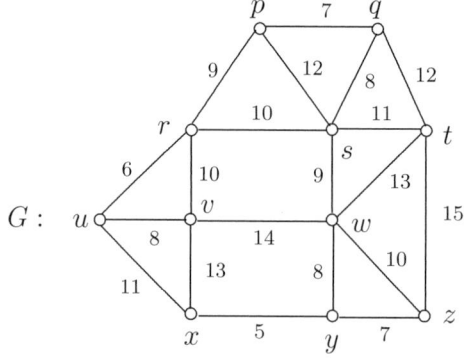

Figure 13.35: The graph G in Exercises 19 and 20

20. Use Prim's algorithm to find a minimum spanning tree T' in the connected weighted graph G shown in Figure 13.35 starting with the vertex t. Indicate the order in which the edges are added to T'.

21. (a) Show that the 3-regular graph K_4 contains two spanning trees T_1 and T_2 such that $\{E(T_1), E(T_2)\}$ is a partition of $E(K_4)$.

   (b) Give an example of a graph G of order 6 such that every vertex of G has degree 3 or 4 and G contains two spanning trees T_1 and T_2 for which $\{E(T_1), E(T_2)\}$ is a partition of $E(G)$.

22. (a) Let G be a graph of even order $n \geq 6$, every vertex of which has degree 3 or 4. If G contains two spanning trees T_1 and T_2 such that $\{E(T_1), E(T_2)\}$ is a partition of $E(G)$, then how many vertices of degree 4 must G contain?

   (b) Show that there exists an even integer $n \geq 6$ and a connected graph G of order n such that (1) every vertex of G has degree 3 or 4, (2) G contains the number of vertices of degree 4 determined in (a) and (3) G does not contain two spanning trees T_1 and T_2 for which $\{E(T_1), E(T_2)\}$ is a partition of $E(G)$.

23. Give an example of three non-isomorphic trees having the same order, same size and whose vertices have the same degrees.

24. Let T be a tree of order n with $V(T) = \{v_1, v_2, \ldots, v_n\}$ and $\deg v_i = d_i$ $(1 \leq i \leq n)$ such that $d_1 \geq d_2 \geq \cdots \geq d_n$. Prove that $d_i \leq \lceil \frac{n-1}{i} \rceil$ for each integer i. [Hint: (1) Use a proof by contradiction and (2) note that if $\deg v > r$, where $v \in V(T)$ and r is an integer, then $\deg v \geq r + 1$.]

25. A tree T has k end-vertices and ℓ vertices of degree 3 where $k \geq 2$ and $\ell \geq 1$. It also contains k vertices of some other degree r (so $r \neq 1$ and $r \neq 3$). What is r?

26. The degrees of the vertices of a graph G are $2, 2, 2, 1, 1$. Which of the following is true?

   (a) G is a tree. (b) G is not a tree.

   (c) There is no way to determine whether G is a tree.

27. A certain nontrivial tree T has order n. Its complement $\overline{T}$ contains two spanning trees T_1 and T_2 such that $E(T_1) \cap E(T_2) = \emptyset$ and $E(T_1) \cup E(T_2) = E(\overline{T})$. What is n?

28. It has been noted that there are seven non-isomorphic nontrivial trees of order 5 or less, namely three of size 4, two of size 3, one of size 2 and one of size 1. The sum of the sizes of these seven trees is 21, which is also the size of K_7. Do there exist seven such pairwise edge-disjoint trees in K_7?

Chapter 14

Planar Graphs and Graph Colorings

In the 1850s a young British mathematician would ask a simple-sounding and curious question which, for well over a century, would have the unexpected consequences of providing a major stimulus for the development of and interest in graph theory: Can the regions of every map be colored with four or fewer colors so that every two regions sharing some common boundary are colored differently? This *Four Color Problem* combines two areas of graph theory, namely, planar graphs and graph colorings. These two areas along with the Four Color Problem are discussed in this chapter.

14.1 Planar Graphs

In the introduction of Chapter 12, the question was asked as to whether utility lines from each of the three utilities water, electricity and natural gas can be connected to three houses without any two of the nine lines crossing each other. This problem is known as the **Three Houses and Three Utilities Problem**.

As one looks at the Three Houses and Three Utilities Problem, perhaps it becomes clear that this problem concerns the graph $K_{3,3}$, where the three vertices of one partite set of $K_{3,3}$ are the three houses and the three vertices of the other partite set are the three utilities. Determining a solution of the Three Houses and Three Utilities Problem is the same as determining whether the graph $K_{3,3}$ can be drawn in the plane without any of its edges crossing. If one attempts to draw $K_{3,3}$ in the plane so that none of its edges cross, then it is quite likely that difficulties arise (see Figure 14.1). It is probably not clear how to draw $K_{3,3}$ in the plane in this manner and also not clear how to convince oneself that any drawing of $K_{3,3}$ in the plane requires some edges to cross.

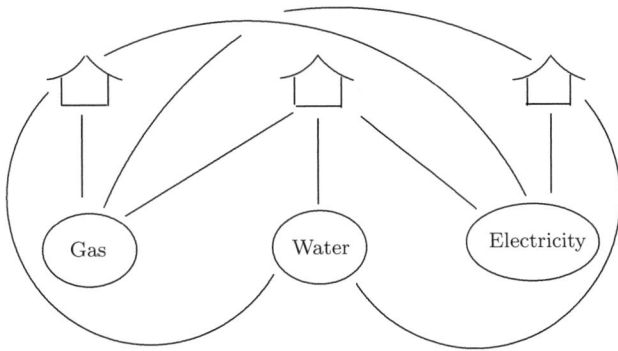

Figure 14.1: The Three Houses and Three Utilities Problem

Another graph that seems to present the same kind of difficulty is K_5 (see Figure 14.2). That is, neither (1) drawing K_5 in the plane without any of its edges crossing nor (2) showing that K_5 cannot be drawn in the plane without any of its edges crossing appears to be an easy task.

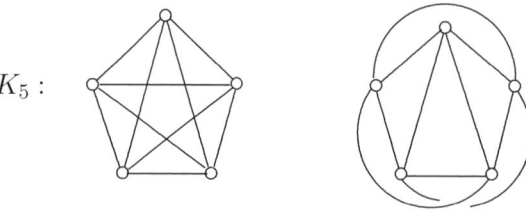

K_5 :

Figure 14.2: Drawing K_5 in the plane

As one would probably expect, there are many graphs in addition to K_5 and $K_{3,3}$ for which it appears to be very difficult to determine whether they can be drawn in the plane without any of its edges crossing. Of course, if we spend enough time trying (unsuccessfully) to draw a graph in the plane without any of its edges crossing, then it becomes increasingly likely to conclude that it can't be done. There is, however, no mathematical basis for such a conclusion.

The Three Houses and Three Utilities Problem suggests the consideration of a certain class of graphs. A graph is a **plane graph** if it is drawn in the plane so that no two of its edges cross. Therefore, the graphs G_1 and G_3 of Figure 14.3 are plane graphs, while the graph G_2 is not a plane graph.

Definition 14.1 *A graph G is a **planar graph** if G can be drawn in the plane so that no two of its edges cross.*

Certainly, the plane graphs G_1 and G_3 of Figure 14.3 are planar graphs as well, but the graph G_2 is also a planar graph since G_2 *can* be drawn as the plane graph G_3 (except for the labeling of the vertices). Consequently, G_2 and G_3 are isomorphic graphs. A graph that is not planar is called, not surprisingly, a **nonplanar graph**. None of the graphs in Figure 14.3 are nonplanar but we will see examples of nonplanar graphs shortly.

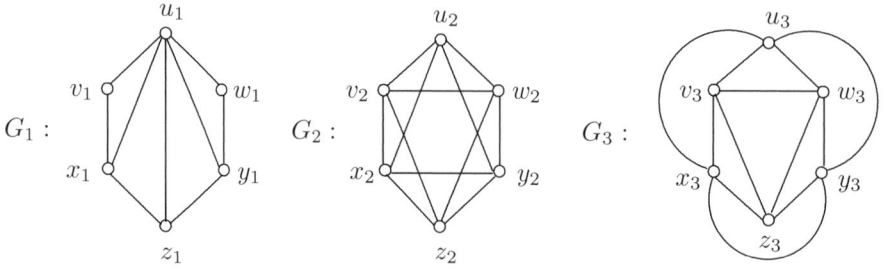

Figure 14.3: Plane and planar graphs

There is a more modern application of planar graphs. Since wire crossing in circuit layouts on silicon chips cause problems, we are interested in those circuits that can be modeled by planar graphs.

Suppose that G is a connected plane graph. Then, of course, G is drawn in the plane without any of its edges crossing. If we interpret a vertex as a point in the plane and an edge as a line segment or curve and remove the vertices and edges of G from the plane, then certain connected pieces of the plane are produced, called the **regions** of G. The vertices and edges of G that are incident with a region R of G form a subgraph of G called the **boundary** of R. There is always one unbounded region, called the **exterior region** of G. If G is a tree (which is a planar graph), then there is only one region. However, if G contains a cycle, then G has at least two regions, as there is at least one region interior to C and at least one region exterior to C. Some observations are useful here.

- An edge lies on the boundaries of two regions if and only if the edge lies on a cycle.

- A bridge lies on the boundary of a single region.

- The boundary of a region is a cycle unless one or more edges of the boundary are bridges.

Example 14.2

The connected plane graph G of Figure 14.4 has four regions denoted by R_1, R_2, R_3 and R_4, where R_4 is the exterior region of G. The boundaries of these four regions are shown in Figure 14.4 as well. The plane graph G of Figure 14.4 has $n = 11$ vertices, $m = 13$ edges and $r = 4$ regions.   ◆

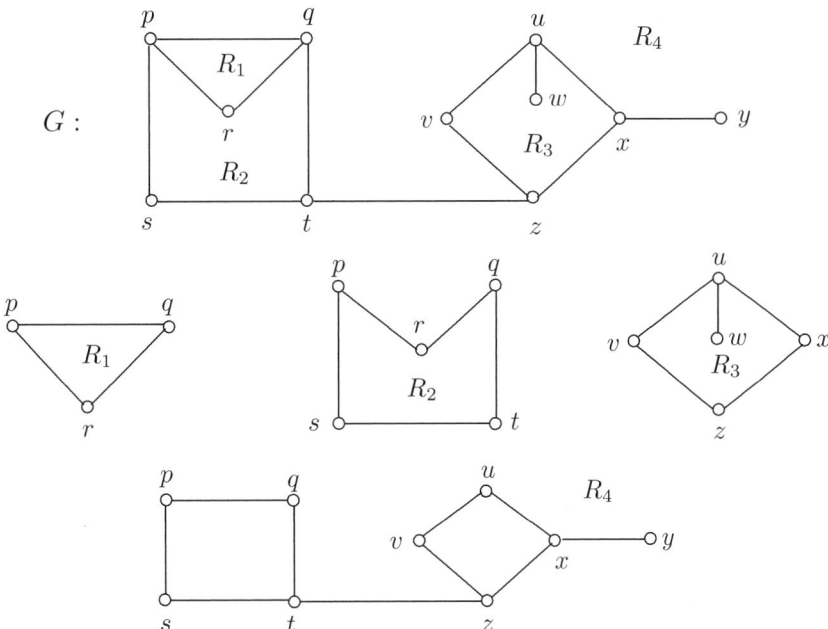

Figure 14.4: The boundaries of the regions of the plane graph in Example 14.2

If the numbers n, m and r are looked at for the graph of Figure 14.4 as well as a few other connected plane graphs, then it might be noticed that $n - m + r = 2$ in every case. This is not a coincidence. In fact, this identity goes back to Euler (the same Euler after whom Eulerian graphs are named).

Euler corresponded with many mathematicians. One of these individuals was Christian Goldbach, perhaps best known for his conjecture that every even integer greater than 2 can be written as the sum of two primes. In a letter from Euler to Goldbach dated November 14, 1750, Euler described an observation he made among the number of vertices, the number of edges and the number of faces of a polyhedron. This eventually became known as the *Euler Polyhedron Formula* or *Euler Polyhedron Identity*.

Theorem 14.3 (**The Euler Polyhedron Identity**) *If a polyhedron has V vertices, E edges and F faces, then*

$$V - E + F = 2.$$

Although the Euler Polyhedron Identity first appeared in print in 1752, the first generally accepted proof of this result was obtained by Adrien-Marie Legendre in 1794.

For every polyhedron, there is associated a plane graph, where the faces of the polyhedron become the regions in the corresponding plane graph. For example, for the two polyhedra cube and dodecahedron, $V - E + F = 8 - 12 + 6 = 2$ and $V - E + F = 20 - 30 + 12 = 2$, respectively, while in the graphs of these two polyhedra, $n - m + r = 8 - 12 + 6 = 2$ and $n - m + r = 20 - 30 + 12 = 2$ (see Figure 14.5).

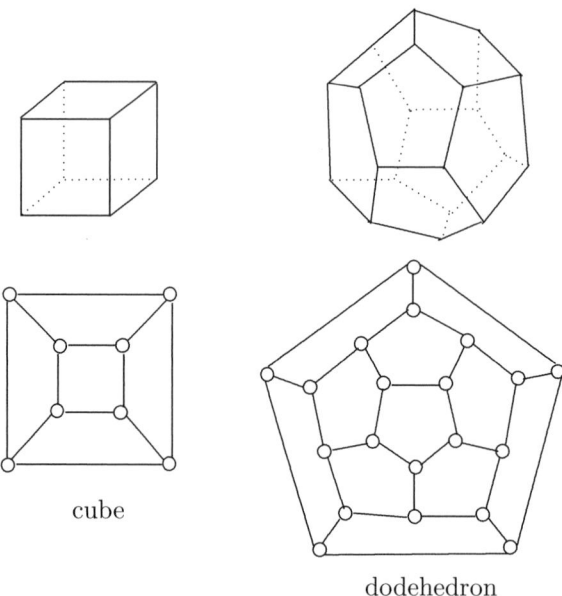

cube   dodehedron

Figure 14.5: The Euler Polyhedron Identity for the cube and dodecahedron

The Euler Identity

The Euler Polyhedron Identity then not only applies to all graphs that are graphs of polyhedra but to all connected plane graphs.

Theorem 14.4 (The Euler Identity) *If G is a connected plane graph with n vertices, m edges and r regions, then*

$$n - m + r = 2.$$

Proof. We proceed by induction on m. For $m = 0$, there is only one connected plane graph, namely K_1. Since K_1 has order $n = 1$, size $m = 0$ and $r = 1$ region, it follows that

$$n - m + r = 1 - 0 + 1 = 2,$$

and the Euler Identity holds for $m = 0$.

Assume that the Euler Identity holds for all connected plane graphs of size $m-1$, where $m-1 \geq 0$. We show that the Euler Identity holds for all connected plane graphs of size m. Let G be a connected plane graph of order n, size m and having r regions. If G is a tree, then there is only one region and so $r = 1$. Since $m = n - 1$, it follows that

$$n - m + r = n - (n - 1) + 1 = 2.$$

We may assume then that G is not a tree. Then G contains a cycle C. Let e be an edge of C. Then $G - e$ is also a connected plane graph of order n, size $m - 1$ and having $r - 1$ regions. By the induction hypothesis,

$$n - (m - 1) + (r - 1) = 2$$

and so $n - m + r = 2$.

By the Principle of Mathematical Induction, it follows that if G is a connected plane graph with n vertices, m edges and having r regions, then $n - m + r = 2$. ∎

There is a class of planar graphs that will be of special interest to us. A graph G is called **maximal planar** if G is planar but the addition of any edge joining two nonadjacent vertices of G produces a nonplanar graph. If G is a maximal planar graph of order $n \geq 3$, then, of course, G can be drawn as a plane graph. The boundary of each region of G is a triangle, for otherwise, an edge could be added to G to produce another planar graph, which is impossible. We state this observation as follows.

Theorem 14.5 *Let G be a plane graph of order 3 or more. Then G is maximal planar if and only if the boundary of every region of G is a triangle.*

The graph G_1 of Figure 14.6 is maximal planar because it can be drawn as a plane graph such that the boundary of every region is a triangle. The graphs G_2 and G_3 are also maximal planar graphs, while G_4 and G_5 are planar graphs that are not maximal planar. Every maximal planar graph of order $n \geq 3$ has a specific size.

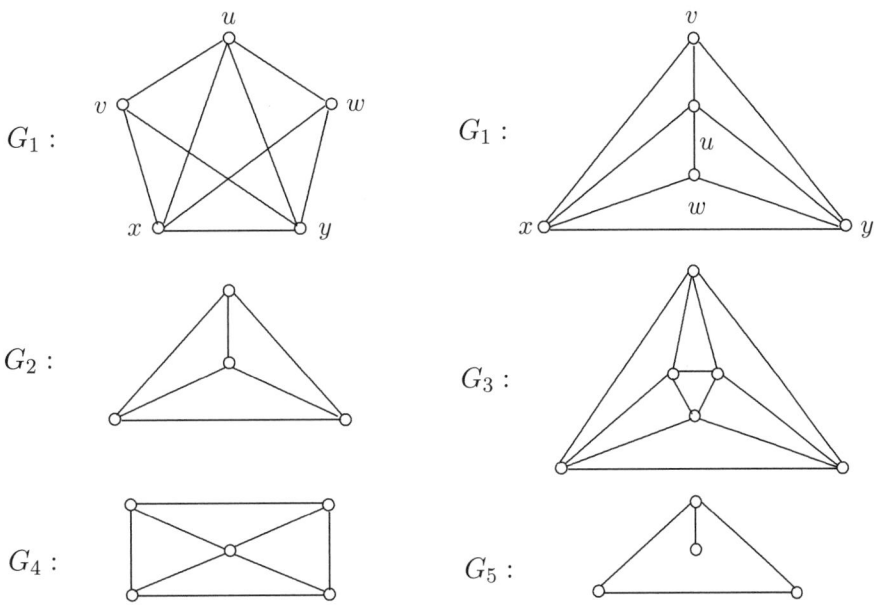

Figure 14.6: Maximal planar graphs and other planar graphs

Theorem 14.6 *If G is a maximal planar graph of order $n \geq 3$ and size m, then*

$$m = 3n - 6.$$

Proof. Draw G as a plane graph, resulting in r regions. By the Euler Identity, $n - m + r = 2$. Since the boundary of every region of G is a triangle, G contains no bridges and every edge of G lies on the boundaries of two regions. Hence if we count the number of edges on the boundary of each region and sum these numbers over all regions of G, then we obtain $3r$. Since this sum counts each edge of G twice, $3r = 2m$. Because $m = n + r - 2$, it follows that

$$3m = 3n + 3r - 6 = 3n + 2m - 6$$

and so $m = 3n - 6$.   ∎

This theorem has an immediate corollary.

Corollary 14.7 *If G is a planar graph of order $n \geq 3$ and size m, then*

$$m \leq 3n - 6.$$

Proof. If G is a maximal planar graph, then $m = 3n - 6$ by Theorem 14.6. If G is not maximal planar, then one or more edges can be added to G to produce a maximal planar graph of order n and size m', where $m' > m$. Hence $m < m' = 3n - 6$.   ∎

There is a consequence of Corollary 14.7 that concerns the degrees of the vertices of a planar graph.

Corollary 14.8 *Every planar graph has a vertex of degree 5 or less.*

Proof. Let G be a planar graph of order n and size m. If $n \leq 6$, then every vertex of G has degree 5 or less; so we may assume that $n \geq 7$. Assume, to the contrary, that there exists a planar graph G of order $n \geq 7$ such that every vertex of G has degree 6 or more. Then

$$\sum_{v \in V(G)} \deg v \geq 6n.$$

Since G is planar, $m \leq 3n - 6$ by Corollary 14.7. It then follows by the First Theorem of Graph Theory that

$$6n \leq \sum_{v \in V(G)} \deg v = 2m \leq 2(3n - 6) = 6n - 12,$$

which is impossible.   ∎

Nonplanar Graphs

According to Corollary 14.7, if G is a planar graph of order $n \geq 3$ and size m, then $m \leq 3n - 6$. Since this is a necessary condition for a graph to be planar, it is most useful when stated in its contrapositive form.

Corollary 14.9 *If G is a graph of order $n \geq 3$ and size m such that*

$$m > 3n - 6,$$

then G is nonplanar.

We are now in a position to prove that the complete graph K_5 of order 5 is nonplanar.

Corollary 14.10 *The complete graph K_5 is nonplanar.*

Proof. The graph K_5 has order $n = 5$ and size $m = 10$. Since $m = 10 > 3n - 6 = 9$, it follows by Corollary 14.9 that K_5 is nonplanar.   ∎

Example 14.11

Determine which of the graphs G_1, G_2 and G_3 of Figure 14.7 are planar.

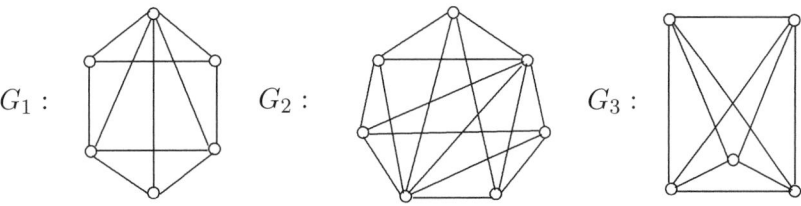

Figure 14.7: Testing three graphs for planarity in Example 14.11

Solution. The graph G_1 of Figure 14.7 is planar as it can be drawn as the plane graph G' in Figure 14.8. The graph G_2 is nonplanar, however, as it has order $n = 7$ and size $m = 16$, where $m > 3n - 6$. Since $G_3 = K_5$, it follows by Corollary 14.10 that G_3 is nonplanar.   ♦

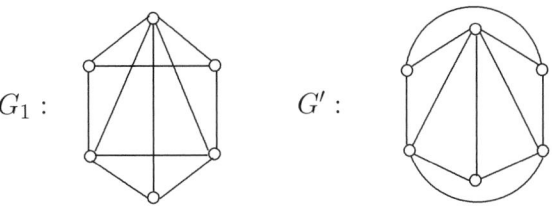

Figure 14.8: A planar graph in Example 14.11

Example 14.12

The vertices of a certain graph G have degrees $2, 4, 4, 4, 5, 5, 7, 7$. Determine whether G is planar.

Solution. The graph G has $n = 8$ vertices. By the First Theorem of Graph Theory, the size of G is

$$m = \frac{1}{2} \sum_{v \in V(G)} \deg v = \frac{1}{2}(2 + 4 + 4 + 4 + 5 + 5 + 7 + 7) = 19.$$

Since $19 = m > 3n - 6 = 3 \cdot 8 - 6 = 18$, it follows by Corollary 14.7 that G is nonplanar.   ♦

Since we have shown an interest in whether $K_{3,3}$ is planar, this suggests attempting to apply Corollary 14.9 to $K_{3,3}$. The order of $K_{3,3}$ is $n = 6$ and its size is $m = 9$. However, $m = 9 < 12 = 3n - 6$. Thus we cannot apply Corollary 14.9 and so no conclusion can be reached. That is, we still do not know whether $K_{3,3}$ is planar. On the other hand, if the proof of Theorem 14.6 is examined more carefully, then an idea may be suggested.

Theorem 14.13   *The graph $K_{3,3}$ is nonplanar.*

Proof. Assume, to the contrary, that $K_{3,3}$ is planar. Therefore, $K_{3,3}$ can be drawn as a plane graph, resulting in r regions, say. By the Euler Identity, $n - m + r = 2$ and so $6 - 9 + r = 2$. Therefore, there are $r = 5$ regions. Since $K_{3,3}$ has no bridges, the boundary of every region of $K_{3,3}$ is a cycle and every edge lies on the boundaries of two regions. Counting the number of edges on the boundary of each region and suming these numbers over all regions, we obtain the number N. Because $K_{3,3}$ is bipartite, it contains no odd cycles and so the boundary of every region contains at least four edges. Hence

$$2m = 18 = N \geq 4r = 4 \cdot 5 = 20,$$

and so $m \geq 10$, which produces a contradiction.   ∎

As a consequence of Theorem 14.13, we now have a solution of the Three Houses and Three Utilities Problem. It is *not* possible to connect the three houses to the three utilities without any of the utility lines crossing.

We also have a great deal of additional information as to which graphs are planar. It should be clear that if a graph G contains a nonplanar subgraph, then G too is nonplanar. Therefore, if a graph G contains K_5 or $K_{3,3}$ as a subgraph, then G is nonplanar. Of course, we also know that if G is a graph of order n and size m such that $m > 3n - 6$, then G is nonplanar. Consider the graph H of Figure 14.9, however. The graph H of Figure 14.9 has order $n = 12$ and size $m = 15$. Since $3n - 6 = 30$, it follows that m is (much) less than $3n - 6$. We have seen that when this occurs, no conclusion can be drawn. The graph H surely does not contain K_5 as a subgraph since K_5 has five vertices of degree 4 and H has no vertices of degree 4 or more. However, H does have six vertices of degree 3; so perhaps H contains a subgraph isomorphic to $K_{3,3}$. Since $K_{3,3}$ contains exactly six vertices of degree 3, each of the two partite sets of $K_{3,3}$ must consist of three of these six vertices. However, $K_{3,3}$ is a 3-regular graph and since each vertex of degree 3 in H is adjacent only to one other vertex of degree 3, it is impossible for H to contain $K_{3,3}$ as a subgraph. We claim, however, that despite all of this, the graph H of Figure 14.9 is nonplanar. Therefore, it must be possible for a graph of order $n \geq 3$ and size m to be nonplanar without having $m > 3n - 6$ and without containing either K_5 or $K_{3,3}$ as a subgraph. Before seeing why H is nonplanar, we consider another idea.

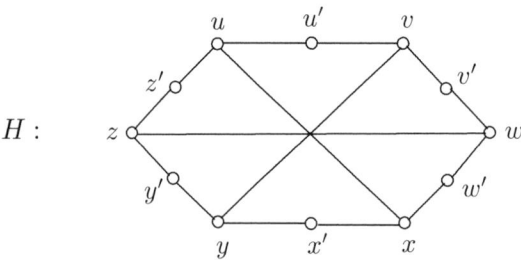

Figure 14.9: A nonplanar graph

A graph G' is called a **subdivision** of a graph G if either $G' = G$ or G' can be obtained from G by inserting vertices of degree 2 into one or more edges of G, that is, by **subdividing** edges of G. All of the graphs G, G_1, G_2 and G_3 of Figure 14.10 are then subdivisions of the graph G. In fact, G_3 is also a subdivision of G_1. The reason that subdivision is important in this subject is because of the following observation.

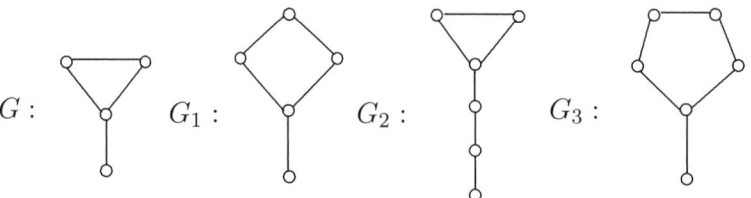

Figure 14.10: Subdivisions of a graph

Theorem 14.14 *Let G' be a subdivision of a graph G. Then G' is planar if and only if G is planar.*

Kuratowski's Theorem

Because K_5 and $K_{3,3}$ are nonplanar, we have the following theorem.

Theorem 14.15 *If G is a graph containing a subdivision of K_5 or $K_{3,3}$ as a subgraph, then G is nonplanar.*

Observe that the graph H of Figure 14.9 is a subdivision of $K_{3,3}$ and is therefore nonplanar. The graph H of Figure 14.9 is redrawn in Figure 14.11 so that it is clear that it is a subdivision of $K_{3,3}$.

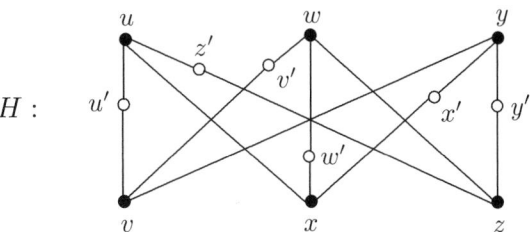

Figure 14.11: A subdivision of $K_{3,3}$

Although Theorem 14.15 may not seem like a particularly remarkable result, what is remarkable is that its converse is also true. This important theorem is known as Kuratowski's theorem.

Kazimierz Kuratowski (1896–1980), well known for his research in the mathematical area of topology, was one of the major contributors to mathematics in Poland during the 20th century, not only because of his research but because he was so influential in mathematics education in his native Poland following World War II.

Theorem 14.16 (Kuratowski's Theorem) *A graph G is planar if and only if G contains no subgraph that is a subdivision of K_5 or $K_{3,3}$.*

Example 14.17

Determine whether the graph F' of Figure 14.12 is planar.

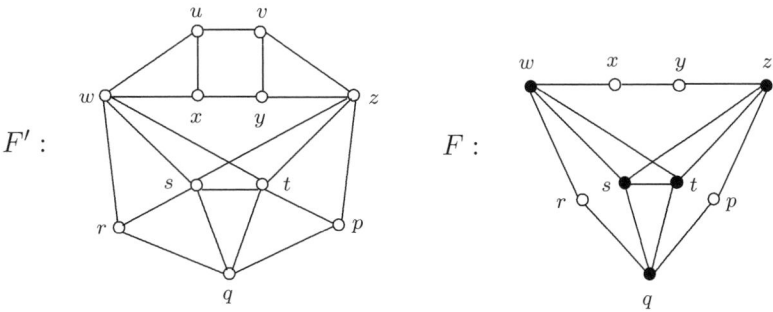

Figure 14.12: A nonplanar graph and a subdivision of K_5

Solution. Since F' contains the graph F as a subgraph and F is a subdivision of K_5, it follows by Kuratowski's theorem that F' is nonplanar.   ♦

Let's summarize then how we can determine whether a given graph is planar or nonplanar. To show that G is planar, we basically have two methods.

- If we can draw G in the plane without any of its edges crossing, then of course G is planar.

- If G contains less than five vertices of degree 4 or more and less than six vertices of degree 3 or more, then it is impossible for G to contain a subgraph that is a subdivision of K_5 or $K_{3,3}$ and so, by Kuratowski's theorem, G is planar.

To show that G is nonplanar, we basically have two methods.

- One method is to identify a subgraph of G that is a subdivision of K_5 or $K_{3,3}$.

- The other method is to show that the size m of G exceeds $3n - 6$, where $n \geq 3$ is the order of G.

This latter method is the preferred method since it is easy to compute m and $3n - 6$ and is so often difficult to locate a subgraph of G that is a subdivision of K_5 or $K_{3,3}$. Of course, the problem with comparing m and $3n - 6$ is that it may be that $m \leq 3n - 6$, in which case we have learned nothing.

Exercises for Section 14.1

1. Show that each of the graphs in Figure 14.13 is planar by drawing it as a plane graph. Verify that the Euler Identity holds for each graph.

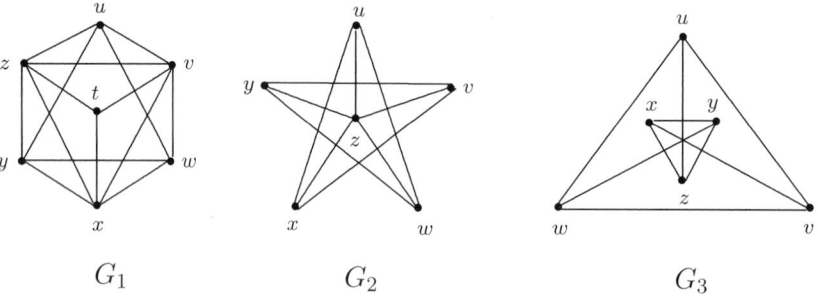

Figure 14.13: Graphs in Exercise 1

2. Determine the boundaries of the regions of the plane graph G in Figure 14.14.

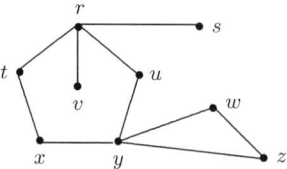

Figure 14.14: The graph G in Exercise 2

3. Let G be a planar graph and let G_1 and G_2 be two plane graphs obtaining by drawing G in the plane. Suppose that G_1 has r_1 regions and G_2 has r_2 regions. What can you say about r_1 and r_2?

4. Let G be a connected 3-regular plane graph of order 8. How many regions does G have?

5. Let G be a connected plane graph of order 8 having five regions. How many edges does G have?

6. The degrees of the vertices of a certain graph G are 3, 4, 4, 4, 5, 6, 6. Determine whether G is planar.

7. The degrees of the vertices of a certain graph G are 2, 2, 2, 3, 5, 5, 6, 7. Indicate which of the following is true.

   (1) The graph G is planar. (2) The graph G is nonplanar.

   (3) It is impossible to determine whether G is planar or nonplanar.

8. A graph G of order $n \geq 3$ has size $m = 3n - 6$. Indicate which of the following is true.

   (1) The graph G is planar. (2) The graph G is nonplanar.

   (3) It is impossible to determine whether G is planar or nonplanar.

9. A planar graph G of order $n \geq 3$ has size $m < 3n - 6$. Let u and v be two nonadjacent vertices of G. Indicate which of the following is true.

   (1) The graph $G + uv$ is planar. (2) The graph $G + uv$ is nonplanar.

   (3) It is impossible to determine whether $G + uv$ is planar or nonplanar.

10. (a) Give an example of a planar graph where no vertex of G has degree 4 or less.

    (b) Give an example of a nonplanar graph where no vertex of G has degree 4 or more.

11. A graph G has a vertex of degree 4 and all other vertices of G have degree 7 or more. Indicate which of the following is true.

    (1) The graph G is planar. (2) The graph G is nonplanar.

    (3) It is impossible to determine whether G is planar or nonplanar.

12. A certain graph G of order 6 has the following properties:

    (a) The graph G does not contain K_5 as a subgraph.

    (b) Exactly five vertices of G have degree 3 or more.

    Indicate which of the following is true.

    (1) The graph G is planar. (2) The graph G is nonplanar.

    (3) It is impossible to determine whether G is planar or nonplanar.

13. State and solve the Five Houses and Two Utilities Problem.

14. Determine all complete graphs K_n that are planar.

15. Determine all complete bipartite graphs $K_{s,t}$ that are planar, where $s \leq t$.

16. A graph G' is a subdivision of a graph G having order n and size m. If the order of G' is $2n$, what is the size of G'?

17. Let U be a subset of the vertex set of a graph G and suppose that $H = G[U]$ is an induced subgraph of G of order k. Let H' be the graph obtained by inserting a vertex of degree 2 into every edge of H. Let G' be the graph obtained by inserting a vertex of degree 2 into every edge of G.

    (a) True or false: If G is planar, then H' is planar.

    (b) What is $G'[U]$?

18. Show that the Petersen graph (see Figure 14.15) is nonplanar by Kuratowski's theorem.

19. Determine, with explanation, which graphs in Figure 14.16 are planar.

20. (a) Prove that if G is a planar graph of order $n \geq 3$ and size m without triangles, then $m \leq 2n - 4$.

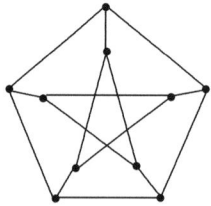

Figure 14.15: The Petersen graph in Exercise 18

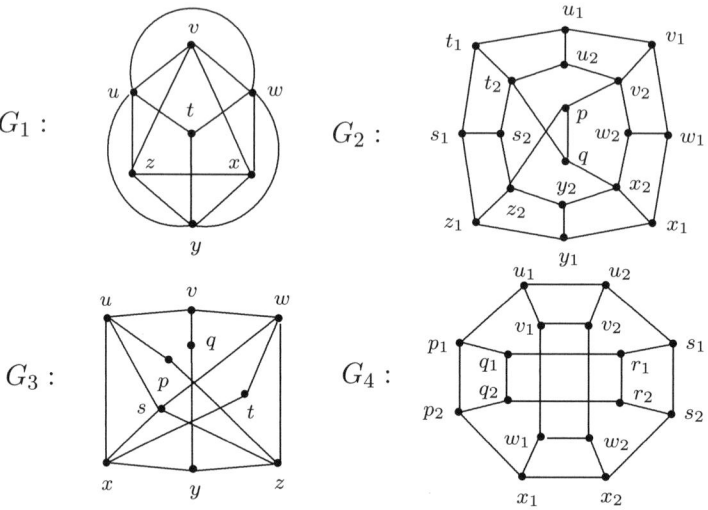

Figure 14.16: The graphs in Exercise 19

(b) Use (a) to show that $K_{3,3}$ is nonplanar.

(c) Prove or disprove: If G is a planar bipartite graph, then G has a vertex of degree 3 or less.

21. Let G be a plane graph of order $n \geq 5$ and size m.

   (a) Prove that if the length of a smallest cycle in G is 5, then $m \leq \frac{5}{3}(n-2)$.

   (b) Use (a) to show that the Petersen graph is nonplanar.

   (c) Prove or disprove: If $n < 20$ and the length of a smallest cycle in G is 5, then G has a vertex of degree 2 or less.

22. Determine all integers $n \geq 3$ such that $\overline{C}_n$ is planar.

14.2   Coloring Graphs

In 1852 a young British mathematician by the name of Francis Guthrie noticed that he could color the counties in a map of England with four colors in such a way that no two counties with a common boundary were assigned the same color. This led Guthrie to wonder whether the regions of any map could be colored with four or fewer colors so that adjacent regions (regions sharing a common boundary) are colored differently. Figure 14.17 shows two maps whose regions can be colored with four or fewer colors, where the colors are the positive integers $1, 2, 3, 4$. Observe that in Map 2, the regions R_1 and R_3 are colored the same. This is acceptable as these two regions share only a common point, not a common boundary.

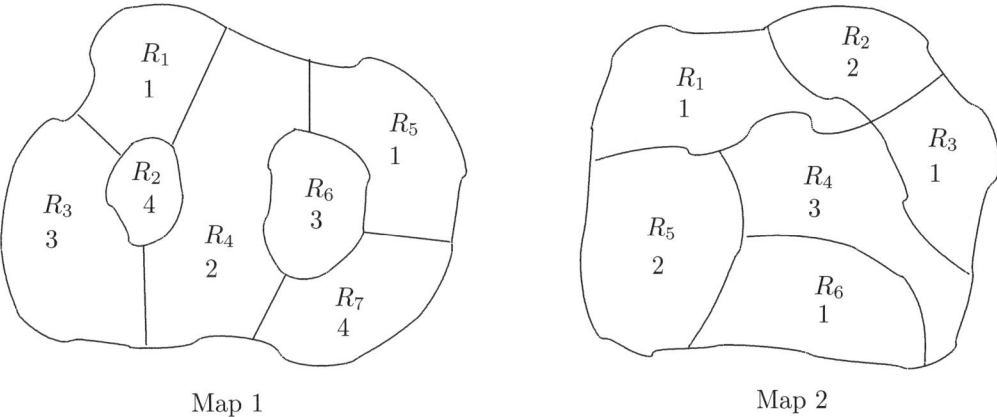

Figure 14.17: Coloring the regions of two maps

Francis Guthrie believed that he may have even proved that every map could be colored with four or fewer colors but he was not satisfied with his argument. Francis mentioned this problem to his younger brother Frederick, who at the time was a student of the famous mathematician Augustus De Morgan (after whom De Morgan's Laws are named). Frederick then mentioned this problem to De Morgan on October 23, 1852. De Morgan was intrigued by this problem and wrote about it that very same day to William Rowan Hamilton (after whom Hamiltonian graphs are named).

> *A student of mine asked me to day to give him a reason for a fact which I did not know was a fact – and do not yet. He says that if a figure be anyhow divided and the compartments differently coloured so that figures with any portion of common boundary line are differently coloured – four colours may be wanted but not more – the following is his case in which four are wanted.*

A B C D are
names of
colours

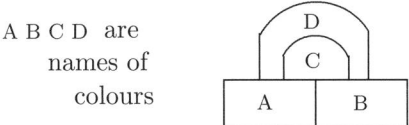

> *Query cannot a necessity for five or more be invented.*

(And, yes, De Morgan wrote *to day* not *today*.) On October 26, 1852 Hamilton gave a prompt, but probably unexpected reply to De Morgan's letter:

> *I am not likely to attempt your "quaternion" of colours very soon.*

Although Hamilton did not appear to show interest in this problem, De Morgan talked about the problem often and to many mathematicians. This problem eventually became known as the Four Color Problem.

There is a natural connection between the Four Color Problem and graphs. With every map M, there is associated a connected graph G, called the **dual graph** of M, where the vertices of G are the regions of M and two vertices of G are adjacent if the corresponding regions share a common boundary. Every dual graph is in fact a connected planar graph. The graphs G_1 and G_2 in Figure 14.18 are the dual graphs of Map 1 and Map 2 of Figure 14.17, where each vertex (region) is labeled by the color of the vertex.

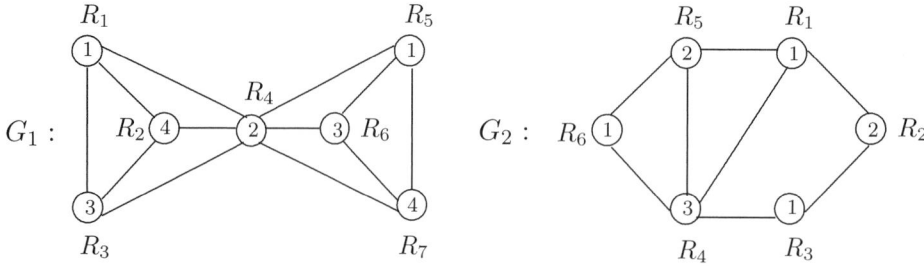

Figure 14.18: The dual graphs of Map 1 and Map 2 of Figure 14.37

Not only is the dual graph of every map a connected planar graph, each connected planar graph is the dual graph of some map. Thus the Four Color Problem, originally stated for maps, can be stated in terms of graphs.

The Four Color Problem Can the vertices of every planar graph be colored with four or fewer colors so that every two adjacent vertices are colored differently?

The Chromatic Number of a Graph

We will return to the Four Color Problem soon, after introducing some terminology on coloring graphs in general.

Definition 14.18 *A **coloring** of a graph G is an assignment of colors to the vertices of G, one color to each vertex of G, such that adjacent vertices of G are colored differently. A k-**coloring** is a coloring that uses k colors. A graph G is k-**colorable** if G can be colored with k or fewer colors.*

Since there are other kinds of "colorings" of graphs, the coloring described in Definition 14.18 is often called a **proper coloring** of a graph. We will assume here, however, that every coloring of (the vertices of) a graph is a proper coloring. We can think of the colors here as being red, blue, green and so on or as positive integers $1, 2, 3, \ldots$ or, indeed, as the elements of any set.

Example 14.19

Using positive integers as colors, several colorings of a graph H are shown in Figure 14.19. If we look at the three colorings of the graph H given in Figure 14.19, we see that the first coloring is a 4-coloring, the second coloring is a 3-coloring and the third coloring is a 2-coloring. Of course, we could assign a different color to each of the six vertices of H and so there are colorings of H that are 6-colorings. It is also easy to find a 5-coloring of H. However, the question of greatest interest here is the following: What is the smallest number of colors in a coloring of H? Actually, this question is quite easy to answer for the graph H of Figure 14.19. We already know that there exists a coloring (namely the third coloring) that uses two colors. Since H contains edges, we certainly need *at least* two colors. Therefore, the minimum number of colors needed to color the vertices of H is 2. ◆

Definition 14.20 *The smallest number of colors needed to color the vertices of a graph G is called the **chromatic number** of G and is denoted by $\chi(G)$.*

Therefore, $\chi(H) = 2$ for the graph H of Figure 14.19. (The symbol χ is the Greek letter *chi*.) The procedure we used to show that the graph H of Figure 14.19 has chromatic number 2 is typical of the method we employ to show that any graph has a specific chromatic number. In particular, to show that a graph G has chromatic number k, say, we must do two things:

(1) Show that the vertices of G can be colored with k colors, say $1, 2, \ldots, k$ (such that adjacent vertices are colored differently).

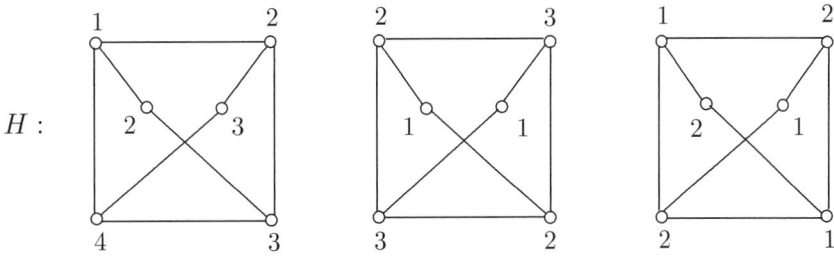

Figure 14.19: Colorings of the graph H in Example 14.19

(2) Show that every coloring of G requires at least k colors.

Actually, (1) establishes the inequality $\chi(G) \leq k$, while (2) establishes the inequality $\chi(G) \geq k$. Combining these two inequalities gives us the desired result that $\chi(G) = k$.

Showing that the vertices of G can be colored with k colors probably seems much easier than showing that we must use at least k colors to color the vertices of G. Although this is often true, there are some observations that will be helpful to us, beginning with a simple one.

Let G be a graph. Then $\chi(G) \geq 2$ if and only if G contains at least one edge.

An immediate consequence of this fact is the following.

Corollary 14.21 *A graph G has chromatic number 1 if and only if $G = \overline{K}_n$ for some positive integer n.*

The graphs with chromatic number 2 are also easy to describe.

Theorem 14.22 *Let G be a nonempty graph. Then $\chi(G) = 2$ if and only if G is bipartite.*

Proof. First, let G be a nonempty bipartite graph with partite sets U and W. Since G is nonempty, $\chi(G) \geq 2$. Assign the color 1 to every vertex of U and the color 2 to every vertex of W. Since every edge of G joins a vertex of U and a vertex of W, adjacent vertices of G are colored differently, as required. Therefore, $\chi(G) = 2$.

For the converse, let G be a graph with chromatic number 2. Hence the vertices of G can be colored with two colors, say 1 and 2, such that adjacent vertices of G are colored differently. Let V_1 be the set of vertices colored 1 and V_2 the set of vertices colored 2. Since every edge of G joins a vertex of V_1 and a vertex of V_2, it follows that G is a bipartite graph with partite sets V_1 and V_2. ∎

Because a graph G is bipartite if and only if G has no odd cycles (Theorem 12.31), there is an immediate corollary.

Corollary 14.23 *Let G be a graph. Then $\chi(G) \geq 3$ if and only if G contains an odd cycle.*

By Corollary 14.23, we can show that the chromatic number of a graph is at least 3 if it contains an odd cycle. Indeed, the chromatic number of each cycle can be determined.

Example 14.24

For every integer $n \geq 3$,

$$\chi(C_n) = \begin{cases} 2 & \text{if } n \text{ is even} \\ 3 & \text{if } n \text{ is odd.} \end{cases}$$

Solution. If n is even, then C_n is a bipartite graph and so $\chi(C_n) = 2$ by Theorem 14.22. If n is odd, then $\chi(C_n) \geq 3$ by Corollary 14.23. To show that $\chi(C_n) \leq 3$, let $C_n = (v_1, v_2, \ldots, v_n, v_1)$. If we assign color 1 to the vertex v_i for each odd integer i with $1 \leq i \leq n - 2$, color 2 to the vertex v_i for each even integer i ($2 \leq i \leq n - 1$), and color 3 to the remaining vertex v_n, then we have a 3-coloring of C_n, implying that $\chi(C_n) \leq 3$. Therefore, $\chi(C_n) = 3$ if n is odd. ♦

Another well-known class of graphs whose chromatic numbers are easy to determine are the complete graphs.

Theorem 14.25 *Let n be a positive integer. A graph G of order n has chromatic number n if and only if $G = K_n$.*

Proof. If we assign the colors 1, 2, ..., n to the vertices of K_n, one color to each vertex, then we have an n-coloring of K_n. Thus $\chi(K_n) \leq n$. On the other hand, because every two vertices of K_n are adjacent, every two vertices of K_n must be colored differently and so $\chi(K_n) \geq n$. Therefore, $\chi(K_n) = n$.

For the converse, let G be a graph of order n that is not complete. Then G contains two nonadjacent vertices u and v. Assigning the color 1 to u and v and distinct colors from $\{2, 3, \ldots, n-1\}$ to the remaining $n - 2$ vertices of G produces an $(n - 1)$-coloring of G. Thus $\chi(G) \neq n$. ∎

We next present a simple but useful observation.

Theorem 14.26 *If H is a subgraph of a graph G, then*

$$\chi(H) \leq \chi(G).$$

Proof. Suppose that $\chi(G) = k$. Then there exists a coloring of the vertices of G with the colors 1, 2, ..., k. Since this also provides a proper coloring of H with some or all of the colors 1, 2, ..., k, it follows that

$$\chi(H) \leq k = \chi(G),$$

as desired. ∎

By Theorem 14.26, it follows that if a graph G contains K_5 as a subgraph, for example, then $\chi(G) \geq \chi(K_5) = 5$. Of course, to show that $\chi(G) = 5$, if this is what we believe, then we are required to find a 5-coloring of G. We now consider an example of this.

Example 14.27

Determine the chromatic number of the graph F of Figure 14.20.

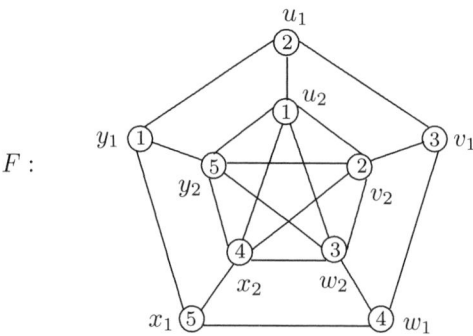

Figure 14.20: The graph F in Example 14.27

Solution. Since the subgraph induced by the set $\{u_2, v_2, w_2, x_2, y_2\}$ of vertices of F is K_5, it follows by Theorem 14.26 that $\chi(F) \geq 5$. However, there is a 5-coloring of F in Figure 14.20 and so $\chi(F) \leq 5$. Therefore, $\chi(F) = 5$. ♦

We mention one other useful fact in coloring graphs. Suppose that G is a graph containing a subgraph H whose chromatic number is known to us, say $\chi(H) = k$. If G contains a vertex $v \notin V(H)$ that is adjacent to every vertex of H, then any color assigned to v must be different from the colors assigned to H and so $\chi(G) \geq k + 1$.

Example 14.28

Determine the chromatic number of the graph G of Figure 14.21.

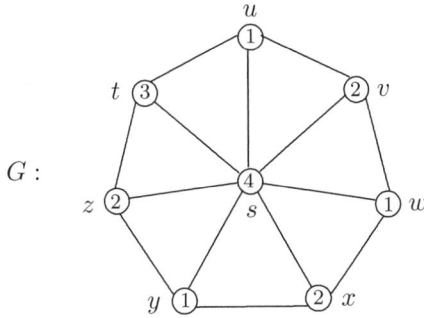

Figure 14.21: The graph G in Example 14.28

Solution. Since G contains an odd cycle (namely the 7-cycle $C = (t, u, v, w, x, y, z, t)$ as well as many triangles), it follows by Corollary 14.23 that $\chi(G) \geq 3$. Since the vertex s is adjacent to every vertex of C, it follows that $\chi(G) \geq 4$. Because the vertices of G can be colored with four colors (namely, 1, 2, 3, 4 in Figure 14.21), $\chi(G) \leq 4$. Therefore, $\chi(G) = 4$. ◆

Example 14.29

Determine the chromatic number of the graph H of Figure 14.22.

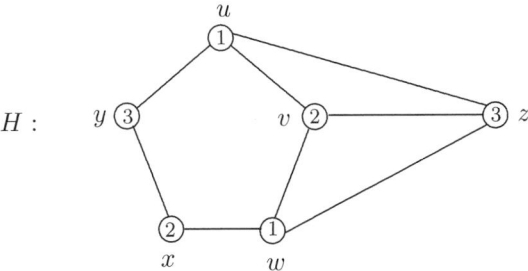

Figure 14.22: The graph H in Example 14.29

Solution. Since the graph H of Figure 14.22 contains an odd cycle (in fact, both triangles and 5-cycles), it follows that $\chi(H) \geq 3$. It might be quite easy to make a mistake at this point. We could start with the 5-cycle and observe that this requires three colors. We might color vertex u with color 1, vertex v with color 2, vertex w with color 3, vertex x with color 1 and vertex y with color 2. However, the vertex z is now adjacent to vertices colored 1, 2, 3 and a different color is required for z, namely color 4. Actually, all this does is establish the inequality $\chi(H) \leq 4$. In fact, the vertices of H can be colored with three colors (as shown in Figure 14.22). Therefore, $\chi(H) \leq 3$ and so $\chi(H) = 3$. ■

Applications of Colorings

Let us return to a question that we posed in the introduction of Chapter 12.

Example 14.30

Eleven chemicals are to be shipped to a chemistry department at a university. Two chemicals that can react with each other should not be placed in the same container during the shipment. The chemicals that cannot be shipped in the same container are listed below:

$$c_1 : \; c_2, c_5, c_7, c_{10} \qquad c_2 : \; c_3, c_6, c_8 \text{ (as well as } c_1) \qquad c_3 : \; c_4, c_7, c_9$$
$$c_4 : \; c_5, c_8, c_{10} \qquad c_5 : \; c_6, c_9 \qquad c_{11} : \; c_6, c_7, c_8, c_9, c_{10}.$$

So, for example, c_1 cannot be shipped in the same container with any of c_2, c_5, c_7, c_{10} and c_2 cannot be shipped in the same container with any of c_3, c_6, c_8 (as well as with c_1 of course). What is the smallest number of containers that can be shipped safely to the university?

Solution. This situation can be modeled by a graph G with $V(G) = \{c_1, c_2, \ldots, c_{11}\}$, where two vertices c_i and c_j ($i \neq j$) are adjacent if they cannot be shipped in the same container. This graph is shown in Figure 14.23(a). The minimum number of containers that can be shipped safely to the university is $\chi(G)$. Since G has an odd cycle, $\chi(G) \geq 3$. Since there is a 4-coloring of G (in Figure 14.23(b)), $\chi(G) \leq 4$. So either $\chi(G) = 3$ or $\chi(G) = 4$.

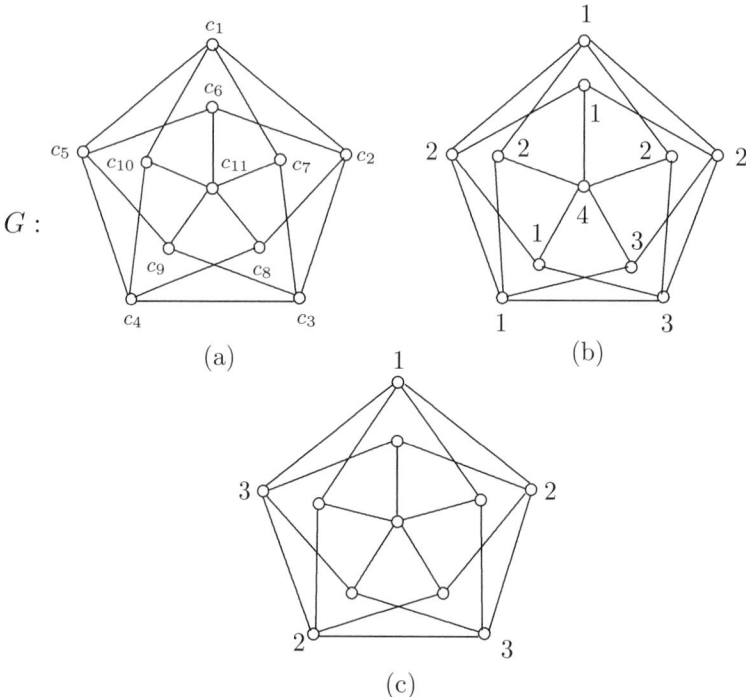

Figure 14.23: The graph G in Example 14.30

We show in fact that $\chi(G) = 4$. Assume, to the contrary, that $\chi(G) = 3$. Then there is a 3-coloring of G using the colors 1, 2, 3, say. Since $C = (c_1, c_2, c_3, c_4, c_5, c_1)$ is a 5-cycle, three colors are required to color the vertices of C. Because of the symmetry of the graph G, we may assume that the color 1 is used to color one of the vertices of C, say vertex c_1, and the colors 2 and 3 are used twice each. (See Figure 14.23(c).) Since c_6 is adjacent to vertices colored 2 and 3, it must be colored 1. Similarly, c_7 is colored 2 and c_{10} is colored 3. However then, c_{11} is adjacent to vertices colored 1, 2 and 3, which makes it impossible to color c_{11} with any of these colors. Since no 3-coloring of G exists, $\chi(G) = 4$. ◆

Observe that we were able to conclude that $\chi(G) = 4$ for the graph G of Figure 14.23(a) and yet G does not contain K_4 as a subgraph. Hence there exists a graph G with $\chi(G) = 4$ that does not contain K_4 as a subgraph. Moreover, this graph G does not even contain K_3 as a subgraph. In fact, for each integer $k \geq 3$, there exists a graph G_k with $\chi(G_k) = k$ such that G_k does not contain K_3 as a subgraph. The graph G of Figure 14.23(a) illustrates this fact for $k = 4$.

Another application of coloring involves questions of scheduling.

Example 14.31

The following eight individuals from eight branch offices of a software company are present at the main office of the company to discuss a number of issues:

Allen (a), Brenda (b), Carlos (c), Dennis (d),
Edith (e), Fay (f), George (g) Harriet (h).

It is the responsibility of Idina, from the main office, to schedule 90-minute meetings of six 3-person committees during one of the days and then meet with all eight people after the six meetings have been taken place. The six committees are

$$C_1 = \{a, c, d\}, \quad C_2 = \{a, e, f\}, \quad C_3 = \{a, b, h\},$$
$$C_4 = \{e, f, g\}, \quad C_5 = \{c, e, g\}, \quad C_6 = \{b, f, g\}.$$

Of course, if the six committee meetings are scheduled one after another, then this will take 9 hours and it will be a long day, to everyone's displeasure. But perhaps these meetings could be scheduled in less than six time periods. For example, committees C_1 and C_4 could be scheduled to meet at the same time since none of the eight individuals are members of both committees. On the other hand, committees C_1 and C_2 must meet during different time periods since Allen is a member of both committees. There is now a natural question: What is the minimum number of time periods for all six committees to meet?

Solution. This situation can be modeled by a graph G whose vertices are the six committees C_1, C_2, ..., C_6. Two vertices C_i and C_j $(i \neq j)$ of G are adjacent if there is an individual who is a member of both of these committees. The graph G is shown in Figure 14.24. The minimum number of time periods for these six committee meetings is $\chi(G)$.

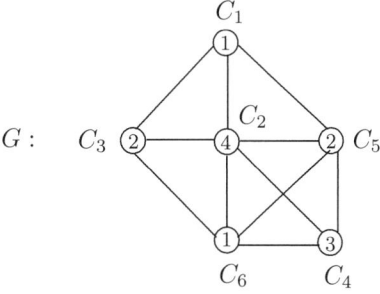

Figure 14.24: The graph G in Example 14.31

Observe that the subgraph H of G induced by $\{C_2, C_4, C_5, C_6\}$ is a complete graph of order 4 and so G contains K_4 as a subgraph. Since $4 = \chi(H) \leq \chi(G)$, it follows that $\chi(G) \geq 4$. On the other hand, the vertices of G can be colored with four colors 1, 2, 3, 4 as shown in Figure 14.24. Therefore, $\chi(G) = 4$.

Since $\chi(G) = 4$, the six meetings can be scheduled in four 90-minute time periods (but no fewer) and within six hours all meetings can be held:

Period 1: $C_1, C_6,$ Period 2: $C_3, C_5,$ Period 3: $C_4,$ Period 4: $C_2.$ ♦

Example 14.32

Figure 14.25 shows six traffic lanes $L_1, L_1, \ldots, L_6$ at the intersection of two streets. A traffic light is located at the intersection. During each phase of the traffic light, those cars in lanes for which the light is green may proceed safely through the intersection into certain permitted lanes. What is the minimum number of phases needed for the traffic light so that all cars may proceed safely through the intersection?

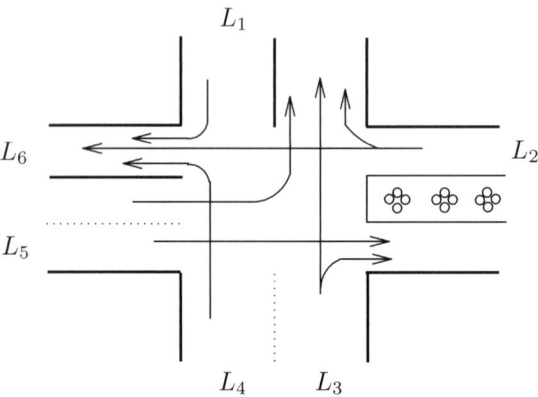

Figure 14.25: Traffic lanes at street intersections in Example 14.32

Solution.   A graph G is constructed with vertex set $V(G) = \{L_1, L_2, \ldots, L_6\}$, where L_i is adjacent to L_j $(i \neq j)$ if cars in lanes L_i and L_j cannot proceed safely through the intersection at the same time. (See Figure 14.26.) The minimum number of phases needed for the traffic light so that all cars may proceed, in time, through the intersection is $\chi(G)$. Since $\{L_1, L_2, L_4\}$ induces a triangle $\chi(G) \geq 3$. Since there is a 3-coloring of G (see Figure 14.26), it follows that $\chi(G) = 3$. For example, since L_1, L_5 and L_6 are colored the same, cars in those three lanes may proceed safely through the intersection at the same time.   ♦

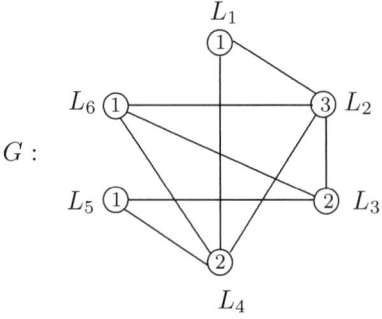

Figure 14.26: The graph of Example 14.32

Upper Bounds for the Chromatic Number

There is no formula that gives the chromatic number of each graph and finding the chromatic number of a graph can be quite challenging. Although we may not be able to determine the chromatic number of a graph G, it can be shown that $\chi(G)$ can never be much larger than its maximum degree $\Delta(G)$.

Theorem 14.33 *For every graph G,*

$$\chi(G) \leq 1 + \Delta(G).$$

Proof. Suppose that the theorem is false. Then there exists some graph G of order n, say, for which $\chi(G) > 1 + \Delta(G)$. Let $\Delta(G) = \Delta$. Let p be the maximum number of vertices of G that can be colored with $1 + \Delta$ colors so that adjacent vertices are colored differently. Let H be the subgraph of G induced by p such vertices and let a coloring of H with $1 + \Delta$ colors be given. By assumption, $p < n$. Let v be one of the vertices of G that does not belong to H. Since v is adjacent to at most Δ vertices of G, it follows that v is adjacent to at most Δ vertices of H. Thus there are at most Δ vertices adjacent to v that have been assigned a color. Hence at least one of the $1 + \Delta$ colors is available for v. Assigning such a color to v produces a contradiction. ∎

For the complete graph K_n, we know that $\chi(K_n) = n$ and $\Delta(K_n) = n - 1$. So

$$\chi(K_n) = 1 + \Delta(K_n).$$

Also, for an odd integer $n \geq 3$, we know that $\chi(C_n) = 3$ and $\Delta(C_n) = 2$. Therefore,

$$\chi(C_n) = 1 + \Delta(C_n).$$

Therefore, there are graphs G, namely C_n, where $n \geq 3$ is odd, and K_n, where $n \geq 1$, for which

$$\chi(G) = 1 + \Delta(G).$$

Rowland Leonard Brooks showed, however, that these are the only two classes of connected graphs for which this is true.

Theorem 14.34 (Brooks' Theorem) *If G is a connected graph that is neither a complete graph nor an odd cycle, then*

$$\chi(G) \leq \Delta(G).$$

The Five Color Theorem and Four Color Theorem

Let us now return to the Four Color Problem, which asks whether every planar graph is 4-colorable. Because the chromatic number of a graph G is the largest chromatic number among the components of G, we can restrict ourselves to *connected* planar graphs only. Since K_4 is a planar graph and $\chi(K_4) = 4$, there are certainly *some* planar graphs that require four colors. But the Four Color Problem suggests that *no* planar graph requires five colors.

A major event in the history of the Four Color Problem occurred in 1879, when the British barrister (lawyer) Alfred Bray Kempe wrote a paper which contained a "proof" that every planar graph is 4-colorable. For the period 1879-1890, the Four Color Problem was considered to be solved. However, in 1890 the British mathematician Percy John Heawood discovered an error in Kempe's argument which neither he nor Kempe could correct. Despite the fact that Kempe's proof was incorrect, his technique was ingenious and Heawood used Kempe's technique to prove the following theorem.

Theorem 14.35 (The Five Color Theorem) *Every planar graph is 5-colorable.*

Proof. We proceed by induction on the order of planar graphs. If G is a planar graph of order 5 or less, then certainly $\chi(G) \leq 5$. This verifies the basis step of induction. We now turn to the inductive step. Assume that the chromatic number of every planar graph of order k, where $k \geq 5$, is at most

5. We show that the chromatic number of every planar graph of order $k+1$ is at most 5. Let G be a planar graph of order $k+1$. By Corollary 14.8, G contains a vertex v with $\deg v \leq 5$. Since $G-v$ is a planar graph of order k, it follows by the induction hypothesis that $\chi(G-v) \leq 5$. Hence there exists a 5-coloring of $G-v$. Let a 5-coloring of $G-v$ be given, using the colors 1, 2, 3, 4, 5. If the neighbors of v in G are colored with four or fewer colors, then at least one of the colors 1, 2, 3, 4, 5 is available for v. Assigning this color to v produces a 5-coloring of G and so $\chi(G) \leq 5$.

Hence we may assume that $\deg v = 5$ and that all five colors 1, 2, 3, 4, 5 are used to color the five vertices adjacent to v. Draw G as a plane graph. We may assume that the vertices adjacent to v are v_1, v_2, v_3, v_4, v_5 located in clockwise order about v. Furthermore, we may assume that v_i is colored i for $1 \leq i \leq 5$. (See Figure 14.27.)

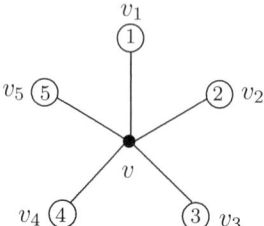

Figure 14.27: The neighbors of v in the planar graph G

Consider two vertices adjacent to v that do not appear consecutively as we proceed about v, say v_1 and v_3. Now there may or may not be a $v_1 - v_3$ path in $G-v$ all of whose vertices are colored 1 or 3. We consider these two cases.

Case 1. There is no $v_1 - v_3$ path in $G-v$, all of whose vertices are colored 1 or 3. Let H be the subgraph of $G-v$ induced by those vertices colored 1 or 3. Since H contains no $v_1 - v_3$ path, v_1 and v_3 belong to different components of H. (See Figure 14.28.) Suppose that v_1 belongs to the component H_1 of H. Interchanging the colors of the vertices in H_1 produces a new 5-coloring of $G-v$ in which v_1 is colored 3. The color of v_3 is not affected, however, and is still colored 3 in this new 5-coloring of $G-v$. Since no vertex adjacent to v is colored 1, the vertex v may now be colored 1, which produces a 5-coloring of G and so $\chi(G) \leq 5$.

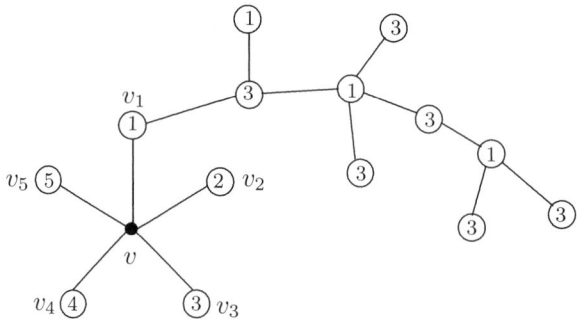

Figure 14.28: The situation in Case 1 in the proof of Theorem 14.35

Case 2. There is a $v_1 - v_3$ path in $G-v$, all of whose vertices are colored 1 or 3. Once again, let H be the subgraph of $G-v$ induced by those vertices colored 1 or 3. Then v_1 and v_3 belong to the same component H_1 of H. (See Figure 14.29.) Therefore, interchanging the colors of the vertices in H_1 has no benefit as the neighbors of v are still colored with five colors. However, since G is a plane graph, there is no $v_2 - v_4$ path in $G-v$, all of whose vertices are colored 2 or 4. Let F be the subgraph of $G-v$ induced by those vertices colored 2 or 4. Then v_2 and v_4 belong to different components of F. Let F_2 be the component of F containing v_2. If we interchange the colors of the vertices in F_2, then no vertex adjacent to v is colored 2. By assigning v the color 2, we have a 5-coloring of G and so $\chi(G) \leq 5$.

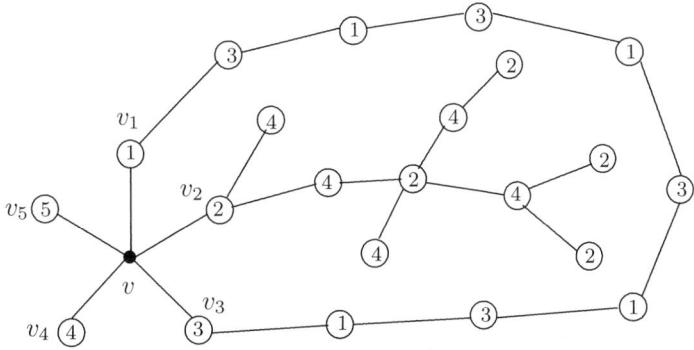

Figure 14.29: The situation in Case 2 in the proof of Theorem 14.35

By the Principle of Mathematical Induction, the chromatic number of every planar graph is at most 5. ∎

By the end of the 19th century, it was known that every planar graph is 5-colorable but no one knew of a single example of a planar graph that required five colors. Many mathematicians worked on the Four Color Problem until 1976 when Kenneth Appel and Wolfgang Haken of the University of Illinois announced that they had constructed a proof of the theorem. This proof, however, involved a large number of cases and required substantial use of computers to verify the details in each case. Consequently and finally, the Four Color Problem had been solved.

The Four Color Theorem Every planar graph is 4-colorable.

Despite the fact that the Four Color Problem had been solved, many mathematicians were dissatisfied with the proof because of the large number of cases and its heavy reliance on computers. Twenty years later, in 1996, a considerably improved proof of the Four Color Theorem was given by Neil Robertson, Daniel Sanders, Paul Seymour and Robin Thomas. Even this proof, however, required extensive use of computers. To this day, no proof of the Four Color Theorem has been given that has not made significant use of computers.

Exercises for Section 14.2

1. For the graph G of order $n = 7$ of Figure 14.30, give an example of a k-coloring, using the colors 1, 2, ..., k, for every integer k with $\chi(G) \leq k \leq n$.

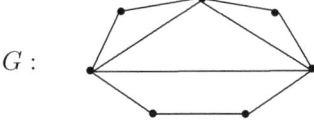

Figure 14.30: The graph in Exercise 1

2. For the map M in Figure 14.31, what is the minimum number of colors needed to color the regions of M so that the regions sharing a common boundary are colored differently.

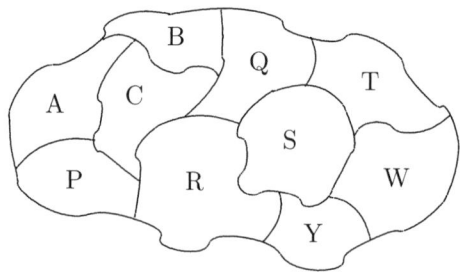

Figure 14.31: The map in Exercise 2

3. (a) Determine the dual graph G of the map M in Figure 14.32.

   (b) Determine $\chi(G)$.

   (c) What information does $\chi(G)$ give us about M?

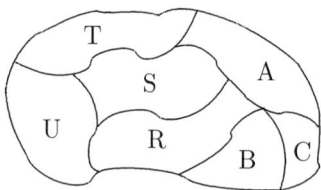

Figure 14.32: The map in Exercise 3

4. Find a map M whose dual graph is isomorphic the connected planar graph in Figure 14.33.

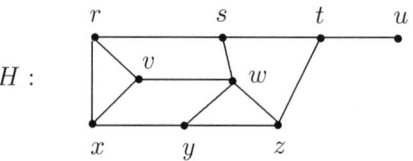

Figure 14.33: The graph in Exercise 4

5. Determine if the following is a proof of the Four Color Theorem.

   Proof. Let G be a planar graph. By Kuratowski's theorem, G does not contain K_5 as a subgraph. Since $\chi(K_5) = 5$ and G does not contain K_5 as a subgraph, it follows that $\chi(G) \leq 4$. ∎

6. Prove or disprove: If G is a graph with an odd cycle for which there exists a 3-coloring, then $\chi(G) = 3$.

7. Prove or disprove: If G is a graph with an odd cycle for which there exists no 3-coloring, then $\chi(G) = 4$.

8. Determine $\chi(Q_3)$.

9. Determine the chromatic number of a nontrivial tree.

10. Determine the chromatic number of the wheel $W_n = C_n + K_1$ for every integer $n \geq 3$.

11. Determine the chromatic number of each of the graphs G and H of Figure 14.34.

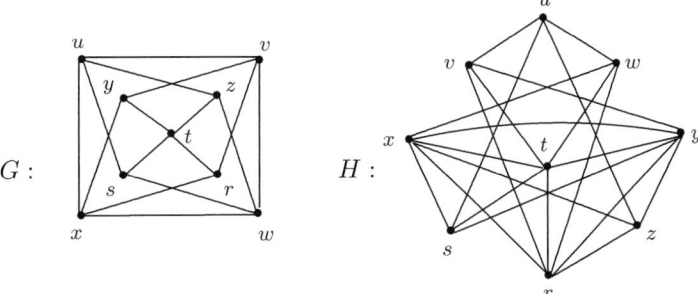

Figure 14.34: Graphs in Exercise 11

12. (a) For the graph G of Figure 14.35, what is the largest complete graph that is a subgraph of G?
    (b) What does (a) say about $\chi(G)$?
    (c) Determine $\chi(G)$.

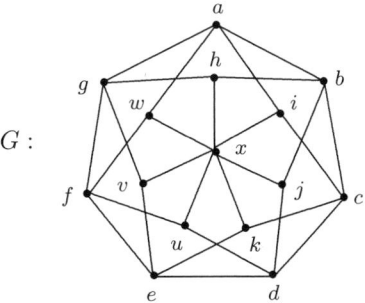

Figure 14.35: Graphs in Exercise 12

13. (a) For the graph G of Figure 14.36, what is the largest complete graph that is a subgraph of G?
    (b) What does (a) say about $\chi(G)$?
    (c) Determine $\chi(G)$.

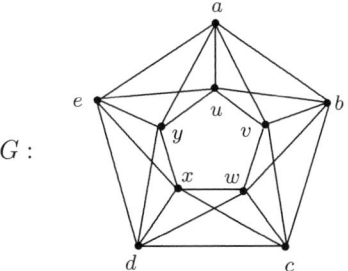

Figure 14.36: Graphs in Exercise 13

14. Let $n \geq 2$ be an integer.

    (a) Show that the graph $\overline{C}_{2n+1}$ contains a complete subgraph of order n but no complete subgraph of order $n + 1$.

(b) According to Theorem 14.26, what does the observation in (a) say about $\chi(\overline{C}_{2n+1})$?

(c) Determine $\chi(\overline{C}_{2n+1})$.

15. Let G be a graph with $V(G) = \{v_1, v_2, \ldots, v_n\}$. A coloring of the vertices of G is given by the following algorithm. The vertex v_1 is assigned the color 1. For an integer k with $1 \leq k < n$, once the vertices v_1, v_2, ..., v_k have been assigned colors, the vertex v_{k+1} is assigned the smallest color such that v_{k+1} and any of its neighbors in $\{v_1, v_2, \ldots, v_k\}$ are colored differently.

   (a) Apply this algorithm to prove Theorem 14.33: *For every graph G, $\chi(G) \leq 1 + \Delta(G)$.*

   (b) Apply this algorithm to the graph G of Figure 14.37.

   (c) Is it true that this algorithm always gives the chromatic number of a graph?

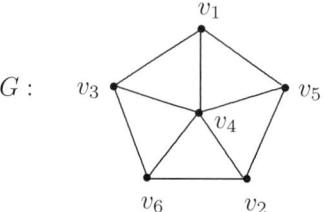

Figure 14.37: The graph in Exercise 15

16. Eleven cities, denoted by $c_1, c_2, \ldots, c_{11}$, have applied for admission to a baseball league. The league is to be divided into a certain number of divisions. Two cities cannot belong to the same division if the distance between them is 60 miles or less. A matrix $\mathbf{D} = [d_{ij}]$ is given where d_{ij} represents the distance between cities c_i and c_j. What is the smallest number of divisions into which the league can be divided? What is one way of dividing the cities into a smallest number of divisions?

$$\mathbf{D} = \begin{bmatrix}
0 & 45 & 50 & 60 & 70 & 85 & 50 & 45 & 65 & 85 & 90 \\
45 & 0 & 60 & 50 & 35 & 55 & 45 & 80 & 105 & 105 & 90 \\
50 & 60 & 0 & 40 & 65 & 105 & 90 & 75 & 60 & 55 & 45 \\
60 & 50 & 40 & 0 & 45 & 90 & 95 & 95 & 100 & 75 & 50 \\
70 & 35 & 65 & 45 & 0 & 50 & 70 & 120 & 115 & 110 & 95 \\
85 & 55 & 105 & 90 & 50 & 0 & 45 & 115 & 135 & 150 & 135 \\
50 & 45 & 90 & 95 & 70 & 45 & 0 & 70 & 105 & 125 & 120 \\
45 & 80 & 75 & 95 & 120 & 115 & 70 & 0 & 45 & 85 & 105 \\
65 & 105 & 60 & 100 & 115 & 135 & 105 & 45 & 0 & 45 & 80 \\
85 & 105 & 55 & 75 & 110 & 150 & 125 & 88 & 45 & 0 & 45 \\
90 & 90 & 45 & 50 & 95 & 135 & 120 & 105 & 80 & 45 & 0
\end{bmatrix} .$$

17. During this coming summer, the Department of Mathematical Sciences at a university plans to offer courses in the following subjects: discrete mathematics (dm), vector calculus (vc), linear algebra (la), data structures (ds), algorithms (al), statistics (st). The following information is known about five students who plan to take courses this summer:

   Alvin plans to take discrete mathematics, vector calculus and linear algebra.

   Beverly plans to take discrete mathematics, data structures and linear algebra.

   Clark plans to take discrete mathematics, algorithms and data structures.

   Donna plans to take vector calculus and statistics.

   Edward plans to take statistics and algorithms.

   The only time periods when courses are taught in the summer are 8:00-9:45, 10:15-12:00, 1:30-3:15. Can all six courses be offered during these time periods so that two courses are not taught at the same time if some student plans to take both courses?

18. Seven chemicals $c_1, c_2, \ldots, c_7$ are to be shipped in a number of packages to a university. Because some pairs of chemicals can react with each other, it is not a good idea to ship such pairs of chemicals in the same package. The chemicals that can react with a given chemical are indicated below:

c_1 : c_2, c_3, c_5, c_6 c_2 : $c_1, c_3, c_4, c_5, c_6, c_7$ c_3 : c_1, c_2, c_6, c_7
c_4 : c_2, c_5, c_6, c_7 c_5 : c_1, c_2, c_4, c_6 c_6 : $c_1, c_2, c_3, c_4, c_5, c_7$
c_7 : c_2, c_3, c_4, c_6

What is the smallest number of packages in which these chemicals can be shipped safely?

19. Figure 14.38 shows eight traffic lanes $L_1, L_1, \ldots, L_8$ at the intersection of two streets. A traffic light is located at the intersection. During each phase of the traffic light, those cars in lanes for which the light is green may proceed safely through the intersection into certain permitted lanes. What is the minimum number of phases needed for the traffic light so that all cars may proceed safely through the intersection?

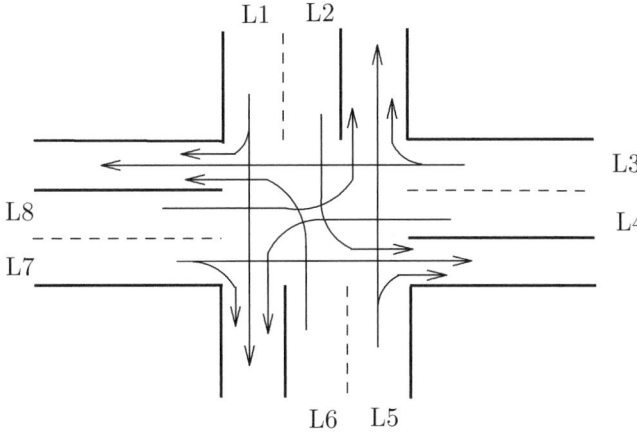

Figure 14.38: Traffic lanes at street intersections in Exercise 19

20. For each possible positive integer k, give an example of a 4-regular graph G_k with $\chi(G_k) = k$.

21. Suppose that G is a graph with chromatic number $k \geq 2$ and that there is given a k-coloring of G using the colors $\{1, 2, \ldots, k\}$. Prove or disprove: For every two distinct colors $i, j \in \{1, 2, \ldots, k\}$, there are adjacent vertices of G colored i and j.

22. Show for every two graphs G and H that $\chi(G + H) = \chi(G) + \chi(H)$.

23. Show for every nontrivial graph G that $\chi(G \times K_2) = \chi(G)$.

Chapter 14 Highlights

Key Concepts

boundary (of a region of a plane graph G): the subgraph whose vertices and edges are incident with the region.

chromatic number of a graph G, $\chi(G)$: the smallest number of colors needed to color the vertices of G (so that every two adjacent vertices are colored differently).

k-colorable graph: a graph that can be colored with k or fewer colors.

coloring (of a graph G): an assignment of colors to the vertices of G, one color to each vertex of G, such that adjacent vertices of G are colored differently.

k-coloring: a coloring that uses k colors.

dual graph (of a map M): a connected graph whose vertices are the regions of M such that two vertices are adjacent if the corresponding regions share a common edge.

exterior region (of a plane graph G): the unbounded region of G.

Four Color Problem: the problem of determining whether the vertices of every planar graph can be colored with four or fewer colors.

maximal planar graph: a planar graph G with the property that the addition of any edge joining two nonadjacent vertices of G produces a nonplanar graph.

nonplanar graph: a graph that is not planar.

planar graph: a graph that can be drawn in the plane so that no two of its edges cross.

plane graph: a graph that is drawn in the plane so that no two of its edges cross.

region (of a plane graph G): a connected piece of the plane resulted from removing the vertices and edges of G from the plane.

subdividing an edge: inserting vertices of degree 2 into an edge.

subdivision (of a graph G): a graph that is either G or obtained from G by inserting vertices of degree 2 into one or more edges of G.

Three Houses and Three Utilities Problem: the problem that asks whether three houses can be connected to three utilities (gas, water and electricity) by water mains and gas and electricity lines without any two of the lines or mains crossing each other.

Key Results

- **Euler Identity:** If G is a connected plane graph with n vertices, m edges and r regions, then $n - m + r = 2$.

- The boundary of every region of a maximal planar graph of order 3 or more is a triangle.

- If G is a maximal planar graph of order $n \geq 3$ and size m, then $m = 3n - 6$.

- If G is a planar graph of order $n \geq 3$ and size m, then $m \leq 3n - 6$.

- Every planar graph has a vertex of degree 5 or less.

- If G is a graph of order $n \geq 3$ and size m such that $m > 3n - 6$, then G is nonplanar.

- The complete graph K_5 is nonplanar.

- The graph $K_{3,3}$ is nonplanar.

- **Kuratowski's Theorem:** A graph G is planar if and only if G does not contain a subdivision of K_5 or $K_{3,3}$ as a subgraph.

- Let G be a graph. Then $\chi(G) \geq 2$ if and only if G contains at least one edge.

- A graph G has chromatic number 1 if and only if $G = \overline{K}_n$ for some positive integer n.

- Let G be a nonempty graph. Then $\chi(G) = 2$ if and only if G is bipartite.

- Let G be a graph. Then $\chi(G) \geq 3$ if and only if G contains an odd cycle.

- If H is a subgraph of a graph G, then $\chi(H) \leq \chi(G)$.

- For every graph G, $\chi(G) \leq 1 + \Delta(G)$.

- **Brooks' Theorem:** If G is a connected graph that is neither a complete graph nor an odd cycle, then $\chi(G) \leq \Delta(G)$.

- **The Five Color Theorem:** If G is a planar graph, then $\chi(G) \leq 5$.

- **The Four Color Theorem:** If G is a planar graph, then $\chi(G) \leq 4$.

Supplementary Exercises for Chapter 14

1. Prove or disprove:

    (a) If G is an Eulerian graph and H is a subdivision of G, then H is Eulerian.

    (b) If G is a Hamiltonian graph and H is a subdivision of G, then H is Hamiltonian.

2. True or False.

    (a) Every 3-regular graph has chromatic number 2.

    (b) Every 3-regular graph is nonplanar.

    (c) Every connected 3-regular graph is Hamiltonian.

    (d) Every connected 3-regular graph is not Eulerian.

3. Let G be a plane graph. For *each* vertex v of G with $\deg v \geq 3$, let $e_1, e_2, \ldots, e_k$ $(k \geq 3)$ be the edges of G incident with v and arranged cyclically about v. Suppose that the vertex u_i (of degree 2) is inserted into the edge e_i $(1 \leq i \leq k)$ and the edges $u_1u_2, u_2u_3, \ldots, u_{k-1}u_k, u_ku_1$ are added and the vertex v is deleted. The resulting graph is called the **cyclic subdivision graph** of G (see Figure 14.39(a)). If a complete graph (rather than a cycle) is constructed on the vertices $u_1, u_2, \ldots, u_k$, then the resulting graph is called the **complete subdivision graph** of G (see Figure 14.39(b)).

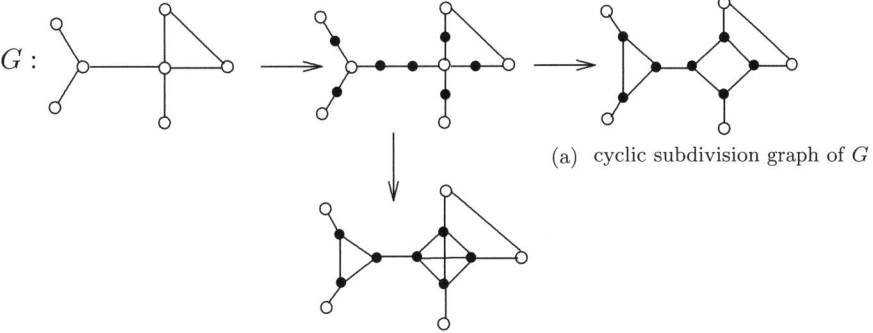

(a) cyclic subdivision graph of G

(b) complete subdivision graph of G

Figure 14.39: The graphs in Exercise 3

Prove or disprove:

    (a) A graph G is Hamiltonian if and only if its cyclic subdivision graph is Hamiltonian.

    (b) A graph G is Eulerian if and only if its complete subdivision graph is Eulerian.

4. The degrees of the ten vertices of a connected graph G are 1, 1, 1, 1, 1, 5, 5, 5, 5, 5. Which of the following is true?

   (a) G is planar. (b) G is nonplanar.

   (c) It is impossible to determine whether G is planar or nonplanar.

5. The degrees of the twelve vertices of a connected graph G are 1, 1, 1, 1, 1, 1, 4, 4, 4, 4, 4, 4. Which of the following is true?

   (a) G is planar. (b) G is nonplanar.

   (c) It is impossible to determine whether G is planar or nonplanar.

6. The minimum degree of a planar graph G with $V(G) = \{v_1, v_2, \ldots, v_8\}$ is 3 and G is known to have exactly one vertex of degree 3, namely v_1. It is also known that $\deg v_2 = 5$, $\deg v_i < 5$ for $i = 3, 4$ and $\deg v_i > 5$ for $i = 5, 6$. What are $\deg v_7$ and $\deg v_8$?

7. Show that the minimum degree of a maximal planar graph of order 4 or more is 3 or more.

8. Determine the minimum number of colors needed to color the regions in the map in Figure 14.40.

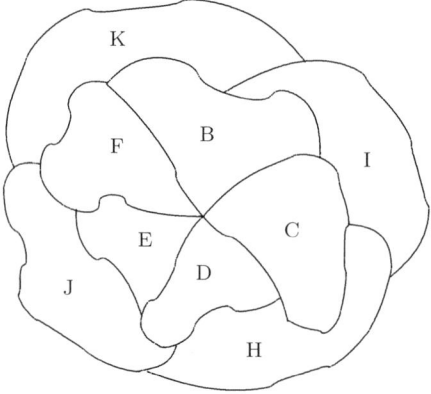

Figure 14.40: The map in Exercise 8

9. If a graph G contains K_4 as a subgraph, then $\chi(G) \geq 4$. Is the converse of this implication true or false?

10. Ten small towns $T_1, T_2, \ldots, T_{10}$ have decided to apply for a channel for an FM radio station. Suppose that two radio stations cannot broadcast on the same channel if the distance between them is 75 miles or less. The distances between various pairs of towns are given below. The distance is not given in the table if this distance exceeds 75 miles. What is the minimum number of channels that can be assigned to the ten radio stations.

	T_1	T_2	T_3	T_4	T_5	T_6	T_7	T_8	T_9	T_{10}
T_1	*	60				50				
T_2	60	*	65	75	55	65				
T_3		65	*	60	70					
T_4		75	60	*	50					65
T_5		55	70	50	*	70		55	60	
T_6	50	65			70	*	50	55		
T_7						50	*	55		
T_8					55	55	55	*	60	
T_9					60			60	*	75
T_{10}				65					75	*

11. Ten juniors from high school were given the opportunity to visit a nearby university with the possibility of attending that university in the future. The day of their visit, however, happened to be one of the days when final exams were given in the high school. Some of the students missed exams in history (H), English (E), government (G), biology (B), mathematics (M) and French (F). The exams the ten students missed are given below.

Alvin: H, E   Brenda: E, F   Charles: M, F   Dina: M, B   Edwin: E, G

Fan: H, F   Ghia: M, H   Howard: H, E, G   Ida: B, G   John: F, B

The next day after the students returned to high school, they were required to take make-up exams for the exams they missed. What is the smallest number of time periods in which all exams can be given if two different exams cannot be given during the same period if some student must take both exams?

12. Among all 5-regular graphs, let G_1 be one with the smallest chromatic number and let G_2 be one with the largest chromatic number. Suppose that $\chi(G_1) = s$ and $\chi(G_2) = t$.

(a) What are s and t?

(b) For each integer k with $s < k < t$, show that there exists a 5-regular graph G_k such that $\chi(G_k) = k$.

13. Let G be a nonempty graph and let H be a graph obtained from G by subdividing a single edge of G exactly once.

(a) Show that $|\chi(G) - \chi(H)| \leq 1$.

(b) Show that it is possible for:

(i) $\chi(H) = \chi(G)$; (ii) $\chi(H) = \chi(G) - 1$; (iii) $\chi(H) = \chi(G) + 1$.

14. Determine the largest positive integer k such that $\chi(H) = \chi(G) = k$, where H is obtained from a nonempty graph G by subdividing each edge of G exactly once.

15. Determine the chromatic number of the graph G of Figure 14.41.

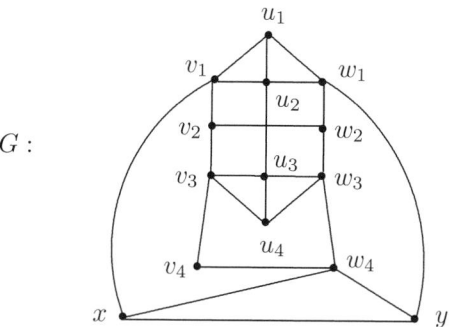

Figure 14.41: The graph in Exercise 15

16. Determine the chromatic number of the graph G of Figure 14.42.

17. A connected plane graph G of order 12 has 10 regions. Suppose that the degrees of 10 of the vertices are 5, 4, 4, 3, 3, 3, 3, 3, 2, 2 and the two remaining vertices have the same degree k. What is k?

18. Prove or disprove: If every edge of a graph G lies on an odd cycle, then $\chi(G) = 3$.

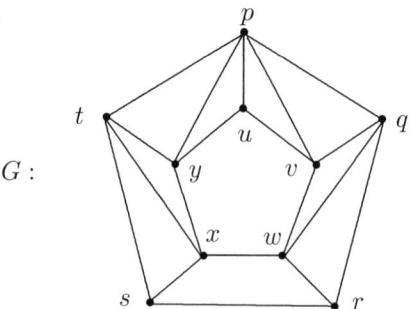

Figure 14.42: The graph in Exercise 16

19. A nonplanar graph G of order 7 has the property that $G - v$ is planar for every vertex v of G.

    (a) Show that G does not contain $K_{3,3}$ as a subgraph.

    (b) Give an example of a graph with this property.

20. Prove or disprove:

    (a) If G is a graph of order 5 that is not K_5, then G is planar.

    (b) If G is a graph of order 6 that does not contain K_5 or $K_{3,3}$ as a subgraph, then G is planar.

21. The degrees of the vertices of a planar graph are 3, 3, 4, 4, 5, 5. Is G a maximal planar graph?

22. Consider the tree T in Figure 14.43.

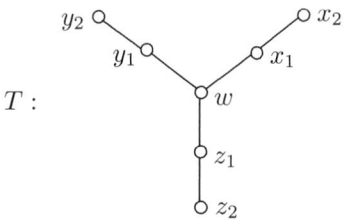

Figure 14.43: The tree T in Exercise 22

    (a) Draw the complement $\overline{T}$.

    (b) Determine whether $\overline{T}$ in (a) is planar or nonplanar. If $\overline{T}$ is nonplanar, then find a subgraph of $\overline{T}$ that is a subdivision of K_5 or $K_{3,3}$. If $\overline{T}$ is planar, then draw $\overline{T}$ as a plane graph and determine whether $\overline{T}$ is maximal planar.

    (c) Find all integers $n \geq 3$ for which there exists a tree T of order n such that $\overline{T}$ is a maximal planar graph.

23. Determine whether the graph G in Figure 14.44 is planar or nonplanar.

24. Determine whether the graphs G_1 and G_2 in Figure 14.45 are planar or nonplanar.

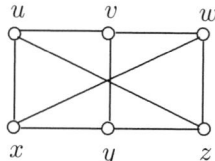

Figure 14.44: The graph in Exercise 23

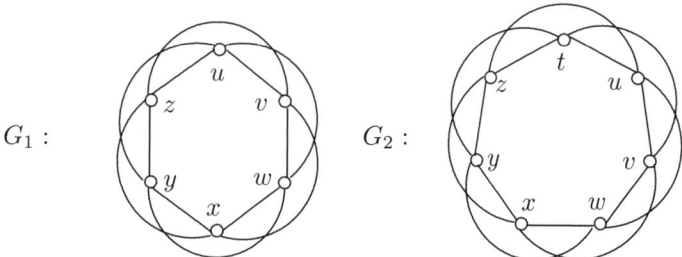

Figure 14.45: The graphs G_1 and G_2 in Exercise 24

25. Let G be a graph with $\chi(G) = k$. Suppose that there is a k-coloring of G such that each vertex is adjacent to vertices, all of which are colored the same. Show that $k \neq 3$.

26. Determine the chromatic number of the graph G in Figure 14.46.

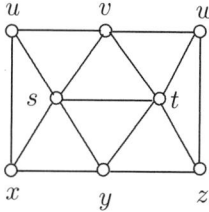

Figure 14.46: The graph G in Exercise 26

27. Let G be a graph with $\chi(G) = k$ and let a k-coloring of G be given. Let H be the graph with $V(H) = V(G)$ such that $uv \in E(H)$ if u and v are assigned different colors in G. Determine $\chi(H)$.

Chapter 15

Directed Graphs

One of the most common applications of graphs is their use in modeling some locations in a city, some pairs of which are connected by streets. We have also noted that graphs do not accurately capture the structure of a street system in which some locations are connected by two or more streets. In this situation, a multigraph can be used. Should some or all of the streets be one-way streets, however, then neither a graph nor a multigraph accurately reflects this street system. In this case, the structure we need is a directed graph, most commonly called a digraph. This concept is introduced in this chapter together with the best known class of digraphs, called tournaments. Digraphs can also be used to model a class of structures known as finite-state machines.

15.1 Fundamental Concepts of Digraph Theory

Suppose that $V = \{2, 4, 8, 12\}$ is a set of vertices and that each vertex is represented in the plane by a point or small circle and that a directed line segment or curve is drawn from vertex i to vertex j $(i \neq j)$ if $i \mid j$. This structure is shown in Figure 15.1 and is called a digraph D.

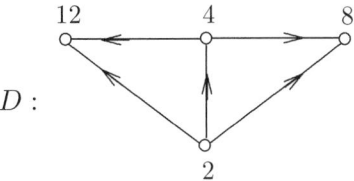

Figure 15.1: A digraph

We now give a formal definition of a digraph.

Definition 15.1 *A* **digraph** (*or* **directed graph**) *D is a finite nonempty set $V(D)$* (*of* **vertices**) *and a set $E(D)$ of ordered pairs of distinct vertices of D, each ordered pair of which is referred to as a* **directed edge** *or an* **arc***.*

A directed edge (u, v) is represented in a diagram of a digraph by drawing a directed line segment or directed curve from u to v. For a directed edge $e = (u, v)$ in a digraph D, the vertex u is called the **initial vertex** of e and v is the **terminal vertex** of e. Unlike uv and vu, which represent the same edge in a graph, the directed edges (u, v) and (v, u) in a digraph are not the same. If u and v are two vertices in a digraph D, then D may contain one, neither or both of the directed edges (u, v) and (v, u). If, in the definition of digraph, for each pair u, v of distinct vertices, at most one of (u, v) and (v, u) is a directed edge, then the resulting digraph is called an **oriented graph**. An oriented graph D can therefore be obtained by assigning a direction to each edge of some *graph G*. In this case, the oriented graph D is called an **orientation** of G. Figure 15.2 shows two digraphs

559

D_1, D_2 and a graph G. The digraph D_1 is not an oriented graph (since both (v, w) and (w, v) are arcs of D_1), while D_2 is an oriented graph. Indeed, D_2 is an orientation of the graph G.

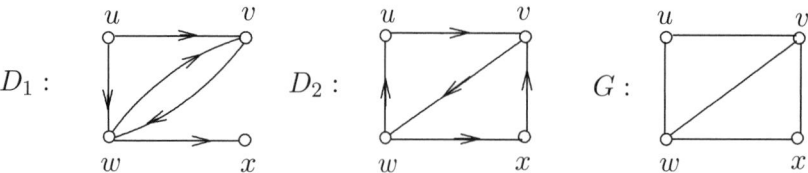

Figure 15.2: Two digraphs and a graph

Example 15.2

Soccer teams from the countries of Brazil, China, Germany and Spain play one another. Brazil defeats China and Spain, while Germany defeats Brazil and Spain. Also, Spain defeats China, which in turn defeats Germany. Model this situation by a digraph.

Solution. Let D be a digraph whose vertices are the four countries. We then draw a directed edge from a country to each country that it defeats. The digraph D is shown in Figure 15.3. Actually, this digraph is an oriented graph.   ♦

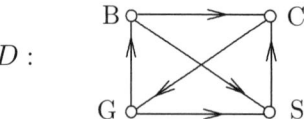

Figure 15.3: The digraph in Example 15.2

Two vertices u and v in a digraph D are **adjacent** if D contains at least one of the directed edges (u, v) and (v, u). More specifically, if (u, v) is a directed edge of D, then we say that u is **adjacent to** v and v is **adjacent from** u. The number of vertices *to which* a vertex v is adjacent is the **outdegree** of v and is denoted by od v or od(v). The number of vertices *from which* v is adjacent is the **indegree** of v and is denoted by id v or id(v). (See Figure 15.4.)

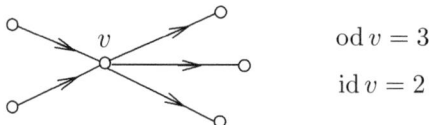

Figure 15.4: The outdegree and indegree of a vertex

As with graphs, the number of vertices in a digraph D is called the **order** of D and the number of directed edges in D is the **size** of D. When the outdegrees of the vertices of a digraph are summed, each directed edge is counted exactly once. The same is true when the indegrees are added. This observation leads us to the digraph counterpart of Theorem 12.10.

Theorem 15.3 (The First Theorem of Digraph Theory) *If D is a digraph of size m, then*

$$\sum_{v \in V(D)} \text{od}\, v = \sum_{v \in V(D)} \text{id}\, v = m.$$

The digraph D of Figure 15.5 has order $n = 4$ and size $m = 5$. Furthermore,

$$\text{od } u = 1, \text{ id } u = 3; \text{ od } v = 1, \text{ id } v = 0;$$
$$\text{od } w = 2, \text{ id } w = 1; \text{ od } x = \text{ id } x = 1.$$

As is guaranteed by Theorem 15.3, the sum of the outdegrees of the vertices of D equals the sum of the indegrees of the vertices of D equals 5.

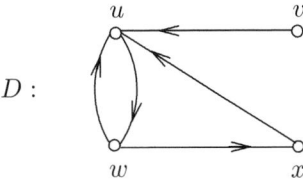

Figure 15.5: A digraph D

For digraphs, the concepts subdigraphs, proper subdigraphs and induced subdigraphs are defined as expected.

Definition 15.4 *The **adjacency matrix** $\mathbf{A} = [a_{ij}]$ of a digraph D with $V(D) = \{v_1, v_2, \ldots, v_n\}$ is defined as expected, namely,*

$$a_{ij} = \begin{cases} 1 & \text{if } (v_i, v_j) \in E(D) \\ 0 & \text{otherwise.} \end{cases}$$

This is illustrated for the digraph D of Figure 15.6.

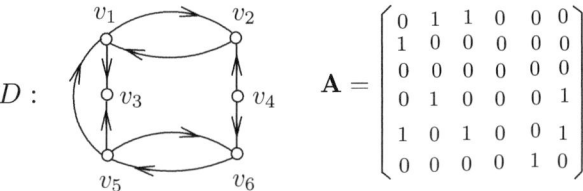

Figure 15.6: The adjacency matrix of a digraph

Corresponding to the concepts of walk, trail, path, circuit and cycle in graphs are expected counterparts in digraphs. The two that will be most useful to us are directed paths and directed cycles. Let $D = (V, E)$ be a digraph. A **directed** $u - v$ **path** in D is a sequence $P = (u = u_0, u_1, \ldots, u_k = v)$ of distinct vertices of D such that $(u_i, u_{i+1}) \in E$ for $0 \le i \le k - 1$. The directed path P has **length** k. A **directed cycle** in D is a sequence $C = (v_0, v_1, \ldots, v_\ell, v_0)$, where $\ell \ge 2$, the vertices $v_0, v_1, \ldots, v_\ell$ are distinct and $(v_i, v_{i+1}) \in E$ for $0 \le i \le \ell - 1$ and $(v_\ell, v_0) \in E$. For the digraph D of Figure 15.7, $P = (v, u, w, y, x)$ is a directed $v - x$ path of length 4, while $C = (u, w, y, v, u)$ is a directed cycle of length 4 and $C' = (u, w, u)$ is a directed cycle of length 2.

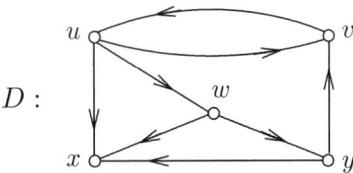

Figure 15.7: Directed paths and directed cycles in a digraph

Connected Digraphs

The **underlying graph** of a digraph D is obtained by removing all directions from the arcs of D and replacing any resulting pair of parallel edges by a single edge. That is, the underlying graph of the digraph D is obtained by replacing each arc (u, v) or pair $(u, v), (v, u)$ of arcs by the edge uv. The graph G of Figure 15.8 is therefore the underlying graph of the digraph D of that figure. A digraph D is **connected** if its underlying graph is connected. Thus the digraph D in Figure 15.8 is connected. A digraph D is **strongly connected** or simply **strong** if D contains both a directed $u - v$ path and a directed $v - u$ path for every pair u, v of distinct vertices of D. While the digraph D_1 of Figure 15.8 is strong, the digraph D_2 is not strong as there is no directed $u - x$ path in D for example. In fact, there is no directed path from any vertex in $\{u, v, w\}$ to any vertex in $\{x, y, z\}$ in D_2.

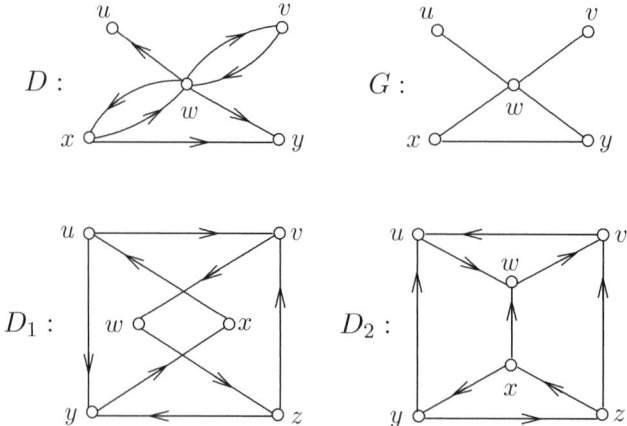

Figure 15.8: Connected digraphs

Distance is defined in digraphs as well as in graphs. If u and v are vertices in a digraph D for which there is a directed $u - v$ path, then the **directed distance** $\vec{d}(u, v)$ **from** u **to** v is the length of a shortest $u - v$ path in D. In order for $\vec{d}(u, v)$ to be defined for every pair u, v of vertices of D, the digraph D must be strong. For example, for the two vertices u and v in the strong digraph D_1 of Figure 15.8, $\vec{d}(u, v) = 1$ and $\vec{d}(v, u) = 5$; while in the digraph D_2 of Figure 15.8 the directed distance $\vec{d}(s, t)$ is not defined if $s \in \{u, v, w\}$ and $t \in \{x, y, z\}$.

In connection with the concept of a strong digraph is the following related problem. A small university town ordinarily has light traffic but the university has a very popular football team, which results in heavy traffic on football Saturdays. To ease the traffic on those days, the town council has decided to convert all streets from two-way streets to one-way streets. Of course, it is essential that after the conversion of the streets to one-way streets it is possible to drive (legally) from any intersection in the town to every other intersection in town. The question here is not only whether this can be done in this town but under what conditions this can be done in any town. This is equivalent to determining which graphs have a strong orientation.

This question was first addressed in 1939 by Herbert Robbins (1915–2001). Robbins was not only a well-known statistician but a co-author (with Richard Courant) of the popular mathematics book *What is Mathematics?*

It is probably clear that in order for a graph G to have a strong orientation, G must be connected and contain no bridges. Robbins showed, however, that these conditions are all that are needed, that is, these conditions are sufficient as well as necessary for G to have a strong orientation.

Before presenting Robbins' theorem, we first make some observations. Let G be a nontrivial connected graph having no bridges. By Theorem 12.36, every edge of G lies on a cycle of G. Therefore, G contains cycles. Let

$$C = (v_1, v_2, \ldots, v_r, v_1)$$

be a cycle of G. Suppose that we direct the edges of C cyclically, obtaining the directed edges

$$(v_1, v_2), (v_2, v_3), \ldots, (v_{r-1}, v_r), (v_r, v)$$

and direct any other edge joining two vertices of C arbitrarily. Then the subgraph H with $V(H) = V(C)$ has a strong orientation. In particular, if C is a Hamiltonian cycle of G, then G itself has a strong orientation.

Theorem 15.5 *A nontrivial graph G has a strong orientation if and only if G is connected and contains no bridges.*

Proof. We have already mentioned that if G has a strong orientation, then G is connected and contains no bridges. It remains therefore only to verify the converse.

Suppose that G is a nontrivial connected graph of order n containing no bridges. Since G contains no bridges, G contains cycles and by our earlier observation, G contains subgraphs having a strong orientation. Let H be a subgraph of maximum order k in G having a strong orientation. If $k = n$, then G itself has a strong orientation. Hence we may assume that $k < n$. Since G is connected, there is a vertex v of G not belonging to H that is adjacent to a vertex u of H. Since $e = uv$ is not a bridge, e lies on a cycle

$$C' = (u = v_1, v = v_2, \ldots, v_r, v_1).$$

Direct the edges of C' cyclically about C' that have not already been directed. Then the subgraph F with

$$V(F) = V(H) \cup V(C') \text{ and } E(F) = E(H) \cup E(C')$$

has a strong orientation. Since the order of F is greater than k, this is a contradiction. ∎

Example 15.6

Since the graph G of Figure 15.9(a) is connected and contains no bridges, it follows by Theorem 15.5 that G has a strong orientation. We now construct a strong orientation of G. Following the proof of Theorem 15.5, we begin with a cycle in G, say

$$C = (v_1, v_2, \ldots, v_{10}, v_1).$$

Directing the edges of C cyclically about C and directing the edges $v_1 v_8$ and $v_2 v_9$ arbitrarily, we obtain a strong orientation of a subgraph H of G shown in Figure 15.9(b). Since H is a proper subgraph of G, the process continues. We seek an edge joining a vertex of H and a vertex not belonging to H. One such edge is $v_3 u_1$ and a cycle containing this edge is

$$C' = (v_3, u_1, u_2, u_3, u_4, u_5, u_7, u_6, v_4, v_3).$$

Direct the edges of C' cyclically about C', except for the edge $v_4 v_3$, which has already been directed as (v_3, v_4). We also direct the edges $u_1 u_4$ and $u_2 u_5$ arbitrarily. This results in a strong orientation D of G (shown in Figure 15.9(c)). ♦

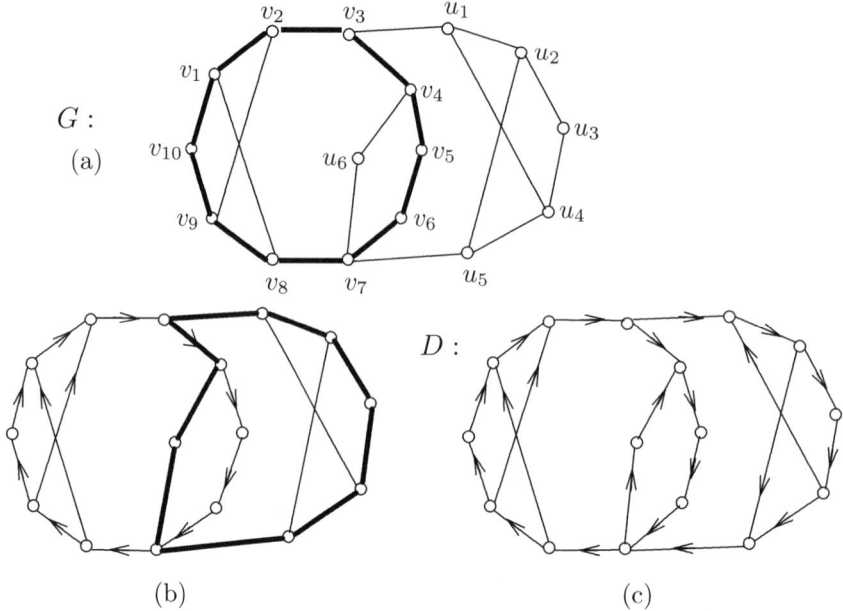

Figure 15.9: A graph with a strong orientation

As with graphs, it is important to know when two digraphs are the same. Two digraphs D_1 and D_2 are **isomorphic** if there exists a one-to-one correspondence $\phi : V(D_1) \to V(D_2)$ such that $(u_1, v_1) \in E(D_1)$ if and only if $(\phi(u_1), \phi(v_1)) \in E(D_2)$. As with isomorphic graphs, two isomorphic digraphs have the same order and same size. Furthermore, the outdegrees (and indegrees) of the vertices of two isomorphic digraphs are the same.

For example, Figure 15.10 shows the two orientations of the 3-cycle C_3 and the four orientations of C_4. Of these orientations, only D' and D_1 are strong. Observe that the digraphs D_2 and D_3 of Figure 15.10 have order 4 and size 4. Also, D_2 and D_3 have two vertices having indegree 1 and outdegree 1, one vertex with indegree 0 and outdegree 2 and one vertex with indegree 2 and outdegree 0. Despite having the same order, same size and the same indegrees and outdegrees of their vertices, the digraphs D_2 and D_3 are not isomorphic. For example, D_2 contains a directed path of length 3 while D_3 does not.

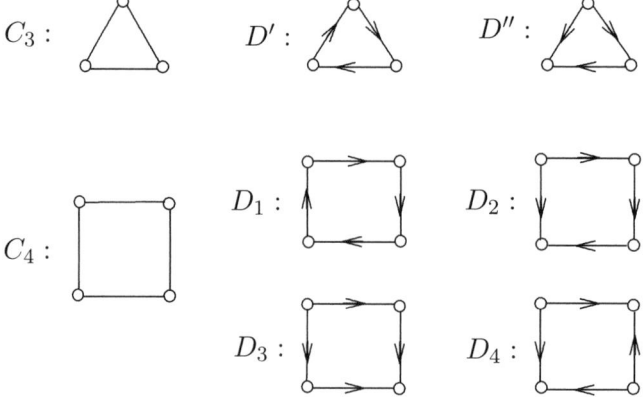

Figure 15.10: The orientations of C_3 and C_4

Exercises for Section 15.1

1. Let $V = \{P(n), Q(n), R(n)\}$ be the set of the following open sentences involving integers n:

$$P(n): \ n \text{ is even.} \quad Q(n): \ 4n + 1 \text{ is even.} \quad R(n): \ n^2 \text{ is even.}$$

Draw a digraph D with vertex set V, where for $A, B \in V$, there is a directed edge from A to B if $\forall n \in \mathbb{Z}, A \Rightarrow B$ is a true statement.

2. Let $S = \{3, 6, 9, 18\}$.

   (a) The vertex set of a graph G is S and two vertices i and j are adjacent if either $i \mid j$ or $j \mid i$. Draw the graph G.

   (b) The vertex set of a digraph D is S and (i, j) is an arc of D if $i \mid j$. Draw the digraph D.

   (c) What can you say about the graph G in (a) and the digraph D in (b)?

3. (a) The vertex set of a digraph D is $V(D) = \{M_1, M_2, M_3, M_4\}$, where M_1, M_2, M_3 and M_4 are the four 2×2 matrices

$$M_1 = \begin{bmatrix} 0 & 1 \\ 0 & 0 \end{bmatrix}, \ M_2 = \begin{bmatrix} 0 & 0 \\ 1 & 0 \end{bmatrix}, \ M_3 = \begin{bmatrix} 1 & 1 \\ 0 & 0 \end{bmatrix} \text{ and } M_4 = \begin{bmatrix} 0 & 0 \\ 1 & 1 \end{bmatrix}.$$

There is a directed edge from the vertex M_i to the vertex M_j $(i \neq j)$ if $M_i M_j = \begin{bmatrix} 0 & 0 \\ 0 & 0 \end{bmatrix}$. Draw the digraph D.

   (b) Use the idea expressed in (a) to define and draw another digraph D'.

4. A digraph D contains three vertices with indegree 0 and three vertices with indegree 2. The remaining two vertices of D have indegree 1. One of these eight vertices has outdegree 0 and another has outdegree 2. The remaining vertices have the same outdegree. What is this outdegree? Draw a digraph having these properties.

5. Let D be a digraph obtained by assigning a direction to each edge of the cycle C_n of order $n \geq 3$. Prove that D either contains an even number of vertices having indegree 0 or an even number of vertices having outdegree 0.

6. Determine the adjacency matrix for each digraph in Figure 15.11.

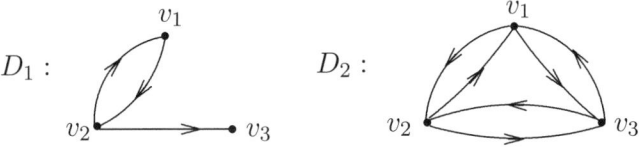

$D_1:$ v_1 v_2 v_3 $D_2:$ v_1 v_2 v_3

Figure 15.11: The digraphs D_1 and D_2 in Exercise 6

7. (a) Prove that if a connected digraph D is strong, then od $v > 0$ and id $v > 0$ for every vertex v of D.

   (b) Is the converse of (a) true?

8. Use the technique employed in the proof of Theorem 15.5 to construct a strong orientation of the graph G of Figure 15.12.

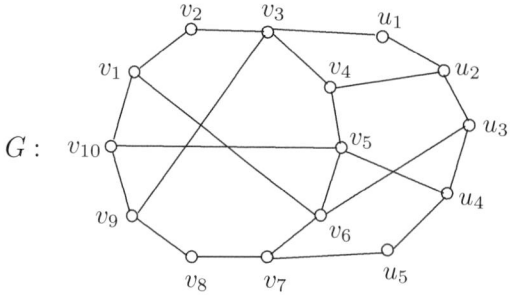

Figure 15.12: The graph G in Exercise 8

9. Determine all (non-isomorphic) orientations of the 5-cycle C_5.

10. Determine the number of orientations of the star $K_{1,t}$ $(t \geq 1)$.

11. Determine the number of orientations of the graph G in Figure 15.13.

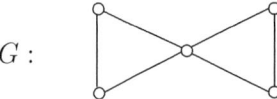

Figure 15.13: The graph G in Exercise 11

12. Let G be a connected graph of order $n \geq 3$. Prove that there is an orientation of G in which no directed path has length 2 if and only if G is bipartite.

13. The **converse** $\vec{D}$ of a digraph D is obtained from D by reversing the direction of every arc of D. Show that a digraph D is strong if and only if its converse $\vec{D}$ is strong.

14. (a) Does there exist a digraph D of order $n \geq 3$ in which no two vertices of D have the same outdegree but every two vertices of D have the same indegree?

    (b) Does there exist a digraph D of order $n \geq 3$ such that no two vertices of D have the same indegree and no two vertices of D have the same outdegree?

15. A digraph D is **Eulerian** if D contains a directed circuit that contains every directed edge of D. Let D be a connected digraph of order $n \geq 3$ with $V(D) = \{v_1, v_2, \cdots, v_n\}$. Prove the following.

    (a) The digraph D is Eulerian if and only if od $v_i = $ id v_i for $1 \leq i \leq n$.

    (b) If od $v_i \geq$ id v_i for $1 \leq i \leq n$, then D is Eulerian.

    (c) If $\sum_{i=1}^{n} |$od $v_i -$ id $v_i| = 0$, then D is Eulerian.

16. Prove that a graph G has an Eulerian orientation (see Exercise 15) if and only if G is Eulerian.

17. According to Theorem 15.5 a nontrivial graph G has a strong orientation if and only if G is connected and contains no bridges.

    (a) Prove that if G is a nontrivial connected graph with at most two bridges, then there exists an orientation D of G having the property that if u and v are any two vertices of D, there is either a directed $u - v$ path or a directed $v - u$ path.

    (b) Show that the statement (a) is false if G contains three bridges.

18. Does there exist a strong orientation D of a graph G and two vertices u and v of G such that $\vec{d}_D(u,v) \neq d_G(u,v)$ and $\vec{d}_D(v,u) \neq d_G(v,u)$?

19. Prove that if every vertex of a digraph D has positive outdegree, then D contains a directed cycle.

20. Let G be a nontrivial connected graph without bridges.

    (a) Show that for every edge e of G and for every orientation of e, there exists an orientation of the remaining edges of G such that the resulting digraph is strong.

    (b) Show that (a) need not be true if we begin with an orientation of two edges of G.

21. Let D be a digraph with $V(D) = \{1, 3, 5, 9, 11, 13\}$ and let $A = \{2, 8\}$ and $B = \{4, 10\}$. A vertex i is adjacent to a vertex j in D if $j - i \in A$ and j is adjacent to i if $j - i \in B$. Is D strong?

22. (a) Two people A and B play a game on the complete graph K_4. These two people alternate assigning a direction to an edge, beginning with A. It is the goal of A is to produce a strong orientation of K_4 while it is the goal of B to stop A from doing this. If both players play perfectly, who will win the game? On which turn will the game end?

    (b) What variation of this game is suggested and what is the outcome of this game?

23. Let T be a nontrivial tree and let r and s be two distinct vertices of T. Let D be the digraph with $V(D) = V(T)$ such that (u,v) is a directed edge of D if the $r - v$ path in T contains u or the $s - v$ path in T contains u. (Thus D may contain both (u,v) and (v,u)). What is the strong subdigraph of D of largest order?

15.2 Tournaments

There are sporting events involving a number of teams or individuals which require every two teams or individuals to compete against each other exactly once. This is referred to as a round robin tournament. Men's soccer has been part of the Summer Olympic Games since 1900. Teams from 16 countries participate, divided into four pools of four teams each. In each pool, a round robin tournament takes place, in which the top two teams in each pool advance to play for Olympic medals. This also occurs during the World Cup for soccer supremacy when 32 countries participate, divided into eight pools of four teams each.

Round robin tournaments can be modeled by a class of digraphs, which are also called tournaments. The vertex set of the tournament (digraph) is the set of teams and there is a directed edge from vertex s to vertex t if team s defeats team t in the round robin tournament. In this case no ties are permitted. The formal definition of a tournament is stated next.

Definition 15.7 *A* **tournament** *is an oriented complete graph.*

Consequently, a tournament is a digraph D with the property that if u and v are any two vertices of D, then exactly one of (u, v) and (v, u) is a directed edge of D. The following theorem is a consequence of the First Theorem of Digraph Theory (see Theorem 15.3).

Theorem 15.8 *Let T be a tournament with $V(T) = \{v_1, v_2, \ldots, v_n\}$. Then*

$$\sum_{i=1}^{n} \operatorname{od} v_i = \sum_{i=1}^{n} \operatorname{id} v_i = \binom{n}{2} = \frac{n(n-1)}{2}.$$

Example 15.9

The digraphs D_1 and D_2 of Figure 15.14 are tournaments. The digraph D_3 is not a tournament since neither (v_3, w_3) nor (w_3, v_3) is a directed edge of D_3; while the digraph D_4 is not a tournament either since both (u_4, w_4) and (w_4, u_4) are directed edges.   ♦

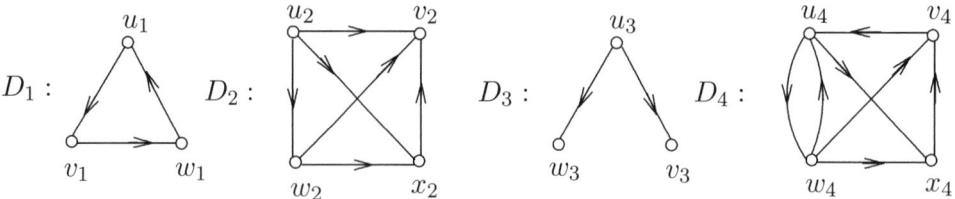

Figure 15.14: The digraphs in Example 15.9

Example 15.10

Suppose that four teams t_1, t_2, t_2, t_4 are participating in a round robin tournament, where t_1 has defeated the other three teams and t_2 has defeated t_3 which has defeated t_4 which has defeated t_2. Then this situation can be modeled by the tournament T of Figure 15.15.

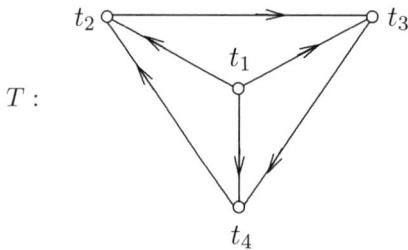

Figure 15.15: A model of the round robin tournament in Example 15.10

Observe that the number of games that a team t_i $(1 \leq i \leq 4)$ wins is its outdegree. Since t_1 has won all three games it has played, od $t_1 = 3$.   ♦

Transitive Tournaments

While it is clear that t_1 is the best team (among the four teams in Example 15.10), it is impossible to tell which is the next best team as each of the three remaining teams has won exactly one game. One might argue that t_2 is better than t_3 since t_2 defeated t_3. However, by the the same argument, t_3 is better than t_4 and t_4 is better than t_2. In fact, as we shall see, the only situation in which there is a clear ranking of all teams is when no two teams have won the same number of games. This brings us to a special class of tournaments.

A tournament T is called a **transitive tournament** if whenever (u, v) and (v, w) are arcs of T, then (u, w) is an arc of T as well (see Figure 15.16).

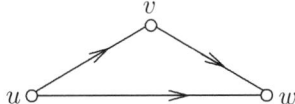

Figure 15.16: Three arcs in a transitive tournament

Figure 15.17 shows transitive tournaments of order n for $n = 3, 4, 5$. The outdegrees of the vertices of every transitive tournament have an interesting property.

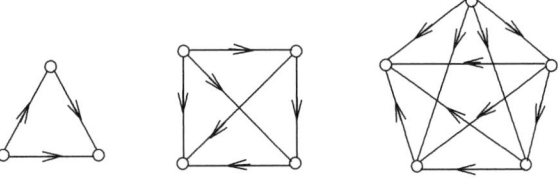

Figure 15.17: Transitive tournaments of orders 3, 4, 5

Theorem 15.11 *A tournament T is transitive if and only if no two vertices of T have the same outdegree.*

Proof. Assume first that T is a tournament of order n in which no two vertices of T have the same outdegree. Since $0 \leq \text{od } v \leq n - 1$ for every vertex v of T, it follows that there is exactly one vertex of outdegree k for every integer k with $0 \leq k \leq n - 1$. We may assume that $V(T) = \{v_1, v_2, \ldots, v_n\}$, where $\text{od } v_i = n - i$ for $i = 1, 2, \ldots, n$. Since $\text{od } v_1 = n - 1$, it follows that v_1 is adjacent to v_i for $i = 2, 3, \ldots, n$. Since $\text{od } v_2 = n - 2$ and v_2 is adjacent from v_1, it follows that v_2 is adjacent to v_i for $i = 3, 4, \ldots, n$. Continuing in this manner, we see that each vertex v_t $(1 \leq t \leq n)$ is adjacent only to those vertices v_i with $t < i \leq n$. We now show that T is transitive. Suppose that u, v and w are vertices of T such that (u, v) and (v, w) are arcs of T. Then $u = v_i$, $v = v_j$ and $w = v_k$ for some integers i, j, k with $1 \leq i < j < k \leq n$. However, since $i < k$, it follows that (v_i, v_k) is an arc of T and so (u, w) is an arc of T. Thus T is transitive.

We now verify the converse. Assume, to the contrary, that there exists a transitive tournament T containing two vertices u and v with the same outdegree, say $\text{od } u = \text{od } v = k$. Since T is a tournament, either (u, v) or (v, u) is an arc of T, say (u, v) is an arc of T. Since $\text{od } v = k$, there is a set S of k vertices to which v is adjacent. Let $w \in S$. So (v, w) is an arc of T. However, (u, v) is an arc of T and T is transitive; so (u, w) is an arc of T. Therefore, u is also adjacent to every vertex of S (see Figure 15.18). Since (u, v) is an arc of T, it follows that $\text{od } u \geq k + 1$, which is a contradiction. ∎

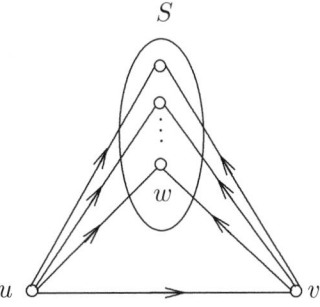

Figure 15.18: A step in the proof of Theorem 15.11

By the proof of Theorem 15.11, a transitive tournament T with vertex set $V(T) = \{v_1, v_2, \ldots,$ $v_n\}$ can be constructed by placing the vertices of T in order in a column with v_1 on the top and v_n on the bottom so that all arcs are directed downward. Perhaps it is now clear that there is only one transitive tournament of each order.

Example 15.12

A transitive tournament T of order 6 is shown in Figure 15.19. If the vertices $v_1, v_2, \ldots, v_6$ represent teams, then v_1 is the best team as it has defeated all other teams, v_2 is the next best team as it has defeated all other teams except v_1 and so on. In other words, there is a clear ranking of all six teams.   ◆

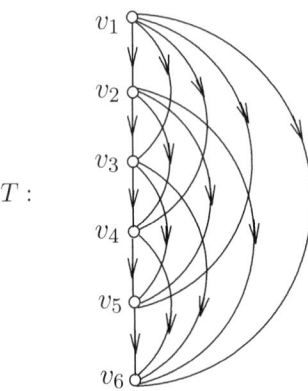

Figure 15.19: The transitive tournament of order 6 in Example 15.12

Recall that a digraph D is strong if for every two vertices u and v, there is both a directed $u - v$ path and a directed $v - u$ path in D. Necessarily, no transitive tournament is strong.

Example 15.13

There are two non-isomorphic tournaments D_1 and D_2 of order 3. These are shown in Figure 15.20. The tournament D_1 is transitive and D_2 is strong. There are four non-isomorphic tournaments of order 4, namely T_1, T_2, T_3 and T_4 of Figure 15.20. While T_1 is transitive and T_2 is strong, T_3 and T_4 are neither. There are 12 non-isomorphic tournaments of order 5 and over 154 billion non-isomorphic tournaments of order 12.   ◆

We now make some observations about the tournaments in Figure 15.20. The transitive tournaments D_1 and T_1 contain no directed cycles. The strong tournament D_2 has a directed 3-cycle, while the strong tournament T_2 has a directed 3-cycle and a directed 4-cycle. Each of the tournaments T_3 and T_4 has a directed 3-cycle but no directed 4-cycle. As the observations above suggest, only transitive tournaments have no directed cycles.

Theorem 15.14 *A tournament T is transitive if and only if T has no directed cycles.*

Proof. Assume first that T is a tournament that contains no directed cycles. We show that T is transitive. Let (u, v) and (v, w) be two arcs of T. Since T contains no directed cycles, T cannot contain the arc (w, u). Thus (u, w) is an arc of T and so T is transitive.

We use a proof by contradiction to verify the converse: If T is a transitive tournament, then T has no directed cycles. Suppose that this statement is false. Then there exists a transitive tournament T that contains a directed cycle C, say

$$C = (v_1, v_2, \ldots, v_k, v_1)$$

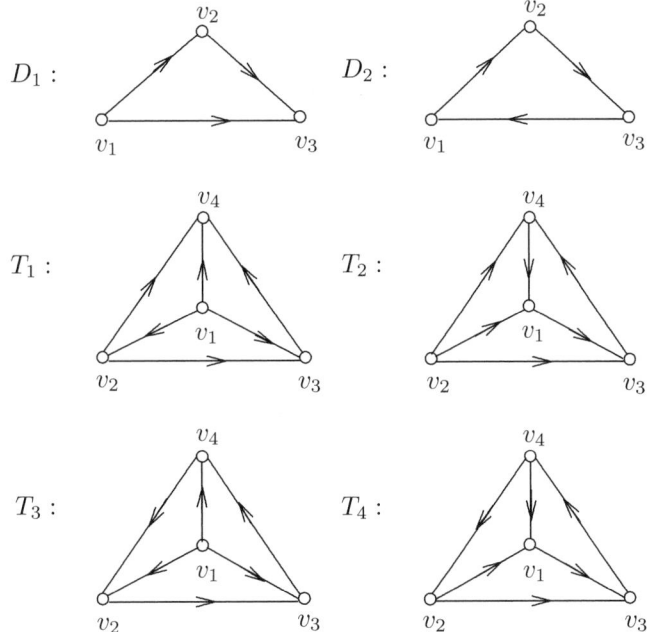

Figure 15.20: The tournaments of order 3 and 4 in Example 15.13

(See Figure 15.21(a)). Since v_1 is adjacent to v_2 and adjacent from v_k, there is a vertex v_i ($1 \leq i \leq k-1$) such that v_1 is adjacent to v_i but adjacent from v_{i+1} (see Figure 15.21(b)). Hence T contains the directed 3-cycle (v_1, v_i, v_{i+1}, v_1). However, (v_1, v_i) and (v_i, v_{i+1}) are arcs of T but (v_1, v_{i+1}) is not an arc of T. This contradicts the assumption that T is transitive.   ∎

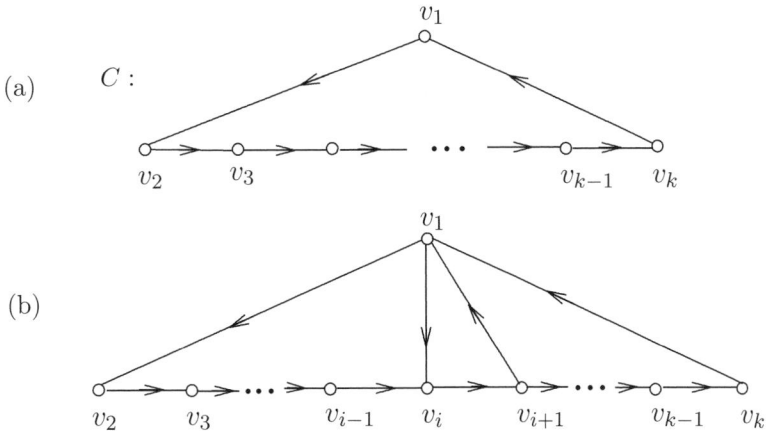

Figure 15.21: A step in the proof of Theorem 15.14

As we have seen, if T is a transitive tournament of order n, then there is a unique vertex u of T having outdegree $n-1$, which is certainly the maximum outdegree of any vertex of T. Therefore, u is adjacent to all other vertices of T and so $\vec{d}(u, v) \leq 1$ for every vertex v of T. This is a special case of the following result.

Theorem 15.15 *Let u be a vertex of maximum outdegree in a tournament T. For every vertex v of T, $\vec{d}(u, v) \leq 2$.*

Proof. Suppose that $\text{od}(u) = k$ and that $u_1, u_2, \ldots, u_k$ are the k vertices of T that are adjacent from u. Then $\vec{d}(u, u_i) = 1$ for $1 \leq i \leq k$. Let v be a vertex of T such that (v, u) is a directed edge of T. Since $\text{od}(v) \leq \text{od}(u)$, the vertex v cannot be adjacent to all of the vertices $u_1, u_2, \ldots, u_k$. Therefore, (u_j, v) is a directed edge of T for some j with $1 \leq j \leq k$. Thus $P = (u, u_j, v)$ is a directed $u - v$ path of length 2 and so $\vec{d}(u, v) = 2$.   ■

Suppose that we have a collection of teams involved in a round robin tournament and the results of the matches in this round robin tournament are modeled by a tournament T. Then the outdegree of a vertex in T is the number of matches won by this team. Let A be a team that has won the greatest number of matches. According to Theorem 15.15, if B is any other team, then either A defeated B or A defeated a team that defeated B.

Hamiltonian Tournaments

Definition 15.16 *A **Hamiltonian path** in a digraph D is a directed path that contains every vertex of D, while a **Hamiltonian cycle** in D is a directed cycle that contains every vertex of D. A digraph D is **Hamiltonian** if it contains a Hamiltonian cycle.*

While it is easy to give an example of a tournament containing a Hamiltonian path, it is impossible to give an example of a tournament that does not contain a Hamiltonian path.

Theorem 15.17 *Every tournament contains a Hamiltonian path.*

Proof. Assume, to the contrary, that there exists a tournament T, of order n say, that contains no Hamiltonian path. Let $P = (v_1, v_2, \ldots, v_k)$ be a directed path of greatest length in T. Necessarily, $k < n$. Hence there exists a vertex v of T that does not lie on P. Since T is a tournament, there is an arc between v and each vertex v_i $(1 \leq i \leq k)$ of P. Since P is a longest path in T, the arcs (v, v_1) and (v_k, v) do not belong to T, for otherwise, a longer path exists. Therefore, both (v_1, v) and (v, v_k) are arcs of T. (see Figure 15.22(a)).

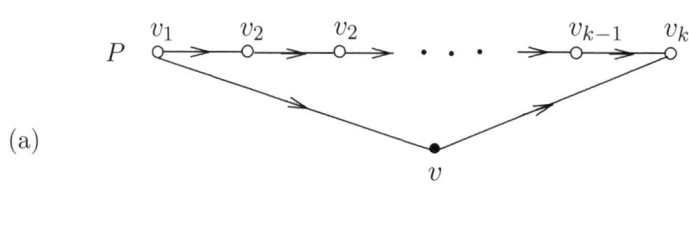

(a)

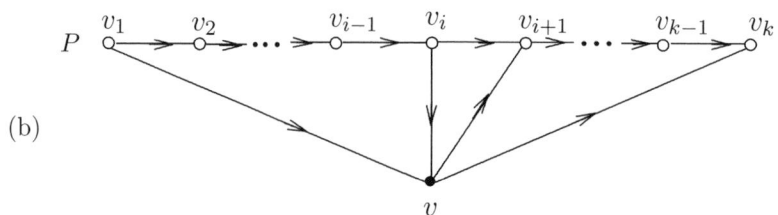

(b)

Figure 15.22: A step in the proof of Theorem 15.17

Since v_1 is adjacent to v and v is adjacent to v_k, there is a vertex v_i $(1 \leq i \leq k-1)$ such that v_i is adjacent to v and v is adjacent to v_{i+1} (see Figure 15.22(b)). Thus

$$(v_1, v_2, \ldots, v_i, v, v_{i+1}, \ldots, v_k)$$

is a directed path whose length exceeds that of P. This is a contradiction. ∎

We have already observed that the strong tournaments of orders 3 and 4 are Hamiltonian. This is true of every strong tournament. Although it is not difficult to show that every Hamiltonian tournament is strong (see Exercise 13), it is a bit challenging to verify the converse. For this reason, we omit the proof of the following result.

Theorem 15.18 *A nontrivial tournament T is Hamiltonian if and only if T is strong.*

Exercises for Section 15.2

1. Is it possible to have a round robin tournament involving six teams where three of the teams win three games and the other three teams win two games?

2. Is it possible to have a round robin tournament involving six teams where two teams win five games each, one team wins two games and the other three teams win one game each?

3. Is it possible to have a round robin tournament with more than three teams where each team wins the same number of games?

4. Give an example of two strong tournaments T' and T'' of the same order that are not isomorphic.

5. Show for every complete graph K_n $(n \geq 3)$ that there exists an orientation of K_n for which there are no directed cycles but the reversal of the direction of one of the edges results in a strong tournament.

6. Does there exist a transitive tournament of order 2?

7. We have seen that there is exactly one transitive tournament of each order. A tournament of order $n \geq 3$ is defined to be **circular** if whenever (u, v) and (v, w) are arcs of T, then (w, u) is an arc of T.

    (a) How many circular tournaments of order 3 are there?

    (b) Show that in a tournament of order 3 or more, every vertex, with at most two exceptions, has positive outdegree and positive indegree.

    (c) How many circular tournaments of order 4 or more are there?

8. Does there exist a round robin tournament containing $2k$ teams, where $k \geq 3$, in which k of these teams win r games and the remaining k teams win s games, where $r \neq s$?

9. A tournament T of order $n \geq 3$ has the property that whenever (u, v) is an arc of T, then $\mathrm{od}\, u \geq \mathrm{od}\, v$. Show that if T is Hamiltonian, then n is odd.

10. (a) Show that if an odd number of teams play in a round robin tournament, then it is possible for all teams to tie for first place.

    (b) Show that if an even number of teams play in a round robin tournament, then it is not possible for all teams to tie for first place.

11. Let u and v be distinct vertices in a tournament such that $\vec{d}(u, v)$ and $\vec{d}(v, u)$ are defined. Show that $\vec{d}(u, v) \neq \vec{d}(v, u)$.

12. Prove that if u and v are vertices of a tournament such that $\vec{d}(u, v) = k$, then $\mathrm{id}\, u \geq k - 1$.

13. Prove that every Hamiltonian tournament is strong.

14. Let T be a tournament of order 3 or more. Prove that the length of a longest cycle in T equals the greatest order of a strong subdigraph in T.

15. (a) Find a Hamiltonian path in each tournament in Figure 15.20.

    (b) Find a Hamiltonian cycle in each strong tournament in Figure 15.20.

16. Prove that every vertex in a nontrivial strong tournament belongs to a directed triangle.

17. Let T be a tournament of order $n \geq 3$ with $V(T) = \{v_1, v_2, \ldots, v_n\}$. Prove that if od $v_i >$ id v_i for $1 \leq i \leq n-1$, then T is not strong.

18. Prove or disprove:

    (a) If every vertex of a tournament T belongs to a cycle in T, then T is strong.

    (b) For every pair u, v of vertices in a strong tournament T, there exists either a Hamiltonian $u - v$ path or Hamiltonian $v - u$ path.

    (c) If (u, v) is a directed edge of a strong tournament T, then (u, v) lies on Hamiltonian cycle of T.

19. Prove that if T is a tournament of order $4r$ with $r \geq 1$, where $2r$ vertices of T have outdegree $2r$ and the other $2r$ vertices have outdegree $2r - 1$, then T is strong.

20. Let T be a tournament of order n. Prove that if every vertex of T has outdegree $\lceil \frac{n-1}{2} \rceil$ or $\lfloor \frac{n-1}{2} \rfloor$, then T is strong.

21. Let T be a tournament of order $n \geq 4$. Prove that if T contains two directed cycles C and C' of order $n - 1$ such that C' contains a vertex not on C, then T is Hamiltonian.

15.3 Finite-State Machines

There is an ever-increasing number of input-output devices. For example,

(1) pushing a sequence of numerical buttons followed by another button (talk) on your cell phone allows you to contact a person at another phone number;

(2) typing in information at a web site and clicking on 'order complete' allows us to order some merchandise over the internet or to make a reservation for a flight to some location;

(3) placing coins in a vending machine permits us to lift the handle to an enclosure and remove a newspaper.

Indeed, a computer is an input-output device. Here, some sort of information is input, which is processed, resulting in an output of some type. The computer must be able to remember past information as it works its way through the input. Devices such as these can be modeled by an abstract structure called a *finite-state machine* or a *sequential circuit*. As the term suggests, a finite-state machine has a finite number of internal states and at each state, the machine has certain information to remember. Before proceeding further with this concept, we give a simple example of a finite-state machine – one involving a vending machine.

Example 15.19

A motel has a laundry room where guests can wash and dry their clothes. There is a vending machine that dispenses small bottles of laundry detergent and small bottles of fabric softener, each bottle costing 75¢. The machine accepts quarters and half-dollars only and returns the change if more than 75¢ is deposited. After returning to the motel at the end of the day, Kevin decides to wash a load of T-shirts. He has no laundry detergent and so decides to use the vending machine to purchase a bottle of laundry detergent. Kevin inserts three quarters into the vending machine. A button labeled LD (laundry detergent) is then pushed. What happens when Kevin purchases the laundry detergent from the vending machine is shown in the table below at various times t_0, t_1, t_2, t_3, t_4, where $t_0 < t_1 < t_2 < t_3 < t_4$ and t_0 is the starting time of this procedure.

Time	t_0	t_1	t_2	t_3	t_4
State	s_0	s_1	s_2	s_3	s_0
Input	25¢	25¢	25¢	LD	
Output	nothing	nothing	nothing	detergent	

The state s_0 is the initial state, where the machine is prepared for a guest to insert a quarter or half-dollar or push the LD or FS (fabric softener) button. At time t_0, Kevin deposits a quarter and nothing is output. Shortly afterwards then (at time t_1), the machine is at state s_1. At state s_1, the machine remembers that 25¢ has already been deposited. Another quarter is then deposited. Again nothing is dispensed. At this point (at time t_2), the machine is at state s_2. At state s_2, the machine remembers that 50¢ has already been deposited. Another quarter is deposited. Again there is no output. Now (at time t_3), the machine is at state s_3, where the machine remembers that 75¢ has already been deposited and that this is the correct amount needed to dispense either of the two products. The LD button is then pushed and a bottle of detergent is released. Soon afterwards, at time t_4, the machine is reset to state s_0. When the next guest chooses to use this vending machine, the machine is at time t_0 and state s_0 and the process repeats itself.

As we mentioned, there are two buttons that can be pushed and the machine will dispense a bottle of detergent or a bottle of fabric softener provided sufficient money has been deposited but does nothing otherwise. That is, if less than 75¢ has been deposited and either button is pushed, then there is no output and the machine remains at its current state. On the other hand, if 75¢ has been deposited, then the machine releases a bottle of detergent or softener (depending on which button is pushed) and the machine is returned to the state s_0 (nothing deposited). If \$1 has been deposited (either two half-dollars or two quarters followed by a half-dollar), then the machine returns 25¢ in change immediately. Thus 75¢ has been deposited. The actions of this vending machine can be modeled by a finite-state machine. This finite-state machine can be represented in two ways. One way is by a digraph or diagram (see Figure 15.23), called the *state digraph* of the finite-state machine. Here LD represents the *laundry detergent button*, FS the *fabric softener button* and N indicates *no output*. For example, there is an arc from state s_0 to state s_1 labeled 25, N. This means that if we deposit 25¢ at state s_0, then the state changes to s_1 but no output is produced. There is an arc from state s_3 to state s_0 labeled FS, softener. This means that if we were to push the fabric softener button at state s_3, then we receive a bottle of fabric softener and the machine returns to state s_0. The directed loop at s_3 labeled 50, 50 indicates that if we deposit 50¢ at state s_3, then 50¢ is returned and the machine remains at state s_3. ♦

We now look at finite-state machines in general. Among the features of a finite-state machine are the following:

(1) The machine can be in only one of finitely many states at any time. These are the internal states of the machine.

(2) The machine can accept as input only a finite number of input values, the entire collection of which is called the *input set* or *input alphabet* (denoted by I). Only these input values are recognized by the machine at each internal state.

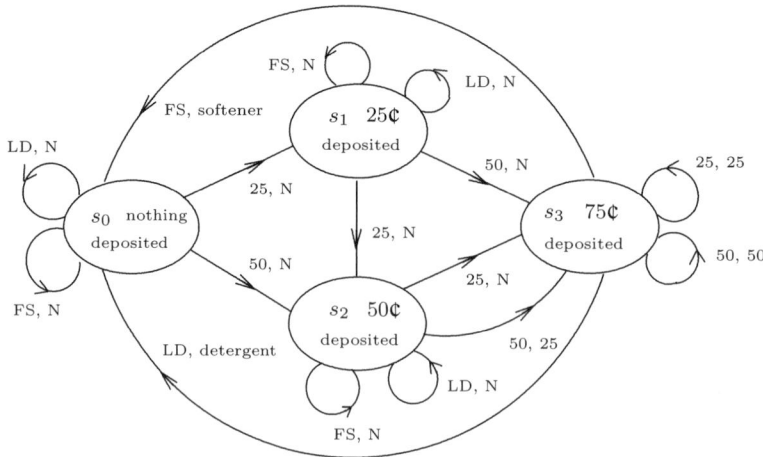

Figure 15.23: The state digraph for the vending machine in Example 15.19

(3) An *output* and *next state* are determined by each state-input pair. The set of all possible outputs is finite and is denoted by O. The set O is called the *output set* of the machine.

(4) We assume that the machine operates in a deterministic manner, where the output is determined by a sequence of input values and the initial state of the machine.

More formally, a **finite-state machine** consists of

(1) a finite set S of internal states,

(2) a finite **input set** I,

(3) a finite **output set** O,

(4) a **next-state** (or **transition**) **function** $f : S \times I \to S$,

(5) an **output function** $g : S \times I \to O$.

Consequently, if the machine is in a certain state s at a particular time and $i \in I$ is input, then $g(s, i)$ is output and there is a transition of the machine to the internal state $f(s, i)$. When the finite-state machine receives its first input, we assume that the machine is in a specified starting state, often denoted by s_0.

The states and functions described above can be represented by a diagram called a **state diagram**. Since the diagram is actually a digraph (possibly with parallel arcs and directed loops), it is also called the **state digraph** or **transition digraph** of the finite-state machine. The vertex set of this digraph is then the set S. If $f(s_i, a) = s_j$ and $g(s_i, a) = b$, then this is represented in the digraph by drawing a directed edge (arc) labeled a, b from vertex s_i to vertex s_j (see Figure 15.24). If there is more than one arc from s_i to s_j, then these arcs can be replaced by a single arc with multiple labels.

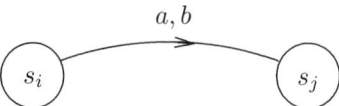

Figure 15.24: An arc (s_i, s_j) in a state digraph

Since the sets S, I and O are finite, the next-state function f and output function g can also be described by means of a table, commonly called the **state table** or **transition table** of the machine. For example, the finite-state machine modeling the vending machine in Example 15.19 can be represented by its state table, which gives the values of the transition function f and the output function g for all state-input pairs. In this case, there are four possible input values, namely (1) deposit 25¢, (2) deposit 50¢, (3) push the laundry detergent button (LD) and (4) push the fabric softener button (FS). The state table for the finite-state machine in Example 15.19 is given below.

	f				g			
	25	50	LD	FS	25	50	LD	FS
s_0	s_1	s_2	s_0	s_0	N	N	N	N
s_1	s_2	s_3	s_1	s_1	N	N	N	N
s_2	s_3	s_3	s_2	s_2	N	25	N	N
s_3	s_3	s_3	s_0	s_0	25	50	det	sof

These ideas are further illustrated with some abstract examples of finite-state machines.

Example 15.20

Construct the state digraph for the finite-state machine whose state table is shown below.

	f		g	
	0	1	0	1
s_0	s_1	s_2	0	0
s_1	s_3	s_1	1	0
s_2	s_2	s_4	1	1
s_3	s_4	s_2	0	0
s_4	s_2	s_0	0	0

Solution. The state digraph for this machine is given in Figure 15.25. For example, since $f(s_0, 0) = s_1$ and $g(s_0, 0) = 0$, there is an arc labeled 0, 0 from s_0 to s_1. Because $f(s_4, 1) = s_0$ and $g(s_4, 1) = 0$, there is an arc labeled 1, 0 from s_4 to s_0. Also, $f(s_2, 0) = s_2$ and $g(s_2, 0) = 1$ and so there is a directed loop at s_2 labeled 0, 1. ◆

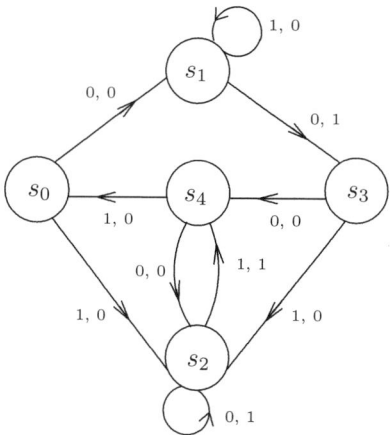

Figure 15.25: The state digraph for the finite-state machine in Example 15.20

Example 15.21

Construct the state table for the finite-state machine whose state digraph is shown in Figure 15.26.

Solution. Since there is an arc from s_0 to s_1 labeled 0, 0 and an arc from s_0 to s_3 labeled 1, 1, it follows that $f(s_0, 0) = s_1$, $f(s_0, 1) = s_3$, $g(s_0, 0) = 0$ and $g(s_0, 1) = 1$. Continuing in this fashion, we obtain the state table shown below for this finite-state machine. ◆

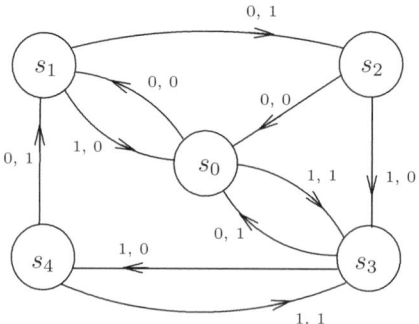

Figure 15.26: The state digraph for the finite-state machine in Example 15.21

	f		g	
	0	1	0	1
s_0	s_1	s_3	0	1
s_1	s_2	s_0	1	0
s_2	s_0	s_3	0	0
s_3	s_0	s_4	1	0
s_4	s_1	s_3	1	1

When dealing with finite-state machines, we are often interested in sequences of inputs. Such a sequence is referred to as an **input string**. An input string takes us from the starting state through a sequence of states, as determined by the transition function. Each input string also produces an output string. For the vending machine in Example 15.19, any of the following input strings places us at state s_3, where we can obtain a bottle of detergent or a bottle of fabric softener by pushing an appropriate button:

(i) 50, 25; (ii) 25, 50; (iii) 25, 25, 25; (iv) 50, 50; (v) 25, 25, 50.

In a finite-state machine, it is often the case that we wish to arrive at some state or states with an appropriate input string. Each such state is called an **accepting state**. We now consider an input string for the finite-state machine described in Example 15.21.

Example 15.22

For the finite-state machine in Figure 15.26 with starting state s_0, find the output string and sequence of states obtained from the input string 1001101.

Solution.   From the state digraph of this finite-state machine in Figure 15.26 (or from its state table in Example 15.21), we see that $f(s_0, 1) = s_3$ and $g(s_0, 1) = 1$; $f(s_3, 0) = s_0$ and $g(s_3, 0) = 1$; and $f(s_0, 0) = s_1$ and $g(s_0, 0) = 0$. Continuing in this way, we obtain the successive states and outputs as shown in the table below. Thus the output string is 1100111 and the resulting sequence of states is $s_0, s_3, s_0, s_1, s_0, s_3, s_0, s_3$. This produces the directed $s_0 - s_3$ walk $(s_0, s_3, s_0, s_1, s_0, s_3, s_0, s_3)$ in the state digraph. ♦

Input	1	0	0	1	1	0	1
State	s_3	s_0	s_1	s_0	s_3	s_0	s_3
Output	1	1	0	0	1	1	1

We now describe a finite-state machine that computes the sum of two positive integers expressed in base 2. Such a finite-state machine is commonly called a **binary adder**.

Example 15.23

Let a and b be two integers expressed in base 2, say $a = a_5a_4a_3a_2a_1a_0 = 011011$ and $b = b_5b_4b_3b_2b_1b_0 = 001110$, where for $0 \leq i \leq 5$, a_i and b_i represent the coefficient of 2^i in a and b, respectively. Therefore, $a = (011011)_2 = 27$ and $b = (001110)_2 = 14$.

The initial bit 0 in the expansion of a and the initial two 0s in the expansion of b are present to have an equal number of bits in a and b and to aid us in computing the sum. Hence we wish to compute $a + b = c = (c_5c_4c_3c_2c_1c_0)_2$.

Let's see how this addition is carried out, bit by bit, from right to left.

$$
\begin{array}{rccccccc}
a & = & 0 & 1 & 1 & 0 & 1 & 1 \\
b & = & 0 & 0 & 1 & 1 & 1 & 0 \\
\hline
c & = & 1 & 0 & 1 & 0 & 0 & 1
\end{array}
$$

In the first addition, $1 + 0 = 1$. In the second addition, we add $1 + 1$, which is 2 (in decimal) and 10 (in binary). So we write down the 0 and carry 1 to the next column, getting $1 + 0 + 1$, which again is 10 in binary. Completing this sum, we obtain $c = (101001)_2 = 41$. In the first addition, when computing $a + b$ for any two integers a and b expressed in base 2, we add the two bits a_0 and b_0, write down 0 if we are adding $0 + 0$ or $1 + 1$ or write down 1 if we are adding $1 + 0$ or $0 + 1$. If $a_0 = b_0 = 1$, then we carry 1 over to the second addition; otherwise we can assume that we are carrying 0 over to the next addition. If we are adding $a_i + b_i$ $(i \geq 1)$, then we proceed in the same manner except we also need to add to $a_i + b_i$ the bit carried over from the preceding addition.

We now decribe how such an addition can be done by a finite-state machine M. The set S of internal states of M is $S = \{s_0, s_1\}$, where s_0 is the state in which 0 is carried over to the next addition and s_1 is the state in which 1 is carried over to the next addition. That is, for $i = 0, 1$, s_i is the state in which i is carried over to the next addition. The only possible additions of bits are $0 + 0$, $0 + 1$, $1 + 0$ and $1 + 1$. We indicate this by letting $I = \{00, 01, 10, 11\}$. When two (or three) bits are added, we either write down 0 or 1. This is indicated by letting $O = \{0, 1\}$. In each case, either 0 or 1 is carried over to the next addition. This is indicated by a transition to a new state.

The state table of this finite-state machine is shown below.

	f				g			
	00	01	10	11	00	01	10	11
s_0	s_0	s_0	s_0	s_1	0	1	1	0
s_1	s_0	s_1	s_1	s_1	1	0	0	1

From this table, we see that $f(s_1, 10) = s_1$ and $g(s_1, 10) = 0$. What this says is that the machine is at the state s_1 and that 10 has been input. Since the machine is at state s_1, the bit 1 has been carried over from the preceding addition. The input 10 tells us that are now adding $1 + 0$. However, since 1 has been carried over, we are actually adding $1 + 1 + 0$, which is 2 in decimal or 10 in binary. Therefore, 0 is written down (which is the output) and 1 is carried over to the next addition, placing us in state s_1. The state digraph for this finite-state machine is shown in Figure 15.27.

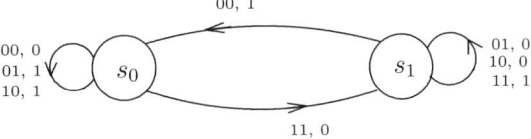

Figure 15.27: The state digraph of the finite-state machine in Example 15.23

Since computing a and b deals with adding $a_0 + b_0$, $a_1 + b_1$, $\ldots$, $a_5 + b_5$, in this order, we are dealing with the input string

$$10, 11, 01, 11, 10, 00.$$

This produces the sequence

$$s_0, \; s_0, \; s_1, \; s_1, \; s_1, \; s_1, \; s_0$$

of states and the output string

$$1 \;\; 0 \;\; 0 \;\; 1 \;\; 0 \;\; 1.$$

Since these are the coeficients $c_0, c_1, c_2, c_3, c_4, c_5$ of $2^0, 2^1, 2^2, 2^3, 2^4, 2^5$, respectively, of the sum c of a and b, it follows that $c = (101001)_2 = 41$.   ♦

Finite-State Automata

There are also finite-state machines that produce no output. These are called *finite-state automata*. The singular of *automata* is *automaton*. More formally, a **finite-state automaton** consists of

(1) a finite set S of states,

(2) a finite set I of input values,

(3) a **transition function** f that associates a next state with each state-input pair.

For example, a finite-state automaton might model the road network of a town (see Figure 15.28). Suppose that the states are the entrances to various street intersections and the input values are *turn left and drive one block* (ℓ), *turn right and drive one block* (r) and go *straight ahead for one block* (s). Whichever state one is in, applying one of the input values will lead us to a new state. An input string, such as $rss\ell ss\ell$ (turn right and drive one block, go straight for two blocks, turn left and drive one block, go straight for two blocks, turn left and drive one block) provides directions on how to drive from a given state (location in this case) to another state (our destination). This is illustrated in the diagram shown in Figure 15.28.

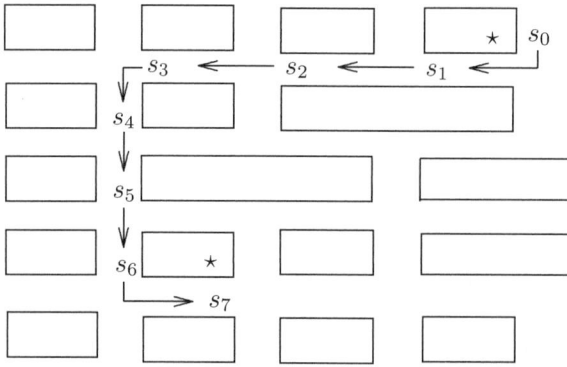

Figure 15.28: The diagram modeling the road network of a town

Here too, there are two common methods used to describe a finite-state automaton. One of these uses a table called a **state table**, while the second uses a digraph (or diagram) called a **state digraph**.

Example 15.24

The state digraph of a finite-state automaton A is shown in Figure 15.29. Thus the states of A are s_0, s_1, s_2, s_3 and the input values of A are 1, 2, 3.

(a) For the transition function f, determine $f(s_2, i)$ for $i = 1, 2, 3$.

(b) Determine the state table for A.

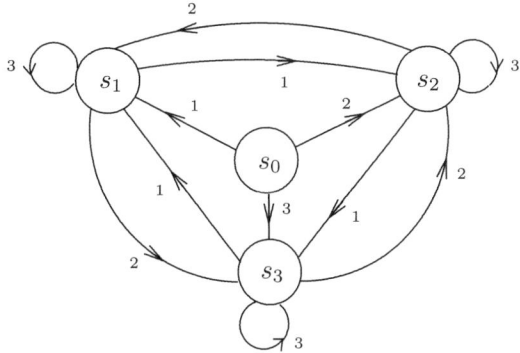

Figure 15.29: The state digraph modeling the finite-state automaton in Example 15.24

Solution.

(a) Since there is an arc labeled 1 from s_2 to s_3, an arc labeled 2 from s_2 to s_1 and a directed loop labeled 3 at s_2, it follows that $f(s_2, 1) = s_3$, $f(s_2, 2) = s_1$ and $f(s_2, 3) = s_2$.

(b) First, consider the starting state s_0. Since there is an arc labeled i from s_0 to s_i for $1 \le i \le 3$, it follows that $f(s_0, i) = s_i$. Next, consider the arcs incident with state s_1. There is an arc labeled 1 from s_1 to s_2, an arc labeled 2 from s_1 to s_3 and a directed loop labeled 3 at s_1. Thus $f(s_1, 1) = s_2$, $f(s_1, 2) = s_3$ and $f(s_1, 3) = s_1$. Continuing in this fashion, we obtain the state table for A given below. ◆

	f		
	1	2	3
s_0	s_1	s_2	s_3
s_1	s_2	s_3	s_1
s_2	s_3	s_1	s_2
s_3	s_1	s_2	s_3

Example 15.25

Determine the state digraph for the finite-state automaton whose state table is shown below.

	f		
	a	b	c
s_0	s_2	s_1	s_3
s_1	s_0	s_2	s_3
s_2	s_1	s_2	s_0
s_3	s_2	s_3	s_2

Solution. The state digraph for this machine is given in Figure 15.30. For example, consider the arcs incident with state s_0. Since $f(s_0, a) = s_2$, $f(s_0, b) = s_1$ and $f(s_0, c) = s_3$, there is an arc labeled a from s_0 to s_2, an arc labeled b from s_0 to s_1 and an arc labeled c from s_0 to s_3. On the other hand, $f(s_1, a) = s_0$ and $f(s_2, c) = s_0$. Thus there is an arc labeled a from s_1 to s_0 and an arc labeled c from s_2 to s_0. ◆

In the following example, there is a clear initial state and a unique accepting state.

Example 15.26

A man has a 6-ounce bottle of medicine. He is directed to take 3 ounces of the medicine in the morning and the remaining 3 ounces at bedtime; however, the bottle has no markings on it. The

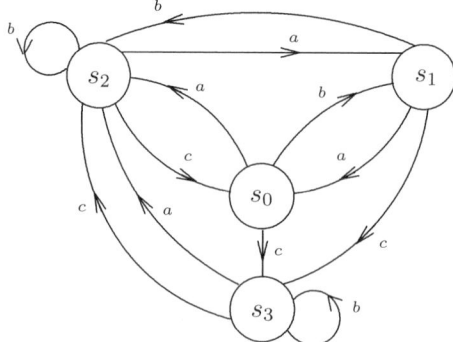

Figure 15.30: The state digraph for the finite-state machine in Example 15.25

man does possess two empty bottles – one with a capacity of 5 ounces and the other with a capacity of 1 ounce, but neither of these has markings either. It is possible to obtain two bottles containing 3 ounces of medicine each by a sequence of pourings. Draw the state digraph of the resulting finite-state automaton. Find a sequence of pourings that will result in two bottles that each contains 3 ounces of medicine.

Solution. Let B_1 denote the bottle containing the 6 ounces of medicine, let B_2 be the empty bottle with the 5-ounce capacity and let B_3 be the empty bottle with the 1-ounce capacity. For distinct integers a and b with $1 \le a, b \le 3$, the label ab on an arc indicates that the contents of bottle a is poured into bottle b, either until bottle b is filled or until bottle a is empty. If at a certain state, bottle a is empty or bottle b is filled, then no arc leaving that state will be labeled ab. Each vertex (state) is labeled as (x_1, x_2, x_3) if bottle i contains x_i ounces of medicine for $i = 1, 2, 3$. Thus $x_1 + x_2 + x_3 = 6$. The initial state is the vertex labeled $(6, 0, 0)$. The accepting state is the vertex labeled $(3, 3, 0)$. The state digraph of this finite-state automaton is shown in Figure 15.31. One sequence of pourings that results in two bottles containing 3 ounces of medicine is 12, 23, 31, 23, 31. ◆

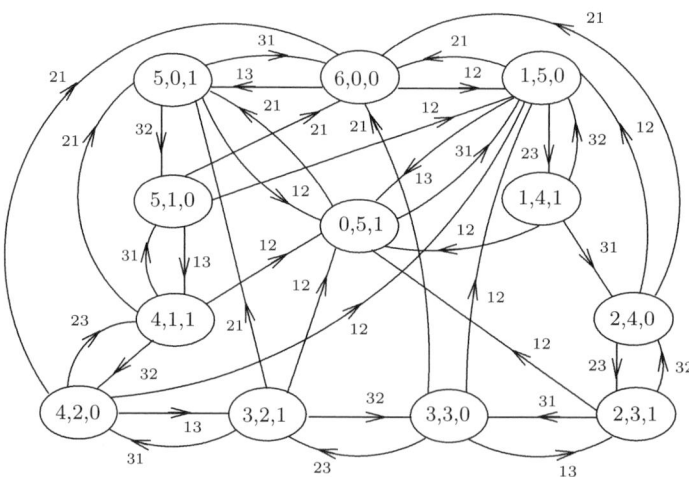

Figure 15.31: The state digraph modeling the finite-state automaton in Example 15.26

The Road Coloring Problem

There is a curious problem concerning finite-state automata called the *Road Coloring Problem*. Suppose that a certain finite-state automaton has n states and the input set consists of d input values. Then, for each state s, applying one of the d input values to s leads us to a new state s'. Thus the resulting state digraph has order n and every vertex has outdegree d.

If the vertices of a digraph D have the same outdegree, then D is said to be **out-regular** or to have **uniform outdegree**. Recall that a digraph D is strong (or strongly connected) if for every two vertices u and v of D, there are both a directed $u-v$ path and a directed $v-u$ path. A digraph D is **periodic** if there is a partition $\{V_1, V_2, \ldots, V_k\}$, $k \geq 2$, of the vertex set $V(D)$ of D such that for every arc (u,v) of D, it follows that $u \in V_i$ and $v \in V_{i+1}$ for some i with $1 \leq i \leq k$, where $V_{k+1} = V_1$. A digraph D is **aperiodic** if it is not periodic. A digraph D is aperiodic if and only if the greatest common divisor of the lengths of its directed cycles is 1.

Suppose, for example, that we have a finite-state automaton A with 8 states, where, this time, we denote the states by $s_1, s_2, \ldots, s_8$. In this instance, we have two input values, which we denote by r and b. The state digraph D of A is shown in Figure 15.32. Since A has two input values, every vertex of D has outdegree 2, that is, D is out-regular (or has uniform outdegree). Since D has directed cycles of length 3 and 4, it follows that D is aperiodic.

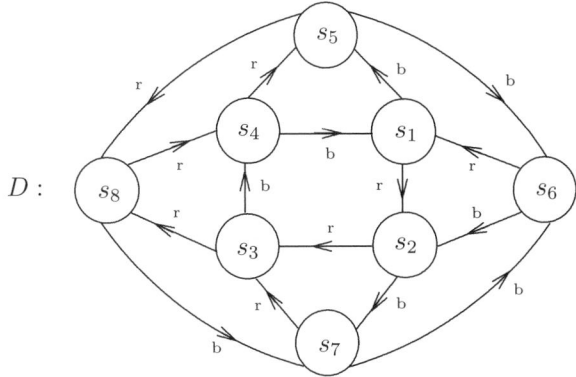

Figure 15.32: The state digraph for a finite-state automaton:
A strong aperiodic digraph with uniform outdegree

Suppose that we were to consider the input string

$$brrbrrbrr \tag{15.1}$$

and we apply this with s_5, say, chosen as the initial state. If f denotes the transition function, then $f(s_5, b) = s_6$. That is, the input value b changes the state s_5 to s_6; so we are now at the state s_6. Since $f(s_6, r) = s_1$, the state s_6 changes to s_1. Because $f(s_1, r) = s_2$, $f(s_2, b) = s_7$ and so on, we arrive at the following directed walk:

$$(s_5, s_6, s_1, s_2, s_7, s_3, s_8, s_7, s_3, s_8).$$

That is, the input string (15.1) results in a directed $s_5 - s_8$ walk. Next, suppose that we apply the same input string with the initial state s_4, say. This results in the directed walk

$$(s_4, s_1, s_2, s_3, s_4, s_5, s_8, s_7, s_3, s_8).$$

In this case, the final state is s_8 once again. In fact, if we were to apply the same input string with any state chosen as the initial state, the final state would always be s_8.

Let us now explain what the input values r and b are meant to designate. These represent colors, namely red and blue, while the states $s_1, s_2, \ldots, s_8$ represent locations. The arcs represent one-way roads. An arc labeled r is a *red road* and an arc labeled b is a *blue road*. Hence if we were to start at the location s_5 and take the blue road, this would lead us to s_6. If we then take the red road (from s_6), then we would be led to the location s_1. Taking the red road out of s_1 moves us to s_2 and so on. Indeed, the string (15.1) can be interpreted as driving directions. Following the driving directions (15.1), we can drive to s_8, regardless of where we begin out trip. Therefore, if we wanted directions on how to drive to s_8, the input string (15.1) serves as driving directions – and it is not necessary to know where we are currently located.

Suppose, however, that we were to consider the new input string

$$bbrbbrbbr \tag{15.2}$$

and apply this, once again using s_5 as the initial state. In this case, we obtain the directed walk

$$(s_5, s_6, s_2, s_3, s_4, s_1, s_2, s_7, s_6, s_1).$$

Hence a directed $s_5 - s_1$ walk is obtained in this case. If we apply the input string (15.2) with the initial state s_6, say, we obtain the directed walk

$$(s_6, s_2, s_7, s_3, s_4, s_1, s_2, s_7, s_6, s_1).$$

So in this case as well, s_1 is the final state of this directed walk. In fact, if we were to apply the input string (15.2) with any initial state, the final state is always s_1.

Perhaps surprisingly, if we were to select any state s in $V(D)$, then there is always an input string (driving directions) such that applying this string to any initial state results in a directed walk having the final state s. Hence the digraph D of Figure 15.32 has a rather unusual property. As we are about to see, this digraph is not so unusual after all.

Suppose that D is a strong aperiodic digraph with uniform outdegree d. A coloring of the arcs of D is called a **proper coloring** if no two arcs leaving the same vertex of D are colored the same. A proper coloring of D with d colors is a **synchronized coloring** if for every vertex v of D there exists a string s_v of these colors such that applying this string to each vertex u of D results in a directed $u - v$ walk. Such a string s_v is called a **synchronized sequence** for v. Thus the string (15.1) is a synchronized sequence for s_8, while the string (15.2) is a synchronized sequence for s_1. In 1970 Benjamin Weiss and Roy Adler conjectured that every strong aperiodic digraph with uniform outdegree has a synchronized coloring. During the 37 years that followed this conjecture, the problem became known as the **Road Coloring Problem**. This problem was solved in 2008 by Avraham Trahtman.

The Road Coloring Theorem *Every strong aperiodic digraph with uniform outdegree r has a synchronized coloring with r colors.*

Exercises for Section 15.3

1. Construct the state table and state digraph of a finite-state machine that models a vending machine in which two cherry cough drops or two lemon cough drops can be purchased for 15¢. The vending machine accepts nickels and dimes only. If more than 15¢ has been deposited, any amount over 15¢ is immediately returned. Once 15¢ has been deposited, one of two buttons (C for cherry and L for lemon) can be pushed to purchase cherry cough drops or lemon cough drops.

2. Construct the state digraph of a finite-state machine that models a vending machine in which a chocolate mint or chocolate cookie can be purchased for 20¢. The vending machine accepts nickels, dimes and quarters only. Whenever more than 20¢ has been deposited, the amount over 20¢ is immediately returned. Once 20¢ has been deposited, the customer can push one of two buttons (M for mint and C for cookie) to purchase either the mint or the cookie. There is a third button RD, which when pushed returns any money currently deposited.

3. Construct the state table for the finite-state machine whose state digraph is shown in Figure 15.33.

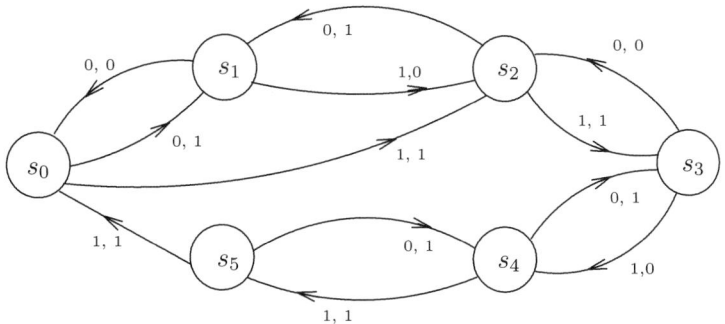

Figure 15.33: The state digraph for the finite-state machine in Exercise 3

4. Suppose that M is a finite-state machine with $S = \{s_0, s_1, s_2, s_3\}$ and $I = O = \{0, 1\}$, where the values of the next-state function f and output function g are given in the state table below.

	f		g	
	0	1	0	1
s_0	s_1	s_2	1	0
s_1	s_2	s_2	0	0
s_2	s_3	s_0	1	0
s_3	s_0	s_1	0	1

Draw the state digraph of a finite-state machine.

5. Draw the state digraphs for the finite-state machines whose state tables are given in (a)-(c). In each case, determine the output produced from the input string 1010011 with initial state s_0.

	f		g	
	0	1	0	1
s_0	s_1	s_2	0	1
s_1	s_2	s_1	0	1
s_2	s_0	s_0	0	1

(a)

	f		g	
	0	1	0	1
s_0	s_1	s_3	0	1
s_1	s_3	s_2	1	1
s_2	s_0	s_1	0	0
s_3	s_2	s_0	1	0

(b)

	f		g	
	0	1	0	1
s_0	s_0	s_3	1	1
s_1	s_2	s_1	1	0
s_2	s_0	s_3	0	0
s_3	s_2	s_1	0	0

(c)

6. For the finite-state machine whose state table is given below, find the output for the input strings in (a)-(d) with initial state s_0.

	f		g	
	0	1	0	1
s_0	s_1	s_2	1	0
s_1	s_2	s_3	1	1
s_2	s_3	s_0	1	1
s_3	s_0	s_1	0	0

(a) 01100   (b) 10101   (c) 111000   (d) 001100.

7. Give the state tables for the finite-state machines whose state digraphs are given in Figures 15.34(a)-(c). In each case, determine the output produced from the input string 01001110 with initial state s_0.

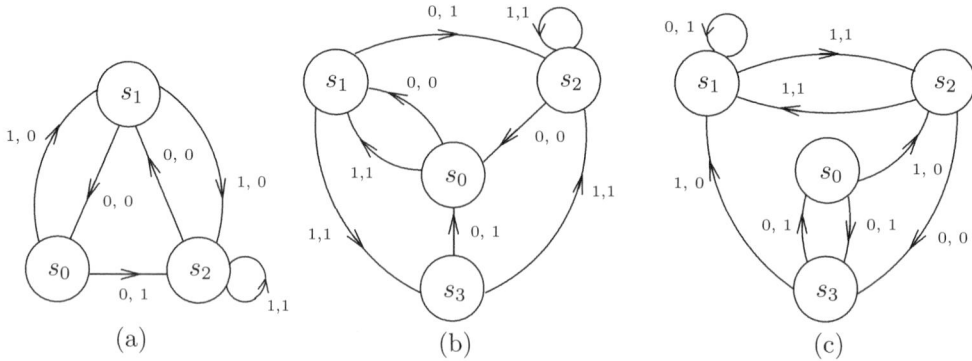

Figure 15.34: State digraphs for the finite-state machines in Exercise 7

8. For the finite-state machine whose state digraph is given in Figure 15.35, find the output for the input strings in (a)-(d) with initial state s_0.

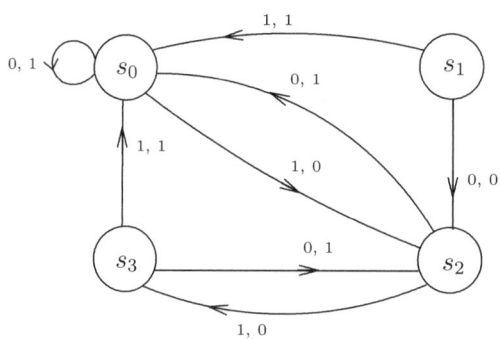

Figure 15.35: The state digraph for the finite-state machine in Exercise 8

(a) 11111   (b) 00100   (c) 101010   (d) 000111.

9. Proceeding as in Example 15.23, use the binary adder to add $a = 55$ and $b = 21$ (expressed in their decimal expansions) to compute $c = a + b$ in base 2. What is the resulting output string (which lists the bits of c in reverse order) and the sequence of states s_0 and s_1, where s_i $(i = 0, 1)$ is the state in which i is carried over to the next addition?

10. Figure 15.36 shows the state digraph D of a finite-state machine with initial state s_0.

   (a) Find an input string that results in a directed path in D whose output string is a number (expressed in base 2) of maximum value.

   (b) Find an input string that results in a directed path in D whose output string in reverse order is a number (expressed in base 2) of maximum value.

11. Suppose that a finite-state automaton has two states s_0 and s_1, two input values 0 and 1 and a transition function f defined by $f(s_0, 0) = s_0$, $f(s_0, 1) = s_1$, $f(s_1, 0) = s_1$, $f(s_1, 1) = s_0$. Construct the state table and state digraph of a finite-state automaton.

12. Determine the state table for the finite-state automaton whose state digraph is shown in Figure 15.37.

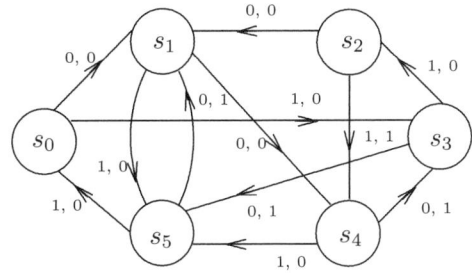

Figure 15.36: The state digraph for the finite-state machine in Exercise 10

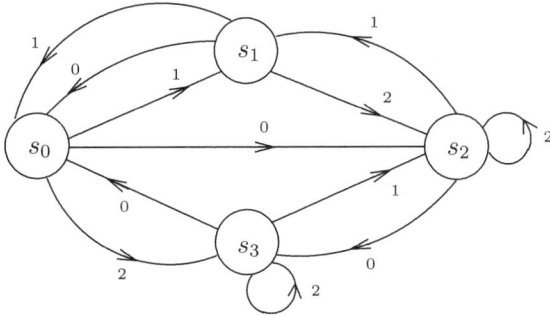

Figure 15.37: The state digraph for the finite-state automaton in Exercise 12

13. A bottle (without markings) contains 4 ounces of liquid. One empty bottle has a capacity of 3 ounces and a second empty bottle has a capacity of 1 ounce. Neither empty bottle has markings. Find a sequence of pourings that results in two bottles both containing 2 ounces of liquid. Construct the state digraph of the finite-state automaton that illustrates these pourings.

14. A bottle (without markings) contains 10 ounces of liquid. One empty bottle has a capacity of 7 ounces and a second empty bottle has a capacity of 3 ounces. Neither empty bottle has markings. Find a sequence of pourings that results in two bottles both containing 5 ounces of liquid. Construct that portion of the state digraph of this finite-state automaton that illustrates these pourings.

15. (a) Explain why there exists a synchronized coloring for the digraph D of Figure 15.38(a). (Do not find such a coloring.)

    (b) Show that the coloring of the arcs of D given in Figure 15.38(b) is not a synchronized coloring.

    (c) For the coloring of the arcs of D given in Figure 15.38(c), determine if any of the following sequences are synchronized sequences for some vertex of D.
    (i) rgb (ii) $rbgrbg$ (iii) $rrbbgg$.

    (d) Is the coloring of the arcs of the digraph in Figure 15.38(c) a synchronized coloring?

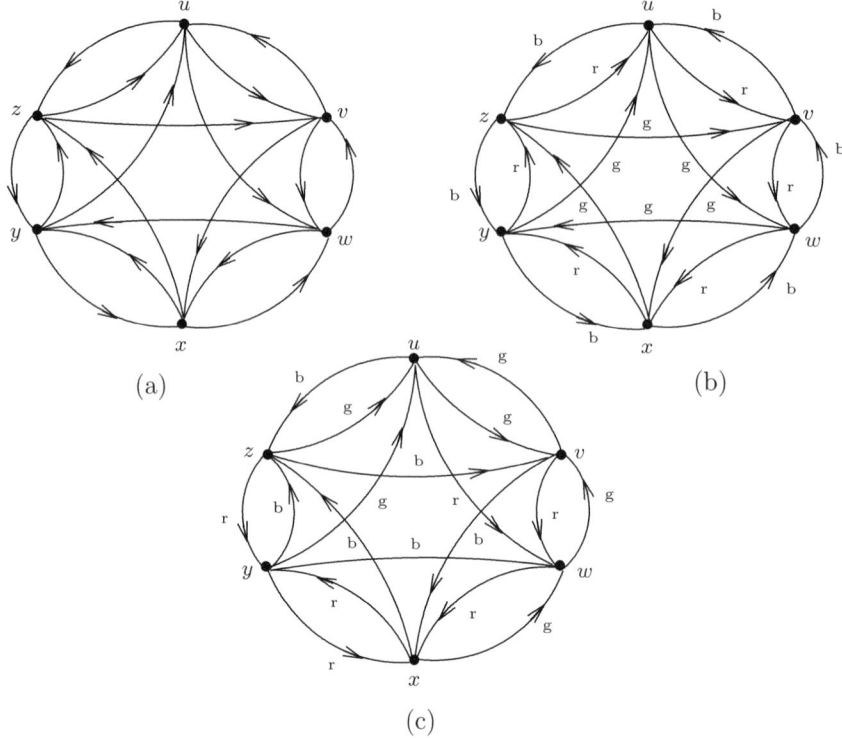

Figure 15.38: The digraphs in Exercise 15

16. Find a synchronized coloring for the digraph D of Figure 15.39.

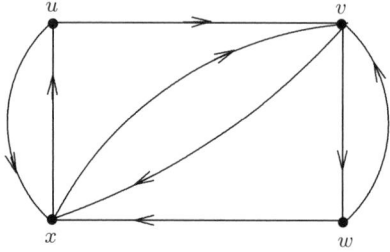

Figure 15.39: The digraph in Exercise 16

17. (a) Explain why there exists a synchronized coloring for the digraph D of Figure 15.40(a). (Do not find such a coloring.)

   (b) For the coloring of the arcs of D given in Figure 15.40(b), determine whether the following sequences are synchronized sequences for some vertex of D.
      (i) *brrbbr*   (ii) *rbrbrb*   (iii) *brbrbb*   (iv) *rrbbbb*.

   (c) Is the coloring of the arcs of D in Figure 15.40(b) a synchronized coloring?

18. Let D be a strong digraph with uniform outdegree. Show that if D is periodic, then D does not have a synchronized coloring.

19. Suppose that there is a coloring of the arcs of a digraph D and that s is a sequence of colors such that when s is applied to every vertex u of D, we arrive at a directed $u - v$ walk for a fixed vertex v of D, except for one vertex x of D, where applying s to x results in a directed $x - y$ walk, where $y \in V(D) - \{x, v\}$. Show that there is synchronized sequence for v.

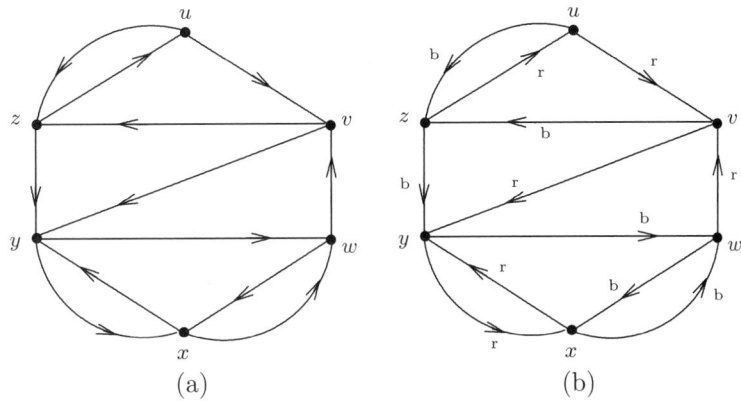

Figure 15.40: The digraph in Exercise 17

20. The digraph D of Figure 15.41(a) is aperiodic, strong and has uniform outdegree 2. Therefore, D has a synchronized coloring.

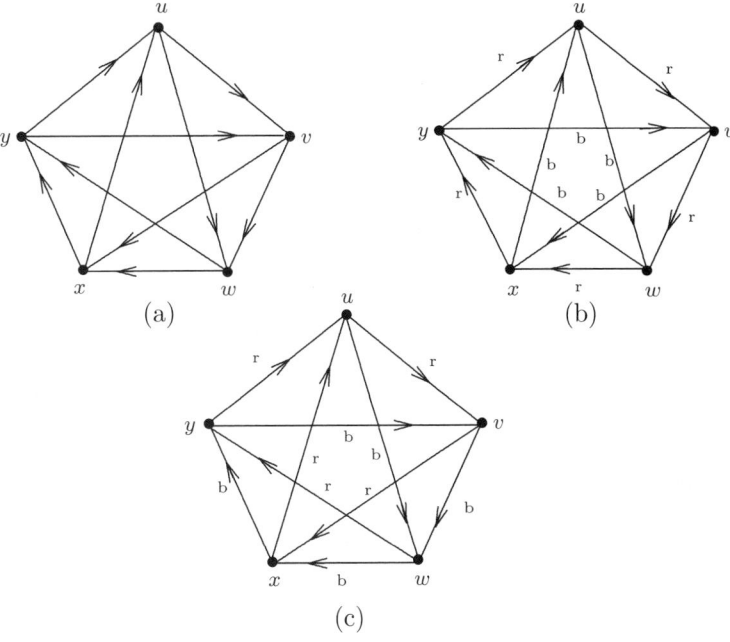

Figure 15.41: The digraph in Exercise 20

(a) Is the coloring of D in Figure 15.41(b) a synchronized coloring?

(b) The following pertains to the coloring of D in Figure 15.41(c).

   (i) Show that $rbbrrbbr$ is a synchronized sequence for some vertex of D.

   (ii) Show that $bbrrbbrr$ is a synchronized sequence for some vertex of D.

   (iii) Show that $brrb$ is a synchronized sequence for some vertex of D.

   (iv) Show that $rrbb$ is not a synchronized sequence for any vertex of D.

   (v) Show that $rrrbbb$ is not a synchronized sequence for any vertex of D.

   (vi) Show that the coloring of D in Figure 15.41(c) is a synchronized coloring of D.

Chapter 15 Highlights

Key Concepts

accepting state: a state in a finite-state machine that is a final state of an appropriate input string.

adjacency matrix (of a digraph D): For $V(D) = \{v_1,\ v_2,\ \ldots,\ v_n\}$, $\mathbf{A} = [a_{ij}]$, where $a_{ij} = 1$ if $(v_i, v_j) \in E(D)$ and $a_{ij} = 0$ if $(v_i, v_j) \notin E(D)$.

aperiodic digraph: a digraph that is not periodic.

binary adder: a finite-state machine that computes the sum of two positive integers expressed in base 2.

directed distance (from a vertex u to a vertex v in a digraph) $\vec{d}(u, v)$: the minimum length of a directed $u - v$ path in the digraph.

finite-state automaton: a structure consisting of (1) a finite set S of states, (2) a finite set I of input values, (3) a transition function f that associates a next state with each state-input pair.

finite-state machine: a structure consisting of (1) a finite set S of internal states, (2) a finite input set I, (3) a finite output set O, (4) a next-state function $f : S \times I \to S$, (5) an output function $g : S \times I \to O$.

Hamiltonian cycle (in a digraph D): a directed cycle containing every vertex of D.

Hamiltonian digraph: a digraph containing a Hamiltonian cycle.

Hamiltonian path (in a digraph D): a directed path containing every vertex of D.

indegree (of a vertex v), id v: the number of vertices from which v is adjacent.

input set: a finite set of inputs.

isomorphic (two digraphs D_1 and D_2 are isomorphic): if there exists a one-to-one correspondence $\phi : V(D_1) \to V(D_2)$ such that $(u_1, v_1) \in E(D_1)$ if and only if $(\phi(u_1), \phi(v_1)) \in E(D_2)$.

length (of a directed path): the number of directed edges in the path.

next-state function (or **transition function**): a function in a finite-state machine or a finite-state automaton that associates a next state with each state-input pair.

orientation (of a graph): an oriented graph obtained by assigning a direction to each edge of the graph.

oriented graph: a digraph such that for each pair u, v of distinct vertices, at most one of (u, v) and (v, u) is a directed edge.

outdegree (of a vertex v), od v: the number of vertices to which v is adjacent.

output function: a function in a finite-state machine that assigns an output to each state-input pair.

output set: a finite set of outputs.

out-regular digraph: a digraph in which all vertices have the same outdegree.

periodic digraph: a digraph D with the property that there is a partition $\{V_1, V_2, \ldots, V_k\}$, $k \geq 2$, of the vertex set $V(D)$ such that for every arc (u, v) of D, it follows that $u \in V_i$ and $v \in V_{i+1}$ for some i with $1 \leq i \leq k$, where $V_{k+1} = V_1$.

proper coloring (in a digraph D): a coloring of the arcs of D such that no two arcs leaving the same vertex of D are colored the same.

Road Coloring Problem: the problem of finding a synchronized coloring in a strong aperiodic digraph with uniform outdegree.

state diagram: a diagram representing the states and functions in a finite-state machine or a finite-state automaton.

state digraph (or **transition digraph**): a digraph representing the states, next-state function and output function in a finite-state machine or a finite-state automaton.

state table (or **transition table**): a table describing the states, next-state function and output function in a finite-state machine or a finite-state automaton.

strong digraph (or **strongly connected digraph**): a digraph D with the property that if for every two vertices u and v, there is both a directed $u - v$ path and a directed $v - u$ path in D.

synchronized coloring: a proper coloring of the arcs of a strong aperiodic digraph D with uniform outdegree d with d colors such that for every vertex v of D there exists a string s_v of these colors such that applying this string to each vertex u of D results in a directed $u - v$ walk.

synchronized sequence: for a given vertex v of a digraph D, a string of colors in a synchronized coloring of D such that applying this string to each vertex u of D results in a directed $u - v$ walk.

tournament: an orientation of a complete graph.

transition digraph (or **state digraph**): a digraph modeling the states and functions in a finite-state machine or a finite-state automaton.

transition function: a function in a finite-state machine or a finite-state automaton that associates a next state with each state-input pair.

transition table (or **state table**): a table describing the states and functions in a finite-state machine or a finite-state automaton.

transitive tournament: a tournament T with the property that if whenever (u, v) and (v, w) are arcs of T, then (u, w) is an arc of T.

uniform outdegree: the property of a digraph having all vertices with the same outdegree.

Key Results

- **The First Theorem of Digraph Theory**: If D is a digraph of size m, then

$$\sum_{v \in V(D)} \operatorname{od} v = \sum_{v \in V(D)} \operatorname{id} v = m.$$

- A nontrivial graph G has a strong orientation if and only if G is connected and contains no bridges.

- Let T be a tournament with $V(T) = \{v_1, v_2, \ldots, v_n\}$. Then $\sum_{i=1}^{n} \operatorname{od} v_i = \sum_{i=1}^{n} \operatorname{id} v_i = \binom{n}{2} = \frac{n(n-1)}{2}$.

- A tournament T is transitive if and only if no two vertices of T have the same outdegree.

- A tournament T is transitive if and only if T has no directed cycles.

- Every tournament contains a Hamiltonian path.

- A nontrivial tournament T is Hamiltonian if and only if T is strong.

- **The Road Coloring Theorem** Every strong aperiodic digraph D with uniform outdegree d has a synchronized coloring of the arcs of D with d colors.

Supplementary Exercises for Chapter 15

1. Let $S = \{1, 2, \ldots, n\}$, where $n \geq 2$ is an integer, and let $f : S \to S$ be a function with the property that $f(i) \neq i$ for every integer i with $1 \leq i \leq n$. Let D be the digraph with $V(D) = S$ such that $(i, j) \in E(D)$ if $f(i) = j$. Suppose that D contains a vertex k with id$(k) \geq 2$. Which of the following is true?

    (a) D is strong.

    (b) D is not strong.

    (c) It is impossible to determine whether D is strong.

2. There are eight students in a graduate computer science class. Each student has a web page and each web page has a link to exactly two web pages of other students in the class. Show that for each student s, there exists a student s^* such that (1) there is no link from the web page of s to the web page of s^* and (2) for each student s' for whom there is a link from the web page of s to the web page of s', there is no link from the web page of s' to the web page of s^*.

3. Let D be a digraph and for each vertex u of D, let $R(u)$ be the set of vertices **reachable** from u and let $r(u) = |R(u)|$. That is, $v \in R(u)$ if D contains a directed $u - v$ path. Since $u \in R(u)$ for every vertex u of D, it follows that $r(u) \geq 1$. Prove that if $r(x) \neq r(y)$ for every two distinct vertices x and y of D, then D contains a Hamiltonian path.

4. (a) Show that the length of a longest directed path in an orientation of a graph with chromatic number 3 is at least 2.

    (b) Show that there exists an orientation D of a graph with chromatic number 3 such that the length of a longest directed path in D is 2.

5. Prove or disprove: There exists a 4-regular graph G of order 7 and an orientation D of G such that for each vertex u of D, there exists either a directed $u - v$ path of length 1 or a directed $u - v$ path of length 2 but not both for every vertex v of D with $v \neq u$.

6. Which of the digraphs in Figure 15.42 are tournaments?

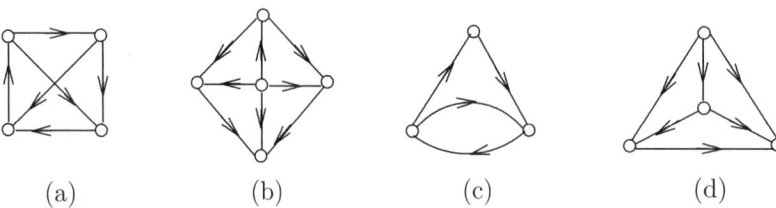

(a)   (b)   (c)   (d)

Figure 15.42: The digraphs in Exercise 6

7. Let D be a digraph with $V(D) = \{1, 3, 5, 7, 9, 11\}$ and let $A = \{4, 6\}$ and $B = \{6, 8\}$. A vertex i is adjacent to a vertex j in D if $j - i \in A$ and j is adjacent to i if $j - i \in B$. Is D strong?

8. Let T be a strong tournament of order 3 or more. Show that if (u, v) is a directed edge of T, then T has a directed cycle that contains (u, v).

9. Let (u, v) be a directed edge of a tournament T. Show that if od $v >$ od u, then (u, v) lies on a directed triangle of T.

10. Show that a tournament can contain three vertices of outdegree 1 but can never contain four vertices of outdegree 1.

11. Determine whether the tournaments T_1 and T_2 in Figure 15.43 are (a) transitive, (b) strong, (c) neither.

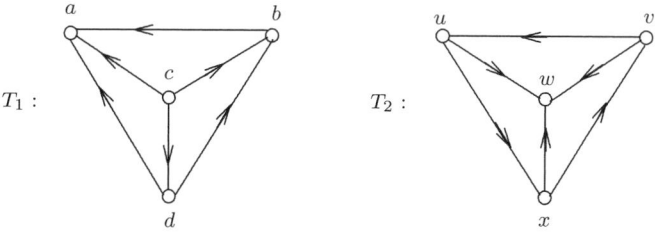

Figure 15.43: The tournaments in Exercise 11

12. Determine the minimum number t of arcs that needs to be reversed in a tournament of order $n \geq 3$ such that the resulting tournament is Hamiltonian.

13. Prove or disprove: If T is a tournament of order $n \geq 4$ that is not transitive, then T is strong.

14. Let T be a tournament of odd order n. Which of the following is true?

    (a) T contains a vertex u such that od $u >$ id u and a vertex v such that od $v <$ id v.

    (b) For every vertex w of T, od $w =$ id w.

    (c) It is impossible to determine whether (a) or (b) occurs.

15. Let T be a tournament of even order n. Which of the following is true?

    (a) T contains a vertex u such that od $u >$ id u and a vertex v such that od $v <$ id v.

    (b) For every vertex w of T, od $w =$ id w.

    (c) It is impossible to determine whether (a) or (b) occurs.

16. Let T be a tournament of order $n \geq 10$. Suppose that T contains two vertices u and v such that when the directed edge joining u and v is removed, the resulting digraph D does not contain a directed $u - v$ path or a directed $v - u$ path. Show that $\text{od}_D u = \text{od}_D v$.

17. What is the largest positive integer k such that for every tournament T with $V(T) = \{v_1, v_2, \ldots, v_n\}$, $\sum_{i=1}^{n} (\text{od } v_i)^k = \sum_{i=1}^{n} (\text{id } v_i)^k$?

18. Let u and v be two vertices in a tournament T. Prove that if $\vec{d}(u, v) = k \geq 2$, then T contains a cycle of order ℓ for each integer i with $3 \leq \ell \leq k + 1$.

19. Let u and v be two vertices in a tournament T. Prove that if u and v do not lie on a common cycle, then od $u \neq$ od v.

20. A high school class of 28 students is to elect a president of the class. Three students Allison (A), Benjamin (B) and Chandra (C) have agreed to be nominated. A ballot consists of six possible choices depending on a student's 1st preference, 2nd preference and 3rd preference for president:

choice 1	choice 2	choice 3	choice 4	choice 5	choice 6
A	A	B	B	C	C
B	C	A	C	A	B
C	B	C	A	B	A

The outcome is as follows: choice 1 (6 votes), choice 2 (5 votes), choice 3 (2 votes), choice 4 (7 votes), choice 5 (4 votes), choice 6 (4 votes). Construct a tournament T with $V(T) = \{A, B, C\}$, where (X, Y) is a directed edge of T if X is preferred over Y by the 28 students. What should the outcome of the election be?

21. Let T be a nontrivial tournament. Let D be a digraph with $V(D) = V(T)$ such that (u, v) is a directed edge in D if T contains a directed $u - v$ path. Under what conditions is $D = T$?

22. A coloring of the arcs of a strong, aperiodic digraph D with uniform outdegree 2 is shown in Figure 15.44.

    (a) Show that $rrrbbb$ is a synchronized sequence for some vertex of D.

    (b) Show that $rrbbrb$ is a synchronized sequence for some vertex of D.

    (c) Show that the coloring of the arcs of D in Figure 15.44 is a synchronized coloring.

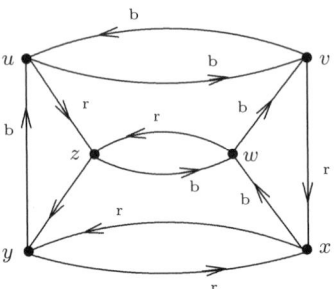

Figure 15.44: The digraph in Exercise 22

23. Figure 15.45 shows a proper coloring of the arcs of a strong, aperiodic digraph D with uniform outdegree 2.

    (a) Show that each of the following is a synchronized sequence for some vertex of D:
       (i) $rrbbrrrrbbrr$ (ii) $bbbrbbbr$ (iii) $brbrrb$
       (iv) $brbbbbrbbb$ (v) $bbrbrrbr$ (vi) $rbbrbb$

    (b) Show that the coloring of D of Figure 15.45 is a synchronized coloring.

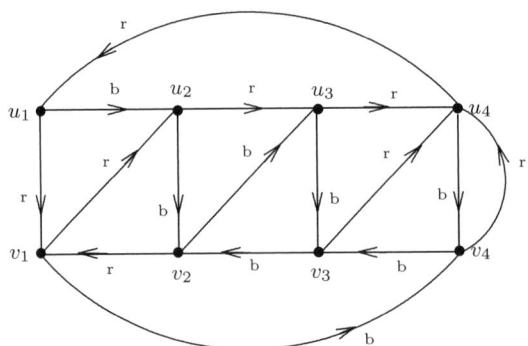

Figure 15.45: The digraph in Exercise 23

Answers and Hints to Odd-Numbered Exercises

Chapter 1: Logic

Section 1.1

1. (a) interrogative. (b) exclamatory. (c) declarative statement (T). (d) imperative (e) declarative, not a statement. (f) declarative statement (F). (g) interrogative . (h) declarative (F).

(i) declarative, not a statement.

3. (a) declarative statement (T). (b) declarative statement (F). (c) interrogative. (d) declarative, not a statement. (e) imperative. (f) declarative, not a statement. (g) exclamatory.

5. (a) $n = 2$. (b) $n = 3$.   **7.** $x = -1, 6$.

9. First, observe that $R(x, y, z)$ can only be true if (i) $x = 0$ or $x = 1$, (ii) $y = 0$ or $y = -1$ and (iii) $z = 0$ or $z = 1$. Consider these possibilities.   **11.** $x = 2$, $y = 1$.

Section 1.2

1. (a)

P	Q	R
T	T	T
T	T	F
T	F	T
T	F	F
F	T	T
F	T	F
F	F	T
F	F	F

(b) Maria prepares for the interview, does well on the interview and gets the job. Maria prepares for the interview, does well on the interview but does not get the job. Maria prepares for the interview, does not do well on the interview but gets the job. Maria prepares for the interview, does not do well on the interview and does not get the job. Maria does not prepare for the interview but does well on the interview and gets the job. Maria does not prepare for the interview, does well on the interview and does not get the job. Maria does not prepare for the interview, does not do well on the interview but gets the job. Maria does not prepare for the interview, does not do well on the interview and does not get the job.

3. (a) −1. (b) −1.   **5.** (a) 0. (b) −1. (c) No. (d) Yes.   **7.** −2.

9. (a) 15 is even. (F) (b) 15 is odd or 21 is even. (T) (c) 15 is odd and 21 is even. (F)

(d) 15 is even or 21 is even. (F) (e) 15 is odd and 21 is odd. (T)

11.

P	Q	$\sim P$	$\sim Q$	$P \wedge Q$	$\sim (P \wedge Q)$	$(\sim P) \vee (\sim Q)$
T	T	F	F	T	**F**	**F**
T	F	F	T	F	**T**	**T**
F	T	T	F	F	**T**	**T**
F	F	T	T	F	**T**	**T**

13. (a) $x \neq 0$ and $y \neq 0$. (b) a or b is negative.

15. (a) $ab > 0$ and $a + b > 0$. (b) $ab > 0$ and $a + b < 0$. (c) $ab < 0$ and $a + b > 0$. (d) $ab < 0$ and $a + b < 0$. (e) It is not the case that $ab > 0$ and $a + b > 0$. (f) $ab < 0$ or $a + b < 0$.

17. (a)

P	Q	$P \wedge Q$	$P \vee (P \wedge Q)$
T	**T**	**T**	**T**
T	**F**	F	**T**
F	**T**	F	**F**
F	**F**	F	**F**

(b) is similar.

19. (T) This can be verified by a truth table.

21. (a) $a \leq 2$ or $a > -2$. $a \leq 2$ and $a > -2$. (b) $a > 2$ and $a \leq -2$. $a > 2$ or $a \leq -2$.

Section 1.3

1. (a) T. (b) F. (c) T. (d) T.

3. (a) If $5n + 3$ is odd, then n is even. (b) If 18 is odd, then 3 is even. (T)
(c) If 13 is odd, then 2 is even. (T)   **5.** (a) T. (b) T. (c) F.

7. (a) ab is odd only if a is odd. ab is odd is sufficient for a to be odd.
(b) $x = -5$ only if $x^2 = 25$. $x = -5$ is sufficient for $x^2 = 25$.

9. Suppose that $P \Rightarrow Q$ is false. Then P is true and Q is false. However then $Q \Rightarrow P$ is true.

11. $P(n)$: $n = 1$; $Q(n)$: $n = 2$; $a = 1$, $b = 2$ and $c = 3$.

13. If $|x| = 2$, then $x = 2$. If $|x| \neq 2$, then $x \neq 2$.

15. (a) If n is even, then $7n - 8$ is odd. (b) If n is odd, then $7n - 8$ is even.

17.

P	Q	$\sim P$	$P \Rightarrow Q$	$(P \Rightarrow Q) \Rightarrow (\sim P)$
T	T	F	T	F
T	F	F	F	T
F	T	T	T	T
F	F	T	T	T

19. R and P are both true.

21. (a) If you drive over 70 miles per hour, then you (will) receive a speeding ticket. (b) If you don't receive a speeding ticket, then you are not driving over 70 miles per hour. (c) Either you don't drive over 70 miles per hour or you receive a speeding ticket. (d) If you receive a speeding ticket, then you were driving over 70 miles per hour. (e) Either you don't receive a speeding ticket or you are driving over 70 miles per hour. (f) If you don't drive over 70 miles per hour, then you won't receive a speeding ticket.

23. (a) $x \neq -2$. $x^2 \neq 4$. (b) If $x = -2$, then $x^2 = 4$. If $x^2 = 4$, then $x = -2$.
(c) If $x \neq -2$, then $x^2 \neq 4$. If $x^2 \neq 4$, then $x \neq -2$.   **25.**

P	Q	$\sim P$	$P \vee Q$	$(\sim P) \Rightarrow Q$
T	T	F	T	T
T	F	F	T	T
F	T	T	T	T
F	F	T	F	F

Section 1.4

1. (a) -5 is odd if and only if $2^3 + 1$ is even. -5 is odd is a necessary and sufficient condition for $2^3 + 1$ to be even. (b) $P \Leftrightarrow Q$ is false.

3. $3n^2$ is even if and only if n^3 is even. $3n^2$ is even is necessary and sufficient for n^3 to be even.

5. $a = 5$ and $b = 3$.   **7.** (a) $a = 0$ and $b = 2$. (b) $a = 0$ and $b = 1$.

9. (a) $a = b = c = 1$. (b) $a = b = 1$ and $c = 2$.

11. The implication $Q \Rightarrow P$ can be expressed as "P is necessary for Q," while $P \Rightarrow Q$ can be expressed as "P is sufficient for Q."

Section 1.5

1. $\sim (P \vee (\sim P)) \equiv (\sim P) \wedge P$.   **3.** (a)

P	Q	$P \Rightarrow Q$	$\sim (P \Rightarrow Q)$	$\sim (P \Rightarrow Q) \Rightarrow P$
T	T	T	F	T
T	F	F	T	T
F	T	T	F	T
F	F	T	F	T

5.

P	Q	$\sim P$	$\sim Q$	$P \Rightarrow Q$	$(P \Rightarrow Q) \wedge (\sim Q)$	$((P \Rightarrow Q) \wedge (\sim Q)) \Rightarrow (\sim P)$
T	T	F	F	T	F	T
T	F	F	T	F	F	T
F	T	T	F	T	F	T
F	F	T	T	T	T	T

7.

P	Q	$\sim P$	$\sim Q$	$P \wedge Q$	$(\sim P) \vee (\sim Q)$	$(P \wedge Q) \Leftrightarrow ((\sim P) \vee (\sim Q))$
T	T	F	F	T	F	**F**
T	F	F	T	F	T	**F**
F	T	T	F	F	T	**F**
F	F	T	T	F	T	**F**

9.

P	Q	$P \vee Q$	$\sim P$	$(P \vee Q) \Rightarrow (\sim P)$
T	T	T	F	**F**
T	F	T	F	**F**
F	T	T	T	**T**
F	F	F	T	**T**

11. If $(\sim R) \Rightarrow S$ is true and S is false, then $\sim R$ is false and R is true.   **13.** $(P \vee Q) \Leftrightarrow R$.

Section 1.6

1. Light both ends of wick #1 and one end of wick #2 at the same time.

3. First, place three coins on each balance pan and consider the cases when they balance or don't.

5. Not necessarily.   **7.** Only Puzzles 1 and 2 have solutions.   **9.** (a) No. (b) Yes.

Supplementary Exercises for Chapter 1

1. (a) true statement. (b) open sentence. (c) true statement. (d) false statement.

(e) open sentence.   **3.** (a) No values of x. (b) All (positive integer) values of x.

5. (a) $n = 1$. (b) n is one of the integers $-1, 0, 1, 2, 3$. (c) n is one of the integers $-1, 0, 2, 3$.

7. (a) $4^2 = 2^4$. (T) (b) $4^2 \neq 2^4$ or $1 + 2 + 3 = 1 \cdot 2 \cdot 3$. (T)

(c) $4^2 \neq 2^4$ and $1 + 2 + 3 = 1 \cdot 2 \cdot 3$. (F) (d) If $4^2 \neq 2^4$, then $1 + 2 + 3 = 1 \cdot 2 \cdot 3$. (T)

9. (a) ab is odd only if a is odd and b is odd. (b) ab is odd is sufficient for a and b to be odd. (c) a is odd and b is odd is necessary for ab to be odd. (d) It is not the case that if ab is odd, then a is odd and b is odd. (e) If a is odd and b is odd, then ab is odd. (f) If a is even or b is even, then ab is even.   **11.** (a) Yes. (b) No. (c) No.

13. (a) If I didn't get the job, then I will go back to college. (b) Either I got the job or I will go back to college. (c) If I get the job, then I will not go back to college. (d) I didn't get the job and I will go back to college.

15. (a) If it is Friday, then I don't go to class. (b) If I go to class, then it is not Friday.

(c) It is Friday and I don't go to class. (d) It is Friday or I go to class.

17. (a) If you get an A in this course, then you do every problem in this book. If you don't get an A in this course, then you don't do every problem in this book.

(b) If a is even and b is odd, then $a + b$ is odd. If a is odd or b is even, then $a + b$ is even.

(c) If $a^2 > 1$, then $a > 1$. If $a^2 \leq 1$, then $a \leq 1$.   **19.** (a) $a = 0$. (b) $b = 1$.

21. Observe that

$$
\begin{aligned}
\sim (P \vee ((\sim P) \wedge Q)) &\equiv (\sim P) \wedge (\sim ((\sim P) \wedge Q)) \equiv (\sim P) \wedge ((\sim (\sim P)) \vee (\sim Q)) \\
&\equiv (\sim P) \wedge (P \vee (\sim Q)) \equiv ((\sim P) \wedge P) \vee ((\sim P) \vee (\sim Q)) \\
&\equiv (\sim P) \vee (\sim Q).
\end{aligned}
$$

23. (a) If the statement is false, then P is true. If the statement is true, then P is false.

(b) If the statement is true, then P is true. If the statement is false, then P is false.

(c) If the statement is false, then P is true. If the statement is true, then P is false.

25. One bottle has three pills remaining and the other has four pills. Take one pill of each type for the next three days and then place the remaining pill with the three pills on the table. Break each of those four pills in half and take half of each pill on day 9 and the remaining four half-pills on day 10.   **27.** 2.

Chapter 2: Sets

Section 2.1

1. (a) $\{0, 1\}$. (b) $\{-1, 0, 1\}$. (c) $\{-3, -2, -1, 0, 1, 2, 3, 4\}$. (d) $\{1, 2, 3, 4\}$. (e) $\emptyset$.

3. (a) $\{n \in \mathbb{Z} : |n| \leq 3\}$. (b) $\{n \in \mathbb{N} : \sqrt{n} \in \mathbb{Z}\}$. (c) $\{n \in \mathbb{Z} : -3 < n < 5\}$.

5. 2 is an element of the sets B and C only.   **7.** (a) 5. (b) 16. (c) 16. (d) 3. (e) 2. (f) 2.

9. (a) $A_1 = A_2$. (b) $B_1 \neq B_2$. (c) $C_1 \neq C_2$.   **11.** (a) T. (b) F. (c) T. (d) F. (e) T. (f) F.

13. See Figure 1.

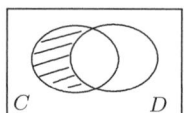

Figure 1: Venn diagram for Exercise 13

15. $A = \{1, 2\}$, $B = \{2, 3\}$, $C = \{1, 3\}$.   **17.** $A = \{1\}$, $B = \{1, 2\}$, $C = \{1, 2\}$.

19. (a) $A = \{1\}$ $B = \{1\}$, $C = \{1, 2\}$. (b) $A = \{1\}$, $B = \{A\}$, $C = \{B\}$.
(c) $A = \{1\}$, $B = \{A\}$, $C = \{1, 2\}$.

21. (a) $\mathcal{P}(A) = \{\emptyset, \{\emptyset\}, \{a\}, A\}$. (b) $\mathcal{P}(B) = \{\emptyset, \{\emptyset\}, \{\{a\}\}, B\}$. (c) $C = \mathcal{P}(B)$ for the set B in (b).

23. $2^5 = 32$.   **25.** (a) $B = \{\{1\}, \{2\}, \{3\}, \{4\}, \{5\}\}$. (b) $B = \{1, 2, 3, 4, 5\}$.

Section 2.2

1. (a) $\{3, 7\}$. (b) $\{1, 2, \ldots, 9\}$. (c) $\{1, 5, 9\}$. (d) $\{2, 4, 6, 8\}$. (e) $\{1, 2, 4, 5, 6, 8, 9\}$. (f) $\{2, 4, 6, 8, 10\}$.
(g) $\{1, 5, 9, 10\}$.   **3.** (a) $\emptyset$. (b) $\mathbb{R}$. (c) $\mathbb{Q}$. (d) $\mathbb{I}$. (e) $\mathbb{R}$. (f) $\mathbb{I}$. (g) $\mathbb{Q}$.

5. See Figure 2.   **7.** See Figure 3.

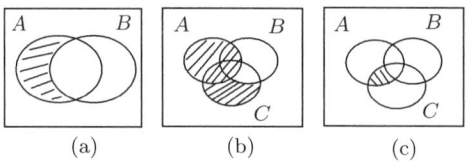

Figure 2: Venn diagrams for Exercise 5

9. (a) $\{3, 4\}$. (b) $\{2, 3, 4, 5, 6\}$. (c) $\{2\}$. (d) $\{5, 6\}$. (e) $\{2, 5, 6\}$. (f) $\{1, 5, 6, 7\}$. (g) $\{1, 2, 7\}$.
(h) $\{1, 2, 5, 6, 7\}$. (i) $\{1, 7\}$. (j) $\{1, 7\}$. (k) $\{1, 2, 5, 6, 7\}$.

11. (a) The set $A \cup B$ consists of those members of the department you have had as an instructor or with whom you have had a conversation. (b) The set $A \cap B$ consists of those members of the department you have had as an instructor and with whom you have had a conversation. (c) The set $A - B$ consists of of those members of the department you have had as an instructor but

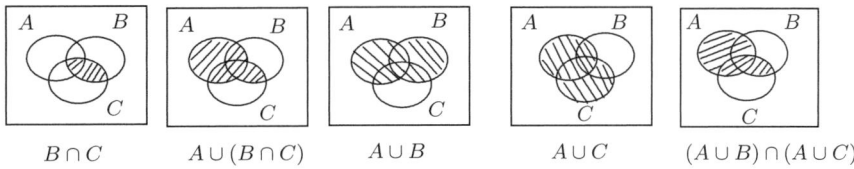

Figure 3: Venn diagrams for Exercise 7

with whom you have never had a conversation. (d) The set $B - A$ consists of those members of the department with whom you have had a conversation but you have never had as an instructor. (e) The set $A \oplus B$ consists of those members of the department with whom you have had a conversation or you have had as an instructor, but not both. (f) The set $\overline{A \cap B}$ consists of those members of the department with whom you have never had a conversation or you have never had as an instructor. (g) The set $\overline{A \cup B}$ consists of those members of the department with whom you have never had a conversation and you have never had as an instructor.

13. (a) See Figure 4. (b) First we show that $A - B \subseteq A \cap \overline{B}$. Let $x \in A - B$. Then $x \in A$ and $x \notin B$. Since $x \notin B$, it follows that $x \in \overline{B}$. Therefore, $x \in A$ and $x \in \overline{B}$; so $x \in A \cap \overline{B}$. Hence $A - B \subseteq A \cap \overline{B}$. Next we show that $A \cap \overline{B} \subseteq A - B$. Let $y \in A \cap \overline{B}$. Then $y \in A$ and $y \in \overline{B}$. Since $y \in \overline{B}$, we have that $y \notin B$. Now because $y \in A$ and $y \notin B$, we conclude that $y \in A - B$. Thus, $A \cap \overline{B} \subseteq A - B$. Therefore, $A - B = A \cap \overline{B}$.

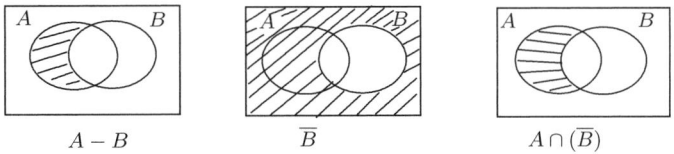

Figure 4: Venn diagrams for Exercise 13(a)

15. (a) (1), (2). (b) (3), (4). (c) (1). (d) (2). (e) (5).

17. Observe that

$$
\begin{aligned}
x \in A \cap (B \cup C) \quad &\equiv \quad x \in A \text{ and } x \in B \cup C \equiv (x \in A) \wedge ((x \in B) \vee (x \in C)) \\
&\equiv \quad ((x \in A) \wedge (x \in B)) \vee ((x \in A) \wedge (x \in C)) \\
&\equiv \quad (x \in A \cap B) \vee (x \in A \cap C) \equiv x \in (A \cap B) \cup (A \cap C).
\end{aligned}
$$

19. No.   **21.** Let $A = \{1, 2, 3, 4\}$, $B = \{1, 2, 3\}$, $C = \{1, 2, 4\}$ and $D = \{1, 3, 4\}$.

Section 2.3

1. $\{(a, a), (a, b), (a, c), (b, a), (b, b), (b, c)\}$. $\{(a, a), (a, b), (b, a), (b, b), (c, a), (c, b)\}$.
$\{(a, a), (a, b), (b, a), (b, b)\}$. $\{(a, a), (a, b), (a, c), (b, a), (b, b), (b, c), (c, a), (c, b), (c, c)\}$.

3. (a) $\{3\}$. (b) $\{(1, 3), (2, 3)\}$. (c) $\{(1, 2), (1, 3), (2, 2), (2, 3)\}$. (d) $\{(1, 3), (1, 4), (2, 3), (2, 4)\}$.
(e) $\{(1, 3), (2, 3)\}$.   **5.** $\{(a, \emptyset), (a, \{a\}), (a, \{b\}), (a, A), (b, \emptyset), (b, \{a\}), (b, \{b\}), (b, A)\}$.

7. $\{\emptyset, \{(1, 1)\}, \{(2, 1)\}, A \times B\}$.

9. (a) $\{(a, 0, x), (a, 0, y), (a, 1, x), (a, 1, y), (b, 0, x), (b, 0, y), (b, 1, x), (b, 1, y), (c, 0, x), (c, 0, y),$ $(c, 1, x), (c, 1, y)\}$. (b) $\{(x, 0, a), (x, 0, b), (x, 0, c), (x, 1, a), (x, 1, b), (x, 1, c), (y, 0, a), (y, 0, b),$ $(y, 0, c), (y, 1, a), (y, 1, b), (y, 1, c)\}$. (c) $\{(0, 0, 0), (0, 0, 1), (0, 1, 0), (0, 1, 1), (1, 0, 0), (1, 0, 1),$ $(1, 1, 0), (1, 1, 1)\}$.   **11.** Let $A = \emptyset$ and $B = \{1\}$.   **13.** $n = 1$.

15. (a) $((0,0),(0,0)),((0,1),(1,0)),\ ((0,0),(1,1)),\ ((1,1),(0,0))$. (b) 16.

Section 2.4

1. (a) Yes. (b) No (since g belongs to no element of $\mathcal{P}_2$). (c) No (since $\emptyset \in \mathcal{P}_3$).
(d) No (since b belongs to two distinct elements of $\mathcal{P}_4$). (e) Yes. (f) Yes.

3. (a) $\{\{(1,3),(1,4)\},\{(2,3)\},\{(2,4)\}\}$. (b) $\{\{\emptyset,\{1\}\},\{\{2\}\},\{A\}\}$.

5. $\mathcal{P}_1 = \{\{a\},\{b\},\{c\},\{d\}\}$. $\mathcal{P}_2 = \{\{a\},\{b\},\{c,d\}\}$. $\mathcal{P}_3 = \{\{a\},\{d\},\{b,c\}\}$. $\mathcal{P}_4 = \{\{a\},\{c\},\{b,d\}\}$.
$\mathcal{P}_5 = \{\{b\},\{d\},\{a,c\}\}$. $\mathcal{P}_6 = \{\{b\},\{c\},\{a,d\}\}$. $\mathcal{P}_7 = \{\{c\},\{d\},\{a,b\}\}$. $\mathcal{P}_8 = \{\{a,b\},\{c,d\}\}$.
$\mathcal{P}_9 = \{\{a,c\},\{b,d\}\}$. $\mathcal{P}_{10} = \{\{a,d\},\{b,c\}\}$. $\mathcal{P}_{11} = \{\{a\},\{b,c,d\}\}$. $\mathcal{P}_{12} = \{\{b\},\{a,c,d\}\}$.
$\mathcal{P}_{13} = \{\{c\},\{a,b,d\}\}$. $\mathcal{P}_{14} = \{\{d\},\{a,b,c\}\}$. $\mathcal{P}_{15} = \{A\}$.

7. $\{\{\emptyset,\{0\},\{1\},\{\{0,1\}\}\},\{\{0,1\},\{0,\{0,1\}\},\{1,\{0,1\}\}\},\{A\}\}$.

9. For $i = 1,2,\ldots k-1$, let $A_i = \{i, n-i\}$ and let $A_k = \{n\}$.

Supplementary Exercises for Chapter 2

1. (a) $\emptyset$. (b) $\{-1\}$. (c) $\emptyset$. (d) $\{0\}$. (e) $\{-2,-1,\ldots,4\}$.

3. If $A \neq B$, then there is an element in one set that is not in the other, say $x \in A$ but $x \notin B$.
Then $\{x\} \in \mathcal{P}(A)$ but $\{x\} \notin \mathcal{P}(B)$. So $\mathcal{P}(A) \neq \mathcal{P}(B)$.

5. $A = \{1,2\}$ and $B = \{1,2,4\}$.   **7.** $\{\emptyset,\{1\},\{3\},\{5\},\{1,3\},\{1,5\},\{3,5\},\{1,3,5\}\}$.

9. (a) $C \cap D$. (b) $C \cup D$. (c) $\overline{C \cup D}$. (d) $C - D$. (e) $D - C$.   **11.** (a) $\{\{x\},\{y\}\}$. (b) $\emptyset$.

13. $\{\{1,2\},\{3,4\},\{5,6\}\}$.

15. (a) $A = \{1\}$, $B = \{1,\{1\}\}$. (b) $A = \{1\}$, $B = \{\{1\}\}$. (c) $A = \{1,2\}$, $B = \{1,\{1,2\}\}$.

17. See Figure 5.   **19.** (a) $\{\{a\},\{b\}\}$. (b) $\{a,b,c\}$. (c) $\{\{a,b\},\{c,d\},\{e,f\},\{g\}\}$.

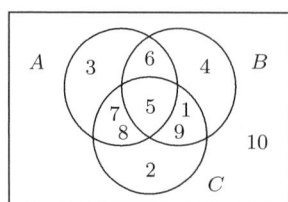

Figure 5: Venn diagram for Exercise 17

21. (a) $\{1,2,3,4,7\}$. (b) $\{1,2,4\}$. (c) $\{3,4\}$. (d) $\{1,4\}$. (e) $\{7\}$. (f) 2.

23. (a) $\{(1,1),(1,2),(1,\{2\}),(\{1\},1),(\{1\},2),(\{1\},\{2\})\}$.
(b) $\{(1,1),(1,\{1\}),(2,1),(2,\{1\}),(\{2\},1),(\{2\},\{1\})\}$. (c) $\{(1,1),(1,\{1\}),(\{1\},1),(\{1\},\{1\})\}$.
(d) $\{\emptyset,\{1\},\{\{1\}\},A\}$. (e) $\{\emptyset,\{1\},\{2\},\{\{2\}\},\{1,2\},\{1,\{2\}\},\{2,\{2\}\},B\}$.

25. (a) T. (b) F. (c) F. (d) F. (e) F. (f) F. (g) F. (h) F.

27. (a) $A_1 = \{1,2\}$, $A_2 = \{1,3\}$, $A_3 = \{2,3\}$.
(b) $B_1 = \{1,2,3\}$, $B_2 = \{1,4,5\}$, $B_3 = \{2,4,6\}$, $B_4 = \{3,5,6\}$.

Chapter 3: Methods of Proof

Section 3.1

1. (a) For every real number x, $(x-1)^2 > 0$.

   (b) There exists a real number x such that $(x-1)^2 > 0$.

3. If n is an odd integer, then $7n + 4$ is odd.

5. For every odd integer n, $n^2 + 1$ is even. There exists an odd integer n such that n^2 is even.

7. (a) There exists some set A such that $A \cap \overline{A} \neq \emptyset$. (b) For every set A, $\overline{A} \not\subseteq A$.

9. (a) There exists an irrational number s such that for every rational number r, $rs \neq 0$.

   (b) There exists a rational number r such that for every irrational number s, $rs \neq 0$.

   (c) For every even integer a there exists an integer b such that ab is odd.

   (d) For every odd integer a there exists an integer b such that ab is even.

11. (a) There exists an even integer a such that either a^2 or $a + 2$ is odd.

    (b) For every real number x, $x + 3 \leq 0$.

13. (a) There was never a time when any man owned a singing frog. (b) No odd integer is the sum of two odd integers. (c) $\sqrt{2}$ is a rational number. (d) There is a smallest positive real number. (e) There are only finitely many primes. (f) There is a rational number that cannot be expressed as the sum of two rational numbers.

15. Let S be the set of even integers, T the set of odd integers and $P(a,b)$: $(a+2)^2 + (b+3)^2 = 0$. (a) $\exists a \in S, \exists b \in T, P(a,b)$. (b) $\forall a \in S, \forall b \in T, \sim P(a,b)$.

    (c) For every even integer a and every odd integer b, $(a+2)^2 + (b+3)^2 \neq 0$.

17. Let $P(a,b)$: $ab \geq 0$. (a) $\exists a \in \mathbb{Z}, \forall b \in \mathbb{Z}, P(a,b)$. (b) $\forall a \in \mathbb{Z}, \exists b \in \mathbb{Z}, \sim P(a,b)$.

    (c) For every integer a, there exists some integer b such that $ab < 0$.

19. There exists an integer a such that for every integer b, $|\frac{2a+1}{2} - b| \geq \frac{1}{2}$.

Section 3.2

1. (a) $R(2)$: 1 is even. (F). $R(4)$: 10 is even. (T). $R(6)$: 35 is even. (F).

   (b) For each $n \in S$, $\dfrac{n^3 - n}{6}$ is even. (F). (c) There exists $n \in S$ such that $\dfrac{n^3 - n}{6}$ is even. (T).

3. **Proof.** Let x be a real number such that $(x-1)^2 = 0$. Thus $x - 1 = 0$ and so $x = 1$. Therefore, $x^3 - 1 = 1^3 - 1 = 0$. ∎

5. **Proof.** Assume that $(x - 2)^4 \leq 0$. Since $(x - 2)^4 \geq 0$, it follows that $(x - 2)^4 = 0$ and so $x - 2 = 0$. Hence $x = 2$. Therefore, $9 - x^2 = 9 - 2^2 = 5 \geq 0$. ∎

7. **Proof.** Let n be an even integer. So $n = 2k$ for some integer k. Then $7n - 2 = 7(2k) - 2 = 14k - 2 = 2(7k - 1)$. Since $7k - 1$ is an integer, $7n - 2$ is even. ∎

9. **Proof.** Let n be an odd integer. Therefore, $n = 2k + 1$ for some integer k. Then

$$7n^2 - 2n + 15 \quad = \quad 7(2k+1)^2 - 2(2k+1) + 15 = 7(4k^2 + 4k + 1) - (4k + 2) + 15$$
$$= \quad 28k^2 + 28k + 7 - 4k - 2 + 15 = 28k^2 + 24k + 20 = 2(14k^2 + 12k + 10).$$

Since $14k^2 + 12k + 10$ is an integer, $7n^2 - 2n + 15$ is even. ∎

11. Proof. Let n be a nonnegative integer in S. Then $n = 0$. Therefore, $2^{-n}(n^2+3) = 2^0(0^2+3) = 3$ is odd. ∎

13. Proof. Let r and s be rational numbers. Then $r = a/b$ and $s = c/d$, where $a, b, c, d \in \mathbb{Z}$ and $b, d \neq 0$. Therefore $5r + 7s = \frac{5a}{b} + \frac{7c}{d} = \frac{5ad+7bc}{bd}$. Since $5ad + 7bc$ and bd are integers and $bd \neq 0$, the number $5r + 7s$ is rational. ∎

15. Proof. Let r and s be rational numbers. Then $r = a/b$ and $s = c/d$, where $a, b, c, d \in \mathbb{Z}$ and $b, d \neq 0$. Therefore $rs = \frac{a}{b} \cdot \frac{c}{d} = \frac{ac}{bd}$. Since ac and bd are integers and $bd \neq 0$, the number rs is rational. ∎

17. Proof. Let P and Q be statements such that $P \wedge Q$ is true. Then P and Q are true and so $P \vee Q$ is true. ∎

19. (a) **Proof.** If T is an equilateral triangle, then its three sides have the same length and so at least two of its sides have the same length. Therefore, T is isosceles. ∎

(b) Yes by (a). (c) No. (d) No.

Section 3.3

1. Proof. Assume that n is an odd integer. Then $n = 2k + 1$ for some integer k. So $5n + 7 = 5(2k + 1) + 7 = 10k + 12 = 2(5k + 6)$. Since $5k + 6$ is an integer, $5n + 7$ is even. ∎

3. Proof. Assume that n is an even integer in $S = \{1, 2, 3\}$. Then $n = 2$. Thus $3n+4 = 3\cdot2+4 = 10$ is even. ∎

5. Proof. First, we prove that if n is an even integer, then n^3 is even. Assume that n is even. Then $n = 2k$ for some integer k. So $n^3 = (2k)^3 = 8k^3 = 2(4k^3)$. Since $4k^3$ is an integer, n^3 is even. Next, we verify the converse, namely if n^3 is even, then n is even. Let n be an odd integer. Then $n = 2\ell+1$ for some integer ℓ. Thus $n^3 = (2\ell+1)^3 = 8\ell^3 + 12\ell^2 + 6\ell + 1 = 2(4\ell^3 + 6\ell^2 + 3\ell) + 1$. Since $4\ell^3 + 6\ell^2 + 3\ell$ is an integer, n^3 is odd. ∎

7. Proof. First, we prove that if $2n^2-n-1 = 0$, then $3n^2-n-2 = 0$. Suppose that $2n^2-n-1 = 0$. Since $2n^2 - n - 1 = (2n + 1)(n - 1) = 0$ and n is an integer, $n = 1$ and so $3n^2 - n - 2 = 3\cdot1^2 - 1 - 2 = 0$. Next, we verify the converse; that is, if $3n^2 - n - 2 = 0$, then $2n^2 - n - 1 = 0$. Suppose that $3n^2 - n - 2 = 0$. Since $3n^2 - n - 2 = (3n + 2)(n - 1) = 0$ and n is an integer, it follows that $n = 1$ and so $2n^2 - n - 1 = 2 \cdot 1^2 - 1 - 1 = 0$. ∎

9. Proof. Assume that $x \geq 0$. Then $x^3 + 5x + 1 \geq 0^3 + 5 \cdot 0 + 1 = 1 > 0$. ∎

11. Proof. Assume that $x < 1$ and $y < 1$. Since x and y are integers, it follows that $x \leq 0$ and $y \leq 0$. Thus $2x + 3y \leq 2 \cdot 0 + 3 \cdot 0 = 0 < 1$. ∎

13. Proof. Suppose that $a \leq 2$, $b \leq 1$ and $c \leq 0$. Thus $c = 0$ and $a + 2b + 3c = a + 2b \leq 2 + 2 \cdot 1 = 4 < 5$. ∎

Section 3.4

1. Proof. Let n be an integer. We consider two cases.

Case 1. n is even. Then $n = 2a$ for some integer a. Therefore, $n^2 - 3n+5 = (2a)^2 - 3(2a)+5 = 4a^2 - 6a + 5 = 2(2a^2 - 3a + 2) + 1$. Since $2a^2 - 3a + 2$ is an integer, $n^2 - 3n + 5$ is odd.

Case 2. n is odd. Then $n = 2b + 1$ for some integer b. Therefore,

$$\begin{aligned} n^2 - 3n + 5 &= (2b + 1)^2 - 3(2b + 1) + 5 = (4b^2 + 4b + 1) - 6b - 3 + 5 \\ &= 4b^2 - 2b + 3 = 2(2b^2 - b + 1) + 1. \end{aligned}$$

Since $2b^2 - b + 1$ is an integer, $n^2 - 3n + 5$ is odd. ∎

3. Proof. Assume that $n^2 + n = n(n + 1) = 0$. Then $n = 0$ and $n = -1$. We consider these two cases.

Case 1. $n = 0$. Then $\frac{2^n + 3^n}{12^n} = \frac{2^0 + 3^0}{12^0} = \frac{1 + 1}{1} = 2$, which is even.

Case 2. $n = -1$. Then $\frac{2^n + 3^n}{12^n} = \frac{2^{-1} + 3^{-1}}{12^{-1}} = \frac{\frac{1}{2} + \frac{1}{3}}{\frac{1}{12}} = 12\left(\frac{1}{2} + \frac{1}{3}\right) = 6 + 4 = 10$, which is even. ∎

5. (a) Proof. Assume that m and n are of the same parity. We consider two cases.

Case 1. m and n are even. Therefore, $m = 2a$ and $n = 2b$ for integers a and b. So $m + n = (2a) + (2b) = 2(a + b)$. Since $a + b$ is an integer, $m + n$ is even.

Case 2. m and n are odd. Therefore, $m = 2a + 1$ and $n = 2b + 1$ for integers a and b. So $m + n = (2a + 1) + (2b + 1) = 2(a + b + 1)$. Since $a + b + 1$ is an integer, $m + n$ is even.

For the converse, assume that m and n are of opposite parity, say m is even and n is odd. Then $m = 2x$ and $n = 2y + 1$ for integers x and y. Thus $m + n = 2x + (2y + 1) = 2(x + y) + 1$. Since $x + y$ is an integer, $m + n$ is odd. ∎

(b) Proof. First, observe that there are six pairs of integers of S, namely, $\{a, b\}$, $\{a, c\}$, $\{a, d\}$, $\{b, c\}$, $\{b, d\}$ and $\{c, d\}$. Thus $m + n = 6$, which is even. By (a), m and n are of the same parity. ∎

7. We first show that if m and n are of opposite parity, then $7m + 3n$ is odd. We consider two cases.

Case 1. m is odd and n is even. Therefore, $m = 2a + 1$ and $n = 2b$ for integers a and b. So $7m + 3n = 7(2a + 1) + 3(2b) = 14a + 7 + 6b = 2(7a + 3b + 3) + 1$. Since $7a + 3b + 3$ is an integer, $7m + 3n$ is odd.

Case 2. m is even and n is odd. (The proof is similar to Case 1.)

Next, we show that if $7m + 3n$ is odd, then m and n are of opposite parity. We use a proof by contrapositive. Assume that m and n are of the same parity. We consider two cases.

Case 1. m and n are even. Case 2. m and n are odd.

9. First, we show that if m and n are odd, then mn^2 is odd. Let m and n be odd integers. Thus $m = 2a + 1$ and $n = 2b + 1$ for integers a and b. Therefore,

$$
\begin{aligned}
mn^2 &= (2a + 1)(2b + 1)^2 = (2a + 1)(4b^2 + 4b + 1) \\
&= 8ab^2 + 8ab + 2a + 4b^2 + 4b + 1 = 2(4ab^2 + 4ab + a + 2b^2 + 2b) + 1.
\end{aligned}
$$

Since $4ab^2 + 4ab + a + 2b^2 + 2b$ is an integer, mn^2 is odd. Next, we verify the converse, that is, if mn^2 is odd, then m and n are odd. Assume that m or n is even. We consider two cases. *Case 1. m is even* and *Case 2. n is even*

11. (a) Proof. Assume that $|2n - 1| \leq 5$. Then $-5 \leq 2n - 1 \leq 5$. Therefore, $-4 \leq 2n \leq 6$. Dividing by 2, we have $-2 \leq n \leq 3$. ∎

(b) Proof. Assume that $n > 3$ or $n < -2$. We consider these two cases.

Case 1. $n > 3$. Then $|2n - 1| = 2n - 1 > 5$.

Case 2. $n < -2$. Then $|2n - 1| = -(2n - 1) = -2n + 1 > 5$. ∎

13. Proof. First, we show that $(A - B) \cup (A - C) \subseteq A - (B \cap C)$. Let $x \in (A - B) \cup (A - C)$. Then $x \in A - B$ or $x \in A - C$. Assume, without loss of generality, that $x \in A - B$. Then $x \in A$ and $x \notin B$. Since $x \notin B$, it follows that $x \notin B \cap C$. Because $x \in A$ and $x \notin B \cap C$, we have $x \in A - (B \cap C)$. Hence $(A - B) \cup (A - C) \subseteq A - (B \cap C)$.

Next, we show that $A - (B \cap C) \subseteq (A - B) \cup (A - C)$. Let $y \in A - (B \cap C)$. Hence $y \in A$ and $y \notin B \cap C$. Since $y \notin B \cap C$, either $y \notin B$ or $y \notin C$. Assume, without loss of generality, that $y \notin B$. Since $y \in A$, we have $y \in A - B$. Therefore, $y \in (A - B) \cup (A - C)$ and so $A - (B \cap C) \subseteq (A - B) \cup (A - C)$. ∎

15. **Proof.** First we show that $(A - B) \cup (B - A) \subseteq (A \cup B) - (A \cap B)$. Let $x \in (A - B) \cup (B - A)$. Then $x \in A - B$ or $x \in B - A$. Assume, without loss of generality, that $x \in A - B$. So $x \in A$ and $x \notin B$. Thus $x \in A \cup B$ and $x \notin A \cap B$. Consequently, $x \in (A \cup B) - (A \cap B)$. Therefore, $(A - B) \cup (B - A) \subseteq (A \cup B) - (A \cap B)$.

Next we show that $(A \cup B) - (A \cap B) \subseteq (A - B) \cup (B - A)$. Let $x \in (A \cup B) - (A \cap B)$. Then $x \in A \cup B$ and $x \notin A \cap B$. Since $x \in A \cup B$, it follows that $x \in A$ or $x \in B$. Without loss of generality, let $x \in A$. Since $x \notin A \cap B$, it follows that $x \notin B$. Therefore, $x \in A - B$ and so $x \in (A - B) \cup (B - A)$. Hence $(A \cup B) - (A \cap B) \subseteq (A - B) \cup (B - A)$. ∎

Section 3.5

1. $n = 2$ is a counterexample.

3. The integers $x = -1$ and $y = -2$ form a counterexample.

5. The sets $A = \{1\}$, $B = \{1, 2\}$, $C = \{1, 3\}$ form a counterexample.

7. The sets $A = \emptyset$ and $B = \{1\}$ form a counterexample.

9. The sets $A = \{1, 2, 3\}$, $B = \{1\}$, $C = \{2\}$ form a counterexample.

11. Consider the positive integer 1.

13. (a) The statement is true. (b) The statement is false. Consider $x = 1/2$.

15. The statement is false ($n = 1$ is a counterexample).

17. The statement is false ($a = 1$ and $b = 2$ form a counterexample).

Section 3.6

1. **Proof.** Let $n = 11$. Then $100 - n^2 = 100 - (11)^2 = 100 - 121 = -21 < 0$. ∎

3. **Proof.** Let $x = \sqrt{5}$. Then $x^2 = (\sqrt{5})^2 = 5$. ∎

5. **Proof.** Let $a = \sqrt{2}$ and $b = 0$. Then $a^b = (\sqrt{2})^0 = 1 \in \mathbb{Q}$. ∎

7. **Proof.** Let $x = \sqrt{2}$. Then $x^4 - x^2 - 2 = (\sqrt{2})^4 - (\sqrt{2})^2 - 2 = 4 - 2 - 2 = 0$. ∎

9. **Proof.** Let $m = n = 3$. Then $(m - 2)^2 - (n - 3)^3 = (3 - 2)^2 - (3 - 3)^3 = 1^2 - 0^2 = 1 - 0 = 1$. ∎

11. Consider the number $\sqrt{3}^{\sqrt{2}}$ and the two cases:

Case 1. $\sqrt{3}^{\sqrt{2}}$ *is rational.* *Case 2.* $\sqrt{3}^{\sqrt{2}}$ *is irrational.*

13. (a) - (b) The statements are true. Let $a = 3$ and $b = \frac{3}{2}$.

15. The statement is false. Let $A = \{1\}$ and let B be an arbitrary set.

Section 3.7

1. **Proof.** Assume, to the contrary, that there is a largest negative rational number r. Then $r = -\frac{a}{b}$, where $a, b \in \mathbb{N}$. Let $s = \frac{r}{2} = -\frac{a}{2b}$. Since s is a negative rational number and $s > r$, this is a contradiction. ∎

3. **Proof.** Assume, to the contrary, that 101 can be expressed as the sum of two even integers, say x and y. Then $x = 2a$ and $y = 2b$ for integers a and b. Furthermore, $101 = x + y = 2a + 2b = 2(a + b)$. Since $a + b$ is an integer, 101 is even, which is a contradiction. ∎

5. Proof. Assume, to the contrary, that there exist a rational number a and an irrational number b such that $a - b$ is rational. Since a and $a - b$ are rational, $a = p/q$ and $a - b = x/y$, where $p, q, x, y \in \mathbb{Z}$ and $q, y \neq 0$. Therefore, $\frac{x}{y} = a - b = \frac{p}{q} - b$; so $b = \frac{p}{q} - \frac{x}{y} = \frac{py - xq}{qy}$. Since $py - xq$ and qy are integers and $qy \neq 0$, it follows that b is a rational number. This is a contradiction. ∎

7. Proof. Assume, to the contrary, that there is a smallest positive irrational number r. Let $s = r/2$. Then $0 < s < r$. We claim that s is irrational. Suppose that s is rational. Then $s = a/b$, where $a, b \in \mathbb{Z}$ and $b \neq 0$. Hence $r = 2s = 2a/b$ and so r is rational, which is impossible. Thus, as claimed, s is irrational. This is a contradiction. ∎

9. Proof. Assume, to the contrary, that $3n + 14$ is and n is also odd. Then $n = 2k + 1$ for some integer k. Therefore, $3n + 14 = 3(2k + 1) + 14 = 6k + 17 = 2(3k + 8) + 1$. Since $3k + 8$ is an integer, $3n + 14$ is odd. This contradicts our assumption that $3n + 14$ is even. ∎

11. Proof. Assume, to the contrary, that $\sqrt{3}$ is rational. Hence $\sqrt{3} = x/y$, where $x, y \in \mathbb{Z}$ and $y \neq 0$. Furthermore, we may assume that x/y has been reduced to lowest terms. Since $\sqrt{3} = x/y$, it follows that $3 = x^2/y^2$ and so $x^2 = 3y^2$. Since y^2 is an integer, $x = 3z$ for some integer z. Hence $(3z)^2 = 3y^2$ and so $9z^2 = 3y^2$. Thus $y^2 = 3z^2$. Since z^2 is an integer, $y = 3w$ for some integer w. Since $x = 3z$ and $y = 3w$ for integers z and w, it follows that x/y is not in lowest terms. This is a contradiction. ∎

13. Proof. Assume, to the contrary, that $\sqrt{6}$ is rational. Then $\sqrt{6} = a/b$ for nonzero integers a and b. We can further assume that a/b has been reduced to lowest terms. Thus $6 = a^2/b^2$; so $a^2 = 6b^2 = 2(3b^2)$. Because $3b^2$ is an integer, a^2 is even. Thus a is even. So $a = 2c$, where $c \in \mathbb{Z}$. Thus $(2c)^2 = 6b^2$ and so $4c^2 = 6b^2$. Therefore, $3b^2 = 2c^2$. Because c^2 is an integer, $3b^2$ is even. Thus either 3 is even or b^2 is even. Since 3 is not even, b^2 is even and so b is even. However, since a and b are both even, each has 2 as a divisor, which contradicts our assumption that a/b has been reduced to lowest terms. ∎

15. Proof. Assume, to the contrary, that two of $\sqrt{a + b}$, $\sqrt{a + c}$ and $\sqrt{b + c}$ are equal, say $\sqrt{a + b} = \sqrt{a + c}$. Squaring both sides, we get $a + b = a + c$ and so $b = c$, which is a contradiction. ∎

Supplementary Exercises for Chapter 3

1. (a) For every positive even integer n, 2^{n-2} is an even integer. (b) There exists a positive even integer n such that 2^{n-2} is an even integer. (c) There exists a positive even integer n such that 2^{n-2} is not an even integer. (d) For every positive even integer n, 2^{n-2} is not an even integer.

3. First, we show that if m is odd and n is even, then $mn + m$ is odd. Assume that m is odd and n is even. Then $m = 2a + 1$ and $n = 2b$ for integers a and b. So $mn + m = (2a + 1)(2b) + (2a + 1) = 4ab + 2b + 2a + 1 = 2(2ab + b + a) + 1$. Since $2ab + b + a$ is an integer, $mn + m$ is odd.

Next, we verify the converse, that is, if $mn + m$ is odd, then m is odd and n is even. Assume that it is not the case that m is odd and n is even. Then either m is even or n is odd. Therefore, either m is even or m and n are both odd. We consider these two cases.

Case 1. m is even. Case 2. m and n are both odd.

5. Proof. Assume that $(x^2 - 1)^2 = 0$. Thus $x^2 - 1 = 0$. Then $x^4 - x^2 = x^2(x^2 - 1) = x^2 \cdot 0 = 0$. ∎

7. Proof. Let r and s be rational numbers, where $s \neq 0$. Then $r = a/b$ and $s = c/d$, where $a, b, c, d \in \mathbb{Z}$ and $b, c, d \neq 0$. Then $\frac{r}{s} = \frac{a/b}{c/d} = \frac{ad}{bc}$. Since ad and bc are integers and $bc \neq 0$, it follows that r/s is rational. ∎

9. Proof. Assume first that n is odd. Then $n = 2a + 1$ for some integer a and so $5n + 1 = 5(2a + 1) + 1 = 10a + 5 + 1 = 10a + 6 = 2(5a + 3)$. Since $5a + 3$ is an integer, $5n + 1$ is even. For the converse, assume that n is even. Then $n = 2b$ for some integer b. Therefore, $5n + 1 = 5(2b) + 1 = 10b + 1 = 2(5b) + 1$. Since $5b$ is an integer, $5n + 1$ is odd. ∎

11. (a) The integers $a = 1$ and $b = 2$ form a counterexample.

(b) The integer $x = -1$ is a counterexample.

13. (a) **Proof.** Assume that n is an even integer. Then $n = 2k$ for some integer k. Then $5n - 7 = 5(2k) - 7 = 10k - 7 = 10k - 8 + 1 = 2(5k - 4) + 1$. Since $5k - 4$ is an integer, $5n - 7$ is odd. ∎

(b) **Proof.** Assume that $5n - 7$ is an even integer. Then $5n - 7 = 2k$ for some integer k. Thus $n = (5n - 7) + (-4n + 7) = 2k - 4n + 7 = 2(k - 2n + 3) + 1$. Since $k - 2n + 3$ is an integer, n is odd. ∎

(c) **Proof.** Assume, to the contrary, that there exists an even integer n such that $5n - 7$ is even. Since n is even, $n = 2k$ for some integer k. Then $5n - 7 = 5(2k) - 7 = 10k - 7 = 10k - 8 + 1 = 2(5k - 4) + 1$. Since $5k - 4$ is an integer, $5n - 7$ is odd, which is a contradiction. ∎

15. (a) **Proof.** Assume that n is odd. Then $n = 2k + 1$ for some integer k and so $7 - n = 7 - (2k + 1) = 6 - 2k = 2(3 - k)$. Since $3 - k$ is an integer, $7 - n$ is even. For the converse, assume that $7 - n$ is even and so $7 - n = 2\ell$ for some integer ℓ. Thus $n = 7 - 2\ell = 2(3 - 2\ell) + 1$. Since $3 - 2\ell$ is an integer, n is odd. ∎

(b) **Proof.** Assume that n is odd. Then $n = 2k + 1$ for some integer k and so $n = 6 - 2k = 2(3 - k)$. Since $3 - k$ is an integer, $7 - n$ is even. For the converse, we show that if $7 - n$ is even, then n is odd. Assume that n is even. Thus $n = 2k$ for some integer k. Hence $7 - n = 7 - 2k = 2(3 - k) + 1$. Since $3 - k$ is an integer, $7 - n$ is odd. ∎

(c) **Proof.** We first show that if n is odd, then $7 - n$ is even. Assume that $7 - n$ is odd. Then $7 - n = 2k + 1$ for some integer k and so $n = 6 - 2k = 2(3 - k)$. Since $3 - k$ is an integer, n is even. For the converse, we show that if $7 - n$ is even, then n is odd. Assume that n is even. Thus $n = 2k$ for some integer k. Hence $7 - n = 7 - 2k = 2(3 - k) + 1$. Since $3 - k$ is an integer, $7 - n$ is odd. ∎

17. **Proof.** Assume, to the contrary, that 10 can be expressed as the sum of an odd integer x and two even integers y and z. Then $x = 2a + 1$, $y = 2b$ and $z = 2c$ for integers a, b and c. Thus $10 = x + y + z = (2a + 1) + 2b + 2c = 2(a + b + c) + 1$. Since $a + b + c$ is an integer, 10 is odd, which is a contradiction. ∎

19. **Proof.** Assume, to the contrary, that there is a positive integer x such that $x < x^2 < 2x$. Dividing by x, we obtain $1 < x < 2$, which is impossible. ∎

21. **Proof.** Assume that $mn = 1$. Then $m \neq 0$ and $n \neq 0$. Hence either $m > 0$ or $m < 0$. We consider these two cases.

Case 1. $m > 0$. Since m is an integer and $mn > 0$, it follows that $m \geq 1$ and $n \geq 1$. If $m > 1$ or $n > 1$, then $mn > 1$, which would produce a contradiction. So $m = n = 1$.

Case 2. $m < 0$. Since m is an integer and $mn > 0$, it follows that $m \leq -1$ and $n \leq -1$. If either $m < -1$ and $n < -1$, then $mn > 1$, which would produce a contradiction. So $m = n = -1$. ∎

23. (a) **Proof.** First, observe that if $a = 0$ or $b = 0$, then the result holds. Thus we may assume that $a \neq 0$ and $b \neq 0$. We consider three cases.

Case 1. $a > 0$ and $b > 0$. Then $ab > 0$. So $|ab| = ab = |a||b|$.

Case 2. $a < 0$ and $b < 0$. Then $ab > 0$. So $|ab| = ab = (-a)(-b) = |a||b|$.

Case 3. Exactly one of a and b is positive and the other is negative, say $a > 0$ and $b < 0$. Then $ab < 0$. So $|ab| = -(ab) = a(-b) = |a||b|$. ∎

(b) Since $|a + b| = |a| + |b|$ if either a or b is 0, we may assume that a and b are nonzero. We proceed by cases.

Case 1. $a > 0$ and $b > 0$. Then $a + b > 0$ and $|a + b| = a + b = |a| + |b|$.

Case 2. $a < 0$ *and* $b < 0$. Since $a + b < 0$, it follows that $|a + b| = -(a + b) = (-a) + (-b) = |a| + |b|$.

Case 3. *Exactly one of a and b is positive and the other is negative, say $a > 0$ and $b < 0$.* We consider two subcases. *Subcase* 3.1. $a + b \geq 0$. *Subcase* 3.2. $a + b < 0$.

25. Assume that a is an odd integer and b is an even integer. Then $a = 2x + 1$ and $b = 2y$ for some integers x and y. Show that $3a + 5b - 4$ is odd.

27. Proof. Let $a, b \in \mathbb{R}$. Assume, without loss of generality, that $a \leq b$. Let $m = \min(a, b)$ and $M = \max(a, b)$. Then $m = a$ and $M = b$. Let $r, s \in \mathbb{R}$ such that $r \leq m$ and $s \geq M$. Then $r \leq m = a \leq b$ and $s \geq M = b \geq a$. ∎

29. Proof. Assume, without loss of generality, that $a \leq b$. Thus $\min(a, b) = a$ and $\max(a, b) = b$. Since $a \leq b$, it follows that $2a = a + a \leq a + b \leq b + b = 2b$ and so $2a \leq a + b \leq 2b$. Dividing by 2, we obtain $\min(a, b) = a \leq \frac{a+b}{2} \leq b = \max(a, b)$. ∎

31. Consider $a = 1$ and $b = \sqrt{2}$.

33. Assume, to the contrary, that $\sqrt[3]{2}$ is rational. Hence $\sqrt[3]{2} = x/y$, where $x, y \in \mathbb{Z}$ and $y \neq 0$. Furthermore, we may assume that x/y has been reduced to lowest terms. Produce a contradiction.

35. Proof. The numbers $\sqrt[3]{2}$ and $\sqrt[3]{3}$ are both irrational. We consider the number $\sqrt[3]{2}^{\sqrt[3]{3}}$. There are two cases, according to whether $\sqrt[3]{2}^{\sqrt[3]{3}}$ is rational or irrational.

Case 1. $\sqrt[3]{2}^{\sqrt[3]{3}}$ *is rational.* Then $a = \sqrt[3]{2}$ and $b = \sqrt[3]{3}$ have the desired properties.

Case 2. $\sqrt[3]{2}^{\sqrt[3]{3}}$ *is irrational.* We consider two subcases.

Subcase 2.1. $\left(\sqrt[3]{2}^{\sqrt[3]{3}} \right)^{\sqrt[3]{3}}$ *is rational.* Then $a = \sqrt[3]{2}^{\sqrt[3]{3}}$ and $b = \sqrt[3]{3}$ have the desired properties.

Subcase 2.2. $\sqrt[3]{2}^{\sqrt[3]{9}}$ *is irrational.* Consider $\left(\sqrt[3]{2}^{\sqrt[3]{9}} \right)^{\sqrt[3]{3}} = 2$. Then $a = \sqrt[3]{2}^{\sqrt[3]{9}}$ and $b = \sqrt[3]{3}$ have the desired properties. ∎

37. Assume, to the contrary, that $\sqrt{3} + \sqrt{5}$ is rational. Then $\sqrt{3} + \sqrt{5} = \frac{a}{b}$, where $a, b \in \mathbb{N}$. Thus $\sqrt{5} = \frac{a}{b} - \sqrt{3}$. Square both sides and obtain a contradiction.

39. Since a and b are of opposite parity, we may assume that a is even and b is odd. For the converse, we use a proof by contrapositive.

Chapter 4: Mathematical Induction

Section 4.1

1. Proof. We proceed by induction. Since $\frac{1}{2\cdot3} = \frac{1}{2\cdot1+4}$, the formula holds for $n = 1$. Assume that $\frac{1}{2\cdot3} + \frac{1}{3\cdot4} + \cdots + \frac{1}{(k+1)(k+2)} = \frac{k}{2k+4}$ where k is a positive integer. We show that $\frac{1}{2\cdot3} + \frac{1}{3\cdot4} + \cdots + \frac{1}{(k+2)(k+3)} = \frac{k+1}{2k+6}$. Observe that

$$
\begin{aligned}
\frac{1}{2 \cdot 3} + \frac{1}{3 \cdot 4} + \cdots + \frac{1}{(k+2)(k+3)} &= \left[\frac{1}{2 \cdot 3} + \frac{1}{3 \cdot 4} + \cdots + \frac{1}{(k+1)(k+2)} \right] + \frac{1}{(k+2)(k+3)} \\
&= \frac{k}{2k+4} + \frac{1}{(k+2)(k+3)} = \frac{k(k+3)+2}{2(k+2)(k+3)} = \frac{k^2+3k+2}{2(k+2)(k+3)} \\
&= \frac{(k+1)(k+2)}{2(k+2)(k+3)} = \frac{k+1}{2k+6}.
\end{aligned}
$$

By the Principle of Mathematical Induction, $\frac{1}{2\cdot3} + \frac{1}{3\cdot4} + \cdots + \frac{1}{(n+1)(n+2)} = \frac{n}{2n+4}$ for every positive integer n. ∎

3. Proof. We proceed by induction. Since $\frac{1}{1\cdot4} = \frac{1}{3+1}$, the formula holds for $n = 1$. Assume that $\frac{1}{1\cdot4} + \frac{1}{4\cdot7} + \cdots + \frac{1}{(3k-2)(3k+1)} = \frac{k}{3k+1}$ where k is an arbitrary positive integer. We show that $\frac{1}{1\cdot4} + \frac{1}{4\cdot7} + \cdots + \frac{1}{(3k+1)(3k+4)} = \frac{k+1}{3k+4}$. Observe that

$$
\begin{aligned}
\frac{1}{1\cdot4} + \frac{1}{4\cdot7} + \cdots + \frac{1}{(3k+1)(3k+4)} &= \left[\frac{1}{1\cdot4} + \frac{1}{4\cdot7} + \cdots + \frac{1}{(3k-2)(3k+1)}\right] + \frac{1}{(3k+1)(3k+4)} \\
&= \frac{k}{3k+1} + \frac{1}{(3k+1)(3k+4)} = \frac{k(3k+4)+1}{(3k+1)(3k+4)} = \frac{3k^2+4k+1}{(3k+1)(3k+4)} \\
&= \frac{(k+1)(3k+1)}{(3k+1)(3k+4)} = \frac{k+1}{3k+4}.
\end{aligned}
$$

By the Principle of Mathematical Induction, $\frac{1}{1\cdot4} + \frac{1}{4\cdot7} + \cdots + \frac{1}{(3n-2)(3n+1)} = \frac{n}{3n+1}$ for every positive integer n. ∎

5. We proceed by induction. Since $1 = 1(2-1)$, the formula holds for $n = 1$. Assume that $1 + 5 + 9 + \cdots + (4k-3) = k(2k-1)$ for an arbitrary positive integer k. Next show that $1 + 5 + 9 + \cdots + (4k+1) = (k+1)(2k+1)$.

7. We proceed by induction. Since $1^2 = \frac{1\cdot2\cdot3}{6}$, the formula holds for $n = 1$. Assume that $1^2 + 2^2 + \cdots + k^2 = \frac{k(k+1)(2k+1)}{6}$ for an arbitrary positive integer k. Next show that $1^2 + 2^2 + \cdots + (k+1)^2 = \frac{(k+1)(k+2)(2k+3)}{6}$.

9. We use induction. Since $1 + 3 = 4 = \frac{3^2-1}{2}$, the formula holds for $n = 1$. Assume that $1 + 3 + 3^2 + \cdots + 3^k = \frac{3^{k+1}-1}{2}$ for an arbitrary positive integer k. Next show that $1 + 3 + 3^2 + \cdots + 3^{k+1} = \frac{3^{k+2}-1}{2}$.

11. We use induction. Since $1 \cdot 2 = \frac{1(1+1)(1+2)}{3}$, the formula holds for $n = 1$. Assume that $1 \cdot 2 + 2 \cdot 3 + 3 \cdot 4 + \cdots + k(k+1) = \frac{k(k+1)(k+2)}{3}$ for a positive integer k. Next show that $1 \cdot 2 + 2 \cdot 3 + 3 \cdot 4 + \cdots + (k+1)(k+2) = \frac{(k+1)(k+2)(k+3)}{3}$.

13. (a) For $n = 1$, $(n+2)(n+3)/2 = 6$, which is even. (b) Assume that $(k+2)(k+3)/2$ is even where $k \in \mathbb{N}$. Then $(k+2)(k+3)/2 = 2x$ for some integer x. Thus $\frac{k^2+5k+6}{2} = 2x$. Hence $\frac{(k+3)(k+4)}{2} = \frac{k^2+7k+12}{2} = \frac{(k^2+5k+6)+(2k+6)}{2} = \frac{k^2+5k+6}{2} + k + 3 = 2x + k + 3$. This is not even when $k = 2$, say. (c) When $k = 2$, $(k+2)(k+3)/2 = 10$ and $(k+3)(k+4)/2 = 15$. Hence the statement in (b) is false, which is the induction step in an induction proof which attempts to prove that $(n+2)(n+3)/2$ is even for every positive integer n.

Section 4.2

1. Proof. We proceed by induction. The inequality holds for $n = 0$ since $2^0 = 1 > 0$. Assume that $2^k > k$, where k is a nonnegative integer. We show that $2^{k+1} > k + 1$. When $k = 0$, we have $2^{k+1} = 2 > 1 = k+1$. We therefore assume that $k \geq 1$. Then $2^{k+1} = 2\cdot2^k > 2k = k+k \geq k+1$. By the Principle of Mathematical Induction, $2^n > n$ for every nonnegative integer n. ∎

3. Proof. We proceed by induction. Since $4! = 24 > 16 = 4^2$, the inequality holds for $n = 4$. Assume that $k! > k^2$ for an arbitrary integer $k \geq 4$. We show that $(k+1)! > (k+1)^2$. Now

$$
\begin{aligned}
(k+1)! &= (k+1)k! > (k+1)k^2 = k^3 + k^2 \\
&= k^2 + k \cdot k^2 \geq k^2 + 16k = k^2 + 2k + 14k > k^2 + 2k + 1 = (k+1)^2.
\end{aligned}
$$

By the Principle of Mathematical Induction, $n! > n^2$ for every integer $n \geq 4$. ∎

5. We proceed by induction. Since $n^2 = 2n + 3$ for $n = 3$, the inequality holds for $n = 3$. Assume that $k^2 \geq 2k + 3$ for an integer $k \geq 3$. Next show that $(k+1)^2 \geq 2k + 5$.

7. We proceed by induction. Since $2^5 = 32 > 25 = 5^2$, the inequality holds for $n = 5$. Assume that $2^k > k^2$ for an integer $k \geq 5$. Next show that $2^{k+1} > (k+1)^2$.

9. We proceed by induction. Since $0^2 + 0 + 1 = 1$ is odd, the statement is true for $n = 0$. Assume that $k^2 + k + 1$ is odd for a nonnegative integer k. Next show that $(k+1)^2 + (k+1) + 1$ is odd.

11. **Proof.** We proceed by induction. We know that x^2 is even if and only if x is even by Theorem 3.23. Thus the statement is true for $n = 2$. Assume that x^k is even if and only if x is even for an arbitrary integer $k \geq 2$. Therefore, x^k is odd if and only if x is odd. We show that x^{k+1} is even if and only if x is even. We first show that if x is even, then x^{k+1} is even. Since x is even, $x = 2a$ for some integer a. Thus $x^{k+1} = x \cdot x^k = (2a) \cdot x^k = 2(a \cdot x^k)$. Since $a \cdot x^k$ is an integer, x^{k+1} is even. For the converse, we show that if x^{k+1} is even, then x is even using a proof by contrapositive. Assume that x is odd. Then x^k is odd by the inductive hypothesis. Let $x = 2s+1$ and $x^k = 2t+1$, where $s, t \in \mathbb{Z}$. Then $x^{k+1} = (2s+1)(2t+1) = 4st+2s+2t+1 = 2(2st+s+t)+1$. Since $2st + s + t$ is an integer, x^{k+1} is odd. ∎

13. We use mathematical induction. By the distributive law for sets, if A, X and Y are any three sets, then $A \cap (X \cup Y) = (A \cap X) \cup (A \cap Y)$. Thus the statement is true for $n = 2$. Assume, for any k sets $D_1, D_2, \ldots, D_k$, that $A \cap (D_1 \cup D_2 \cup \cdots \cup D_k) = (A \cap D_1) \cup (A \cap D_2) \cup \cdots \cup (A \cap D_k)$. Let $C_1, C_2, \ldots, C_{k+1}$ be any $k+1$ sets. Next show that $A \cap (C_1 \cup C_2 \cup \cdots \cup C_{k+1}) = (A \cap C_1) \cup (A \cap C_2) \cup \cdots \cup (A \cap C_{k+1})$.

15. **Proof.** We proceed by induction. Since $1 \leq 2\sqrt{1}$, the inequality holds for $n = 1$. Assume that $1 + \frac{1}{\sqrt{2}} + \frac{1}{\sqrt{3}} + \cdots + \frac{1}{\sqrt{k}} \leq 2\sqrt{k}$ for some positive integer k. We show that $1 + \frac{1}{\sqrt{2}} + \frac{1}{\sqrt{3}} + \cdots + \frac{1}{\sqrt{k+1}} \leq 2\sqrt{k+1}$. Now

$$
\begin{aligned}
1 + \frac{1}{\sqrt{2}} + \frac{1}{\sqrt{3}} + \cdots + \frac{1}{\sqrt{k+1}} &= \left(1 + \frac{1}{\sqrt{2}} + \frac{1}{\sqrt{3}} + \cdots + \frac{1}{\sqrt{k}}\right) + \frac{1}{\sqrt{k+1}} \leq 2\sqrt{k} + \frac{1}{\sqrt{k+1}} \\
&= \frac{2\sqrt{k^2+k}+1}{\sqrt{k+1}} \leq \frac{2\sqrt{k^2+k+\frac{1}{4}}+1}{\sqrt{k+1}} = \frac{2(k+\frac{1}{2})+1}{\sqrt{k+1}} \\
&= \frac{2k+2}{\sqrt{k+1}} = \frac{2(k+1)}{\sqrt{k+1}} = 2\sqrt{k+1}.
\end{aligned}
$$

By the Principle of Mathematical Induction, the inequality holds for every positive integer n. ∎

17. **Proof.** We proceed by induction. Since the sum of the interior angles of each triangle is $180° = (3-2) \cdot 180°$, the result holds for $n = 3$. Assume that the sum of the interior angles of every k-gon is $(k-2) \cdot 180°$ for an arbitrary integer $k \geq 3$. We show that the sum of the interior angles of every $(k+1)$-gon is $(k-1) \cdot 180°$. Let P_{k+1} be a $(k+1)$-gon whose $k+1$ vertices are $v_1, v_2, \ldots, v_{k+1}$ and whose edges are $v_1 v_2, v_2 v_3, \ldots, v_k v_{k+1}, v_{k+1} v_1$. Now let P_k be the k-gon whose vertices are $v_1, v_2, \ldots, v_k$ and whose edges are $v_1 v_2, v_2 v_3, \ldots, v_{k-1} v_k, v_k v_1$ and let P_3 be the 3-gon whose vertices are v_k, v_{k+1}, v_1 and whose edges are $v_k v_{k+1}, v_{k+1} v_1, v_1 v_k$. Observe that the sum of the interior angles of P_{k+1} is the sum of interior angles of P_k and the interior angles of P_3. By the induction hypothesis, the sum of the interior angles of P_k is $(k-2) \cdot 180°$ and the sum of the interior angles of P_3 is $180°$. Therefore, the sum of the interior angles of P_{k+1} is $(k-2) \cdot 180° + 180° = (k-1) \cdot 180°$. By the Principle of Mathematical Induction, the sum of the interior angles of every n-gon is $(n-2) \cdot 180°$. ∎

Section 4.3

1. $a_1 = 5/3, a_2 = -7/6, a_3 = 3/4, a_4 = -11/24$.

3. $a_1 = -1/2^2 = -1/4, a_2 = 1/2^4 = 1/16, a_3 = -1/2^6 = -1/64, a_4 = 1/256$.

5. (a) 3, 3, 3, 3. (b) 0, 8, 0, 32. (c) 1, 5, 19, 65.

7. (1) $a_n = 2^{n-1}$. (2) $b_n = \frac{n^2 - n + 2}{2}$. (3) $c_n = 3^{n-1} - n! + 1$.

9. (a) (a, b, c, d, e). (b) $(0, 1, 0, 1, 0, 1)$.

(c) any bit string whose first term is 1 and alternates 0, 1 thereafter.

11. Each placement of coins corresponds to the 4-bit string $a_1 a_2 a_3 a_4$, where $a_i = 1$ $(1 \le i \le 4)$ if a black coin is placed on square i and $a_i = 0$ if a red coin is placed on square i (or interchange red and black).

13. (a) $a_2 = 2a_1 - 1 = 3$, $a_3 = 5$, $a_4 = 9$, $a_5 = 17$. (b) For every positive integer, $a_n = 2^{n-1} + 1$. **Proof.** We proceed by induction. Since $a_1 = 2 = 2^{1-1} + 1$, the formula holds for $n = 1$. Assume that $a_k = 2^{k-1} + 1$ for a positive integer k. Now $a_{k+1} = 2a_k - 1 = 2(2^{k-1} + 1) - 1 = 2^k + 1$. By the Principle of Mathematical Induction, $a_n = 2^{n-1} + 1$ for every positive integer n. ∎

15. $a_i = i$ for $3 \le i \le 6$.   **17.** (a) $s_1 = 2$, $s_2 = 4$ and $s_3 = 7$. (b) $s_1 = 2$, $s_2 = 4$, $s_3 = 7$ and $s_n = s_{n-1} + s_{n-2} + s_{n-3}$ for $n \ge 4$.

19. (a) $s_1 = 0$, $s_2 = 1$ and $s_3 = 3$. (b) $s_1 = 0$. $s_n = s_{n-1} + (n-1)$ for $n \ge 2$.

(c) $s_4 = s_3 + (4 - 1) = 3 + 3 = 6$.

21. We proceed by induction. Since $F_1 = 1 = 2 - 1 = F_3 - 1$, the statement is true for $n = 1$. Assume that $F_1 + F_2 + \cdots + F_k = F_{k+2} - 1$ for a positive integer k. Next show that $F_1 + F_2 + \cdots + F_{k+1} = F_{k+3} - 1$.

23. Proof. We proceed by induction. When $n = 2$, we have $F_{n-1}F_{n+2} = F_1 F_4 = 1 \cdot 3 = 3$ and $F_n F_{n+1} = F_2 F_3 = 1 \cdot 2 = 2$. Thus when $n = 2$, $F_{n-1}F_{n+2} = F_n F_{n+1} + (-1)^n$ is a true statement. Assume that $F_{k-1}F_{k+2} = F_k F_{k+1} + (-1)^k$ for an integer $k \ge 2$. We show that $F_k F_{k+3} = F_{k+1}F_{k+2} + (-1)^{k+1}$. Observe that

$$
\begin{aligned}
F_k F_{k+3} &= F_k(F_{k+1} + F_{k+2}) = F_k F_{k+1} + F_k F_{k+2} = F_{k-1}F_{k+2} - (-1)^k + F_k F_{k+2} \\
&= F_{k-1}F_{k+2} + F_k F_{k+2} + (-1)^{k+1} = (F_{k-1} + F_k)F_{k+2} + (-1)^{k+1} \\
&= F_{k+1}F_{k+2} + (-1)^{k+1}.
\end{aligned}
$$

By the Principle of Mathematical Induction, $F_{n-1}F_{n+2} = F_n F_{n+1} + (-1)^n$ for every integer $n \ge 2$. ∎

Section 4.4

1. (a) $a_2 = 2$, $a_3 = 4$, $a_4 = 8$ and $a_5 = 16$. (b) Claim: $a_n = 2^{n-1}$ for every positive integer n.

(c) **Proof.** We proceed by induction. Since $a_1 = 1 = 2^{1-1} = 2^0$, the formula holds for $n = 1$. Assume that $a_k = 2^{k-1}$ for a positive integer k. We show that $a_{k+1} = 2^k$. Observe that $a_{k+1} = 2a_k = 2 \cdot 2^{k-1} = 2^k$. By the Principle of Mathematical Induction, $a_n = 2^{n-1}$ for every positive integer n. ∎

3. Proof. We apply the Strong Principle of Mathematical Induction. Since $a_1 = 4 = 1^2 + 3$, the formula holds for $n = 1$. Assume for an arbitrary positive integer k that $a_i = i^2 + 3$ for every integer i with $1 \le i \le k$. We show that $a_{k+1} = (k+1)^2 + 3$. Since $a_2 = 7 = (1+1)^2 + 3$, it follows that $a_{k+1} = (k+1)^2 + 3$ when $k = 1$. Hence we may assume that $k \ge 2$. Observe that

$$
\begin{aligned}
a_{k+1} &= 2a_k - a_{k-1} + 2 = 2(k^2 + 3) - [(k-1)^2 + 3] + 2 = 2k^2 + 6 - (k^2 - 2k + 1 + 3) + 2 \\
&= 2k^2 + 6 - k^2 + 2k - 4 + 2 = k^2 + 2k + 4 = (k+1)^2 + 3.
\end{aligned}
$$

By the Strong Principle of Mathematical Induction, $a_n = n^2 + 3$ for every positive integer n. ∎

5. (a) $a_4 = a_3 - a_2 + a_1 + 2(8 - 3) = 9 - 4 + 1 + 10 = 16$, $a_5 = 25$ and $a_6 = 36$.

(b) Claim: $a_n = n^2$ for every positive integer n.

(c) **Proof.** We apply the Strong Principle of Mathematical Induction. Since $a_1 = 1^2$, the formula holds for $n = 1$. Assume for an arbitrary positive integer k that $a_i = i^2$ for every integer i with $1 \le i \le k$. We show that $a_{k+1} = (k+1)^2$. Since $a_2 = (1+1)^2 = 4$ and $a_3 = (2+1)^2 = 9$, $a_{k+1} = (k+1)^2$ when $k = 1$ and $k = 2$. Hence we may assume that $k \ge 3$. Observe that

$$
\begin{aligned}
a_{k+1} &= a_k - a_{k-1} + a_{k-2} + 2[2(k+1) - 3] = k^2 - (k-1)^2 + (k-2)^2 + 2(2k-1) \\
&= k^2 - (k^2 - 2k + 1) + (k^2 - 4k + 4) + 4k - 2 = k^2 + 2k + 1 = (k+1)^2.
\end{aligned}
$$

By the Strong Principle of Mathematical Induction, $a_n = n^2$ for each positive integer n. ■

7. Observe that $a_3 = a_2 + 2a_1 = -1 + 2 = 1$ and $a_4 = a_3 + 2a_2 = 1 + 2(-1) = -1$. We conjecture that $a_n = (-1)^{n+1}$.

9. If $(1+c) \le c^2$, then $c^2 - c - 1 \ge 0$. Using the quadratic formula on $c^2 - c - 1 = 0$, we see that $c = \frac{1+\sqrt{5}}{2}$. Observe that $\frac{1+\sqrt{5}}{2} < \frac{5}{3}$. We show that $F_n \le \left(\frac{1+\sqrt{5}}{2}\right)^n$ for every positive integer n.

Proof. We apply the Strong Principle of Mathematical Induction. Since $F_1 = 1 < \frac{1+\sqrt{5}}{2}$, the inequality holds for $n = 1$. Assume for an arbitrary positive integer k that $F_i \le \left(\frac{1+\sqrt{5}}{2}\right)^i$ for every integer i with $1 \le i \le k$. We show that $F_{k+1} \le \left(\frac{1+\sqrt{5}}{2}\right)^{k+1}$. Since $F_2 = 1 < \left(\frac{1+\sqrt{5}}{2}\right)^2$, it follows that $F_{k+1} \le \left(\frac{1+\sqrt{5}}{2}\right)^{k+1}$ when $k = 1$. Hence we may assume that $k \ge 2$. Observe that

$$
\begin{aligned}
F_{k+1} &= F_{k-1} + F_k \le \left(\frac{1+\sqrt{5}}{2}\right)^{k-1} + \left(\frac{1+\sqrt{5}}{2}\right)^k = \left(\frac{1+\sqrt{5}}{2}\right)^{k-1}\left(1 + \frac{1+\sqrt{5}}{2}\right) \\
&= \left(\frac{1+\sqrt{5}}{2}\right)^{k-1}\left(\frac{3+\sqrt{5}}{2}\right) = \left(\frac{1+\sqrt{5}}{2}\right)^{k-1}\left(\frac{1+\sqrt{5}}{2}\right)^2 = \left(\frac{1+\sqrt{5}}{2}\right)^{k+1}.
\end{aligned}
$$

By the Strong Principle of Mathematical Induction, $F_n \le \left(\frac{1+\sqrt{5}}{2}\right)^n$ for every positive integer n. ■

11. The primary error is not being careful when attempting to give a proof. The second line of the proof should read: Assume, for an arbitrary *nonnegative* integer k, that $2i = 0$ for every integer i with $0 \le i \le k$. Suppose that $k = 0$. Then we must show that $2(k+1) = 0$ where $k = 0$. The next sentence states: Let i and j be integers such that $0 \le i \le k$ and $0 \le j \le k$ and $i + j = k + 1$. When $k = 0$ (which is a possible value of k), the only nonnegative integers i and j for which $i + j = k + 1 = 1$ are $i = 0$ and $j = 1$ (or $i = 1$ and $j = 0$). However, there is no integer j satisfying $0 \le j \le k$ when $j = 1$ and $k = 0$. So no such integers i and j exist. Hence we cannot show that $2(k+1) = 0$ for an arbitrary nonnegative integer k (because it's not true).

Supplementary Exercises for Chapter 4

1. We proceed by induction. Since $1 = \frac{1 \cdot (1+1)(1+2)}{6}$, the formula holds for $n = 1$. Assume that $1 + 3 + 6 + \cdots + \frac{k(k+1)}{2} = \frac{k(k+1)(k+2)}{6}$ for an arbitrary positive integer k. Next show that $1 + 3 + 6 + \cdots + \frac{(k+1)(k+2)}{2} = \frac{(k+1)(k+2)(k+3)}{6}$.

3. We verify this formula by induction. Since $1^2 = 1 = \frac{1(2 \cdot 1 - 1)(2 \cdot 1 + 1)}{3}$, the formula holds for $n = 1$. Assume that $1^2 + 3^2 + \cdots + (2k-1)^2 = \frac{k(2k-1)(2k+1)}{3}$, where k is an arbitrary positive integer. Next show that $1^2 + 3^2 + \cdots + (2k+1)^2 = \frac{(k+1)(2k+1)(2k+3)}{3}$.

5. Proof. We use induction. Since $1(1!) = 1 = (1+1)! - 1 = 2! - 1$, the formula holds for $n = 1$. Assume that $1(1!) + 2(2!) + \cdots + k(k!) = (k+1)! - 1$ for an arbitrary positive integer k. We show that $1(1!) + 2(2!) + \cdots + (k+1)[(k+1)!] = (k+2)! - 1$. Observe that

$$
\begin{aligned}
1(1!) + 2(2!) + \cdots + (k+1)[k+1)!] &= [1(1!) + 2(2!) + \cdots + k(k!)] + (k+1)[k+1)!] \\
&= [(k+1)! - 1] + (k+1)[(k+1)!] \\
&= (k+2)[(k+1)!] - 1 = (k+2)! - 1.
\end{aligned}
$$

By the Principle of Mathematical Induction, $1(1!) + 2(2!) + \cdots + n(n!) = (n+1)! - 1$ for every positive integer n. ∎

7. Observe that for $n \geq 1$, $2^{n+1} < 1 + 2 \cdot 2^n \leq 1 + (n+1)2^n$. Also, this statement can be verified by induction. **Proof.** We use induction. Since $2^2 = 4 < 5 = 1 + (1+1) \cdot 2^1$, the inequality holds for $n = 1$. Assume that $2^{k+1} < 1 + (k+1)2^k$ for an arbitrary positive integer k. We show that $2^{k+2} < 1 + (k+2)2^{k+1}$. Observe that

$$
\begin{aligned}
2^{k+2} &= 2 \cdot 2^{k+1} < 2\left[1 + (k+1)2^k\right] = 2 + (k+1)2^{k+1} = 2 + (k+2)2^{k+1} - 2^{k+1} \\
&= 1 + (k+2)2^{k+1} + \left(1 - 2^{k+1}\right) < 1 + (k+2)2^{k+1}.
\end{aligned}
$$

By the Principle of Mathematical Induction, $2^{n+1} < 1 + (n+1)2^n$ for every positive integer n. ∎

9. Proof. Since $1024 = 2^{10} > 10^3 = 1000$, the inequality holds when $n = 10$. Assume that $2^k > k^3$ where $k \geq 10$ is an arbitrary integer. We show that $2^{k+1} > (k+1)^3$. Observe that

$$
\begin{aligned}
2^{k+1} &= 2 \cdot 2^k > 2k^3 = k^3 + k^3 \geq k^3 + 10k^2 = k^3 + 3k^2 + 7k^2 \\
&> k^3 + 3k^2 + 7k = k^3 + 3k^2 + 3k + 4k > k^3 + 3k^2 + 3k + 1 = (k+1)^3.
\end{aligned}
$$

By the Principle of Mathematical Induction, $2^n > n^3$ for every integer $n \geq 10$. ∎

11. Proof. We proceed by induction. Since $2^2 = 4 > 2! = 2$, the statement is true for $n = 2$. Assume that $k^k > k!$ for an arbitrary integer $k \geq 2$. We show that $(k+1)^{k+1} > (k+1)!$. Observe that $(k+1)^{k+1} = (k+1)(k+1)^k > (k+1)k^k > (k+1)k! = (k+1)!$. By the Principle of Mathematical Induction, $n^n > n!$ for every integer $n \geq 2$. ∎

13. We use induction. Since

$$
\begin{aligned}
\frac{1}{\sqrt[3]{1}} + \frac{1}{\sqrt[3]{2}} + \frac{1}{\sqrt[3]{3}} + \frac{1}{\sqrt[3]{4}} &= 1 + \frac{1}{2}\sqrt[3]{4} + \frac{1}{3}\sqrt[3]{9} + \frac{1}{4}\sqrt[3]{16} > 1 + \frac{1}{2}\sqrt[3]{\frac{27}{8}} + \frac{1}{3}\sqrt[3]{8} + \frac{1}{4}\sqrt[3]{\frac{125}{8}} \\
&= 1 + \frac{1}{2} \cdot \frac{3}{2} + \frac{2}{3} + \frac{1}{4} \cdot \frac{5}{2} = 1 + \frac{3}{4} + \frac{2}{3} + \frac{5}{8} = \frac{73}{24} > 3 = \sqrt[3]{27} > \sqrt[3]{25} = 5^{2/3},
\end{aligned}
$$

the inequality holds for $n = 4$. Assume that $\frac{1}{\sqrt[3]{1}} + \frac{1}{\sqrt[3]{2}} + \cdots + \frac{1}{\sqrt[3]{k}} > (k+1)^{2/3}$ for an arbitrary integer $k \geq 4$. Next show that $\frac{1}{\sqrt[3]{1}} + \frac{1}{\sqrt[3]{2}} + \cdots + \frac{1}{\sqrt[3]{k+1}} > (k+2)^{2/3}$.

15. Proof. We proceed by induction. Since $(1+x)^1 = 1 + 1x$, the inequality holds when $n = 1$. Assume that $(1+x)^k \geq 1 + kx$, where k is an arbitrary positive integer. We show that $(1+x)^{k+1} \geq 1 + (k+1)x$. Observe that $(1+x)^{k+1} = (1+x)(1+x)^k \geq (1+x)(1+kx)$ since $1 + x > 0$. Thus $(1+x)^{k+1} \geq (1+x)(1+kx) = 1 + (k+1)x + kx^2 \geq 1 + (k+1)x$ since $kx^2 \geq 0$. By the Principle of Mathematical Induction, $(1+x)^n \geq 1 + nx$ for every positive integer n. ∎

17. Proof. We proceed by induction. Since $a + 0 \cdot b = a = \frac{1}{2}(0+1)(2a + 0 \cdot b)$, the formula holds for $n = 0$. Assume that $\sum_{i=0}^{k}(a + ib) = \frac{(k+1)(2a+kb)}{2}$ for some nonnegative integer k. We show that $\sum_{i=0}^{k+1}(a + ib) = \frac{(k+2)[2a+(k+1)b]}{2} = \frac{bk^2 + 2ak + 3bk + 4a + 2b}{2}$. Observe that

$$
\begin{aligned}
\sum_{i=0}^{k+1}(a + ib) &= \left[\sum_{i=0}^{k}(a+i)\right] + [a + (k+1)b] = \frac{(k+1)(2a+kb)}{2} + [a + (k+1)b] \\
&= \frac{(k+1)(2a+kb) + 2(a + kb + b)}{2} = \frac{bk^2 + 2ak + 3bk + 4a + 2b}{2}.
\end{aligned}
$$

By the Principle of Mathematical Induction, $\sum_{i=0}^{n}(a+ib) = \frac{1}{2}(n+1)(2a+nb)$ for every nonnegative integer n. ∎

19. **Proof.** We proceed by induction. Since $a_0 = 0 = 0 \cdot 3^{-1}$, the formula holds for $n = 0$. Assume that $a_k = k3^{k-1}$ for a nonnegative integer k. We show that $a_{k+1} = (k+1)3^k$. Note that $a_{k+1} = 3a_k + 3^k = 3(k3^{k-1}) + 3^k = k3^k + 3^k = (k+1)3^k$. By the Principle of Mathematical Induction, $a_n = n3^{n-1}$ for every nonnegative integer n. ∎

21. We proceed by the Strong Principle of Mathematical Induction. Since $a_1 = 2 = 1 \cdot 2$, the formula holds for $n = 1$. Assume for a positive integer k that $a_i = i(i+1)$ for every integer i with $1 \le i \le k$. Next show that $a_{k+1} = (k+1)(k+2)$.

23. We proceed by the Strong Principle of Mathematical Induction. Since $F_3 = 2 > \left(1+\sqrt{5}\right)/2$, the inequality holds for $n = 1$. Assume for an integer $k \ge 3$ that $F_i > \left(\frac{1+\sqrt{5}}{2}\right)^{i-2}$ for every integer i with $3 \le i \le k$. We show that $F_{k+1} > \left(\frac{1+\sqrt{5}}{2}\right)^{k-1}$. Since $F_4 = 3 > \left(\frac{1+\sqrt{5}}{2}\right)^2$, it follows that $F_{k+1} > \left(\frac{1+\sqrt{5}}{2}\right)^{k-1}$ when $k = 3$. Hence we may assume that $k \ge 4$. Observe that

$$F_{k+1} = F_{k-1} + F_k > \left(\frac{1+\sqrt{5}}{2}\right)^{k-3} + \left(\frac{1+\sqrt{5}}{2}\right)^{k-2}.$$

25. (a) Observe that S is a subset of itself.

(b) Observe that $(0,1] \subseteq [0,1]$ and $(0,1]$ does not have a least element.

27. For $1 \le n \le 210$, let $P(n)$: There exists a subset of S, the sum of whose elements is n. We prove that $P(n)$ is true for every integer n with $1 \le n \le 210$ by applying the Principle of Finite Induction. Since the sum of the element(s) of the subset $\{1\}$ is 1, the basis step is established. Assume, for an integer k with $1 \le k < 210$, that there exists a subset T of S, the sum of whose elements is k. We show that there is a subset of S, the sum of whose elements is $k+1$. If $1 \notin T$, then $T \cup \{1\}$ has the desired property. Hence we may assume that $1 \in T$. Since $k < 210$, it follows that $T \subset S$. Thus there exists a smallest integer $m \ge 2$ such that $m \notin T$.

29. We proceed by induction. Since $2\cos\frac{\pi}{4} = 2\frac{\sqrt{2}}{2} = \sqrt{2}$, the statement holds for $n = 1$. Assume that $\sqrt{2 + \sqrt{2 + \sqrt{2 + \cdots + \sqrt{2}}}} = 2\cos\frac{\pi}{2^{k+1}}$, where the number 2 occurs k times in the expression on the left. Next show that $\sqrt{2 + \sqrt{2 + \sqrt{2 + \cdots + \sqrt{2}}}} = 2\cos\frac{\pi}{2^{k+2}}$, where the number 2 occurs $k+1$ times in the expression on the left. Since $\cos 2x = 2\cos^2 x - 1$ for every real number x, it follows that $4\cos^2 x = 2 + 2\cos 2x$ and so $4\cos^2\frac{\pi}{2^{k+2}} = 2 + 2\cos\frac{\pi}{2^{k+1}}$.

Chapter 5: Relations and Functions

Section 5.1

1. $R = \{(1,5), (2,5), (2,6)\}$ 3. See Figure 6.   5. $R = \{(1,2), (2,3), (2,5), (4,3), (5,2)\}$.

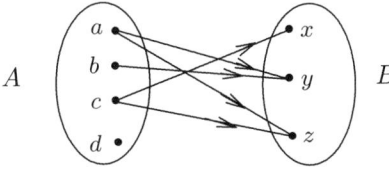

Figure 6: The relation R in Exercise 3

7. (a) symmetric. (b) transitive. (c) reflexive. (d) transitive. (e) reflexive, transitive.

9. (a) $3\ R\ 1$ but $1\ \not R\ 3$. (b) reflexive, transitive.

11. (a) reflexive, transitive. (b) symmetric (c) reflexive, transitive. (d) symmetric.

Section 5.2

1. (a) **Proof.** Let $(a, b) \in \mathbb{N} \times \mathbb{N}$. Since $a + b = b + a$, it follows that $(a, b)\ R\ (a, b)$. Thus R is reflexive. Next, assume that $(a, b)\ R\ (c, d)$, where $(a, b), (c, d) \in \mathbb{N} \times \mathbb{N}$. Then $a + d = b + c$ and so $c + b = d + a$. Thus $(c, d)\ R\ (a, b)$ and R is symmetric. Finally, assume that $(a, b)\ R\ (c, d)$ and $(c, d)\ R\ (e, f)$, where $(a, b), (c, d), (e, f) \in \mathbb{N} \times \mathbb{N}$. Thus $a + d = b + c$ and $c + f = d + e$. Adding these two equations, we obtain $(a + d) + (c + f) = (b + c) + (d + e)$. Thus $a + f = b + e$ and so $(a, b)\ R\ (e, f)$. Therefore, R is transitive. ∎

(b) $[(3, 1)] = \{(3, 1), (4, 2), (5, 3), \ldots\}$. $[(5, 5)] = \{(1, 1), (2, 2), (3, 3), \ldots\}$.
$[(4, 7)] = \{(1, 4), (2, 5), (3, 6), \ldots\}$.

3. $\{(a, a), (a, d), (b, b), (b, f), (c, c), (c, e), (d, a), (d, d), (e, c), (e, e), (f, b), (f, f)\}$.

5. $R = \{(1, 1), (1, 3), (1, 4), (1, 5), (2, 2), (3, 1), (3, 3), (3, 4), (3, 5), (4, 1), (4, 3), (4, 4), (4, 5), (5, 1), (5, 3), (5, 4), (5, 5), (6, 6)\}$.

7. (a) R is an equivalence relation. Let $a \in \mathbb{Z}$. Then $a - a = 0$ and so $a\ R\ a$. Thus R is reflexive. Assume that $a\ R\ b$, where $a, b \in \mathbb{Z}$. Then $a + b = 0$ or $a - b = 0$. If $a + b = 0$, then $b + a = a + b = 0$; while if $a - b = 0$, then $b - a = -(a - b) = 0$. Thus $b + a = 0$ or $b - a = 0$. Therefore, $b\ R\ a$ and R is symmetric. We now assume that $a\ R\ b$ and $b\ R\ c$, where $a, b, c \in \mathbb{Z}$. Hence (1) $a + b = 0$ or $a - b = 0$ and (2) $b + c = 0$ or $b - c = 0$. Next show that $a + c = 0$ or $a - c = 0$. (b) For each $a \in \mathbb{Z}$, $[a] = \{a, -a\}$.

9. (a) Let $a \in \mathbb{Z}$. Then $11a - 5a = 6a = 2(3a)$ is an even integer and so $a\ R\ a$. Thus R is reflexive. Assume that $a\ R\ b$, where $a, b \in \mathbb{Z}$. Then $11a - 5b$ is even. Thus $11a - 5b = 2k$ for some integer k. Observe that $11b - 5a = (11a - 5b) - 16a + 16b = 2k - 16a + 16b = 2(k - 8a + 8b)$. Since $k - 8a + 8b \in \mathbb{Z}$, it follows that $11b - 5a$ is even. Therefore, $b\ R\ a$ and R is symmetric. Finally, assume that $a\ R\ b$ and $b\ R\ c$, where $a, b, c \in \mathbb{Z}$. Show that $a\ R\ c$ and R is transitive. (b) $[0] = \{x \in \mathbb{Z} : x \text{ is even}\}$ and $[1] = \{x \in \mathbb{Z} : x \text{ is odd}\}$.

11. No (R is not reflexive).   **13.** Yes.

15. (a) **Proof.** Let $x \in S$. Since $\mathcal{P}$ is a partition of S, it follows that $x \in S_i$ for some integer i with $1 \leq i \leq k$. Thus, $x\ R\ x$ and R is reflexive. Assume that $x\ R\ y$, where $x, y \in S$. Then $x, y \in S_i$ for some i with $1 \leq i \leq k$. Thus $y, x \in S_i$ for some i with $1 \leq i \leq k$. Therefore, $y\ R\ x$ and R is symmetric. Finally, we show that R is transitive. Assume that $x\ R\ y$ and $y\ R\ z$, where $x, y, z \in S$. Thus there exist integers i, j with $1 \leq i, j \leq k$ such that $x, y \in S_i$ and $y, z \in S_j$. Since $y \in S_i$ and $y \in S_j$, it follows that $i = j$ because $\mathcal{P}$ is a partition of S. Thus $x, z \in S_i$ for some i with $1 \leq i \leq k$. Hence $x\ R\ z$ and R is transitive. ∎

(b) The distinct equivalence classes are $S_1, S_2, \ldots, S_k$.   **17.** The statement is true.

Section 5.3

1. (a) Yes. (b) No. (c) No. (d) No. (e) Yes. (f) No.   **3.** (a) The domain of f is A, the codomain of f is B and the range of f is $\{x, z\}$. (b) x. (c) No. (d) $\{x, z\}$. (e) $g = \{(x, d), (y, c), (z, b)\}$.

5. Proof. We first show that $f(A_1 \cup A_2) \subseteq f(A_1) \cup f(A_2)$. Let $b \in f(A_1 \cup A_2)$. Then there exists $a \in A_1 \cup A_2$ such that $f(a) = b$. Since $a \in A_1 \cup A_2$, it follows that $a \in A_1$ or $a \in A_2$, say the former. Thus $b = f(a) \in f(A_1)$. Since $f(A_1)$ is a subset of $f(A_1) \cup f(A_2)$, we have $b \in f(A_1) \cup f(A_2)$ and so $f(A_1 \cup A_2) \subseteq f(A_1) \cup f(A_2)$. Next, we show that $f(A_1) \cup f(A_2) \subseteq$

$f(A_1 \cup A_2)$. Let $b \in f(A_1) \cup f(A_2)$. Then $b \in f(A_1)$ or $b \in f(A_2)$, say the former. Thus there exists $a \in A_1$ such that $f(a) = b$. Since $a \in A_1$, it follows that $a \in A_1 \cup A_2$ and so $b = f(a) \in f(A_1 \cup A_2)$. Hence $f(A_1) \cup f(A_2) \subseteq f(A_1 \cup A_2)$. ∎

7. Proof. Let $A = \{1,2,3\}$, $B = \{a,b\}$ and let a function $f : A \to B$ be defined by $f = \{(1,a),(2,b),(3,a)\}$. Let $A_1 = \{1,2\}$ and $A_2 = \{2,3\}$. Since $A_1 \cap A_2 = \{2\}$, it follows that $f(A_1 \cap A_2) = f(\{2\}) = \{b\}$. On the other hand, $f(A_1) = f(\{1,2\}) = \{a,b\}$ and $f(A_2) = f(\{2,3\}) = \{a,b\}$; so $f(A_1) \cap f(A_2) = \{a,b\}$. Therefore, $f(A_1 \cap A_2) \neq f(A_1) \cap f(A_2)$. ∎

9. $f_1 = \{(a,0),(b,0)\}$. $f_2 = \{(a,0),(b,1)\}$. $f_3 = \{(a,0),(b,2)\}$. $f_4 = \{(a,1),(b,0)\}$. $f_5 = \{(a,1),(b,1)\}$. $f_6 = \{(a,1),(b,2)\}$. $f_7 = \{(a,2),(b,0)\}$. $f_8 = \{(a,2),(b,1)\}$. $f_9 = \{(a,2),(b,2)\}$.

11. No. $((1,1)$ and $(1,-1)$ both lie on the parabola.)

13. The domain and codomain of f are $\mathbb{R}$. The range of f is $\{y \in \mathbb{R} : y \le 4\}$.

15. (a) $\{(-3,0),(-2,2),(-1,1),(0,0),(1,1),(2,-2),(3,0)\}$. (b) The domain of f is A, the codomain of f is B and the range of f is $\{-2,0,1,2\}$. (c) $\{0,1,2\}$.   **17.** See Figure 7.

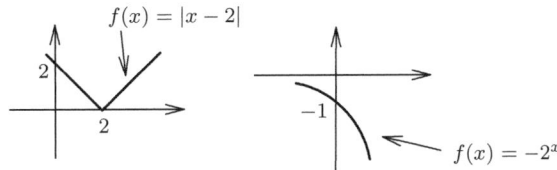

$f(x) = |x - 2|$

$f(x) = -2^x$

Figure 7: The graphs of the functions in Exercise 17

19. (a) $\{0,1,3\}$. (b) $\{0,1,3\}$. (c) $\{0,3,8,16\}$. (d) $\{1,5,12,21\}$.

21. (a) $f\left(\frac{1}{2}\right) = 1$ and $f(\sqrt{2}) = 2$. (b) $g\left(\frac{1}{2}\right) = 0$ and $g(\sqrt{2}) = 1$. (c) $h\left(\frac{1}{2}\right) = -1$ and $h(\sqrt{2}) = \frac{1}{2}$.

23. Proof. First assume that $x \in \mathbb{Z}$. Then $\lceil x \rceil + \lfloor x \rfloor = x + x = 2x$. Next, assume that $2x \in \mathbb{Z}$ but $x \notin \mathbb{Z}$. Then $x = n + \frac{1}{2}$ for some $n \in \mathbb{Z}$ and so $2x = 2n + 1$. Hence $\lceil x \rceil + \lfloor x \rfloor = (n + 1) + n = 2n + 1 = 2x$. For the converse, suppose that $x \notin \mathbb{Z}$ and $2x \notin \mathbb{Z}$. Then $n < x < n + 1$ for some integer n but $x \neq n + \frac{1}{2}$. Thus $\lceil x \rceil + \lfloor x \rfloor = (n + 1) + n = 2n + 1 \neq 2x$. ∎

25. (a) - (c) No.

27. Proof. First, suppose that e is an even integer. Then $e = 2d$ for some integer d. Assume that $f(d) = k$. Then $f(e) = f(2d) = f(d + d) = f(d) + f(d) = 2f(d) = 2k$, which is even. Assume that c is an odd integer such that $f(c)$ is even. Then $c = 2x + 1$ for some integer x. Then $f(c) = f(2x + 1) = f(2x) + f(1)$. Hence $f(1) = f(c) - f(2x)$. Since $f(c)$ and $f(2x)$ are even, so is $f(1)$, say $f(1) = 2\ell$. Now let y be an odd integer. Then $y = 2w + 1$ for some integer w. Hence $f(y) = f(2w) + f(1) = 2f(w) + 2\ell = 2(f(w) + \ell)$. Since $f(w) + \ell$ is an integer, $f(y)$ is even. ∎

29. (a) w. (b) $\{(a,y), (b,z), (c,z), (d,w)\}$. (c) $\{(a,a), (b,c), (c,c), (d,a)\}$.

31. (a) $(g \circ f)(0) = 2$, $(f \circ g)(0) = 28$. (b) $(g \circ f)(x) = 15x^2 + 2$, $(f \circ g)(x) = 75x^2 - 90x + 28$.

Section 5.4

1. (a) $f = \{(1,1), (2,2), (3,3), (4,3)\}$. (b) $g = \{(1,1), (2,2), (3,3), (4,4)\}$.
(c) $h = \{(1,2), (2,3), (3,4), (4,5)\}$.

3. (a) $f = \{(a,a), (b,b), (c,a), (d,a)\}$. (b) $g = \{(a,a), (b,a), (c,b), (d,c)\}$.
(c) $h = \{(a,b), (b,c), (c,a), (d,c)\}$.

5. (a) The function f is not onto since there is no n, for example, such that $f(n) = 0$.

(b) The function f is onto. (c) The function f is onto. (d) The function f is not onto.

7. (a)-(b) No such function exists.

(c) $f = \{(a, q), (b, r), (c, s), (d, t)\}$. (d) $f = \{(a, q), (b, q), (c, q), (d, q)\}$.

9. Since $|A| < |B|$, there exists no function from A to B that is onto.

11. (a) The function f is not onto. (b) The function f is one-to-one.

Proof. Assume that $f(a) = f(b)$, where $a, b \in \mathbb{Z}$. Then $2a = 2b$. Thus $a = b$.   ∎

13. (a) The function f is one-to-one. **Proof.** Let $a, b \in \mathbb{Z}$ such that $f(a) = f(b)$. Then $5a + 1 = 5b + 1$. Subtracting 1 from both sides and dividing by 5, we obtain $a = b$.   ∎

(b) The function f is not onto since, for example, there is no integer n such that $f(n) = 0$.

15. The function f is one-to-one and onto. **Proof.** First, we show that f is one-to-one. Assume that $f(a) = f(b)$, where $a, b \in \mathbb{R} - \{0\}$. Then $\frac{1}{2a} = \frac{1}{2b}$ and so $2a = 2b$. Hence $a = b$. Next we show that f is onto. Let $r \in \mathbb{R} - \{0\}$. Consider $\frac{1}{2r} \in \mathbb{R} - \{0\}$. Since $f\left(\frac{1}{2r}\right) = \frac{1}{2\left(\frac{1}{2r}\right)} = r$, it follows that f is onto.   ∎

17. **Proof.** Suppose that there exists no positive integer n such that $f(n) < n$ and that f is not the identity function on $\mathbb{N}$. Since f is not the identity function on $\mathbb{N}$, there are integers n such that $f(n) > n$. Let m be the smallest positive integer such that $f(m) > m$. Since there is a positive integer a with $f(a) = 1$, it follows that $m > 1$. Therefore, $f(k) = k$ for every integer k with $1 \le k < m$. Since there is an integer ℓ such that $f(\ell) = m$, it follows that $\ell > m$. Hence $f(\ell) < \ell$, a contradiction.   ∎

19. (a)-(d) not one-to-one. (e) one-to-one.

21. (a) The statement is true. **Proof.** Assume, to the contrary, that f is not one-to-one. Then there exist $a_1, a_2 \in A$ such that $a_1 \ne a_2$ and $f(a_1) = f(a_2)$. Let $A_1 = \{a_1\}$ and $A_2 = \{a_2\}$. Then $A_1 \ne A_2$. On the other hand, since $f(a_1) = f(a_2)$, it follows that $f(A_1) = \{f(a_1)\} = \{f(a_2)\} = f(A_2)$. This implies that $A_1 = A_2$, which is a contradiction.   ∎

(b) The statement is true. **Proof.** Let $b \in B$ and let $B_1 = \{b\} \subseteq B$. By assumption, there exists a subset A_1 of A such that $f(A_1) = B_1 = \{b\}$. Let $a \in A_1$. Then $f(a) = b$ and so f is onto.   ∎

23. (a) Let $A = \{a, b\}$, $B = \{1, 2, 3\}$, $C = \{x, y\}$, $f = \{(a, 1), (b, 2)\}$ and $g = \{(1, x), (2, y), (3, y)\}$.

(b) Let $A = \{a, b\}$, $B = \{1, 2, 3\}$, $C = \{x, y, z\}$, $f = \{(a, 1), (b, 2)\}$ and $g = \{(1, x), (2, y), (3, z)\}$.

(c) Let $A = \{a, b, c\}$, $B = \{1, 2\}$, $C = \{x, y\}$, $f = \{(a, 1), (b, 2), (c, 2)\}$ and $g = \{(1, x), (2, y)\}$.

25. **Proof.** Assume that f is not injective. Then there exist distinct integers a and b such that $f(a) = f(b)$, say $f(a) = f(b) = c$. Then $(f \circ f)(a) = f(f(a)) = f(c)$, while $(f \circ f)(b) = f(f(b)) = f(c)$. Thus $f \circ f$ is not injective.   ∎

27. (a) c_3. (b) c_1. (c) c_i for $i = 1, 2, 3, 4$.   **29.** (a) one-to-one. (b) No (since f is not onto).

31. The function f is bijective. **Proof.** First, we show that f is one-to-one. Assume that $f(a) = f(b)$, where $a, b \in \mathbb{R}$. Then $5a - 7 = 5b - 7$. Adding 7 to both sides and dividing by 5, we obtain $a = b$. Next we show that f is onto. Let $r \in \mathbb{R}$. Then $(r + 7)/5 \in \mathbb{R}$. Therefore, $f\left(\frac{r+7}{5}\right) = 5\left(\frac{r+7}{5}\right) - 7 = r$.   ∎

$f^{-1}(x) = (x + 7)/5$ for $x \in \mathbb{R}$.

33. Let $C = \{a, b, c, e\}$ and $g = \{(a, 2), (b, 3), (c, 1), (e, 4)\}$.

35. (a) $g \circ f = \{(1, x), (2, y), (3, z)\}$, $(g \circ f)^{-1} = \{(x, 1), (y, 2), (z, 3)\}$.

(b) $f^{-1} = \{(a, 2), (b, 3), (c, 1)\}$, $g^{-1} = \{(x, c), (y, a), (z, b)\}$, $f^{-1} \circ g^{-1} = \{(x, 1), (y, 2), (z, 3)\}$.

37. Let $S = \{1, 2, 3, 4\}$ and let f be defined by $f(1) = 2$, $f(2) = 3$, $f(3) = 4$ and $f(4) = 1$. For $a = 1$, $f(1) = 2$, $(f \circ f)(1) = f(f(1)) = f(2) = 3$ and $f^{-1}(1) = 4$ are distinct as are all of $f(a)$, $(f \circ f)(a)$ and $f^{-1}(a)$ for each $a \in S$.

Section 5.5

1. The function $f : E \to O$ defined by $f(n) = n + 1$ is bijective.

3. Proof. Let $A = \{a_1, a_2, a_3, \ldots\}$ and $B = \{b_1, b_2, \ldots, b_n\}$, where $n \in \mathbb{N}$. Then the function $f : \mathbb{N} \to A \cup B$ defined $f(i) = b_i$ for $1 \le i \le n$ and $f(i) = a_{i-n}$ for $i \ge n + 1$ is bijective and so $A \cup B = \{b_1, b_2, \ldots, b_n, a_1, a_2, \ldots\}$ is denumerable. ∎

5. Since $\mathbb{Q}$ is denumerable, $\mathbb{Q} \times \mathbb{Q}$ is denumerable. Thus, there is a bijection f from $\mathbb{N}$ to $\mathbb{Q} \times \mathbb{Q}$. Define the function $g : \mathbb{Q} \times \mathbb{Q} \to C$ by $g((a, b)) = a + b\sqrt{2}$ for all $(a, b) \in \mathbb{Q} \times \mathbb{Q}$.

7. (a) T. (b) T. (c) F. (d) F.

9. The statement is false. If A is finite, then A is not denumerable.   **11.** (a)-(c) T. (d) F.

13. The statement is false. The set $\mathbb{Q}$ of rational numbers is denumerable and the set $\mathbb{C}$ of complex numbers is uncountable. However, $\mathbb{Q} \subseteq \mathbb{R} \subset \mathbb{C}$ but $\mathbb{R}$ is uncountable.

15. The statement is true. Let $f : S - \{a, b\} \to S - \{c, d\}$ be defined by $f(x) = x$ if $x \in S - \{a, b, c, d\}$, $f(c) = a$ and $f(d) = b$. Then f is a bijection.

Supplementary Exercises for Chapter 5

1. $\{(\{1, 2\}, \{3, 4\}), (\{1, 3\}, \{2, 4\}), (\{1, 4\}, \{2, 3\}), (\{2, 4\}, \{1, 3\}), (\{2, 3\}, \{1, 4\}), (\{3, 4\}, \{1, 2\})\}$.   **3.** symmetric.   **5.** (a) $1 \ R \ 1$, $2 \ R \ 1$, $1 \ \not R \ 3$, $2 \ \not R \ 5$. (b) reflexive.

7. (a) $0 \ R \ -2$, $-2 \ R \ 2$, $2 \ \not R \ 2$, $-7 \ \not R \ -2$. (b) symmetric.

9. (a) Let $A_1 = \{1, 2\}$, $A_2 = \{3\}$, $A_3 = \{4\}$ and $A_4 = \{5\}$. Define an equivalence relation R on S by $R = \{(1, 1), (1, 2), (2, 1), (2, 2), (3, 3), (4, 4), (5, 5)\}$. Then the equivalence classes resulting from R are A_1, A_2, A_3, A_4.

(b) Let $B_1 = \{1, 2\}$, $B_2 = \{3, 4\}$ and $B_3 = \{5\}$. Define an equivalence relation R' on S by $R' = \{(1, 1), (1, 2), (2, 1), (2, 2), (3, 3), (3, 4), (4, 3), (4, 4), (5, 5)\}$. Then the equivalence classes resulting from R' are B_1, B_2, B_3.

11. First assume that R is an equivalence relation on A. Thus R is reflexive. It remains only to show that R is circular. Assume that $x \ R \ y$ and $y \ R \ z$, where $x, y, z \in A$. Since R is transitive, $x \ R \ z$. Since R is symmetric, $z \ R \ x$. Thus R is circular. It remains to verify the converse.

13. (a) Let $a \in \mathbb{N}$. Then $a^2 + a^2 = 2(a^2)$ is an even integer and so $a \ R \ a$. Thus R is reflexive. It remains to show that R is symmetric and transitive.

(b) $[0] = \{0, \pm 2, \pm 4, \ldots\}$, $[1] = \{\pm 1, \pm 3, \pm 5, \ldots\}$.

15. Let $R = R_1 \cap R_2$. First, we show that R is reflexive. Let $a \in S$. Since R_1 and R_2 are equivalence relations on S, it follows that $(a, a) \in R_1$ and $(a, a) \in R_2$. Thus $(a, a) \in R$ and so R is reflexive. It remains to show that R is symmetric and transitive.

17. (a) $f = \{(0, -2), (1, -1), (3, 0), (4, 1)\}$. (b) $\{-2, -1, 0, 1\}$ **19.** No.   **21.** No.

23. (a) No. (b) The function f is onto.

Proof. Let $m \in \mathbb{Z}$. Then $3m \in \mathbb{Z}$ and $f(3m) = \lceil (3m)/3 \rceil = \lceil m \rceil = m$. Thus f is onto. ∎

25. (a) **Proof.** We first show that $f^{-1}(B_1) \cup f^{-1}(B_2) \subseteq f^{-1}(B_1 \cup B_2)$. Let $a \in f^{-1}(B_1) \cup f^{-1}(B_2)$. Then $a \in f^{-1}(B_1)$ or $a \in f^{-1}(B_2)$, say the former. Thus $f(a) \in B_1$. Since $B_1 \subseteq B_1 \cup B_2$, it follows that $f(a) \in B_1 \cup B_2$ and so $a \in f^{-1}(B_1 \cup B_2)$. Thus $f^{-1}(B_1) \cup f^{-1}(B_2) \subseteq f^{-1}(B_1 \cup B_2)$. Next, we show that $f^{-1}(B_1 \cup B_2) \subseteq f^{-1}(B_1) \cup f^{-1}(B_2)$. Let $a \in f^{-1}(B_1 \cup B_2)$. Then $f(a) \in B_1 \cup B_2$ and so $f(a) \in B_1$ or $f(a) \in B_2$, say the former. Hence $a \in f^{-1}(B_1)$. Since $f^{-1}(B_1) \subseteq f^{-1}(B_1) \cup f^{-1}(B_2)$, we have $a \in f^{-1}(B_1) \cup f^{-1}(B_2)$, which implies that $f^{-1}(B_1 \cup B_2) \subseteq f^{-1}(B_1) \cup f^{-1}(B_2)$. ∎

(b) **Proof.** We first show that $f^{-1}(B_1 \cap B_2) \subseteq f^{-1}(B_1) \cap f^{-1}(B_2)$. Let $a \in f^{-1}(B_1 \cap B_2)$. Then $f(a) \in B_1 \cap B_2$ and so $f(a) \in B_1$ and $f(a) \in B_2$. Hence $a \in f^{-1}(B_1)$ and $a \in f^{-1}(B_2)$. Thus $a \in f^{-1}(B_1) \cap f^{-1}(B_2)$, implying that $f^{-1}(B_1 \cap B_2) \subseteq f^{-1}(B_1) \cap f^{-1}(B_2)$. Next, we show that $f^{-1}(B_1) \cap f^{-1}(B_2) \subseteq f^{-1}(B_1 \cap B_2)$. Let $a \in f^{-1}(B_1) \cap f^{-1}(B_2)$. Then $a \in f^{-1}(B_1)$ and $a \in f^{-1}(B_2)$. Hence $f(a) \in B_1$ and $f(a) \in B_2$, implying that $f(a) \in B_1 \cap B_2$ and so $a \in f^{-1}(B_1 \cap B_2)$. Thus $f^{-1}(B_1) \cap f^{-1}(B_2) \subseteq f^{-1}(B_1 \cap B_2)$. ∎

27. (a) bijective. (b)-(d) not bijective.   **29.** (a) $f(x) = x^3 - x$. (b) $f(x) = 2^x$.

31. Proof. Assume that $f(c) = f(d)$, where $c, d \in \mathbb{R}$. Then $ac + b = ad + b$. Subtracting b from both sides and dividing by a, we have $c = d$ and so f is one-to-one. To show that f is onto, let $r \in \mathbb{R}$. Then $(r - b)/a \in \mathbb{R}$. Since $f\left(\frac{r-b}{a}\right) = a\left(\frac{r-b}{a}\right) + b = r$, f is onto. ∎

33. Assume that $f(a) = f(b)$, where $a, b \in \mathbb{R} - \{5\}$. Then $\frac{a+1}{a-5} = \frac{b+1}{b-5}$ and so $(a+1)(b-5) = (a-5)(b+1)$. Simplifying, we obtain $ab + b - 5a - 5 = ab + a - 5b - 5$. Therefore $6a = 6b$ and so $a = b$. Thus f is one-to-one. It remains to show that f is onto.

35. Proof. We proceed by induction. The derivative of $f(x) = \ln x$ is $f'(x) = f^{(1)}(x) = 1/x$. For $n = 1$, $\frac{(-1)^{n+1}(n-1)!}{x^n} = \frac{(-1)^2 0!}{x} = \frac{1}{x}$ and so the result holds for $n = 1$. Assume that the kth derivative of $f(x)$ is $f^{(k)}(x) = \frac{(-1)^{k+1}(k-1)!}{x^k} = (-1)^{k+1}(k-1)!x^{-k}$, where k is an arbitrary positive integer. We show that $f^{(k+1)}(x) = \frac{(-1)^{k+2}k!}{x^{k+1}}$. Observe that

$$
\begin{aligned}
f^{(k+1)}(x) &= \frac{d}{dx}f^{(k)}(x) = \frac{d}{dx}\left[(-1)^{k+1}(k-1)!x^{-k}\right] \\
&= (-1)^{k+1}(k-1)!(-k)x^{-(k+1)} = (-1)^{k+2}k(k-1)!x^{-(k+1)} = \frac{(-1)^{k+2}k!}{x^{k+1}}.
\end{aligned}
$$

By the Principle of Mathematical Induction, $f^{(n)}(x) = \frac{(-1)^{n+1}(n-1)!}{x^n} = (-1)^{n+1}(n-1)!x^{-n}$ for every positive integer n. ∎

37. The function $f : \mathbb{N} \to A$ defined by $f(n) = -n$ is bijective.

39. Divide the proof into cases, according to whether A is finite or denumerable and B is finite or denumerable.

41. False. Let A be the set of even integers and B the set of odd integers.

43. (a) **Proof.** Assume, to the contrary, that nr is rational. Then $nr = a/b$ for some integers a and b and $b \neq 0$. Thus $r = a/(nb)$. Since a and nb are integers and $nb \neq 0$, it follows that r is rational, a contradiction. ∎

(b) False. Consider $S = \{nr : n \in \mathbb{N}\}$ where $r \in \mathbb{I}$.

Chapter 6: Algorithms and Complexity

Section 6.1

1. This is illustrated in the table:

i	a_i	x	output
		3	
2	5	5	
3	4	5	
4	5	5	$x = 5$

3. (a) **Algorithm:** *Find the Minimum of Three Numbers a, b, c.*

Input: Three numbers a, b, c.

Output: $x = \min(a, b, c)$.

    1. $x := a$ [x is assigned the value a]

    2. **if** $b < x$ **then** $x := b$ [if $b < x$, then x is assigned the value b]

    3. **if** $c < x$ **then** $x := c$ [if $c < x$, then x is assigned the value c]

    4. **output** x

(b) In this case $a = 8$, $b = 6$, $c = 4$. In Step 1, x is assigned the value 8. Since $6 < 8$, x is assigned the value 6 in Step 2. Since $4 < 6$, x is assigned the value 4 in Step 3. Step 4 then outputs x, namely 4.

(c) In this case, $a = 7$, $b = 5$, $c = 5$. In Step 1, x is assigned the value 7. Since $5 < 7$, x is assigned the value 5 in Step 2. In Step 3, we check whether $c < x$, that is, whether $5 < 5$. Since this is not true, we move on to Step 4, where $x = 5$ is output.

5. (a) **Algorithm:** *Find the Next-to-Largest Number in a List $s : a_1, a_2, \ldots, a_n$ of $n \geq 2$ Distinct Numbers.*

Input: An integer $n \geq 2$ and a sequence $s : a_1, a_2, \ldots, a_n$ of n distinct numbers.

Output: The next-to-largest number y in s.

    1. $x := a_1$ [x is assigned the first number in the sequence]

    2. **for** $i := 2$ **to** n **do** [i is assigned the integers from 2 to n]

    3.   **if** $a_i > x$ **then** $x := a_i$

         [if $a_i > x$ then a value larger than x has been found and x is replaced by a_i]

    4. $y := a_1$ [y is assigned the first number in the sequence]

    5. **if** $y = x$ **then** $y := a_2$

         [if y is the largest number in the sequence, then y is replaced by a_2]

    6. **for** $i := 1$ **to** n **do**

    7.   **if** $y < a_i < x$ **then** $y := a_i$

         [if $a_i > y$ but $a_i \neq x$, then a value larger than y, but not

         the largest number in s, has been found and y is replaced by a_i]

    8. **output** y [the next-to-largest number in s is output]

(b) In this case, $a_1 = 7$, $a_2 = 10$, $a_3 = 12$, $a_4 = 15$. The numbers in this sequence s are input, as is the length $n = 4$ of the sequence. In Step 1, x is assigned the value 7. Step 2 is a for loop and i is initially assigned the value 2. Because $a_2 = 10 > 7 = x$, the value 10 is assigned to x in Step 3. We repeat Step 3 for $i = 3$. Because $a_3 = 12 > 10$, x is assigned the value 12 in Step 3. We repeat Step 3 for $i = 4$. Since $a_4 = 15 > 12$, x is assigned the value 15 in Step 3. Since $i = n = 4$, we move on to Step 4, where y is given the initial value $a_1 = 7$. In Step 5, since $y \neq x$, the conclusion of Step 5 is not executed. Step 6 is a for loop where i is initially assigned the value 1. Because $y = a_1 = 7$, the conclusion of Step 7 is not executed when $i = 1$. Step 7 is repeated for $i = 2$. Since $a_2 = 10$ and $7 = y < 10 < x = 15$, y is assigned the value 10 in Step 7. Step 7 is repeated for $i = 3$. Since $a_3 = 12$ and $10 = y < 12 < x = 15$, y is assigned the value 12 in Step 7. Since $x = a_4$, Step 7 is not executed when $i = 4 = n$. Thus $y = 12$, which is output in Step 8.

7. (a) 19. (b) This algorithm finds the middle number in a list $s : a_1, a_2, \ldots, a_n$ of n distinct integers, where n is odd.

9. Here $k = 11$, $n = 4$ and $a_1 = 9$, $a_2 = 10$, $a_3 = 14$, $a_4 = 11$. In Step 1, i is assigned the value 1. Step 2 is a while loop. Since $1 = i \leq n = 4$, Steps 3 and 4 are performed. Since $a_1 \neq 11$, the conclusion of Step 3 is not executed. In Step 4, i is increased to 2. Because $i = 2 \leq 4$, Steps 3 and 4 are performed again. Since $a_2 \neq 11$, we proceed to Step 4, where i is increased to 3. Because $i = 3 \leq 4$, Steps 3 and 4 are executed again. Since $a_3 \neq 11$, we proceed to Step 4, where i is increased to 4. Because $i = 4 \leq 4$, Steps 3 and 4 are executed once again. Now $a_4 = 11$ and so 11 *is in the sequence* is output. Also, i is assigned the valued $n + 1 = 5$. In Step 4, i is increased to 6. Since $6 \leq 4$ is not true, the while loop is exited and we proceed to Step 5. Because $i \neq 5$, the algorithm ends and we are left with the output: 11 *is in the sequence.*

11. Algorithm: *Determine Whether a Sequence $s : a_1, a_2, \ldots, a_n$ of n Numbers Contains a Negative Number.*

Input: A positive integer n and a sequence $s : a_1, a_2, \ldots, a_n$ of n numbers.

Output: "the term" a_i "is a negative number in s" or "there is no negative number in s", as appropriate.

    1. $i := 0$

    2. **while** $i < n$ **do**
        begin

    3. $i := i + 1$

    4.   **if** $a_i < 0$ **then** $i := n + 1$ and
            output "the term" a_i "is a negative number in s"
        end

    5. **if** $i = n$ **then output** "there is no negative number in s".

13. Algorithm: *Determine Whether and Where a Given Number Appears in a Sequence*

This algorithm determines whether a given number k appears in a sequence $s : a_1, a_2, \ldots, a_n$ of n distinct numbers and, if so, where in the sequence it appears.

Input: A positive integer n, a sequence $s : a_1, a_2, \ldots, a_n$ of n distinct numbers and a number k.

Output: "the number" k "is term" i "in the sequence" or k "is not in the sequence", as appropriate.

    1. $i := 1$

    2. **while** $i \leq n$ **do**
        begin

    3.   **if** $a_i = k$ **then output** "the number" k "is term" i "in the sequence"
            and $i := n + 1$.

    4. $i := i + 1$
        end

    5. **if** $i = n + 1$ **then output** k "is not in the sequence".

15. The integer $n = 3$ is input as is the sequence $s : 1, 2, 3$. So $a_1 = 1$, $a_2 = 2$, $a_3 = 3$. In Step 1, i is initialized to 0. Since $0 \leq 2$ in Step 2, Steps 3–5 are performed. In Step 3, i is increased to 1 and j is defined to be 2 in Step 4. Since $2 = j \leq n = 3$ in Step 5, Steps 6 and 7 are performed. Since $a_1 \neq a_2$, the conclusion of Step 6 is not executed. In Step 7, j is increased to 3. We return to Step 5. Since $3 = j \leq n = 3$ in Step 5, Steps 6 and 7 are performed. Since $a_1 \neq a_3$, the conclusion of Step 6 is not executed. In Step 7, j is increased to 4. We return to Step 5. Since $4 = j \leq n = 3$ is false, this completes Step 5 and we return to Step 2. Continue from here.

Section 6.2

1. Let $C' = \max(C_1, C_2, \ldots, C_{k-1}, C)$.

3. Since $5n + 7 \leq 5n + 7n = 12n \leq 12n^2$ for $n \geq 1$, it follows that $f = O(g)$, where $C = 12$ and $k = 1$ in the definition. Next we show that $g \neq O(f)$. Assume, to the contrary, that $g = O(f)$. Then there exist a positive constant C and a positive integer k such that $n^2 \leq C(5n + 7)$ for every integer $n \geq k$. Dividing by n^2, we obtain $1 \leq \frac{5C}{n} + \frac{7C}{n^2}$.

   Let n be an integer greater than k and $14C$. So $n > \max(k, 14C)$. Therefore, $\frac{5C}{n} < \frac{5C}{14C} = \frac{5}{14} < \frac{1}{2}$ and $\frac{7C}{n^2} < \frac{7C}{14Cn} = \frac{1}{2n} \leq \frac{1}{2}$, and so $1 \leq \frac{5C}{n} + \frac{7C}{n^2} < \frac{1}{2} + \frac{1}{2} = 1$, which is a contradiction.

5. Since $2^n \geq n$ for every positive integer n, it follows that $\log(2^n) \geq \log n$; so $n \geq \log n$ for $n \geq 1$. Therefore, $\log n = O(n)$, where $C = k = 1$ in the definition.

7. $f(n) = O(n^2)$ for the functions f in (a)–(d).

9. First, we show that $f = O(g)$. Since $2n + 1 \leq 2n + n = 3n$ for $n \geq 1$, it follows that $f = O(g)$, where $C = 3$ and $k = 1$ in the definition. Next, we show that $g = O(f)$. Since $n \leq 2n + 1$ for $n \geq 1$, we have $g = O(f)$, where $C = k = 1$ in the definition.

11. First, we show that $f = O(g)$. Since $n^2 + 4n + 1 \leq n^2 + 4n^2 + 1 = 5n^2 + 1 \leq 5n^2 + 20 = 5(n^2 + 4)$ for $n \geq 1$, we have $f = O(g)$, where $C = 5$ and $k = 1$ in the definition. Next, we show that $g = O(f)$. Since $n^2 + 4 \leq n^2 + 4n + 1$, it follows that $g = O(f)$, where $C = k = 1$ in the definition.

13. Assume that $f = \Theta(g)$. Then $f = O(g)$ and $g = O(f)$ by Theorem 6.20. Also, by Theorem 6.20, $g = \Theta(f)$. ∎

15. Observe that $f(n) = \frac{1}{2}n^2 + 5n + 1 \leq \frac{1}{2}n^2 + 5n^2 + \frac{33}{4} = \frac{11}{2}n^2 + \frac{33}{4} = \frac{11}{4}(2n^2 + 3) = \frac{11}{4}g(n)$ for every positive integer n. Therefore, $f = O(g)$, where $C = \frac{11}{4}$ and $k = 1$ in the definition. Next, $g(n) = 2n^2 + 3 \leq 2n^2 + 20n + 4 = 4(\frac{1}{2}n^2 + 5n + 1) = 4f(n)$ for every positive integer n. Thus $g = O(f)$, where $C = 4$ and $k = 1$ in the definition.

Section 6.3

1. **Algorithm:** *Find a Number in a List $s : a_1, a_2, \ldots, a_n$ of $n \geq 3$ Distinct Numbers That is Neither the Maximum nor the Minimum Number in the List.*

   Input: An integer $n \geq 3$ and a sequence $s : a_1, a_2, \ldots, a_n$ of n distinct numbers.

   Output: A number y in s that is neither the maximum nor the minimum number in s.

   > 1. $x := a_1$
   > 2. $z := a_2$
   > 3. **if** $x > z$ **then** $t := x$, $x := z$ and $z := t$
   >   [If $x > z$, then the values of x and z are interchanged so that in every case, $x < z$.]
   > 4. **if** $x < a_3 < z$ **then** $y := a_3$
   > 5. **if** $a_3 < x$ **then** $y := x$
   > 6. **if** $a_3 > z$ **then** $y := z$
   > 7. **output** y

   The time complexity of this algorithm is $5 = O(1)$.

3. **Algorithm:** *Compute $(a_1 + a_2 + \cdots + a_n)(a_1 + a_2 + \cdots + a_{n-1}) \cdots (a_1 + a_2)a_1$ for the Sequence $s : a_1, a_2, \ldots, a_n$ of n Numbers*

   Input: A positive integer n and a sequence $s : a_1, a_2, \ldots, a_n$ of n numbers.

   Output: $x = (a_1 + a_2 + \cdots + a_n)(a_1 + a_2 + \cdots + a_{n-1}) \cdots (a_1 + a_2)a_1$

1. $i := 1$
2. $x := a_1$
3. $s := a_1$
4. **while** $i < n$ **do**
    begin
        $i := i + 1$
        $s := s + a_i$
        $x := sx$
    end
5. **output** x

The time complexity of this algorithm is $n = \Theta(n)$.

5. Algorithm: *Compute* $|a_1 - a_2| + |a_1 - a_3| + \cdots + |a_1 - a_n| + |a_2 - a_3| + \cdots + |a_{n-1} - a_n|$ *for the Sequence* $s : a_1, a_2, \ldots, a_n$ *of* $n \geq 2$ *Numbers*

Input: An integer $n \geq 2$ and a sequence $s : a_1, a_2, \ldots, a_n$ of n numbers.

Output: $s = |a_1 - a_2| + |a_1 - a_3| + \cdots + |a_1 - a_n| + |a_2 - a_3| + \cdots + |a_{n-1} - a_n|$.

1. $s := 0$
2. **for** $i := 1$ **to** $n - 1$ **do**
    begin
3.   **for** $j := i + 1$ **to** n **do**
        begin
4. $x := |a_i - a_j|$
5. $s := s + x$
        end
    end
6. **output** s

The time complexity of this algorithm is $\Theta(n^2)$.

7. (a)**Algorithm:** *Compute* a^n *For a Real Number* a *and a Nonnegative Integer* n.

Input: A real number a and a nonnegative integer n.

Output: $p = a^n$

1. $p := 1$
2. **for** $i := 1$ **to** n **do**
3. $p := ap$
4. **output** y

(b) The time complexity of this algorithm is $n = \Theta(n)$.

9. (a) $3n = \Theta(n)$. (b) Horner's algorithm is more efficient than the algorithm in Exercise 8.

Section 6.4

1. In this case, $k = 17$, $n = 9$ and $a_1 = 3$, $a_2 = 6$, $a_3 = 7$, $a_4 = 8$, $a_5 = 15$, $a_6 = 17$, $a_7 = 19$, $a_8 = 23$, $a_9 = 24$. By Steps 1 and 2, $a = 1$ and $b = 9$. Since $1 = a \leq b = 9$, Steps 4–8 are executed. By Step 4, $p = 5$. Among Steps 5–7, only $a_5 < k$ is true and so only the conclusion of Step 7 is performed, where a is assigned the value $5 + 1 = 6$. The subsequence a_6, a_7, a_8, a_9 is now considered. Now continue.

3. In this sequence, $n = 4$ and $a_1 = 11$, $a_2 = 9$, $a_3 = 10$, $a_4 = 8$. See the table below.

original sequence	$i = 1$ $j = 1$	$i = 2$ $j = 1$	$i = 3$ $j = 1$	$i = 1$ $j = 2$	$i = 2$ $j = 2$	$i = 1$ $j = 3$	final sequence
11	**9**	9	9	**9**	9	**8**	8
9	**11**	**10**	10	10	**8**	**9**	9
10	10	**11**	**8**	8	**10**	10	10
8	8	8	**11**	11	11	11	11

5. Algorithm: This algorithm sorts the terms of a sequence $s : a_1, a_2, \ldots, a_n$ of n numbers so that the terms of s are reordered from largest to smallest.

Input: A positive integer n and a sequence $s : a_1, a_2, \ldots, a_n$ of n numbers.

Output: A sequence of n numbers whose terms are those of s in nonincreasing order.

1. **for** $j := 1$ **to** $n - 1$ **do**
2.   **for** $i := 1$ **to** $n - j$ **do**
3.   **if** $a_i < a_{i+1}$ **then swap**(a_i, a_{i+1})
4. **output** s

7. See Figure 8.

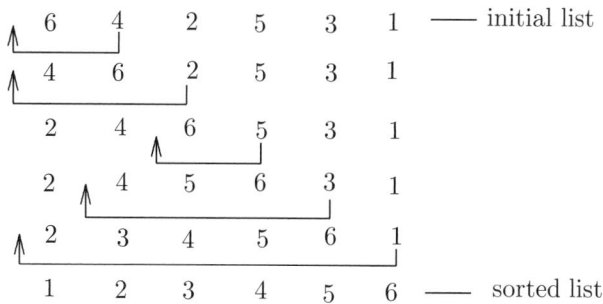

Figure 8: Sorting a list by the Insertion Sort Algorithm in Exercise

9. The steps performed on the lists are given in the following table.

List 1	List 2	Comparison	Merged List
1 3 4 7 8 9	2 5 6	$1 < 2$	1
3 4 7 8 9	2 5 6	$2 < 3$	1 2
3 4 7 8 9	5 6	$3 < 5$	1 2 3
4 7 8 9	5 6	$4 < 5$	1 2 3 4
7 8 9	5 6	$5 < 7$	1 2 3 4 5
7 8 9	6	$6 < 7$	1 2 3 4 5 6
7 8 9	$\emptyset$		1 2 3 4 5 6 7 8 9

Supplementary Exercises for Chapter 6

1. (a) **Algorithm:** *Determine Whether Some Number Appears Twice in a Sequence and, If So, Where in the Sequence It Occurs*

Input: A positive integer n and a sequence $s : a_1, a_2, \ldots, a_n$ of n numbers.

Output: a_i "appears in s as terms" i "and" j or "no term appears twice in s", as appropriate.

1. $i := 0$
2. **while** $i \leq n - 1$ **do**
   begin
3. $i := i + 1.$
4. $j := i + 1$

5.   **while** $j \leq n$ **do**
                begin
6.   **if** $a_i = a_j$ **then output** a_i "appears in s as terms" i "and" j,
                $i = n + 1$ and $j := n$
7. $j := j + 1$
                end
        end

8. **if** $i = n$ **then output** "no term appears twice in s"

(b) For this sequence, $n = 4$ and $a_1 = 6$, $a_2 = 7$, $a_3 = 5$, $a_4 = 7$. By Step 1, i is assigned the value 0. Since $0 = i \leq n - 1 = 3$ in Step 2, Steps 3–5 are executed. In Steps 3 and 4, we have $i = 1$ and $j = 2$. Since $2 = j \leq n = 4$ in Step 5, Steps 6 and 7 are executed. Since $6 = a_1 \neq a_2 = 7$ in Step 6, the conclusion of Step 6 is not executed. In Step 7, j is increased to 3. Since $3 = j \leq n = 4$ in Step 5, Steps 6 and 7 are executed. Since $6 = a_1 \neq a_3 = 5$, we move on to Step 7 and j is increased to 4. Since $4 = j \leq n = 4$ in Step 5, we execute Steps 6 and 7. Since $6 = a_1 \neq a_4 = 7$, we move on to Step 7 and j is increased to 5. However, $5 = j \leq n = 4$ is false; so we have completed Step 5. We now return to Step 2. Now continue.

3. (a) Some possibilities: (1) Determine Whether Some Number Appears Three Times in s. (2) Determine Whether Some Number Appears Exactly Twice in s. (3) Determine Whether s Contains Two Terms Whose Absolute Values Are Equal. (4) Determine Whether s Contains Two Terms a_i and a_j Whose Difference $a_i - a_j$ is a Fixed Constant k.

5. Since $2^n \geq n$ for every positive integer n, it follows that $\log n \leq n$ for every positive integer n. Multiplying by n, we have that $n \log n \leq n^2$ for every positive integer n. Hence $n \log n = O(n^2)$, where $C = k = 1$ in the definition.

7. For each positive integer n, observe that $f(n) = \frac{2n^3 + n}{n + 2} \leq \frac{2n^3 + 4n^2}{n + 2} = \frac{2n^2(n + 2)}{n + 2} = 2n^2$. Thus $f = O(n^2)$, where $C = 2$ and $k = 1$ in the definition.

9. Observe, for every positive integer n, that $f(n) = n^2 + 3n + 1 \leq n^3 + 3n^3 + n^3 = 5n^3 = 5g(n)$. Thus $f = O(g)$, where $C = 5$ and $k = 1$ in the definition. Next, we show that $g \neq O(f)$. Assume, to the contrary, that $g = O(f)$. Then there exist a positive constant C and a positive integer k such that $n^3 \leq C(n^2 + 3n + 1)$ for every integer $n \geq k$. Dividing by n^3, we have $1 \leq C\left(\frac{1}{n} + \frac{3}{n^2} + \frac{1}{n^3}\right) \leq C\left(\frac{1}{n} + \frac{3}{n} + \frac{1}{n}\right) = \frac{5C}{n}$. Let $n > \max(k, 5C)$. So $\frac{1}{n} < \frac{1}{5C}$ and so $1 \leq \frac{5C}{n} < (5C)\frac{1}{5C} = 1$ which is impossible. Since $g \neq O(f)$, it follows that $g \neq \Theta(f)$

11. Since $2^n \leq 3^n$ for every positive integer n, it follows that $f(n) = O(g(n))$. Now we show that $f(n) \neq \Theta(g(n))$. Assume, to the contrary, that there exist a positive constant C and a positive integer k such that $g(n) \leq Cf(n)$ for every integer $n \geq k$. Thus $3^n \leq C \cdot 2^n$ and so $\frac{3^n}{2^n} = (1.5)^n \leq C$. Since $\sqrt{2} < 1.5$, it follows that $2^{\frac{n}{2}} = (\sqrt{2})^n < (1.5)^n \leq C$. Taking logarithms of both sides, we have $n/2 \leq \log C$ and so $n \leq 2 \log C$. Let n be an integer greater than $\max(k, 2 \log C)$. Then $n > 2 \log C$. This is a contradiction.

13. There are several possible solutions. Suppose first that we use the Bubble Sort Algorithm, say, to sort the sequence. The time complexity of this algorithm is $\Theta(n^2)$. Hence we may assume that s is a nondecreasing sequence. Thus the minimum value of $|a_i - a_j|$ among all pairs a_i, a_j of distinct terms of s occurs when $j = i + 1$ for some integer i. An algorithm that does the following can be written.

Algorithm: *This algorithm determines the minimum value of* $|a_i - a_{i+1}|$ *where* $1 \leq i \leq n - 1$ *for a sequence* $s : a_1, a_2, \ldots, a_n$ *of n numbers listed in nondecreasing order.*

Input: A positive integer n and a nonincreasing sequence $s : a_1, a_2, \ldots, a_n$ of n numbers.

Output: $m = \min\{|a_i - a_{i+1}| : 1 \leq i \leq n - 1\}$

15. **Algorithm:** *This algorithm inserts an integer a in an appropriate position in an increasing sequence $s : a_1, a_2, \ldots, a_n$ of n (distinct) numbers resulting in a nondecreasing sequence of $n+1$ numbers.*

    Input: An integer a and an increasing sequence $s : a_1, a_2, \ldots, a_n$ of n numbers.

    Output: A nondecreasing sequence of $n + 1$ whose terms are a and those of s.

    1. $j := n$
    2. **for** $i := 1$ **to** n **do**
    3.   **if** $a < a_i$ **then** $j := i - 1$ and $i := n + 1$,
    4. **if** $j := 0$ **then**
       begin
    5.   **output** a
    6.   **for** $i := 1$ **to** n **do**
    7.   **output** a_i
       end
    8. **if** $j := n$ **then**
       Then continue.

17. **Algorithm:** This algorithm determines the number of 1s in an n-bit string.

    Input: A positive integer n and an n-bit string $a_1, a_2, \ldots, a_n$.

    Output: The number k of 1s in the string.

    1. $k := 0$
    2. **for** $i := 1$ **to** n **do**
       if $a_i = 1$ **then** $k := k + 1$
    3. **output** k

19. **Algorithm:** This algorithm determines the location of each term in a sequence $s : a_1, a_2, \ldots, a_n$ of integers that is greater than the sum of all previous terms in the sequence or determines that s contains no such term.

    Input: A positive integer n and a sequence $s : a_1, a_2, \ldots, a_n$ of n integers.

    Output: The location of each term of s that is greater than the sum of all previous terms in s or "The sequence s contains no such term", as appropriate.

    1. $k := 0$
    2. $s := 0$
    3. **for** $i := 2$ **to** n **do**
       begin
    4. $s := s + a_{i-1}$
    5.   **if** $a_i > s$ **then** $k := k + 1$ and **output** i
       end
    6. **if** $k = 0$ **then output** "the sequence s contains no such term"

Chapter 7: Integers

Section 7.1

1. (a) $7 \mid -70$, $c = -10$. (b) $16 \nmid -40$. (c) $1 \mid 10$, $c = 10$.
   (d) $8 \mid -8$, $c = -1$. (e) $14 \mid 0$, $c = 0$. (f) $0 \nmid 14$ (not defined).

3. **Proof.** Assume that $a \mid b$. Then $b = ax$ for some integer x. Thus $-b = -(ax) = a(-x)$ and $b = (-a)(-x)$. Since $-x$ is an integer, $a \mid (-b)$ and $(-a) \mid b$.   ∎

5. **Proof.** First, assume that $ac \mid bc$. Then $bc = (ac)x = c(ax)$ for some integer x. Since $c \neq 0$, we can divide by c, obtaining $b = ax$. So $a \mid b$. For the converse, assume that $a \mid b$. Then $b = ax$ for some integer x. Multiplying by c, we have $bc = (ax)c = (ac)x$. Since x is an integer, $ac \mid bc$.   ∎

7. $a = 1$ and $b = -1$ is a counterexample.

9. **Proof.** Assume that $a \mid b$ or $a \mid c$, say $a \mid b$. Then $b = ax$ for some integer x. Thus $bc = (ax)c = a(xc)$. Since xc is an integer, $a \mid bc$.   ∎

11. We proceed by induction. Since $3 \mid (4 \cdot 0^3 + 5 \cdot 0)$, the statement is true for $n = 0$. Assume that $3 \mid (4k^3 + 5k)$, where k is a nonnegative integer. Then $4k^3 + 5k = 3\ell$ for some integer ℓ. Next show that 3 divides $4(k+1)^3 + 5(k+1)$. Observe that $4(k+1)^3 + 5(k+1) = 3(\ell + 4k^2 + 4k + 3)$.

13. We proceed by induction. Since $7 \mid (2^{3 \cdot 0} - 1)$, the statement is true for $n = 0$. Assume that $7 \mid (2^{3k} - 1)$, where k is a nonnegative integer. Then $2^{3k} - 1 = 7\ell$ for some integer ℓ. Thus $2^{3k} = 7\ell + 1$. Next show that 7 divides $2^{3(k+1)} - 1$. Observe that $2^{3(k+1)} - 1 = 7(8\ell + 1)$.

15. If $n < 0$, then $n = -k$ for some positive integer k. Then apply Result 7.6 and Exercise 3.

17. We proceed by induction. For $n = 1$, $11^{n+1} + 12^{2n-1} = 11^2 + 12 = 121 + 12 = 133$. Since $133 \mid 133$, the statement is true for $n = 1$. Assume that $133 \mid (11^{k+1} + 12^{2k-1})$ for a positive integer k. Then $11^{k+1} + 12^{2k-1} = 133x$ for some integer x and so $12^{2k-1} = 133x - 11^{k+1}$. Next show that $133 \mid (11^{k+2} + 12^{2k+1})$. Observe that $11^{k+2} + 12^{2k+1} = 133(144x - 11^{k-1})$.

Section 7.2

1. (a) $2 \cdot 5^3$. (b) $3^3 \cdot 11$. (c) $2 \cdot 11^3$. (d) $5^2 \cdot 7^2$. (e) $3^4 \cdot 11$.   3. 1009.

5. **Proof.** Assume, to the contrary, that there is a finite number of primes greater than $1,000,000$. Since there is certainly a finite number of primes less than $1,000,000$, it follows that there is a finite number of primes. This is a contradiction.   ∎

7. (a) $111 = 3 \cdot 37$. (b) $1111 = 11 \cdot 101$. (c) $111,111 = 111 \cdot 1001$. (d) $11,111 = 41 \cdot 271$.

9. First, ab may be prime, such as 31. However, no other integer in the list can be prime. If $abab \cdots ab$ contains $2k$ digits, where $k \geq 2$, then $abab \cdots ab = ab\,(1010 \cdots 01)$.

11. **Proof.** If $n = 1$, then $n^3 + 1 = 2$ is a prime. Assume, to the contrary, that there exists a prime $p = n^3 + 1$ for some integer $n \geq 2$. Then $p = (n+1)(n^2 - n + 1)$. Therefore, either $n + 1 = 1$ or $n^2 - n + 1 = 1$. If $n + 1 = 1$, then $n = 0$, which is impossible; while if $n^2 - n + 1 = 1$, then $n^2 - n = 0$, implying that $n = 1$ or $n = 0$, both of which are impossible.   ∎

13. **Proof.** Assume that Goldbach's conjecture is true. Then every even integer that is 4 or greater can be expressed as the sum of two primes. Let $n \geq 3$ be an integer. We show that n can be expressed as the sum of three integers, each of which is either 1 or a prime. Since $3 = 1 + 1 + 1$ and $4 = 1 + 1 + 2$, the conjecture is true if $n = 3$ and $n = 4$. Hence we may assume that $n \geq 5$. We consider two cases.

Case 1. n is odd. Then $n - 1$ is even and $n - 1 \geq 4$. Thus $n - 1$ can be expressed as the sum of two primes p_1 and p_2 and so $n = 1 + p_1 + p_2$.

Case 2. n is even. Then $n - 2$ is even and $n - 2 \geq 4$. Thus $n - 2$ can be expressed as the sum of two primes p_1 and p_2 and so $n = 2 + p_1 + p_2$. ∎

15. It's not clear if the conjecture is true.

17. Since $3^k - 1$ is even and $3 \mid (4^k - 1)$ for every integer $k \geq 2$, these numbers are not primes.

Section 7.3

1. (a) $q = r = 4$. $48 = 11 \cdot 4 + 4$. (b) $q = r = 0$. $0 = 11 \cdot 0 + 0$.

  (c) $q = -5$ and $r = 7$. $-48 = 11 \cdot (-5) + 7$. (d) $q = 0$ and $r = 9$. $9 = 11 \cdot 0 + 9$.

3. (a) 3 and 5. (b) 3 and 15. (c) 4 and 0. (d) 0 and 24. (e) 0 and 0. (f) −3 and 26.

5. Let $n = 2k + 1$ for some integer k. Then $n^2 = (2k + 1)^2 = 4(k^2 + k) + 1$.   **7.** No.

9. (a) Let $n \in \mathbb{Z}$. Then $n = 3q$, $n = 3q + 1$ or $n = 3q + 2$ for some integer q. Consider these three cases. (b) The proof is similar to (a).

11. Let a, b, c, d and e be five consecutive integers. Hence we may assume that $a = n$, $b = n + 1$, $c = n + 2$, $d = n + 3$ and $e = n + 4$ for some integer n. Thus $n = 5q$, $n = 5q + 1$, $n = 5q + 2$, $n = 5q + 3$ or $n = 5q + 4$ for some integer q. Consider these five cases.

13. (a) Let $n + 1, n + 2, \ldots, n + p$ be p consecutive integers. Then $n = pq + r$ for some integers q and r, where $0 \leq r \leq p - 1$. Since $1 \leq p - r \leq p$, one of the integers $1, 2, \ldots, p$ is $p - r$. Hence one of the integers $n + 1, n + 2, \ldots, n + p$ is $n + (p - r)$. Now continue.

  (b) Let $n + 1, n + 2, \ldots, n + p$ be p consecutive integers. Then

$$(n + 1) + (n + 2) + \cdots + (n + p) = pn + (1 + 2 + \cdots + p) = pn + \frac{p(p + 1)}{2} = p(n + \frac{p + 1}{2}).$$

Now continue.

15. Assume first that a and b are of the same parity and show that $4 \mid (a^2 - b^2)$. Consider two cases. *Case 1. a and b are even. Case 2. a and b are odd.*

For the converse, assume that a and b are of opposite parity and show that $4 \nmid (a^2 - b^2)$. Consider two cases. *Case 1. a is odd and b is even. Case 2. a is even and b is odd.*

17. Assume that $3 \nmid a$ and $3 \nmid b$. Then one of the following four situations occurs: (1) $a = 3s + 1$ and $b = 3t + 1$; (2) $a = 3s + 1$ and $b = 3t + 2$; (3) $a = 3s + 2$ and $b = 3t + 1$; (4) $a = 3s + 2$ and $b = 3t + 2$, where $s, t \in \mathbb{Z}$. We consider these four cases.

Case 1. $a = 3s + 1$ and $b = 3t + 1$. Then

$$\begin{aligned} a^2 - b^2 &= (3s + 1)^2 - (3t + 1)^2 = (9s^2 + 6s + 1) - (9t^2 + 6t + 1) \\ &= 9s^2 + 6s - 9t^2 - 6t = 3(3s^2 + 2s - 3t^2 - 2t). \end{aligned}$$

Since $3s^2 + 2s - 3t^2 - 2t$ is an integer, $3 \mid (a^2 - b^2)$.

[The proofs of the other cases are similar.]

19. Let p be a prime different from 2 and 5. Dividing p by 10, we obtain $p = 10k + r$ for some integers k and r, where $0 \leq r \leq 9$. If $r = 0$, then $10 \mid p$, which is impossible. If $r = 2$, then $p = 10k + 2 = 2(5k + 1)$. Since $5k + 1$ is an integer, $2 \mid p$, again an impossibility since $p \neq 2$. If $r = 4$, then $p = 10k + 4 = 2(5k + 2)$. Since $5k + 1$ is an integer, $2 \mid p$, which is a contradiction. Now continue.

21. Let $p \geq 5$ be a prime. Dividing p by 6, we obtain $p = 6k + r$ for some integers k and r, where $0 \leq r \leq 5$. If $r = 0$, then $6 \mid p$, which is impossible. If $r = 2$, then $p = 6k + 2 = 2(3k + 1)$. Since $3k + 1$ is an integer, $2 \mid p$. Since $p \geq 5$, we have a contradiction. Now continue.

23. Let $n = 3q + r$ where $r = 0, 1, 2$ and consider the three cases:

Case 1. $n = 3q$. *Case 2.* $n = 3q + 1$. *Case 3.* $n = 3q + 2$.

Section 7.4

1. (a) $47 \equiv 23 \pmod{8}$. (b) $18 \equiv 38 \pmod{5}$. (c) $20 \not\equiv 10 \pmod{3}$. (d) $12 \equiv 12 \pmod{13}$. (e) $37 \equiv 35 \pmod{2}$.   **3.** $-9, 4, 17, 30$.

5. First, we prove that if $a \equiv b \pmod{n}$, then $b = a + \ell n$ for some integer ℓ. Assume that $a \equiv b \pmod{n}$. By Theorem 7.25, $a = b + kn$ for some integer k. Then $b = a - kn$. Thus $b = a + \ell n$, where $\ell = -k$. Now verify the converse: If $b = a + \ell n$ for some integer ℓ, then $a \equiv b \pmod{n}$.

7. Proof. Since $m \mid n$, it follows that $n = mx$ for some integer x. Assume that $a \equiv b \pmod{n}$. Thus $n \mid (a - b)$ and so $a - b = ny$ for some integer y. Hence $a - b = ny = (mx)y = m(xy)$. Since xy is an integer, $m \mid (a - b)$ and so $a \equiv b \pmod{m}$.   ∎

9. Proof. Assume that $a \equiv b \pmod{n}$ and $c \equiv d \pmod{n}$. By Theorem 7.25, $a = b + nx$ and $c = d + ny$ for some integers x and y. Hence $a - c = (b - d) + (nx - ny) = (b - d) + n(x - y)$. Since $x - y$ is an integer, $a - c \equiv b - d \pmod{n}$ by Theorem 7.25.   ∎

11. Proof. We proceed by induction. By Exercise 10, the statement is true for $r = 2$. Assume that if $a \equiv b \pmod{n}$, then $a^k \equiv b^k \pmod{n}$ for an integer $k \geq 2$. Again, it then follows by Exercise 10 that $a^{k+1} \equiv b^{k+1} \pmod{n}$. By the Principle of Mathematical Induction, if $a \equiv b \pmod{n}$, then $a^r \equiv b^r \pmod{n}$ for every integer $r \geq 2$.   ∎

13. Proof. Assume that $a \equiv 0 \pmod{5}$ and $b \equiv 2 \pmod{5}$. Then $a = 5x$ and $b = 5y + 2$ for integers x and y. Observe that $a^2 + b^2 = 5(5x^2 + 5y^2 + 2y) + 4$. Since $5x^2 + 5y^2 + 2y$ is an integer, $a^2 + b^2 \equiv 4 \pmod{5}$.   ∎

15. Assume that one of a and b is congruent to 0 modulo 3 and that the other is not congruent to 0 modulo 3. We show that $a^2 + 2b^2 \not\equiv 0 \pmod{3}$. We consider two cases.

Case 1. $a \equiv 0 \pmod{3}$ *and* $b \not\equiv 0 \pmod{3}$. Since $a \equiv 0 \pmod{3}$, it follows that $a = 3p$ for some integer p. Since $b \not\equiv 0 \pmod{3}$, either $b = 3q + 1$ or $b = 3q + 2$ for some integer q. There are two subcases.

Subcase 1.1. $b = 3q + 1$. Then $a^2 + 2b^2 = (3p)^2 + 2(3q + 1)^2 = 9p^2 + 2(9q^2 + 6q + 1) = 3(3p^2 + 6q^2 + 4q) + 2$. Since $3p^2 + 6q^2 + 4q$ is an integer, $3 \nmid (a^2 + 2b^2)$ and so $a^2 + 2b^2 \not\equiv 0 \pmod{3}$.

Subcase 1.2. $b = 3q + 2$. (The proof is similar to that of Subcase 1.1.)

Now consider *Case 2.* $a \not\equiv 0 \pmod{3}$ *and* $b \equiv 0 \pmod{3}$.

17. (a) 6, 6, Yes. (b) 1, 1, Yes. (c) 6, 6, Yes. (d) 11, 1, No. (e) 0, 0, Yes. (f) 7, 2, No.

Section 7.5

1. DTZ. FWJ. HTWWJHY. XNW.   **3.** The likely message is HELP IS HERE.

5. (a) SIK. (b) TWIN. (c) DAN.

7. Both A and N are transformed into the same letter A, for example. Since this function is not one-to-one, two different words can be transformed into the same word.

9. (a) ABOVE. (b) $f^{-1}(x) = f(x) = x + (-1)^x$.

11. Possible values for a and b are $a = 5$ and $b = 0$. In this case, the word WING would correspond to the actual word GONE.

Section 7.6

1. (a) 2. (b) 1. (c) 10. (d) 2. (e) 17.   **3.** (a) $1 = 4 \cdot (-1) + 5 \cdot 1$. (b) $5 = 15 \cdot (-2) + 35 \cdot 1$.
(c) $12 = 12 \cdot 1 + 36 \cdot 0$. (d) $3 = 9 \cdot (-1) + 12 \cdot 1$. (e) $6 = 3 \cdot 30 + (-2) \cdot 42$.

5. Proof. Let a and b be integers, not both 0, and let $d = \gcd(a, b)$. Assume first that n is a linear combination of a and b. Then $n = ax + by$ for some integers x and y. Since $d \mid a$ and $d \mid b$, it follows that $a = ds$ and $b = dt$ for some integers s and t. Thus $n = ax + by = (ds)x + (dt)y = d(sx + ty)$. Since $sx + ty$ is an integer, $d \mid n$. For the converse, assume that $d \mid n$. Then $n = dk$ for some integer k. Since $d = \gcd(a, b)$, it follows that d is a linear combination of a and b and so $d = ax + by$ for some integers x and y. Therefore, $n = dk = (ax + by)k = a(xk) + b(yk)$. Since xk and yk are integers, n is a linear combination of a and b.   ■

7. Proof. Let k be a positive integer and $d = \gcd(a, b)$. Since $d = \gcd(a, b)$, it follows that $d = ax + by$ for some integers x and y. Thus $kd = k(ax + by) = (ka)x + (kb)y$. Since kd is a linear combination of ka and kb and $\gcd(ka, kb)$ is the smallest integer that is a linear combination of ka and kb, it follows that $\gcd(ka, kb) \le k \gcd(a, b)$. On the other hand, $\gcd(ka, kb)$ is a linear combination of ka and kb and so $\gcd(ka, kb) = (ka)s + (kb)t$ for some integers s and t. Since $d \mid a$ and $d \mid b$, it follows that $kd \mid ka$ and $kd \mid kb$. Thus $ka = (kd)x$ and $kb = (kd)y$ for some integers x and y. Therefore, $\gcd(ka, kb) = (ka)s + (kb)t = (kdx)s + (kdy)t = (kd)(xs + yt)$. Since $xs + yt$ is an integer, $kd \mid \gcd(ka, kb)$, implying that $k \gcd(a, b) \le \gcd(ka, kb)$. Therefore, $k \gcd(a, b) = \gcd(ka, kb)$.   ■

9. Proof. Let $a = p_1^{a_1} p_2^{a_2} \dots p_k^{a_k}$ and $b = p_1^{b_1} p_2^{b_2} \dots p_k^{b_k}$, where $p_1 < p_2 < \dots < p_k$ are primes. For each i ($1 \le i \le k$), the prime p_i divides exactly one of a and b and so exactly one of a_i and b_i is 0. Therefore, $\gcd(a, b) = p_1^{\min(a_1, b_1)} p_2^{\min(a_2, b_2)} \dots p_k^{\min(a_k, b_k)} = p_1^0 p_2^0 \dots p_k^0 = 1$. Since $ab = \gcd(a, b) \operatorname{lcm}(a, b)$, it follows that $\operatorname{lcm}(a, b) = ab$.   ■

11. $2^4 \cdot 3^4 \cdot 5^2 \cdot 7^{10}$.   **13.** $a = 1$ and $b = p^3$ is the only pair.

15. Proof. Assume that $\gcd(a^n, b^n) = 1$, where $a, b, n \in \mathbb{N}$ with $n \ge 2$. Then there exist integers x and y such that $a^n x + b^n y = 1$. Thus $a(a^{n-1}x) + b(b^{n-1}y) = 1$. Since $a^{n-1}x, b^{n-1}y \in \mathbb{Z}$, it follows that $\gcd(a, b) = 1$.   ■

17. $1, 5, 7, 11, 13, 17$.

19. Let a and b be integers that are not multiples of 3 but differ by 3. Then $a = 3q + r$ for some integers q and r, where $r \in \{1, 2\}$. We may assume, without loss of generality, that $b = a + 3 = 3q + 3 + r$. Consider the cases where $r = 1$ and $r = 2$.

21. Proof. Since $a \mid c$ and $b \mid c$, there exist integers x and y such that $c = ax$ and $c = by$. Furthermore, since a and b are relatively prime, there exist integers s and t such that $1 = as + bt$. Multiplying by c and substituting, we obtain $c = c \cdot 1 = c(as + bt) = c(as) + c(bt) = (by)(as) + (ax)(bt) = ab(sy + xt)$. Since $sy + xt$ is an integer, $ab \mid c$.   ■

23. Proof. Since a and b are relatively prime, there exist integers x and y such that $ax + by = 1$. Then $a(x + y) + (b - a)y = ax + ay + by - ay = ax + by = 1$. Since 1 is a linear combination of a and $b - a$, it follows that a and $b - a$ are relatively prime.   ■

25. We use the Strong Principle of Mathematical Induction. Since $\gcd(a_1, a_2) = \gcd(1, 2) = 1$, the statement is true for $n = 1$. Assume for a positive integer k, that $\gcd(a_i, a_{i+1}) = 1$ for every integer i with $1 \le i \le k$. We show that $\gcd(a_{k+1}, a_{k+2}) = 1$. When $k = 1$, $\gcd(a_{k+1}, a_{k+2}) = \gcd(a_2, a_3) = \gcd(2, 3) = 1$. Since $\gcd(a_{k+1}, a_{k+2}) = 1$ for $k = 1$, we can assume that $k \ge 2$. We consider two cases, according to whether $k + 1$ is even or $k + 1$ is odd.

Case 1. $k + 1$ *is odd.* Then $k + 1 = 2\ell + 1$ for some positive integer ℓ. Then $a_{k+1} = a_{2\ell+1} = a_\ell + a_{\ell+1}$, while $a_{k+2} = a_{2\ell+2} = a_{\ell+1}$. By the induction hypothesis, $\gcd(a_\ell, a_{\ell+1}) = 1$. Hence there exist integers x and y such that $a_\ell x + a_{\ell+1} y = 1$. Since $a_{k+1} x + a_{k+2}(y - x) = (a_\ell + a_{\ell+1})x + a_{\ell+1}(y - x) = a_\ell x + a_{\ell+1} x + a_{\ell+1} y - a_{\ell+1} x = a_\ell x + a_{\ell+1} y = 1$, it follows that $\gcd(a_{k+1}, a_{k+2}) = 1$.

Case 2. $k + 1$ *is even.* (The proof is similar to case 1).

Section 7.7

1. (a) 43. (b) 119. (c) 34.   **3.** (a) 13. (b) 153. (c) 2748.

5. (a) $(133)_8$. (b) $(323)_8$. (c) $(632)_8$.   **7.** (a) 135. (b) 193. (c) 143.

9. (a) $(011110011011)_2$. (b) $(467)_8$. (c) $(9D9)_{16}$.   **11.** $b = 1023 - a$.

13. Multiplying $(1011)_2$ and $(1011)_2$, we obtain $(1111001)_2$.

Supplementary Exercises for Chapter 7

1. (a) 6. (b) 9. (c) 1. (d) No. (e) Yes.

3. **Proof.** Assume that $a \mid (b + c + d)$ and that a divides any two of b, c and d, say $a \mid b$ and $a \mid c$. Thus $b + c + d = ax$, $b = ay$ and $c = az$, where $x, y, z \in \mathbb{Z}$. Then $ax = b + c + d = ay + az + d$. Therefore, $d = ax - ay - az = a(x - y - z)$. Since $x - y - z$ is an integer, $a \mid d$.   ■

5. **Proof.** We proceed by induction. The statement is true for $n = 1$. Assume that if $a \mid b$, then $a^k \mid b^k$ for a positive integer k. We show that if $a \mid b$, then $a^{k+1} \mid b^{k+1}$. Assume that $a \mid b$. By the induction hypothesis, $a^k \mid b^k$. Thus $b^k = a^k x$ for some integer x. Since $a \mid b$, it follows that $b = ay$ for some integer y. Then $b^{k+1} = b \cdot b^k = (ay)(a^k x) = a^{k+1} xy$. Since xy is an integer, $a^{k+1} \mid b^{k+1}$. By the Principle of Mathematical Induction, if $a \mid b$, then $a^n \mid b^n$ for every positive integer n.   ■

7. We proceed by induction. Since $3 \mid (1^3 + 2 \cdot 1)$, the statement is true for $n = 1$. Assume that $3 \mid (k^3 + 2k)$, where k is a positive integer. Then $k^3 + 2k = 3\ell$ for some integer ℓ. We show that 3 divides $(k + 1)^3 + 2(k + 1)$. Observe that $(k + 1)^3 + 2(k + 1) = 3(\ell + k^2 + k + 1)$.

9. $2 \cdot 3^2 \cdot 13$.   **11.** Yes.   **13.** (a) True. (b) False. (c) True.

15. False. Consider $a = 2$ and $b = 3$.

17. Let $d = \gcd(a, b)$ and $d' = \gcd(a, b + ak)$. Since $d \mid a$ and $d \mid b$, it can be shown that $d \mid (b + ak)$. This implies that $d \mid d'$ and so $d \leq d'$. For the converse, since $d' \mid a$ and $d' \mid (b + ak)$, it can be shown that $d' \mid b$. Then $d' \mid d$, implying that $d' \leq d$.   **19.** $a = 2^3$ and $b = 2^6$.

21. (a) **Proof.** We first assume that $\gcd(a, b) = \gcd(a, c) = 1$. Then $as + bt = 1$ and $ax + cy = 1$ for some integers s, t, x, y. Thus

$$1 = (as + bt)(ax + cy) = a^2 sx + acsy + abtx + bcty = a(asx + csy + btx) + bc(ty).$$

Since $asx + csy + btx$ and ty are integers, $\gcd(a, bc) = 1$. For the converse, assume that $\gcd(a, bc) = 1$. Then $ak + (bc)\ell = 1$ for some integers k, ℓ. Thus $ak + b(c\ell) = 1$ and $ak + c(b\ell) = 1$. Since k, $c\ell$ and $b\ell$ are integers, $\gcd(a, b) = \gcd(a, c) = 1$.   ■

(b) **Proof.** Since a and b are relatively prime, $\gcd(a, b) = 1$. By (a), $\gcd(a, b^2) = 1$ (letting $c = b$). Since $\gcd(a, b^2) = \gcd(b^2, a)$, it follows that $\gcd(b^2, a) = 1$. Again by (a), $\gcd(b^2, a^2) = 1$ (letting $c = a$) and so a^2 and b^2 are relatively prime.   ■

23. $1 \cdot 8^2 + 2 \cdot 8^1 + 3 \cdot 8^0 = 64 + 16 + 3 = 83$.

25. We proceed by the Strong Principle of Mathematical Induction. First, consider $n = 1$. Since $a_1 \equiv 2 \pmod{3}$, it follows that $a_1 \not\equiv 0 \pmod{3}$. Since $1 \not\equiv 0 \pmod{4}$, we conclude that $a_n \equiv 0 \pmod{3}$ if and only if $n \equiv 0 \pmod{4}$ is true for $n = 1$. Assume for a positive integer k that $a_i \equiv 0 \pmod{3}$ if and only if $i \equiv 0 \pmod{4}$ for all integers i with $1 \le i \le k$. We show that $a_{k+1} \equiv 0 \pmod{3}$ if and only if $k + 1 \equiv 0 \pmod{4}$. Since $a_2 \equiv 2 \pmod{3}$ and from the recurrence relation $a_3 \equiv 1 \pmod{3}$ and $a_4 \equiv 0 \pmod{3}$, it follows that $a_{k+1} \equiv 0 \pmod{3}$ if and only if $k + 1 \equiv 0 \pmod{4}$ when $1 \le k \le 3$. Hence we may assume that $k \ge 4$. We now consider four cases, depending on whether $k + 1$ is congruent to 0, 1, 2 or 3 modulo 4.

Case 1. $k + 1 \equiv 0 \pmod{4}$. Since $k - 3 \in \mathbb{N}$ and $k - 3 \equiv 0 \pmod{4}$, it follows that $a_{k-3} \equiv 0 \pmod{3}$. Since $k - 2 \not\equiv 0 \pmod{4}$, it follows that $a_{k-2} \not\equiv 0 \pmod{3}$. Therefore, either $a_{k-2} \equiv 1 \pmod{3}$ or $a_{k-2} \equiv 2 \pmod{3}$. By the recurrence relation, we either have (1) $a_{k-2} \equiv 1 \pmod{3}$, $a_{k-1} \equiv 1 \pmod{3}$, and $a_k \equiv 2 \pmod{3}$ or (2) $a_{k-2} \equiv 2 \pmod{3}$, $a_{k-1} \equiv 2 \pmod{3}$, and $a_k \equiv 1 \pmod{3}$. If either (1) or (2) occurs, then $a_{k+1} \equiv 0 \pmod{3}$, as desired.

(The proofs of other cases are similar.)

27. Proof. Assume that k is not a prime. Then there exist integers a and b such that $2 \le a \le k-1$, $2 \le b \le k - 1$ and $k = ab$. Thus $2^k - 1 = 2^{ab} - 1 = (2^a)^b - 1 = (2^a - 1)(2^{(b-1)a} + 2^{(b-2)a} + \cdots + 2^a + 1)$. Since $2^a - 1 > 1$ and $2^{(b-1)a} + 2^{(b-2)a} + \cdots + 2^a + 1 > 1$, it follows that $2^k - 1$ is not a prime. ∎

29. We proceed by the Strong Principle of Mathematical Induction. Since $F_1 = 1$, the statement is true for $n = 1$. Assume for a positive integer k that $2 \mid F_i$ if and only if $3 \mid i$ for all integers i with $1 \le i \le k$. We show that $2 \mid F_{k+1}$ if and only if $3 \mid (k + 1)$. Since $F_2 = 1$, it follows that $2 \mid F_{k+1}$ if and only if $3 \mid (k + 1)$ when $k = 1$. Hence we may assume that $k \ge 2$. Now consider the three cases, according to whether $k + 1 = 3q$, $k + 1 = 3q + 1$ or $k + 1 = 3q + 2$ for some integer q.

31. Algorithm: This algorithm determines in a sequence $s : a_1, a_2, \ldots, a_n$ of integers the number of times that the terms in the sequence change from even to odd or from odd to even.

Input: A positive integer n and a sequence $s : a_1, a_2, \ldots, a_n$ of n integers.

Output: The number k of times that the terms in the sequence change from even to odd or from odd to even.

1. $k := 0$

2. **for** $i := 1$ **to** $n - 1$ **do**

3.   **if** $(a_i + a_{i+1})$ **mod** $2 = 1$ **then** $k := k + 1$

4. **output** k.

33. $20 \cdot 5 + 33 \cdot (-3) = 1$.

35. Proof. By Theorem 7.41, $\gcd(a, b) \, \text{lcm}(a, b) = ab$. Assume that $\gcd(a, b) = 1$. So $\text{lcm}(a, b) = ab$. For the converse, assume that $\text{lcm}(a, b) = ab$. Then $\gcd(a, b) = 1$. ∎

37. Proof. Since $b = a(-1) + (-c)$, it follows by Theorem 7.35 that $\gcd(a, b) = \gcd(-c, a) = \gcd(c, a) = \gcd(a, c)$. Similarly, $\gcd(b, c) = \gcd(a, b)$. ∎

39. $(444)_5 = (1330)_4$.

41. (a) $a = 1$, $x = 1$; $a = 3$, $x = 7$; $a = 7$, $x = 3$; $a = 9$, $x = 9$. (b) $\gcd(a, 10) = 1$.

43. Note that $a = 2b$ and so one possibility is $a = 10$ and $b = 5$.

Chapter 8: The Basic Principles of Counting

Section 8.1

1. $3 \cdot 4 = 12$.   **3.** $26 \cdot 50 = 1300$.   **5.** $3 \cdot 4 \cdot 2 = 24$.

7. (a) 6^5. (b) 5^6. (c) $6 \cdot 5 \cdot 4 \cdot 3 \cdot 2 = 720$. (d) none.   **9.** $2 \cdot 4! = 48$.

11. (a) $6^4 = 1296$. (b) $6 \cdot 5 \cdot 4 \cdot 3 = P(6, 4)$.   **13.** $3 \cdot 5 \cdot 3 = 45$.

15. $7 + 5 + 4 = 16$.   **17.** (a) $2 \cdot 10^4 = 20,000$. (b) $7 \cdot 6 \cdot 5 \cdot 4 = 840$.

19. $2^5 = 32$.   **21.** $2^5 + 2^3 = 40$.   **23.** $2^6 + 2^4 = 80$.

25. $26^3 \cdot 10^3 + 10^3 \cdot 26^3 = 35,152,000$.   **27.** (a) $5 \cdot 6 = 30$. (b) $5 + 6 = 11$.

Section 8.2

1. (a) $2^5 + 2^6 = 96$. (b) $2^5 + 2^6 - 2^3 = 88$.

3. (a) $2^6 + 2^6 = 128$. (b) $2^6 + 2^6 - 2^2 = 124$. (c) $2^2 = 4$.

5. $n^{2n} + n^{2n} - n^n = 2n^{2n} - n^n = n^n(2n^n - 1)$.   **7.** 175.   **9.** 8.

11. (a) 10. (b) 18. (c) 25.   **13.** 21.   **15.** 26.

17. How many different 9-digit numbers are there where each digit is 1, 2 or 3 and such that either the first three digits are 111, the middle three digits are 222 or the final three digits are 333?

Section 8.3

1. **Proof.** Suppose first that n is even. Then $n = 2k$ for some integer k. Thus $\lceil n/2 \rceil = \lfloor n/2 \rfloor = k$ and so $\lceil n/2 \rceil + \lfloor n/2 \rfloor = k + k = n$. Suppose next that n is odd. Then $n = 2k + 1$ for some integer k. Therefore, $\lceil n/2 \rceil = k + 1$ and $\lfloor n/2 \rfloor = k$ and so $\lceil n/2 \rceil + \lfloor n/2 \rfloor = (k+1) + k = 2k + 1 = n$. ∎

3. 11   **5.** 2.

7. If at most 2 of these 9 marbles are red, at most 2 are blue and at most 4 are green, then there are at most 8 marbles in the bowl. This is a contradiction.

9. 13.   **11.** 14.   **13.** 25.   **15.** 9.   **17.** 7.   **19.** 11.

21. Assume, to the contrary, that the lady does not have at least four pairs of black shoes, at least four pairs of brown shoes or at least five pairs of white shoes. Then she has at most three pairs of black shoes, at most three pairs of brown shoes and at most four pairs of white shoes. But this says that she has at most ten pairs of shoes, which is not true.

23. A die is thrown three times and the result is recorded each time. For example, if 3 is obtained on the first two throws and 5 on the third throw, we record this sequence by 335. By the Multiplication Principle, there are $6 \cdot 6 \cdot 6 = 6^3 = 216$ different outcomes. What is the smallest number of times we would have to do this to be guaranteed that we obtain the same sequence 7 times?

Section 8.4

1. (a) 120. (b) 56. (c) 720. (d) 1/4. (e) n.

3. (a) 42 and 21. (b) 336 and 56. (c) 3024 and 126. (d) 10! and 1.

5. 6!.   **7.** 2520.   **9.** 384.   **11.** 20.   **13.** 120.   **15.** 57.

17. 10.   **19.** 120.   **21.** (a) 79. (b) $2^{79} - 1$. (c) $2^{79} - 1 - 79$.

23. (a) $125,970$. (b) 56. (c) 60.   **25.** 90.   **27.** $\binom{10}{3} = 120$.

29. $rC(n,r) = r\frac{n!}{r!(n-r)!} = \frac{n!}{(r-1)!(n-r)!} = n\frac{(n-1)!}{(r-1)![(n-1)-(r-1)]!} = nC(n-1, r-1)$.

31. How many subsets of a 7-element set contain an odd number of elements?

33. Let S be a set of cardinality $3n$. Divide S into three subsets A, B and C, each of cardinality n. There are $\binom{3n}{3}$ 3-element subsets of S. Since there are $3\binom{n}{3}$ subsets of S that are subsets of exactly one of A, B and C, there are $6n\binom{n}{2}$ subsets of S such that exactly two elements belong to one of A, B and C and n^3 subsets of S such that exactly one element belongs to each of A, B and C, it follows that $\binom{3n}{3} = 3\binom{n}{3} + 6n\binom{n}{2} + n^3$.

35. $\binom{n^2}{3} = n\binom{n}{3} + n^2(n-1)\binom{n}{2} + n^3\binom{n}{3}$.

Section 8.5

1. 66.   **3.** 4500.   **5.** 11760.   **7.** 1200.   **9.** (a) 3003. (b) 6. (c) 2877. (d) 950. (e) 1410.

11. (a) 220. (b) 660. (c) 1320.   **13.** $\frac{(10n)!}{n!(2n)!(3n)!(4n)!}$.

Supplementary Exercises for Chapter 8

1. 16.   **3.** 576.   **5.** (a) $30,000$. (b) 1680.   **7.** 12.

9. Suppose that $|A| = n$ and $|B| = k$, where $n > k$. Let $B = \{b_1, b_2, \ldots, b_k\}$. For $1 \leq i \leq k$, let A_i be the set of elements of A whose image in B is b_i. Thus A is divided into the subsets $A_1, A_2, \ldots, A_k$. By the Pigeonhole Principle, at least one of the subsets $A_1, A_2, \ldots, A_k$, say A_j, has at least $\lceil n/k \rceil \geq 2$ elements. Let a_1 and a_2 be two distinct elements in A_j. Then $f(a_1) = f(a_2) = b_j$ and so f is not one-to-one.

11. 137.   **13.** (a) 1. (b) none. (c) 8. (d) 7.   **15.** 504.   **17.** (a) 1000. (b) 270.

19. (a) 64. (b) 36.   **21.** August.   **23.** 17.   **25.** 24.

27. By the Pigeonhole Principle, there is at least one element $b \in B$ that is the image of at least $\lceil \frac{17}{5} \rceil = 4$ elements of A.

29. 66.   **31.** (a) 55. (b) 45.   **33.** 968.   **35.** 1296.   **37.** (a) 219. (b) 270.

39. For n lockers and n students, there will be $\lfloor \sqrt{n} \rfloor$ open lockers at the end.

Chapter 9: Advanced Counting Methods

Section 9.1

1. (a) The sums are 9, 16 and 25. The sum of the four numbers inside such a rhombus whose lower left number is $C(n, 1)$, $n \geq 3$, is n^2. If four numbers lie inside the rhombus, then the four numbers are $C(n-1, 1), C(n-1, 2), C(n, 1)$ and $C(n, 2)$. Observe that $C(n-1, 1) + C(n-1, 2) + C(n, 1) + C(n, 2) = n^2$.

(b) For a rhombus whose lower left number is $C(n, 1)$, $n \geq 3$, the value of the number computed is $C(n, 2)$. Observe that $C(n-1, 1) \cdot C(n, 2) - C(n, 1) \cdot C(n-1, 2) = C(n, 2)$.

3. Yes. $\binom{22}{1} = 22$.   **5.** $x^6 - 6x^5y + 15x^4y^2 - 20x^3y^3 + 15x^2y^4 - 6xy^5 + y^6$.

7. $x^8 + 8x^7y + 28x^6y^2 + 56x^5y^3 + 70x^4y^4 + 56x^3y^5 + 28x^2y^6 + 8xy^7 + y^8$.

9. $210x^4y^6$.   **11.** $14x^6y^5$.   **13.** $-15,120$.   **15.** $-448x^6y^5$ **17.** -1920.

19. By Theorem 9.1, $\binom{s}{r} = \binom{s-1}{r-1} + \binom{s-1}{r}$ for integers r and s with $1 \le r \le s-1$. Thus $1 \cdot \binom{4}{0} + 4 \cdot \binom{4}{1} + 6 \cdot \binom{4}{2} + 4 \cdot \binom{4}{3} + 1 \cdot \binom{4}{4} = \left[\binom{4}{0} + \binom{4}{1}\right] + 3\left[\binom{4}{1} + \binom{4}{2}\right] + 3\left[\binom{4}{2} + \binom{4}{3}\right] + \left[\binom{4}{3} + \binom{4}{4}\right] = 1 \cdot \binom{5}{1} + 3 \cdot \binom{5}{2} + 3 \cdot \binom{5}{3} + 1 \cdot \binom{5}{4}$.

21. (a) 78. (b) Carlos: 78; Dina: 75.8; Eric: 80.   **23.** Allyson: 90.4; Brian: 73.9.

25. Since $\sin^2 x + \cos^2 x = 1$, it follows that $(\sin^2 x + \cos^2 x)^4 = 1^4 = 1$. By the Binomial Theorem, $(\sin^2 x + \cos^2 x)^4 = \sin^8 x + 4\sin^6 x \cos^2 x + 6\sin^4 x \cos^4 x + 4\sin^2 x \cos^6 x + \cos^8 x$. Now simplify.

27. Observe that $5^n = (4+1)^n = \sum_{r=0}^{n} \binom{n}{r} 4^{n-r} \cdot 1^r = \sum_{r=0}^{n} \binom{n}{r} 2^{2(n-r)} = \sum_{r=0}^{n} \binom{n}{r} 2^{2n-2r}$.

Section 9.2

1. (a) 30. (b) 120. (c) 60. (d) 720. (e) 3360. (f) 34,650.   **3.** (a) 210. (b) 35.

5. 35.   **7.** 4200.   **9.** 210.   **11.** 420.   **13.** 28.   **15.** 2520.   **17.** 13,860.

19. 36.   **21.** 2376.   **23.** 54.   **25.** (a) 56. (b) 83. (c) 115.

Section 9.3

1. (a) $(1 + x + x^2 + x^3)^3 = 1 + 3x + 6x^2 + 10x^3 + 12x^4 + 12x^5 + 10x^6 + 6x^7 + 3x^8 + x^9$.

(b) 10. (c) 12.

3. (a) Let r, b and g denote the number of red, blue and green balls selected, respectively. There

r	b	g
3	1	4
4	0	4
4	2	2
5	1	2
6	0	2
6	2	0
7	1	0
8	0	0

are 8 possibilities.

(b) The coefficient of x^8 in $(x^3 + x^4 + x^5 + x^6 + x^7 + x^8)(1 + x + x^2)(1 + x^2 + x^4 + x^6 + x^8)$ is 8.

5. (a) 1. (b) $\frac{e(e-1)}{2!}$. (c) $\frac{\frac{1}{2}(-\frac{1}{2})(-\frac{3}{2})}{3!}$. (d) 0. (e) 35. (f) -10. (g) 0. (h) 1. (i) $\frac{(-\frac{1}{2})(-\frac{3}{2})}{2!}$. (j) -1.

7. $\sum_{r=0}^{\infty} \binom{-3}{r} x^r = 1 - 3x + 6x^2 - 10x^3 + \cdots$.   **9.** $f(x) = \frac{1}{(1+x)^2} = \sum_{k=0}^{\infty} (-1)^k (k+1) x^k$.

11. $f(x) = \sum_{k=0}^{\infty} [(-1)^k + (k+1)] x^k = 2 + x + 4x^2 + 3x^3 + 6x^4 + 5x^5 + 8x^6 + \cdots$, the generating function of the sequence $2, 1, 4, 3, 6, 5, 8, \ldots$.

13. (a) $f(x) = (1 + x + x^2 + \cdots)^5 = \frac{1}{(1-x)^5}$. The coefficient of x^{12} in $f(x) = (1-x)^{-5}$ is 1820.

(b) $\frac{1}{(1-x)^5} = \sum_{r=0}^{\infty} \binom{r+4}{r} x^r$.   **15.** $f(x) = \frac{1}{1-2x} = \sum_{n=0}^{\infty} 2^n x^n$ and $a_n = 2^n$ for $n \ge 0$.

17. $f(x) = (-1 + 4x)(1 + 2x + 3x^2 + 4x^3 + \cdots)$ and $a_n = 3n - 1$ for $n \ge 0$.

19. $f(x) = \frac{\frac{1}{5}}{1-2x} + \frac{\frac{4}{5}}{1+3x} = \sum_{n=0}^{\infty} \left[\frac{1}{5} \cdot 2^n + \frac{4}{5} \cdot (-1)^n 3^n\right] x^n$ and $a_n = \frac{1}{5} \cdot 2^n + \frac{4}{5} \cdot (-1)^n 3^n$ for $n \ge 0$.

Supplementary Exercises for Chapter 9

1. 1260.   **3.** 739,201.   **5.** -70.   **7.** 1260.   **9.** 5040.

11. 365.   **13.** 84.   **15.** 2520.   **17.** 1287.

19. There are 15 ways. (a) Let r, b and g denote the number of red, blue and green balls selected.

r	b	g
2	2	6
2	3	5
2	4	4
2	5	3
2	6	2
3	2	5
3	3	4
3	4	3
3	5	2
4	2	4
4	3	3
4	4	2
5	2	3
5	3	2
6	2	2

(b) The coefficient of x^{10} in $(x^2+x^3+x^4+x^5+x^6)(x^2+x^3+x^4+x^5+x^6)(x^2+x^3+x^4+x^5+x^6) = (x^2+x^3+x^4+x^5+x^6)^3$ is 15. (c) $s = 4$, $t = 3$ and $\binom{s+t-1}{s} = 15$.

21. (a) $\frac{(\frac{1}{3})(\frac{-2}{3})}{2!}$. (b) 0. (c) 56. (d) -35. (e) $\sqrt{2}$.

23. $1 + 2 \cdot 3x + 3 \cdot 3^2 x^2 + 4 \cdot 3^3 x^3 + \cdots + (n+1) \cdot 3^n x^n + \cdots$.

25. $f(x) = \frac{-x}{(1+x)^2} = -x(1 - 2x + 3x^2 - 4x^3 + \cdots) = -x + 2x^2 - 3x^3 + 4x^4 - 5x^5 + \cdots$ and $a_n = (-1)^n n$ for $n \geq 0$.   **27.** 1001.

Chapter 10: Discrete Probability

Section 10.1

1. 1/3.   **3.** 5/9.   **5.** (a) 1/216. (b) 1/8.   **7.** $1 - 9/2^8 = 247/256$.

9. 1/6.   **11.** $\frac{\binom{13}{5} \cdot 4^5}{\binom{52}{5}} = \frac{1,317,888}{2,598,960}$.

13. The strategy is very bad. First, it will cost him $120,526,770 to purchase these lottery tickets. Even if he could purchase one lottery ticket per second, he could only purchase 604,800 tickets in a week, which would take him too much time. Furthermore, even if he could make arrangements to purchase all tickets in the appropriate amount of time, he would have to locate the winning ticket among the 120,526,770 tickets he bought. Even worse, it is quite possible that there are other winning tickets, in which case the grand prize would be shared and he has lost more than $45,000,000.

15. 17/24.   **17.** $\frac{638}{1024}$.   **19.** $330/500 = 0.66$.

21. (a) $\frac{1}{3^{10}}$. (b) $\frac{2^{10}}{3^{10}}$. (c) $\frac{126,000}{3^{10}}$.   **23.** 1/32.   **25.** 1/2.

Section 10.2

1. (a) 3/8. (b) 3/8.   **3.** (a) 1/2. (b) 1/2.   **5.** Yes.   **7.** (a) 13/25. (b) 3/5. (c) 2/5.

9. No.   **11.** (a) Twice. (b) 7 times. (c) 4 times.   **13.** 0.75.   **15.** 0.6.

17. (a) $S = \{(C_1, H), (C_1, T), (C_2, H)\}$. (b) 1/3.   **19.** $\frac{0.045}{0.235} \approx 0.19$.   **21.** 3/5.   **23.** 1/9.

25. (a) **Proof.** Since E and F are independent, $p(E \mid F) = p(E) \cdot p(F)$. Let S be the sample space. Then

$$\begin{aligned} p(E) &= p(E \cap S) = p(E \cap (F \cup \overline{F})) = p((E \cap F) \cup (E \cap \overline{F})) \\ &= p(E \cap F) + p(E \cap \overline{F}) = p(E) \cdot p(F) + p(E \cap \overline{F}). \end{aligned}$$

Therefore, $p(E \cap \overline{F}) = p(E) - p(E) \cdot p(F) = p(E)(1 - p(F)) = p(E) \cdot p(\overline{F})$ and so E and $\overline{F}$ are independent. ∎

(b) **Proof.** Since E and F are independent, E and $\overline{F}$ are independent by (a). Thus $\overline{E}$ and $\overline{F}$ are independent by (a). ∎

Section 10.3

1. Let s be the color of the first ball and t the color of the second ball. (a) $1/12$ if $(x, y) = (b, g)$ or (g, b); otherwise, $1/6$. (b) If $x, y \neq g$, then $X(x, y) = 0$ and $X(x, y) = 1$ if one of x and y is g. (c) $\{0, 1\}$. (d) $p(X = 0) = 1/2$, $p(X = 1) = 1/2$.   **3.** 2.

5. (a) The player would expect to lose about \$20.63. (b) The player's actual loss or gain will not be close to the amount in (a). If he/she does not lose \$100, then he/she will win at least \$900.

7. $\frac{35}{18} \approx 2$.   **9.** You would expect to lose \$10 − \$7.50 = \$2.50 each time you play the game.

11. (a) $4/3$. (b) $4/9$.   **13.** about 53.   **15.** 110 dollars.

Supplementary Exercises for Chapter 10

1. $1/12$.   **3.** (a) 0.5. (b) 0.8. (c) 0.36. (d) $105/1225 \approx 0.0086$. (e) $510/1225 \approx 0.416$.

5. The probability is approximately $1 - 0.883 = 0.117$.

7. (a) $1 - \frac{380}{650} \approx 0.42$. (b) $\frac{6 \cdot 5}{650} \approx 0.046$. (c) 0.24. (d) 0.2.   **9.** Yes.

11. (a) $\binom{5}{3}(0.49)^3(0.51)^2 \approx 0.306$. (b) $\binom{5}{3}(0.51)^3(0.49)^2 \approx 0.318$. (c) $1 - (0.51)^5 \approx 0.965$.
(d) $1 - (0.49)^5 \approx 0.972$. (e) $(0.51)^5 + (0.49)^5 \approx 0.063$.

13. 86.   **15.** $1/10$.   **17.** $1/24$.   **19.** $5/6$.   **21.** 4th place.

Chapter 11: Partially Ordered Sets and Boolean Algebras

Section 11.1

1. (a) $\{(a, b)\}$. (b) $\{(a, b), (b, a)\}$. (c) $\emptyset$. (d) $\{(a, b), (b, a), (c, d)\}$.

3. (a) No. For example, $(-2) \mid 2$ and $2 \mid (-2)$, but $2 \neq -2$. (b) Yes.

5. (a) No, for example, $7 \not{R} 7$. (b) No, for example, $3 R 5$ and $5 R 3$ but $3 \neq 5$. (c) Yes.
(d) No, for example, $5 R 3$ and $3 R 5$ but $5 \not{R} 5$.   **7.** antisymmetric, transitive.

9. The relation R on $\mathbb{Z}$ defined by $a R b$ if $a = b$.   **11.** No. R is not transitive.

13. Let $S = \{a, b, c, d, e, f, g\}$ and

$$\preceq \ = \ \{(a, a), (b, b), (c, c), (d, a), (d, b), (d, c), (d, d), (e, a), (e, b), (e, c), (e, e), (f, a),$$
$$(f, b), (f, c), (f, d), (f, e), (f, f), (g, a), (g, b), (g, c), (g, d), (g, e), (g, g)\}$$

15. Let $S = \{a, b\}$ and $R = \{(a, a), (a, b), (b, b)\}$. Then (S, R) is a poset.

17. Proof. Let (S, R) be a poset. Let $a \in S$. Since $a R a$, it follows that $a R^{-1} a$. Hence R^{-1} is reflexive. Assume that $a R^{-1} b$ and $b R^{-1} a$, where $a, b \in S$. Then $b R a$ and $a R b$. Because R is antisymmetric, $b = a$. Thus R^{-1} is antisymmetric. Next assume that $a R^{-1} b$ and $b R^{-1} c$, where $a, b, c \in S$. Then $b R a$ and $c R b$. Since R is transitive, $c R a$. Thus $a R^{-1} c$ and so R^{-1} is transitive. Therefore, (S, R^{-1}) is a poset. ∎

19. (a) and (d).

21. Proof. Let $(S, \preceq)$ be a finite nonempty poset. Among the chains in S, let $x_1 \prec x_2 \prec \cdots \prec x_k$ be one containing a maximum number of elements. There can be no element $x \in S$ such that $x_k \prec x$, for otherwise, $x_1 \prec x_2 \prec \cdots \prec x_k \prec x$ is a chain with more than k elements. Hence x_k is a maximal element. ∎

23. Proof. Assume, to the contrary, that there exists a finite poset $(S, \preceq)$ containing a unique minimal element m such that m is not a least element. Since m is not a least element, there exists an element $a \in S$ such that $m \not\preceq a$. Since m is a minimal element, it is impossible for $a \prec m$. Thus a and m are incomparable. If there is no element a' such that $a' \prec a$, then a is a minimal element of $(S, \preceq)$, contradicting our assumption that m is the only minimal element of $(S, \preceq)$. Hence there must be some element a' such that $a' \prec a$. Since S is finite, there is a chain $a_1 \prec a_2 \prec \cdots \prec a_k = a$ with a maximum number of elements. However, then a_1 is a minimal element. Because $a_1 \prec a$ and a and m are incomparable, $a_1 \neq m$, which contradicts the assumption that m is the unique minimal element of $(S, \preceq)$. ∎

25. (a) See Figure 9. (b) 6, 10, 14, 15, 21 and 35 are maximal elements, while 2, 3, 5 and 7 are minimal elements. (c) There is no greatest or least element.

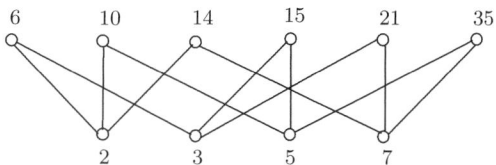

Figure 9: A Hasse diagram for the poset in Exercise 25

27. Algorithm Find a maximal element in a finite poset.

This algorithm finds a maximal element in a finite poset $(S, \preceq)$, where $s : a_1, a_2, \ldots, a_n$ is a list of the n elements of S and $s' : e_1, e_2, \ldots, e_m$ is a list of the m elements (a, b) of $\preceq$ for which $a \neq b$.

> **Input**: Two positive integers n and m, a list $s : a_1, a_2, \ldots, a_n$ of the n elements of S and a list $s' : e_1, e_2, \ldots, e_m$ of the m elements (a, b) of $\preceq$ for which $a \neq b$.
>
> **Output**: a maximal element y of $(S, \preceq)$.

1. $y := a_1$
2. **for** $i := 2$ **to** n **do**
3.   **for** $j := 1$ **to** m **do**
4.   **if** $e_j = (y, a_i)$ **then** $y := a_i$ and $j := m + 1$.
5. **output** y

29. Proof. We proceed by induction. If $n = 2$, then the result follows from Theorem 11.19. Assume that the result is true for $k \geq 2$ totally ordered sets and let $(A_1, \preceq_1), (A_2, \preceq_2), \ldots,$ $(A_{k+1}, \preceq_{k+1})$ be $k + 1$ totally ordered sets. We show that the lexicographic order $\preceq$ defined on $A_1 \times A_2 \times A_{k+1}$ is a total order. By the induction hypothesis, the lexicographic order $\preceq$ defined on $A = A_1 \times A_2 \times \cdots A_k$ is a total order and so $(A, \preceq)$ is a totally ordered set. Since $A_1 \times A_2 \times \cdots A_{k+1} = A \times A_{k+1}$, it follows by Theorem 11.19 that $(A \times A_{k+1}, \preceq)$ is a totally ordered set. ∎

31. $a_4 \prec a_6 \prec a_8 \prec a_1 \prec a_2 \prec a_3 \prec a_5 \prec a_7$.

33. (a) See Figure 10. (b) (i) $2 \prec 4 \prec 8 \prec 24 \prec 72 \prec 40 \prec 200 \prec 1800$; (ii) $2 \prec 4 \prec 8 \prec 24 \prec 40 \prec 72 \prec 200 \prec 1800$.

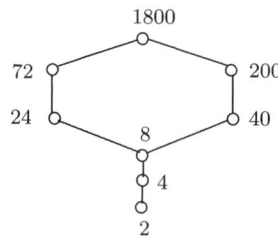

Figure 10: A Hasse diagram for the poset in Exercise 33

35. Proof. Let $q = r/s$, where r and s are integers with $1 \leq r < s$. Let $p_1, p_2, \ldots, p_{2s-1}$ be $2s - 1$ distinct primes. Let $a = 1$, $b_i = p_i$ for $1 \leq i \leq 2r$, $c = b_1 b_2 \cdots b_{2r}$ and $d_j = p_{2r+j} c$ for $1 \leq j \leq 2s - 2r - 1$. Let $S = \{a, b_1, b_2, \ldots, b_{2r}, d_1, d_2, \ldots, d_{2s-2r-1}\}$. Then $(S, \mid)$ is a poset with $|S| = 2s$. The set $I = \{b_1, b_2, \ldots, b_{2r}\}$ is the set of intermediate elements of S. Then $\frac{|I|}{|S|} = \frac{2r}{2s} = \frac{r}{s} = q$. ∎

Section 11.2

1. Proof. Assume, to the contrary, that there exists a poset $(S, \preceq)$ containing a nonempty subset A having two distinct least upper bounds u_1 and u_2. Since u_1 is a least upper bound, $u_1 \preceq u_2$. Also, since u_2 is a least upper bound, $u_2 \preceq u_1$. Since $\preceq$ is antisymmetric, $u_1 = u_2$, contradicting the assumption that u_1 and u_2 are distinct. ∎

3. No, a_2 and a_3, for example, do not have a unique least upper bound.

5. (a) **Proof.** Let $d = a \vee c$ and $e = b \vee c$. Then $b \preceq e$ and $c \preceq e$. Since $a \preceq b$, it follows that $a \preceq e$. Thus e is an upper bound of $\{a, c\}$. Because $d = a \vee c$, we have $d \preceq e$, that is, $a \vee c \preceq b \vee c$. ∎

(b) **Proof.** Let $d = a \wedge c$ and $e = b \wedge c$. Thus $d \preceq a$ and $d \preceq c$. Since $a \preceq b$, it follows that $d \preceq b$. Thus d is a lower bound of $\{b, c\}$. Since $e = b \wedge c$, we have $d \preceq e$ and so $a \wedge c \preceq b \wedge c$. ∎

7. Proof. First, assume that $c \preceq a$ and $c \preceq b$. Hence c is a lower bound of $\{a, b\}$ and so $c \preceq a \wedge b$. For the converse, assume that $c \preceq a \wedge b$. Let $d = a \wedge b$. Then $d \preceq a$ and $d \preceq b$. Since $c \preceq d$, it follows that $c \preceq a$ and $c \preceq b$. ∎

9. (a) **Proof.** Suppose first that $a \vee b = b$. Since $a \vee b$ is the least upper bound of a and b, it follows that $a \preceq a \vee b = b$. Thus $a \preceq b$. For the converse, suppose that $a \preceq b$. Thus b is an upper bound of a and b. Since $a \vee b$ is the least upper bound of a and b, it follows that $a \vee b \preceq b$. Since $a \vee b$ is an upper bound of a and b, we have $b \preceq a \vee b$. Because $\preceq$ is antisymmetric, $a \vee b = b$. ∎

(b) is similar to (a). (c) **Proof.** By (b), $a \wedge b = a$ if and only if $a \preceq b$. By (a), $a \preceq b$ if and only if $a \vee b = a$. Consequently, $a \wedge b = a$ if and only if $a \vee b = a$. ∎

11. (a) 6, 6. (b) 30, 30. (c) Parts (a) and (b) illustrates that $(S, \mid)$ is a distributive lattice, where S, say, is the set of positive integer divisors of 300.

13. The lattice $(S, \leq)$ is complemented if $n = 2$; while $(S, \leq)$ is not complemented for $n \geq 3$.

15. Proof. Observe that $b \wedge (a \vee b) = (b \wedge a) \vee (b \wedge b) = (b \wedge a) \vee b = b$ and that

$$\begin{aligned} b \wedge (a \vee b) &= b \wedge (a \vee c) = (b \wedge a) \vee (b \wedge c) = (a \wedge b) \vee (b \wedge c) = (a \wedge c) \vee (b \wedge c) \\ &= (c \wedge a) \vee (c \wedge b) = c \wedge (a \vee b) = c \wedge (a \vee c) = (c \wedge a) \vee (c \wedge c) = (c \wedge a) \vee c = c. \end{aligned}$$

Thus $b = c$. ∎

17. (a) **Proof.** Let $(L, \preceq)$ be a distributive lattice and let $a, b, c \in L$, where $a \preceq c$. We show that $a \vee (b \wedge c) = (a \vee b) \wedge c$. First, observe that since $a \preceq c$, we have $c \wedge a = a$. Now $(a \vee b) \wedge c = c \wedge (a \vee b) = (c \wedge a) \vee (c \wedge b) = a \vee (b \wedge c)$. ∎

(b) First observe that $b \preceq d$. Now $b \vee (e \wedge d) = b \vee a = b$, while $(b \vee e) \wedge d = f \wedge d = d$. Thus the lattice is not modular.

19. Proof. Suppose that there is an element $a \in L$ that has two complements, say b and c. Then $a \vee b = a \vee c = 1$ and $a \wedge b = a \wedge c = 0$. Therefore,

$$
\begin{aligned}
b &= b \wedge 1 = b \wedge (a \vee c) = (b \wedge a) \vee (b \wedge c) = 0 \vee (b \wedge c) = b \wedge c = c \wedge b = 0 \vee (c \wedge b) \\
&= (c \wedge a) \vee (c \wedge b) = c \wedge (a \vee b) = c \wedge 1 = c
\end{aligned}
$$

and so $b = c$. ∎

21. If $a \in S$ such that a and n/a are relatively prime, then $\bar{a} = n/a$.

Section 11.3

1. Proof. Let $a, b \in S$. To see that $\overline{a \cdot b} = \bar{a} + \bar{b}$, we show that $\bar{a} + \bar{b}$ is the complement of $a \cdot b$. Hence we must verify that $a \cdot b + (\bar{a} + \bar{b}) = 1$ and $(a \cdot b) \cdot (\bar{a} + \bar{b}) = 0$. Observe by a distributive law that

$$
\begin{aligned}
a \cdot b + (\bar{a} + \bar{b}) &= (\bar{a} + \bar{b}) + (a \cdot b) = ((\bar{a} + \bar{b}) + a) \cdot ((\bar{a} + \bar{b}) + b) = (a + (\bar{a} + \bar{b})) \cdot ((\bar{a} + \bar{b}) + b) \\
&= ((a + \bar{a}) + \bar{b}) \cdot (\bar{a} + (\bar{b} + b)) = (1 + \bar{b}) \cdot (\bar{a} + 1) = 1 \cdot 1 = 1. \\
(a \cdot b) \cdot (\bar{a} + \bar{b}) &= ((a \cdot b) \cdot \bar{a}) + ((a \cdot b) \cdot \bar{b}) = (\bar{a} \cdot (a \cdot b)) + (a \cdot (b \cdot \bar{b})) \\
&= ((\bar{a} \cdot a) \cdot b) + (a \cdot 0) = (0 \cdot b) + 0 = 0 + 0 = 0.
\end{aligned}
$$

Therefore, $\overline{a \cdot b} = \bar{a} + \bar{b}$. ∎

3. (a) 1. (b) 0. (c) 1.   **5.** (a) $w + x$. (b) y. (c) $\overline{(w + x)}$.

7. (a)

x	y	$\bar{x}$	$\bar{y}$	$x + \bar{y}$	$(x + \bar{y}) \cdot y$	$f = (x + \bar{y}) \cdot y + \bar{x}$
1	1	0	0	1	1	1
1	0	0	1	1	0	0
0	1	1	0	0	0	1
0	0	1	1	1	0	1

(b) and (c) Proceed as with (a).

9. $xyz + xy\bar{z} + x\bar{y}\,\bar{z} + \bar{x}yz + \bar{x}y\bar{z} + \bar{x}\,\bar{y}z + \bar{x}\,\bar{y}\,\bar{z}$.   **11.** $xyz + xy\bar{z} + x\bar{y}\,\bar{z} + \bar{x}yz + \bar{x}y\bar{z} + \bar{x}\,\bar{y}\,\bar{z}$.

13. See Figure 11.   **15.** See Figure 12.

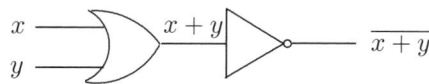

Figure 11: The combinatorial circuit in Exercise 13

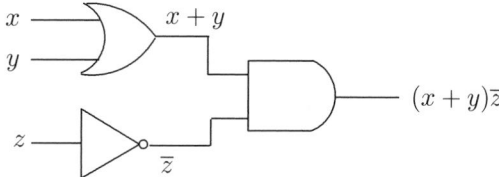

Figure 12: The combinatorial circuit in Exercise 15

Supplementary Exercises for Chapter 11

1. No.   **3.** $a = 1$, $b = 2$, $c = 3$ and $d = 6$.

5. (a) **Proof.** Let $(a, b) \in S \times S$. Since $a \le a$ and $b \le b$, it follows that $(a, b)\, R\, (a, b)$ and so R is reflexive. Next, assume that $(a, b)\, R\, (c, d)$ and $(c, d)\, R\, (a, b)$, where $(a, b), (c, d) \in S \times S$. Since $(a, b)\, R\, (c, d)$, we have $a \le c$ and $b \le d$. Furthermore, since $(c, d)\, R\, (a, b)$, it follows that $c \le a$ and $d \le b$. Thus $a = c$ and $b = d$ and so $(a, b) = (c, d)$. Therefore, R is antisymmetric. Finally, assume that $(a, b)\, R\, (c, d)$ and $(c, d)\, R\, (e, f)$, where $(a, b), (c, d), (e, f) \in S \times S$. Therefore, $a \le c$, $b \le d$, $c \le e$ and $d \le f$. Thus $a \le e$ and $b \le f$. This implies that $(a, b)\, R\, (e, f)$ and so R is transitive. ∎

(b) See Figure 13. (c) $(1,1) \prec (1,2) \prec (2,1) \prec (1,3) \prec (2,2) \prec (3,1) \prec (2,3) \prec (3,2) \prec (3,3)$.
(d) Yes.

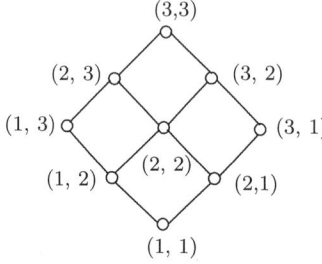

Figure 13: A Hasse diagram of the poset in Exercise 5(b)

7. (a) The maximal elements are h and i and the minimal elements are a, b and c.

(b) The set A_1 does not have a least upper bound or a greatest lower bound.

(c) The set A_2 does not have a least upper bound or a greatest lower bound.

9. (a) is true. Since R is symmetric, it suffices to show that R is reflexive and transitive. Let $a \in S$. Since one of the ordered pairs (a, a) and (a, a) belongs to R, it follows that $(a, a) \in R$ and R is reflexive. Next show that R is transitive.

11. **Proof.** Let $(S, \preceq)$ be a totally ordered set of cardinality $n \ge 2$. Hence we may assume that $S = \{a_1, a_2, \ldots, a_n\}$, where $a_1 \prec a_2 \prec \cdots \prec a_n$. For $1 \le i \le n$, let $S_{a_i} = \{a_i, a_{i+1}, \ldots, a_n\}$ and so $|S_{a_i}| = n - i + 1$. Thus $f(a_i) = n - i + 1$ and so f is bijective. For the converse, assume that f is a bijective function. Hence for each integer i with $1 \le i \le n$, there exists $x_i \in S$ such that $f(x_i) = i$. Since $f(x_n) = n$, it follows that $x_n \preceq x_i$ for all i with $1 \le i \le n$. Since $\preceq$ is antisymmetric, it follows that $x_{n-1} \not\preceq x_n$. Furthermore, since $f(x_{n-1}) = n - 1$, it follows that $x_{n-1} \preceq x_i$ for all i with $1 \le i \le n - 1$. Thus $x_n \prec x_{n-1} \prec \cdots \prec x_1$ and so $(S, \preceq)$ is a totally ordered set. ∎

13. See Figure 14.   **15.** See Figure 15.

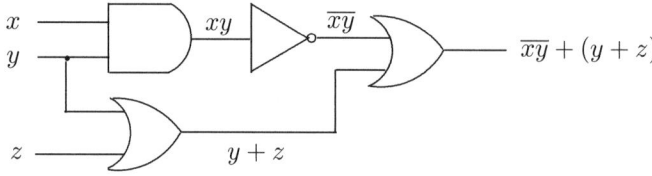

Figure 14: The combinatorial circuit in Exercise 13

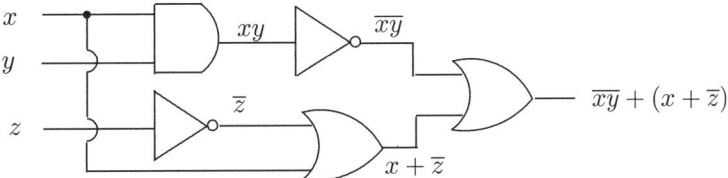

Figure 15: The combinatorial circuit in Exercise 15

Chapter 12: Introduction to Graphs

Section 12.1

1. Yes. **3-9.** See Figure 16

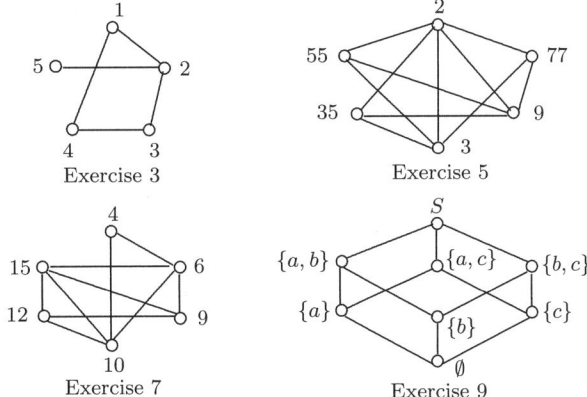

Figure 16: The graphs in Exercises 3-9

11. $\mathbf{A} = \begin{bmatrix} 0 & 1 & 1 & 0 & 0 & 0 \\ 1 & 0 & 1 & 1 & 0 & 0 \\ 1 & 1 & 0 & 1 & 1 & 0 \\ 0 & 1 & 1 & 1 & 1 & 1 \\ 0 & 0 & 1 & 1 & 1 & 1 \\ 0 & 0 & 0 & 0 & 1 & 1 \end{bmatrix}$
$\quad\quad u : v, w \quad\quad v : u, w, x \quad w : u, v, x, y$
$\quad\quad x : v, w, y, z \quad y : w, x, z \quad z : x, y.$

13. The statement is false. The number of 1's in row i is the degree of v_i. The total number of 1's in the adjacency matrix is the sum of the degrees of the vertices of the graph, which is twice the size of the graph. **15.** 10. **17.** 4.

19. For $k = 3$, let G be the graph with $V(G) = \{v_1, v_2, \ldots, v_6\}$ and $E(G) = \{v_1v_2, v_1v_5, v_1v_6, v_2v_3, v_2v_5, v_3v_4, v_4v_5\}$.

21. Proof. Suppose that G contains a vertices of degree $2k$, b vertices of degree $2k + 1$ and c vertices of degree $2k + 2$. Assume, to the contrary, that $a \leq 2k$, $b \leq 2k + 1$ and $c \leq 2k$. Thus $a + b + c \leq 2k + (2k + 1) + 2k = 6k + 1$. Since G has $6k + 1$ vertices, $a = 2k$, $b = 2k + 1$ and $c = 2k$. However then, G contains an odd number b of odd vertices, which is impossible. ∎

23. (a) P_4. (b) C_5. (c) If $G \cong \overline{G}$, then G and $\overline{G}$ have the same size m. Thus $2m = \binom{6}{2} = 15$ and so $m = 15/2$. This is impossible.

25. The statement is false. There is no 5-regular graph of order $2n + 1$.

27. No. **29.** The statement is false.

31. Since $\deg_G v = \deg_{\overline{G}} v = r$ and $\deg_G v + \deg_{\overline{G}} v = n - 1$, it follows that $2r = n - 1$ and so $n = 2r + 1$ is odd. **33.** See Figure 17.

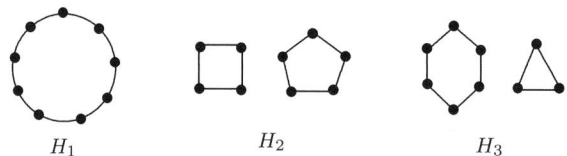

$$H_1 \qquad H_2 \qquad H_3$$

Figure 17: The graphs in Exercises 33

35. Define $f : V(G_1) \to V(G_2)$ by $f(u_1) = v_2$, $f(v_1) = w_2$, $f(w_1) = z_2$, $f(x_1) = u_2$, $f(y_1) = x_2$, $f(z_1) = y_2$ and $g : V(H_1) \to V(H_2)$ by $g(p_1) = r_2$, $g(q_1) = s_2$, $g(r_1) = p_2$, $g(s_1) = q_2$, $g(t_1) = t_2$.

37. (a) The statement is true. **Proof.** Suppose that $A_1 = A_2$. Then A_1 and A_2 are both $n \times n$ matrices for some positive integer n. Hence the orders of G_1 and G_2 are n. Let $u_1, u_2, \ldots, u_n$ be a labeling of the vertices of G_1 that gives the adjacency matrix A_1 and let $v_1, v_2, \ldots, v_n$ be a labeling of the vertices of G_2 that gives the adjacency matrix A_2. Define $f : V(G_1) \to V(G_2)$ by $f(u_i) = v_i$ for $1 \le i \le n$. Since $A_1 = A_2$, two vertices u_i and u_j are adjacent in G_1 if and only if v_i and v_j are adjacent in G_2. Hence $G_1 \cong G_2$. ∎

(b) The statement is false. Let G_2 and G_3 be the paths P_3 where $V(G_2) = \{u_1, u_2, u_3\}$ and $V(G_3) = \{v_1, v_3, v_2\}$ such that u_2 and v_3 have degree 2.

39. (a) $E(G) = \{A_1 A_3, A_3 A_2, A_2 A_4, A_4 A_1\}$. (b) If $n = 2k$ is even, then let $S = \{(i - 1, 1 - i) : 0 \le i \le k - 1\} \cup \{(i + 1, 2 - i) : 0 \le i \le k\}$. If $n = 2k + 1$ is odd, then let $S = \{(i - 1, 1 - i) : 0 \le i \le k\} \cup \{(i + 1, 2 - i) : 0 \le i \le k\}$. (c) No.

Section 12.2

1. Let G be the star $K_{1,3}$ with $V(G) = \{u, v, w, x\}$, where $\deg x = 3$ and $W = (u, x, v, x, w, x, v)$.

3. (a) (q, u, y, v, r, u, x). (b) (t, x, u, y). (c) $(u, q, t, x, u, r, v, y, u)$. (d) (s, v, z, w, s).

5. See Figure 18.

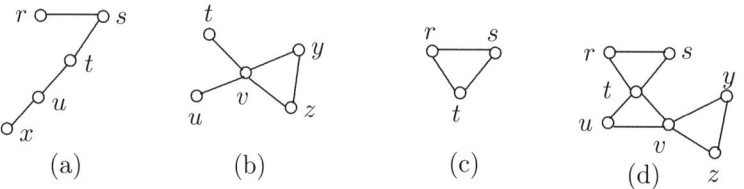

Figure 18: The graphs in Exercise 5

7. (a) **Proof.** Suppose that $P = (u = v_0, v_1, v_2, v_3, v_4 = v)$ is a $u - v$ path of length 4 in G. If $uv \in E(G)$, then P together with the edge uv forms a 5-cycle in G, which is impossible since G is bipartite and contains no odd cycle. Therefore, $uv \notin E(G)$. ∎

(b) If u and v are two vertices in a connected bipartite graph G such that there is a $u - v$ path of length $2k$ in G for some integer k, then $uv \notin E(G)$.

9. To compute $\mathbf{A}^2$, observe that $\deg v_1 = \deg v_4 = 2$ and $\deg v_2 = \deg v_3 = 3$. For $i \ne j$, the (i, j)-entry of $\mathbf{A}^2$ is the number of different $v_i - v_j$ paths of length 2. Thus $\mathbf{A}^2 = \begin{bmatrix} 2 & 1 & 1 & 2 \\ 1 & 3 & 2 & 1 \\ 1 & 2 & 3 & 1 \\ 2 & 1 & 1 & 2 \end{bmatrix}$.

To compute $\mathbf{A}^3$, observe that v_1 and v_4 belong to one triangle and v_2 and v_3 belong to two triangles each. For $i \neq j$, the (i,j)-entry of $\mathbf{A}^3$ is the number of different $v_i - v_j$ paths of length 3. Thus $\mathbf{A}^3 = \begin{bmatrix} 2 & 5 & 5 & 2 \\ 5 & 4 & 5 & 5 \\ 5 & 5 & 4 & 5 \\ 2 & 5 & 5 & 2 \end{bmatrix}$.   **11.** Theorem 12.22 holds for $k = 0$.

13. Consider the graph consisting of the two components C_3 and C_4.

15. Suppose that the order of G_1 is at most 3, the order of G_2 is at most 3 and the order of G_3 is at most 5. Then the order of G is at most $3 + 3 + 5 = 11$, which is impossible.

17. Let $x, y \in V(G)$. If one of x and y is v, say $x = v$, then x is adjacent to y and so (x, y) is a path. If neither x nor y is v, then v is adjacent to x and y and so (x, v, y) is an $x - y$ path.

19. Yes.   **21.** (a) $ug(G_1) = ug(G_2) = K_6$ and $ug(G_3) = G_3$. (b) $G = K_{2,3}$. (c) $G = C_5$. (d) Yes.

23. Proof. Assume, to the contrary, that K_n, $n \geq 3$, is bipartite. Then there is partition of the vertex set of K_n into two subsets U and W such that no two vertices of U are adjacent and no two vertices of W are adjacent. By the Pigeonhole Principle, at least one of U and W contains at least two vertices, say U contains at least two vertices. Since every two vertices of U are adjacent, this is a contradiction. ∎

25. The size of $\overline{C}_n$ is $n(n-3)/2$.

27. Proof. If G is not bipartite, then we have the desired result. Thus, we may assume that G is bipartite. Let V_1 and V_2 be the two partite sets of G. Since the order of G is at least 5, at least one of V_1 and V_2 contains 3 or more vertices, say $|V_1| = k \geq 3$. Since the subgraph of $\overline{G}$ induced by V_1 is the complete graph K_k and $k \geq 3$, it follows that $\overline{G}$ contains a triangle. By Theorem 12.31, $\overline{G}$ is not bipartite. ∎

29. (a) Note that if $G - X$ is 2-regular, then $G - Y$ is 1-regular, where $Y = E(G) - X$. (b) No.

31. If v is a cut-vertex of G, then $G - v$ is disconnected. Show that $\overline{G - v} = \overline{G} - v$ is connected.

33. Proof. Since $e = uv$ is an edge, G contains the $u - v$ path $P = (u, v)$. We show that P is the only $u - v$ path in G. Assume, to the contrary, that G contains another $u - v$ path P'. Observe that $P' \neq P$ and so the path P' contains at least three vertices. Since u and v are the end-vertices of P' and P' is a path, it follows that $u, v \notin V(P') - \{u, v\}$ and so $e \notin E(P')$. Thus P' together with $e = uv$ form a cycle containing e. Since e is a bridge in G, a contradiction to Theorem 12.36 is produced. ∎

35. Proof. If v is a cut-vertex in a connected graph G, then, of course, $G - v$ contains two or more components. If u and w are vertices in distinct components of $G - v$, then u and w are not connected in $G - v$. On the other hand, u and w are necessarily connected in G. Thus v lies on every $u - w$ path in G. For the converse, suppose that there are two vertices u and w distinct from v such that v lies on every $u - w$ path in G. Then there is no $u - w$ path in $G - v$. Thus u and w are not connected in $G - v$ and so $G - v$ is disconnected. Therefore, v is a cut-vertex of G. ∎

37. (a) Suppose that G contains distinct vertices u and v such that $G - u$ and $G - v$ are connected. To show that G itself is connected, show that every two vertices of G are connected. Let x and y be two vertices of G. Consider the two cases.

Case 1. $\{x, y\} \neq \{u, v\}$, say $u \notin \{x, y\}$. *Case 2.* $\{x, y\} = \{u, v\}$, say $x = u$ and $y = v$.

For the converse, let u and v be two vertices of G such that $d(u, v) = \text{diam}(G)$. Show, by contradiction, that $G - u$ and $G - v$ are both connected.

(b) The statement is false. Consider $G = P_5$.

Section 12.3

1. (a)-(b) No. (c) This provides a somewhat different solution of the Königsberg Bridge Problem.

3. Proof. Assume first that G contains an Eulerian trail T. Thus T is a $u - v$ trail for some distinct vertices u and v. We now construct a new connected graph H from G by adding a new vertex x of degree 2 and joining it to u and v. Then $C = (T, x, u)$ is an Eulerian circuit in H. Thus, every vertex of H is even and so only u and v have odd degrees in $G = H - x$. For the converse, we proceed in a similar manner. Let G be a connected graph containing exactly two vertices u and v of odd degree. We show that G contains an Eulerian trail T, where T is either a $u - v$ trail or a $v - u$ trail. Add a new vertex x of degree 2 to G and join it to u and v, calling the resulting graph H. Therefore, H is a connected graph all of whose vertices are even. Thus H is an Eulerian graph containing an Eulerian circuit C. Since it is irrelevant which vertex of C is the initial (and terminal) vertex, we assume that C is an $x - x$ circuit. Since x is incident only with the edges ux and vx, one of these is the first edge of C and the other is the final edge of C. Deleting x from C results in an Eulerian trail T of G that begins either at u or v and ends at the other. ∎

5. See Figure 19.

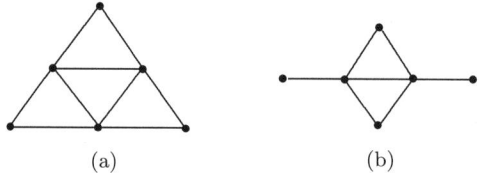

Figure 19: The graphs in Exercises 5

7. Note that C_n is a 2-regular graph and that for $n \geq 5$, $\overline{C}_n$ is a connected $(n-3)$-regular graph.

9. Note that Q_n is connected and n-regular.

11. Suppose that there is an Eulerian graph of order $n \geq 3$, where at most two vertices have the same degree. Consider the two cases: *Case 1. n is even, say $n = 2k$ for some $k \geq 2$. Case 2. n is odd, say $n = 2k + 1$ for some $k \geq 1$.*

13. The statement is true.   **15.** See Figure 20.   **17.** 2^n, $n \cdot 2^{n-1}$.

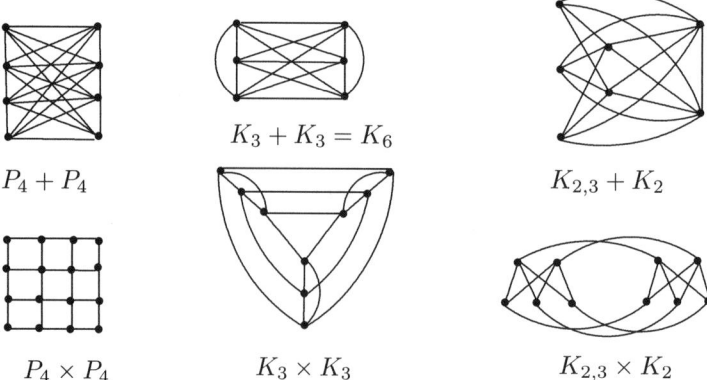

Figure 20: The graphs in Exercise 15

19. Let G_i be an r_i-regular graph of order n_i $(i = 1, 2, 3)$. Note that G is connected and (1) every vertex of G_1 in G has degree $r_1 + n_2 + n_3$, which is even, (2) every vertex of G_2 in G has degree $r_2 + n_1 + n_3$, which is even and (3) every vertex of G_3 in G has degree $r_3 + n_1 + n_2$, which is even.

21. The statement is false. If G and H are not Eulerian, then $G + H$ is not Eulerian.

23. Floyd is lying as the multigraph modeling this situation has 4 odd vertices.

Section 12.4

1. (a)-(b) No.   **3.** (a) Nothing. (b) Yes.

5. If $S = \{u, v\}$, then $G - S$ has at least three components.

7. Not necessarily.   **9.** $C_3 \times C_3$ is Hamiltonian.

11. If $s = t$, then $K_{s,t}$ is an s-regular graph of order $n = 2s$. Since $s = \frac{n}{2}$, it follows by Theorem 12.59 that $K_{s,t}$ is Hamiltonian. For the converse, assume that $s \neq t$, say $s < t$, and apply Theorem 12.57.

13. Proof. Since G is Hamiltonian, G contains a Hamiltonian cycle $C_n = (v_1, v_2, \ldots, v_n, v_1)$. Assume, to the contrary, that $G \neq C_n$, then there is an edge $e = v_i v_j$ where $|i - j| \geq 2$ and $G - e$ is still Hamiltonian. This is a contradiction. ∎

15. The statement is false. Consider $G = C_n$ for $n \geq 5$.

Section 12.5

1. (a) See Figure 21. (b) Since all cycles are even, G is bipartite. Let U and W be the partite sets of G. Necessarily, u and v belong to the same partite set of G and so every $u - v$ path must have even length.

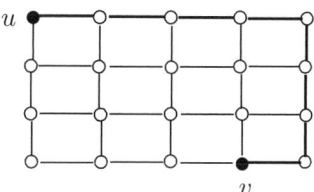

Figure 21: The graph in Exercise 1(a)

3. The following table gives the directed distances from v_1 to the other vertices of D.

v_1	v_2	v_3	v_4	v_5	v_6	S		
0	∞	∞	∞	∞	∞	v_1	$d(v_1, v_2) = 14$	(v_1, v_2)
	$(14, v_1)$	$(16, v_1)$	$(18, v_1)$	$(19, v_1)$	$(42, v_1)$	v_2	$d(v_1, v_3) = 16$	(v_1, v_3)
		$(16, v_1)$	$(18, v_1)$	$(19, v_1)$	$(40, v_2)$	v_3	$d(v_1, v_4) = 18$	(v_1, v_4)
			$(18, v_1)$	$(19, v_1)$	$(33, v_3)$	v_4	$d(v_1, v_5) = 19$	(v_1, v_5)
				$(19, v_1)$	$(32, v_4)$	v_5	$d(v_1, v_6) = 31$	(v_1, v_5, v_6)
					$(31, v_5)$	v_6		

5. 42.   **7.** (a) The least expensive trip will cost \$1384 if it is taken in the order NYC, LOS, MIN, HOU, MIA, NYC. (b) The least expensive trip will cost \$1254 if it is taken in the order NYC, HOU, MIA, LOS, NYC (or NYC, LOS, MIA, HOU, NYC) (c) If Minneapolis is simply removed from the trip in (a), giving NYC, MIA, HOU, LOS, NYC, then the cost of this trip is \$1384, which is identical to the cost with Minneapolis included.

Supplementary Exercises for Chapter 12

1-5. See Figure 22. The vertices of the graph of Exercise 3 are labeled with their degrees.

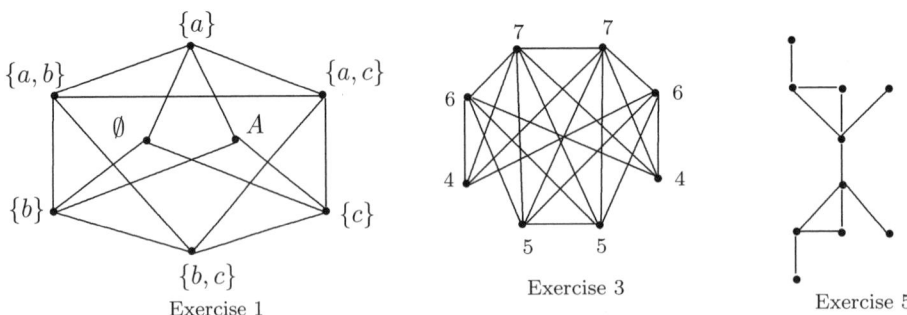

Figure 22: The graphs in Exercise 1, 3, 5

7. $r = 0$ or $r = n - 1$.   **9.** 9.   **11.** The order is 10 and the size is 34.

13. Proof. Assume, to the contrary, that there is a graph G of order $2k + 1$ such that k vertices of G have degree k and the remaining $k + 1$ vertices have degree $k + 1$. Observe that for any integer k, either k or $k + 1$ is odd. In each case, G has an odd number of odd vertices, which is impossible.   ∎

15. 5.   **17.** (a) The maximum size of a graph of order 10 is 45. (b) No graph has an odd number of odd vertices. (c) No bipartite graph contains an odd cycle. (d) Two isomorphic graphs must have the same size.

19. The graphs G and H are isomorphic.

21. $E(G) = \{v_1v_2, v_1v_3, v_1v_4, v_1v_5, v_2v_4, v_4v_5, v_5v_3\}$.   **23.** 8.

25. Proof. Assume, to the contrary, that there is a graph G of order n that contains a double odd vertex u and a mixed vertex v. Then $\deg_G u + \deg_{\overline{G}} u = n - 1$. Since $\deg_G u + \deg_{\overline{G}} u$ is even, it follows that n is odd. On the other hand, $\deg_G v + \deg_{\overline{G}} v = n - 1$. Since $\deg_G v + \deg_{\overline{G}} v$ is odd, it follows that n is even, which is impossible.   ∎

27. See Figure 23.

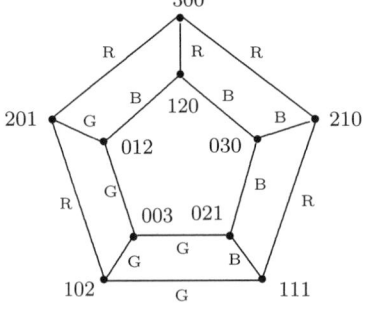

Figure 23: The graph in Exercise 27

29. Proof. Let G be a graph of order $n \geq 2$ such that $\deg v \geq \frac{n-2}{3}$ for every vertex v of G. Assume, to the contrary, that G contains at least three components. Then G contains a component G_1 of order at most $n/3$. Let $v \in V(G_1)$. Then $\deg v \leq (n/3) - 1 = (n - 3)/3$, contradicting the fact that that $\delta(G) \geq (n - 2)/3$.   ∎

31 (a) See Figure 24. (b) Let G_n be the graph of order n. Necessarily, G must contain a vertex of each of the degrees $1, 2, \ldots, n-1$ and one other vertex having one of these degrees. There can be only one vertex of degree $n-1$, for otherwise, there are no vertices of degree 1. Hence $\overline{G}_n$ has exactly one isolated vertex. In fact, the degrees of $\overline{G}_n$ are $0, 1, 2, \ldots, n-2$. If we delete the isolated vertex from $\overline{G}_n$, then we obtain the graph H_n of order $n-1$ containing a vertex of each of the degrees $1, 2, \ldots, n-2$ and one vertex of one of these degrees. This is G_{n-1}.

$G_3:$ $G_4:$ $G_5:$

Figure 24: The graphs G_n $(3 \leq n \leq 5)$ in Exercise 31

33. (a) **Proof.** Let P be a longest path in G. Suppose that P is a $u-v$ path. Since P is a longest path in G, every vertex adjacent to u must lie on P. Hence P contains at least $\delta + 1$ vertices and so its length is at least δ. ∎

(b) **Proof.** Let P be a longest path in G, say $P = (u = v_1, v_2, \ldots, v_k = v)$. Let j be the largest integer such $uv_j \in E(G)$. Then $j \geq \delta$. The $u - v_j$ subpath P' of P together with the edge uv_j produces a cycle of length at least $\delta + 1$. ∎

35. **Proof.** Assume, to the contrary, that u is a cut-vertex of G. Then $G - u$ has two or more components. Let G_1 and G_2 be two components of $G-u$ such that $v \in V(G_1)$. Let $w \in V(G_2)$. Then every $v - w$ path in G contains u. This implies that every $v - w$ path is longer than P, which is a contradiction. ∎

37. $k = 1$ and G is a complete graph.   39. No.

41. (a) See Figure 25. (b) $G = C_5$. (c) Let $G = C_p$, where p is prime and $p \geq 2k + 1$.

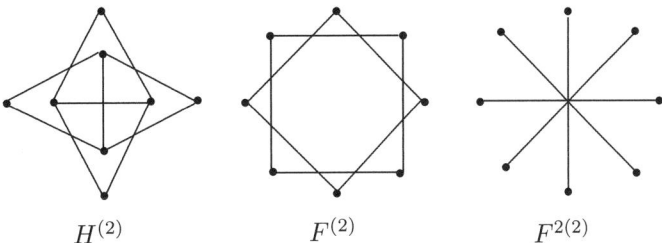

$H^{(2)}$ $F^{(2)}$ $F^2(2)$

Figure 25: The graphs $H^{(2)}$, $F^{(2)}$ and $F^2(2)$ in Exercise 41(a)

43. Suppose that $G_1, G_2, \ldots, G_\ell$ are the components of $G - u$, where $\ell \geq 2$. We may assume that $v \in V(G_1)$. Let $w \in V(G_2)$. Since every $v - w$ path in G contains u, it follows that $d(v, w) > d(u, v) = k$.

45. The graph G is Eulerian.

47. (a) **Proof.** Since $\deg v \geq 1$, it follows that $\deg u - \deg v \leq n - 2$. Thus $\deg u - \deg v = n - 2$, which implies that $\deg u = n - 1$ and $\deg v = 1$. Since $\deg u$ is odd, $n = \deg u + 1$ is even.

(b) Let G be the graph obtained from K_5 by adding a new vertex v and joining v to a vertex u of K_5.

49. (a) **Proof.** Let G be an Eulerian graph. For each vertex v in G, $\deg v = 2k$ for some integer k. If $k = 2\ell$ is even, then $\deg v = 4\ell$ and $\deg v \equiv 0 \pmod{4}$; while if $k = 2\ell + 1$ is odd, then $\deg v = 4\ell + 2$ and $\deg v \equiv 2 \pmod 4$. ∎

(b) **Proof.** Let $V = V_1 \cup V_2$ such that each vertex v in V_1 has $\deg v \equiv 0 \pmod 4$ while each vertex v in V_2 has $\deg v \equiv 2 \pmod 4$. Suppose that $V_1 = \{u_1, u_2, \ldots, u_a\}$ and $V_2 = \{v_1, v_2, \ldots, v_b\}$, where $\deg u_i = 4k_i$ for $1 \le i \le a$ while $\deg v_j = 4\ell_j + 2$ for $1 \le j \le b$. Let m be the size of G. By the First Theorem of Graph Theory, $\sum_{v \in V} \deg v = \sum_{v \in V_1} \deg v + \sum_{v \in V_2} \deg v = 4 \sum_{i=1}^{a} k_i + 2 \sum_{j=1}^{b} (2\ell_j + 1) = 2m$. This implies that $2 \sum_{i=1}^{a} k_i + \sum_{j=1}^{b} (2\ell_j + 1) = m$. Since m is even, $\sum_{j=1}^{b} (2\ell_j + 1) = m - 2 \sum_{i=1}^{a} k_i$ is even. Since $\sum_{j=1}^{b} (2\ell_j + 1)$ is the sum of odd integers, b is even. ∎

51. (a) $\binom{7}{3} = 35$. (b) Yes. (c) 3. (d) 6.

Chapter 13: Trees

Section 13.1

1. See Figure 26.

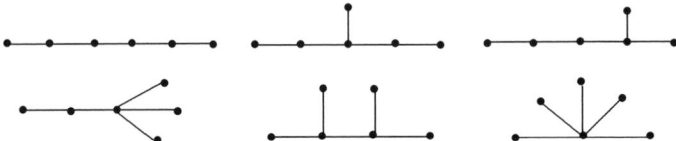

Figure 26: The trees in Exercise 1

3. They are all paths of order 4: (x, u, v, w), (u, x, w, v), (u, x, v, w), (u, v, x, w).

5. See the two trees with diameter 4 in Exercise 1.   **7.** Only K_1 and P_4.

9. Not necessarily.   **11.** one.   **13.** 25.   **15.** 64.

17. **Proof.** Assume, to the contrary, that there is such a tree T. Let x be the number of vertices of degree 1 in T. Then there are $100 - x$ vertices of degree 5. By Theorem 13.12, $x = 2 + 3(100 - x)$. Solving for x, we obtain $x = 75.5$, which is impossible. ∎

19. (a) Let $\Delta(T) = k \ge 3$ and let n_i be the number of vertices of degree i in T, where $2 \le i \le k$. Since $\Delta(T) = k$, it follows that $n_k \ge 1$. Therefore, by Theorem 13.12, $n_1 = 2 + n_3 + 2n_4 + \cdots + (k-2)n_k \ge 2 + (k-2)n_k \ge 2 + n_k \ge 3$.

(b) If T has exactly two end-vertices, then $\Delta(T) \le 2$ by (a) and so T is a path.

21. **Proof.** It is clear that if $G = P_n = (u = v_1, v_2, \ldots, v_n = v)$, then $d(u, v) = n - 1$. For the converse, suppose that G contains two vertices u and v with $d(u, v) = n - 1$. Then G contains a $u - v$ path Q of length $n - 1$, say $Q = (u = u_1, u_2, \ldots, u_n = v)$. Since the order of G is n, it follows that $V(Q) = V(G)$. Assume, to the contrary, that $G \ne Q$. Then there is $e \in E(G) - E(Q)$, say $e = u_i u_j$ where $1 \le i < j$ and $j - i \ge 2$. However then, $Q' = (u = u_1, u_2, \ldots, u_i, u_j, \ldots, u_n = v)$ is a $u - v$ path of length at most $n - 2$. This implies that $d(u, v) \le n - 2$, which is a contradiction. ∎

23. $G = T$.

Section 13.2

1. No.   **3.** (a) $h_1 = \text{rad}(T)$ and $h_2 = \text{diam}(T)$. (b) r_1 is a central vertex of T.

5. See Figure 27.   **7.** See Figure 28. The label of each leaf indicates the counterfeit coin.

9. (a) The counterfeit coins can be identified in two weighings. (b) The counterfeit coins cannot be identified.   **11-15.** See Figure 29.

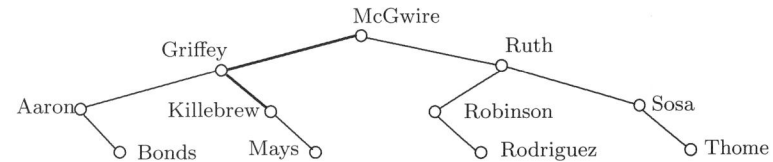

Figure 27: The tree in Exercise 5

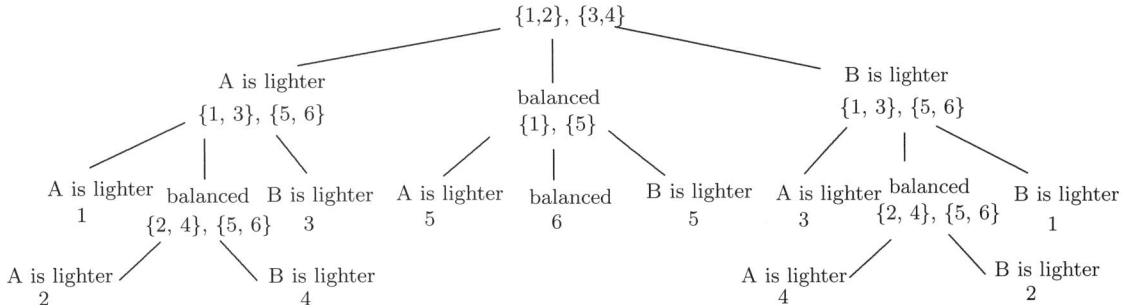

Figure 28: The decision tree in Exercise 7

Section 13.3

1. See Figure 30(a).   **3.** A unique minimum spanning tree.

5. See Figure 30(b), where we start with the vertex u.   **7.** Not necessarily.

9. Suppose that the maximum weight of an edge of G is k. Construct a new weighted graph G' from G by assigning each edge e of G the weight $w'(e) = k + 1 - w(e)$. Then a minimum spanning tree of G' is a maximum spanning tree of G.

11. 60 for G_1 and 512 for G_2.

13. Proof. Let T be a minimum spanning tree of G. First, assume that T is the unique minimum spanning tree of G. Assume, to the contrary, that there exist edges e and f of G such that (1) $e \in E(G) - E(T)$, (2) f is on the cycle of $T + e$ and (3) $w(f) \geq w(e)$. Then $T' = T + e - f$ is a spanning tree of G with $w(T') = w(T) + w(e) - w(f) \leq w(T)$, which implies that either T' is another minimum spanning tree of G or T is not a minimum spanning tree of G, a contradiction. For the converse, assume that the weight of each edge $e \in E(G) - E(T)$ exceeds the weight of every edge on the cycle of $T + e$. Then by Kruskal's algorithm, T be a minimum spanning tree of G. We claim that T is unique, for otherwise, let $T' \neq T$ be another minimum spanning tree of G. So $w(T') = w(T)$. Let $e' \in E(T') - E(T)$ and let C' be the cycle in $T + e'$. By our assumption, $w(e') > w(f')$ for every edge f' in C'. Thus let $f' \in E(C)$ and so $w(T') = w(T) + w(e') - w(f') > w(T) = w(T')$, which is a contradiction. ∎

Supplementary Exercises for Chapter 13

1. 20.   **3.** (a) C_n. (b) $K_4 - e$. (c) Yes.

5. Suppose that the order of G is n and the size is m. Since $G - u$ and $G - v$ are both trees of order $n - 1$, the sizes of $G - u$ and $G - v$ are both $n - 2$. Observe that $m - \deg u = n - 2$ and $m - \deg v = n - 2$.   **7.** 485.   **9.** $n/2$.

11. Let G be the graph obtained from the 4-cycle (u, v, w, x, u) by (1) adding two vertices u_1 and u_2 and joining each u_i $(1 \leq i \leq 2)$ to u and (2) adding two vertices w_1 and w_2 and joining each w_i $(1 \leq i \leq 2)$ to w.

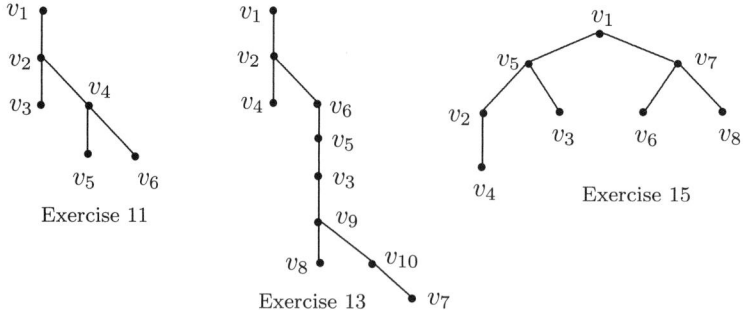

Figure 29: The binary search trees in Exercises 11, 13 and 15

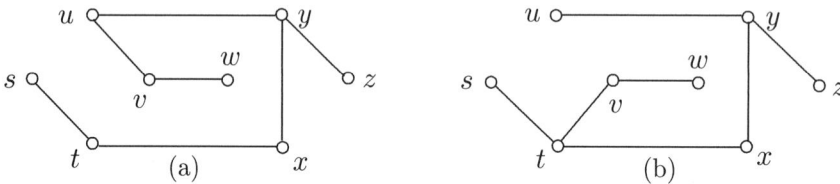

Figure 30: Minimum spanning trees in Exercises 1 and 5

13. 9.   **15.** 62. Let n_i be the number of vertices of degree i in T. Then $n_{10} = 3$ and $n_{20} = 2$ and $n_i = 0$ for $i \neq 1, 10, 20$. Thus $n_1 = 2 + 8n_{10} + 18n_{20} = 2 + 8 \cdot 3 + 18 \cdot 2 = 2 + 24 + 36 = 62$.

17. 98.   **19.** See Figure 31.

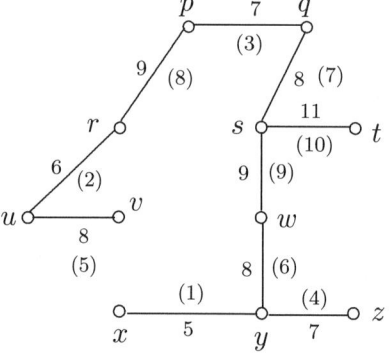

Figure 31: A minimum spanning tree T in Exercise 19

21. (a) Let $V(K_4) = \{u, v, w, x\}$. Suppose that T_1 and T_2 are spanning trees of K_4, each of which is isomorphic to P_4, say $T_1 = (w, u, v, x)$ and $T_2 = (u, x, w, v)$. Then $\{E(T_1), E(T_2)\}$ is a partition of $E(K_4)$. (b) See Figure 32 .

23. Let $P_7 = (v_1, v_2, \ldots, v_7)$ be a path of order 7. Let T_1 be the tree obtained from P_7 by adding a new vertex x and joining x to v_2, let T_2 be the tree obtained from P_7 by adding a new vertex y and joining y to v_3 and let T_3 be the tree obtained from P_7 by adding a new vertex z and joining y to v_4.   **25.** 2.   **27.** 6.

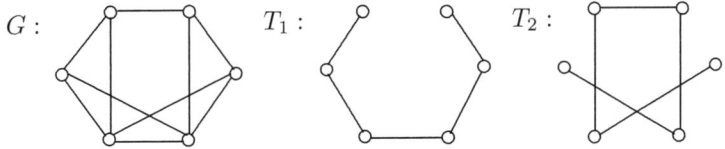

Figure 32: The graphs in Exercise 21(b)

Chapter 14: Planar Graphs and Graph Colorings

Section 14.1

1. See Figure 33.   **3.** $r_1 = r_2$.   **5.** 11.   **7.** (1).   **9.** (3).   **11.** (2).

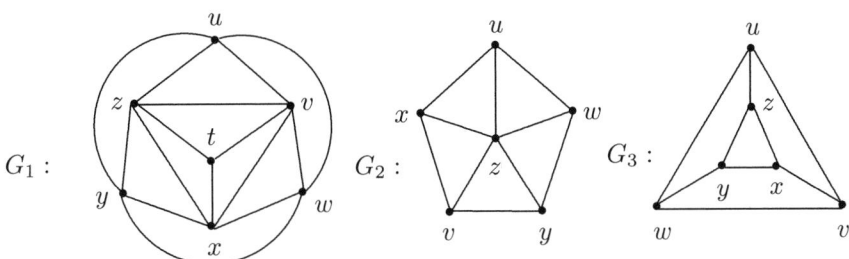

Figure 33: The graphs in Exercise 1

13. Can five houses be connected to two utilities without any of the utility lines crossing? The answer is yes as $K_{2,5}$ is planar.

15. $K_{1,t}$ ($t \geq 1$) and $K_{2,t}$ ($t \geq 2$).   **17.** (a) True. (b) $\overline{K}_k$.

19. Only G_1 and G_4 are planar. Each of G_2 and G_3 contains a subdivision of $K_{3,3}$.

21. (a) **Proof.** Let $R_1, R_2, \ldots, R_r$ be the r regions of G. So if m_i is the number of edges on the boundary of R_i ($1 \leq i \leq r$), then $m_i \geq 5$. Let $M = \sum_{i=1}^{r} m_i \geq 5r$. The number M counts an edge once if the edge is a bridge and counts it twice if the edge is not a bridge. So $M \leq 2m$. Therefore, $5r \leq M \leq 2m$ and so $5r \leq 2m$. Applying the Euler Identity $2 = n - m + r$ to G, we have $10 = 5n - 5m + 5r \leq 5n - 5m + 2m = 5n - 3m$. Solving for m, we get $m \leq \frac{5}{3}(n-2)$. ∎

(b) The smallest cycle in the Petersen graph is a 5-cycle.

(c) The statement is true. **Proof.** Assume, to the contrary, that G has no vertex of degree 2 or less and so $\delta(G) \geq 3$. Then $2m \geq 3n$ or $m \geq \frac{3}{2}n$. On the other hand, $m \leq \frac{5}{3}(n-2)$ by (a). Therefore, $\frac{3}{2}n \leq m \leq \frac{5}{3}(n-2)$ and $\frac{3}{2}n \leq \frac{5}{3}(n-2)$, which implies that $n \geq 20$, a contradiction. ∎

Section 14.2

1. $\chi(G) = 3$. Examples of k-colorings of G for $3 \leq k \leq 7$ are shown in Figures 34(a)-(e).

3. (a) See Figure 35. (b) $\chi(G) = 3$.

(c) This tells us that the smallest number of colors needed to color the regions of M is 3.

5. This is not a proof. The fact that a planar graph G does not contain K_5 as a subgraph and $\chi(K_5) = 5$ does not mean that $\chi(G) \leq 4$.   **7.** False. Consider $G = K_5$.

9. 2.   **11.** $\chi(G) = 3$ and $\chi(H) = 4$.   **13.** (a) K_4. (b) $\chi(G) \geq 4$. (c) $\chi(G) = 5$.

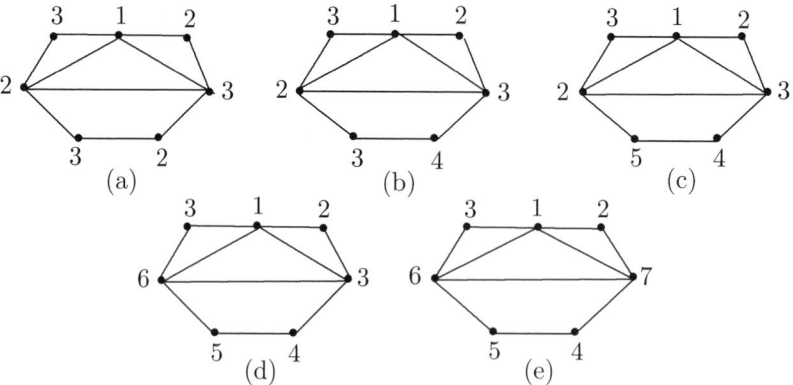

Figure 34: The colorings in Exercise 1

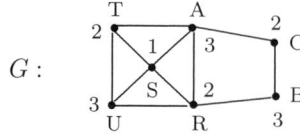

Figure 35: The graph G in Exercise 3(a)

15. (a) We employ finite induction (see Exercise 27 in Supplementary Exercises for Chapter 4). We show, by this algorithm, that every vertex of a graph G of order n can be colored with one of the colors $1, 2, \ldots, 1 + \Delta$, where $\Delta = \Delta(G)$. The vertex v_1 is colored 1; so the basis step is true. We now verify the inductive step. Assume, for an integer k with $1 \leq k < n$, that each of the vertices $v_1, v_2, \ldots, v_k$ can be assigned one of the colors $1, 2, \ldots, 1 + \Delta$ such that adjacent vertices are colored differently. Consider the vertex v_{k+1}. Since $\deg v_{k+1} \leq \Delta$, the vertex v_{k+1} is adjacent to at most Δ vertices. Even if all of these vertices have been assigned distinct colors, there is at least one color among $1, 2, \ldots, 1 + \Delta$ that has not been used to color the vertices adjacent to v_{k+1}. Select the smallest unused color and assign this color to v_{k+1}. Hence all of the vertices $v_1, v_2, \ldots, v_{k+1}$ are colored with $1, 2, \ldots, 1 + \Delta$ in such a way that adjacent vertices are colored differently. This verifies the inductive step. By the Principle of Finite Induction, all of the vertices of G can be colored with $1, 2, \ldots, 1 + \Delta$ so that adjacent vertices are colored differently. Therefore, $\chi(G) \leq 1 + \Delta(G)$. ∎

(b) $\chi(G) = 4$. This algorithm produces a 4-coloring c of G defined by $c(v_1) = 1$, $c(v_2) = 1$, $c(v_3) = 2$, $c(v_4) = 3$, $c(v_5) = 2$, $c(v_6) = 4$. (c) No.

17. Yes. This situation can be modeled by a graph whose chromatic number equals to the minimum number of time periods needed.   **19.** 4.

21. The statement is true. **Proof.** Let c be a k-coloring of G and let V_i be the set of vertices colored i for $1 \leq i \leq k$. Assume, to the contrary, that there are $i, j \in \{1, 2, \ldots, k\}$ such that no two adjacent vertices of G are colored i and j, say $i = k - 1$ and $j = k$. Then the $(k-1)$-coloring c' of G defined by $c'(v) = c(v)$ if $v \in V_i$ and $1 \leq i \leq k - 1$ and $c'(v) = k - 1$ if $v \in V_k$ is a proper $(k-1)$-coloring of G, which is impossible. ∎

23. Since G is a subgraph of $G \times K_2$, it follows that $\chi(G) \leq \chi(G \times K_2)$. To show that $\chi(G \times K_2) \leq \chi(G)$, suppose that $\chi(G) = k$ and let c be a k-coloring of G, using the colors $1, 2, \ldots, k$. The graph $G \times K_2$ is constructed from two copies G and G' of G, where uu' is an edge of $G \times K_2$ if u' corresponds to u in G. Let u' be colored $i + 1$ if u is colored i $(1 \leq i \leq k - 1)$ and color u' with the color 1 if u is colored k. This produces a k-coloring of $G \times K_2$ and so $\chi(G \times K_2) \leq \chi(G)$. Therefore, $\chi(G \times K_2) = \chi(G)$.

Supplementary Exercises for Chapter 14

1. (a) True. (b) False.   **3.** (a) False. (b) True.   **5.** (c).

7. Proof. Let G be a maximal planar graph of order $n \geq 4$ and size m. Then $m = 3n - 6$. Assume, to the contrary, that $\delta(G) \leq 2$. Let $v \in V(G)$ with $\deg v = \delta(G)$. Then $G - v$ has order $n - 1$ and size $m' \geq m - 2$. Furthermore, $G - v$ is planar. Thus $m - 2 \leq m' \leq 3(n-1) - 6$. Therefore, $m \leq 3n - 7$, a contradiction. ∎

9. False.   **11.** 3.

13. (a) Let G be a nonempty graph and let H be a graph obtained from G by subdividing a single edge xy of G. Suppose that xy is replaced by two edges xz and zy in H. Suppose that $\chi(G) = k$ and let there be a k-coloring of G with the colors $1, 2, \ldots, k$. By assigning z the color $k + 1$, a $(k + 1)$-coloring of H is obtained, which implies that $\chi(H) \leq k + 1$. Next, suppose that $\chi(H) = k$ and let c be a k-coloring of H with the colors $1, 2, \ldots, k$. If the edges xz and zy are replaced by xy and x is assigned the color $k + 1$, then a $(k + 1)$-coloring of G is produced, which implies that $\chi(G) \leq k + 1$. Therefore, $|\chi(G) - \chi(H)| \leq 1$.

(b) (i) Let $G = P_2$ and $H = P_3$. (ii) Let $G = C_3$ and $H = C_4$. (iii) Let $G = C_4$ and $H = C_5$.

15. 3.   **17.** 4.

19. (a) **Proof.** Assume, to the contrary, that G contains $K_{3,3}$ as a subgraph. Let v be the vertex of G that is not a vertex of $K_{3,3}$. Then $G - v$ contains $K_{3,3}$ as a subgraph and so $G - v$ is nonplanar, which is a contradiction. ∎

(b) Let G be a subdivision of K_5 or a subdivision of $K_{3,3}$ of order 7.

21. Yes.   **23.** Nonplanar.

25. Proof. Assume, to the contrary, that $\chi(G) = 3$. Then each vertex that is colored 3 is adjacent only to vertices colored 1 or is adjacent only to vertices colored 2. Let V_1 be the set of vertices colored 3 and are adjacent only to vertices colored 1 and let V_2 be the set of vertices colored 3 and are adjacent only to vertices colored 2. Now change the color of each vertex of V_1 to 2 and change the color of each vertex of V_2 to 1. The resulting coloring is a 2-coloring, contradicting our assumption that $\chi(G) = 3$. ∎

27. $\chi(H) = k$.

Chapter 15. Directed Graphs

Section 15.1

1. $V(D) = \{P, Q, R\}$ and $E(D) = \{(P, R), (Q, P), (Q, R), (R, P)\}$.

3. (a) $V(D) = \{M_1, M_2, M_3, M_4\}$ and $E(D) = \{(M_1, M_3), (M_2, M_4)\}$.

5. Proof.   Let a be the number of vertices having indegree 0 and let b be the number of vertices having outdegree 0. Then there are a vertices having outdegree 2 and $n - a - b$ vertices having outdegree 1. By the First Theorem of Digraph Theory, $\sum_{v \in V(D)} \operatorname{od} v = a \cdot 2 + (n - a - b) \cdot 1 + b \cdot 0 = n + a - b = n$. This implies that $a = b$ and so $a + b = 2a$ is even. ∎

7. (a) Suppose that D is a digraph such that $\operatorname{od} v = 0$ or $\operatorname{id} v = 0$ for some vertex v of D. Let u be a vertex in $V(D) - \{v\}$. If $\operatorname{od} v = 0$, then there is no $v - u$ directed path in D; while if $\operatorname{id} v = 0$, then there is no $u - v$ directed path in D. In either case, D is not strong.

(b) The converse of (a) is false.

9. There are 4 orientations.   **11.** There are 10 orientations.

13. If $(u = v_0, v_1, v_2, \ldots, v_r = v)$ is a $u - v$ path in D, then $(v = v_r, v_{r-1}, \ldots, v_1, v_0 = u)$ is a $v - u$ path in $\vec{D}$.

15. (a) **Proof.** First, let D be an Eulerian digraph. Then D contains an Eulerian circuit C. Let v be a vertex of C. Assume first that v is not the initial vertex of C (and so v is not the terminal vertex either). Whenever v is encountered on C, an arc is used to enter v and another is used to exit v. This contributes 1 to both the indegree and outdegree of v. If v is encountered k times on C, then $\operatorname{od} v = \operatorname{id} v = k$. If v is the initial vertex of C, then an arc is used to exit v. The final arc of C enters v. Any other occurrences of v on C contribute 1 to both the indegree and outdegree of v and so $\operatorname{od} v = \operatorname{id} v$ in this case as well.

For the converse, let D be a nontrivial connected digraph for which $\operatorname{od} v = \operatorname{id} v$ for every vertex v of D. For a vertex u of D, let T be a trail of maximum length with initial vertex u. Suppose that T is a $u - v$ trail. Assume first that $u \neq v$ and that v is encountered k times on T, where $k \geq 1$. Then T contains k arcs directed towards v and $k - 1$ arcs directed away from v. However, since $\operatorname{od} v = \operatorname{id} v$, there is an arc directed away from v that does not belong to T. This means, however, that T can be extended to a longer trail with initial vertex u. Since this is impossible, $u = v$ and T is a circuit C in D. Consequently, C is a circuit of maximum length in D containing u.

We claim that C contains all of the arcs of D and that C is therefore an Eulerian circuit. Assume, to the contrary, that C does not contain all of the arcs of D. Since D is connected, there is a vertex w on C that is incident with arcs not on C. Let $D' = D - E(C)$ be the spanning subdigraph of D whose arcs are those not belonging to C. Since $\operatorname{od}_D v = \operatorname{id}_D v$ and $\operatorname{od}_C v = \operatorname{id}_C v$ for every vertex v on C, it follows that $\operatorname{od}_{D'} v = \operatorname{id}_{D'} v$ for every vertex of D'. Let T' be a trail of maximum length in D' with initial vertex w. As before, T' is a circuit C' in D''. If we attach C' to C at w, then we produce a circuit C'' in D containing u and containing more arcs than C, which is impossible. Hence C is an Eulerian circuit. ∎

(b) **Proof.** Let m be the size of D. By the First Theorem of Digraph Theory, $\sum_{i=1}^n \operatorname{od} v_i = \sum_{i=1}^n \operatorname{id} v_i = m$. Thus $\sum_{i=1}^n (\operatorname{od} v_i - \operatorname{id} v_i) = 0$. Since $\operatorname{od} v_i \geq \operatorname{id} v_i$ for $1 \leq i \leq n$, it follows that $\operatorname{od} v_i - \operatorname{id} v_i \geq 0$ for $1 \leq i \leq n$. Thus $\operatorname{od} v_i - \operatorname{id} v_i = 0$ or $\operatorname{od} v_i = \operatorname{id} v_i$ for $1 \leq i \leq n$, implying that D is Eulerian. ∎

(c) **Proof.** If $\sum_{i=1}^n |\operatorname{od} v_i - \operatorname{id} v_i| = 0$, then $|\operatorname{od} v_i - \operatorname{id} v_i| = 0$ for $1 \leq i \leq n$. Thus $\operatorname{od} v_i - \operatorname{id} v_i = 0$ or $\operatorname{od} v_i = \operatorname{id} v_i$ for $1 \leq i \leq n$. Therefore, D is Eulerian by (a). ∎

17. (a) **Proof.** The result is true if G contains no bridges. Thus we may assume that G contains at least one bridge. We consider two cases.

Case 1. G contains exactly one bridge, say $e = uv$. Then $G - e$ has exactly two components G_1 and G_2, each of which has a strong orientation. Let D_i be a strong orientation of G_i $(i = 1, 2)$. Then the orientation D of G obtained from D_1 and D_2 by adding the arc (u, v) has the desired property.

Case 2. G contains exactly two bridges, say $e = uv$ and $f = xy$. We may assume that $d(v, x) = \min\{d(z, z'), z \in \{u, v\}, z' \in \{x, y\}\}$. Then $G - e - f$ has exactly three components G_1, G_2 and G_3, each of which has a strong orientation. Let D_i be a strong orientation of G_i $(i = 1, 2, 3)$. Furthermore, v and x belong to the same component. Suppose that $u \in V(G_1)$, $v, x \in V(G_2)$ and $y \in V(G_3)$. Then the orientation D of G obtained from D_1, D_2 and D_3 by adding the arcs (u, v) and (x, y) has the desired property. ∎

(b) For example, no orientation of $G = K_{1,3}$ has the desired property.

19. Proof. Let P be a longest path in D, say P is a $u - v$ path in D. Since $\operatorname{od} v > 0$ and P is a longest path, it follows that v is adjacent to a vertex of P, producing a cycle. ∎

21. Yes.   **23.** The subdigraph of D induced by the vertices on the $r - s$ path in T.

Section 15.2

1. Yes. **3.** Yes.

5. Let D be the transitive orientation of K_n. Then D contains no cycles but D has a Hamiltonian path P, say P is a $u - v$ path. Then $(u, v) \in E(D)$. Then replacing (u, v) by (v, u) results in a strong digraph.

7. (a) 1. (b) Let T be a tournament of order 3 or more and let u and v be two vertices of T such that $\operatorname{od} u = 0$ and $\operatorname{id} v = 0$. Let w be any other vertex of T. Thus (w, u) and (v, w) are arcs in T. Hence $\operatorname{od} w \geq 1$ and $\operatorname{id} w \geq 1$. (c) None.

9. Suppose that T contains the Hamiltonian cycle $C = (v_1, v_2, \ldots, v_n, v_1)$. So $\operatorname{od} v_1 \geq \operatorname{od} v_2 \geq \cdots \geq \operatorname{od} v_n \geq \operatorname{od} v_1$. This implies that all vertices of T have the same outdegree x. Since $nx = \binom{n}{2}$, it follows that $x = (n-1)/2$ and so n is odd.

11. If $(u, v) \in E(T)$, then $(v, u) \notin E(T)$. Thus $\vec{d}(u, v) = 1$ and $\vec{d}(v, u) \neq 1$.

13. **Proof.** Suppose that a tournament T contains a Hamiltonian cycle $C = (v_1, v_2, \ldots, v_n, v_1)$. We show that T is strong. Let u and v be any two distinct vertices of T. Then $u = v_i$ and $v = v_j$, where $1 \leq i \neq j \leq n$. We may assume that $i < j$. Then $P = (u = v_i, v_{i+1}, \ldots, v_j = v)$ and $P' = (v = v_j, v_{j+1}, \ldots, v_n, v_1, v_2, \ldots, v_i = u)$ are $u - v$ and $v - u$ paths and so T is strong. ∎

15. (a) In D_1, (v_1, v_2, v_3). In D_2, (v_1, v_2, v_3). In T_1, (v_1, v_2, v_3, v_4). In T_2, (v_2, v_1, v_3, v_4). In T_3, (v_1, v_2, v_3, v_4). In T_4, (v_1, v_3, v_4, v_2). (b) In D_2, (v_1, v_2, v_3, v_1). In T_4, $(v_1, v_3, v_4, v_2, v_1)$.

17. **Proof.** Recall that $\sum_{i=1}^n \operatorname{od} v_i = \sum_{i=1}^n \operatorname{id} v_i$; so $\sum_{i=1}^n (\operatorname{od} v_i - \operatorname{id} v_i) = 0$. Now

$$0 = \sum_{i=1}^n (\operatorname{od} v_i - \operatorname{id} v_i) = \sum_{i=1}^{n-1} (\operatorname{od} v_i - \operatorname{id} v_i) + (\operatorname{od} v_n - \operatorname{id} v_n) \geq (n-1) + \operatorname{od} v_n - \operatorname{id} v_n.$$

Therefore, $\operatorname{id} v_n - \operatorname{od} v_n \geq n - 1$, which implies that $\operatorname{id} v_n = n - 1$ and $\operatorname{od} v_n = 0$. Since $\operatorname{od} v_n = 0$, it follows that T is not strong. ∎

19. **Proof.** Observe that the indegree of a vertex in T is at most $2r$. Let $u, v \in V(T)$. We may assume, without loss of generality, that $(u, v) \in E(T)$. It remains to show that T contains a directed $v - u$ path. Let A be the set of vertices from which u is adjacent and let B the set of vertices to which v is adjacent. Then $2r - 1 \leq |A| \leq 2r$ and $2r - 1 \leq |B| \leq 2r$. If $A \cap B \neq \emptyset$, then there is $w \in A \cap B$ and so (v, w, u) is a directed $v - u$ path in T. Hence we may assume that $A \cap B = \emptyset$ and so $|A| = |B| = 2r - 1$. If there is an $x \in B$ and $y \in A \cup \{u\}$ such that $(x, y) \in E(T)$, then either (v, x, y, u) or $(v, x, y = u)$ is a directed $v - u$ path in T; for otherwise, every vertex in $A \cup \{u, v\}$ is adjacent to every vertex in B. Thus, each vertex in B has indegree at least $2r + 1$, which is a contradiction. ∎

21. **Proof.** Let $C = (v_1, v_2, \ldots, v_{n-1}, v_1)$ be a cycle of order $n - 1$ in T and let $v \in V(C')$ such that $v \notin V(C)$. Since $v \in V(C')$, it follows that $\operatorname{od} v > 0$ and $\operatorname{id} v > 0$. Thus v is adjacent to some vertex of C and is adjacent from another vertex of C. Assume, without loss of generality, that $(v_1, v) \in E(T)$. Since $\operatorname{od} v > 0$, there exists a vertex $v_i \in V(C)$ $(i \geq 2)$ such that $(v, v_i) \in E(T)$. Thus there exists j with $1 \leq j \leq n - 2$ such that $(v_j, v), (v, v_{j+1}) \in E(T)$. Therefore, $(v_1, v_2, \ldots, v_j, v, v_{j+1}, \ldots, v_{n-1}, v_1)$ is an Hamiltonian cycle of T and so T is Hamiltonian. ∎

Section 15.3

1. The state table is shown below (where 'ch' means that two cherry cough drops are output and 'le' means that two lemon cough drops are output).

	f				g			
	5	10	C	L	5	10	C	L
s_0	s_1	s_2	s_0	s_0	N	N	N	N
s_1	s_2	s_3	s_1	s_1	N	N	N	N
s_2	s_3	s_3	s_2	s_2	N	5	N	N
s_3	s_3	s_3	s_0	s_0	5	10	ch	le

The state digraph is shown in Figure 36.

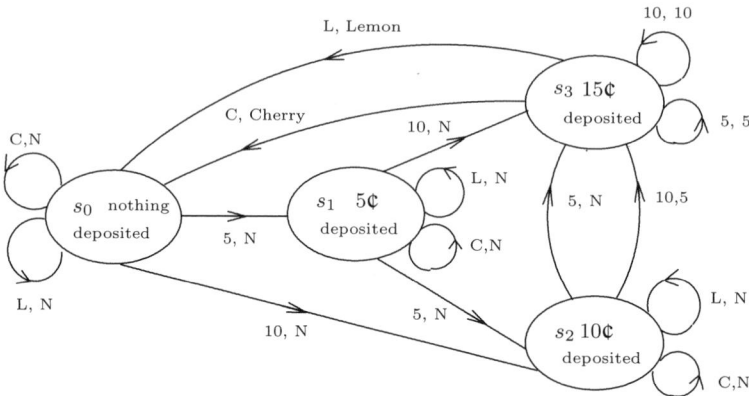

Figure 36: The state digraph for the finite-state machine in Exercise 1

3.

	f		g	
	0	1	0	1
s_0	s_1	s_2	1	1
s_1	s_0	s_2	0	0
s_2	s_1	s_3	1	1
s_3	s_2	s_4	0	0
s_4	s_3	s_5	1	1
s_5	s_4	s_0	1	1

5. See Figure 37. (a) The output is 1010011. (b) The output is 1101100.

(c) The output is 1000010.

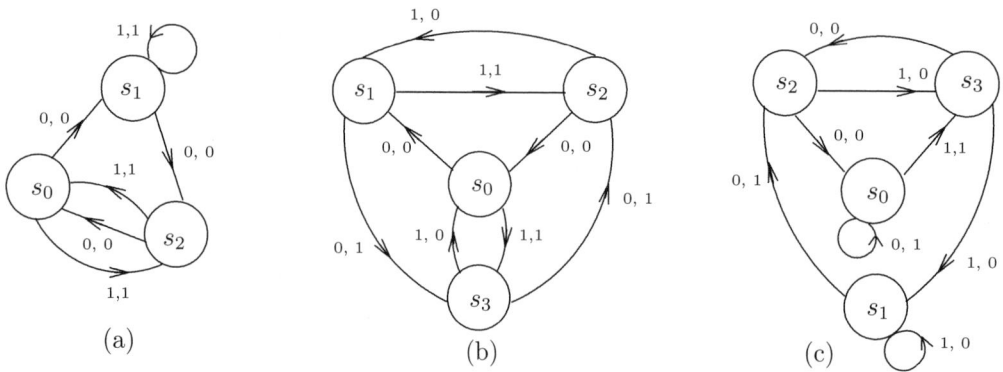

Figure 37: State digraphs for the finite-state machines in Exercise 5

7. (a)

	f		g	
	0	1	0	1
s_0	s_2	s_1	1	0
s_1	s_0	s_2	0	0
s_2	s_1	s_2	0	1

The output is 11000010.

(b)

		f		g	
		0	1	0	1
	s_0	s_1	s_1	0	1
	s_1	s_2	s_3	1	1
	s_2	s_0	s_2	0	1
	s_3	s_0	s_2	1	1

The output is 01101110.

(c)

		f		g	
		0	1	0	1
	s_0	s_3	s_1	1	0
	s_1	s_1	s_2	1	1
	s_2	s_3	s_1	0	1
	s_3	s_0	s_1	1	0

The output is 10111110.

9. $a = (0110111)_2$, $b = (0010101)_2$, $c = (1001100)_2 = 76$ and the resulting output string is 0011001. Choosing s_0 as the initial state, the resulting sequence of states is $s_0, s_1, s_1, s_1, s_0, s_1, s_1$.

11.

	f	
	0	1
s_0	s_0	s_1
s_1	s_1	s_0

The state digraph of this machine is shown in Figure 38.

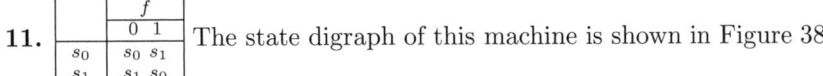

Figure 38: The state digraph for the finite-state automaton in Exercise 11

13. See Figure 39.

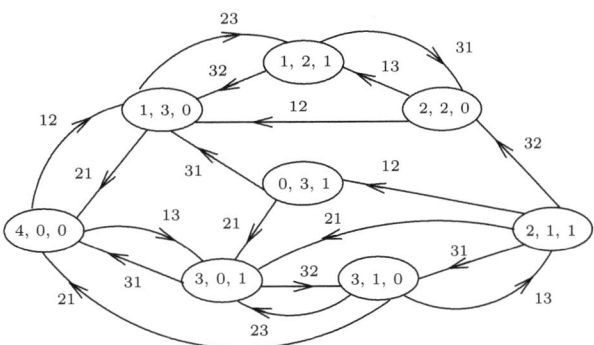

Figure 39: The state digraph for the finite-state machine in Exercise 13

15. (a) Since the digraph D is strong, aperiodic (it has directed cycles of lengths 2 and 3) and has uniform outdegree, it follows by the Road Coloring Theorem that D has a synchronized coloring. (b) Suppose that the the coloring of D is synchronized. Then there is a synchronized sequence s for u. Applying this sequence to u results in a directed $u - u$ walk. Because of the symmetry of this coloring, when s is applied to v, a directed $v - v$ walk is produced, contradicting the assumption that s is a synchronized sequence for u. (c) (i). (d) Yes.

17. (a) Since the digraph D is a strong, aperiodic digraph and has uniform outdegree, it follows by the Road Coloring Theorem that D has a synchronized coloring. (b) (i) is not. (c) Yes.

19. The sequence s' constructed by following s by s is a synchronized sequence for v.

Supplementary Exercises for Chapter 15

1. (b).

3. Proof. Suppose that the order of D is n. The vertices of D can be labeled as $v_1, v_2, \ldots, v_n$ so that $r(v_i) = i$ for $1 \le i \le n$. Necessarily, D contains no directed cycle, for otherwise, if u and v belong to a cycle, then $r(u) = r(v)$, which is impossible. We show that D contains the Hamiltonian path $(v_n, v_{n-1}, \ldots, v_2, v_1)$. Suppose that it does not. First, we claim that $(v_n, v_{n-1}) \in E(D)$. Suppose that this is not the case. Since $r(v_n) = n$, there is a directed $v_n - v_{n-1}$ path P in D such that P has an internal vertex x. Since x can reach every vertex that v_{n-1} can reach, $r(x) \ge r(v_{n-1})$, which is impossible. Thus, as claimed, v_n is adjacent to v_{n-1} in D. Let k be the largest positive integer such that $(v_k, v_{k-1}) \notin E(D)$. Then $k \le n - 1$. Since D contains no directed cycle, D contains no directed path from v_k to any of the vertices $v_n, v_{n-1}, \ldots, v_{k+1}$. Thus v_k cannot reach any of $v_n, v_{n-1}, \ldots, v_{k+1}$ and must reach all of $v_1, v_2, \ldots, v_{k-1}$. Hence D contains a $v_k - v_{k-1}$ path with an internal vertex v_j such that (v_k, v_j) is a directed edge and $1 \le j \le k - 2$. Since D contains a directed $v_j - v_{k-1}$ path and $r(v_{k-1}) = k - 1$, it follows $r(v_j) \ge k - 1$, which is a contradiction. Thus D contains the Hamiltonian path $(v_n, v_{n-1}, \ldots, v_2, v_1)$. ∎

5. The statement is false.   **7.** Yes.

9. Proof. Let A be the set of vertices of T to which v is adjacent. Then $|A| = \text{od } v$. We claim that there is $x \in A$ such that $(x, u) \in E(T)$ and so (u, v, x, u) is a triangle. If $(u, w) \in E(T)$ for all $w \in E(T)$, then $\text{od } u \ge |A| + 1 = \text{od } v + 1 > \text{od } v$, which is impossible. ∎

11. T_1 is transitive, T_2 is neither.   **13.** The statement is false.   **15.** (a) is true.

17. $k = 2$. Observe that $\text{od } v_i = n - 1 - \text{id } v_i$ for $i = 1, 2, \ldots, n - 1$.

19. Suppose that u and v do not lie on a common cycle. Show that if $(u, v) \in E(T)$, then $\text{od } u > \text{od } v$.

21. T is transitive.

23. (a) (i) u_1. (ii) v_1. (iii) u_2. (iv) v_2. (v) u_3. (vi) v_3. (b) The sequence $brbr$ is a synchronized sequence for u_4. The sequence $rbrb$ is a synchronized sequence for v_4.

Index

List of Symbols